RUSSELL'S
SOIL CONDITIONS AND PLANT GROWTH

ELEVENTH EDITION

Frontispiece Rice growing in West Java, Indonesia, with incipient soil erosion on the upper slope. Stream beds run through the terraces; further erosion on the upper slope may cause increased run-off and erosion of the terraces (P. J. Gregory)

Russell's Soil Conditions and Plant Growth

ELEVENTH EDITION

edited by
ALAN WILD

*Department of Soil Science,
University of Reading*

Longman
Scientific &
Technical

Copublished in the United States with
John Wiley & Sons, Inc., New York

Longman Scientific & Technical,
Longman Group UK Limited,
Longman House, Burnt Mill, Ha
Essex CM20 2JE, England
and Associated Companies throu

Copublished in the United States with
John Wiley & Sons, Inc., 605 Third Avenue, New York, NY 10158

First published (in Monographs in Biochemistry) by E. J. Russell	*1912*
Second Edition	*1915*
Third Edition	*1917*
Fourth Edition (in Rothamsted Monographs on Agricultural Science)	*1921*
Fifth Edition	*1927*
Sixth Edition	*1932*
Seventh Edition	*1937*
Eighth Edition rewritten and revised by E. W. Russell	*1950*
Ninth Edition	*1961*
Tenth Edition	*1973*
Eleventh Edition edited by Alan Wild	*1988*

British Library Cataloguing in Publication Data
Russell, E. W.
  Russell's Soil conditions and plant growth. – 11th ed.
  1. Soils 2. Field crops
  I. Title   II. Wild, Alan
  631.4   S592.5

ISBN 0-582-44677-5

Library of Congress Cataloging-in-Publication Data
Russell, Edward J. (Edward John), Sir, 1872–1965.
  Russell's soil conditions and plant growth.

  Rev. ed. of: Soil conditions and plant growth.
10th ed. 1973.
  Bibliography: p.
  Includes indexes.
  1. Soil science. 2. Crops and soils. I. Wild, Alan. II. Russell, Edward
J. (Edward John), Sir, 1872–1965. Soil conditions and plant growth.
III. Title. IV. Title: Soil conditions and plant growth. S591.R84 1988
631.4   86–27391

ISBN 0-470-20796-5 (Wiley, USA only).

Printed and bound in Great Britain at the Bath Press, Avon.

Contents

Foreword

It was 75 years ago that my father brought out the first edition of this book, then a slim volume of 168 pages. It was written fairly quickly, both because his time was very fully occupied with building up Rothamsted to which he had recently been appointed Director and, because there was little active research in agricultural science, he was familiar with most of the research in progress. Further, Rothamsted had a long history of agricultural research behind it and as it was already internationally known its Director had contacts with the relatively few research stations in Europe and America. The book obviously fulfilled a need because a fourth edition, already considerably enlarged, was published only nine years after the first.

Subsequently the problem of revision became increasingly time consuming as the volume of research expanded in the 1920s. As the pace of agricultural research in the world increased, due in part to the success of the First International Congress of Soil Science held in Washington, and as the task of revision became greater, my father asked me to assist him in the production of the seventh edition and to take full responsibility for the book thereafter. This gave me the opportunity to recast it entirely for the eighth edition. The next two editions, although involving very considerable revision and rewriting, followed the same pattern.

After I had completed the revision for the tenth edition I realized that I would be unable to continue its subsequent revision so it was with particular pleasure that I accepted the suggestion of the staff of the Department of Soil Science of the University of Reading, of which I was the first Professor, that they should be responsible for undertaking the revision of this book under the editorship of Professor Wild, the present Professor. The result is a book in which each chapter has either been rewritten or suitably revised by a member of the Department or by outside colleagues with specific expertise.

I am sure that this new edition will be a worthy successor to the earlier ones and that it will enjoy the confidence so many research scientists and students have placed in it.

E. W. Russell

Preface to the eleventh edition

Since its first appearance in 1912 *Soil Conditions and Plant Growth* has achieved worldwide recognition as one of the most influential books in agricultural science. Its success has been largely due to the encyclopaedic knowledge of Professor Walter Russell and his father Sir John Russell. With the rapid growth of the subject no one person would now be able to write a similar book which is as authoritative. This edition is the work of several authors, mostly from the Department of Soil Science, University of Reading, of which Walter Russell was the first Head.

It will be seen that the degree of rewriting differs between chapters. Some are entirely new whereas others, on topics in which there has been less development, have been brought up to date, but much of Professor Russell's material remains. In the Contents no distinction is made between chapters which are essentially new and those where little revision has been required. Authors of all chapters recognize the debt to Professor Russell.

The book itself is about soils and soil/crop relations, which are subjects of interest to a wide range of scientists and to agriculturists, and it deals with soil properties that affect crop growth. After the first chapter on the historical development of the subject, there are chapters on crop growth, soil formation, chemical and physical properties of soil, soil organisms and the transformations they bring about, plant nutrients in soil and, finally, the application of soil science to important problems in soil management. Each chapter gives references to enable readers to study each topic in more detail although the selection becomes ever more difficult. As with previous editions the objective has been to present topics as comprehensively as space allowed and to show the cross-linkages between them.

It is hoped that the book will contribute to the solution of practical problems. To this end we have presented principles as well as applications. This is justified because principles are universal and need to be understood before considering individual problems in a particular environment. For example, new farming systems, especially in the Third World, will only be successful in the long term if they are based on sound methods of soil conservation and the maintenance of fertility. We can now formulate general guidelines, but these need to be properly understood and developed if they are to be applied successfully because each region of the world has its own particular soil problems, and these inter-relate with the local climate and social system. Soil science has now reached the point where transfer of its technology can be made from one region to another, but for this to be successful it is necessary for the subject to be properly understood. Our hope is that the book will help to provide this understanding and thereby help with the appropriate transfer of the technology of soil science.

In preparing this edition it is a great pleasure to acknowledge the generous help of many colleagues. Particular thanks are due to Dr W. A. Adams, Dr P. B. Barraclough, Dr M. D. Dennett, Professor J. Elston, Dr J. K. R. Gasser, Dr K. W. T. Goulding, Dr J. M. Grainger, Professor T. R. G. Gray, Dr D. J. Greenland, Dr R. M. Jackson, Dr S. C. Jarvis, Dr. Angela A. Jones, Mr E. A. Kirkby, Mr R. Lawton, Dr P. H. Le Mare, Dr B. Marshall, Dr G. E. G. Mattingly, Professor J. L. Monteith. Dr J. Moorby, Dr J. Oster, the late Dr J. C. Ryden, Dr R. Summerfield and Dr P. B. Tinker. I also thank the secretarial staff for typing the manuscripts and my wife for assistance in preparing the Author and Subject Indexes.

Alan Wild
University of Reading
January 1987

Contributors to the eleventh edition

Dr G. W. Cooke, Rothamsted Experimental Station, Harpenden, Herts.
Dr P. J. M. Cooper, ICARDA, Aleppo, Syria.
Professor B. E. Davies, Department of Environmental Science, University of Bradford.
Dr D. B. Davies, ADAS, Cambridge.
Dr P. J. Gregory, Department of Soil Science, University of Reading.
Dr P. J. Harris, Department of Soil Science, University of Reading.
Dr D. S. Jenkinson, Rothamsted Experimental Station, Harpenden, Herts.
Dr L. H. P. Jones, Department of Soil Science, University of Reading.
Dr J. M. Lynch, Glasshouse Crops Research Institute, Littlehampton, West Sussex.
Dr C. J. B. Mott, Department of Soil Science, University of Reading.
Mr J. Newman, ICI Jealott's Hill Research Station, Bracknell, Berks.
Dr S. Nortcliff, Department of Soil Science, University of Reading.
Mr D. Payne, Department of Soil Science, University of Reading.
Dr D. L. Rowell, Department of Soil Science, University of Reading.
Professor A. Wild, Department of Soil Science, University of Reading.
Dr M. Wood, Department of Soil Science, University of Reading.

1

Historical

In all ages the growth of plants has interested thoughtful men. The mystery of the change of an apparently lifeless seed to a vigorous growing plant never loses its freshness, and constitutes, indeed, no small part of the charm of gardening. The economic problems are of vital importance, and become more and more urgent as time goes on and populations increase and their needs become more complex.

There was an extensive literature on agriculture in Roman times which maintained a pre-eminent position until comparatively recently In this we find collected many of the facts which it has subsequently been the business of agricultural experts to classify and explain. The Roman literature was collected and condensed into one volume about the year 1240 by a senator of Bologna, Petrus Crescentius, whose book[1] was one of the most popular treatises on agriculture of any time, being frequently copied, and in the early days of printing, passing through many editions – some of them very handsome, and ultimately giving rise to the large standard European treatises of

---

1 *Ruralium commodorum libri duodecim*, Augsburg, 1471, and many subsequent editions.

the sixteenth and seventeenth centuries. Many other agricultural books appeared in the fifteenth and early sixteenth centuries, notably in Italy, and later in France. In some of these are found certain ingenious speculations that have been justified by later work. Such, for instance, is Palissy's remarkable statement in 1563: 'You will admit that when you bring dung into the field it is to return to the soil something that has been taken away. . . . When a plant is burned it is reduced to a salty ash called alcaly by apothecaries and philosophers. . . . Every sort of plant without exception contains some kind of salt. Have you not seen certain labourers when sowing a field with wheat for the second year in succession, burn the unused wheat straw which had been taken from the field? In the ashes will be found the salt that the straw took out of the soil; if this is put back the soil is improved. Being burnt on the ground it serves as manure because it returns to the soil those substances that had been taken away.' But for every speculation that has been confirmed will be found many that have not, and the beginnings of agricultural chemistry must be sought later, when men had learnt the necessity for carrying on experiments.

The search for the 'principle' of vegetation, 1630–1750[2]

It was probably very early discovered that manures, composts, dead animal bodies, and parts of animals, such as blood, all increased the fertility of the land; and this was the basis of the ancient saying that 'corruption is the mother of vegetation'. Yet the early investigators consistently ignored this ancient wisdom when they sought for the 'principle' of vegetation to account for the phenomena of soil fertility and plant growth. Thus the great Francis Bacon, Lord Verulam, believed that water formed the 'principal nourishment' of plants, the purpose of the soil being to keep them upright and protect them from excessive cold or heat, though he also considered that each plant drew a 'particular juyce' from the soil for its sustenance, thereby impoverishing the soil for that particular plant and similar ones, but not necessarily for other plants. Van Helmont (1577–1644) regarded water as the sole nutrient for plants, and his son thus records his famous Brussels experiment: 'I took an earthen vessel in which I put 200 pounds of soil dried in an oven, then I moistened with rain-water and pressed hard into it a shoot of willow weighing 5 pounds. After exactly five years the tree that had grown up weighed 169 pounds and about three ounces. But the vessel had never

---

2 A more detailed account of the British contribution to the development of agricultural science, particularly as it affects crop production, is given by Russell E. J., *A History of Agricultural Science in Great Britain*, Allen and Unwin, 1966, which takes the story up to 1955. For a general account see Browne, C. A., *A Source Book of Agricultural Chemistry*, Chronica Botanica, Waltham, Mass., 1944.

received anything but rain-water or distilled water to moisten the soil when this was necessary, and it remained full of soil, which was still tightly packed, and, lest any dust from outside should get into the soil, it was covered with a sheet of iron coated with tin but perforated with many holes. I did not take the weight of the leaves that fell in the autumn. In the end I dried the soil once more and got the same 200 pounds that I started with, less about two ounces. Therefore the 164 pounds of wood, bark, and root arose from the water alone.'[3]

The experiment is simple and convincing, and satisfied Robert Boyle,[4] who repeated it with 'squash, a kind of Italian pompion' and obtained similar results. Boyle further distilled the plants and concluded, quite justifiably from his premises, that the products obtained, 'salt, spirit, earth, and even oil (though that be thought of all bodies the most opposite to water), may be produced out of water'. Nevertheless, the conclusion is incorrect, because two factors had escaped van Helmont's notice – the parts played by the air and by the missing two ounces of soil. But the history of this experiment is thoroughly typical of experiments in agricultural chemistry generally: in no other subject is it so easy to overlook a vital factor and draw from good experiments a conclusion that appears to be absolutely sound, but is in reality entirely wrong.

Some years later J. R. Glauber[5] set up the hypothesis that saltpetre is the 'principle' of vegetation. Having obtained saltpetre from the earth cleared out from cattle sheds, he argued that it must have come from the urine or droppings of the animals, and must, therefore, be contained in the animal's food, i.e. in plants. He also found that additions of saltpetre to the soil produced enormous increases in crop. He connected these two observations and supposed that saltpetre is the essential principle of vegetation. The fertility of the soil and the value of manures (he mentions dung, feathers, hair, horn, bones, cloth cuttings) are entirely due to saltpetre.

This view was supported by John Mayow's experiments.[6] He estimated the amounts of nitre in the soil at different times of the year, and showed that it occurs in greatest quantity in spring when plants are just beginning to grow, but is not to be found 'in soil on which plants grow abundantly, the reason being that all the nitre of the soil is sucked out by the plants'. J. A. Külbel,[7] on the other hand, regarded a *magma unguinosum* obtainable from humus as the 'principle' sought for.

The most accurate work in this period was published by John Woodward[8] in a remarkable paper. Setting out from the experiments of van Helmont and

3 *Ortus medicinae*, pp. 84–90. *Complexionum atque mistionum elementalium figmentum*, Amsterdam, 1652.
4 *The Sceptical Chymist*, Pt. II, 1661.
5 *Des Teutschlandts Wohlfart* (*Erster Theil*), *das dritte Capittel. De concentratione Vegetabilium, Miraculum Mundi*, Amsterdam, 1656.
6 *Tractatus quinque medico-physici*, 1674 (Alembic Club reprint, Edinburgh, 1907).
7 *Cause de la fertilité des terres*, Bordeaux, 1741.
8 *Phil. Trans. R. Soc.*, 1699, **21**, 382.

of Boyle, but apparently knowing nothing of the work of Glauber and of Mayow, he grew spearmint in water obtained from various sources with the following results among others:

Source of water	Weight of plants		Gained in 77 days	Expense of water (i.e. trans- piration)	Proportion of increase of plant to expense of water
	When put in	When taken out			
	grains*	grains	grains	grains	
Rain water	$28\frac{1}{4}$	$45\frac{3}{4}$	$17\frac{1}{2}$	3004	1 to $171\frac{23}{35}$
River Thames	28	54	26	2493	1 to $95\frac{23}{26}$
Hyde Park conduit	110	249	139	13 140	1 to $94\frac{74}{139}$
Hyde Park conduit plus $1\frac{1}{2}$ oz. garden mould	92	376	284	14 950	1 to $52\frac{182}{284}$

* 1000 grains = 64.8 grams

Now all these plants had abundance of water, therefore all should have made equal growth had nothing more been needed. The amount of growth, however, increased with the impurity of the water. 'Vegetables', he concludes, 'are not formed of water, but of a certain peculiar terrestrial matter. It has been shown that there is a considerable quantity of this matter contained in rain, spring and river water, that the greatest part of the fluid mass that ascends up into plants does not settle there but passes through their pores and exhales up into the atmosphere: that a great part of the terrestrial matter, mixed with the water, passes up into the plant along with it, and that the plant is more or less augmented in proportion as the water contains a greater or less quantity of that matter; from all of which we may reasonably infer, that earth, and not water, is the matter that constitutes vegetables.'

He discusses the use of manures and the fertility of the soil from this point of view, attributing the well-known falling off in crop yield when plants are grown for successive years on unmanured land to the circumstance that 'the vegetable matter that it at first abounded in being extracted from it by those successive crops, is most of it borne off. . . . The land may be brought to produce another series of the same vegetables, but not until it is supplied with a new fund of matter, of like sort with that it at first contained; which supply is made several ways, either by the ground's being fallow some time, until the rain has poured down a fresh stock upon it; or by the tiller's care in manuring it.' The best manures, he continues, are parts either of vegetables or of animals, which ultimately are derived from vegetables.

In his celebrated textbook of chemistry, H. Boerhaave[9] taught that plants

9 *A New Method of Chemistry*, London, 1727.

absorb the juices of the earth and then work them up into food. The raw material, the 'prime radical juice of vegetables, is a compound from all the three kingdoms, viz. *fossil* bodies and putrified parts of *animals* and *vegetables*'. This 'we look upon as the *chyle of the plant*; being chiefly found in the first order of vessels, viz. in the roots and body of the plant, which answers to the stomach and intestines of an animal'.

For many years no such outstanding work as that of Glauber and Woodward was published, if we except Stephen Hales's *Vegetables Staticks* in 1727, the interest of which is physiological rather than agricultural.[10] Advances were, however, being made in agricultural practice. One of the most important was the introduction of the drill and the horse-hoe by Jethro Tull, an Oxford man of a strongly practical turn of mind, who insisted on the vital importance of getting the soil into a fine, crumbly state for plant growth. Tull was more than an inventor; he discussed in most picturesque language the sources of fertility in the soil.[11] In his view it was not the juices of the earth, but the very minute particles of soil loosened by the action of moisture, that constituted the 'proper pabulum' of plants. The pressure caused by the swelling of the growing roots forced these particles into the 'lacteal mouths of the roots, where they entered the circulatory system. All plants lived on these particles, i.e. on the same kind of food; it was incorrect to assert, as some had done, that different kinds of plants fed as differently as horses and dogs, each taking its appropriate food and no other. Plants will take in anything that comes their way, good or bad. A rotation of crops is not a necessity, but only a convenience. Conversely, any soil will nourish any plant if the temperature and water supply are properly regulated. Hoeing increased the surface of the soil or the 'pasture of the plant', and also enabled the soil better to absorb the nutritious vapours condensed from the air. Dung acted in the same way, but was more costly and less efficient.

So much were Tull's writings esteemed, Cobbett tells us, that they were 'plundered by English writers not a few and by Scotch in whole bandittis'.

The position at the end of this period cannot better be summed up than in Tull's own words: 'It is agreed that all the following materials contribute in some manner to the increase of plants, but it is disputed which of them is that very increase or food: (1) nitre, (2) water, (3) air, (4) fire, (5) earth.'

The search for plant nutrients

The phlogistic period, 1750–1800

Great interest was taken in agriculture in this country during the latter half of the eighteenth century. 'The farming tribe', writes Arthur Young during

10 He shows, however, that air is 'wrought into the composition' of plants.
11 *Horse Hoeing Husbandry*, London, 1731.

this period, 'is now made up of all ranks, from a duke to an apprentice'. Many experiments were conducted, facts were accumulated, books written, and societies formed for promoting agriculture. The Edinburgh Society, established in 1755 for the improvement of arts and manufactures, induced Francis Home[12] 'to try how far chymistry will go in settling the principles of agriculture'. The whole art of agriculture, he says, centres in one point: the nourishing of plants. Investigation of fertile soils showed that they contain oil, which is therefore a food of plants. But when a soil has been exhausted by cropping, it recovers its fertility on exposure to air,[13] which therefore supplies another food. Home made pot experiments to ascertain the effect of various substances on plant growth. 'The more they [i.e. farmers] know of the effects of different bodies on plants, the greater chance they have to discover the nourishment of plants, at least this is the only road.' Saltpetre, epsom salt, vitriolated tartar (i.e. potassium sulphate) all lead to increased plant growth, yet they are three distinct salts. Olive oil was also useful. It is thus clear that plant food is not one thing only, but several; he enumerates six: air, water, earth, salts of different kinds, oil, and fire in a fixed state. As further proof he shows that 'all vegetables and vegetable juices afford those very principles, and no other, by all the chymical experiments which have yet been made on them with or without fire'.

The book is a great advance on anything that had gone before it, not only because it recognizes that plant nutrition depends on several factors, but because it indicates so clearly the two methods to be followed in studying the problem – pot cultures and plant analysis. Subsequent investigators, J. G. Wallerius,[14] the Earl of Dundonald[15] and R. Kirwan[16] added new details but no new principles. The problem, indeed, was carried as far as was possible until further advances were made in plant physiology and in chemistry. The writers just mentioned are, however, too important to be passed over completely. Wallerius, in 1761, professor of chemistry at Uppsala, after analysing plants to discover the materials on which they live, and arguing that *Nutritio non fieri potest a rebus heterogeneis, sed homogeneis*, concludes that humus, being *homogeneous*, is the source of their food – the *nutritiva* – while the other soil constituents are *instrumentalia*, making the proper food mixture, dissolving and attenuating it, till it can enter the plant root. Thus chalk and probably salts help in dissolving the 'fatness' of the humus. Clay helps to retain the 'fatness' and prevent it being washed away by rain: sand keeps the soil open and pervious to air. The Earl of Dundonald, in 1795, adds

12 *The Principles of Agriculture and Vegetation*, Edinburgh, 1757.
13 Recorded by most early writers, e.g. Evelyn (*Terra, a philosophical discourse of earth*, 1674).
14 *Agriculturae Fundamenta Chemica: Akerbrukets Chemiska Grunder*, Uppsala, 1761.
15 *A Treatise Showing the Intimate Connection that Subsists between Agriculture and Chemistry, etc.*, London, 1795.
16 *The Manures most Advantageously Applicable to the Various Sorts of Soils and the Cause of their Beneficial Effects in each Particular Instance* (4th edn), London, 1796.

alkaline phosphates to the list of nutritive salts, but he attaches chief importance to humus as plant food. The 'oxygenation' process going on in the soil
makes the organic matter insoluble and therefore useless for the plant; lime,
'alkalis and other saline substances' dissolve it and change it to plant food;
hence these substances should be used alternately with dung as manure.
Manures were thus divided, as by Wallerius, into two classes: those that
afford plant food, and those that have some indirect effect.

Throughout this period it was believed that plants could generate alkalis.
'Alkalis', wrote Kirwan in 1796, 'seem to be the product of the vegetable
process, for either none, or scarce any, is found in the soils, or in rain water.'
In like manner Lampadius thought he had proved that plants could generate
silica. The theory that plants agreed in all essentials with animals was still
accepted by many men of science; some interesting developments were made
by Erasmus Darwin.[17]

Between 1770 and 1800 work was done on the effect of vegetation on air
that was destined to revolutionize the ideas of the function of plants in the
economy of nature, but its agricultural significance was not recognized until
later. Joseph Priestley,[18] knowing that the atmosphere becomes vitiated by
animal respiration, combustion, putrefaction, etc., and realizing that some
natural purification must go on, or life would no longer be possible, was led
to try the effect of sprigs of living mint on vitiated air. He found that the mint
made the air purer, and concludes 'that plants, instead of affecting the air
in the same manner with animal respiration, reverse the effects of breathing,
and tend to keep the atmosphere pure and wholesome, when it is become
noxious in consequence of animals either living, or breathing, or dying, and
putrefying in it'. But he had not yet discovered oxygen, and so could not give
precision to his discovery: and when, later on, he did discover oxygen and
learn how to estimate it, he unfortunately failed to confirm his earlier results
because he overlooked a vital factor, the necessity of light. He was therefore
unable to answer Scheele, who had insisted that plants, like animals, vitiate
the air. It was Jan Ingen-Housz[19] who reconciled both views and showed that
purification goes on in light only, while vitiation takes place in the darkness.
Jean Senebier at Geneva had also arrived at the same result. He also studied
the converse problem – the effect of air on the plant, and in 1782[20] argued
that the increased weight of the tree in van Helmont's experiment (p. 2) came
from the fixed air. 'Si donc l'air fixe, dissous dans l'eau de l'atmosphère, se
combine dans la parenchyme avec la lumière et tous les autres éléments de
la plante; si le phlogistique de cet air fixe est sûrement précipité dans les
organes de la plante, si ce précipité reste, comme on le voit, puisque cet air

17 *Phytologia, or the Philosophy of Agriculture and Gardening*, London, 1800.
18 *Experiments and Observations on Different Kinds of Air*, London, 1775.
19 *Experiments upon Vegetables, Discovering their Great Power of Purifying Common Air
   in the Sunshine and of Injuring in the Shade and at Night*, London, 1779.
20 *Mémoires Physico-chimiques*, 1782.

fixe sort des plantes sous la forme d'air déphlogistiqué, il est clair que l'air fixe, combiné dans la plante avec la lumière, y laisse une matière qui n'y seroit pas, et mes expériences sur l'étoilement suffisent pour le démontrer.' Later on Senebier translated his work into the modern terms of Lavoisier's system.

The period 1800–60

The foundation of plant physiology

We have seen that Home in 1757 pushed his enquiries as far as the methods in vogue would permit, and in consequence no marked advance was made for 40 years. A new method was wanted before further progress could be made, or before the idea introduced by Senebier could be developed. Fortunately, this was soon forthcoming, in 1804. To Théodore de Saussure,[21] son of the well-known de Saussure of Geneva, is due the quantitative experimental method which more than anything else has made modern agricultural chemistry possible; which formed the basis of subsequent work by Boussingault, Liebig, Lawes and Gilbert, and, indeed, still remains our safest method of investigation. Senebier tells us that the elder de Saussure was well acquainted with his work, and it is therefore not surprising that the son attacked two problems that Senebier had also studied – the effect of air on plants and the nature and origin of salts in plants. De Saussure grew plants in air or in known mixtures of air and carbon dioxide, and measured the gas changes by eudiometric analysis and the changes in the plant by 'carbonization'. He was thus able to demonstrate the central fact of plant respiration – the absorption of oxygen and the evolution of carbon dioxide, and further to show the decomposition of carbon dioxide and evolution of oxygen in light. Carbon dioxide in small quantities was a vital necessity for plants, and they perished if it was artificially removed from the air. It furnished them not only with carbon, but also with some oxygen. Water is also decomposed and fixed by plants. On comparing the amount of dry matter gained from these sources with the amount of material that can enter through the roots even under the most favourable conditions, he concludes that the soil furnished only a very small part of the plant food. Small as it is, however, this part is indispensable: it supplies nitrogen – *une partie essentielle des végétaux* – which, as he had shown, was not assimilated direct from the air; and also ash constituents, *qui peuvent contribuer à former, comme dans les animaux, leur parties solides ou osseuses*. Further, he shows that the root is not a mere filter allowing any and every liquid to enter the plant; it has a special action and

21 *Recherches chimiques sur la végétation*, Paris, 1804.

takes in water more readily than dissolved matter, thus effecting a concentration of the solution surrounding it; different salts, also, are absorbed to a different extent. Passing next to the composition of the plant ash, he shows that it is not constant, but varies with the nature of the soil and the age of the plant; it consists mainly, however, of alkalis and phosphates. All the constituents of the ash occur in humus. If a plant is grown from seed in water there is no gain in ash: the amount found at the end of the plant's growth is the same as was present in the seed excepting for a relatively small amount falling on the plant as dust. Thus he disposes finally of the idea that the plant *generated* potash.

After the somewhat lengthy and often wearisome works of the earlier writers it is very refreshing to turn to de Saussure's concise and logical arguments and the ample verification he gives at every stage. But for years his teachings were not accepted, nor were his methods followed.

The two great books on agricultural chemistry then current still belonged to the old period. A. von Thaer and Humphry Davy, while much in advance of Wallerius, the textbook writer of 1761, nevertheless did not realize the fundamental change introduced by de Saussure; it has always been the fate of agricultural science to lag behind pure science. Thaer published his *Grundsätze der rationellen Landwirtschaft* in 1809–12: it had a great success on the Continent as a good, practical handbook, and was translated into English as late as 1844 by Cuthbert Johnson. In it he adopted the prevailing view that plants draw their carbon and other nutrients from the soil humus. 'Die Fruchtbarkeit des Bodens', he says, 'hängt eigentlich ganz vom Humus ab. Denn ausser Wasser ist er es allein, der den Pflanzen Nahrung gibt. So wie der Humus eine Erzeugung des Lebens ist, so ist er auch eine Bedingung des Lebens. Er gibt den Organismen die Nahrung. Ohne ihn lässt sich kein individuelles Leben denken.' Humphry Davy's book[22] grew out of the lectures which he gave annually at the Royal Institution on agricultural chemistry between 1802 and 1812; it forms the last textbook of the older period. While no great advance was made by Davy himself he carefully sifted the facts and hypotheses of previous writers, and gives us an account, which, however defective in places, represents the best accepted knowledge of the time, set out in the new chemical language. His great name gave the subject an importance it would not otherwise have had.[23] He did not accept de Saussure's conclusion that plants obtain their carbon chiefly from the carbonic acid of the air: some plants, he says, appear to be supplied with carbon chiefly from this source, but in general he supposes the carbon to be taken in through the roots. Oils are good manures because of the carbon and

22 *Elements of Agricultural Chemistry*, London, 1813.
23 Thus Charles Lamb, *Essays of Elia* (1820–23) in the 'Old and New Schoolmaster', writes: 'The modern schoolmaster is required to know a little of everything because his pupil is required not to be entirely ignorant of anything. He is to know something of pneumatics, of chemistry, the quality of soils, etc. . . . .'

hydrogen they contain; soot is valuable, because its carbon is 'in a state in which it is capable of being rendered soluble by the action of oxygen and water'. Lime is useful because it dissolves hard vegetable matter. Once the organic matter has dissolved there is no advantage in letting it decompose further: putrid urine is less useful as manure than fresh urine, while it is quite wrong to cause farmyard manure to ferment before it is applied to the land. All these ideas have been given up, and, indeed, there never was any sound experimental evidence to support them. It is even arguable that they would not have persisted so long as they did had it not been for Davy's high reputation. His insistence on the importance of the physical properties of soils – their relationship to heat and to water – was more fortunate and marks the beginning of soil physics, afterwards developed considerably by Gustav Schübler.[24] On the Continent, to an even greater extent than in England, it was held that plants drew their carbon and other nutrients from the soil humus, a view supported by the very high authority of J. J. Berzelius.[25]

The foundation of agricultural science

Hitherto experiments had been conducted either in the laboratory or in small pots: about 1834, however, J. B. Boussingault, who was already known as an adventurous traveller in South America, began a series of field experiments on his farm at Bechelbronn in Alsace.[26] These were the first of their kind: to Boussingault, therefore, belongs the honour of having introduced the method by which the new agricultural science was to be developed. He reintroduced the quantitative methods of de Saussure, weighed and analysed the manures used and the crop obtained, and at the end of the rotation drew up a balance sheet, showing how far the manures had satisfied the needs of the crop and how far other sources of supply – air, rain and soil – had been drawn upon. The results of one experiment are given in Table 1.1.[27] At the end of the period the soil had returned to its original state of productiveness, hence the dry matter, carbon, hydrogen and oxygen not accounted for by the manure must have been supplied by the air and rain, and not by the soil. On the other hand, the manure afforded more mineral matter than the crop took off, the balance remaining in the soil. Other things being equal, he argued that the best rotation is one which yields the greatest amount of organic matter over and above what is present in the manure. No fewer than five rotations were studied, but it will suffice to set out only the nitrogen statistics

24 *Grundsätze der Agrikulturchemie in Näherer Beziehung auf Land- und Fortswirtschaftliche Gewerbe*, Leipzig, 1838.
25 *Traité de Chimie*, Brussels, 1838.
26 See McCosh F. W. J., *Boussingault Chemist and Agriculturist*, Reidel, Dordrecht, 1984 for a biography.
27 *Ann. Chim. Phys.* (III), 1841, **1**, 208.

Table 1.1   Statistics of a rotation

	Weight in kg ha⁻¹ of					
	Dry matter	*Carbon*	*Hydrogen*	*Oxygen*	*Nitrogen*	*Mineral matter*
1 Beets	3 172	1 357.7	184.0	1 376.7	53.9	199.8
2 Wheat	3 006	1 431.6	164.4	1 214.9	31.3	163.8
3 Clover hay	4 029	1 909.7	201.5	1 523.0	84.6	310.2
4 Wheat	4 208	2 004.2	230.0	1 700.7	43.8	229.3
Turnips (catch crop)	716	307.2	39.3	302.9	12.2	54.4
5 Oats	2 347	1 182.3	137.3	890.9	28.4	108.0
Total during rotation	17 478	8 192.7	956.5	7 009.0	254.2	1 065.5
Added in manure	10 161	3 637.6	426.8	2 621.5	203.2	3 271.9
Difference not accounted for taken from air, rain or soil	+7 317	+4 555.1	+529.7	+4 387.5	+51.0	−2 206.4

Table 1.2   Nitrogen statistics of various rotations

	kg ha⁻¹			
Rotation	*Nitrogen in manure*	*Nitrogen in crop*	*Excess in crop over that supplied in manure*	
			Per rotation	*Per annum*
(1) Potatoes, (2) wheat, (3) clover, (4) wheat, turnips,* (5) oats	203.2	250.7	47.5	9.5
(1) Beets, (2) wheat, (3) clover, (4) wheat, turnips,* (5) oats	203.2	254.2	51.0	10.2
(1) Potatoes, (2) wheat, (3) clover, (4) wheat, turnips,* (5) peas, (6) rye	243.8	353.6	109.8	18.3
Jerusalem artichokes, two years	188.2	274.2	86.0	43.0 †
(1) Dunged fallow, (2) wheat, (3) wheat	82.8	87.4	4.6	1.5
Lucerne, five years	224.0	1078.0	854.0	170.8

* Catch crop, i.e. taken in autumn after the wheat.
† This crop does not belong to the Leguminosae; but it is possible that the nitrogen came from the soil, and that impoverishment was going on.

(Table 1.2), which show a marked gain of nitrogen when the newer rotations are adopted, but not where wheat only is grown.

Now the rotation has not impoverished the soil, hence he concludes that 'l'azote peut entrer directement dans l'organisme des plantes, si leur parties vertes sont aptes à le fixer'. Boussingault's work covers the whole range of agriculture and deals with the composition of crops at different stages of their growth, with soils, and with problems in animal nutrition. Unfortunately the classic farm of Bechelbronn did not remain a centre of agricultural research and the experiments came to an end after the war of 1870. Some of the work was summarized by J. B. A. Dumas and Boussingault[28] in a very striking essay that has been curiously overlooked by agricultural chemists.

During this period (1830–40) Carl Sprengel was studying the ash constituents of plants, which he considered were probably essential to nutrition.[29] Schübler was working at soil physics, and a good deal of other work was quietly being done. No particularly important discoveries were being made, no controversies were going on, and no great amount of interest was taken in the subject.

But all this was changed in 1840 when Liebig's famous report to the British Association[30] upon the state of organic chemistry, published as *Chemistry in its Application to Agriculture and Physiology* in 1840, came like a thunderbolt upon the world of science. With polished invective and a fine sarcasm he holds up to scorn the plant physiologists of his day for their continued adhesion, in spite of accumulated evidence, to the view that plants derive their carbon from the soil and not from the carbonic acid of the air. 'All explanations of chemists must remain without fruit, and useless, because, even to the great leaders in physiology, carbonic acid, ammonia, acids and bases are sounds without meaning, words without sense, terms of an unknown language, which awake no thoughts and no associations.' The experiments quoted by the physiologists in support of their view are all 'valueless for the decision of any question'. 'These experiments are considered by them as convincing proof, while they are fitted only to awake pity.' Liebig's ridicule did what neither de Saussure's nor Boussingault's logic had done: it finally killed the humus theory. Only the boldest would have ventured after this to assert that plants derive their carbon from any source other than carbon dioxide, although it must be admitted that we have no proof that plants really do obtain all their carbon in this way. Thirty years later, in fact, L. Grandeau[31] adduced evidence that humus may, after all, contribute something to the carbon supply, and his view found some accept-

---

28 *Essai de Statique Chimique des Êtres Organisés*, Paris, 1841.
29 *Chemie für Landwirthe, Forstmänner und Cameralisten*, Göttingen, 1832.
30 There is no record of this Report ever having been presented to the Association.
31 *Comp. Rend.*, 1872, **74**, 988; *Publication de la Station Agronomique de l'Est*, 1872.

ance in France;[32] for this also, however, convincing proof is lacking. But for
the time carbon dioxide was considered to be the sole source of the carbon
of plants. Hydrogen and oxygen came from water, and nitrogen from
ammonia. Certain mineral substances were essential: alkalis were needed for
neutralization of the acids made by plants in the course of their vital
processes, phosphates were necessary for seed formation, and potassium
silicates for the development of grasses and cereals. The evidence lay in the
composition of the ash: plants might absorb anything soluble from the soil,
but they excreted from their roots whatever was non-essential. The fact of
a substance being present was therefore sufficient proof of its necessity.

Plants, Liebig argued, have an inexhaustible supply of carbonic acid in the
air. But time is saved in the early stages of plant growth if carbonic acid is
being generated in the soil, for it enters the plant roots and affords extra
nutrient over and above what the small leaves are taking in. Hence a supply
of humus, which continuously yields carbonic acid, is advantageous. Further,
the carbonic acid attacks and dissolves some of the alkali compounds of the
soil and thus increases the mineral food supply. The true function of humus
is to evolve carbonic acid.

The alkali compounds of the soil are not all equally soluble. A weathering
process has to go on, which is facilitated by liming and cultivation, whereby
the comparatively insoluble compounds are broken down to a more soluble
state. The final solution is effected by acetic acid excreted by the plant roots,
and the dissolved material now enters the plant.

The nitrogen is taken up as ammonia, which may come from the soil, from
added manure, or from the air. In order that a soil may remain fertile it is
necessary and sufficient to return in the form of manure the mineral constituents and the nitrogen that have been taken away. When sufficient crop
analyses have been made it will be possible to draw up tables showing the
farmer precisely what he must add in any particular case.

An artificial manure known as Liebig's patent manure was made up on
these lines and placed on the market.

Liebig's book was meant to attract attention to the subject, and it did; it
rapidly went through several editions, and as time went on Liebig developed
his thesis, and gave it a quantitative form: 'The crops on a field diminish or
increase in exact proportion to the diminution or increase of the mineral
substances conveyed to it in manure.' He further adds what afterwards
became known as the Law of the Minimum, 'by the deficiency or absence
of *one* necessary constituent, all the others being present, the soil is rendered
barren for all those crops to the life of which *that one* constituent is indispensable'. These and other amplifications in the third edition, 1843, gave rise

32 See for example Cailletet L., *Comp. Rend.*, 1911, **152**, 1215, Lefèvre J., *op. cit.* 1905,
   141, 211, and Laurent J., *Rev. gén. bot.*, 1904, **16**, 14.

to much controversy. So much did Liebig insist, and quite rightly, on the necessity for alkalis and phosphates, and so impressed was he by the gain of nitrogen in meadow land supplied with alkalis and phosphates alone, and by the continued fertility of some of the fields of Virginia and Hungary and the meadows of Holland, that he began more and more to regard the atmosphere as the source of nitrogen for plants. Some of the passages of the first and second editions urging the necessity of ammoniacal manures were deleted from the third and later editions. 'If the soil be suitable, if it contain a sufficient quantity of alkalis, phosphates, and sulphates, nothing will be wanting. The plants will derive their ammonia from the atmosphere as they do carbonic acid', he writes in the *Farmer's Magazine*.[33] Ash analysis led him to consider the turnip as one of the plants 'which contain the least amount of phosphates and therefore require the smallest quantity for their development'. These and other practical deductions were seized upon and shown to be erroneous by Lawes and Gilbert,[34] who had for some years been conducting vegetation experiments. Lawes does not discuss the theory as such, but tests the deductions Liebig himself draws and finds them wrong. Further trouble was in store for Liebig; his patent manure when tried in practice *had failed*. This was unfortunate, and the impression in England at any rate was, in Philip Pusey's words: 'The mineral theory, too hastily adopted by Liebig, namely, that crops rise and fall in direct proportion to the quantity of mineral substances present in the soil, or to the addition or abstraction of these substances which are added in the manure, has received its death-blow from the experiments of Mr Lawes.'

And yet the failure of the patent manure was not entirely the fault of the theory, but only affords further proof of the numerous pitfalls of the subject. The manure was sound in that it contained potassium compounds and phosphates (it ought, of course, to have contained nitrogen compounds), but it was unfortunately rendered insoluble by fusion with lime and calcium phosphate so that it should not too readily wash out in the drainage water. Not till Way had shown in 1850[35] that *soil precipitates soluble salts of ammonium, potassium and phosphates* was the futility of the fusion process discovered, and Liebig[36] saw the error he had made.

Meanwhile the great field experiments at Rothamsted had been started by Lawes and Gilbert in 1843. These experiments were conducted on the same general lines as those begun earlier by Boussingault, but they have the advantage that they are still going on, having been continued year after year on the same ground without alteration, except in occasional details, since

---

33 *Farmer's Magazine*, 1847, **16**, 511. A good summary of Liebig's position is given in his *Familiar Letters on Chemistry* (3rd edn), 1851, 34th letter, p. 519.
34 Lawes, J. B. and Gilbert, J. H., *J. R. agric. Soc. Eng.*, 1847, **8**, 226; 1851, **12**, 1; 1855, **16**, 411.
35 Way, J. T., *J. R. agric. Soc. Eng.*, 1850, **11**, 313; 1852, **13**, 123.
36 *Familiar Letters on Chemistry* (3rd edn), London, 1851.

1852. The mass of data now accumulated is considerable and it is being treated by modern statistical methods. Certain conclusions are so obvious, however, that they can be drawn on mere inspection of the data. By 1855 the following points were definitely settled:[37]

1 Crops require phosphates and salts of the alkalis, but the composition of the ash does not afford reliable information as to the amounts of each constituent needed, e.g. turnips require large amounts of phosphates, although only little is present in their ash. Some of the results are:

Composition of ash, per cent (1860 crop)		Yield of turnips, tons per acre* (1843)	
K_2O	44.8	Unmanured	4.5 (11.3)
P_2O_5	7.9	Superphosphate	12.8 (32.1)
		Superphosphate + potassic salts	11.9 (29.9)

* In parentheses: t ha^{-1}

2 Non-leguminous crops require a supply of some nitrogenous compounds, nitrates and ammonium salts being almost equally good. Without an adequate supply no increases of growth are obtained, even when ash constituents are added. The amount of ammonia obtainable from the atmosphere is insufficient for the needs of crops. Leguminous crops behave abnormally.
3 Soil fertility may be maintained for some years at least by means of artificial manures.
4 The beneficial effect of fallowing lies in the increase brought about in the available nitrogen compounds in the soil.

Although many of Liebig's statements were shown to be wrong, the main outline of his theory as first enunciated stands. It is no detraction that de Saussure had earlier published a somewhat similar, but less definite view of nutrition: Liebig had brought matters to a head and made men look at their cherished, but unexamined, convictions. The effect of the stimulus he gave can hardly be over-estimated, and before he had finished, the essential facts of plant nutrition were settled and the lines were laid down along which scientific manuring was to be developed. The water cultures of Knop and other plant physiologists showed conclusively that potassium, magnesium, calcium, iron, phosphorus, along with sulphur, carbon, nitrogen, hydrogen

37 Lawes and Gilbert's papers are collected in ten volumes of *Rothamsted Memoirs*, and the general results of their experiments are summarized by Hall, A. D. in *The Book of the Rothamsted Experiments*. A detailed investigation of the early experiments of Lawes in their relation to the discovery of superphosphate has been made by Max Speter in *Superphosphate*, 1935, **8**.

and oxygen are all necessary for plant life. The list differs from Liebig's only in the addition of iron and the withdrawal of silica; but even silica, although not strictly essential, is advantageous for the nutrition of cereals.

In two directions, however, the controversies went on for many years. Farmers were slow to believe that 'chemical manures' could ever do more than stimulate the crop, and declared they must ultimately exhaust the ground. The Rothamsted plots falsified this prediction; manured year after year with the same substances and sown always with the same crops, they even now, after more than a hundred years of chemical manuring, continue to produce good crops, although secondary effects have sometimes set in. In France the great missionary was Georges Ville,[38] whose lectures were given at the experimental farm at Vincennes during 1867 and 1874–75. He went even further than Lawes and Gilbert, and maintained that artificial manures were not only more remunerative than dung, but were the only way of keeping up fertility. In recommending mixtures of salts for manure he was not guided by ash analysis but by field trials. For each crop one of the four constituents, nitrogen compounds, phosphates, lime and potassium compounds (he did not consider it necessary to add any others to his manures) was found by trial to be more wanted than the others and was therefore called the 'dominant' constituent. Thus for wheat he obtained the following results, and therefore concluded that on his soil wheat required a good supply of nitrogen, less phosphate, and still less potassium:

Constituent	Crop per acre bushels*	
Normal manure	43	(2.98)
Manure without lime	41	(2.84)
Manure without potash	31	(2.14)
Manure without phosphate	26.5	(1.83)
Manure without nitrogen	14	(0.97)
Soil without manure	12	(0.83)

* In parentheses: t ha^{-1}, for 1 bushel = 28 kg

Other experiments of the same kind showed that nitrogen was dominant for all cereals and beetroot, potassium for potatoes and vines, phosphate for turnips and swedes. An excess of the dominant constituent was always added to the crop manure. The composition of the soil had to be taken into account, but soil analysis was no good for the purpose. Instead he drew up a simple scheme of plot trials to enable farmers to determine for themselves

38 *On Artificial Manures, Their Chemical Selection and Scientific Application to Agriculture.* Trans. by Crookes, W., London, 1879.

just what nutrient was lacking in their soil. His method was thus essentially empirical, but it still remains the best we have; his view that chemical manures are always better and cheaper than dung is, however, too narrow and has not survived.

The second controversy dealt with the source of nitrogen in plants. Priestley had stated that a plant of *Epilobium hirsutum* placed in a small vessel absorbed during the course of the month seven-eighths of the air present. De Saussure, however, denied that plants assimilated gaseous nitrogen. J. B. Boussingault's pot experiments[39] showed that peas and clover could get nitrogen from the air while wheat could not, and his rotation experiments emphasized this distinction. He himself did not make as much of this discovery as he might have done, but later[40] fully realized its importance.

Liebig, as we have seen, maintained that ammonia, but not gaseous nitrogen, was taken up by plants, a view confirmed by Lawes, Gilbert and Pugh[41] in the most rigid demonstration that had yet been attempted. Plants of several natural orders, including the Leguminosae, were grown in surroundings free from ammonia or any other nitrogen compound. The soil was burnt to remove all traces of nitrogen compounds, while the plants were kept throughout the experiment under glass shades, but supplied with washed and purified air and with pure water. In spite of the ample supply of mineral food the plants languished and died: the conclusion seemed irresistible that plants could not utilize gaseous nitrogen. For all non-leguminous crops this conclusion agreed with the results of field trials. But there remained the very troublesome fact that leguminous crops required no nitrogeous manure and yet they contained large quantities of nitrogen, and also enriched the soil considerably in this element. Where then had the nitrogen come from? The amount of combined nitrogen brought down by the rain was found to be far too small to account for the result. For years experiments were carried on, but the problem remained unsolved. Looking back over the papers[42] one can see how very close some of the older investigators were to the discovery of the cause of the mystery: in particular Lachmann[43] carefully examined the structure of the nodules, which he associated with the nutrition of the plant, and showed that they contained 'vibrionenartige' organisms. His paper, however, was published in an obscure journal and attracted little attention. Atwater in 1881 and 1882 showed that peas acquired large quantities of nitrogen from the air, and later suggested that they might 'favour the action

39 Boussingault, J. B., *Ann. Chim. Phys.*, 1838 (II), **67**, 5; **69**, 353; 1856 (III), **46**, 5.
40 Dumas J. B. A. and Boussingault, J. B., *Essai de Statique Chimique des Êtres Organisés*, Paris, 1841.
41 Lawes, J. B. *et al.*, *Phil. Trans.*, 1861, **151**, 431; 1889, **180**A, 1; *J. R. agric. Soc. Eng.*, 1891, ser. 3, **2**, 657.
42 A summary of the voluminous literature is contained in Löhnis's *Handbuch der landw. Bakteriologie*, pp. 646 *et seq.*
43 Lachmann, J., *Mitt. Landw. Lehranst.*, Poppelsdorf, 1858, **1**, Reprinted in *Zbl. Agrik. Chem.*, 1891, **20**, 837.

of nitrogen-fixing organisms'.[44] But he was too busily engaged to follow the matter up, and once again an investigation in agricultural chemistry had been brought to a standstill for want of new methods of attack.

The beginnings of soil bacteriology

It had been a maxim with the older agricultural chemists that 'corruption is the mother of vegetation'. Animal and vegetable matter had long been known to decompose with formation of nitrates: indeed nitre beds made up from such decaying matter were the recognized source of nitrates for the manufacture of gunpowder during the European wars of the seventeenth and eighteenth centuries.[45] No satisfactory explanation of the process had been offered, although the discussion of rival hypotheses continued up till 1860, but the conditions under which it worked were known and on the whole fairly accurately described.

No connection was at first observed between nitrate formation and soil productiveness. Liebig[46] rather diverted attention from the possibility of tracing what now seems an obvious relationship by regarding ammonia as the essential nitrogenous plant nutrient, though he admitted the possible suitability of nitrates. Way came much nearer to the truth. In 1856 he showed that nitrates were formed in soils to which nitrogenous fertilizers were added. Unfortunately he failed to realize the significance of this discovery. He was still obsessed with the idea that ammonia was essential to the plant, and he believed that ammonia, unlike other nitrogen compounds, could not change to nitrate in the soil, but was absorbed by the soil by the change he had already described (p. 14). But he only narrowly missed making an important advance in the subject, for after pointing out that nitrates are comparable with ammonium salts as fertilizers he writes: 'Indeed the French chemists are going further, several of them now advocating the view that it is in the form of nitric acid that plants make use of compounds of nitrogen. With this view I do not myself at present concur: and it is sufficient here to admit that nitric acid in the form of nitrates has at least a very high value as a manure.'

It was not till ten years later, and as a result of work by plant physiologists, that the French view prevailed over Liebig's, and agricultural investigators recognized the importance of nitrates to the plant and of nitrification to soil fertility. It then became necessary to discover the cause of nitrification.

During the 1860s and 1870s great advances were being made in bacteri-

44 Atwater, W. O., *Am. Chem. J.*, 1885, **6**, 365; **8**, 327.
45 *Instructions sur l'Établissement des Nitrières, Publié par les Régisseurs Généraux des Poudres et Salpètre*, Paris, 1777.
46 *Principles of Agricultural Chemistry with Special Reference to the Late Researches Made in England*, London, 1855.

ology, and it was definitely established that bacteria bring about putrefaction, decomposition and other changes; it was therefore conceivable that they were the active agents in the soil, and that the process of decomposition there taking place was not the purely chemical 'eremacausis' Liebig had postulated. Pasteur himself had expressed the opinion that nitrification was a bacterial process. The new knowledge was first brought to bear on agricultural problems by Th. Schloesing and A. Müntz[47] during a study of the purification of sewage water by land filters. A continuous stream of sewage was allowed to trickle down a column of sand and limestone so slowly that it took eight days to pass. For the first 20 days the ammonia in the sewage was not affected, then it began to be converted into nitrate; finally all the ammonia was converted during its passage through the column, and nitrates alone were found in the issuing liquid. Why, asked the authors, was there a delay of 20 days before nitrification began? If the process were simply chemical, oxidation should begin at once. They therefore examined the possibility of bacterial action and found that the process was entirely stopped by a little chloroform vapour, but could be started again after the chloroform was removed by adding a little turbid extract of dry soil. Nitrification was thus shown to be due to micro-organisms – 'organized ferments', to use their own expression.

Warington[48] had been investigating the nitrates in the Rothamsted soils, and at once applied the new discovery to soil processes. He showed that nitrification in the soil is stopped by chloroform and carbon disulphide; further, that solutions of ammonium salts could be nitrified by adding a trace of soil. By a careful series of experiments described in his four papers to the Chemical Society he found that there were two stages in the process and two distinct organisms: the ammonia was first converted into nitrite and then to nitrate. But he failed altogether to obtain the organisms, in spite of some years of study, by the gelatin methods then in vogue. However, Winogradsky,[49] in a brilliant investigation, isolated these two groups of organisms, showing they were bacteria. He succeeded where Warington failed because he realized that carbon dioxide should be a sufficient source of carbon for them, so that they ought to grow on silica gel plates carefully freed from all organic matter; and it was on this medium that he isolated them in 1890.

Warington also established definitely the fact that nitrogen compounds rapidly change to nitrate in the soil, so that whatever compound is supplied as manure, plants get practically nothing but nitrate as food. This closed the long discussion as to the nitrogenous food of non-leguminous plants; in natural conditions they take up nitrate only (or at any rate chiefly), because the activities of the nitrifying organisms leave them no option. The view that

47 Schloesing, Th. and Müntz, A., *Comp. Rend.*, 1877, **84**, 301; **85**, 1018; 1878, **86**, 892.
48 Warington, R., *J. Chem. Soc.*, 1878, **33**, 44; 1879, **35**, 429; 1884, **45**, 637; 1891, **59**, 484.
49 Winogradsky, S., *Ann. Inst. Pasteur*, 1890, **4**, 213, 257, 760.

plants assimilate gaseous nitrogen has from time to time been revived,[50] but it is not generally accepted.

The apparently hopeless problem of the nitrogen nutrition of leguminous plants was soon to be solved. In a striking series of sand cultures Hellriegel and Wilfarth[51] showed that the growth of non-leguminous plants, barley, oats, etc., was directly proportional to the amount of nitrate supplied, the duplicate pots agreeing satisfactorily; while in the case of leguminous plants no sort of relationship existed and duplicate pots failed to agree. After the seedling stage was passed the leguminous plants grown without nitrate made no further progress for a time, then some of them started to grow and did well, while others failed. This stagnant period was not seen where nitrate was supplied. Two of their experiments are given in Table 1.3.

Table 1.3   Relation between nitrogen supply and plant growth

Nitrogen in the calcium nitrate supplied per pot, g	none	0.056	0.112	0.168	0.224	0.336
Weight of oats obtained (grain and straw)	0.361 0.419	5.902 5.851 5.287	10.981 10.941	15.997	21.273 21.441	30.175
Weight of peas obtained (grain and straw)	0.551 3.496 5.233	0.978 1.304 4.128	4.915 9.767 8.497	5.619	9.725 6.646	11.352

Analysis showed that the nitrogen contained in the oat crop and sand at the end of the experiment was always a little less than that originally supplied, but was distinctly greater in the case of peas; the gain in three cases amounted to 0.910, 1.242 and 0.789 g per pot respectively. They drew two conclusions: (1) the peas took their nitrogen from the air; (2) the process of nitrogen assimilation was conditioned by some factor that did not come into their experiment except by chance. In trying to frame an explanation they connected two facts that were already known. Berthelot[52] had made experiments to show that certain micro-organisms in the soil can assimilate gaseous nitrogen. It was known to botanists that the nodules on the roots of Leguminosae contained bacteria.[53] Hellriegel and Wilfarth, therefore, supposed that

50 For example, Pfeiffer, Th., Franke, E., *Landw. Vers.-Stat.*, 1896, **46**, 117; Jamieson, T., *Aberdeen Res. Assoc. Repts.*, 1905–08; Lipman C. B. and Taylor, J. K., *J. Franklin Inst., Calif.*, 1924, p. 475.
51 Hellriegel, H. and Wilfarth, H., *Ztschr. Rübenzucker-Ind.*, Beilageheft, 1888.
52 Bertholet, M., *Comp. Rend.*, 1885, **101**, 775.
53 This had been demonstrated by Lachmann (p. 17) and by Woronin, M., *Mem. Acad. Sci.*, St Petersburg, 1866, ser. 7, **10**, No. 6. Eriksson, J., in 1874 Doctor's dissertation, abs. in *Botan. Ztg.*, 1874, **32**, 381 carried on the investigation, while Brunchorst, G., in 1885 *Ber. Deut. Bot. Ges.*, **3**, 241 gave the name 'bacteroids'.

the bacteria in the nodules assimilated gaseous nitrogen, and then handed on some of the resulting nitrogenous compounds to the plant. This hypothesis was shown to be well founded by the following facts:

1 In absence of nitrate peas made only small growth and developed no nodules in sterilized sand; when calcium nitrate was added they behaved like oats and barley, giving regular increases in crop for each increment of nitrate (the discordant results of Table 1.3 were obtained on unsterilized sand).

2 They grew well and developed nodules in sterilized sand watered with an extract of arable soil.

3 They sometimes did well and sometimes failed when grown without soil extract and without nitrate in *unsterilized* sand, which might or might not contain the necessary organisms. An extract that worked well for peas might be without effect on lupins or serradella. In other words, the organism is specific.

Hellriegel and Wilfarth read their paper and exhibited some of their plants at the Naturforscher-Versammlung at Berlin in 1886. Gilbert was present at the meeting, and on returning to Rothamsted repeated and confirmed the experiments. At a later date Schloesing *fils* and Laurent[54] showed that the weight of nitrogen absorbed from the air was approximately equal to the gain by the plant and the soil, and thus finally clinched the evidence:

	Control	Peas	Mustard	Cress	Spurge
Nitrogen lost from the air, mg	1.0	134.6	−2.6	−3.8	−2.4
Nitrogen gained by crop and soil, mg	4.0	142.4	−2.5	2.0	3.2

The organism was isolated by Beijerinck[55] and called *Bacillus radicicola*, but is now known as *Rhizobium*.

Thus another great controversy came to an end, and the discrepancy between the field trials and the laboratory experiments of Lawes, Gilbert and Pugh was cleared up. The laboratory experiments gave the correct conclusion that leguminous plants, like non-leguminous plants, have themselves no power of assimilating gaseous nitrogen; this power belongs to the bacteria associated with them. But so carefully was all organic matter removed from the soil, the apparatus and the air in endeavouring to exclude all trace of ammonia, that there was no chance of infection with the necessary bacteria.

54 Schloesing, Th. and Laurent, E., *Ann. Inst. Pasteur*, 1892, **6**, 65.
55 Beijerinck, M. W., *Bot. Ztg.*, 1888, **46**, 725, 741, 757; 1890, **48**, 837.

Hence no assimilation could go on. In the field trials the bacteria were active, and here there was a gain of nitrogen.

The general conclusion that bacteria are the real makers of plant food in the soil, and are, therefore, essential to the growth of all plants, was developed by Wollny[56] and Berthelot.[57] It was supposed to be proved by Laurent's[58] experiments. He grew buckwheat on humus, obtained from well-rotted dung, and found that plants grew well on the untreated humus, but only badly on the humus sterilized by heat. When, however, soil bacteria were added to the sterilized humus (by adding an aqueous extract of unsterilized soil) good growth took place. The experiment looks convincing, but is really unsound. When a rich soil is heated some substance is formed toxic to plants. The failure of the plants on the sterilized humus was, therefore, not due to absence of bacteria, but to the presence of a toxin. No one has yet succeeded in carrying out this fundamental experiment of growing plants in two soils differing only in that one contains bacteria while the other does not.

The development and application of modern knowledge of soils*

Our understanding of the physical, chemical and biological factors that control the fertility of soils has advanced greatly in this century. The application of this knowledge has resulted in great increases in productivity. The most useful gains have been from new understanding of the storage and movement of water in soils, the value and valuation of reserves of plant nutrients, the physical, chemical and biological conditions in the rhizosphere and from the role of cultivations in modern production systems. Both research and its application have been greatly aided by developments in the basic sciences, in optical and electronic instruments, and in computers. Experimental work in the field with crops always provides the practical assessment of advances in basic soil studies. Here the new experimental designs initiated by R. A. Fisher have been invaluable by providing methods of solving problems in soil management and cropping systems which could not be attempted before. The development of statistical methods of assessing error in biological experiments was also a major advance. Field experimentation became a major technique in research on soil fertility and, because the

56 Wollny, E., *Bied. Zbl. Agric. Chem.*, 1884, **13**, 796.
57 Berthelot, M., *Comp. Rend.*, 1888, **106**, 569.
58 Laurent, E., *Bull. Acad. R. Belgique*, 1886, **2**, 128. See also Duclaux, E., *Comp. Rend.*, 1885, **100**, 66.

* Contributed by G. W. Cooke.

precision of the results could be estimated, they were readily accepted by other scientists. Designs in which several factors were tested simultaneously and their interactions were measured, were an important advance. They opened the way to multidisciplinary research, for example the effects of biological, physical and management factors on crop nutrition can now be investigated.

The development of research on soils, and its application, has been aided in many countries by organizational advances. In Britain the Agricultural Research Council was established in 1931 to provide guidance and support for the various branches of research in agriculture and to coordinate programmes. The application of this research to develop agriculture was the responsibility of scientists attached to universities and agricultural colleges until 1946; in that year the Ministry of Agriculture and Fisheries established the National Agricultural Advisory Service (NAAS) – which later became the Agricultural Development and Advisory Service (ADAS).

The close collaboration established between the research and the development scientists has advanced the work of both services and led to considerable improvements in agricultural productivity. Similar developments involving the setting up of national institutions have occurred in many other countries. On the international scale the most important advance has been the establishment of the chain of institutes sponsored and supervised by the Consultative Group for International Agricultural Research (CGIAR) to aid food production in developing countries. An example of a well-known institute whose workers have added much to our scientific knowledge of soils that are managed for flooded rice production is the International Rice Research Institute (IRRI) established at Manila in the Philippines in 1960.

The Food and Agriculture Organization of the United Nations (FAO) was established in 1945 with the high ideal of freeing the people of the world from hunger by improving both agriculture and the distribution of the food produced. It has had a vital role in assessing and reporting on the food situation in various parts of the world and indicating how progress may be made by research and development work on agricultural systems, and by solving the social and economic problems which often lead to shortages. FAO has concentrated on developing natural resources and improving their utilization to increase food production in developing countries. It has sponsored scientific work on soils and on their improvement and conservation by the use of fertilizers and better management. A notable activity has been the Fertilizer Programme established in 1960; experiments made in many regions which measured the effects of fertilizers on farmers' crop yields have led to improved recommendations for efficient and economic use of the fertilizers that are available.

Advances in soil science

The development of our knowledge of various aspects of soil science will be clear from the succeeding chapters of this volume. Important advances made in this century have resulted from the application of new methodologies such as X-ray analysis which led to our understanding of the crystalline structures of layered aluminosilicate minerals, and from spectrographic analyses which provided much information on the distribution of micronutrients.

Work on the forms of nutrient reserves in soils, their mobilities and availabilities, and the fate of nutrients applied in fertilizers, has continued through this century. Work on phosphorus was advanced when the radioisotope ^{32}P became available. Other advances were made by applying thermodynamic concepts to the solubilities of nutrient ions. Understanding of cation relationships took a notable step forward when Schofield's Ratio Law was proposed in 1947; later advances came from the use of the Quantity (Q) and Intensity (I) factors. Concepts of nutrient ion mobility, first developed in the USA, have been used in the mathematical modelling of processes of nutrient uptake by Nye and Tinker.[59] These models of nutrient processes in soil will in future become the basis for management of nutrients in farming systems.

Much research has been done on the role of nitrogen in soil/crop systems. This is justifiable as, in many conditions, the supply of nitrogen has a greater effect on crop performance than do the supplies of other nutrients, therefore on the world scale this nutrient dominates fertilizer markets. In years past it was generally admitted that no more than 30 per cent of the nitrogen applied was, on average, taken up by the crops grown. The 70 per cent that was not recovered represented a serious loss to farmers and was also the cause of environmental pollution – nitrate leached into waters used for public supply and nitrous oxide formed by denitrification was thought to damage the ozone layer of the stratosphere. Improved technologies, including the use of the stable isotope ^{15}N, have made it possible to show in experiments that totals of 90 per cent or more of the nitrogen applied as fertilizer can be accounted for in uptake by the crop plus that nitrogen stored in the soil which may benefit future crops. It does seem that we are in sight of securing much higher efficiency of fertilizer nitrogen applied to crops, ranging from wheat in Europe to rice in Southeast Asia. The investigations described above of the roles of micro-organisms in the fixation of nitrogen by leguminous crops have led to methods of preparing cultures of the organisms (*Rhizobium* spp.) which are specifically associated with particular crops. These cultures have been made available to farmers for inoculating crops which are to be grown on soils where the appropriate species of *Rhizobium* is lacking.

The management of other major nutrients, notably phosphorus and potas-

59 Nye, P. H. and Tinker, P. B., *Solute Movement in the Soil-Root System*, Oxford, Blackwell, 1977.

sium but also calcium and magnesium, has been greatly aided by studies of the soils and crops in long-term experiments. The subject was well reviewed in an international conference in 1976.[60] The classical experiments at Rothamsted begun by Lawes in the last century have been invaluable in these studies; they have shown that reserves of phosphate and potassium accumulated in soils from fertilizer additions have considerable value in crop production. Long-term experiments also provide the best basis for relating soluble nutrients in soils to crop performance and to the need for fertilizers; they also lead to calculations of nutrient cycles which are essential for the efficient management of crop nutrition.

The identification of nutrient deficiencies and the need for fertilizers were considerably advanced by studies of deficiency symptoms in plants; this work was pioneered in the 1930s at Long Ashton Research Station. It is still of much value to workers who are concerned with intensifying farming in areas of the world where there is little information on soils or on the requirements of crops for both macro- and micronutrients.

The management of soil has been much improved as a result of scientific work. The cultivations necessary for good crop growth have been defined; minimum cultivation systems lessen the energy required and conserve soil structure. Erosion of the cultivated layer of soil by water, and by wind, is a serious threat to efficient agriculture in many parts of the world. Studies of the mode and extent of losses of soil under practical conditions have led to recommendations for improved management of cropping systems which avoid these losses. Another serious hazard is the damage done to plant growth, and also to soil structure, by the salts which accumulate in saline soils. Salinity may occur naturally, or it may be the result of irrigating with unsuitable water. Investigations of this problem have led to definitions of the water quality that is required to remove soluble salts in drainage and to the use of gypsum for reclaiming saline soils. With good quality irrigation water and correctly managed drainage salinity need not now be a problem for efficient crop production.

Soil surveys

The first proposals for the classification and mapping of soils were made by Russian workers about 100 years ago. Surveys were made in Southeast England at the end of last century, other local surveys were made early in this century. Now soil maps on the scale of 1 : 250 000 are available for all of Great Britain and land capability maps have been derived.

The importance of soil surveys for 'technology transfer' cannot be overemphasized. We require correct identification of the nature of soil

60 Institut National de la Recherche Agronomique, *Annal. Agron.*, 1976, **27** (5–6).

profiles and their physical and chemical properties, as affected by the nature of the parent materials and the genetic processes which formed them.

With such information the results of experimental work on one site may be transferred to a distant site where the soil classification is the same. In addition to the soil maps advisers require computer-based information services giving the capability for cropping and physical, chemical and biological properties of the soil. Such information is becoming essential to modern management systems which aim to control productivity of soil. On a world scale the most important tool for technology transfer is undoubtedly the *Soil Taxonomy* system developed in the USA.[61]

Fertilizers

The use of fertilizers has expanded greatly in this century in Britain and in most other countries and this has resulted in large increases in crop yield. These increases have resulted from the application by farmers' advisers of the research work which has identified the deficiencies of the main nutrients in soils and crops. The amounts used at intervals during the period from 1913 to 1984 are shown in Table 1.4.

Table 1.4   The fertilizers used in the world and the UK in specified years from 1913 to 1984 (thousands of tonnes)

	N	*P$_2$O$_5$*	*K$_2$O*
World use			
1913	1 300	2 000	900
1939	2 600	3 600	2 800
1960	9 700	9 700	8 600
1970	28 700	18 800	15 500
1983	61 021	30 833	22 844
1984	66 907	32 856	25 408
United Kingdom use			
1913	29	183	23
1939	61	173	76
1950	229	468	238
1969	803	484	465
1982	1 416	446	483
1984	1 588	488	559

61 Soil Survey Staff. *Soil Taxonomy: a basic system of soil classification for making and interpreting soil surveys*, Agricultural Handbook No. 436, USDA, US Government Printing Office, Washington DC, 1975.

Both in the world as a whole, and in the UK, the amounts of nitrogen used dominate the totals; this nutrient is responsible for the major part of the cost of fertilizers to the world's farmers (in the UK about three-quarters of the total spent on fertilizers is for nitrogen). Table 1.4 also shows the very great increases in the nutrients applied which have occurred in the last 45 years. The extent to which farmers change their fertilizer practice on particular crops is shown by *Surveys of Fertiliser Practice* which were initiated in the 1940s and still continue. Similar surveys are made in a few other countries. Some results of surveys in England and Wales made during the last 40 years are given in Table 1.5.

Table 1.5   Average amount of fertilizers used by farmers in England and Wales in selected years on winter wheat, maincrop potatoes, and temporary grass leys (kilograms per hectare)
(Data from *Surveys of Fertiliser Practice*)

	Winter wheat			*Maincrop potatoes*			*Temporary grassland*		
	N	P_2O_5	K_2O	N	P_2O_5	K_2O	N	P_2O_5	K_2O
1943/5	19	30	2	79	92	100	—	—	—
1950/2	33	28	15	117	124	166	16	35	15
1966	90	44	44	161	173	241	66	50	30
1979	135	46	38	193	195	257	173	33	37
1984	187	56	53	214	228	278	190	33	51

The great increases in the amounts of nitrogen used on all three crops are very striking; relatively the increases in the phosphorus used are much less marked; there have also been considerable increases in the potassium applied. Such surveys are very useful in indicating whether current fertilizer practices are deficient or adequate; the amounts farmers apply are assessed against the background of current information on crop needs for each nutrient, on soil properties, and on the farming system followed, as this determines the return of nutrients to the soil in crop wastes or in manures made on the farm. The data in Table 1.5 emphasize the large expenditure on fertilizers incurred by farmers in developed countries such as the UK. This cost to crop production justifies good advisory services supported by current research on the farming systems based on the results of long-term experiments sited on the important soil types in a country.

It is essential to ensure that all recommended inputs to farming systems are cost effective, the increase in yield that each produces being more than sufficient to pay for the input. This is particularly important with fertilizers which usually are responsible for the major part of the variable costs of production.

Improvements in agricultural productivity

The average yields of crops have increased in most countries in this century and particularly in the last 40 years. The recent increases have resulted from the breeding of improved varieties, from measures to control pests, diseases and weeds, and from improved soil management practices (including drainage and cultivations); but the most important improvements have certainly resulted from the increased use of fertilizers which have rectified deficiencies of nutrients in soils and have supported the increased growth that has been made possible by the improvement in varieties and practices noted above. Examples of the increased productivity achieved during the last century are given in Table 1.6.

Table 1.6   Average yields of crops in the UK for three-year periods from 1885–87 to 1981–83 (*All yields are in tonnes per hectare*)

	1885–87	*1911–13*	*1937–39*	*1951–53*	*1961–63*	*1971–73*	*1981–83*
Wheat, grain	2.0	2.2	2.3	2.9	3.8	4.3	6.1
Barley, grain	2.0	1.9	2.1	2.7	3.4	3.9	4.6
Oats, grain	1.6	1.7	2.0	2.4	2.7	3.8	4.4
Potatoes, tubers	15	15	18	20	22	29	33
Sugar beet, roots	—	—	21	28	31	39	42
Tomatoes (under glass)	—	—	84	84	86	109	152

Yields of the four crops for which we have records changed little between 1885–87 and 1937–39; since this latter date there have been very large increases; roughly we can say that yields have, on average, doubled in the last 45 years. The most marked increases have been in wheat yields which have benefited from much research in several branches of agricultural science. But it is fair to claim that much credit for all these increases in yields must be given to the application of the results of research on the constraints which soils have imposed on crop growth and to the development of materials and practices which overcome these constraints.

The needs for future research on soil conditions in relation to plant growth

All agricultural production has its origin in plants which grow in soils. The ultimate potential yield is set by the genetic capacity of the crop which is grown and the solar radiation which it receives at the site. These potentials have been established by models based on the physiology of plants and many have now been confirmed by field experiments on farms. Average yields throughout the world, as recorded by FAO, are far below the recognized potentials because a variety of constraints limit crop growth. Some constraints result from attacks by pests and diseases but most are associated with soil conditions. Chemical constraints may be due to shortages of nutrients or the presence of toxins; physical constraints arise from shortages or excesses of water or from mechanical impedance to root growth. Scientific work to improve further agricultural production must therefore concentrate on the identification of constraints and the further development of inputs and/or management practices which overcome these constraints. The multifactorial field experiments which test these means of overcoming constraints must involve all the disciplines concerned with plant growth, and the objective must be to produce reliably the recognized potential yield at all sites that are investigated by the best combination of treatments. Successful work on wheat along these lines has been done in a project to investigate the causes of yield variability sponsored by the Agricultural Research Council; the experimental work has been described by Lester and Prew.[62] Extensions of this work to other crops, to other areas, and to other countries, are essential if the growing population of the world is to be assured of an adequate supply of food at prices people can afford. In these extensions soil science will always have a major role in first identifying, and then achieving, ideal physical conditions for root growth together with beneficial rhizosphere populations, while ensuring that nutrient and water supplies are adequate while toxins are absent.

In considering the nature of future experimental work we should note the advice given by Sir Humphrey Davy in the 1821 edition of his book, for this is still relevant. He discussed the need to substitute sound and rational principles for 'vague, popular prejudices . . . Nothing is more wanting in agriculture than experiments in which all the circumstances are minutely and scientifically detailed. The results of truly philosophical experiments in agricultural chemistry would be of more value in enlightening and benefiting the farmer than the greatest possible accumulation of imperfect trials conducted merely in the empirical spirit'.

The multidisciplinary experimentation in which soil science has a fundamental basic role requires adequate support from those responsible for agricultural improvement. It is therefore appropriate to quote from the writings of another giant of the last century: John Bennett Lawes concluded his paper[63] published in 1850 to report the first work ever done on the effect of plant nutrients on water use efficiency, with these words:

'The interest and progress of agriculture would be more surely and permanently served if its great patron Societies were to permit to their

62 Lester, E. and Prew, R. D., in *The Yield of Cereals*, p. 79, R. agric. Soc., Stoneleigh, 1983.
63 Lawes, J. B., *J. hort. Soc. Lond.*, 1850, **5**, 38.

scientific officers a wider range of discretion, and more liberal means for the selection and carrying out of definite questions of research. Results of this kind promise, it is true, but little prospect of immediate and practical application, but by their aid the uncertain dictates, whether of common experience, theory, or speculation, may, ere long, be replaced by the unerring guidance of principles; and then alone can it reasonably be anticipated that miscellaneous and departmental analyses may find their true interpretation, and acquire a due and practical value.'

2

Crop growth and development

The advances in crop nutrition outlined in the previous chapter were paralleled by growth in other aspects of crop science. At the same time as the main nutrients essential for plant growth were being identified, other scientists were looking for explanations as to how weather influenced growth and, particularly in the USA, they were making observations on the relation between growth and water supply; all this at a time when the essential features of genetic inheritance were also being determined. This chapter draws together some of these fields of study and describes the basic relations underlying crop responses to their edaphic and climatic environments. It deals almost entirely with whole plants or stands of crops and includes little about individual enzymes, hormones or the molecular basis of the phenomena described. For these, other texts should be consulted.[1]

1 For more comprehensive accounts of crop physiology, see the following texts: Evans, L. T., *Crop Physiology: some case histories*, Cambridge Univ. Press, 1975; Milthorpe, F. L. and Moorby, J., *An Introduction to Crop Physiology* (2nd edn), Cambridge Univ. Press, 1979; Street, H. E. and Opik, H., *The Physiology of Flowering Plants: their growth and development* (3rd edn), Arnold, 1984; Wilkins, M. B. (ed.), *Advanced Plant Physiology*, Pitman, 1984.

Plants are grown by man for a variety of products; some are for fuel and building materials, some for eating fresh, and others for eating after conversion into other products either by animals or by manufacturing processes. A wide range of plant types and species is used in these endeavours but this book refers predominantly to annual crops. In this chapter, the general principles of crop growth and development are described principally in relation to the aerial environment and the constraints that that environment places upon crops. Subsequent chapters will examine the constraints imposed by the soil.

Overall pattern of growth

The dry weight per unit area of an annual crop, or of individual plants, usually follows an approximately S-shaped course with time as shown in Fig. 2.1. As growth proceeds beyond the so-called 'lag' period, during which seedlings emerge from the soil, there is first a period of accelerating growth (sometimes referred to as the 'exponential' growth phase), followed by a period when the rate of growth is almost constant (the linear growth phase), and then a period when growth rates are close to zero, and ultimately, because of senescence or leaf abscission or both, growth declines and 'negative' growth is recorded. The change from exponential to linear growth is commonly associated with the closure of the crop canopy so that the interception of light then determines the rate of growth.

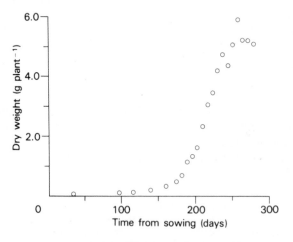

Fig. 2.1 Changes in shoot dry weight of field-grown winter wheat.

Rates of growth can be expressed in several ways[2] but the most common are the absolute growth rate, G, and the relative growth rate, R (sometimes termed the specific growth rate). The instantaneous absolute growth rate, G, is given by the slope of a graph of crop dry weight, W, against time t:

$$G = \frac{dW}{dt} \qquad [2.1]$$

or measured over a time interval (t_1 to t_2):

$$\bar{G} = \frac{W_2 - W_1}{t_2 - t_1} \qquad [2.2]$$

Absolute growth rates are most frequently used to compare treatments or species during the linear phase of growth. In many cases, though, and particularly during early growth, it is more logical to express the rate of growth relative to the amount of dry matter present. At any given instant, the relative growth rate, R, is given by:

$$R = \frac{1}{W} \cdot \frac{dW}{dt} = \frac{d \ln W}{dt} \qquad [2.3]$$

or averaged over a time interval (t_1 to t_2):

$$\bar{R} = \frac{\ln W_2 - \ln W_1}{t_2 - t_1} \qquad [2.4]$$

The concept of relative growth rate may be regarded as a special case of G when rate is proportional to size. If the physical and chemical conditions within the plant are kept constant then exponential rate of growth can be maintained for extended periods. Commonly, though, R decreases in plants because the necessary resources for growth (e.g. carbohydrate and nutrients) are not available at rates proportional to plant size. In crops, R also generally decreases but the decrease may sometimes be compensated for by the environment becoming more favourable for growth. Blackman,[3] almost 70 years ago, conducted one of the first reviews of differences between species in relative growth rate and showed substantial differences between crop species during early growth (Table 2.1). Differences in $\bar{R}$ between species are also important in natural ecosystems and may contribute to the success of different plant communities in different environments. For example, a study of 132 species of flowering plants[4] showed that woody species and fine-leaved grasses had a bias towards low values of maximum $\bar{R}$ (assessed between two and five weeks after germination) whereas annual plants were most

2 See Hunt, R., *Plant Growth Curves: the functional approach to plant growth analysis*, Arnold, 1982.
3 Blackman, V. H., *Ann. Bot.*, 1919, **33**, 353.
4 Grime, J. P. and Hunt, R., *J. Ecol.*, 1975, **63**, 393.

Table 2.1   Mean relative growth rates of young plants grown under favourable conditions. (From Blackman, V. H., *Ann. Bot.*, 1919, **33**, 353)

Species		$R(day^{-1})$
Hemp	*Cannabis sativa*	0.13
Sunflower	*Helianthus annuus*	0.18
Tobacco	*Nicotiana tabacum*	0.21
Maize	*Zea mays*	0.07

frequently in the high maximum $\bar{R}$ category. It was also found that fast-growing species were dominant, and slow-growing species absent, in disturbed or productive habitats, while the reverse pertained in several more stable, less productive habitats.

Crop and weather relations

From simple observation, it is clear that the growth of plants is substantially affected by climatic differences between locations and that growth at a specific location varies from year to year in response to variations in the local weather. However, the ways in which growth, and ultimately yield, are related to weather are generally less well understood than climatic effects, so that the importance of changes in weather during a growing season for the ultimate yield is frequently speculative. In their survey of 1880, Lawes and Gilbert[5] determined that the years of highest wheat yields in England were characterized by higher than average temperatures during most of the winter and early spring but could reach no firm conclusions about the importance of summer temperatures. They also drew attention to the interactions between weather and soil factors by observing that good yields were frequently correlated with lower than average rainfall in the winter and spring and by showing experimentally how this observation was probably related to the loss of nitrate by leaching.

Statistical procedures have sometimes been useful tools for analysing seasonal variations in yield particularly in those countries where a few weather factors, such as water or temperature, dominate all others. For example, a multiple correlation technique with rainfall and temperature averaged over various parts of the growing season was used to analyse the variations in wheat yields of six mid-western states of the USA.[6] The regressions accounted for 80–92 per cent of the variation after removing trends

5 Lawes, J. B. and Gilbert, J. H., *J. R. agric. Soc. Eng.*, 1880, **16**, 173.
6 Thompson, L. M., *J. Soil Wat. Conserv.*, 1969, **24**, 219.

with time caused by improvements in soil management and the introduction of new varieties.

Similarly, variations throughout Europe in wheat, sugar beet and tobacco yields in relation to weather have been analysed using multiple regression analysis having first defined homogeneous regions of yield and weather fluctuations by principal component analysis.[7] The analysis shows that the weather factors most closely associated with yield variations were different in different areas. For example, tobacco yields were positively correlated with high summer temperature and high summer rainfall in northern and southern Europe, respectively.

While such statistical analyses have their uses, they have limitations if there is no physiological basis for the terms used in the correlation. Moreover, the direct effects of weather on the rate of dry matter production by a crop are confounded with such indirect effects as the incidence and severity of pests and diseases, and the timeliness of farm operations. Similarly, false conclusions will often result if the analysis is extrapolated from one climatic region to another. For example, yields of cereals in many semi-arid regions are frequently positively correlated with rainfall because low rainfall limits growth, whereas in England grain yield and rainfall are negatively related, largely because years of above average rainfall are often years with intense summer storms that may induce lodging or make harvesting difficult, or favour pathogens.

A further disadvantage of statistical procedures for describing crop and weather relations is that the interactions between physical and physiological processes are often ignored. However, the development of micrometeorological techniques[8] has allowed measurements on an hourly basis of many environmental variables together with the growth and transpiration of undisturbed crops and forests. This has paved the way for the analysis of many crop and weather interactions. These techniques have considerable advantages over the older methods of growth analysis involving measurement of leaf areas and dry weights of crops because, while the latter can only detect changes in growth over intervals of a week or more (during which time weather may have varied considerably), the former can determine hour by hour changes in both growth and weather. During the 1970s the combination of long-term micrometeorological measurements with measurements of dry weight obtained at regular intervals during a whole growing season led to substantial improvements in the understanding of how crop plants respond to their aerial environment. This has, in turn, allowed the development of predictive models based on sound physiological principles.

7 Dennett, M. D. *et al.*, *Agric. Meteorol.*, 1980, **21**, 249.
8 See for example Biscoe, P. V. *et al.*, *J. appl. Ecol.*, 1975, **12**, 227;   Baldocchi, D. D. *et al.*, *Agron. J.*, 1981, **73**, 706.

Growth and development

To understand the importance of both edaphic and climatic variables to plant physiology, it is important to distinguish between the processes of growth and development. Growth is defined as the increase in size of a plant. The increase may be in dry weight or in dimensions, and arises as a consequence of the formation of new cells, the expansion of the constituent cells, and the production of assimilates. The rate of growth is therefore expressed as the increase in weight, volume, area or length per unit time. Development is the progress of a plant from germination to maturity through a series of stages which, in many crops, are often well defined. For example, the normal development of wheat always proceeds through germination to emergence, the production of leaves and tillers, ear initiation and stem elongation followed by flowering and grain filling, and finally to physiological maturity and death. Such well-defined stages form the basis for several systems of describing unambiguously the morphological development of both cereals and legumes.[9] Differences in the phenology of genotypes within a crop species are often exploited in plant breeding programmes to produce varieties better suited to the duration of the growing season. Some developmental stages only happen once for each plant (e.g. germination and flowering) so that the rate of development can be expressed in terms of the reciprocal of the time taken to reach that particular stage. In contrast, at other stages similar sets of organs may be produced (e.g. leaf appearance and root axis appearance) so that the rate of development is then a number per unit time.

Development

In general, the principal factors influencing the rate of development are temperature and photoperiod whereas those determining the rate of growth are more numerous and include light, carbon dioxide, nutrients and water.

Temperature has a marked and obvious effect on all rates of development. For most plants growing in temperate regions there is no development and not much growth up to a temperature of about 0–5 °C (the base temperature). The rate of development increases rapidly to a temperature of about 20–25 °C (the optimum temperature), decreases at higher temperatures and ceases at about 35–40 °C (the maximum temperature). These observations can be explained biochemically in terms of the effects of temperature on individual biological processes but the processes involved are complex.[10]

---

9 For cereals see Tottman, D. R. *et al.*, *Ann. appl. Biol.*, 1979, **93**, 221.   For soyabeans see Fehr, W. R. and Caviness, C. E., *Special Report* No. 80, 1977, Co-operative Extension Service, Iowa State University.

10 For a review of the temperature dependence of crop processes see Johnson, I. R. and Thornley, J. H. M., *Ann. Bot.*, 1985, **55**, 1.

Rates of development often increase linearly with temperature (T) in the range between the base and optimum temperatures. This may be expressed by the equation:

$$1/t = (T - T_b)/\theta \qquad [2.5]$$

where $1/t$ is effectively a rate of development, T is the daily mean temperature, T_b is the base temperature at which the rate is zero, and θ is the accumulated temperature (sometimes referred to as the 'thermal time') required to complete the developmental process and is commonly expressed in day-degrees. Figure 2.2 shows that rates of appearance of both millet and groundnut leaves are linearly related to the mean temperature experienced by the meristem of the shoot; in millet, the meristem is initially below the soil surface but in groundnut it is above ground. Above the optimum temperature, thermal time may also be used to calculate rates of development by using a negative relationship between development and increasing temperature; the magnitude of the slope is, though, different on either side of the optimum.[11]

The base temperature, T_b, varies from species to species and may also vary for different phases of development within the same species. Observations on the emergence of 44 species of field crops sown at monthly intervals at

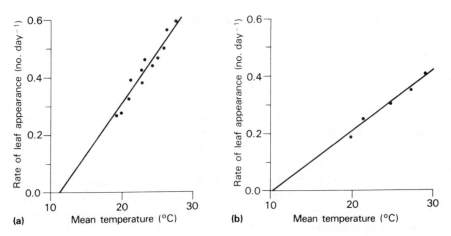

Fig. 2.2   Rate of leaf appearance as a function of shoot temperature for (a) pearl millet (Ong, C. K., *J. exp. Bot.*, 1983, **34**, 322) and (b) groundnut (Leong, S. K. and Ong, C. K., *op. cit.*, 1983, **34**, 1551). The base temperature, T_b, for these species is the temperature where the line crosses the temperature axis.

11 For an example see the germination of millet seeds, Garcia-Huidobro, J. *et al.*, *J. exp. Bot.*, 1982, **33**, 287.

Table 2.2   Values of base temperature (T_b) and the number of day degrees (thermal time) required for emergence (θ_e) for some common crops. Standard errors are in brackets[12]

Species		T_b (°C)	θ_e
Barley	*Hordeum vulgare*	2.6 (0.3)	79.3 (2.5)
Wheat	*Triticum aestivum*	2.6 (0.2)	77.9 (2.5)
Maize	*Zea mays*	9.8 (0.3)	60.8 (2.6)
Field pea	*Pisum sativum*	1.4 (0.5)	110.3 (5.3)
Lentil	*Lens esculenta*	1.4 (1.3)	90.1 (8.0)
Soyabean	*Glycine max*	9.9 (0.6)	70.5 (7.0)
Cowpea	*Vigna unguiculata*	11.0 (0.2)	43.0 (1.9)
Groundnut	*Arachis hypogaea*	13.3 (0.2)	76.3 (5.7)
Rape	*Brassica napus*	2.6 (0.3)	79.0 (2.8)
Sesame	*Sesamum indicum*	15.9 (2.3)	21.3 (15.9)

three sites in eastern Australia[12] showed that the base temperature ranged from 1.4 °C for field peas to 15.9 °C for sesame (Table 2.2). The variation of base temperature for different phases of development within a crop species has received considerably less attention, although for pearl millet only a small range of base temperature (between 10 and 12 °C) has been found for processes as diverse as germination, emergence, leaf appearance, root axis appearance, floral initiation and tiller development.[13] Eleven different genotypes of cowpea examined for variation in base temperature for flowering showed that, although there was some slight variation, a base temperature of 8 °C accurately predicted development in all genotypes except for one accession originating in a hotter environment.[14]

Thermal time is often calculated using temperatures obtained from meteorological screens. However, the site of cellular differentiation may vary from deep in the soil when considering the branching of a root to some distance above the soil surface when considering the initiation of flowers. In many cereals the position of the meristem producing the main shoot remains just below the soil surface until around the time that the reproductive organs are initiated so that both air and soil temperatures play a role in the development of the shoot. The importance of soil temperature in determining rates of development is discussed on pp. 293–297.

Photoperiod also plays a crucial role in the development of plants. This is seen, for example, in the important role that daylength plays in the induction of flowering. Some plants only flower, or flower most rapidly, with fewer than

12 Angus, J. F. *et al.*, *Field Crops Res.*, 1980, **3**, 365.
13 For germination see ref. 11.   For leaves see Ong, C. K., *J. exp. Bot.*, 1983, **34**, 322. For panicle emergence *ibid.*, **34**, 337.   For root development see Gregory, P. J., *op. cit.*, **34**, 744.
14 Hadley, P. *et al.*, *Ann. Bot.*, 1983, **51**, 531.

a certain number of hours of light in each day (short-day plants); others only flower, or do so most rapidly, with more than a certain number of hours of light in each day (long-day plants), while others flower at more or less the same time irrespective of photoperiod (daylength indifferent).[15] Modifying the daylength that a plant receives can result in bizarre consequences (Plate 2.1).

Although leaf appearance of winter wheat is a linear function of temperature, it has also been shown that the mean rate of leaf appearance per unit of thermal time is faster for spring-sown than autumn-sown crops.[16] When

Plate 2.1 Effects of photoperiod on the panicle formation of pearl millet. Right, normal head induced by 12 short days; left, distorted head induced by 4 short days. (From Ong, C. K. and Everard, A., *Expl. Agric.*, 1979, **15**, 401.)

15 Vince-Prue, D., *Photoperiodism in Plants*, McGraw-Hill, 1975.
16 Baker, C. K. *et al.*, *Pl. Cell Environ.*, 1980, **3**, 285.

the mean rate of leaf appearance per unit of thermal time was plotted against the rate of change of daylength at crop emergence for eight emergence dates (ranging from mid-October to mid-June), a strong linear correlation was found. This suggests that the rate of change of daylength has a marked effect on leaf appearance. Similar relations have also been found in barley although, in contrast with the previous study cited above, the base temperature was also affected by the date of sowing.[17]

Sensitivity to daylength is an important criterion in determining the suitability of different genotypes of a crop species when cultivated outside the range of latitudes to which they were initially adapted. In cowpeas and soyabeans, genotypes have been identified as either photoperiod insensitive or photoperiod sensitive (Fig. 2.3).[18] In the insensitive genotypes the time to flowering is predicted solely from equation [2.5], while in the sensitive genotypes there is a progressive delay in the onset of flowering as photoperiod increases beyond a critical value which changes with temperature. This response to photoperiod can be described by an equation similar to equation [2.5] (substituting photoperiod for temperature) so that the time to flowering is determined by either photoperiod or mean temperature, whichever causes the greater delay in flowering.

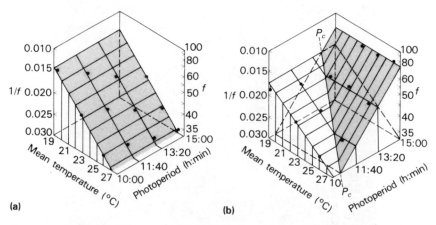

Fig. 2.3 Effects of photo-thermal regime on the number of days from sowing to flowering (f) in cowpea accessions (a) TV_u 1009 and (b) TV_u 1188. The solid circles denote the experimental mean values. Note that TV_u 1188 has a critical period (the line p_c) so that flowering is determined by either photoperiod or mean temperature, whichever causes the greater delay.[14]

17 For barley see Kirby, E. J. M. *et al.*, *Pl. Cell Environ.*, 1982, **5**, 477. For barley and wheat, *idem*, *Agronomie*, 1985, **5**, 117.
18 For cowpeas see ref. 14. For soyabean see Hadley, P. *et al.*, *Ann. Bot.*, 1984, **53**, 669.

Development, then, is determined principally by the photo-thermal environment although in specific circumstances other environmental factors may have a second-order effect. A slight shortage of water may accelerate crop development by raising the temperature of the plant thereby decreasing the calendar time required to reach a particular thermal time.[19] Application of phosphate fertilizer to phosphate deficient soils has also been shown to decrease the calendar time to flowering and maturity of barley in both Australia and northern Syria[20]; the mechanism of this response is not known.

Growth

Once the seed reserves are exhausted, further plant growth depends upon the interception of sunlight by green tissues and the interception of nutrients by roots. Chapter 4 considers the growth of roots and the soil processes contributing to uptake, so that this chapter will consider mainly the shoots and the aerial environment. For most of a crop's life, the leaves are the main plant organs intercepting sunlight and converting it to chemical energy and then to dry matter by the process of photosynthesis. Maximum crop growth is therefore highly dependent on the expansion of leaf area to intercept the maximum amount of radiation. For many crops a leaf area index (area of leaves per unit area of ground) of about 3–5 is necessary to intercept 80–90 per cent of the incident, photosynthetically active radiation.[21] At the start of the growing season, any factor which restricts the rate of expansion of the leaf surface will directly limit the dry matter production until a leaf area index of 3–5 is achieved. Similarly, factors causing premature leaf senescence at the end of the growing season will also adversely affect dry matter production.

Expansion of leaf area depends on the number of leaves, the rate at which they expand and their final size; these are primarily affected by temperature, plant water stress, and nutrient availability. Results from several experiments with cereals and grasses[22] show that, in the absence of water stress, the rate of leaf extension is a linear function of mean temperature (Fig. 2.4) and there is no difference in the response to day and night temperatures. However,

19 See Angus, J. F. and Moncur, M. W., *Aust. J. agric. Res.*, 1977, **28**, 177. Deficiency of N fertilizer can have similar effects; see Seligman, N. G. *et al.*, *J. agric. Sci.*, 1983, **101**, 691.
20 See Woodroffe, K. and Williams, C. H., *Aust. J. agric. Res.*, 1953, **4**, 127 for example in Australia;   Cooper, P. J. M. in *Proc. 17th Coll. Int. Potash Inst. Berne*, 1983, p. 63 for example in Syria.
21 LAI of 5 for winter wheat, see Gallagher, J. N. and Biscoe, P. V., *J. agric. Sci.*, 1978, **91**, 47;   LAI of 3 for potato, see MacKerron, D. K. L. and Waister, P. D., *Agric. For. Meteorol.*, 1985, **34**, 241;   LAI of about 4 for cassava, see Fukai, S. *et al.*, *Field Crops Res.*, 1984, **9**, 347.
22 For wheat see Gallagher, J. N., *J. exp. Bot.*, 1979, **30**, 625.   For millet see Ong C. K., *J. exp. Bot.*, 1983, **34**, 1731.   For perennial ryegrass see Keatinge, J. D. H. *et al.*, *J. agric. Sci.*, 1979, **92**, 175.

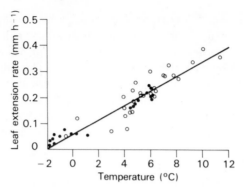

Fig. 2.4   The rate of leaf extension for a winter wheat leaf in relation to temperature. The open circles are day-time and the closed circles night-time measurements. (From Biscoe, P. V. and Gallagher, J. N., *Agric. Prog.*, 1978, **53**, 34.)

during periods of water or nutrient stress, rates of leaf extension are slower. For example, on sunny days, extension rates of barley leaves were slower in the afternoon (1.4 mm h^{-1}) than in the morning (1.9 mm h^{-1}) at similar temperatures of 12 °C.[23] The slower rates of leaf extension in the afternoon were associated with a larger difference between soil and leaf water potentials (0.9 and 0.4 MPa for the afternoon and morning respectively) indicating greater plant water stress. The rate of extension of leaves may also be reduced by lowering the atmospheric humidity.[24] Shortage of nutrients has a similar effect to shortage of water by reducing the area of individual leaves through effects on both length and width. Table 2.3 shows the effects of applying fertilizer nitrogen on the leaf area produced by sugar beet crops. Crops where nitrogen was not applied had smaller leaves resulting from decreased rates of expansion; the duration of the period of leaf expansion was largely unaffected by fertilizer application. Similarly in crops of wheat and barley, nitrogen fertilizers have only a small effect on the number of leaves produced on an individual main shoot or tiller but their main effect is to produce a larger leaf area by increasing the size of individual leaves and ensuring that a larger number of tillers survive.[25]

For many determinate crops (those that produce their seeds at almost the same time), leaf growth ceases shortly before or soon after flowering and thereafter photosynthesis depends mainly on the persistence of the existing leaves. Moreover, once fully extended a leaf does not remain photosynthetically active for long and in barley was found to decrease rapidly only five

23 Biscoe, P. V. and Gallagher, J. N. in *Environmental Effects on Crop Physiology* (eds J. J. Lansberg and C. V. Cutting), p. 75, Acad. Press, 1977.
24 Squire, G. R. and Ong, C. K., *J. exp. Bot.*, 1983, **34**, 846; and with Black, C. R. and Ong, C. K., *op. cit.*, 1983, **34**, 856.
25 Dale, J. E. and Wilson, R. G., *J. agric. Sci.*, 1978, **90**, 503.

Table 2.3 The influence of nitrogen fertilizer (0 and 125 kg N ha^{-1}) on the growth of sugar beet leaves in 1979. (Taken from Milford, G. F. J. *et al.*, *Ann. appl. Biol.*, 1985, **106**, 187)

Leaf number	Final area (cm²)		Expansion rate (cm² day⁻¹)	
	0	*125*	*0*	*125*
4	123	123	6.7	5.9
6	225	259	9.9	10.1
8	229	353	11.5	14.3
10	173	378	8.2	14.6
12	110	372	4.7	13.9
14	81	337	2.7	13.3
16	49	294	1.7	11.8
18	33	256	0.9	8.5
20	25	206	0.7	6.4

days after complete expansion.[26] In some plants (e.g. tomato) the photosynthetic activity of leaves starts to decrease even before full expansion. During reproductive growth nearly all of the assimilate produced is used to fill the seeds or fruits so that the persistence of green leaves can have important effects on crop yield.

Water stress accelerates leaf senescence and when it occurs late in the growing season is almost always associated with nutrient stress. For example, the high demand for nitrogen by growing grains (grains of wheat frequently contain almost 70 per cent and seeds of soyabeans 70–80 per cent of total plant nitrogen at final harvest[27]) means that if the demand cannot be met by uptake from the soil, nitrogen will be translocated from the green stems and leaves resulting in early senescence. In winter wheat, nitrogen translocation from leaves to the grain was accompanied by decreasing rates of photosynthesis,[28] implying that the continued production of dry matter by cereals during grain growth depends on the availability and uptake of sufficient nitrogen to delay leaf senescence.

Photosynthesis

For a green plant to increase in dry matter, atmospheric CO_2 and water within the plant must be chemically combined to form sugars and then a wide

26 Littleton, E. J. cited by Biscoe, P. V. and Gallagher, J. N., *Agric. Progress*, 1978, **53**, 34.
27 For wheat see Austin, R. B. *et al.*, *J. agric. Sci.*, 1977, **88**, 159. For soyabeans see Sinclair, T. R. and de Wit, C. T., *Agron. J.*, 1976, **68**, 319.
28 Gregory, P. J. *et al.*, *J. agric. Sci.*, 1981, **96**, 539.

range of more complex molecules. This process uses energy derived from sunlight and is known as photosynthesis. Its essential features are summarized in the equation:

$$CO_2 + 2H_2O \xrightarrow[\text{chlorophyll}]{\text{light}} CH_2O + O_2 + H_2O$$

During photosynthesis, water is oxidized by the removal of hydrogen and oxygen is released (a process requiring energy to break the chemical bonds); carbon dioxide is then reduced by the hydrogen to carbohydrate.[29]

Electromagnetic radiation with wavelengths between 400 and 700 nm is visible to the human eye and part of it (sometimes called photosynthetically active radiation, PAR) is used by plants during photosynthesis. Light, like all radiant energy, is emitted and absorbed in discrete units of energy called quanta. The energy of a quantum (E) depends on the frequency (v) of the radiation and is equal to the product of frequency and Planck's constant (h):

$$E = h v \qquad \qquad [2.6]$$

The frequency is given by the speed of light (c) divided by the wavelength (λ), so that:

$$E = \frac{hc}{\lambda} \qquad \qquad [2.7]$$

Chlorophyll absorbs mainly in the blue and red parts of the spectrum but the energy available for photosynthesis is derived almost entirely from red light, with a maximum near the red absorption peak at a wavelength of 680 nm, i.e. an energy of 2.9×10^{-19} J for each quantum. Measurements show that about 8 to 10 quanta of wavelength 680 nm reduce one molecule of CO_2 to CH_2O. Multiplying by Avogadro's number to convert from molecules to gram molecules, this gives 1.4 MJ of light energy used to produce one gram mole of CH_2O. One gram mole of CH_2O burned in a calorimeter yields aproximately 0.47 MJ of heat. Therefore on a molecular basis, the maximum conversion efficiency of light energy to chemical energy in photosynthesis is about 0.33 (0.47/1.4). In practice, the efficiency is frequently much smaller because light comprising a range of wavelengths is absorbed by plants, of which only about 50 per cent is photosynthetically active radiation. This reduces the efficiency of conversion to about 11–12 per cent.

The efficiency of conversion by a crop of the incident solar energy into the chemical energy of carbohydrates (sometimes called the photosynthetic efficiency) is smaller than the efficiency at the molecular level for two further reasons. First, crops rarely intercept the whole of the incident light and

29 For a brief review see Hall, D. O. and Rao, K. K., *Photosynthesis*, IOB Studies in Biology, No. **37**, Arnold, 1974.

second, some of that which is intercepted is lost by transmission and reflection; these factors combine to reduce the efficiency from 11–12 per cent to about 8 per cent. Furthermore, plants lose accumulated carbohydrate by respiration and photorespiration (light-stimulated respiration converting newly fixed carbon to CO_2). This loss is typically about 40 per cent of the carbon assimilated in photosynthesis so that the apparent overall efficiency is generally no greater than 5 per cent.[30] However, crop species differ in their efficiency of conversion and to understand why, it is necessary to examine briefly the various mechanisms involved.

The basic biochemical pathway whereby CO_2 is converted to organic carbon is the reductive pentose phosphate cycle (RPP cycle, sometimes called the Calvin cycle). Figure 2.5 illustrates the cycle in which the pentose sugar phosphate, ribulose 1, 5-diphosphate, is converted to glucose 6-phosphate which, after further transformation, gives starch. The cycle is maintained because of the 6 C_3 molecules produced in the initial combination of 3 CO_2 and 3 pentose (C_5) molecules, 5 are rearranged in a variety of reactions to regenerate 3 C_5 molecules. Light energy is used in the production of ATP and NADPH, and the cycle requires 9 molecules of ATP and 6 molecules of NADPH to produce 1 molecule of glyceraldehyde 3-phosphate.[31]. The whole process occurs inside the chloroplasts and is catalysed by enzymes located in the stroma region. Plants using this cycle of photosynthesis are commonly referred to as C_3 plants and comprise most of the common, temperate cereals, legumes and trees.

In some plants, commonly referred to as C_4 plants, there is an additional pathway whereby atmospheric CO_2 is first converted to malate and aspartate (molecules with 4 carbon atoms) and is later released to be assimilated via the RPP cycle.[32] Associated with this pathway are anatomical distinctions between C_3 and C_4 plants. In both, the mesophyll cells contain chloroplasts, but in C_4 plants there is an additional layer of cells containing chloroplasts around the vascular tissue. These cells, known as Kranz or bundle sheath cells, generally have thick walls and contain chloroplasts larger than those typically present in the mesophyll cell.[33] The essential features of the cycle are shown in Fig. 2.6. Atmospheric CO_2 carboxylates phosphoenolpyruvate (using the enzyme PEP carboxylase) to form oxaloacetate, which is rapidly

30 Edwards, G. and Walker, D., C_3, C_4:*mechanisms, and cellular and environmental regulation of photosynthesis*, Blackwell, 1983.
31 For an account of the energy costs see Bassham, J. A. in *Encycl. Pl. Physiol.*, Vol. 6 (eds M. Gibbs and E. Latzko), p. 9, Springer-Verlag, 1979.
32 Much of the early work on this cycle was performed with sugar cane. See Hatch, M. D. and Slack, C. R., *Biochem. J.*, 1966, **101**, 103 and *Ann. Rev. Pl. Physiol.*, 1970, **21**, 141. For a more recent review see Edwards, G. E. and Huber, S. C. in *Biochemistry of Plants: a comprehensive treatise*, Vol. III (eds M. D. Hatch and N. K. Boardman), Acad. Press, 1981.
33 Brown, W. V., *Am. J. Bot.*, 1975, **62**, 395; Kennedy, R. A., *Pl. Physiol.*, 1976, **58**, 573.

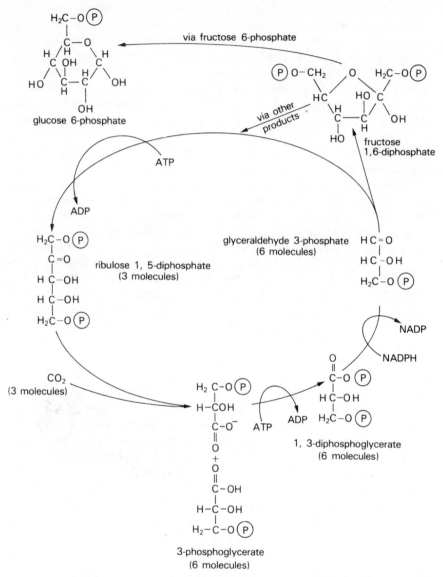

Fig. 2.5 Outline of the reductive pentose phosphate cycle for the conversion of CO_2 to organic carbon. Abbreviations: ADP, adenosine diphosphate, ATP, adenosine triphosphate, NADP, NADPH, oxidized and reduced nicotinamide adenine dinucleotide phosphate.

converted to malate and aspartate. These products are transferred to the bundle sheath cells where decarboxylation occurs to produce CO_2 for use in the RPP cycle and to regenerate phosphoenolpyruvate. Only the Kranz cells

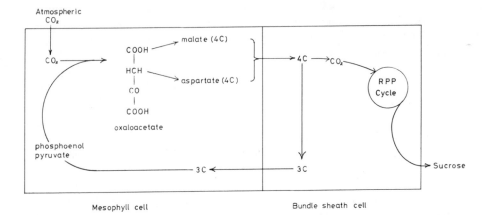

Fig. 2.6 Outline of the C_4 pathway for photosynthesis emphasizing the association of mesophyll and bundle sheath cells. 3C is the carbon unit remaining after decarboxylation of the 4C acid.

contain the enzymes able to convert phosphoglycerate to hexose phosphate, so there is no net fixation of CO_2 in the mesophyll cells. The precise mechanism allowing the rapid transport of metabolites between mesophyll and bundle sheath cells is not yet known but plasmodesmata are known to connect the two cell types in some species.[34] There are also known to be differences between C_4 species in the mode of C_4 decarboxylation.[35]

These anatomical and biochemical differences have important physiological consequences. One consequence of the C_4 pathway is to maintain a higher concentration of CO_2 in the bundle sheath cells than would normally exist by diffusion of CO_2 into the cells from the atmosphere (i.e. there is a CO_2 'pump'). This means that CO_2 concentration is enhanced compared with a C_3 plant and is less limiting to rates of photosynthesis. Furthermore the increased CO_2 concentration diminishes the activity of enzymes involved in the process of photorespiration. In C_3 plants, photorespiration results in 20 to 50 per cent of the fixed carbon being immediately respired but in C_4 plants there is no detectable evolution of CO_2 in the light.[36] It is uncertain at present whether photorespiration is completely inhibited in the bundle sheath cells of C_4 plants or whether such plants refix CO_2 produced by photorespiration in the mesophyll cells.[37]

34 Laetsch, W. M., *Sci. Prog. Oxford*, 1971, **57**, 323.
35 Rathnam, C. K. and Edwards, G. E., *Arch. Biochem. Biophys.*, 1977, **182**, 1.
36 See Hofstra, G. and Hesketh, J. D., *Planta*, 1969, **85**, 228. For review of photorespiration see Ogren, W. L., *Ann. Rev. Pl. Physiol.*, 1984, **35**, 415.
37 Tolbert, N. E., *Ann. Rev. Pl. Physiol.*, 1971, **22**, 45.

The conversion of CO_2 to carbohydrate by the C_4 and RPP cycles requires the use of energy from two molecules of ATP per molecule of CO_2 more than the RPP cycle alone. However, in terms of dry matter production, this additional energy cost is offset by the lack of photorespiration. Many C_4 species are native to tropical and subtropical regions where high irradiance and high temperature tend to bring about CO_2 limitation of photosynthesis in the RPP cycle alone; the C_4 cycle, by enhancing the CO_2 concentration at the sites of conversion to carbohydrate, is advantageous where CO_2 rather than light is limiting.

There are several important consequences of this CO_2-concentrating mechanism for other aspects of crop physiology.[38] C_4 crops (e.g. sugar cane, maize, sorghum and millet) are more efficient in their use of water, being able to synthesize about 2 to 3 times as much dry matter per unit mass of water transpired as C_3 plants. This occurs in C_4 plants because in addition to high CO_2 concentrations being maintained in the bundle sheath cells, stomatal regulation maintains the intercellular CO_2 concentration at about 100–150 ppm. In C_3 plants it is maintained at around 220–260 ppm.[39] Since the ratio of photosynthesis to transpiration is inversely related to the intercellular CO_2 concentration (assuming the ambient CO_2 concentration and the saturation deficit of the air are constant), then the dry matter production per unit of water transpired will be greater in C_4 than in C_3 plants (see Ch. 11). This greater water use 'efficiency' does not mean, however, that C_4 plants are more drought tolerant or better able to withstand low plant water potentials. Tolerance of low water potentials depends on many other factors (e.g. osmotic adjustment and leaf morphology) not simply related to the photosynthetic pathway.

A third mechanism for photosynthesis, known as crassulacean acid metabolism (CAM), is common in the family Crassulacea (which contains many succulents) and also among cacti (family Cactaceae). A crop species having the CAM mechanism is pineapple. Frequently, only a few species of a plant family are CAM plants and some can exhibit either C_3 or CAM photosynthesis depending on plant age or environmental conditions.[40] Essentially, the normal pattern of stomatal opening during the day and closure at night is reversed in CAM plants, although the biochemical pathway is similar initially to the C_4 pathway. In CAM plants, CO_2 is fixed during the night by PEP (phosphoenolpyruvate) carboxylase activity with the eventual formation of malate, which is stored in vacuoles. The cells where photosynthesis occurs typically possess large vacuoles and the pH of the vacuole solution decreases during the night as the malic acid content increases. During the day, the malate is decarboxylated, the CO_2 enters the RPP cycle as a substrate for ribulose 1, 5-bisphosphate carboxylase and starch is produced. The starch

38 For a review see Pearcy, R. W. and Ehleringer, J., *Pl. Cell Environ.*, 1984, **7**, 1.
39 Wong, S. C. et al., *Nature, Lond.*, 1979, **282**, 424.
40 Szarek, S. R. and Ting, I. P., *Photosynthetica*, 1977, **11**, 330.

produced in the day provides the energy source for PEP formation the following night thereby allowing CO_2 fixation to occur without a direct requirement for light energy. CAM is therefore biochemically similar to C_4 photosynthesis but whereas C_4 photosynthesis occurs by two contemporaneous processes spatially separated, CAM photosynthesis occurs by two spatially contiguous processes, temporally separated.[41]

This temporal separation of the processes is of significance as an adaptational mechanism facilitating photosynthesis in arid environments; it enables plants to conserve water during the day when transpiration would normally be most rapid. This can be seen from typical values of water transpired per unit of dry matter accumulated, which are around 50–100 g g^{-1} in CAM plants compared to 250–300 in C_4 plants and 400–500 in C_3 plants. The conservation of water is accompanied by conservation of carbon, and respiratory losses of CO_2 are minimized so that a plant may pass through an entire day without any net efflux of CO_2. However, the conservation of water also results in very slow growth rates for such plants. In the presence of adequate water supplies, some CAM plants can revert to the C_3 cycle alone and increase their rate of growth.[42]

Production of dry matter

The presence of green leaves and the process of photosynthesis together result in the production of dry matter by plants. The rate of photosynthesis of individual leaves depends upon a number of environmental factors but principally upon the irradiance (light intensity), CO_2 concentration, temperature and water status. Figure 2.7 shows the response of photosynthesis to light for leaves of winter wheat. In this C_3 plant, the rate of photosynthesis is directly proportional to irradiance until a point is reached beyond which the rate does not increase even in very bright light; the leaf is then said to be light saturated. These leaves were grown outdoors; leaves of plants grown in growth rooms (usually at low irradiance) frequently show a different response to irradiance that is more closely described by a form of rectangular hyperbola.[43] The shape of the light response curve also varies with the type of photosynthetic pathway used by the crop. For C_3 plants the rate of photosynthesis generally increases more rapidly at low irradiance than in C_4 plants but they become light saturated at much lower irradiance than C_4

41 This diurnal change in acidity can be tasted. A letter written by B. Heyune to the Vice-President of the Linnean Society in 1813 describes how leaves of *Bryophyllum calycinum* were 'as acid as sorrel' in the morning and 'tasteless' around noon. It was 150 years later that the processes underlying this observation were deduced. For reviews see Osmond, G. B., *Ann. Rev. Pl. Physiol.*, 1978, **11**, 81; Kluge, M. in *Encycl. Pl. Physiol.*, Vol. 6 (eds M. Gibb and E. Latzko), p. 113, Springer-Verlag, 1979.
42 Hartsock, T. L. and Nobel, P. S., *Nature, Lond.*, 1976, **262**, 574.
43 See Marshall, B. and Biscoe, P. V., *J. exp. Bot.*, 1977, **28**, 1008.

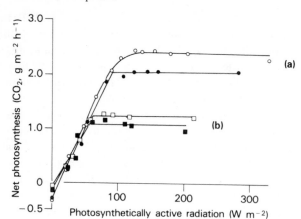

Fig. 2.7   The photosynthetic response of flag leaves of winter wheat to irradiance. Measurements made (a) at anthesis and (b) about 3 weeks
later in crops given no fertilizer (open symbols) and 151 kg N ha^{-1}
(closed symbols).[28]

plants. This means that in cool, dull conditions C_3 plants will photosynthesize
more than C_4 plants, but in hot, light conditions, the reverse is true.

Figure 2.7 also shows that nitrogen fertilizer has no effect on the shape of
the light response curve although there was a 40 per cent difference in the
nitrogen concentration of the leaves. Similar results have been obtained by
Pearman at Rothamsted[44] indicating that, over a wide range of applications
of nitrogen fertilizer, the primary effect of nitrogen is to increase leaf area
but the photosynthetic rate per unit area of leaf is unaffected. This contrasts
with the results reported by other workers in which nitrogen has been shown
to affect the rate of photosynthesis per unit area of leaf.[45] However, such
experiments have usually been performed with plants grown in media
containing concentrations of nitrate considerably lower than those commonly
present in soils so that the concentration of nitrogen in the leaves may be
lower than that of soil-grown plants.

As leaves age, the irradiance at which photosynthesis is light saturated
decreases although the response in dull light remains the same. This means
that the maximum rate of photosynthesis decreases with time and that the
rate responds much less to changes in irradiance.

The rate of photosynthesis is directly related to stomatal conductance so
that water stress resulting in stomatal closure will also slow leaf photosynthesis. This can be seen in the measurements of net photosynthesis for a crop
of barley recorded at regular intervals through the day (Fig. 2.8). At the start
of the period of dry weather, net photosynthesis responded to irradiance in
the same way in the morning and afternoon. Three days later, when the crop
was more water stressed, the crop response to irradiance was similar in the

44 See Pearman, I., *et al.*, *Ann. Bot.*, 1977, **41**, 93; *ibid.*, 1979, **43**, 613.
45 Natr, L., *Photosynthetica*, 1972, **6**, 80.

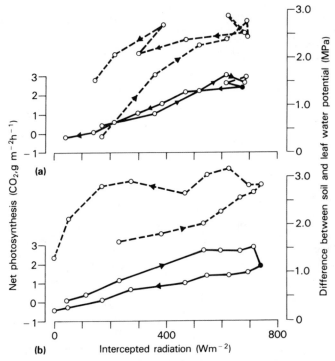

Fig. 2.8 Changes in the rate of net photosynthesis (●) and the difference between soil and leaf water potential (○) during one day (a) at the start and (b) at the end of a period of warm, dry weather. (From Biscoe, P. V. and Gallagher, J. N. in *Environmental Effects on Crop Physiology* (eds J. J. Landsberg and C. V. Cutting), p. 84, Acad. press, 1977).

morning to that previously but in the afternoon, net photosynthesis at any given irradiance was only about one-half that of the morning.

Crop photosynthesis depends, of course, upon the photosynthetic contribution of all of the leaves present in the canopy and is best considered in terms of its response to irradiance. For a cereal crop, there are four broad responses determined by the developmental and growth stages (Table 2.4).[46] Similar divisions.could be made for other crops although the precise shape of the light-response curve would depend on the photosynthetic pathway and the energy requirement for nitrogen fixation in legumes.

Carbohydrate produced by photosynthesis undergoes further reactions to form the structural and cellular materials comprising the dry matter of a crop. This synthesis and the maintenance of existing tissues requires energy, and this energy is derived from respiration. Respiration, R_t, then, may be conveniently divided into two components, synthesis respiration, R_s, and maintenance respiration, R_m. Measurements on a daily basis showed that for

46 Biscoe, P. V. and Gallagher, J. N., *Agric. Progress*, 1978, **53**, 34.

Table 2.4 Summary of response of crop photosynthesis to irradiance in cereal crops

Stage	Condition	Response of crop photosynthesis
Young crop	Row structure still distinct	Crop photosynthesis saturated at irradiances only slightly greater than those of individual leaves
Tillering, stem extension	Canopy of young leaves sufficient to intercept most of the sunlight	Little tendency to saturate even in bright light; irradiance is the major determinant of crop photosynthetic rate
Flag leaf emergence, flowering, early grain filling	Young leaves at top of canopy, leaf area index >3	As previous stage but increasing likelihood of water stress during bright days. If dry, photosynthesis slows in proportion to degree of stomatal closure
Grain filling	Canopy ageing	Rate of canopy photosynthesis slows in bright light. Drought exaggerates ageing effect by closing stomata

white clover and sorghum plants, R_s was a constant fraction, a, of gross photosynthesis, P_g, and R_m was proportional to the total dry weight of the plant, W, expressed as g CO_2, i.e.

$$R_t = aP_g - bW \qquad [2.8]$$

where b is the fraction of W respired daily for the maintenance of existing structures.[47] Values of a and b were found to be 0.25 and 0.015 respectively, and similar values have been determined for maize, barley and cotton.[48]

During early growth, R_m is small compared with R_s, but in later growth R_m may become large although the rate is strongly dependent on temperature. Because R_s is related to P_g, losses by respiration are greater during bright days. Figure 2.9 illustrates that CO_2 uptake by a crop proceeds by a series of accretions and losses depending upon the balance between photosynthesis and respiration. During the day photosynthesis is faster than respiration while at night photosynthesis ceases and CO_2 is lost by respiration. Total loss of CO_2 is greater following bright days when photosynthesis is fast. Overall, over periods of a week or more, crops lose by respiration

47 McCree, K. J., *Crop Sci.*, 1974, **14**, 509.
48 For cotton see Baker, D. N. *et al.*, *Crop Sci.*, 1972, **12**, 431.   For barley see Biscoe, P. V. *et al.*, *J. appl. Ecol.*, 1975, **12**, 269.

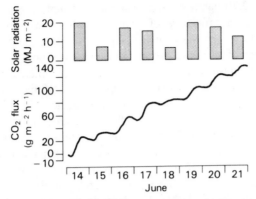

Fig. 2.9 Hourly rates of CO_2 fixation by a spring barley crop (cv. Proctor) summed over the period of 14–21 June 1972. The histograms show the total solar radiation for each day. (From Biscoe, P. V. *et al., J. appl. Ecol.*, 1975, **12**, 269.)

an almost constant fraction (0.4) of the CO_2 fixed by photosynthesis.[49]

In terms of the production of crop dry matter, the result of these processes is that for well-fertilized crops that are not subjected to water stress, crop growth is linearly related to the amount of radiation intercepted during the period of growth (Fig. 2.10). The gradient of a composite line drawn through

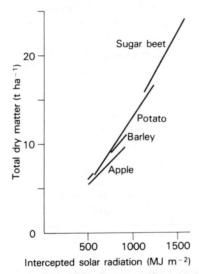

Fig. 2.10 Relation between total dry matter at harvest and radiation intercepted by foliage during the growing season. (From Monteith, J. L., *Phil. Trans. R. Soc. Lond.*, 1977, **B281**, 277.)

49 For barley see Biscoe, P. V. *et al., J. appl. Ecol.*, 1975, **12**, 269. For rice see Takeda, T. *et al., Proc. Crop Sci. Soc. Japan*, 1976, **45**, 139.

all lines on Fig. 2.10 has a slope of 1.4 g MJ^{-1} (total solar radiation) which is equivalent to an efficiency of conversion of energy of 2.4 per cent (assuming 17.5 kJ g^{-1} as the heat of combustion of dry matter). This linear relation has been determined for many crops over both a whole growing season and much shorter periods of growth.[50] The efficiency of conversion of energy varies between crops (each crop in Fig. 2.10 has a different slope) and there can be variations within a crop species. C_4 plants would be expected to have a better conversion efficiency than C_3 plants because of the considerably smaller loss of carbon by photorespiration, and this should be reflected in crop growth rates. Table 2.5 shows that maximum short-term growth rates are indeed faster in C_4 plants than in C_3 plants though part of this difference may arise because C_4 plants grown in the tropics may experience more intense sunshine than C_3 plants in temperate regions. However, the difference between C_3 and C_4 crops is not confined to maximum growth rates alone. Monteith[51] determined that the mean growth rate for the season (the above-ground dry weight at harvest divided by the duration of the growing season) for a range of C_4 crops was 22.0 $\pm$ 3.6 g m^{-2} day^{-1} while that for C_3 crops was 13.0 $\pm$ 1.6 g m^{-2} day^{-1} giving conversion efficiencies of total solar radiation of 2.0 and 1.4 per cent respectively.

Legumes (C_3 plants) may have smaller conversion efficiencies than cereals because of the energy requirement for the nitrogen fixation process in the nodules. There are currently few measurements to determine whether such a difference exists, and although nitrate reduction is less demanding of energy than fixation of nitrogen gas, the difference is marginal.[52] For well-watered

Table 2.5 Maximum short-term growth rates of C_3 and C_4 plants[51]

Group	Species		Crop growth rate $(g\ m^{-2}\ day^{-1})$
C_4	Maize	*Zea mays*	52
	Pearl millet	*Pennisetum typhoides*	54
	Sorghum	*Sorghum vulgare*	51
C_3	Bulrush	*Typha latifolia*	34
	Corn cockle	*Agrostemma githago*	39
	Potato	*Solanum tuberosum*	37
	Rice	*Oryza sativa*	36

50 For barley and wheat see Gallagher, J. N. and Eiscoe, P. V., *J. agric. Sci.*, 1978, **91**, 47.   For sorghum see Sivakumar, M. V. K. and Huda, A. K. S., *Agric. For. Meteorol.*, 1985, **35**, 47.   For willow trees, see Cannell, M. G. R. *et al.*, *J. appl. Ecol.*, 1987, **24**, 261.
51 Monteith, J. L., *Expl Agric.*, 1978, **14**, 1.
52 See Postgate, J. R. *The Fundamentals of Nitrogen Fixation*, pp. 184–7, Cambridge Univ. Press, 1982; Phillips, D. A., *Ann. Rev. Pl. Physiol.*, 1980, **31**, 29.

crops of pigeonpea,[53] dry matter production and the amount of radiation intercepted were linearly related and gave a conversion of 1.23 ± 0.12 g dry matter per MJ total solar radiation and had an efficiency of conversion of 2.1 per cent, which is comparable with the values calculated from Fig. 2.10; the crop was, however, given 40 kg ha^{-1} of nitrogen. In contrast, dry matter production by monocropped and intercropped groundnuts was not linearly related to the amount of intercepted radiation during pod filling; the reduced dry matter production during pod filling was possibly related to the requirement for energy for nitrogen fixation which occurred at a fast rate during this period.[54]

The effect of water stress on the relation shown in Fig. 2.10 is to reduce the efficiency of conversion. Results[55] show that when pigeonpea was water stressed, not only was the amount of radiation intercepted by the crop reduced, but the conversion rate fell to 0.65 ± 0.07 g MJ^{-1} as did the efficiency of conversion to about 1.1 per cent. Similarly, persistent water stress during the growth of winter wheat in England in the exceptionally hot, dry summer of 1976, also reduced the conversion efficiency.[56]

Partitioning of dry matter and components of yield

Our understanding of the control of partitioning of photosynthetic assimilates is currently very poor. This is evident from the number of terms available to describe partitioning (for example 'functional equilibrium' and 'sink' concepts) most of which are conceptually acceptable but difficult to quantify. Another problem lies in finding a mechanism of control operating at all of the various levels within the plant where transfer and partitioning of assimilate occurs. In a growing leaf, assimilate is partitioned between further leaf growth, temporary storage and exported material. Once outside the leaf, the exported material is partitioned both between other growing organs (sinks) and between different chemical constituents within an organ. It is unlikely that there is one mechanism controlling all of these processes, but loading and unloading of the sieve tubes within the phloem appears to play a major role in the transfer of assimilates and, together with the sinks themselves, may form the basis for regulating carbon flow. Gradients of sucrose concentration developing in the sieve tubes between sources and sinks may create gradients of hydrostatic pressure sufficient to enable mass flow to equalize supply and demand.[57]

53 Hughes, G. and Keatinge, J. D. H., *Field Crops Res.*, 1983, **6**, 171.
54 Marshall, B. and Willey, R. W., *Field Crops Res.*, 1983, **7**, 141.
55 See ref. 53.
56 See ref. 50.
57 Gifford, R. M. and Evans, L. T., *Ann. Rev. Pl. Physiol.*, 1981, **32**, 485.

Partitioning is usually discussed in relation to the distribution of dry weight. The distribution between roots and shoots will be discussed in more detail in Chapter 4, but is summarized by:

root mass $\times$ specific root activity $\propto$ shoot mass $\times$ specific shoot activity.[58]

Changes in the partitioning of new dry matter between roots and shoots will therefore result if the specific activity of shoots or roots changes with time or if the elemental composition of new plant dry matter changes.

For agricultural crops with a clearly defined time of maturity, partitioning is frequently assessed at harvest in terms of the harvest index, H, defined as:

$$H = \frac{\text{Wt of grain (or other economic yield)}}{\text{Wt of shoot, including grain (above-ground biological yield)}}$$

This ratio can be estimated by a variety of techniques but the current standard method is to cut mature plants at ground level, weigh to give total yield, and thresh to give the yield of grain.[59] Oven dry weights should be used whenever possible to allow different sites and seasons to be compared without variations in water content of the tissues. Harvest index varies between varieties of a crop species and may also vary from site to site depending upon the availability of resources and weather during the grain filling period. In experiments in a glasshouse, the harvest index of wheat increased linearly (from 0.15 to 0.55) as the fraction of total water used after anthesis increased from 0.02 to 0.32.[60] Although such large variation has not been determined in field experiments, there are frequently smaller effects of water shortage.

The increased grain yields in many countries during the last two to three decades have been associated with increasing use of fertilizer (thereby increasing above-ground biological yield) and with new varieties in which the harvest index is higher. In the UK, an examination of the genetic improvements in yields of winter wheat since 1900 concluded that the increased grain yield of the newer varieties was associated with a shorter length of straw and a greater harvest index.[61] However, using harvest index as a physiological indicator can be misleading since the newer varieties also reach anthesis earlier than the older varieties and intercept more light during grain filling.[62] Calculations show that a linear relation is obtained when grain yield of the old and new wheat varieties is plotted against the amount of incident radiation intercepted during grain filling. Similarly, the shorter-strawed varieties of rice which have increased yields are responsive to fertilizer because lodging

58 Davidson, R. L., *Ann. Bot.*, 1969, **33**, 561.
59 Donald, C. M. and Hamblin, J., *Adv. Agron.*, 1976, **28**, 361.
60 Passioura, J. B., *J. Aust. Inst. agric. Sci.*, 1977, **43**, 117.
61 Austin, R. B. *et al.*, *J. agric. Sci.*, 1980, **94**, 675. For a review of partitioning in crops see Synder, F. W. and Carlson, G. E., *Adv. Agron.*, 1984, **37**, 47.
62 Charles-Edwards, D. A., *Physiological Determinants of Crop Growth*, p. 110, Acad. Press, 1982.

is less; fertilizer promotes leaf growth and increases the time for which the leaf area is green, thereby increasing the amount of radiation intercepted and dry matter produced. Incidentally, harvest index is increased. Harvest index is, then, dependent on phenological, physiological and environmental effects, and grain yield is better analysed in terms of the main factors determining yield.

Grain yield per unit area, Y, can be considered as consisting of a number of components such that:

Y = plants/unit area × ears/plant × grains/ear × weight/grain for cereals,

or:

Y = plants/unit area × pods/plant × seeds/pod × weight/seed for legumes.

All environmental effects can now be examined in relation to the number of grains per unit area of ground and the average size of each grain. Numerous experiments have been performed in an attempt to elucidate and predict these effects. By assuming that each pod (or harvestable unit) requires assimilate at some critical rate during its initiation if it is not to be aborted, it can be predicted that the number of pods per plant will be inversely related to the planting density. Figure 2.11 shows this to be so for mung bean crops that were shaded or thinned to modify plant density and assimilate supply.[63]

The principal factors affecting the number of ears and grains per ear in wheat and barley crops and the effects of temperature on grain growth have been reviewed by Gallagher and Biscoe.[64] Generally, increasing temperature

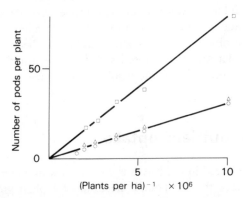

Fig. 2.11 Relation between the average number of pods per plant and the reciprocal of plant density for crops of mung beans; cv. Berken (o), CES-ID-21 (△) and Regur (□).[63]

63 Muchow, R. C. and Charles-Edwards, D. A., *Aust. J. agric. Res.*, 1982, **33**, 53.
64 Gallagher, J. N. and Biscoe, P. V., *Agric. Progress*, 1978, **53**, 51.

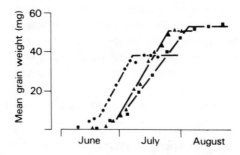

Fig. 2.12   Changes in the mean weight of grains of winter wheat (cv. Maris Huntsman) in 1974 (■), 1975 (▲) and the hot, dry summer of 1976 (●). (From Gallagher, J. N. *et al., Nature, Lond.*, 1976, **264**, 541.)

increases the rate of grain filling but reduces its duration, as shown in Fig. 2.12. In 1974 and 1975, grains grew at 1.4 mg day^{-1} and 1.8 mg day^{-1} respectively during the linear phase of growth but 1975 was warmer and duration was reduced from 33 to 25 days. This gave grains of approximately 50 mg in both years; the size being limited by the structure of the ears. In the hot, dry summer of 1976, however, the rate of grain growth was faster (2.1 mg day^{-1}) but drought reduced the duration of the linear phase of grain filling to 15 days and the mean weight per grain to 37 mg. Similar patterns of grain growth have also been observed in legumes.

One additional aspect of partitioning during grain filling is the ability of many plant species to translocate assimilates stored in the stem before anthesis to the growing grain or storage organs; this may be of considerable importance in maintaining yields in drought-prone environments. In the UK it was found that about 15 per cent of pre-anthesis assimilate was translocated to grains of wheat although much higher values have been estimated in dry years; in pearl millet grown on stored soil moisture, translocated materials were apparently responsible for the whole of grain yield.[65]

Growth and nutrient uptake

Photosynthate produced in the leaves is combined with minerals absorbed by the roots to produce the compounds necessary for plant growth. Although seeds contain appreciable quantities of mineral nutrients, uptake from the soil starts soon after germination and rapidly contributes most of the minerals. The availability of mineral ions in the soil and their transport to the root are

65 For wheat see Austin, R. B. *et al., Ann. Bot.*, 1977, **41**, 1309; Gallagher, J. N. *et al., Nature, Lond.*, 1976, **264**, 541.   For millet see Gregory, P. J. and Squire, G. R., *Expl Agric.*, 1979, **15**, 161.

considered in detail in other chapters of this book, so this section will concentrate on the observed relations between growth and nutrient uptake.

It has been observed in many studies that growth is usually increased if the supply of nutrients to a plant is increased from a level that is low. Liebig was one of the first workers to attempt to express the relation between growth and nutrient supply in what is now generally called the 'Law of the Minimum'; namely, the amount of plant growth is regulated by the factor present in the minimum amount and rises or falls according as this is increased or decreased in amount. According to this hypothesis, growth increases with additions of the limiting factor until it ceases to be limiting. Growth then becomes independent of this factor as it increases still more until a point is reached when it becomes toxic and causes growth to decrease. The relation between plant growth and the amount of the limiting nutrient present in the soil can, therefore, be represented as a curve such as is shown in Fig. 2.13. However, this 'law' has limited validity because nutrients frequently have interactive effects on growth so that if several factors are low, but not excessively so, then increasing any one will increase growth.

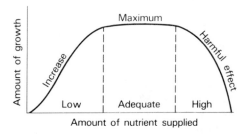

Fig. 2.13　General relation between any particular nutrient or growth factor and the amount of growth made by the plant.

Experiments conducted in the late 1800s and early 1900s suggested that the relation between growth and nutrient supply was predictable and could be described by simple mathematical equations; E. A. Mitscherlich was among the first to do this. He assumed that a plant should produce a certain maximum yield if all conditions were ideal, but if an essential factor were deficient, there would be a corresponding decrease in the yield. Moreover, he assumed that the increase in yield produced by unit increment of the lacking factor is proportional to the decrement from the maximum, or expressed mathematically:

$$\frac{dy}{dx} = (Y-y)\,C \qquad\qquad [2.9]$$

where y is the yield obtained when x is the amount of the factor present, Y is the maximum yield obtainable if the factor were present in excess and C

is a constant. On integration, and assuming that $y = 0$ when $x = 0$,

$$y = Y(1 - e^{-Cx}) \qquad [2.10]$$

The form of this equation implies that as each unit of factor x is added, the response to each diminishes. Mitscherlich's experiments were made with plants grown in sand culture supplied with excess of all nutrients except the one under investigation. Table 2.6 shows some typical results obtained with oats supplied with increasing amounts of monocalcium phosphate fertilizer.[66] Mitscherlich claimed to show that the proportionality factor, C, was a constant for each fertilizer, and independent of the crop, site or other conditions. If this were so, an experimenter knowing its value could predict the yields obtainable from any given quantity of fertilizer and it would be possible to estimate directly by pot experiment the amount of available plant nutrients in a soil.

Table 2.6   Growth of oats given different amounts of phosphate fertilizer

P_2O_5 applied (g)	Dry matter produced (g)	Dry matter calculated from Mitscherlich equation (g)	Increase in growth per 0.05 g increment of P_2O_5 added
0	9.8 ± 0.5	9.8	—
0.05	19.3 ± 0.5	18.9	9.5
0.10	27.2 ± 2.0	26.6	7.9
0.20	41.0 ± 0.9	38.6	6.9
0.30	43.9 ± 1.1	47.1	1.5
0.50	54.9 ± 3.7	57.4	2.8
2.00	61.0 ± 2.2	67.6	0.2

Using this approach, and collating the results of many experiments, workers were able to calculate appropriate values of C for the UK and formulate a fertilizer policy.[67] They determined that C had values of 1.1 for N, 0.8 for P and K, and 0.04 for farmyard manure and were then able to predict the optimal, economic application rates. However, despite this success, it was rapidly realized that C was not a constant and depended on many other factors during growth.[68] Moreover, the response curve is sometimes sigmoidal, especially if the nutrient under consideration is present at a very low level; fertilizer in excess frequently decreases the crop yield, and the calculated maximum yield of the crop is sometimes far in excess of anything that can be obtained.

66 *Landw. Jahrb.*, 1909, **38**, 537.
67 Crowther, E. M. and Yates, F., *Emp. J. expl Agric.*, 1941, **9**, 77.
68 See for examples Bullen, E. R. and Lessels, W. J., *J. agric. Sci.*, 1957, **49**, 319; Boyd, D. A., *J. Sci. Fd. Agric.*, 1961, **12**, 493.

Attempts have been made to improve Mitscherlich's equation,[69] and have had some success. As amounts of fertilizers applied to crops increased in the UK during the 1960s and 1970s, it was found that, for practical purposes, the response curve could be broken down into two parts.[70] First, a linear response to increasing amounts up to a certain point followed by, second, a sharp break above which additional fertilizer either had no effect on yield, or caused only a small increase or sometimes a small decrease in yield. This is illustrated by Fig. 2.14 which shows the response of wheat and barley to nitrogen on three fields of low nitrogen status at three experimental farms in England. These observed responses have implications for the fertilizer policies adopted by different farmers. If fertilizer is readily available and the price obtained for the crop far exceeds the cost of fertilizer, then many farmers may add more fertilizer than is strictly necessary and thereby ensure maximum yields even though this is not the maximum economic return. If, however, fertilizers are in short supply or expensive relative to the price obtained for the crop, greater economic returns will come from using small amounts over as wide an area as possible rather than large amounts over a restricted area.

More recently, attempts have been made to develop simple equations that allow for the mineralization of organic nitrogen and the release of slowly exchangeable forms of phosphate and potassium in the interpretation of conventional fertilizer trials. For example, responses of sugar beet to nitrogen fertilizer frequently vary considerably from season to season,[71] but allowing for the distribution and mineralization of organic nitrogen in the soil profile and appropriate growth and nitrogen concentrations in the plant, a simple

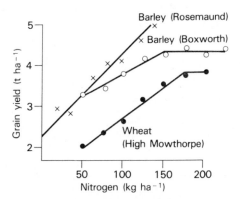

Fig. 2.14   Responses of wheat and barley to nitrogen on soils low in nitrogen at three experimental husbandry farms in England.

69 For a review see Cooke, G. W., *Fertilizing for Maximum Yield* (3rd ed.), Granada, 1982.
70 Boyd, D. A., *Proc. 9th Congr. Int. Potash Inst.*, 1971, 461.
71 Draycott, A. P. *et al.*, *Rothamsted. Ann. Rep. for 1972*, Part 2, p. 5.

model accurately predicted the observed responses in 8 out of 9 trials conducted at 2 sites over 5 years.[72]

Attention has also focused more sharply on the relation between growth and the uptake of nutrients by plants. For example, Table 2.7 shows that phosphate fertilizer added to oats grown in sand culture increased the growth of the plants and the amount of phosphate taken up, but the concentration of phosphate in the tissues (expressed on the basis of dry weight) remained almost constant.[73] Typically, the dry weight of a plant varies by several orders of magnitude during growth whereas the concentration of nutrients in plant tissues varies at most by an order of magnitude and more commonly by a factor of 2 to 5. For example, in nodulated soyabeans (without fertilizer) the concentrations of P and K typically varied from 0.3 and 1.4, to 0.2 and 1.0 respectively,[74] during growth, and similarly in winter wheat their concentrations ranged from 0.4 and 4.2, to 0.2 and 1.0.[75] The basic aspects of the relation between growth and nutrient uptake can be expressed by:

$$\frac{dU}{dt} = \frac{d(WX)}{dt} = W.\frac{dX}{dt} + X.\frac{dW}{dt} \qquad [2.11]$$

where U is the total nutrient content of the plant, W is the total plant dry weight, X is the concentration of nutrient U, and t is time.[76] Since X is normally constant relative to W (i.e. if $dX/dt \simeq 0$) then:

Table 2.7   Uptake and concentration of phosphate in oats[73]

	Initial external supply of P (mg pot⁻¹)	
	8	*90*
Uptake of P (mg pot⁻¹)		
leaves	0.47	2.8
stems	0.38	4.1
roots	0.55	2.1
Uptake of P per g increment in plant weight		
leaves	0.6	0.5
stems	1.0	0.9
roots	0.9	0.7

72 Greenwood, D. J., *Fertil. Res.*, 1984, **5**, 355.
73 Williams, R. F., *Aust. J. Sci. Res. B*, 1948, **1**, 333.
74 Hanway, J. J. and Weber, C. R., *Agron. J.*, 1971, **63**, 286.
75 Page, M. B. *et al.*, *Pl. Soil*, 1978, **49**, 149.
76 Nye, P. H. and Tinker, P. B., *J. appl. Ecol.*, 1969, **6**, 293;   see also Brewster, J. L. and Tinker, P. B., *Soils Fertil.*, 1972, **35**, 355.

$$\frac{X}{W}\frac{dW}{dt} = XR \qquad\qquad [2.12]$$

where R is the relative growth rate. This analysis indicates that plant growth and nutrient uptake are directly linked and that continued uptake of nutrients only occurs if the size of the plant increases. To test this idea, young barley seedlings were grown with different photoperiods to obtain different rates of growth.[77] Figure 2.15 shows that after three days, the transport of the potassium into the shoot was proportional to the relative rate of growth of the shoot. Similarly the relation between R and the accumulation of a range of nutrients was investigated in tobacco over a longer period of time (25 days).[78] For P and K, dX/dt was 0 so that the relative accumulation rates equalled R for the whole plant, but for N, Ca and Mg, X decreased during growth and the relative accumulation rates were less than R for the whole plant. However, this approach is limited in its scope because the concentration of nutrients within a plant does generally fall with time because as a plant grows it generally produces tissues to support the structure of the plant and these have lower nutrient content than, for examples, leaves. The relationship may be applied to individual meristems over short periods but as a plant ages, nutrients are supplied to growing tissues by both uptake and redistribution of nutrients previously absorbed and stored in older tissues.

To quantify the relation between uptake of nutrients by roots and the

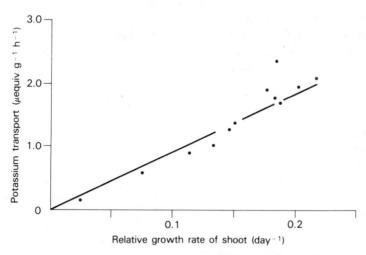

Fig. 2.15   Relation between the rate of potassium transport relative to shoot weight and the relative growth rate of the shoot for barley plants. (From Pitman, M. G., *Aust. J. biol. Sci.*, 1972, **25**, 905.)

77 Pitman, M. G., *Aust. J. biol. Sci.*, 1972, **25**, 909.
78 Raper, C. D. *et al.*, *Pl. Soil*, 1977, **46**, 473.

concentration of ions in the external solution at the root surface (C_l), it is convenient to define a coefficient, α, the root absorbing power such that[79]:

$$\text{Flux}(F) = \alpha C_l \qquad [2.13]$$

or $\qquad\qquad \text{inflow } (I) = 2\pi\alpha r C_l \qquad [2.14]$

where F is the uptake per unit surface area of root, I is the uptake per unit length of root, and r is the root radius. The mean αr over all roots in the root system, $\overline{\alpha r}$, is termed the root demand coefficient.

If the initial concentration of ions at the root surface is C_{li}, and L is the total root length

$$\frac{dU}{dt} = 2\pi\, \overline{\alpha r}\, L\, C_{li} \qquad [2.15]$$

Combining equations [2.11] and [2.15]

$$2\pi\, \overline{\alpha r} = \frac{1}{LC_{li}}\left[W.\frac{dX}{dt} + X.\frac{dW}{dt} \right] \qquad [2.16]$$

$$= \frac{W.X}{LC_{li}}\left[\frac{dX}{dt}.\frac{1}{X} + R \right] \qquad [2.17]$$

If the specific root length (L/W), the internal : external concentrations (X/C_{li}) and the relative rate of concentration change are constant, then $\overline{\alpha r}$ will change only with R and is dependent only upon the nature of the plant and the nutrient absorbing properties of the root.

Plants grown in flowing nutrient solutions[80] have been used to measure values of $\overline{\alpha}$ and $\overline{\alpha r}$ and to examine differences between species. Typically values of $\overline{\alpha}$ for plants grown in static solution culture at concentrations giving maximum growth rates are about 10^{-5} cm s^{-1}.[81] When grown in flowing solution culture, very low solution concentrations give maximum rates of growth and values of $\overline{\alpha}$ are two to three magnitudes higher.[82] Table 2.8 compares the potassium flux of four grass species, and values of $\overline{\alpha}$ and $\overline{\alpha r}$, and shows that the minimum concentration of potassium in solution necessary for maximum growth varied between species but that at that concentration, $\overline{\alpha}$ varied only by a factor of 2. Experiments to determine $\overline{\alpha}$ in soils generally give values that are 10 to 100 times smaller than those found in solution culture.[83]

79 Passioura, J. B., *Pl. Soil*, 1963, **18**, 225.
80 For details see Clement, C. R. *et al.*, *J. exp. Bot.*, 1974, **25**, 81.
81 Nye, P. H. and Tinker, P. B., *Solute Movement in the Soil–Root System*, p.118, Blackwell, 1977.
82 Wild, A. *et al.*, *J. appl. Ecol.*, 1974, **11**, 801.
83 See for example Brewster, J. L. and Tinker, P. B., *Soil Sci. Soc. Am. Proc.*, 1970, **34**, 421.

Table 2.8 Potassium flux (F), root absorbing power ($\bar{\alpha}$), and root demand coefficient ($\bar{\alpha}r$) for four plant species at concentrations of K giving maximum growth rates in solution culture[82]

Species	Concentration of K in solution (M)	F (mol cm^{-2} s^{-1})	$\bar{\alpha}$ (cm s^{-1})	$\bar{\alpha}r$ (cm^2 s^{-1})
Cocksfoot *Dactylis glomerata*	0.3×10^{-5}	2.6×10^{-12}	86×10^{-5}	8.0×10^{-6}
Scented vernal grass *Anthoxanthum odoratum*	0.1×10^{-5}	1.0×10^{-12}	97×10^{-5}	7.9×10^{-6}
Red clover *Trifolium pratense*	0.3×10^{-5}	1.7×10^{-12}	57×10^{-5}	5.2×10^{-6}
Black medick *Medicago lupulina*	0.3×10^{-5}	3.4×10^{-12}	112×10^{-5}	19.9×10^{-6}

Potential productivity of crops and crop modelling

The increasing understanding of the basic relations between growth and radiation interception, nutrient uptake and water use over the last decade has allowed the development of predictive equations to describe both the potential productivity of crops and to simulate growth for the whole growing season. Most have as their basis the relation between dry matter production and intercepted radiation discussed earlier. In conditions where water and nutrients are not limiting, and pests and diseases are assumed to be absent, the potential grain yield of a crop (Y) can be calculated from:

$$Y = I \cdot e \cdot \frac{H}{C} \qquad [2.18]$$

where I is the amount of light intercepted throughout the life of the canopy (MJ of PAR), e is the efficiency of conversion of light energy to dry matter (expressed as a fraction), H is the harvest index and C is the calorific value of the crop (MJ g^{-1}). Assigning appropriate values to this equation ($I = 756$ MJ PAR; $e = 0.045$; $H = 0.53$, and $C = 0.017$ MJ g^{-1}) indicates that the potential grain yield of winter wheat in England is about 10.3 t ha^{-1} (12.1 t ha^{-1} at 15 per cent moisture).[84] To achieve this yield, full crop cover would be required from the beginning of April until the end of July in a typical English growing season. Estimates of potential yield[85] using slightly

84 Gallagher, J. N. and Biscoe, P. V., *J. agric. Sci.*, 1978, **91**, 47.
85 Austin, R. B., *J. agric. Sci.*, 1982, **98**, 447.

more complex calculations, explored the effects of canopy characteristics on potential yield of wheat assuming that all carbohydrate for grain growth originates from photosynthesis during the grain filling period. Such calculations show that the potential yield of the best varieties of winter wheat grown in eastern England is in the range 12 to 14 t ha^{-1} (15 per cent moisture). The importance of growing and maintaining a large leaf area is highlighted by these calculations.

Leaf development and the growth of the canopy to achieve maximum interception of light is frequently the next process simulated by such models. The winter wheat model (ARCWHEAT 1)[86] developed jointly by research institutes in the UK uses photo-thermal time as the basis for leaf production and temperature as the prime determinant of leaf and canopy expansion. In this model the growth of individual leaves is summed to give the size of the canopy; other models attempt to simulate canopy growth. For example in modelling potato growth,[87] the date after planting for 50 per cent emergence of sprouted potatoes is estimated from planting date and depth of planting using soil temperature measured at 10 cm depth (base temperature (T_b) 2 °C, extension rate of shoot 1 mm per degree day). The expansion of the canopy was found to be dependent on mean air temperature and occurred exponentially to a leaf area index of 3. The relative leaf growth rate, R (day^{-1}), was linearly dependent (slope 0.013 °C^{-1} day^{-1}) on the mean air temperature above a base of 2.4 °C so that the leaf area (L) at time t (days) is

$$L_t = L_E \, e^{Rt} \qquad [2.19]$$

where L_E is the initial leaf area at emergence. During canopy expansion the amount of intercepted radiation was taken to be proportional to the leaf area index until a value of 3 was reached. Using a conversion efficiency of 1.84 g MJ^{-1} (total radiation), a 'harvest index' (a partitioning coefficient expressing the proportion of the total dry matter in the tuber) of 0.75, and a dry matter content of the tubers of 20 per cent, potential yield could be estimated for crops not limited by water, nutrients or disease. Table 2.9 compares predicted and observed yields of crops of potatoes grown under commercial conditions on farms in the east of Scotland. All the crops were well fertilized and irrigated and the results indicate the usefulness of such simple calculations to predict potential yield.

In all the approaches to modelling discussed so far, nutrients and water are assumed to be present in adequate amounts, pests and diseases absent, and, essentially, the interception of radiation and its conversion efficiency form the only constraints to yield. For practical purposes, however, it is necessary to

86 See Weir, A. H. *et al.*, *J. agric. Sci.*, 1984, **102**, 371; Porter, J. R., *op. cit.*, 1984, **102**, 383.
87 MacKerron, D. K. L. and Waister, P. D., *Agric. For. Meteorol.*, 1985, **34**, 241; MacKerron, D. K. L., *op. cit.*, 1985, **34**, 285.

Table 2.9 Comparison between predicted and observed yields (t ha^{-1}) of potatoes grown on commercial farms with adequate water and fertilizer. (From MacKerron, D. K. L., *Agric. For. Meteorol.*, 1985, **34**, 285)

Farm:	Balhungie	Balcathie	Auchrennie	East Scryne
Cultivar:	Maris Piper	Maris Piper	Desiree	Maris Piper
Total intercepted solar radiation (MJ m^{-2})	1141	1111	929	1084
Predicted tuber yield	78.7	76.7	64.1	74.8
Observed tuber yield	80.3	72.1	65.1	63.7

examine the relation between yields and nutrient and water supply. Nitrogen supply is frequently a major limitation to yields and many studies of ecosystems in Western Europe have suggested that typically around 30–50 kg N ha^{-1} is available to plants via atmospheric deposition, mineralization and non-symbiotic fixation.[88] Assuming radiation and water supply are not limiting, potential yield (Y in g m^{-2}) can be estimated from

$$Y = \frac{N_a}{N_c} \cdot H \qquad [2.20]$$

where N_a is the amount of nitrogen available to the crop (g m^{-2}), N_c is the average concentration of nitrogen in plant tissues at harvest (g g^{-1} dry matter) and H is the harvest index. Assigning appropriate values to this equation ($N_a = 3.0$ g m^{-2}; $N_c = 0.015$ g g^{-1}; $H = 0.4$) gives yields of grain of 80 g m^{-2} (0.8 t ha^{-1}); this is close to the long-term yields of wheat at Rothamsted where no fertilizer has been applied.[89] Similar calculations could be made for other areas and other crops to indicate the yields attainable where fertilizers are unavailable and agriculture is sustained only through nutrient cycling.

In practice, yields most frequently lie somewhere between the maximum determined by radiation interception and the minimum determined by nutrient and water supply (assuming no pests and diseases). The effects of nutrient and water supply on the production of leaves, their growth and maintenence, and the final yield of crops are not yet fully understood and most of the models developed to date reflect this. The Department of Theoretical Production Ecology and the Centre for Agrobiological Research at Wageningen have been at the forefront in the development of models in environ-

88 Tinker, P. B. in *Nitrogen Assimilation of Plants* (eds E. J. Hewitt and C. V. Cutting), Acad. Press, 1979.
89 Garner, H. V. and Dyke, G. V., *Rothamsted Ann. Rept. for 1968*, Part 2, p. 26.

ments where water and nutrients are limiting.[90] The basis of many models to predict the effects of water shortage is first to calculate the amount of water in the soil available to roots, second, to compare the atmospheric demand for water with the availability of water in the soil, and finally to accumulate dry matter at a rate proportional to the rate of transpiration divided by the saturation deficit. A general problem with such models is the considerable difference in time steps of the various processes operating within the models and much remains to be learned. An alternative approach has been to assume that, for example, canopy expansion occurs at the potential rate determined by temperature until a critical fraction of the available water in the soil has been depleted, when the rate of expansion then decreases linearly as more water is depleted. The size of the canopy is then used to calculate the amount of radiation intercepted and, using an appropriate value for the conversion to dry matter, crop growth is estimated.[91] The use of such relations provides a simple link between the soil scientist and the crop physiologist but their general validity has yet to be assessed.

Predicting the availability of nutrients in soils, their supply to the root, and the consequences for crop growth is presently even less well developed than the effects of water. The concepts underlying these processes are slowly being elucidated and models of nutrient uptake by plants have been developed.[92] The link, though, between uptake and growth is more difficult to quantify not least because of the difficulties of defining both the demand for nutrients by crops and the consequent relations to other physiological processes (e.g. leaf expansion and rates of photosynthesis).

In summary, in the absence of soil-related constraints and of pests and diseases, the potential productivity of many crops and ecosystems can be predicted with reasonable accuracy. The challenge to soil science, in its agricultural context, is to raise current production nearer to the potential limits imposed by the weather by understanding the individual processes operating within the soil and communicating and integrating that work with that of the other branches of crop science. The remainder of this book is, in great part, directed to that end.

90 See Penning de Vries, F. W. T. and van Laar, H. H. (eds), *Simulation of Plant Growth and Crop Production*, Pudoc, Wageningen 1982.
91 Sinclair, T. R., *Field Crops Res.*, 1986, **15**, 125;   Muchow, R. C. and Sinclair, T. R., *op. cit.*, 1986, **15**, 143.
92 See for example Nye, P. H. *et al.*, *Pl. Soil*, 1975, **42**, 161;   Claassen, N. and Barber, S. A., *Agron. J.*, 1976, **68**, 961;   Baldwin, J. P., *J. Soil Sci.*, 1975, **26**, 195.

3

Mineral nutrition of crop plants[1]

Elements required by crops

As described in the previous chapter, higher plants synthesize their tissues from simple substances which they take from the atmosphere and the soil. They absorb carbon dioxide and oxygen from the air for photosynthesis and respiration, respectively, and take up water and mineral nutrients from the soil. There are some exceptions to these general statements; for example, most plants need to have some oxygen supplied to their roots from the soil;

---

1 For general accounts of mineral nutrition of plants see Epstein, E., *Mineral Nutrition of Plants: principles and perspectives*, Wiley, 1972; Gauch, H. G., *Inorganic Plant Nutrition*, Dowden, Hutchinson & Ross, 1972; Hewitt, E. J. and Smith, T. A., *Plant Mineral Nutrition*, English Univ. Press, 1975; Mengel, K. and Kirkby, E. A., *Principles of Plant Nutrition* (3rd edn). Int. Potash Inst., 1982; *Inorganic Plant Nutrition* (eds A. Läuchli and R. L. Bieleski); *Encycl. Pl. Physiol.* (New Series), Vol. 15A, B, Springer-Verlag, Berlin, 1983; *Diagnosis of Mineral Disorders in Plants*, Vol. 1, *Principles* (eds C. Bould, E. J. Hewitt and P. Needham), London, HMSO, 1983; Marschner, H., *Mineral Nutrition of Higher Plants*, Acad. Press, 1986. A comprehensive review of the amounts of 68 elements in soils has been given by Ure, A. M. and Berrow, M. L. in *Environmental Chemistry*, Vol. 2 (Sen. Reporter H. J. M. Bowen), R. Soc. Chem., Burlington House, London, 1982.

second, nitrogen used in the metabolism of leguminous plants (and some others) can be obtained from the atmosphere through biological fixation of dinitrogen gas in their root nodules; and third, plants can absorb mineral nutrients through their leaves, with the absorption of sulphur as sulphur dioxide as perhaps the most important example (see Ch. 22).

In general, all higher plants require the same nutrients, as far as this has been tested, although there are a small number of exceptions as mentioned below. For crop plants grown under intensive conditions the aim is usually to prevent nutrient supply from limiting yields. To do this it is necessary for the plants to have access to all the essential mineral nutrients, and for the rate of supply of each to at least equal the rate of demand by the crop.

Following Arnon and Stout[2] an element is essential to a plant if: (i) a deficiency of it makes it impossible for the plant to complete the vegetative or reproductive stage of its life cycle; (ii) such deficiency is specific to the element in question and can be prevented or corrected only by supplying this element; and (iii) the element is directly involved in the nutrition of the plant quite apart from its possible effects in correcting some unfavourable micro-biological or chemical condition of the soil or other culture medium. In order to satisfy criterion (iii) the physiological and/or biochemical functions should be known. The elements now considered essential for higher plants are as follows:

1 Macronutrients: carbon, hydrogen, oxygen; nitrogen, phosphorus, potassium; calcium, magnesium, sulphur.
2 Micronutrients: iron, manganese, copper, zinc, boron, molybdenum, chlorine.
3 Beneficial elements, as discussed below, include cobalt, sodium and silicon.

Carbon, hydrogen and oxygen are obtained from the atmosphere and water as described in Chapter 2; they account for about 90–95 per cent of plant dry weight. For agronomic purposes the mineral macronutrients required from soil are separated arbitrarily into two groups[3]: (i) nitrogen, phosphorus and potassium are taken up by plants in moderate or large amounts, deficiencies are common, and they are the major constituents of commercial fertilizers; (ii) calcium, magnesium and sulphur are taken up by plants in moderate amounts and, although deficiencies are generally less common, they can be very important regionally. The micronutrients, iron, manganese, copper, zinc, boron, molybdenum and chlorine, also known as trace elements, are believed to be required by all higher plants, whereas a requirement for cobalt has only been established for biological nitrogen fixation systems, as in

2 Arnon, D. I. and Stout, P. R., *Plant Physiol.*, 1939, **14**, 371 and discussion by Hewitt, E. J. in *Plant Physiology*, Vol. III (ed. F. C. Steward), p. 97, Acad. Press, 1963.
3 For a classification of the nutrients on a physiological basis, see Mengel, K. and Kirkby, E. A. (ref. 1).

legume nodules. Sodium is not an essential element for crop plants, although it is beneficial for crops such as sugar beet and it is considered to be essential for the salt bush *Atriplex vesicaria* and other species.[4] Similarly, silicon is essential for *Equisetum arvense*[5] but for crop plants it should at present be considered beneficial rather than essential. Plants often take up large amounts of sodium and silicon and their effects are considered later in this chapter. There are also claims, as yet unsubstantiated, that nickel, tungsten and vanadium are essential.[6] The list of elements required by animals is the same except that boron is apparently not required, whereas selenium, fluorine, iodine, silicon, chromium, vanadium, tin, arsenic and nickel are now listed as essential.[7]

The list of essential elements given above for higher plants should not be regarded as final. Some of the enzyme systems for which micronutrients are required in animal metabolism might also be required by plants. In order to create a deficiency experimentally, the concentrations of any such elements in the external solution or in plant tissues will, however, be very small indeed, and there are severe difficulties in achieving and measuring these very low concentrations.[8]

The distinction between macronutrients and micronutrients is based on the relative amounts required by plants, as established from solution culture experiments. Large amounts of the micronutrients are sometimes found in plants, but this reflects the supply, or sometimes contamination by soil, rather than a requirement.

Amounts of nutrients taken up by crops

In Tables 3.1 and 3.2 the amounts of some of the elements found at harvest in the common agricultural and horticultural crops are expressed in kg t^{-1} dry matter (DM). Crop composition is variable and the amounts given in the tables are those commonly found. They are higher where uptake considerably exceeds the requirement, and they are lower where the nutrient supply limits

4 Brownell, P. F. and Wood, J. G., *Nature, Lond.*, 1957, **179**, 635; Brownell, P. F., *Adv. bot. Res.*, 1979, **7**, 117.
5 Chen, C.-H. and Lewin, J., *Canad. J. Bot.*, 1969, **47**, 125.
6 Nickel is required as a component of the urease enzyme, at least in some species, see Eskew, D. L., Welch, R. M. and Cary, E. E., *Science*, 1983, **222**, 621. Tungsten has a beneficial effect under conditions of severe molybdenum deficiency, see Fido, R. J. *et al.*, *Aust. J. Pl. Physiol.*, 1977, **4**, 675. Vanadium is required by *Chlorella* and other algae and bromine is required by some marine algae, see Hewitt, E. J. in *Diagnosis of Mineral Disorders in Plants*, Vol. 1, *Principles* (eds C. Bould, E. J. Hewitt and P. Needham), p. 7, London, HMSO, 1983. These effects do not meet the requirements of essentiality as defined by Arnon and Stout.
7 Underwood, E. J., *Trace Elements in Human and Animal Nutrition*, Acad. Press, 1977.
8 Hewitt, E. J. in *Metals and Micronutrients: uptake and utilization by plants* (eds D. A. Robb and W. S. Pierpoint), p. 277, Acad. Press, 1983.

Table 3.1 Approximate amounts of various elements present in the harvested parts of agricultural crops expressed in kg t^{-1} of crop dry matter

	N	P	K	Ca	Mg	S
Wheat – grain	20	4	5.5	0.6	1.5	1.5
– straw	7	0.8	8	3.5	0.9	1
Maize – grain	20	4	10	2	1.5	—
– straw	10	2	12	2	2	—
Rice – grain	18	4	5	1	0.2	—
– straw	4	1	10	3	1.6	—
Soyabeans	30	4	7	—	—	—
Sugar beet – roots	7.5	1.5	8	2.5	1	2
– tops	20	3	30	9	5	4
Potatoes, tubers	14	2	22	0.9	0.8	1.5
Oilseed rape	36	7	10	4	2.5	10
Ryegrass hay	16	3	18	4	1.2	1.2
Kale	24	3.5	28	20	2	6

Data selected from Archer, J., *Crop Nutrition and Fertiliser Use*, Farming Press, 1985 and Sanchez, P. A., *Properties and Management of Soils in the Tropics*, Wiley-Intersci., NY, 1976, who give analyses for other crops. Other analyses are given by Malavolta, E. *et al.*, *On the Mineral Nutrition of some Tropical Crops*, Intern. Potash Inst., Berne, 1962, and by Jollans, J. L., *Fertilizers in UK Farming*, Centre for Agricultural Strategy, Rept. No. 9, 1985, Reading Univ. For early analyses which include other elements see Warington, R., *The Chemistry of the Farm* (7th edn), Vinton & Co., Lond., 1891; Wolff, E., *Aschen Analysen*, Berlin, 1871; Wilfarth, H. *et al.*, *Landw. Vers-Stat.*, 1905, **63**, 1.

yield. In using the data in Tables 3.1 and 3.2, it is also necessary to take into account differences between crop varieties and, for vegetative crops, their age and the part of the crop that is harvested, as referred to in later chapters.

The concentrations of micronutrients in crop plants appear to vary more than those of macronutrients and Table 3.3 gives a range of values commonly found in crop plants neither deficient nor suffering from toxicity. The amounts taken up by crops are small. In 1 t of crop DM containing 0.5 μg Mo g^{-1} there is only 0.5 g molybdenum, and for copper using the concentrations in Table 3.3, the range is 5 to 20 g in 1 t DM.

The data given in Tables 3.1, 3.2 and 3.3 provide a guide to the nutrient loss from the soil due to crop removal. With wheat, for example, and using the data in Table 3.1, 59 per cent of the potassium and 85 per cent of the calcium in the crop at harvest are present in the straw whereas only 17 per cent of the phosphorus is in the straw. Returning straw, whether burned or unburned, to the soil therefore conserves potassium and calcium relatively more than phosphorus. This kind of information is needed to compare nutrient inputs and outputs from a field or some other land area, although outputs in drainage water and in gaseous form, e.g. nitrogen and nitrous oxide by denitrification and sulphur dioxide from straw burning, and inputs

Table 3.2 Content of six elements in a range of vegetable crops at final harvest. The yield is that obtained with standard fertilizer addition at the National Vegetable Research Station, Wellesbourne, England, and chemical analysis was of the whole crop excluding fibrous roots. (From D. J. Greenwood *et al.*, *J. Sci. Fd. Agric.*, 1980, **31**, 1343 and private communication)

Crop	Yield total DM ($t\ ha^{-1}$)	Amounts of nutrient ($kg\ t^{-1}\ DM$)					
		N	P	K	Ca	Mg	Na
Amaryllidaceae							
Leek	14.5	16	2.5	14	7.1	2.0	1.5
Chenopodiaceae							
Red beet	11.6	19	4.6	25	6.3	5.5	6.7
Spinach	2.0	43	5.7	56	15.8	9.4	2.0
Compositae							
Lettuce	2.3	25	4.2	33	1.1	5.6	6.9
Cruciferae							
Radish	1.0	38	4.2	40	16.3	4.2	4.1
Summer cabbage	5.3	40	4.1	26	24.8	4.7	4.6
Swede	9.8	24	4.4	25	8.9	2.5	1.7
Turnip	8.7	30	5.6	34	17.0	3.3	3.4
Winter cabbage	5.6	34	5.2	27	—	—	—
Leguminosae							
Broad bean	6.5	29	3.4	18	8.3	3.1	2.1
French bean	4.2	22	3.6	22	10.1	4.0	1.6
Pea	6.5	21	4.0	19	11.2	2.9	0.2
Umbelliferae							
Carrot	12.9	16	2.9	21	11.7	2.9	9.9
Parsnip	8.2	18	4.7	17	4.8	2.5	0.4

from the atmosphere, are also needed in order to arrive at a true nutrient balance.

Nutrient uptake usually increases throughout the course of crop growth, as shown in Fig. 3.1 for winter wheat. For each nutrient the content in roots and shoots was determined at a succession of harvests until the final grain harvest ($6.45\ t\ ha^{-1}$). Although there were some variations over the season, the general pattern was for an increase in the content of each nutrient towards the final harvest. Potassium and sulphur were exceptions, about half of each being lost after anthesis. Leaching by rain of potassium and other nutrients from leaves, especially as they mature, has commonly been observed.[9] The high loss of sulphur shown in Fig. 3.1 might be due to high

9 Tukey, H. B. Jr., *Ann. Rev. Pl. Physiol.*, 1970, **21**, 305;   Clement, C. R. *et al.*, *J. appl. Ecol.*, 1972, **9**, 249;   Cole, D. W. and Rapp, M. in *Dynamic Properties of Forest Ecosystems* (ed D. E. Reichle), p. 341, 1981.

Table 3.3   Range of concentration of micronutrients commonly found in mature leaves with a sufficient supply. (Adapted from J. Benton Jones, Jr. in *Micronutrients in Agriculture* (eds J. J. Mortvedt *et al.*), p. 319, Soil Sci. Soc. Am., Madison, 1972. See also Chapman, H. D., *Diagnostic Criteria for Plants and Soils*, Univ. Calif., 1966)

Micronutrient	Concentration $\mu g \ g^{-1}$ DM, except for Cl
B	20–100
Cu	5–20
Fe	50–250
Mn	20–500
Mo	0.2–1
Zn	25–150
Cl	0.2–1.8%

Note: For comparison with values in Tables 3.1 and 3.2, concentrations given as $\mu g \ g^{-1}$ should be divided by 1000 to give kg t^{-1} DM

sulphate concentrations on the leaves from a nearby industrial source, and other measurements are needed from non-polluted sites. Nutrients are also returned to the soil from a standing crop by leaf fall and by loss from roots. Gaseous loss of nitrogen as ammonia and of sulphur as volatile organic compounds can also occur.[10] The total uptake of nutrients by a crop is therefore usually underestimated if the content at the final harvest is used. The distinction between nutrient uptake and removal in the crop is important in assessing nutrient requirements and nutrient balance, respectively.

As mentioned above, nutrients taken up by crop plants are distributed at the final harvest between the part of the plant for which the crop is mainly grown, e.g. cereal grain, and the rest of the plant which is often of less economic value. The harvest index of a nutrient is the fraction of the nutrient taken up by the crop which is present in the economically important part of the crop at final harvest. For nitrogen in a cereal crop it is nitrogen in the grain/nitrogen in the whole crop (kg kg^{-1}). The harvest index of nitrogen and phosphate is higher with newer, short-strawed varieties of cereals than with older varieties,[11] which would be expected because the grain/straw ratio is higher. The harvest index of a nutrient is sometimes used to describe its efficiency of use. Another index is the physiological efficiency, expressed as

10 For nitrogen losses from plants see Wetselaar, R. and Farquhar, G. D., *Adv. Agron.*, 1980, **33**, 263.   The soil sulphur cycle is reviewed by Freney, J. R. and Williams, C. H. in *The Global Biochemical Sulphur Cycle* (eds M. V. Ivanov and J. R. Freney), Wiley, p. 129, 1983.
11 For nitrogen see Fischer, R. A., *Pl. Soil*, 1981, **58**, 249.   For phosphorus see Jessop, R.S. *et al.*, *3rd Int. Congr. on Phosphorus Compounds*, p. 445, 1983.   Earlier work by Jessop, R. S., *Aust. J. exp. Agric. Anim. Husb.*, 1974, **14**, 387 showed that the response of wheat to phosphate fertilizer depended on the variety.

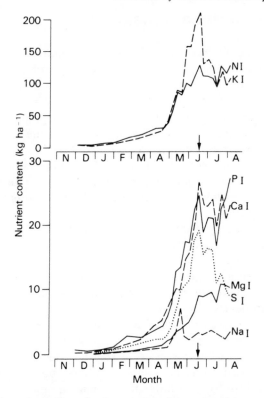

Fig. 3.1   Weights of nutrients in the winter wheat crop (shoot and root). The
standard error (I) shown for each nutrient is a mean for the whole
season and is an overestimate early in the season and slight underestimate later. The time of anthesis is shown by the arrow. (From
Gregory, P. J. *et al.*, *J. agric. Sci.*, 1979, **93**, 485.)

useful yield per unit uptake of nutrient, e.g. grain yield/nitrogen uptake
(kg kg^{-1}).

There are appreciable differences between crops and between varieties in
the amounts of each element they take up from the same soil, the differences
being greatest with the micronutrients. Because of the big genotype differences, especially in the uptake of micronutrients, there are good prospects
for breeding crop varieties that can grow well under conditions where there
would otherwise be nutrient deficiencies or toxicities.[12]

12 Gerloff, G. C. and Gabelman, W. H. in *Inorganic Plant Nutrition* (eds A. Läuchli and
R. L. Bieleski), *Encycl. Pl. Physiol.* (New Series), Vol. 15B, p. 453, Springer-Verlag,
Berlin, 1983;   Clark, R. B. in *Genetic Aspects of Plant Nutrition* (eds M. R. Sarić and
B. C. Loughman), p. 49, Nijhoff, The Hague, 1983;   Graham, R. D. in *Adv. Plant
Nutrition*, Vol. 1 (eds P. B. Tinker and A. Läuchli), p. 57, Praeger, NY, 1984.

Principles of nutrient uptake

The principles that are currently used to describe nutrient uptake by plants have been given in Chapter 2. In essence, the uptake of a nutrient by plant roots is described by the product of the concentration of the nutrient in the soil solution and the root absorbing power, that is, $I = 2\pi\bar{\alpha}\bar{r}C_l$ where I is uptake per unit length of root, $\bar{\alpha}$ is the root absorbing power, $\bar{r}$ is root radius and C_l is the concentration of nutrient in solution at the root surface. The root absorbing power is expressed in terms of: (i) the ratio of total root length (or root surface area) to plant weight; (ii) the ratio of external to internal nutrient concentration; (iii) the relative growth rate (RGR) of the plant; and (iv) the relative rate of change of internal nutrient concentration (see equation [2.16]).

Nutrients taken up by plants are used for their growth and development or are stored, and the concentration at the root surface, C_l, plays a key role in meeting these requirements. This is illustrated in Fig. 3.2 which shows the potassium content and growth of fodder radish (*Raphanus sativus*) and perennial ryegrass (*Lolium perenne*) over a range of low potassium concentrations. The nutrient solutions were circulated through the root systems of the plants and the concentrations of potassium were kept constant by continuous monitoring and replenishment. Other nutrients were non-limiting. Uptake of potassium by both species, as shown by the potassium content of the plants, increased with the solution concentration, which was up to 4 mg l^{-1} K (about 10^{-4}M) and lower than commonly found in soil solutions (see p. 751). The effects of the external concentration on growth differed, however, between species. Whereas radish gave a growth response over the whole concentration range, perennial ryegrass gave none above the lowest concentration tested, 0.05 mg l^{-1} K, indicating a requirement of less than 1.2 $\times$ 10^{-6} M at the root surface. The explanation for the difference between the two species was that ryegrass was able to meet its requirements for growth at a very low potassium concentration because of its low RGR (0.19 g g^{-1} day^{-1}) and high ratio of root surface area to plant fresh weight (0.08 cm^2 mg^{-1}). Radish had a higher RGR (0.28 g g^{-1} day^{-1}) and a lower ratio of root surface area to plant fresh weight (0.04 cm^2 mg^{-1}). This example illustrates two points. First, at relatively low external nutrient concentrations, uptake and plant growth increase with increased concentration in the external solution. Second, the relation between growth and nutrient concentration in the external solution differs between species. In other words, in considering the requirement and availability of nutrients for plants, the properties of the plant, and as shown below also of the soil, need to be taken into account. In this experiment the nutrient solution was circulated through the root system, thus maintaining constant concentrations of potassium at, or close to, the root surface. Systems such as sand culture and soil require higher nutrient concentrations in the bulk solution because of depletion close to the root.

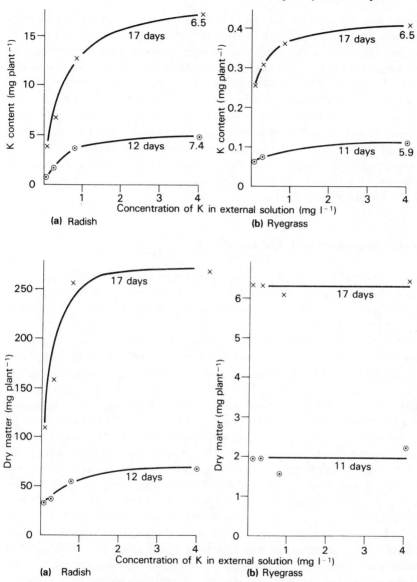

Fig. 3.2 Uptake of potassium and growth of two plant species in flowing nutrient solutions containing 0.05 to 4 mg K l⁻¹. (Adapted from Woodhouse, P. J. *et al.*, *J. exp. Bot.*, 1978, **29**, 885, and unpublished.)

For some nutrient ions, including the major anions and potassium, uptake into the root is usually against the electrical potential gradient and therefore requires energy which is provided by oxidation of photosynthate. The oxygen is provided either from the soil solution or through aerenchyma (see p. 310).

Uptake also depends on pH, and for plants grown in nutrient solutions is generally highest at about pH 5 to 6.[13] A high concentration of one nutrient ion can also depress the uptake of another, a phenomenon known as ion antagonism. It applies generally between cations but is less common between anions. Potassium competes strongly with other cations, and in excess can cause deficiency of magnesium if this supply is marginal. Concentration of any element, if sufficiently high in the external solution, can also lead directly to reduced growth and death of the plant. The elements that most commonly harm plants include aluminium, manganese, copper, nickel and zinc although this group includes three (manganese, copper and zinc) which are essential nutrients. Metal toxicity is discussed elsewhere (see Ch. 25 for aluminium and Ch. 23 for other metal toxicities).

Plant requirement for nutrients

If the concentration in the soil solution is high enough nutrient uptake is largely in response to plant demand. The requirements are for vegetative growth and for the production of storage organs (grain, roots, tubers), and each will now be considered.

The highest recorded rates of dry matter increase over short periods of time are 510–540 kg ha^{-1} day^{-1} for C_4 plants and 340–390 kg ha^{-1} day^{-1} for C_3 plants (Table 2.5). In Britain and parts of Western Europe the highest average value is about 280 kg ha^{-1} day^{-1}. In order to achieve these growth rates factors other than solar radiation must be non-limiting. The requirement for nutrients is that their concentrations in plant tissues be maintained above a critical value. The critical concentration required for maximum growth should refer to particular plant tissues,[14] but most data are for whole plants. For nitrogen, phosphorus and potassium during the period of rapid growth the critical values are about 2, 0.4 and 2 per cent respectively of the dry matter (DM) of the whole plant. It follows that to maintain a growth rate of 280 kg DM ha^{-1} day^{-1}, the required nutrient uptake rates are 5.6, 1.1 and 5.6 kg ha^{-1} day^{-1} for nitrogen, phosphorus and potassium, respectively. The highest rates in experiments with four field grown crops are of this order of magnitude (Table 3.4). These high rates of uptake were maintained for only short periods and showed considerable fluctuation, probably because of variation of solar radiation and the restriction of nutrient availability imposed by low soil water contents. In general, relatively high amounts of nutrients are taken up during the early stages of vegetative growth as shown in Fig. 3.3

13 Blamey, F. P. C. *et al., Proc. 9th Int. Pl. Nutr. Coll.*, 1982, **1**, 66.
14 Loneragan, J. F., *Nature, Lond.*, 1968, **220**, 1307; Clarkson, D. T. and Hanson, J. B., *Ann. Rev. Pl. Physiol.*, 1980, **31**, 239.

Table 3.4 Highest rates of nutrient uptake observed for three crops during vegetative growth

Crop	Period	Highest uptake rates (kg ha^{-1} day^{-1})		
	(days)	N	P	K
Maize (1)	10	9.0	1.2	10.6
Winter wheat (2)	7	3.7	0.7	10.2
Potatoes (3)	14	4.6	0.7	7.3
Ryegrass (4)	7	7.5	—	—

(1) Mengel, D. B. and Barber, S. A., *Agron. J.*, 1974, **46**, 399.
(2) Gregory, P. J. *et al., J. agric. Sci.*, 1979, **93**, 485.
(3) Asfary, A. F. *et al., J. agric. Sci.*, 1983, **100**, 87.
(4) From Whitehead, D C., *Commonw. Bur. Past. Field Crops Bull.*, No. 48, 1970.

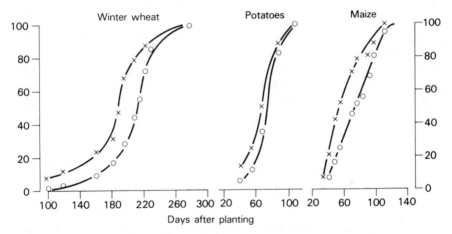

Fig. 3.3 Accumulation of dry matter (o) and nitrogen (x) in winter wheat, potatoes, and maize, as per cent of final harvest. (Gregory, P. J. *et al., J. agric. Sci.*, 1979, **93**, 485 for wheat; Asfary, A. F. *et al., J. agric. Sci.*, 1983, **100**, 87 for potatoes; and Mengel, D. B. and Barber, S. A., *Agron. J.*, 1974, **66**, 399 for maize.)

for the uptake of nitrogen by winter wheat, potatoes and maize. Because the period of the greatest uptake of nitrogen precedes the period of highest growth rate, the percentage of nitrogen in the DM of the whole crop decreases during the growing season. The pattern of uptake of other nutrients is similar at the vegetative stage to that shown for nitrogen, and in general the percentage of all nutrients in the crop DM decreases towards harvest.

During the development of the storage organs (seed, root, tuber), their nutrient requirements are met partly by continued uptake of nutrient and partly by transfer from stems and leaves. A distinction is needed between elements, which include nitrogen, phosphorus and potassium, which are mobile in the phloem and those which are comparatively immobile, for example, calcium, boron and iron, and are transferred only slowly to the developing organ. An example of the uptake and transfer of nitrogen and phosphorus during the period of grain filling of winter wheat is given in Table 3.5. Nitrogen in the grain at harvest (110–200 kg N ha^{-1}) came partly from uptake after anthesis (51–106 kg N ha^{-1}), and partly from the straw (up to 70–105 kg N ha^{-1}). A small amount of nitrogen was present in the inflorescence, which is not given in Table 3.5, and some was lost from the vegetative parts and not transferred to the grain. On average, 48 per cent of the nitrogen in the grain was taken up after anthesis and about 52 per cent was transferred from the vegetative parts. With phosphorus the pattern was similar: 45 per cent of the phosphorus in the grain was taken up after anthesis and about 55 per cent was transferred from the vegetative parts. During the five-week period from anthesis to harvest the net uptake of nitrogen corresponded to an average rate of between 1.5 and 3.0 kg ha^{-1} day^{-1}, and for phosphorus 0.36 to 0.41 kg ha^{-1} day^{-1}. More potassium was lost from the vegetative parts than was found in the grain, so there was loss of potassium from the whole plant, as discussed earlier. In this experiment the high uptake of nitrogen and phosphorus after anthesis might have been due to the good supply of water and nutrients in the subsoil. Late application of fertilizer nitrogen increased

Table 3.5 Uptake of nitrogen and phosphorus by winter wheat during the period of grain filling. (From Spiertz, J. H. J. and Ellen, J., *Neth. J. agric. Sci.*, 1978, **26**, 210)

Fertilizer N* (kg ha^{-1})	Grain (at harvest) (t ha^{-1})	(%N)	In grain at harvest	Net uptake after anthesis	Loss from straw after anthesis
			Nitrogen (kg ha^{-1})		
50	6.29	1.75	110	51	70
100	6.91	2.07	143	72	82
100 + 50	8.56	2.09	179	93	101
100 + 100	8.58	2.33	200	106	105
			Phosphorus (kg ha^{-1})		
50	6.29	1.75	27.4	12.8	15.6
100	6.91	2.07	28.1	12.9	16.6
100 + 50	8.56	2.09	30.6	13.7	17.3
100 + 100	8.58	2.33	32.0	14.3	18.3

* Applications of 50 or 100 kg N ha^{-1} early (16 March) and 0, 50 or 100 kg N ha^{-1} late (20 May).

the uptake rate of nitrogen (kg ha^{-1} day^{-1}) during grain filling from 2.1 (no late N) to 2.6 (50 kg late N) to 3.0 (100 kg late N). Not all experiments show this late uptake (see Fig. 3.1 for example).

The high uptake rates of nitrogen and phosphorus after anthesis are a response to the nutrient demand during grain filling. At this period the soil is often dry, which will slow nutrient transport to the root. For high grain yields it is therefore necessary for roots to be able to take up nutrients at a high rate after anthesis, which may explain the benefit from deep incorporation of nutrients in dry seasons[15] and for split applications.

Soil as a source of nutrients

In studies of the mineral nutrition of plants and crops grown in soil it is assumed that nutrients enter the root from the soil solution. Although there can be no proof that this is always true there is a mass of self-consistent evidence to support it, which comes mainly from solution culture work.

There have been claims that nutrient cations in general can be taken up directly from the solid phase of the soil without passing through the solution. This was the 'contact exchange' concept of Jenny and Overstreet,[16] according to which hydrogen ions on the root surface exchange with nutrient cations held on soil surfaces because of an overlapping of their oscillation volumes. However, the evidence from later experimental work,[17] and the application of diffusion theory to nutrients in soil (see pp. 155–163), make it improbable that contact exchange is a general phenomenon.

The evidence from solution culture work is that an actively growing plant takes up nutrient ions at rates which are related to their concentration or activity in solution.[18] An apparent difficulty with the solution theory was that in soils able to sustain good plant growth the solution concentration of some nutrients, e.g. potassium[19] and phosphate might be up to 100 times lower than that required to sustain comparable growth rates in conventional solution culture. Experiments have shown, however, that plants can obtain their nutrients at sufficient rates to maintain rapid growth at very low concentrations if the concentrations at the root surface are maintained by frequent addition and mixing. For example, using flowing nutrient solutions, concentrations of phosphate of about 0.03 mg l^{-1} P (10^{-6} M) and below have been found sufficient to maintain high growth rates of ryegrass as long as signifi-

15 McEwan, J. and Johnston, A. E., *J. agric. Sci.*, 1979, **92**, 695.

16 Jenny, H. and Overstreet, R., *Soil Sci.*, 1939, **47**, 257; *J. phys. chem.*, 1939, **43**, 1185.

17 Olsen, R. A. and Peech, M., *Soil Sci. Soc. Am. Proc.*, 1960, **24**, 257.

18 For discussion of the use of ionic concentrations and ionic activities in the mineral nutrition of plants see Khasawneh, F. E., *Soil Sci. Soc. Am. Proc.*, 1971, **35**, 426; Sparks, D. L., *Soil Sci. Soc. Am. J.*, 1984, **48**, 514.

19 Wild, A. *et al.*, *J. appl. Ecol.*, 1974, **11**, 801.

cant nutrient depletion of the solution does not occur. Similar results have been reported for potassium, nitrogen and calcium.[20] These experiments have also shown that the relationships between the concentration of a nutrient ion in the external solution and the rates of uptake and growth are not unique but depend on the plant species, its growth rate and stage of development. Thus, the parameters used by plant physiologists to describe nutrient uptake kinetics (V_{max} = maximum uptake rate, and K_m = external concentration of ion when the uptake rate is half maximum) can give the false impression that uptake rates are linked only with ion concentrations in the external solution.

Using flowing nutrient solutions, concentrations of ions can be maintained constant at, or very close to, the root surface. For plants growing in soil, however, the concentration of a nutrient at the root surface depends on the relative rates of uptake by the root and of transport to the root surface from the bulk soil solution. In contrast to flowing nutrient solutions, this introduces the requirement for diffusion of the nutrient, or mass flow of the soil solution, to carry the nutrient to the root surface. There is also a requirement to replenish the soil solution because the concentrations of many of the nutrients in the solution are low. This applies, for example, to phosphate and potassium, for which the amounts in solution at any time may represent only about 1–10 per cent of the amounts taken up by an annual crop. Replenishment occurs by desorption from the surfaces of the mineral and organic components of soil and by mineralization of soil organic matter. The amount of

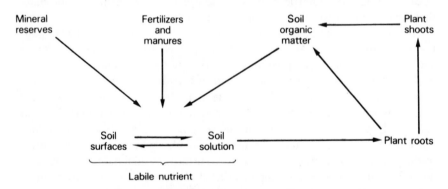

Fig. 3.4   Schematic representation of the readily available nutrients in soil, their replenishment, and their transfer to plant roots and shoots.

20 For phosphate see Breeze, V. G. *et al.*, *J. exp. Bot.*, 1984, **35**, 1210; for earlier work Asher, C. J. and Loneragan, J. F., *Soil Sci.*, 1967, **103**, 225 and Loneragan, J. F. and Asher, C. J., *op. cit.*, 311.   For potassium see Woodhouse, P. J. *et al.*, *J. exp. Bot.*, 1978, **29**, 885; for earlier work Asher, C. J. and Ozanne, P. G., *Soil Sci.*, 1967, **103**, 155.   For nitrate see Clement, C. R. *et al.*, *J. exp. Bot.*, 1978, **29**, 453. For calcium see Loneragan, J. F. *et al.*, *Aust. J. agric. Res.*, 1968, **19**, 845.

nutrient which can rapidly replenish the soil solution is, together with the amount in solution, described as labile (Fig. 3.4). The determination of the labile fraction, and the relation between concentration in solution and the amount held on surfaces are described in later chapters. The labile fraction can be replenished by addition of fertilizer, by mineralization of soil organic matter which provides mainly nitrogen, phosphorus and sulphur, and by slow release from within the structure of the soil minerals. The supply system in soil for most nutrients is therefore usually well buffered. An important implication is that no single measure of nutrient availability is generally valid. A complete description of nutrient availability should include: (i) the concentration in the soil solution; (ii) the rate of replenishment of the solution at the root surface; (iii) the amount of nutrient which can replenish the solution; and (iv) plant factors referred to below.

Which of the three soil factors limits uptake differs between the nutrients.[21] For example, with phosphate the concentration at the root surface will generally limit uptake because of its low rate of movement through the soil. With nitrogen in a moist soil it is the amount in the labile fraction which limits uptake rather than rate of transport. The amount which is available to a crop is therefore the amount of mineral nitrogen (NH_4^+-N and NO_3^--N) in the soil profile at the start of the growing season and the amount which is mineralized while the crop is growing. No measure of nutrient availability will, however, suit all plant species because, as discussed above, they differ in root length, growth rate and nutrient requirement. Further, plants themselves can have a profound effect on nutrient availability through mycorrhizal associations and rhizosphere processes (see Ch. 17).

Because nitrate is readily transported through moist soil the total length and distribution of roots are not prime determinants of uptake. Phosphate and potassium are, however, comparatively immobile in the soil, and for these nutrients availability depends on their being close to the root. Substantial work since about 1960 has shown that diffusion to the root surface is the step that usually limits the rate of uptake of phosphate and potassium. Under carefully controlled conditions in pot experiments, the rate of uptake of phosphate has been predicted moderately well for some crops using soil solution concentrations, diffusion coefficients and growth rates (see Ch. 4). It is more difficult to make similar predictions for field grown crops and is likely to prove too complex a method for providing advice on fertilizer use. For most nutrients we therefore resort to the use of various extractants which rank soils according to an index of availability of a nutrient as discussed later in this chapter.

21 Barber, S. A. and Silberbush, M. in *Roots, Nutrient and Water Flux and Plant Growth*, p. 65, Wiley, 1984.

Functions of nutrient elements in plants

The nutrients in a germinating seed meet the requirements of the embryonic plant for only the first few days of its life. Plant roots grow before the shoot has emerged from the soil, and the nutrients that are absorbed by them, together with photosynthesis when the shoot has emerged, allow the seedling to continue to grow. Plant growth involves a large number of highly integrated enzyme reactions and physiological or metabolic processes, all of which are dependent on adequate supplies of nutrients. However, knowledge of the functional roles of the essential nutrients varies from element to element, and most of them have several functions. A full account of these is given in the books listed under ref. 1, and what follows is an outline of their main functions as they relate to crop yields. There is further discussion of the macro- and micronutrients in later chapters.

Carbon, hydrogen, oxygen

About 90–95 per cent of the dry weight of most higher plants consists of carbon, hydrogen and oxygen which occur in about the same ratio as in carbohydrates. This reflects the fact that carbohydrates are the main repository of photosynthetic energy; they comprise: (i) simple sugars and their conjugates active in intermediary metabolism; (ii) storage or reserve compounds, e.g. starch, sucrose, fructans; and (iii) structural polysaccharides of plant cell walls, principally cellulose, hemicelluloses and pectins. Also associated with the structural polysaccharides are phenolic compounds, including lignin, and silicon which is taken up in large quantitites by cereals and grasses and deposited as $SiO_2.nH_2O$ in the cell walls.

Macronutrient elements

Nitrogen

Nitrogen is usually the fourth most abundant element in plants, following carbon, oxygen and hydrogen. It has an essential role as a constituent of proteins, nucleic acids, chlorophyll and growth hormones. Most plants depend entirely for growth on inorganic nitrogen taken up from the soil as nitrate or as ammonium ions, but legumes can also utilize dinitrogen which is used via the mediation of symbiotic bacteria which reduce it to ammonium. Most of the absorbed nitrate is reduced sequentially by enzymes to nitrite then ammonium. Irrespective of whether nitrate, ammonium or dinitrogen is the source of nitrogen, ammonium is the intermediate which serves in the formation of amino acids and amides and subsequently proteins, as well as other, high molecular weight compounds containing reduced nitrogen. The synthesis of the various amino acids and amides needed to make proteins depends on an adequate supply of the appropriate carbon compounds synthesized in the leaves. With increases in nitrogen supply there are increases in both soluble amino compounds and proteins; the additional protein allows the leaves to grow larger and hence to have a larger area for photosynthesis.

Increasing the nitrogen supply has other effects on the leaf in addition to increasing growth. The higher the supply the greater the demand for carbon compounds and the smaller the proportion of carbohydrate left available for cell wall material. Excessive supplies of nitrogen produce leaves with large, thin-walled cells that are susceptible to attack by insects and fungi[22] and are harmed by unfavourable weather such as droughts and frosts. A very low nitrogen supply on the other hand gives yellow or pale green leaves with small cells and thick walls, and the leaves are in consequence harsh and fibrous. The leaves become darker as the nitrogen supply increases and become very dark green when it is excessive. Increasing the nitrogen supply to the leaves also tends to keep them green for a longer time, and in many cereals an excessive supply increases the length of the growing season and delays the onset of maturity.

Crops grown for their carbohydrates, such as root and tuber crops and sugar cane, thus benefit from nitrogen manuring mainly through the increased leaf area and its duration, so that the additional yield of carbohydrate is usually proportionately less than the increase in leaf area. Root crops show this effect most strikingly when the length of the growing season is varied. Thus, a high addition of nitrogen fertilizer mainly affects the shoots of a crop which is only in the ground for a short time, such as white turnips, but has a very considerable effect on both shoots and roots of a crop that is in the ground for a long time, such as sugar beet and mangolds. Late-sown sugar beet sometimes responds in leaf growth to nitrogen fertilizer, but this might be accompanied by a relatively low increase in yield of root and sugar.

With cereals the main effect of raising the nitrogen supply is to increase tiller survival and growth, and hence leaf area. This is shown by the effects of applications of fertilizer nitrogen early in the growing season, which increase grain yields by increasing the number of ears per unit area, and the number of grains per ear. The addition of nitrogen fertilizer, however, usually increases the yield of straw more than the yield of grain. The harvest index, expressed as yield of grain/yield of grain plus straw is therefore decreased, as shown by results from the continuous wheat plots on the Broadbalk experiment at Rothamsted Experimental Station (Table 20.7).

The addition of nitrogen fertilizer usually increases the nitrogen concentration in the crop. If, however, the fertilizer greatly stimulates crop growth,

22 For the relationship between plant disease and the supply of nitrogen (and other nutrients) see Last, F. T., *Pl. Path.*, 1962, **11**, 133 and comprehensive review by Graham, R. D. in *Adv. bot. Res.* (ed H. W. Woolhouse), 1983, **10**, 221.

as may happen where nitrogen is a severely limiting factor, the nitrogen content of the crop, expressed as a percentage of its dry or fresh weight, will sometimes decrease, although the total uptake will increase. Nitrogen fertilizers can often be made to increase the nitrogen content of the crop more economically if they are applied sufficiently near harvest for the crop to take up much of the added nitrogen, but for the synthesized proteins not to have time to increase growth appreciably. Thus, by adding nitrogen fertilizer to grassland one to three weeks before the grass is cut, the protein content of the hay will be increased though the yield will scarcely be affected.[23] With cereals on the other hand, nitrogen fertilizer appears to increase the protein content of the grain whenever it is applied during vegetative growth,[24] unless there is gaseous or leaching loss of nitrogen.

Crops that respond to nitrogen commonly take up between one-third and two-thirds of the amount added as fertilizer. The remainder is lost to the crop, being either denitrified, volatilized as ammonia, leached out of the soil during wet weather, or immobilized within the soil (see (Ch. 20).

Phosphorus

Phosphorus is present in the soil solution as $H_2PO_4^-$ and HPO_4^{2-} ions and is generally believed to be taken up by plants mainly as $H_2PO_4^-$. After absorption, much of the phosphate reacts very quickly to form organic compounds,[25] and it plays an important role in the very large number of enzyme reactions that depend on phosphorylation. These include the incorporation of phosphate in nucleotides, notably adenosine di- and triphosphates (ADP and ATP) and analogous nucleotide triphosphates, and hence it has a key role in the conservation and transfer of energy for a wide range of biochemical processes. Phosphate is also a constituent of nucleic acids and of phospholipids, including those of cytoplasmic membranes. Phytic acid, the hexaphosphate ester of inositol, or its calcium or magnesium salt, is formed in seeds and provides an important source of phosphate during germination.

Phosphate and the other nutrient ions are essential for cell division and for the development of meristem tissue. The higher concentration of phosphorus in these tissues can be demonstrated if radioactive phosphorus (^{32}P) is mixed with the main supply of phosphorus. An autoradiograph of the phosphorus in a young barley plant is reproduced in Plate 3.1. It shows very clearly that the actively growing meristematic leaf and root cells contain far more

23 Ferguson W. S., *J. agric. Sci.*, 1948, **38**, 33.
24 Holbrook, J. R., *et al.*, *J. agric. Sci.*, 1982, **99**, 163.
25 Loughman, B. C. and Russell, R. S., *J. exp. Bot.*, 1957, **8**, 280; Loughman, B. C., *New Phytol.*, 1966, **65**, 389. For a review of phosphorus in plant metabolism see Bieleski, R. L. and Ferguson, I. B. in *Inorganic Plant Nutrition* (eds A. Läuchli and R. L. Bieleski), *Encycl. Pl. Physiol.* (New Series), Vol. 15A, p. 422, Springer-Verlag, Berlin, 1983.

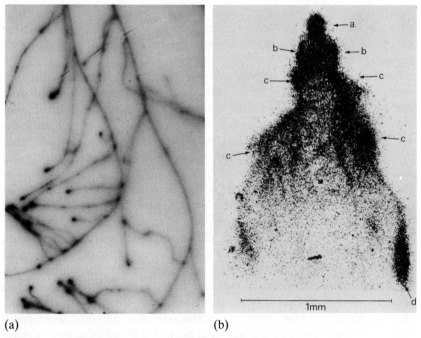

(a)   (b)

Plate 3.1 Autoradiographs of barley plants containing radioactive phosphorus:
(a) Showing the concentration of phosphorus in the tips of the growing roots (natural size).
(b) Showing the concentration of phosphorus in the apical meristem: (a) the leaf primordia; (b) the bases of leaf sheaves; (c) and (d) the initials of adventitious roots (× 53).

phosphorus, in fact from several hundred to several thousand times more, than the cells that have ceased to divide.[26]

Cereals suffering from phosphorus deficiency are retarded at every stage of their life history, from the emergence of the second leaf to the time of ripening. They have a stunted root system and an even more stunted leaf and stem; the leaf colour is a dull greyish-green, a red pigment due to anthocyanins is often produced in the leaf bases and the dying leaves, and tillering and the number of tillers bearing grain are depressed. On the other hand, except in extreme cases, the ratio of grain to straw is not affected. On soils with low supplies of phosphate, phosphatic fertilizers hasten the ripening processes, which can be of crucial importance in getting a grain yield in semi-arid regions of the world. This effect on ripening is well shown on the barley plots at Rothamsted Experimental Station where plants receiving phosphate are golden yellow in colour while those on the phosphate-starved plots

26 Russell, R. S. and Martin, R. P., *Nature, Lond.*, 1949, **163**, 71.

remain green. Certain indirect effects also follow; the ears of barley emerge a few days in advance of those receiving insufficient phosphate, and therefore have a better chance of escaping attack by the larvae of the gout fly (*Chlorops toeniopus* Meig.).

Root crops suffering from severe phosphorus deficiency are also very stunted, and the effect of added phosphate can be spectacular. The early workers were so impressed with the great increase in the yield of roots obtained by phosphatic fertilizers that they assumed that phosphate had a specific action in encouraging root development. Doubt has, however, been cast on this assumption by later experimentation. For example, using flowing nutrient solutions the proportion of plant dry weight present in the roots of several plant species has been found to be greatest at very low concentrations of phosphate in solution, and to decrease with increased concentrations.[27] In conventional nutrient solutions it has also been shown that the number of second-order laterals of barley roots may be higher at low concentrations of phosphate.[28] On the other hand, localized placement of phosphate in the root zone produces a localized proliferation of lateral roots.[29] The reason for this difference in effect between a localized and a general, high concentration in the root zone is not yet known. As with other nutrients, however, the main effect of phosphate arises from the feedback between root and shoot development. The addition of phosphate to a plant that would otherwise be phosphate deficient leads to more rapid leaf growth, and some of the extra carbohydrate which is synthesized is used for further root growth.

Potassium

Potassium is one of the three or four elements that are most commonly in sufficiently short supply in the soil to limit crop yield. The content in plants is about the same as that of nitrogen and it is the most abundant cellular cation. Although not known to be a constituent of any essential organic compound in plants, potassium has several physiological and biochemical roles, for example in protein synthesis. It is required in large concentrations to activate many enzymes and to neutralize anions, and anionic groups of macromolecules, thereby contributing to the osmotic potential. Potassium plays a specific role in the mechanism of opening and closing of stomata; it accumulates, with malate, in the guard cells when stomata are open and is released when they are closed, thereby contributing to the osmotic potential on which the turgor of the guard cells depends. Potassium is also involved in the transport of photosynthate from the leaves and as a result it can have an indirect effect on photosynthesis. Potassium can be substituted to some degree by sodium in its physiological roles, particularly in the maintenance

27 Loneragan, J. F. and Asher, C. J., *Soil Sci.*, 1967, **103**, 311.
28 Hackett, C., *New Phytol.*, 1968, **67**, 287.
29 Drew, M. C., *New Phytol.*, 1975, **75**, 479.

of cell turgor; the sparing effect of sodium on potassium deficiency is usually attributed to an osmotic role.

The potassium supply in the soil may be adequate for a crop under conditions of low nitrogen and phosphorus supplies, but becomes inadequate if these are increased. Hence, signs of potassium deficiency are often seen only when nitrogen and phosphate fertilizers are given to a crop. The most characteristic sign is the premature death of the older leaves as potassium is highly mobile and readily transported to the young leaves. When nitrogen and potassium are simultaneously in short supply, the plants are stunted, their leaves are small and rather ashy-grey in colour, dying prematurely, first at the tips and then along the outer edges, and the fruit and seed are small in quantity and size. These effects are general, and are seen in crops on all soils, but deficiency is most common on sandy or chalky soils and on some peats, and it is on these soils that potassium fertilizers are most likely to benefit all crops.

With large supplies of nitrogen relative to potassium the leaves are large but relatively inefficient photosynthesizers. Hence abnormal concentrations of soluble nitrogen compounds, particularly amines, compared with soluble carbohydrates accumulate in the leaf, leading to various undesirable effects, including a greater liability to fungal and bacterial diseases, and to a reduced resistance against damage by drought.

Crops differ greatly in their responsiveness to potassium. Many fruit trees, including apples, gooseberries, red currants, need ample supplies for good cropping and quality: *Vicia* beans and potatoes among field crops, and tomatoes among the glasshouse crops, are all very responsive. Clovers and lucerne also seem to need good supplies of potassium, particularly if they are to compete successfully with grasses; and for lucerne potassium also increases winter hardiness, possibly because it encourages the plant to store more carbohydrate and protein in its root system.[30]

Excess potassium in the soil, as brought about by too high a level of potassium manuring, reduces very considerably the amounts of other cations, especially magnesium, the crop can take up. This may lead to growth being depressed by these induced deficiencies of other cations.

Calcium

The concentration of calcium in the dry weight of plants can vary over the wide range of about 0.1 to 2.5 per cent (Tables 3.1 and 3.2) but, as it has low mobility and cannot be readily redistributed from older to new leaves, the higher concentrations are probably greater than the metabolic requirement. Calcium is essential for the growth of meristems, and particularly for the proper growth and functioning of root tips. It plays a key role in the

30 Graber, L. F. *et al.*, *Wisc. Agric. Expt. Stn. Res. Bull.*, No. 80, 1927.

maintenance of the integrity of membranes, protecting them against leakiness, and with roots in particular, against ion imbalance, low pH, and toxic ions such as aluminium. Calcium is required by a few enzymes, including amylase and some nucleases. It is the main cation associated with the middle lamella of cell walls and, in seeds, is present as the salt of phytic acid. There is recent evidence that the Ca^{2+} protein calmodulin plays a key role in cellular activity in plants by moderating enzyme activity.[31]

Calcium deficiency causes the growth of the root system to be stunted and leaf margins, especially of young leaves, to curl backwards and the tip forwards. Later, the leaves may show marginal chlorosis and these areas eventually become chlorotic. Degeneration at the apex of young fruit ('blossom end rot') is a common symptom of deficiency in tomatoes, as is 'bitter pit' in apples. Calcium deficiency occurs rarely as a field problem. Crop yields on acid soils are often improved by liming, but the effects are usually due to lower aluminium concentrations in the soil solution, as described in Chapter 25.

Magnesium

The concentration of magnesium in the dry matter of plants is variable, but generally lower than calcium (Tables 3.1 and 3.2). Unlike calcium, however, magnesium is mobile and a large proportion of the total magnesium is associated with organic anions, e.g. malate. Magnesium is a specific constituent of chlorophyll (magnesium porphyrin), in which one atom of magnesium is bound to four pyrrole rings. Magnesium also plays a major role in numerous enzyme reactions. As a cofactor of most of the enzymes that act on phosphorylated substrates it is of great importance in the transfer of energy. It also activates certain other enzymes, including carboxylases and some dehydrogenases. Because of its role as a cofactor of enzymes that act on phosphorylated substrates, the distribution of magnesium in plants is often similar to that of phosphorus.

The symptoms of magnesium deficiency vary among species, but the first symptom is always interveinal yellowing, or chlorosis, of the older leaves due to diminished levels of chlorophyll. This is accompanied by a decreased rate of photosynthesis. Biosynthetic pathways are disrupted due to the inhibition of phosphorylation processes, and soluble nitrogenous compounds accumulate with concomitant decreases in the proportion of protein. There are reports of deficiency from many countries, especially in sugar beet, potatoes and fruit crops, and with many crops following heavy applications of lime to acidic, sandy soils. This is due to a restriction of magnesium uptake by

31 For reviews of the functions of calcium in plants see Hanson, J. B. in *Adv. Plant Nutrition*, Vol. 1 (eds P. B. Tinker and A. Läuchli), p. 149, Praeger, 1984; Marmé, D. in Inorganic Plant Nutrition (eds A. Läuchli and R. L. Bieleski), *Encycl. Pl. Physiol.* (New Series), Vol. 15B, p. 599, Berlin, Springer-Verlag, 1983.

calcium ions, sometimes called 'antagonism', or 'interaction'. High levels of potassium or ammonium ions in the root medium similarly restrict the uptake of magnesium.

Sulphur

Sulphur is absorbed from the soil solution as SO_4^{2-}, and can also be assimilated from the sulphur dioxide present in the atmosphere. A large proportion of the sulphur is then reduced from sulphate to sulphide, which is incorporated in the amino acids: cysteine, cystine and methionine. Normally these amino acids are present in the free state in very small quantities as they are rapidly converted to proteins. Reduced sulphur is also present in some coenzymes, including biotin, thiamine and coenzyme A, which are essential for metabolism when attached to appropriate apoenzymes. The ferrodoxins, which are non-haem proteins involved in photosynthesis and other electron transfers, contain reduced sulphur (as sulphide) and iron in equivalent amounts. Some species, notably brassicas, synthesize mustard oils which contain sulphur, both as sulphate and in reduced form.

Proteins are the form in which most of the sulphur and nitrogen are stored in plants. When the supply of sulphur is ample, the rate of uptake may exceed the rates of reduction and assimilation into proteins and other organic compounds with the result that sulphate accumulates in the tissues. The atomic ratio of protein-nitrogen to protein-sulphur is fairly constant with an average of about 36 : 1, which on a weight basis is 15.7 : 1. This is close to the ratio of total nitrogen to total sulphur in plants grown under conditions where there is no excess uptake of either element. If the supply of sulphate from the soil is inadequate, the ratio of total nitrogen to total sulphur will increase above 15 or 16 : 1, and this relationship is sometimes used to assess the adequacy of supply of sulphur from soil (see also Ch. 22).

As sulphur is essential for the synthesis of the sulphur-containing amino acids and hence of proteins including enzymes, a deficiency in supply disrupts many biochemical reactions. In deficient plants there are increased levels of amines, amides and nitrate due to a restricted use of nitrogenous substrates for the synthesis of protein. The level of carbohydrates is often reduced as a result of diminished photosynthesis. Sulphur and nitrogen deficiencies have profound and similar effects on growth: deficient plants are stunted and weak with thin stems while the leaves remain small and are sometimes distorted. A deficiency of either sulphur or nitrogen causes chlorosis, or yellowing, of the leaves due to diminished levels of chlorophyll. Sulphur is relatively immobile within the plant with the result that the newer leaves usually show chlorosis before the older leaves. Because nitrogen is highly mobile chlorosis appears first on the older leaves.

Sulphur deficiency is most common in those regions of the world where accessions from the atmosphere to the crop canopy and soil are less than

$10 \text{ kg ha}^{-1} \text{ a}^{-1}$. The sulphur compounds in the atmosphere comprise mainly sulphur dioxide derived from the combustion of sulphur containing fuels, and sulphate from sea spray. As a result, deficiency usually occurs in crops grown at great distances from these sources, notable examples being in New Zealand, in the eastern states of Australia, in the western provinces of Canada and along the Pacific coast of the USA.

Micronutrient elements[32]

Iron

The functions and physiological effects of iron in plants depend on changes in its oxidation state between Fe(II) and Fe(III) and the formation of complexes with organic and inorganic ligands. Plant roots have been shown to reduce Fe(III) to Fe(II) and this appears to be an essential prerequisite to absorption from solution.[33] After absorption, Fe(II) is oxidized and translocated to the shoots in anionic form as ferric citrate.[34] Some of the iron can be stored in the leaves as a ferric phosphoprotein, phytoferritin, which serves as a reserve for developing plastids and hence for photosynthesis.[35]

The active iron occurs in numerous enzyme systems either as a structural component of prosthetic groups or as a constituent of the protein itself. The best known prosthetic groups are the iron porphyrins which, when attached to specific proteins, are known as haem proteins.[36] These include peroxidase, catalase and some dehydrogenases, as well as the cytochromes which function prominently as electron carriers in photosynthesis and respiration. In addition, a cytochrome is one of three prosthetic groups which contribute to the transfer of electrons in nitrate reductase. The reduction of nitrite to ammonia also depends on iron, as nitrite reductase itself comprises a haem protein, called sirohaem and a non-haem component containing iron and sulphur. Leghaemoglobin, another haem protein, is required for the fixation of dinitrogen in legume root nodules where, through reversible oxygenation it regulates the supply to nitrogenase of oxygen, to which it is highly sensitive. The most important non-haem protein is ferredoxin which has a very high

---

32 General references: *Diagnosis of Mineral Disorders in Plants*, Vol. 1, *Principles* (eds C. Bould, E. J. Hewitt and P. Needham), London, HMSO, 1983; *Micronutrients in Agriculture* (eds J. J. Mortvedt, P. M. Giordano and W. L. Lindsay), Soil Sci. Soc. Am., Madison, 1972; Kabata-Pendias, A. and Pendias, H., *Trace Elements in Soils and Plants*, CRC Press, Boca Raton, Florida, 1984.

33 Brown, J. C., *Pl. Cell Environ.*, 1978, **1**, 249; Chaney, R. L. *et al.*, *Pl. Physiol.*, 1972, **50**, 208; Römheld, V. and Marschner, H. in *The Soil–Root Interface* (eds J. L. Harley and R. Scott Russell), London, Acad. Press, 1979.

34 Tiffin, L. O., *Pl. Physiol.*, 1970, **45**, 280, and in *Micronutrients in Agriculture* (ref. 32).

35 Hyde, B. B. *et al.*, *J. Ultrastruct. Res.*, 1963, **9**, 248.

36 Rains, D. W. in *Plant Biochemistry* (eds J. Bonner and J. E. Varner), p. 561, Acad. Press, 1976.

negative redox potential and acts as an electron carrier in photophosphorylation, in the photosynthetic reduction of NADP (nicotinamide adenine dinucleotide phosphate) and in the reduction of nitrite. In legume root nodules and other biological nitrogen fixing systems, either ferredoxin or flavodoxin acts as an electron carrier to nitrogenase, which comprises a non-haem protein with molybdenum as an essential constituent and an iron–sulphur protein. It is generally believed that iron plays a role in the synthesis of chlorophyll but this is still not understood.

The symptoms of iron deficiency resemble those of magnesium deficiency as both cause extensive interveinal chlorosis of the leaves. However, the younger leaves are affected most by a deficiency of iron as it does not move readily out of the older leaves. Excess supplies of manganese, zinc, copper and some other heavy metal cations may induce symptoms similar to those of iron deficiency,[37] possibly by competing with iron for absorption or for functional sites within the plant. In plants suffering from iron chlorosis there is a relatively high ratio of phosphorus to iron but it is not clear if this is a cause or result of the chlorosis. Lime induced chlorosis is frequently found in some plants growing on calcareous soils. Although high pH and competitive effects of calcium have been suggested as causes, it appears that the chlorosis is attributable to the bicarbonate ion which reduces the absorption or mobility of iron within the plant.[38]

Manganese

Manganese is absorbed from solution as $Mn(\text{II})$ ions* and translocated to the shoots primarily as the free ion. Although its content in plant dry matter is typically 50 $\mu g\ g^{-1}$ (Table 3.3) it can approach 1000 $\mu g\ g^{-1}$ in some plants growing on acid soils.[39] Manganese, like magnesium, can act as a cofactor of many enzymes that act on phosphorylated substrates. Some of the enzyme reactions in the tricarboxylic cycle, notably decarboxylases and dehydrogenases, are also activated by manganese, but as manganese can be substituted by magnesium, the relative importance of these two ions is uncertain. The main role identified for manganese is the one it plays in the evolution of oxygen in photosynthesis. Manganese is present in the chloroplasts in a complex which oxidizes water to produce molecular oxygen, hydrogen ions and electrons (photolysis). The structure and functioning of the complex are

37 Wallihan, E. F. in *Diagnostic Criteria for Plants and Soils* (ed. H. D. Chapman), p. 203, Univ. Calif., 1966; Needham, P. in *Diagnosis of Mineral Disorders in Plants* Vol. 1, *Principles* (eds C. Bould, E. J. Hewitt and P. Needham), p. 137, London, HMSO, 1983.
38 Rutland, R. B. and Bukovac, M. J., *Pl. Soil*, 1971, **35**, 225.
39 Labanauskas, C. K. in *Diagnostic Criteria for Plants and Soils* (ed. H. D. Chapman), p. 264, Univ. Calif., 1966.

* In specifying ionic forms here and elsewhere, hydrolysed and hydrated species are not excluded.

not known, but it has been suggested that the manganese may have a redox role by changing its oxidation state.[40] The role of manganese in the photosynthetic evolution of oxygen is reflected in gross changes in structure of the chloroplasts in manganese deficient plants.[41] Manganese also plays a role in regulating the levels of auxin in plant tissues by activating the auxin oxidase system.

The visual effects of manganese deficiency vary considerably between species and cultivars. Symptoms may appear first in either younger or older leaves and comprise a wide variety of chlorotic and necrotic patterns. Among the cereals, oat is the most susceptible, the deficiency being known as 'grey speck' because the symptoms appear as small necrotic areas on the leaves; these areas tend to coalesce while the apical tissues remain green. In peas, beans, sugar beet and the brassicas, symptoms appear first on the older leaves, generally as interveinal yellowing or chlorosis in a mottled pattern. In potato the young leaves turn pale and in some cultivars small brown spots appear on the leaves.

On some acid soils, especially when poorly drained, plants can take up excessive amounts of manganese with various results including reduced growth and lesions in the leaves, often appearing as a chlorosis.

Copper

Copper is absorbed by plant roots as $Cu(\mathrm{II})$ ions and translocated to the shoots predominantly in anionic form and possibly some as the free ion. Copper is an essential constituent of a group of enzymes known as oxidases in which molecular oxygen is used directly in the oxidation of substrate; these include cytochrome oxidase, phenol oxidase, laccase, ascorbic acid oxidase, and amine oxidase. It is believed that the catalytic activity of these enzymes depends on the ability of their copper to undergo reversible change in oxidation state between $Cu(\mathrm{II})$ and $Cu(\mathrm{I})$. In addition to the oxidases which catalyse the reduction of molecular oxygen, plants have superoxide dismutase, an enzyme containing both copper and zinc, which reacts with the superoxide ion O_2^- to produce molecular oxygen and hydrogen peroxide. As the superoxide ion is toxic and readily produced from molecular oxygen, superoxide dismutase plays an important, protective role in plant metabolism. Copper also plays a role in photosynthesis as an essential constituent of plastocyanin. This enzyme is located in the chloroplasts and forms part of the electron transport chain between the two photochemical systems of photosynthesis. The precise role of copper in plastocyanin is not known, but it has been attributed to reversible change in oxidation state.

40 Cheniae, G. M., *Ann. Rev., Pl. Physiol.*, 1970, **21**, 467; Radmer, R. and Cheniae, G. M. in *Primary Processes of Photosynthesis* (ed. J. Barber) p. 303, Elsevier, 1977.
41 Possingham, J. V. *et al.*, *J. Ultrastruct. Res.*, 1964, **11**, 63.

There is some evidence from work with clover (*Trifolium subterraneum*) that copper plays a role in the symbiotic fixation of nitrogen.[42] Apparently when copper is in short supply, both nodulation by *Rhizobium* and the fixation of nitrogen are depressed. The role of copper in nodulation is not known, but as it is essential for cytochrome oxidase a deficiency in supply may cause the oxygen in the nodules to reach levels that reduce the activity of nitrogenase.[43]

Crops vary in their response to low supplies of copper. Cereals and fruit trees can be seriously affected with substantial losses in yield while, for example, sugar beet shows few visible symptoms and yield losses are only small. In cereal crops the symptoms of deficiency appear during tillering when the leaves become twisted or rolled and their tips turn grey or white; ear emergence and grain filling are seriously affected. In fruit trees the leaves of the terminal shoots become dark green and curled and may then develop brown or necrotic areas. This is usually followed by withering of the leaves, defoliation and death, or 'die-back', of the shoot.

Zinc

Zinc is absorbed as Zn(II) and translocated to the shoots primarily as the free ion; its concentration in plant dry matter is at least three- or four-fold greater than copper. In some metalloenzymes, which are not involved in oxidation-reduction reactions, zinc can be substituted by magnesium or manganese and it is perhaps significant that none of these metal ions is known to be oxidized or reduced in plants. Zinc is known to be an essential constituent of only three plant enzymes, namely carbonic anhydrase, alcohol dehydrogenase and superoxide dismutase.[44] However, studies of zinc-requiring enzymes isolated from other organisms, and also with zinc deficient plants, suggest that it is specifically required for many other plant enzymes including additional dehydrogenases, DNA and RNA nucleotidyltransferases (polymerases) and some peptidases and proteinases. It has long been known that zinc has a marked effect on the level of auxin.[45] There is uncertainty about where zinc is involved in the synthesis of auxin, but it appears to be required in the synthesis of intermediates in the metabolic pathway, through tryptophan to auxin.

There is some evidence that the synthesis of protein, which is mediated by RNA (ribonucleic acid), is regulated by the concentration of zinc. Thus when zinc is deficient in *Euglena gracilis* the ribosomes become unstable, and both the levels of RNA in the cells and the rates of synthesis of protein are mark-

42 Hallsworth, E. G. *et al.*, *Nature, Lond.*, 1960, **187**, 79.
43 Cartwright, B. and Hallsworth, E. G., *Pl. Soil*, 1970, **33**, 685.
44 Vallee, B. L. and Wacker, W. E. C. in *CRC Handbook of Biochemistry and Molecular Biology: proteins*, Vol. 2 (ed. G. D. Fasman), CRC Press, p. 276, 1976.
45 Skoog, F., *Am. J. Bot.*, 1940, **27**, 939.

edly reduced.[46] With citrus it has been found that the levels of both RNA and protein were lower in zinc deficient than in healthy leaves, and the levels of ribonuclease and soluble nitrogen compounds were higher in the deficient leaves.[47] The synthesis of protein from amino acids is, however, vastly complicated and although it has been suggested that zinc has a specific role in inhibiting ribonuclease it is most probable that zinc and other metals are essential for several, as yet unknown, pathways both in the synthesis and breakdown of proteins.

There are numerous reports that zinc deficiency is induced by the application of phosphate fertilizers, the best known example being with citrus trees. This effect is unlikely to be due to an interaction between phosphorus and zinc in the soil or to a dilution of zinc in the plant tissues owing to the growth response to phosphorus. There is, however, evidence that phosphate may interfere with the translocation of zinc or with its utilization, but no mechanisms have been suggested for either effect.[48] With potato it has been found that phosphate does not affect the translocation of zinc, but that a ratio of phosphorus to zinc in the leaves in excess of 400 to 1 is associated with zinc deficiency.[49]

The characteristic symptom of zinc deficiency is a failure of leaves to expand and stems to elongate, giving a terminal rosette effect. These effects are believed to be associated with a disturbance of auxin metabolism. Fruit trees, particularly citrus, are frequently affected by zinc deficiency and maize, tomato and cotton are also especially sensitive. In fruit trees such as citrus, apple and peach, the mature leaves may show the first symptoms as pale green to yellow interveinal mottling, while more severe effects on growth and rosetting often occur in the terminal leaflets of young shoots. The older leaves of maize may have purple tints while yellow or white interveinal chlorotic stripes develop in the younger leaves. In tomato and cotton the symptoms appear as irregular areas of interveinal chlorosis which become necrotic.

Boron

It is generally assumed that boron is absorbed as undissociated boric acid $(B(OH)_3)$ and follows the flow of water into the roots. Although boron contents of $20–100 \ \mu g \ g^{-1}$ dry weight are considered adequate for normal growth, certain species will show toxicity symptoms[50] if the level exceeds $200 \ \mu g \ g^{-1}$. There is no evidence that boron is a specific activator or constitu-

46 Schneider, E. and Price, C. A., *Biochim. Biophys. Acta*, 1962, **55**, 406; Praske, J. A. and Flocke, D. J., *Pl. Physiol.*, 1971, **48**, 150.
47 Kessler, B. and Monselise, S. P., *Physiol. Plant.*, 1959, **12**, 1.
48 Olsen, S. R. in *Micronutrients in Agriculture* (eds J. J. Mortvedt, P. M. Giordano and W. L. Lindsay), p. 243, Soil Sci. Soc. Am., Madison, 1972.
49 Boawn, L. C. and Leggett, G. E., *Soil Sci. Soc. Am. Proc.*, 1964, **28**, 229.
50 Bradford, G. R. in *Diagnostic Criteria for Plants and Soils* (ed. H. D. Chapman), p. 33, Univ. Calif., 1966.

ent of any enzyme and its biochemical functions in plants are still not ident-ified.[51] However, some indications of the roles of boron have arisen from studies based on deficient plants. Other indications come from the knowledge that the borate ion forms complexes with polyhydroxy compounds having two hydroxy groups (diols) with appropriate configuration; suitable compounds include several sugars, phosphate esters and o-diphenols. One hypothesis is that boron facilitates the translocation of sugars across cell membranes. Another suggests that as boron complexes with 6-phosphogluconate it slows down the formation of compounds formed in the 'pentose shunt', including phenolic acids which accumulate when boron is deficient. A very rapid effect of withholding boron is a reduction in the level of RNA and cessation of cell division in root tip meristems; the roots subsequently become shortened and appear bumpy. Auxins are also known to stop meristematic cell division but no direct relationship between boron deficiency and auxin activity has been established.

In general, dicotyledons are more susceptible to boron deficiency than monocotyledons with the exception of maize and sorghum. Boron deficient plants show disorganized meristems often leading to early death of stem tips. Leaves may become crinkled and misshapen while stems and petioles are thickened and cracked. Storage organs are often affected as, for example, in 'water core' of turnip and 'heart rot' in sugar beet. Flowering may be totally suppressed while fruit and seed formation, if it occurs, is abnormal.

Molybdenum

Molybdenum is absorbed from soils predominantly as MoO_4^{2-} but increasingly as $HMoO_4^-$ as pH decreases to about 4. The content of molybdenum in plants that is considered to be adequate for healthy growth is only 0.1 $\mu g\ g^{-1}$ dry weight, but it is commonly up to ten-fold greater than this. It is now well established that molybdenum is an essential constituent of two important enzymes involved in the assimilation of nitrogen, namely nitrate reductase and nitrogenase. Nitrate reductase is a molybdoflavo-protein and is present in leaves and roots, probably in the cytoplasm.[52] Nitrate reductases from higher plants and lower organisms are complex in structure and mechanism. However, present evidence indicates that the enzyme of higher plants uses NADH (reduced nicotinamide adenine dinucleotide) as reductant from which electrons are transferred to nitrate through the three prosthetic groups of the enzyme functioning in series, these being FAD (flavin adenine dinucleotide), a cytochrome and a molybdenum protein. It has been suggested that the final step in the reaction of molybdenum with nitrate involves a reversible change

51 Dugger, W. M. in *Inorganic Plant Nutrition* (eds A. Läuchli and R. L. Bieleski), *Encycl. Pl. Physiol.* (New Series), Vol. 15B, p. 628, Springer-Verlag, 1983.
52 Oaks, A. in *Nitrogen Assimilation in Plants* (eds E. J. Hewitt and C. V. Cutting), p. 216, Acad. Press, 1979; Hageman, R. H., *op. cit.*, p. 591.

between Mo(IV) and Mo(V).[53] The activity of nitrate reductase depends on the flux of nitrate into the cell with the result that leaves adequately supplied with nitrate have a high nitrate reductase activity and vice versa. Nitrite, the immediate product of the nitrate reductase reaction, is further reduced by nitrite reductase to ammonia which is the form of nitrogen metabolized to produce organic nitrogen compounds. The fact that the concentration of nitrite in plant tissues is very low indicates the close link between the activities of nitrite and nitrate reductases.

Molybdenum is specifically required for the biological fixation of dinitrogen both in the symbiotic systems of legumes and certain non-legumes, and in free-living organisms. The main biochemical pathways of biological fixation have been established using cell-free extracts from free-living organisms but are probably the same in symbiotic fixation. Nitrogenase, the enzyme responsible for the reduction of dinitrogen, consists of two iron–sulphur proteins, the larger one of which contains molybdenum and is an iron–molybdenum protein, and the smaller one is an iron protein.

The reduction of dinitrogen to ammonia is a complicated, chain of reactions which requires, in addition to the two enzymes, a source of electrons, ATP, magnesium and an environment low in oxygen.[54] The following is a simplified account of the nitrogenase reaction: electrons are supplied by NADH and carried by ferredoxin to the iron protein; ATP from the respiratory chain is bound through magnesium to the iron protein which, together with the iron–molybdenum protein, forms a ternary complex. The formation of this complex reduces the redox potential and enables the transfer of electrons from the iron protein to the iron–molybdenum protein where the bound dinitrogen is reduced. The reduction is accompanied by hydrolysis of ATP at a rate of about 20 molecules for every 2 molecules of ammonia produced.

Molybdenum, albeit at very low concentrations compared with the other micronutrients, plays a key role in the assimilation of nitrogen in both the legume–*Rhizobium* system and in non-legumes. When legumes rely entirely on atmospheric nitrogen a deficiency of molybdenum is manifested as a deficiency of nitrogen, with characteristic symptoms including a proliferation of small nodules on the roots. However, where the host plant is grown with nitrate as the sole or principal source of nitrogen, molybdenum is still required for the nitrate reductase reaction, as in the non-legumes. The effect of molybdenum deficiency is generally to reduce the levels of amino acids, amides and amines concomitantly with an accumulation of nitrate which may reach 20 per cent of the dry weight.

53 Notton, B. A. and Hewitt, E. J. in *Nitrogen Assimilation in Plants* (eds E. J. Hewitt and C. V. Cutting), p. 227, London, Acad. Press, 1979; Beevers, L. and Hageman, R. H. in *Inorganic Plant Nutrition* (eds A. Läuchli and R. L. Bieleski), *Encycl. Pl. Physiol.* (New Series), Vol. 15A, p. 351, Berlin, Springer-Verlag, 1983.
54 Bothe, H. *et al.* in *Inorganic Plant Nutrition* (eds A. Läuchli and R. L. Bieleski), *Encycl. Pl. Physiol.* (New Series), Vol. 15A, p. 241, Berlin, Springer-Verlag, 1983.

Plants differ in susceptibility to molybdenum deficiency due to their different absorptive capacities, requirements, seed reserves and nitrogen supply. Visible symptoms usually appear first in the oldest leaves, and progress to the apex of the plant; leaves show marked, yellow interveinal mottling and become distorted with their margins often curling upwards. In addition to the legumes, brassicas are particularly susceptible. With cauliflower, which shows characteristic symptoms known as 'whiptail', older leaves may remain unaffected while young leaves develop chlorotic interveinal areas and their margins break down leaving intact, narrow green areas along an elongated and distorted midrib. However, in the absence of molybdenum cauliflower remains healthy when grown under sterile conditions with ammonium as the sole source of nitrogen.[55]

Chlorine

Chlorine is absorbed by plants as the chloride ion, the concentration in the dry matter usually being in the range 0.2–1.8 per cent, but can be very much higher in halophytes (salt tolerant species). Despite its accumulation by plants, chlorine is required in relatively low concentrations; thus, with several species grown in nutrient solutions deficiency symptoms occurred when the content was in the range 50 to 180 μg g^{-1} dry weight.[56]

From studies with isolated chloroplasts it was originally concluded that chlorine was required for the photosynthetic evolution of oxygen.[57] However, an absolute requirement for chlorine in photosynthesis remains in doubt as it has been shown that deficiency in sugar beet leads to a large reduction in growth through effects on cell division and leaf area and not on the rate of photosynthesis per unit leaf area.[58] As chloride is mobile and tolerated in relatively high concentration in plant tissues, it may also play a role as counter ion for potassium and thereby contribute to osmotic potential.

Deficiency symptoms have been reported for several crop species grown under controlled conditions in nutrient solutions, the common features being chlorosis and wilting of the plant. Chloride deficiency has never been encountered in field grown crops, due in large part to accessions from the atmosphere but also to additions in fertilizers. However, excess chloride poses serious problems in salt affected soils which occur over extensive areas of the world (Ch. 27).

55 Hewitt, E. J. and Gundry, C. S., *J. hort. Sci.*, 1970, **45**, 351.
56 Johnson, C.M. *et al.*, *Pl. Soil*, 1957, **8**, 337.
57 Warburg, O. and Luttgens, W., *Biokhimiya*, 1946, **11**, 303; Bishop, N. I., *Ann. Rev. Biochem.*, 1971, **40**, 197.
58 Terry, N., *Pl. Physiol.*, 1977, **60**, 69.

Beneficial elements (Co, Na, Si)

Cobalt

Although cobalt is not known to be essential for higher plants it is specifically required by *Rhizobium* in legume root nodules.[59] In the absence of cobalt the growth of the host plant is drastically reduced with symptoms of nitrogen deficiency, and nodulation is poor with low levels of leghaemoglobin. The role of cobalt is as the metal component of coenzyme B_{12}; this is required for leghaemoglobin formation and other important reactions in nitrogen fixation which are dependent on either vitamin B_{12} or the coenzyme form. There are several reports of field responses by nodulated legumes, including lupins, groundnuts and subterranean clover, to applications of cobalt.[32]

Sodium

Sodium is absorbed by plants to a varying degree depending on the species. It is not known to be essential except for some salt tolerant *Atriplex* species. Sodium can, however, sometimes have beneficial effects on plant growth. It can partially replace potassium in osmotic roles and the maintenance of turgor; this effect is greatest when the supply of potassium is inadequate. With some plants, especially members of the beet family, sodium can increase growth irrespective of the supply of potassium.

Silicon

Silicon is present in the soil solution as monosilicic acid $(Si(OH)_4)$ in which form it is absorbed by plants. Although silicon has not been shown to be essential for plants other than *Equisetum arvense* ('horsetail') it can have beneficial effects on the growth of some crop plants, notably rice. There are marked differences between species in the extent to which they accumulate silicon: rice commonly contains 4 to 7 per cent in the foliage dry matter and other cereals and grasses contain 1 to 2 per cent, in contrast to dicotyledonous species, including legumes, which generally contain only 0.05 to 0.20 per cent. Following absorption, silicic acid is translocated to the shoots and, as water is lost by transpiration, opaline silica $(SiO_2.nH_2O)$ is deposited in cell walls, mainly in epidermal and vascular tissues. In cereal plants the bulk of the total silicon is present as opaline silica, but in dicotyledons a relatively high proportion remains in solution as silicic acid. As a constituent of the cell walls of epidermal and vascular tissues, silica plays an important structural role in strengthening these tissues in the stems and leaves. With rice in particular it changes the leaf angle and hence improves light interception, with the result that growth and grain yields are increased. The silica of cell

59 Ahmed, S., and Evans, H. J., *Soil Sci.*, 1960, **90**, 205;   Reisenauer, H. M., *Nature, Lond.*, 1960, **186**, 375; Delwiche, C. C. *et al.*, *Pl. Physiol.*, 1961, **36**, 73.

walls inhibits fungal infection of rice plants, e.g. by *Piricularia oryzae* (blast disease) and of other cereals by *Erysiphe graminis* (powdery mildew). It also reduces the susceptibility of rice to attack by certain insect pests.[60]

Fertilizers and organic manures

Organic

Farmyard manure, which is the traditional method of returning nutrients to the land, is made by composting faeces and urine of farm animals, usually cattle, with straw or other bedding material. Its plant nutrient content depends on the animal from which it was derived, the fodder the animals received, the composition of the bedding material; its nitrogen content also depends on the loss of ammonia by volatilization and of soluble nutrients in drainage. The composition is therefore variable, and the dry material in Britain commonly contains 1.7–2.5 per cent N, 0.2–0.5 per cent P, and 1.3–2.0 per cent K. An application of 25 t ha^{-1} of the fresh manure returns to the soil about 160 kg N, 22 kg P and 110 kg K. It also supplies Ca, Mg, S and micronutrients.

Field experiments have shown that in the year of application most of the phosphate and potassium is as effective as fertilizer phosphate and potassium, whereas only about one-quarter to one-third of the nitrogen is as effective as that present in fertilizers.[61] Part of the nitrogen that is present in organic substances, which only slowly decompose in soil, becomes available to crops in subsequent years. Farmyard manure is therefore a source of all the plant nutrients. It is valued as a source of nitrogen where fertilizers are expensive or unavailable but, compared with a balanced fertilizer, it supplies less nitrogen and more potassium. Responses to applications of 2 to 5 t ha^{-1} on phosphate deficient soils have been attributed to its phosphate content[62]; and probably because of its content of Ca, Mg, K and Na it can maintain or increase the soil pH where the use of ammonium fertilizers or urea causes acidification.[63] It may also reduce phosphate adsorption and thereby increase the concentration of phosphate in the soil solution.[64] As shown by the example in Table 20.7 crop yields may be higher on plots treated liberally with farmyard manure than on plots receiving liberal applications of fertil-

60 Jones, L. H. P. and Handreck, K. A., *Adv. Agron.*, 1967, **19**, 107.
61 Cooke, G. W.,*The Control of Soil Fertility*, Crosby Lockwood, 1967.
62 Hartley, K. T., *Emp. J. expl. Agric.*, 1933, **1**, 113; and with Greenwood, M., op. cit., 1937, **5**, 254.
63 Bache, B. W. and Heathcote, R. G., *Expl. Agric.*, 1969, **5**, 241.
64 Johnston, A. E. and Warren, R. G., *Rothamsted Ann Rept. for 1963*, p. 56.

izers. A suggestion by Cooke,[65] who gives other examples, is that organic treatments are associated with the provision of nitrogen for crops in ways and at times that are not fully imitated by fertilizers applied in conventional ways. Aside from these effects on nutrient supply, long continual use of farmyard manure at high rates of application increases the organic matter content of the soil (Fig. 20.2), and this can improve soil structure (see p. 437).

It is now common for the faeces and urine from cattle and pigs to be collected in tanks, usually diluted with wash water. The liquid is known as slurry, and its nutrient content depends on its degree of dilution as well as on the factors given above for farmyard manure. If it is not diluted, cattle slurry has a dry matter content of about 10 per cent and a nutrient content of about 0.5 per cent N, 0.1 per cent P and 0.4 per cent K. The availability to crops of its phosphate and potassium is similar to that of farmyard manure, but for nitrogen the availability depends on the time of its application in relation to crop growth and on the conditions in the soil. About half its total nitrogen is present as ammonium which may be quickly volatilized after the slurry has been spread.[66] Because slurry is a liquid there can be large denitrification losses, especially if applied to soils with poor drainage which are already wet. If applied at the optimum time in spring under optimum conditions the nitrogen in slurry is about half as effective as nitrogen in inorganic fertilizers.[67]

Domestic sewage has about the same nitrogen and phosphorus composition (on a dry matter basis) as farmyard manure but has a much lower content of potassium because after treatment the liquid is allowed to run to waste. Sewage from industrial areas may contain elements in quantities which are harmful to crops (see pp. 812–814). Poultry manure and wastes from the processing of animal and plant products are used mainly in horticulture where they provide nitrogen and phosphate and small amounts of other nutrients. They do not add to the organic matter content of soil and are often described as organic fertilizers rather than as organic manures.[68]

Inorganic

Until the introduction of inorganic fertilizers the maintenance of nutrient supplies depended largely on effective conservation through the use of farmyard manure. The landmarks in the development of fertilizer use were: (i) the

65 Cooke, G. W., *J. Soil Sci.*, 1979, **30**, 187.
66 Ryden, J. C. and Lockyer, D. R., *The Grassland Res. Inst. Ann. Rept. for 1984–85*, p. 19.
67 Gostick, K. G., *Phil. Trans. R. Soc. Lond.*, 1982, **B296**, 329.
68 For a fuller discussion of organic manures and fertilizers see Cooke, G. W., *Fertilizing for Maximum Yields*, Granada, 1982.

manufacture of superphosphate from deposits of rock phosphate which started in the 1840s; (ii) the introduction in the 1870s of potassium salts as fertilizers using mineral deposits in France and Germany; and (iii) the introduction in the decade 1910–20 of the Haber-Bosch process of ammonia synthesis which quickly led to the use of synthetic nitrogen fertilizers. Various industrial by-products were also used, of which the most important were basic slag, a by-product of the steel industry, and ammonium sulphate which was produced from the ammonia separated from town gas.

The most commonly used nitrogen fertilizers are now urea, ammonium nitrate, ammonium phosphates and ammonia gas, and for phosphate the more concentrated form, triple superphosphate, has largely replaced superphosphate (or single superphosphate). Potassium is supplied mainly as the chloride. One effect of the changes has been to increase the concentration of nitrogen and phosphorus in fertilizers, which has reduced transport costs (see Table 3.6 for composition of fertilizers in common use). By comparison with farmyard manure and with fertilizers which are industrial by-products, modern nitrogen, phosphate and potassium fertilizers supply little calcium, magnesium, sulphur and micronutrients unless designed to do so, and these other nutrients must be added separately if they are required. The outstanding benefit from the introduction of synthetic fertilizers is that their use has made possible the big increase in yields of crops which has occurred in many countries since about 1950.

Table 3.6   Composition of fertilizers in most common use

	Main nutrient (%)	*Other nutrients (%)*
Nitrogen fertilizers	N	
Ammonium sulphate	21	S(24)
Ammonium nitrate	32–34.5	
Calcium ammonium nitrate	25	Ca(11)
Urea	46	
Anhydrous ammonia	82	
Aqueous ammonia	27	
Phosphorus fertilizers	*Total P and water solubility*	
Single superphosphate	8–9 sol	S(12–14), Ca(20)
Triple superphosphate	20 sol	Ca(14)
Basic slag	4–7 insol	Ca and others (variable)
Rock phosphate	12.5–15.5 insol	Ca(31–35)
Monoammonium phosphate	26 sol and 12% N	
Diammonium phosphate	23 sol and 21% N	
Potassium fertilizers	K	
Potassium chloride	50	
Potassium sulphate	42	S(17)

Trends in the use of fertilizers are shown in Table 1.4. The trends reflect food prices and fertilizer costs, and hence the cost/benefit of their use. The use of phosphate fertilizers in the United Kingdom is now almost constant from year to year because of the value of residues from previous applications. It has become the usual practice to use only maintenance applications of phosphate and potassium, keeping crop removal and nutrient addition roughly in balance. This does not apply to developing countries where there is little or no history of fertilizer use.

Assessment of fertilizer requirements

Field experiments and soil chemical analysis are the two general methods of investigation for advising farmers how much fertilizer to use. In field experiments the aim is to find the yield response curve to increments of a nutrient, or in a factorial experiment to measure the response to two or more nutrients. Properly conducted, a field experiment can be used to assess the potential yield at the site and also the optimum fertilizer rate under the conditions of the experiment; but, as discussed below, it is uncertain how widely the results can be applied. Together with the results from field experiments, soil analysis gives an index of nutrient availability which will indicate whether fertilizer is required, and how much is required at sites additional to those on which the experiments were conducted. The two methods are therefore complementary.

Soil analysis

As discussed in later chapters, no single method of extracting a nutrient from soil can simulate the complex dynamic processes which supply nutrients to plants. Instead, the aim of the soil chemist in analysing soils is to provide indices of nutrient availability which correlate with crop responses in field or glasshouse experiments. A good method of analysis is the one that gives a large correlation coefficient, and the method chosen will vary with the nutrient and the properties of the soils under investigation. For example, a commonly used extractant for phosphate is 0.5 M sodium bicarbonate at pH 8.5, and for potassium and magnesium a 1 M solution of ammonium nitrate. Using these extractants under standard conditions soils can be grouped and each group given an index of availability. These are the methods used in the United Kingdom and the analyses are grouped as shown in Table 3.7. With a nutrient index of 0 all crops can be expected to respond to applications of fertilizer, but because crops differ in their ability to take up nutrients, and potential yields differ between sites, interpretation of the index depends on experience and the results from field experiments. The methods and interpretation of soil chemical analysis for individual nutrients are discussed in later chapters.

Table 3.7 Classification of soil analysis results as used in England and Wales, concentrations as μg cm^{-3} soil. (From *MAFF Reference Book*, **209**, London, HMSO, 1985, which also gives the interpretation for fertilizer use. Phosphorus is extracted with 0.5 M sodium bicarbonate (1 cm^3 soil : 20 cm^3 solution), potassium and magnesium are extracted with 1 M ammonium nitrate (1 cm^3 soil : 5 cm^3 solution). For methods see *The Analysis of Agricultural Materials, MAFF Reference Book*, **RB427** (3rd edn), London, HMSO, 1986)

Index	Phosphorus	Potassium	Magnesium
0	0–9	0–60	0–25
1	10–15	61–120	26–50
2	16–25	121–240	51–100
3	26–45	241–400	101–175
4	46–70	401–600	176–250
5	71–100	601–900	251–350
6	101–140	901–1500	351–600
7	141–200	1501–2400	601–1000
8	201–280	2401–3600	1001–1500
9	over 280	over 3600	over 1500

Field experiments

An example of a crop response curve for increasing applications of fertilizer nitrogen is given in Fig. 3.5. Further examples are given later (Figs 20.6 and 20.8). The equation used to describe a response curve is important because it is used to determine the optimum fertilizer application. Fig. 3.5 shows experimental points for the yield of barley grain with increasing applications of fertilizer nitrogen. Following the Mitscherlich equation (see pp. 59–61) and using a quadratic equation, the curve fits the data rather less well than two straight lines. The importance of the equation is shown by the optimum application of nitrogen being assessed as 138 and 95 kg ha^{-1} using the curve and straight lines respectively. The evidence of Boyd[69] is that two intersecting straight lines generally give the best fit of the response of wheat to nitrogen applications, but this may not always be so.

A serious limitation of response curves is that they are site and year specific, that is, they relate only to the growing conditions at the site of the experiment and in the year it was conducted. For advising farmers on fertilizer use, field experiments are usually carried out at several sites over several seasons and the results are grouped for similar growing conditions before calculating an average response curve. The severe limitation of response curves is, however, that their use as a basis of advice to farmers is valid only for the same conditions as in the experiments. Information is therefore needed of the conditions for crop growth in the experiments and on the farm, which

69 Boyd, D. A. *et al.*, *J. agric. Sci.*, 1976, **89**, 149.

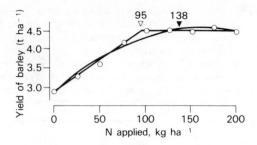

Fig. 3.5   Results of an experiment on barley interpreted by fitting either a quadratic equation, or two straight lines. The marked points (▼ and ▽)
show the differences in optimal applications of N fertilizer determined
by the two methods. (From Boyd *et al.*[69])

requires adequate monitoring. As discussed throughout this book, the
conditions that usually have to be taken into account include not only the
supply of available soil nutrients, but also soil physical conditions, the
weather during the growing season, control of pests and diseases, and such
crop characteristics as the species, variety and seed rate.

Field experiments also need to take account of interactions between two
or more nutrients, and between nutrients and other factors that determine
crop growth. According to the Law of the Minimum, introduced by Liebig,
growth increases as the factor present in minimum amount is increased (see
p. 59). It makes no allowance for interaction between factors, yet interactions
are important in most agronomic work. Sometimes the effects of increasing
two factors are simply additive and there is no interaction, that is, the yield
response of the crop with two factors together is equal to the sum of the
response to each factor separately. There is said to be a positive interaction
if the two factors together give a response which is greater than the sum of
the two factors separately. The interaction is negative if the response to the
two factors together is less than the sum of the two factors separately.

Table 3.8 illustrates the interaction between nitrogen, potassium and farmyard manure in an experiment on potatoes. There is a positive interaction
between nitrogen and potassium of 1.8 t, arrived at as 3.6/2 where 3.6 t is
the difference between the response of the two nutrients together (10.6 t) and
the separate responses to nitrogen (0.9 t) and potassium (6.1 t). The division
by 2 allows the main effects and interactions to be added together, which is
required for statistical reasons. The meaning of the positive interaction is that
each nutrient only has its full effect in the presence of the other. The negative
interaction (−4.4 t) between farmyard manure and potassium would be
expected because farmyard manure is a good source of potassium. The
negligible interaction (0.2 t) between farmyard manure and nitrogen indicates
that they act almost independently of each other. Books on statistical

Table 3.8   Yields of potatoes and responses (t ha^{-1}) to farmyard manure (FYM), nitrogen and potassium. For statistical analysis of interactions see Snedecor, G. W. and Cochran, W. G., *Statistical Methods* (6th edn), Ch. 12, Iowa State Univ. Press, 1974)

Yields	*Without N and K*	*N alone*	*K alone*	*N + K*
Without FYM	10.0	10.9	16.1	20.6
With FYM	23.1	27.6	23.6	28.5
Response to FYM	13.1	16.7	7.5	7.9

	Response to nitrogen			*Response to potassium*			*Interaction*
	without K	*with K*	*mean*	*without N*	*with N*	*mean*	
Without FYM	0.9	4.5	2.7	6.1	9.7	7.9	1.8
With FYM	4.5	4.9	4.7	0.5	0.9	0.7	0.2

	With adequate K		
Main effect of		Response to	
FYM	7.7	FYM alone	7.5
N	4.7	N alone	4.5
Interaction	0.2	FYM + N	12.4

	With adequate N		
Main effect of		Response to	
FYM	12.3	FYM alone	16.7
K	5.3	K alone	9.7
Interaction	−4.4	FYM + K	17.6

methods should be consulted for detailed analysis of multifactorial experiments.

The example that has been given is for the interaction of nitrogen, potassium and an organic manure. There may, however, be an interaction between any two or more agronomic factors. An important example is that yield potentials of new crop varieties are usually reached only if they are adequately fertilized. Similarly, irrigation may give a higher yield potential and require more fertilizer. These, and other factors, may act independently of each other, but often there is an interaction which does not fall off with increasing rates of input, as main effects do. Thus the best returns from one input are often obtained at high levels of a second input.

Optimum rates

The optimum rate of application of fertilizer is determined by the difference

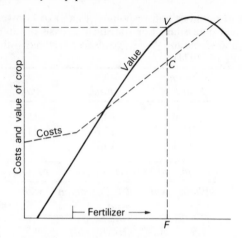

Fig. 3.6 Fertilizer use and maximum profit. (Adapted from Jollans, J. L., *Fertilizers in UK Farming, Centre for Agricultural Strategy Rept.*, 9, 1985, Univ. Reading.)

between the cost of fertilizer and the saleable value of the crop. This is shown schematically in Fig. 3.6 where the most profitable application, *F*, gives the biggest difference between costs, *C*, and the value of the crop, *V*. Data in Table 3.9 are of averaged linear responses and compare the value of the product with the cost of fertilizer. They show with barley, for example, that with existing prices (about 10 pence kg^{-1} grain and 35 pence kg^{-1} fertilizer nitrogen) the breakeven response is 3.5 kg grain per kg fertilizer N, and over the approximately linear part of the response curve 20 kg grain per kg fertilizer N can be expected. In developing countries the value/cost ratio is often much less favourable because of higher costs of fertilizer and lower farm prices, although the response to fertilizer under good growing conditions may be similar. The example in Table 3.9 is for the use of fertilizer nitrogen, but the same principles apply to the other nutrients.

Aside from the evaluation of fertilizers on economic grounds there are environmental considerations. The early criticism that fertilizers would lead to a decline of soil fertility has been discredited by the long-term experiments at Rothamsted Experimental Station which were established in the 1840s. They showed that high crop yields could be maintained with fertilizers alone (see for example Table 20.7), and experiments in other countries have led to the same conclusion as long as the fertilizers are correctly used. More recently there has been concern that higher energy costs would make fertilizer use uneconomic, and that pollution problems were making high rates of fertilizer use unacceptable.

In Britain the energy input to fertilizer manufacture accounts for about 25

Table 3.9 Typical crop responses in the U.K. to fertilizer nitrogen and comparison between value of product and cost of fertilizer. (From Eagle, D. J., *et al.*, *MAFF Tech. Bull.*, No. 32, 355, London, HMSO, 1976)

	Crop response kg kg⁻¹ N	kg crop product of equal value to 1 kgN	*Value : cost ratio*
Barley			
(grain at 85% DM)			
Chalk and limestone soils	20	3.5	6
Silt and clay soils	10	3.5	3
Potatoes	64	8	8
Sugar beet (total sugar)	16	2.5	6
Grass for silage (DM)	20	7–8	2.5–3

Note: Columns 2 and 3 assume a linear relation between yield and nitrogen application.

per cent of the energy used in agriculture, or about 1 per cent of the total energy consumption throughout the country.[70] Energy costs have risen (in real money terms) since the early 1970s, but fertilizer use remains highly cost effective.

The pollution effects attributed to fertilizers include eutrophication of surface waters and high nitrate concentrations in drinking water. Whatever the source, be it fertilizers, soil, sewage or industrial effluent, high concentrations of nutrients, particularly phosphorus, may increase the growth of algae and weeds in fresh waters (eutrophication). When these plant materials decompose, removal of oxygen from the water may lead to the death of fish, and also give the water an unpleasant taste which has to be removed if it is to be used as a source for drinking. There has been more concern, however, over the possible hazards to human health of the higher concentrations of nitrate now generally found in drinking water in many countries. There is convincing evidence[71] that nitrate concentrations in river water have been increasing for several years. Fertilizer nitrogen has contributed to this increase but the size of its contribution is uncertain. The possible health hazards associated with high nitrate intake through vegetables and water have been the subject of several public inquiries, without any clear conclusion emerging that there is any problem at present, if the water is free of bacterial contamination. There is further reference to pollution in later chapters.

70 Lewis, D. A. and Tatchell, J. A., *J. Sci. Fd Agric.*, 1979, **30**, 449.
71 Wild, A. and Cameron, K. C. in *Soils and Agriculture* (ed P. B. Tinker), p. 35, Blackwell, 1980.

Nutrient balances

Where there is no population pressure on the land, cropping periods can alternate with resting periods under a fallow. In the humid tropics the land might be allowed to revert to secondary forest, and in drier regions to a grass or bush fallow. The fallow period confers several benefits, of which nutrient accumulation is one, as shown in Table 3.10 for forested and savanna regions in West Africa. The forest fallows accumulate large amounts of nutrients in the vegetation, part of which is returned annually to the soil surface in rain wash and in timber and litter fall (Table 3.11). When the land is brought into cultivation the traditional practice is to burn the ground vegetation and the smaller trees. Although much of the nitrogen is lost as ammonia and as oxides of nitrogen and some of the sulphur as sulphur dioxide, the mineral nutrients, including phosphorus, potassium, calcium and magnesium are retained in the ash and become available to crops. Nutrients which have accumulated in the soil by the cycling process through vegetation, and the nitrogen, phosphorus and sulphur which have accumulated in the soil organic matter under the fallow, also become available to crops. After a few years of cultivation the land is allowed to revert to a fallow. As practised tradi-

Table 3.10   Nutrient content of some tropical fallows and in underlying soils. (From Nye, P. H. and Greenland, D. J., *The Soil Under Shifting Cultivation*, Commonw. Bur. Soils, Harpenden, Tech. Comm., No. 51, 1960)

Site	Annual rainfall (*mm*)	Fallow	Nutrients stored in fallow crop (*kg ha^{-1}*)				
			N	P	K	Ca	Mg
Kade, Ghana	1650	40-year mature secondary forest	1830	125	819	2524	346
Yangambi, Zaïre	1850	5-year secondary forest	391	24	344	293	
Ejura, Ghana	1500	High grass savanna, undisturbed over 20 years	127	23	192	270	89
			Nutrients stored in soils under fallows* (*kg ha^{-1} to about 30 cm depth*)				
Kade, Ghana			4600	12	650	2600	370
Yangambi, Zaïre			2200–3400	19	360	100	53
Ejura, Ghana			1800	22	190	2900	380

* Nutrients are reported as 'total' for N, dilute acid extractable for P, exch. for K, Ca, Mg.

Table 3.11 Nutrient cycle in mature secondary forest, Kade, Ghana. (From Nye, P. H. and Greenland, D. J., *The Soil Under Shifting Cultivation*, Commonw. Bur. Soils, Harpenden, Tech. Comm. No. 51, 1960)

	Wt. of material (kg ha⁻¹ a⁻¹)	*Nutrient elements (kg ha⁻¹ a⁻¹)*				
	$(kg\ ha^{-1}\ a^{-1})$	*N*	*P*	*K*	*Ca*	*Mg*
Rain wash from leaves		12	3.7	220	29	18
Timber fall	11 000	36	2.9	6	82	8
Litter fall	10 500	199	7.3	68	206	45
Total addition to soil surface		247	13.9	294	317	71

tionally the only nutrient input is of nitrogen through biological fixation; for other nutrients replenishment depends on rock and mineral weathering. The system ceases to be viable when population pressure requires more land to be used for cultivation. The fallow period then becomes too short for a sufficient build-up of nutrients.[72]

Table 3.12 Annual inputs and outputs of N on dairy farms under intensive and extensive management in the Netherlands. (From Van der Meer, H. G., *Proc. 9th General Meeting of European Grassland Federation (Reading)*, 61–88, Brit. Grassland Soc., 1983)

	Intensive	*Extensive*
	$(kg\ N\ ha^{-1})$	
Inputs		
Fertilizer	383	—
Symbiotic fixation	—	65
Purchased feed	127	24
Precipitation	23	23
	533	112
Outputs		
Milk	72	31
Liveweight	12	7
	84	38
N not accounted for	449	74

72 For accounts of systems used in the tropics see Allan, W., *The African Husbandman*, Oliver and Boyd, 1965; Ruthenberg, H., *Farming Systems in the Tropics*, Oxford Univ. Press, 1971. For nutrient cycling through forest ecosystems see Reichle, D. E. (ed.), *Dynamic Properties of Forest Ecosystems*, Cambridge Univ. Press, 1981.

The system described above is suitable for an environment where animals are largely excluded because of diseases or insect pests. In other environments a rotation of grass or grass/clover swards, i.e. leys, with arable cropping, has been used mainly where high quality herbage is needed for dairy cattle. Most of the benefit conveyed by the ley to the growth of the arable crop is due to the nitrogen that accumulates under the sward and is subsequently mineralized.[73] The amount of nitrogen mineralized after a grazed ley has been estimated[74] as 140 kg ha^{-1} a^{-1}. Other practices of adding nitrogen to the soil/plant system, for example use of the alga *Anabaena azolla* in symbiotic association with the freshwater form *Azolla* in paddy fields, are described in Chapter 19.

Nutrients are conserved if plant residues and animal wastes are returned to the land. The amount of nitrogen potentially available in Britain from animal manures and slurries has been estimated at 455 000 t per year.[75] However, losses occur during storage and spreading, and the efficiency of use of the nitrogen by a crop is often about 30 to 50 per cent. There are also large losses when faeces and urine are voided on to grassland (see Ch. 20). The net utilization of nitrogen to produce animal products is therefore low,[76] as shown by Table 3.12. On the intensively managed dairy farm 16 per cent of the nitrogen input was recovered in the milk and liveweight, compared with 34 per cent on the extensively managed farm.[77]

73 Clement, C. R. in *Residual Value of Applied Nutrients, MAFF Tech. Bull.*, No. 20, 166, 1971.
74 Cooke, G. W., *J. Sci. Fd. Agric.*, 1975, **26**, 1065.
75 Gostick, K. G., *Phil. Trans. R. Soc. Lond.*, 1982, **B296**, 329.
76 Ryden, J. C., *Fertil. Soc. Proc.*, No. 229, 1984, has reviewed the evidence.
77 For further examples see Frissel, M. J. (ed), *Cycling of Mineral Nutrients in Agricultural Ecosystems*, Elsevier, 1978.

4

Growth and functioning of plant roots

The soil affects the plant primarily through its root system but because of the problem of measuring root systems of crops growing in the field, our knowledge of the interaction between soil and roots is incomplete. Roots growing in soil are very much more inaccessible than shoots growing in the air so that many of the methods available for their study are both time consuming and involve the destruction of the root environment itself. Numerous techniques have been employed to study root growth,[1] the method selected depending upon the attribute of the root system that is to be measured. A common procedure involves taking either disturbed or undisturbed cores of soil, washing out the roots, and then separating the 'live' roots (usually pale brown) from decaying roots and other organic matter.[2] The

---

1 For a review see Böhm, W., *Methods of Studying Root Systems*, Vol. 33, *Ecological Studies*, Springer-Verlag, 1979.
2 See for example Welbank, P. J. *et al.*, *Rothamsted Ann. Rept. for 1973*, Part 2, p. 26.

isolated roots can then be used to determine both total root length and dry weight per unit ground area, and the distribution of length and weight with soil depth. Considerable impetus was given to the study of roots by Newman's[3] work showing that root length (*L*) could be estimated by counting intersections of roots (number *N*) with randomly placed straight lines (length *H*) such that:

$$L = \frac{\pi N A}{2H}$$ [4.1]

where *A* is the area over which the roots were spread. Subsequently, other workers have modified equation [4.1] to make the estimation more rapid and susceptible to automation.[4] While the estimation of length has been speeded up considerably by this innovation, the overall time requirement is long, especially if samples have to be cleaned to remove organic debris; the technique is also destructive of both the soil and the above-ground crop.

Non-destructive techniques for measuring root growth have also been developed including the use of underground root laboratories, or rhizotrons,[5] to study the distribution of root length and the persistence of roots; and the injection of radioisotopes (especially ^{32}P) into soils for the measurement of rooting depth.[6] Rooting depth alone is rarely a good indicator of root growth and function so that radioisotopes are currently little used. However, although large rhizotrons are expensive to build and maintain, the use of mini-rhizotrons (glass tubes) with either an endoscope, or fibre-optics, or a video camera to observe the root growth at the soil/glass interface is a practical technique in field studies. This approach has the advantages of being quick, thereby allowing comparison of several experimental treatments, and non-destructive, thereby allowing temporal comparisons at the same spot. There are, though, problems in achieving good glass/soil contact although these can be reduced if tubes are inserted at an angle of 45° when the soil is moist.

Typically, roots rarely occupy more than about 5 per cent of the soil volume even in the upper 100–150 mm of soil where they are most abundant. For many crops the volume occupied decreases rapidly with depth and is often no more than one-hundredth to one-thousandth of 1 per cent at 0.5 m. It follows, then, that only a small fraction of the soil within the rooting zone is in direct contact with roots. Thus, root function in soils is determined not

3 Newman, E. I., *J. appl. Ecol.*, 1966, **3**, 139.
4 See Tennant, D., *J. Ecol.*, 1975, **63**, 995;   for an automated method see Rowse, H. R. and Phillips, D. A., *J. appl. Ecol.*, 1974, **11**, 309.
5 Browning, V. D. *et al.*, *Agric. Exp. Stn. Auburn Univ. Alabama, Bull.* No. 467, 1975; Gregory, P. J., *J. exp. Bot.*, 1979, **30**, 205;   Upchurch, D. R. and Ritchie, J. T., *Agron. J.*, 1983, **75**, 1009.
6 See for example Bassett, D. M. *et al.*, *Agron. J.*, 1970, **62**, 200.

only by the physiological characteristics of the root system but also by those soil factors that determine the rate at which nutrients and water can move to the soil/root interface.

This chapter deals mainly with the growth of roots of crops but also includes some information on forest trees. The principal soil factors influencing the rates of growth and the size of root systems are summarized together with the principal soil processes and root physiology determining the uptake sites and rates of water and nutrients. Other texts and reviews should be consulted for more specialized information.[7]

Structure and function of roots

Roots have a number of functions; they absorb nutrients and water from the soil, and transport these from where they were absorbed to the stems; they are the site of synthesis of a number of plant hormones and growth regulators; they may act as storage organs, as, for example, in the root vegetables; and they anchor the plant into the soil. A typical young root (Fig. 4.1) consists of a growing tip, behind which are the meristem cells which later differentiate into the cells forming the epidermis, the cortex and the stele. During differentiation the cells increase in size and elongate to push the root through the soil. Behind this elongating zone, the cells of the stele differentiate into the endodermal and phloem cells and the xylem vessels.

As a given part of the root ages, it changes colour from white to pale brown, and suberization occurs in the endodermal cells as a band of a waxy substance, suberin, is laid down within their radial and transverse walls. Later still the roots of many plants, excluding the grasses and cereals, may start to have secondary thickening due to the development of a layer of cambium behind the endodermis and the laying down of an outer corky layer over the root surface. The cell arrangement in the epidermal and cortical layer is such that there is an appreciable proportion of space between the individual cells and within the cell walls. This space, known as the root free space, contains solution which forms a continuum with the soil solution, that is, there is no intervening membrane.

---

7 For more comprehensive accounts of roots and root/soil interactions and processes see Barber, S. A. and Bouldin D. R. (eds), *Roots, Nutrient and Water Influx, and Plant Growth*, ASA Special Publication, No. 49, 1984;   Carson E. W. (ed.), *The Plant Root and Its Environment*, Univ. Virginia Press, 1974;   Gregory, P. J., Lake, J. V. and Rose, D. (eds), *Root Development and Function: effects of the physical environment*, Cambridge Univ. Press, 1987;   Harley, J. L. and Russell, R. Scott (eds), *The Soil–Root Interface*, Acad. Press, 1979;   Nye, P. H. and Tinker, P. B., *Solute Movement in The Soil-Root System*, Blackwell, 1977;   Russell, R. Scott *Plant Root Systems: their function and interaction with the soil*, McGraw-Hill, 1977;   Torrey, J. G. and Clarkson, D. T. (eds), *The Development and Function of Roots*, Acad. Press, 1975.

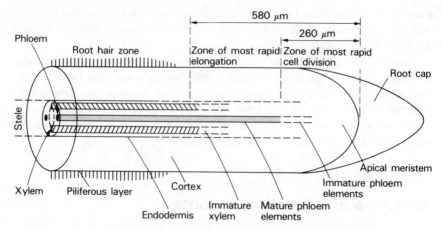

Fig. 4.1   Diagrammatic reresentation of a longitudinal section through a tobacco root tip showing the spatial relations and order of maturation of different tissues. (Redrawn from Esau, K., *Plant Anatomy*, Wiley, 1953; reproduced with permission from Sutcliffe, J. F., *Mineral Salts Absorption in Plants*, p. 15, Pergamon Press, 1962.)

The cell walls of the epidermal and cortical cells contain polysaccharide and polyuronide gums, composed of hexose, pentose and amino sugars, and carboxylic acid groups which are dissociated at the pH prevailing outside the cells. Thus the cell walls are negatively charged and have a diffuse double layer outside them in which cations are concentrated and anions excluded. The water volume affected by this double layer is known as the Donnan free space. The cell walls thus have an appreciable cation exchange capacity. Values of cation exchange capacity ranging from 5.3 to 23.6 meq/100 g dry matter and 20.2 to 53.2 meq/100 g dry matter have been reported for the roots of several monocotyledons and dicotyledons respectively.[8] The outer cells of the root cap may also excrete gums which coat the young root with a mucilaginous layer up to 1 μm thick. This mucigel production is particularly abundant in the region of the root cap but smaller amounts may also extend over the root hair region and further back along the root.[9] Table 4.1 shows the composition of mucigel produced by *Zea mays* (maize). During drought, the production of mucigel is particularly noticeable and substantial quantities of soil particles may adhere to the roots[10]; the significance of this for water uptake is not known with certainty.

The cells in the meristematic region of the root cap have also been impli-cated in the synthesis of plant hormones and growth regulating substances.

8 Crooke, W. M., *Pl. Soil*, 1964, **21**, 43.
9 Greaves, M. P. and Darbyshire, J. F., *Soil Biol. Biochem.*, 1972, **4**, 443.
10 See for example Sprent, J. I., *New Phytol.*, 1975, **74**, 461.

Table 4.1 The composition of mucigel secreted by cells of *Zea mays*. (Adapted from Barlow, P. W. in *The Development and Function of Roots* (eds J. G. Torrey and D. T. Clarkson), p. 44, Acad. Press, 1975)

Variety/hybrid	Orla 266	WF-9X M-14	WF9 X38-11
Cell type	root cap	root cap	lst or 2nd nodal roots 0.5–2.5 cm long
Surface sterilized	Yes	Yes	No
Analytical method	hydrolysis, g.l.c	ethanol, g.l.c.	hydrolysis, paper chromatography
% by weight of constituent sugars			
Galacturonic acid	Present	12	
Glucuronic acid	Present		
Galactose	21	39	18.4
Glucose	22	22	27.3
Mannose	6	6	
Fucose	32	8	16.6
Rhamnose		1	
Arabinose	15	7	15.1
Xylose	4	5	6.2
Fructose			16.4

Plant hormones have been found in both xylem and phloem sap in roots and there is some evidence that the cytokinins and gibberellins moving upwards to the shoot are important in controlling shoot growth. While the sites of production and the mode of action of hormones are still far from clear, the root cap constitutes an important region of root capable of perceiving a variety of stimuli and translating them into a growth or a regulating response.[11]

Root hairs

Some of the epidermal cells behind the meristem, in the so-called 'piliferous layer', elongate to produce root hairs which grow into the soil. Root hairs have a diameter of about 10 to 20 μm and pores of this size are emptied of water by suctions of about 15 to 30 kPa. Since field capacity corresponds to a suction of about 5 to 10 kPa depending upon the soil type (see p. 353), much of the available water is in pores that are inaccessible to roots or root hairs. It is not obvious, therefore, how the root cells maintain contact with

11 For reviews of the roles of root hormones see Feldman, L. J., *Ann. Rev. Plant Physiol.*, 1984, **35**, 223; Torrey, J. G., *Ann. Rev. Physiol.*, 1976, **27**, 435.

water in pores <10 μm diameter. Root hairs may play a major role in maintaining root/soil contact and in freely drained soils they are likely to be of major importance in water and nutrient absorption. Root hairs may increase the surface area of the root system 5 to 18 times[12] but their production varies from species to species. For example, young roots of lucerne, maize and perennial ryegrass grown in a sandy loam produced 105, 161 and 88 root hairs per mm of root respectively.[13] The total length of root hairs per mm length

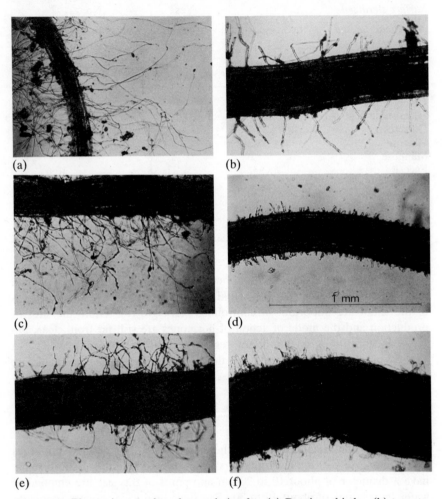

(a)

(b)

(c)

(d)

$\overline{1}$ mm

(e)

(f)

Plate 4.1   Photomicrographs of root hairs for (a) Russian thistle, (b) tomato, (c) lettuce, (d) wheat, (e) carrot, and (f) onion. (From Itoh, S. and Barber, S. A., *Agron. J.*, 1983, **75**, 457.)

12 Dittmer, H. J., *Am. J. Bot.*, 1937, **24**, 417.
13 Reid, J. B., *Pl. Soil*, 1981, **62**, 319.

of root was estimated as 37, 146 and 99 mm for lucerne, maize and perennial ryegrass respectively.

Root hairs are important for the uptake of ions moving to the root mainly by diffusion (e.g. phosphate) because they effectively extend outwards the surface of the root[14] (see later section). In a comparative study of Russian thistle, tomato, lettuce, wheat, carrot and onion, which vary considerably in their degree of root hair growth (Plate 4.1),[15] it was shown that root hairs contributed substantially to uptake of phosphate by Russian thistle, tomato and to a lesser degree by lettuce, but not by wheat. The density of root hairs for wheat was low compared with the other species and they contributed only very slowly to the development of an additional depletion zone. In practice, many environmental factors (e.g. the mechanical properties of the soil) influence the growth of root hairs. Generally they persist for only a few days although for longer in some grasses.

The rhizosphere

The soil just outside the root surface contains the rhizosphere population and is a zone of great microbial activity. Its primary energy source is derived from the plant, possibly in part from the mucigel and in part from soluble organic compounds excreted by the roots, and from sloughed root cells. Shown later in Plate 17.1 is the bacterial colonization of the roots of cereals grown in solution, and other micrographs show the rhizosphere of wheat grown in the field at about the time of flowering.[16] There is considerable growth of microorganisms in both the rhizosphere and on the outer cortical cells and cell walls. Decortication of roots during the growing season is a common occurrence and, for example, about half the dry matter of the white extension roots of apple trees may be lost during the growing season.[17] The losses of root material, the consequences for microbial activity, and the associations between plant roots and micro-organisms are more thoroughly discussed on pp. 526 to 530.

Growth and development of root systems

A great variety of names has been used to distinguish one type of root from another; some are merely a description of the site or origin while others imply

14 Bhat, K. K. S. and Nye, P. H., *Pl. Soil*, 1973, **38**, 161.
15 Itoh, S. and Barber, S. A., *Agron. J.*, 1983, **75**, 457.
16 Foster, R. C. and Rovira, A. D., *New Phytol.*, 1976, **76**, 343.
17 Rogers, W. S. and Head, G. C. in *Root Growth* (ed. W. J. Whittington), Butterworths, p. 280, 1968.

some functional significance. For convenience, the system of description employed for cereals by Hackett[18] is adopted here.

An individual cylinder of root tissue developing from the seed or stem is called an axis and branches from this are known as primary laterals. Branches from primary laterals are termed secondary laterals and these, in turn, give rise to tertiary laterals. Each axis plus laterals is a root and each root is a member of the root system. Seminal axes are derived from differentiated cells present in the embryo and nodal axes develop from the growing shoot.

For non-cereal crops, the same system of description of laterals can be employed but there are usually far fewer axes and frequently only one, which is termed the taproot. In trees, a framework of older roots frequently has a mass of finer roots which grow and decay quite rapidly making detailed description difficult.[19]

Most root systems are branched (the root systems of some bulbs, corms and vegetables such as leeks and onions frequently are not) and the lateral roots generally form some distance behind the main root apices. The meristems of the new laterals originate within the root and must therefore traverse other tissues in order to emerge; this disruption of the cortex and epidermis may allow sites for micro-organisms to enter the root. There is some predictability as to the position in which lateral roots arise since it is related to the internal vascular pattern of the root.[20]

The genetic control of the number of root axes together with the genetic/mechanical determination of the position of laterals means that although many factors may influence the size of root systems, their general morphology may be specified and is characteristic of the species. Pictorial representations of the morphology of root systems enable some broad generalizations to be made about rooting depth and the relative distribution of roots.[21] Plate 4.2 shows root systems of sugar beet, maize, wheat and clover near to maturity all of which (except clover) penetrate to at least 1 metre. Nearly all such photographs reveal that roots grow initially at some angle relative to the plant shoot but then become more vertical. The quality of light may play a role in determining the angle at which roots initially grow in soil but the geotropic response together, perhaps, with the predominance of vertical cracks in deeper soil layers ensures that most roots grow almost vertically. These photographs also enable clear distinctions to be drawn between certain types of root system, notably the cereals, legumes and root crops.

Dicotyledons differ from monocotyledons in that the primary root system

18 Hackett, C., *New Phytol.*, 1968, **67**, 288.
19 Reynolds, E. R. C., *op. cit.*, 1987, **98**, 397.
20 McCully, M. E. in *The Development and Function of Roots* (eds J. G. Torrey and D. T. Clarkson), p. 105 Acad. Press, 1975.
21 See Weaver, J. E., *Root Development of Field Crops*, McGraw-Hill, 1926; Kutschera, L., *Wurzelatlas mitteleuropäischer Ackerunkräuter und Kulturpflanzen*, 1960.

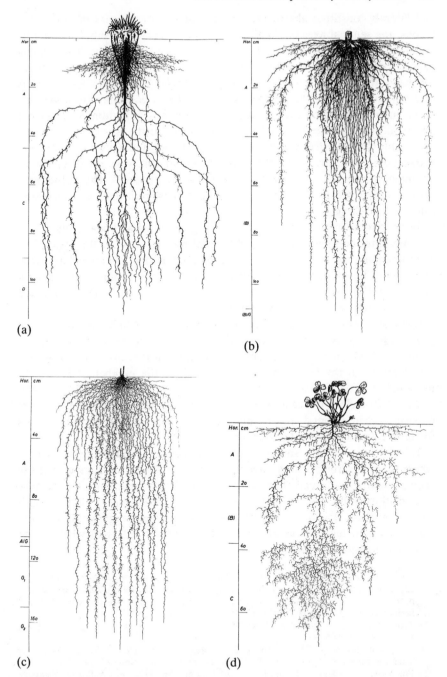

(a)

(b)

(c)

(d)

Plate 4.2 Root systems of (a) sugar beet, (b) maize, (c) wheat and (d) clover. (From Kutschera, L., *Wurzelatlas Mitteleuropäishcer Ackerunkräuter und Kulturpflanzen*, 1960.)

and laterals constitute almost the entire root system whereas in monocotyledons, the seminal axes originating in the seed embryo are supplemented by
nodal axes produced from nodes at the base of the leaf sheaths. The seminal
axes of cereals vary in number from about 4 to 7 for winter wheat, 1 to 11
for maize (commonly 3 to 5) and only 1 for millet,[22] but for a given variety
of a particular cereal, the number is usually almost constant. More variable
is the number of nodal axes produced because their initiation is dependent
on favourable conditions for tillering and their growth is prevented by unfavourable soil physical conditions such as drying and crusting of the soil
surface. For example, when the upper layers of soil were dried, clones of
perennial ryegrass ceased to produce nodal root axes.[23] The subsequent
rewetting resulted in the rapid production of new root axes and increased
rates of leaf elongation.

Seminal roots of some tropical cereals (e.g. sorghum and pearl millet) may
die soon after germination leaving the nodal roots to support the plant. This
phenomenon may also occur in many grasses but usually at a much later stage
of growth. In most temperate cereals, though, the seminal root system
persists and crops of winter wheat have yielded grain while maintained
entirely by the seminal root system.[24]

Because nodal root development in cereals is closely related to the
production of leaves and tillers, leaf number can be used to predict the
number of nodal axes. For example, the number of root axes, RA, of winter
wheat grown in the field can be calculated from the number of leaves, L,
using $RA = 1.95L - 3.06$, as they can also for pearl millet using $RA = 1.42L
- 2.26$.[25] Branching of the axes is more difficult to predict because it depends
on many environmental factors, but the relations described provide a basis
for future predictions of development and growth.

Table 4.2 presents a summary of the principal geometric features of a
wheat or barley root system growing in topsoil. Similar data show that the
radii of maize roots vary from about 0.3 to 3 mm depending on the type
and position of the root; most are in the range 0.4 to 1.0 mm.[26] There is much
less information available for dicotyledonous plants although soyabean roots
were shown to have diameters in the range 0.12 to 0.50 mm with 0.15 to
0.30 mm the most common.[27]

22 For wheat see Troughton, A., *Mimeo. Pub.*, **21**, Commonw. Bur. Past. Field Crops,
   1962.   For maize see Hector, J. M., *Introduction to the Botany of Field Crops*, Vol. 1,
   p. 92. Central News Agency, Johannesburg, 1936.   For millet see Gregory, P. J. and
   Squire, G. R., *Expl Agric.*, 1979, **15**, 161.
23 Troughton, A., *J. agric. Sci.*, 1980, **95**, 533; see also Cornish, P. S., *Aust. J. agric.
   Res.*, 1982, **33**, 665.
24 For a pot experiment see Passioura, J. B., *Aust. J. agric. Res.*, 1972, **23**, 745.   For
   a field experiment see Gregory, P. J., Ph. D thesis, Univ. Nottingham, 1976.
25 For wheat see Klepper, B. *et al.*, *Agron. J.*, 1984, **76**, 117.   For millet see Gregory,
   P. J., *J. exp. Bot.*, 1983, **34**, 744.
26 Weihing, R. M., *J. Am. Soc. Agron.*, 1935, **27**, 526.
27 Arya, L. M. *et al.*, *Soil Sci. Soc. Am. Proc.*, 1975, **39**, 437.

Table 4.2 General features of the geometric relations of wheat and barley roots. (Taken from Barley, K. P., *Adv. Agron.*, 1970, **22**, 159.)

	Order				
	Main axis	Primary lateral	Secondary lateral	Tertiary lateral	Root hair
Diameter (cm)	5×10^{-2}	2×10^{-2}	1×10^{-2}	5×10^{-3}	1×10^{-3}
Number per cm of root of next higher order	—	2	1	5×10^{-1}	1×10^3
Length (cm) per cm³ of soil	1	5	2	5×10^{-1}	1×10^3

Clearly, a root system changes throughout the life of a crop and many researchers have attempted to relate such changes to phases of shoot growth. Flowering appears as a particularly important stage of growth in many crops, as after this assimilates are required pre-eminently to fill the growing reproductive structures. As Fig. 4.2 shows, the weight of cereal roots frequently reaches a maximum at around flowering and may even decline thereafter.[28]

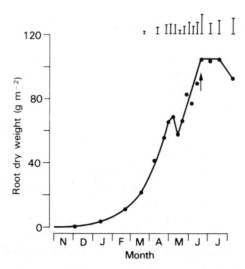

Fig. 4.2 Changes in the dry weight of roots of winter wheat with time. The standard error for each sample is shown at the top of the figure and the time of flowering is indicated by the arrow. (From Gregory, P. J. *et al.*, *J. agric. Sci.*, 1978, **91**, 91.)

28 For grass see also Troughton, A., *J. Brit. Grassland. Soc.*, 1960, **15**, 41. For maize see Mengel, D. B. and Barber, S. A., *Agron. J.*, 1974, **66**, 341.

This does not necessarily mean that all root growth has ceased since most of the techniques currently used to assess growth really measure the balance between growth and decay; indeed as Fig. 4.4 later illustrates, growth of deep roots after flowering may occur while decay occurs elsewhere. In other types of plant, notably those legumes where the growth of the pods is indeterminate, the root system may continue to increase in weight throughout growth. For example, in one study substantial increases in the depth of soyabean root systems and the production of new roots was observed during pod filling.[29] In contrast other studies with soyabeans show that the size of the root system decreases during pod filling[30]; the pattern of root growth in legumes is, then, uncertain.

Little is known about the stages of root growth for crops whose storage organs are located below ground. Although total length of fibrous roots of potato grown on a loamy sand near Reading did not increase after about 2 weeks after emergence (6 weeks from planting), workers in the western USA found the longest total root length occurred between 5 and 12 weeks after emergence but decreased thereafter.[31] Growth of fibrous roots has been reported throughout the whole growing period for cassava and for most of the period for sugar beet.[32]

Root length and weight

Table 4.3 draws together information from a number of experiments to show the maximum recorded values of root dry weight, root length and depth of rooting of various crops. These data were obtained using a variety of methods for extracting roots all of which are liable to criticism on the grounds that some fine roots may have been lost during the extraction procedure. Table 4.3 shows differences between crops, with the cereals generally possessing greater weights and lengths. The lengths fall within the range 2.5 to 23.5 km m^{-2}, again with cereals at the upper end of the range. Unfortunately some sources do not give the developmental stages of the crop when sampling occurred, but it appears that in many temperate cereals, about 10 per cent of the total plant dry weight is in the roots at maturity.

In coniferous forests, estimates of root dry weight are about 15 to 25 per cent of total tree weight, although individual estimates range from 9 to 44

29 Kaspar, T. C. *et al.*, *Agron. J.*, 1978, **70**, 1105.
30 Mayaki, W. C. *et al.*, *Crop Sci.*, 1976, **16**, 92.
31 Asfary, A. F. *et al.*, *J. agric. Sci.*., 1983, **100**, 87 at Reading; Lesczynski, D. B. and Tanner, C. B., *Am. Pot. J.*, 1976, **53**, 69 in USA.
32 For cassava see Aresta, R. B. and Fukai, S., *Field Crops Res.*, 1984, **9**, 361; for sugar beet see Brown, K. F. and Biscoe, P. V., *J. agric. Sci.*, 1985, **105**, 679.

Table 4.3 Maximum length, dry weight and depth of rooting of some common crops

Crop	Soil	Maximum root			Source
		dry weight ($g\ m^{-2}$)	length ($km\ m^{-2}$)	depth (m)	
Winter wheat	Batcombe series, England	170	32	>1.0	1
	Blithe series, England	150	27	>1.0	
Soyabean	Australia	—	11.5	>0.95	2
Sugar beet ⎫				0.9	3
Potato ⎬	Stretham series, England			0.8	
Barley ⎭				>1.0	
Winter wheat	Denchworth series, England				4
	drained		13.8	1.0	
	undrained		8.5	1.0	
Winter wheat	Astley-Hall series, England	105	23.5	2.0	5
Groundnut ⎫	Alfisol, ICRISAT, India	—	2.5	0.7	6
Millet ⎭		—	3.5	0.9	
Clover ⎫	psammentic	—	9.2	0.4–1.0	7
Pea ⎪	Entisols,	—	1.3	1.5–0.9	
Medic ⎪	W. Australia	—	5.7	0.6–1.3	
Wheat ⎭		—	4.0	1.0–1.4	
Maize	Typic Argiaquoll, USA	160	15.1	>0.75	8
Sorghum	Tindall clayloam, Australia	100	26.5	1.35	9
Soyabean	Typic Udorthents, USA	58	5.5	1.8	10
Paddy rice ⎫	Aquic Tropudalf, IRRI	—	32.8	0.5	11
Dryland rice ⎭	Philippines	—	24.7	>0.8	

1 Barraclough, P. B. and Leigh, R. A., *J. agric. Sci.*, 1984, **103**, 59.
2 Burch, G. J. *et al.*, *Aust. J. Pl. Physiol.*, 1978, **5**, 169.
3 Durrant, M. J. *et al.*, *Ann. appl. Biol.*, 1973, **4**, 387.
4 Ellis, F. B. *et al.*, *J. agric. Sci.*, 1984, **102**, 583.
5 Gregory, P. J. *et al.*, *J. agric. Sci.*, 1978, **91**, 91.
6 Gregory, P. J. and Reddy, M. S., *Field Crops Res.*, 1981, **5**, 241.
7 Hamblin, A. P. and Hamblin, J., *Aust. J. agric. Res.*, 1985, **36**, 63.
8 Mengel, D. B. and Barber, S. A., *Agron. J.*, 1974, **66**, 341.
9 Myers, R. J. K., *Field Crops Res.*, 1980, **3**, 53.
10 Sivakumar, M. V. K. *et al.*, *Agron. J.*, 1977, **69**, 470.
11 Yoshida, S. and Hasegawa, S. in *Drought Resistance in Crops*, p. 97, IRRI, 1982.

per cent.[33] The dry weight of roots in conifer stands less than 200 years old has been variously estimated at 3 to 85 t ha⁻¹ with fine roots ranging from 1 to 12.6 t ha⁻¹ with a mean of 5 t ha⁻¹. Endotrophic mycorrhizas may account for an additional 8 per cent of total tree weight in young Douglas

33 Santantonio, D. *et al.*, *Pedobiologia*, 1977, **17**, 1. For root dynamics in Scots pine see Persson, H., *Oikos*, 1978, **30**, 508.

fir stands, and roots and mycorrhizas together may constitute 63 to 70 per cent of total net primary production in Douglas fir and Pacific silver fir stands.[34]

Partitioning between roots and shoots

Root weight as a fraction of the total plant weight changes throughout growth and in many temperate cereals ranges from around 0.4 to 0.1 depending upon growth stage (Fig. 4.3). Nitrogen supply can have a marked effect on the ratio of root : plant weight and the considerable range of values shown in the figure during the spring probably reflects different combinations of timing of nitrogen fertilizer application and soil nitrogen levels. In more arid environments where total plant weights are frequently much lower than those of the crops used to calculate Fig. 4.3, root : plant weight ratios are generally higher throughout growth.

The environmental and physiological factors determining the partitioning of assimilates between roots and shoots are little understood although some researchers have suggested that many plants are wasteful in diverting too much assimilate to root growth at the expense of shoot growth and useful yield. Such statements imply that scope exists for changing the balance between root and shoot growth in a controlled manner.

Exact relations between root and shoot are currently difficult to define but it is generally assumed that a balance must exist between the activities of the two systems. The experimental work of Brouwer[35] and others led a number of workers to postulate the existence of an equilibrium between the two systems under a range of conditions. The experimental results can be described quantitatively by defining the size and activity of the shoot and root systems, i.e.:

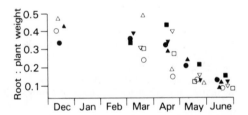

Fig. 4.3 Changes in the ratio of root : plant weight of winter wheat crops with time. Solid symbols, 1980 crops; open symbols, 1981 crops. (Adapted from Barraclough, P. B., *J. agric. Sci.*, 1984, **103**, 439.)

34 Fogel, R., *Pl. Soil*, 1983, **71**, 75.
35 Brouwer, R., *Jaarb. Inst. biol. scheik. Onderz. LandbGewass.*, 1963, 31.

(root mass × specific absorption rate) ∝ (shoot mass × specific photosynthetic rate).

Following from this, a simple model can be based on the vegetative growth of a two-component system (root and shoot), one supplying nitrogen, the other carbon.[36] The model shows that plants possessing a number of chemical activities each essential for growth will, when undergoing steady state growth in a constant environment, adjust their growth so that the individual activities bear a constant ratio to each other independent of the environment. However, crops rarely experience constant environments, and sink sizes and demand vary during growth so that such simple models generally only apply when plants are young and growing vegetatively.

In crops with storage roots, the partitioning between roots and shoots is crucial to the production of utilizable yields. Based on the concept of competing sinks, a simple empirical relationship for assimilate distribution in storage roots has been developed.[37] The relation states that the shoot dry matter (S_w) and the weight of the storage root (R_w) are related by:

$$\ln S_w = \alpha + \gamma \ln R_w - \eta t,$$

where α, γ and η are constants for respectively, the shoot : root ratio at time zero, the ratio of rates of assimilation of dry matter (activities × efficiencies) for shoot and root, and the sum of rates of losses due to leaf death and maintenance respiration rates of shoot and root. Measurements with carrots, sugar beet and a range of root vegetables[38] support this general relation although γ may not be constant in young plants or at high plant densities. Differences in γ between species reflect both anatomical differences in the storage tissues and biochemical differences in the form of sugars produced and stored.

Depth of rooting

The depth of rooting is of considerable agricultural importance especially if water or nutrients are likely to be in short supply. In an early, descriptive study, Weaver[39] found that the roots of winter wheat had grown to a depth of 2 m and many laterals were present to 1.4 m. Table 4.3 shows that under favourable conditions most crops will root to at least 1 m though this is, of course, highly dependent on the water supply and other factors affecting

36 Thornley, J. H. M., *Ann. Bot.*, 1972, **36**, 431; see also Ch. 2.
37 Barnes, A., *Ann. Bot.*, 1979, **43**, 487.
38 For carrots see Hole, C. C. *et al.*, *Ann. Bot.*, 1983, **51**, 175. For sugar beet see McLaren, J. S., *Ann. Bot.*, 1984, **54**, 383. For carrot, parsnip, radish and red beet see Hole, C. C. *et al.*, *Ann. Bot.*, 1984, **53**, 625.
39 Weaver, J. E., *Root Development of Field Crops*, McGraw-Hill, 1926.

growth. For example, in Western Australia, rooting depth of wheat varied considerably with soil type and in three out of four soils was related to the depth of water penetration.[40] Maximum depths of root penetration in the sand, sandy loam, grey clay and sand over clay soils were 169, 173, 31 and 73 cm respectively and were reached 10 to 14 weeks after planting (about 7 to 3 weeks before anthesis).

Besides measuring the maximum depth which a root system may attain, it is also important to know the rate at which it becomes established and thereby secures an uninterrupted supply of water and nutrients. For example, on the sandy soils of Gleadthorpe Experimental Husbandry Farm, Nottinghamshire, England, late spring-sown cereals can suffer badly during drought because the potential rate of evaporation is typically about 2.5 mm day^{-1} while the available water is only 1 mm cm^{-1} depth. Therefore, to maintain evaporation close to the potential rate in the absence of rain, roots must elongate by at least 25 mm day^{-1}, a rate of extension that is rarely attained. In general, the axes of wheat elongate at 5 to 30 mm day^{-1} and a value of 10 mm day^{-1} is typical.[41] Rates of extension for soyabeans are typically[42] 20 mm day^{-1} but faster rates of penetration of up to 60 mm day^{-1} have been reported for maize grown at high temperatures.[43] Extension is highly dependent on both temperature and assimilate supply, factors that were discussed in Chapter 2. While rapid rates of downward extension may guarantee supplies of water, nutrients are frequently concentrated in the upper soil layers so that the effects of drought are also often inextricably linked with shortages of nutrients.

Distribution of roots in the soil profile

In general, the length of roots (cm root cm^{-3} soil) decreases with depth in the soil but considerable variation exists in the distribution of roots at different stages of growth. Figure 4.4 shows how the distribution of winter wheat and soyabean roots changed at various stages of growth; the soyabean roots appear more uniformly distributed within the soil profile. Typically root lengths (cm root cm^{-3} soil) in the top 10–15 cm are 20–25 for temperate grass swards (though up to 100), 20–30 for paddy rice, 8–10 for temperate cereals, 5–10 for dryland rice, 1–2 for cereals such as millet and sorghum, and about 1 for many legumes (groundnut, chickpea and pigeonpea). Beneath this top layer, length per unit volume rapidly decreases although the presence of compact layers or zones of nutrient accumulation can substantially change this general pattern. Since many theories assume the uptake of water and

40 Tennant, D., *Aust. J. exp. Agric. Anim. Husb.*, 1976, **16**, 570.
41 Welbank, P. J. *et al.*, *Rothamsted Ann. Rept. for 1973*, Part 2, p. 26.
42 Stone, J. A. *et al.*, *Agron. J.*, 1983, **75**, 1050.
43 Taylor, H. M. *et al.*, *Agron. J.*, 1970, **62**, 807.

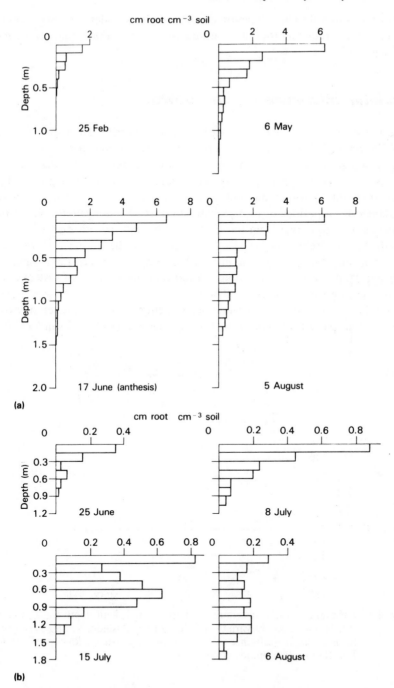

Fig. 4.4 Distribution of root length with depth during the growing season for (a) winter wheat and (b) soyabeans (Adapted from: (a) Gregory, P. J. *et al.*, *J. agric. Sci. Camb.*, 1978, **91**, 91; (b) Sivakumar, M. V. K. *et al.*, *Agron. J.*, 1977, **69**, 470.)

nutrients from soil to be a function of root length, considerable effort is being expended on predicting root distributions (see later section on modelling root growth).

Varietal differences in root growth

The foregoing account has dealt with comparisons between crops but there are known to be differences in root growth between varieties of a crop. Figure 4.5 shows that in rice there are obvious differences between upland and paddy varieties in respect of total length of roots, of depth, and of their distribution within the soil profile. Similarly, Russell,[44] in discussing possible critical periods of root growth, contrasts the elongation rates of two varieties of barley, one from the UK (Proctor) and the other grown on the South African Veldt (Swanneck). The UK variety had a much slower rate of extension than that grown in South Africa where the rapid onset of drought would ensure an advantage to those plants whose root system could secure an uninterrupted supply of water.

In wheat, differences in the number of seminal axes, diameter of xylem vessels, and pattern of root growth have been observed, and similar differ-

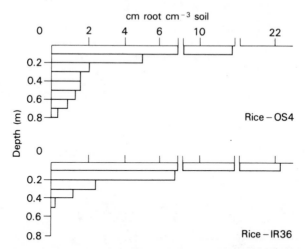

Fig. 4.5   Differences in the distribution of root length with depth for an upland (OS4) and a paddy variety of rice. (From Yoshida, S. and Hasegawa, S. in *Drought Resistance in Crops with Emphasis on Rice*, p. 97, IRRI, Los Banos, Philippines, 1982.)

44 Russell, R. S. in *Potential Crop Production* (eds P. F. Wareing and J. P. Cooper), p. 100, Heinemann, 1971.

ences are known to exist in other species.[45] However, there have been only few quantitative studies showing that these differences in rooting are important either for competition between plant species or for allowing one genotype to grow better than another at a specific location. In rice, some progress has been made in establishing the genetic control of root elongation and this opens possibilities for modifying root growth in the future.[46]

Environmental effects on root growth

Clearly, many factors affect root growth, e.g. pH, aluminium toxicity, specific nutrient deficiencies, but while many of these are of over-riding importance in specific circumstances, more generally the principal factors affecting growth are: the soil moisture content; the soil temperature; the system of pores into which roots can grow together with the shear strength and compressibility of the soil; the oxygen supply; the level of toxins and pathogens in the soil; and the nutrient supply. To some extent, many of these factors interact with each other so that, for example, compacting a soil may not only reduce the mean pore size but also change the aeration and water status of the soil. Nevertheless, a harmful effect of one unfavourable factor can often be diminished by improving some of the others, as, for example, in the application of fertilizer to crops suffering from soil compaction. Because soil properties vary spatially, restricted root growth in one region may be compensated for by improved growth in more favourable regions. When roots of young barley plants were excised, desiccated or exposed to unfavourable temperatures, compensatory growth of other roots occurred.[47]

This section is concerned mainly with the effects of mechanical impedance, soil water, and nutrient supply on root growth. Other environmental effects are dealt with in other chapters and, in particular, Chapters 8, 9 and 25 consider temperature, aeration and aluminium toxicity respectively.

Effect of soil physical conditions

The ability of the plant root to find space in which to grow, or to force its way into the soil, is often an important factor limiting plant growth. There is not a great deal of information on the minimum size of pore into which a root can grow without having to enlarge it but few crops have roots with

45 For number see Robertson, B. M. *et al.*, *Crop Sci.*, 1979, **19**, 843. For diameter see Richards, R. A. and Passioura, J. B., *Crop Sci.*, 1981, **21**, 253. For pattern see Hurd, E. A., *Agric. Meteorol.*, 1974, **14**, 39.
46 Armenta-Soto, J. *et al.*, *SABRAO Journal*, 1983, **15**, 103.
47 Crossett, R. N. *et al.*, *Pl. Soil*, 1975, **42**, 673.

a diameter of less than 0.1 mm and the young growing roots of most trees and herbaceous plants are considerably larger than this. Roots are therefore larger, and usually very much larger, than the pores drained at field capacity (diameter 60 μm) so that transmission pores are the major routes for growing roots; poor soil structure will, then, inhibit root growth. As a consequence, roots must force their way into a soil and can therefore only grow in soils which are compressible.

The mechanism of root growth is that the meristem cells behind the root tip elongate longitudinally and radially pushing the root tip forward. Growth occurs when the turgor pressure inside the elongating cells is sufficient to overcome the constraint imposed by the cell walls and any external constraint caused by the soil matrix. The difference between the turgor pressure and the cell wall constraint is termed the root growth pressure. For several plant species, maximum root growth pressures ranged from 700 to 2500 kPa for beans (*Vicia faba*), maize (*Zea mays*), vetch (*Vicia sativa*) and horse chestnut (*Aesculus hippocastanum*).[48] Similarly, root growth pressures in cotton, peas and peanuts averaged 940, 1300 and 1150 kPa, respectively, although there was considerable variation within each plant type.[49]

Goss[50] grew barley in a bed of glass beads 1 mm in diameter, so the minimum size of pore was 0.16 mm, and found that if the bed was subjected to external pressures of 20 kPa and 50 kPa, then root elongation was reduced by 50 per cent and 80 per cent respectively. The anatomy of the roots was also changed; increasing the external pressure to 20 kPa increased the diameter of the roots (principally because the thickness of the cortex increased), increased the number of cells in transverse section, and increased the diameter of the outer cells (though decreased that of the inner ones).[51] The very substantial difference in the pressures required to reduce growth in these experiments and those commented upon earlier is probably a consequence of arching or locking of the beads so that external pressures should not be equated with the pressures experienced by the root.[52] In most soils, roots grow partly through existing pores and partly by moving aside soil particles, and assessments of root elongation in relation to penetrometer resistance give pressures very much greater than those suggested by Goss. For example, Fig. 4.6 shows that penetration of cotton roots into 2.5 cm thick cores of four soils was reduced by 50 per cent when penetrometer resistance was about 700 kPa, by 80 per cent when it reached 1400 kPa, and effectively ceased at about 2000 kPa; these values are of the same magnitude as the maximum root growth pressures cited earlier.

48 Gill, W. R. and Bolt, G. H., *Agron. J.*, 1955, **47**, 166.
49 Taylor, H. M. and Ratcliffe, L. F., *Agron. J.*, 1969, **61**, 398.
50 Goss, M. J., *J. exp. Bot.*, 1977, **28**, 96.
51 Wilson, A. J. *et al.*, *J. exp. Bot.*, 1977, **28**, 1216.
52 See Barley, K. P., *Soil Sci.*, 1963, **96**, 175; Richards, B. G. and Greacen, E. L., *Aust. J. Soil Res.*, 1986, **24**, 393.

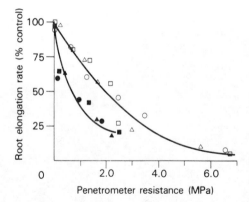

Fig. 4.6 Effect of penetrometer resistance on the rate of root elongation of groundnut (open symbols) and cotton (closed symbols). Approximate water contents (g g^{-1}) of the soil were 7 per cent (circles), 5 per cent (triangles) and 4 per cent (squares). (Redrawn from Taylor, H. M. and Ratcliff, L. F., *Soil Sci.*, 1969, **108**, 113.)

Roots growing in clay soils are sometimes concentrated on the surfaces of cracks separating larger blocks of clay (Plate 4.3). Although the largest roots may not penetrate the blocks, some fine secondary or tertiary laterals, or root hairs may be able to do so. The roots on the clod surfaces are often flattened by the pressure exerted by the adjacent clay clods when they swell on wetting in the autumn or winter, after they have shrunk by drying in the summer. This same effect can be seen when roots grow between a ped and a stone.

As a general indication, roots will be severely impeded if bulk densities exceed 1.55, 1.65, 1.80 and 1.85 g cm^{-3} on clay loams, silt loams, fine sandy loams and loamy fine sands, respectively.[53] However, although root growth may be restricted by a mechanically impeded layer produced, for example, by machinery or treading, the effect may be short-lived and not translated into an effect on yield. This is particularly so on soils with a rich fauna and on clay soils that shrink and crack, for if a zone of weakness is present a root may grow around an obstacle. For example, manually created pans in two soils were readily lost on drying and the pans had no effect on yields of ryegrass, spring barley or faba beans although there were transient effects on water uptake by roots.[54]

Under controlled conditions the behaviour of pea, rape and safflower roots growing in cracks was dependent on the strength of the peds, the width of the crack, and the orientation of the crack relative to the preferred geotropic

53 Bowen, H. D. in *Modifying the Root Environment to Reduce Crop Stress* (eds G. F. Arkin and H. M. Taylor), p. 21, ASAE Monog., No. 4, 1981.
54 McGowan, M. *et al.*, *J. Soil Sci.*, 1983, **34**, 233.

Plate 4.3   Plant roots growing through cracks and on the outside of blocks of a clay soil. (From Goss, M. J., Rothamsted Experimental Station.)

growth direction.[55] Experiments using field clods from a fine sandy loam soil showed that even when the soil was near to saturation with a penetrometer resistance of about 1.5 MPa, ped surfaces were still a substantial barrier to root penetration. However, there were differences between species in the mean length of root hairs, the zones of active root hair development and the mean length of the roots before they penetrated the ped, all of which contributed to differences in the ability of species to penetrate peds.

55 Whiteley, G. M. and Dexter, A. R., *Pl. Soil*, 1983, **74**, 153; *ibid*, 1984, **77**, 141.

If an annual crop is growing on a deep, well-drained soil with a suitable range of pore sizes, the depth and distribution of rooting are very dependent on the water supply. In general, frequent light showers of rain, or frequent irrigation, by keeping the surface soil moist, encourages the proliferation of shallow roots. Thus lucerne roots only penetrated to about 1 m when frequently irrigated with 50 mm water, but to 1.5 m when irrigated less frequently with 120 mm.[56] This result is not always found and the depth of rooting of ryegrass on the loam soil of the Institute for Grassland and Animal Production, England, was unaffected by supplementary irrigation.[57] Similarly, root penetration of sorghum to the base of a 1.08 m deep lysimeter was unaffected by the frequency of irrigation which ranged from daily watering to watering when the matric potential at 25 cm depth was −500 kPa followed by a second irrigation when the matric potential at 40 cm depth was −1.0 MPa.[58] However, whether or not the total depth of rooting of annual crops is affected by the frequency of wetting, the length and distribution of roots most certainly are. In the sorghum study cited, root length and dry weight were 20 per cent to 30 per cent greater under daily irrigation than under the least frequent irrigation and the root system was also relatively larger in the surface layers. Similarly, the root length of mature field-grown lettuces was 75 per cent less in plants protected from rainfall after establishment compared to plants that had been frequently irrigated, although the root weights were very similar. Non-irrigated plants had thicker roots and greater length and weight at depth.[59]

In annual crops, the period of most active root growth is limited to the time before the filling of the reproductive organs but in perennial crops and pastures a longer period of growth is often possible and differences in rooting depth are more apparent between wet and dry years. In such circumstances, severe drought restricts the production of dry matter and in turn limits the rooting depth. This is illustrated in Fig. 4.7 for a range of prairie species at Hays, Kansas,[60] after a period of moist years, and after a period of dry years. In the wetter years 65 per cent of the species sent their roots to below 1.5 m, many reached 2.7 m and a few reached 6 m, whereas in the dry years most of the grass roots were in the top 0.6 m and only a few herbaceous roots penetrated deeper than 1.5 m. This dependence of depth of rooting on water supply is affected also by the availability of carbohydrates to the root; and this becomes of great importance in the correct management of pastures. Hard grazing of pastures reduces surplus carbohydrate production, and all

56 Thompson, C. A. and Burrows, E. L., *New Mexico Agric. Exp. Stn. Bull.* No. 123, 1920.
57 Garwood, E. A., *J. Brit. Grassland Soc.*, 1967, **22**, 176.
58 Merrill, S. D. and Rawlins, S. L., *Agron. J.*, 1979, **71**, 738.
59 Rowse, H. R., *Pl. Soil*, 1974, **40**, 381.
60 Albertson, F. W., *Ecol. Monog.*, 1937, **7**, 481; with Weaver, J. E., *op cit.*, 1943, **13**, 1.

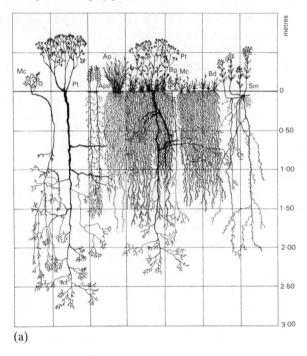

(a)

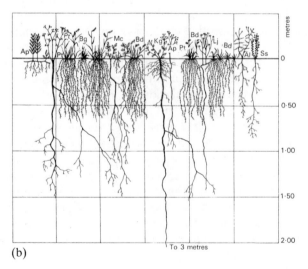

(b)

Fig. 4.7 The root systems of plants in a typical short-grass prairie at Hays, Kansas after (a) several years with average rainfall and (b) several years of drought. Al *Allionia linearis*; Ap *Aristi da purpurea*; Aps *Ambrosia psilostachya*; Bd *Buchloe dactyloides*; Bg *Bouteloua gracilis*; Kg *Kuhnia glutinosa*; Lj *Lygodesmia juncea*; Mc *Malvastrum coccineum*; Pt *Psoralia tenuiflora*; Sm *Solidago mollis*; Ss *Sideranthus spinulosus*.

carbohydrates produced are needed to make good the loss of leaf; hence such pastures are typically shallow rooted. Hard grazing is therefore a practice only suited to moist climates; as the climate becomes more arid, the herbage must be allowed to grow longer and the grazing must be lighter. As examples, on the semi-arid ranges of Utah, the roots of bunch wheat grass (*Agropyron inerme*) only penetrated to about 0.4 m on overgrazed range and to about 0.6 m on protected range, and on New Mexico ranges, the roots of black and blue grama grasses (*Bouteloua eripoda and B. gracilis*) penetrated to about 1.2 m on properly grazed, to about 0.6 m on overgrazed, and to less than 0.3 m on badly overgrazed ranges.[61]

Drying of soils affects both soil strength and plant turgor so that it is difficult to state with certainty the lowest water content beyond which roots will not grow. In an early study, it was found that bean and sunflower roots grew only a few millimetres into soil at or below the permanent wilting point.[62] More recently, experiments[63] with maize and tomato roots grown in sandy loam and clay loam soils respectively have shown that roots can continue to grow in soils with water potentials less than −4.0 MPa. There was more root growth at higher potentials and growth appeared to cease when potentials were −5.0 MPa to −10.0 MPa. However, some root tips remained alive even at these potentials and were able to elongate rapidly when the soil was again wetted.

Effect of nutrient supply

The amount, form and distribution of nutrients in soils will also influence the amount of root growth and its distribution. Generally, a nutrient deficient soil will result in reduced growth although different components of the root system may be affected differently. For example, both phosphorus and potassium deficiency reduced the size of the nodal root system of barley, but had little effect on the seminal root system.[64] The numbers of axes and the lengths of primary laterals were reduced but the mean length of axes, the numbers of primary laterals per axis, and the diameter of root members were little affected. Similarly, the length and number of nodal roots together with the length of primary lateral roots arising from seminal axes were also reduced

61 Hanson, W, R. and Stoddart, L. A., *J. Am. Soc. Agron.*, 1940, **32**, 278. For further examples see Stoddart, L. A. and Smith, A. D., *Range Management*, New York, 1943; Schuster, J. L., *Ecology*, 1964, **45**, 63. For the effect of different grazing methods on the depth of the root zone of pastures in West European conditions see Klapp, E., *Pflanzenbau*, **19**, 221.
62 Hendrickson, A. H. and Veihmeyer, F. J., *Pl. Physiol.*, 1931, **6**, 567.
63 Portas, C. A. M. and Taylor, H. M., *Soil Sci.*, 1976, **121**, 170.
64 For barley see Hackett, C., *New Phytol.*, 1968, **67**, 287. For grasses see Christie, E. K., *Aust. J. agric. Res.*, 1975, **26**, 437.

by phosphorus deficiency in studies with three semi-arid grass species in Australia.[64]

More detailed experiments to investigate the influence of localized concentrations of nutrients on root morphology have been conducted and Plate 4.4 shows the results of applying localized concentrations of a single nutrient to the root system of barley grown for 21 days in sand culture continuously irrigated with nutrient solution.[65] For phosphate, nitrate and ammonium,

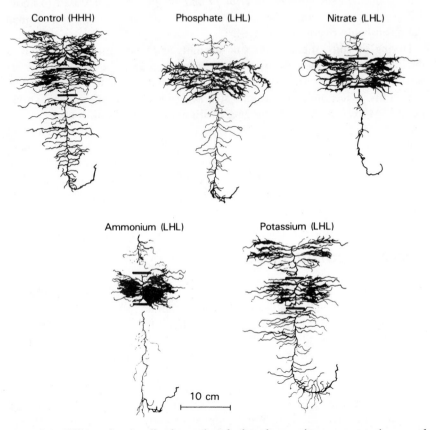

Plate 4.4   Effect of a localized supply of phosphate, nitrate, ammonium, and potassium on root form. Control plants (HHH) received the complete nutrient solution in all parts of the root system. The other roots (LHL) received the complete nutrient solution only in the middle zone, the top and bottom being supplied with a solution deficient in the specified nutrient. Note similarity between control and potassium treated roots. (From Drew, M. C., *New Phytol.*, 1975, **75**, 479.)

65 Drew, M. C., *New Phytol.*, 1975, **75**, 479;   with Saker, L. R. and Ashley, T. W., *J. exp. Bot.*, 1973, **24**, 1189;   with Saker, L. R., *J. exp. Bot.*, 1975, **26**, 79;   *ibid*, 1978, **29**, 435.

exposure to high concentrations of parts of the seminal roots resulted in considerable morphological changes to the root system, which were caused by the initiation and extension of primary and secondary lateral roots within the exposed zone. Exposure to high concentrations of potassium, though, resulted in growth of laterals to almost the same extent as the controls throughout the entire root system. The contrasting behaviour of nitrogen and phosphorus to that of potassium in promoting the growth of laterals remains unexplained. Potassium ions might be more readily translocated from zones of high concentration to zones of low concentration within the root system although there is no evidence for this. Alternatively, there might be a threshold concentration of all nutrients within the plant above which the initiation and extension of laterals is not limited; the threshold may not have been reached for nitrogen and phosphorus. The localized stimulation of root growth by concentrations of nutrients is also evident in trees such as Sitka spruce.[66]

Increased root branching in soil volumes where nutrients are concentrated can be seen in the field in, for example, the vicinity of bands of nitrogen and phosphate fertilizer placed close to seed during drilling.[67] However, phosphate alone or nitrogen (as ammonium sulphate) alone did not result in proliferation, and the form of nitrogen fertilizer may also influence root proliferation. For example, when wheat was grown in both sterile and non-sterile soil for a period of 8 weeks after banded applications of urea or ammonium sulphate, the amount of root growth was different in the two soils. There was a marked proliferation of roots in the fertilizer-affected zone between 4 and 8 weeks in the non-sterile soil treated with ammonium sulphate but not in the urea-treated soil, while in the sterile soil there was no proliferation of roots with ammonium sulphate but an intense proliferation surrounding a small zone (diameter 2 cm) which was devoid of roots in the urea treatment.[68] Form of fertilizer and microbial activity will, therefore, influence the precise extent and timing of root proliferation.

The effect of fertilizers on the root system of a crop depends on the fertility of the soil, but if the soil has a low nutrient status then suitable manures or fertilizers will increase both shoot growth and root growth, though generally root growth increases relatively less. In field experiments on sandy silt loams at Woburn, nitrogen fertilizer produced larger cereal crops with relatively smaller root systems than plants not given fertilizer (Table 4.4). Nitrogen also generally decreased root weight in the soil below 150 mm and decreased the specific root length. Applications of phosphate and potassium in this experiment did not affect root growth independently of their effects on the plants as a whole.

66 Coutts, M. P. and Philipson, J. J., *J. exp. Bot.*, 1976, **27**, 1102.
67 Miller, M. H. and Ohlrogge, A. J., *Agron. J.*, 1958, **50**, 95; Miller, M. H. *et al.*, *op. cit.*, 1958, **50**, 605.
68 Passioura, J. B. and Wetselaar, R., *Pl. Soil*, 1972, **36**, 461.

Table 4.4 Effects of N fertilizer on root weight as a fraction of total plant dry weight for cereals. (From Welbank, P. J. *et al., Rothamsted Ann. Rept. for 1973*, Part 2, 26)

Date	Winter wheat		Summer wheat		Oats		Barley		SE
	−N	+N	−N	+N	−N	+N	−N	+N	
5 May	0.381	0.222	0.349	0.297	0.268	0.218	0.251	0.210	0.024
2 June	0.188	0.115	0.252	0.134	0.297	0.158	0.242	0.109	0.018
30 June	0.129	0.085	0.145	0.094	0.184	0.115	0.125	0.084	0.009

In drier environments, the response to fertilizers may interact with water supply. A study of root growth of barley in Syria at two sites of contrasting soils and rainfall[69] showed that fertilizer application increased the size of the root system at both sites, although the crops at the wetter site (rainfall 320 mm) generally had heavier and longer root systems than those at the drier site (rainfall about 260 mm). Applications of phosphate alone significantly increased the weight of root systems but not the length, while applications of nitrogen and phosphate together increased both weight and length. As in the experiments at Woburn, applications of fertilizer reduced the proportion of total plant weight in the root system.

The interaction between water supply, fertilizer and root growth was first demonstrated by Lawes and Gilbert[70] who showed that in a drought, unmanured grass at Rothamsted could only take water from the top 0.67 m, but if suitably manured from 0.9 m. Deep rooting also allows a crop to take up additional nutrients if the nutrient supply in the surface soil is inadequate. Thus, savanna grasses and their associated herbs in Ghana took up 30 per cent of their phosphate (about 6 to 12 kg ha^{-1} P annually) from below 0.3 m depth; whereas annual crops, such as maize and millet, only took between 7 and 15 per cent from below this depth, probably because of their more restricted rooting depth.[71]

The effect of fertilizers on the root system is also illustrated in Plate 4.5 which shows the stunted root system of an eight-year-old gooseberry bush growing on a sandy soil without added nutrients, and the much deeper and more extensive root system of a similar bush given a good dressing of farmyard manure before planting. Soils that are permeable and of low nutrient status often carry crops with a sparse but well-distributed root system which is often well supplied with mycorrhizas. Under such conditions the roots will ramify in any pocket of soil well supplied with nutrients, such as rotting frag-

69 Gregory, P. J. *et al., J. agric. Sci.*, 1984, **103**, 429.
70 Lawes, J. B. and Gilbert, J. H., *J. R. agric. Soc. England*, 1871, **7**, 1.
71 Nye, P. H. and Foster, W. N. M., *J. agric. Sci.*, 1961, **56**, 299.

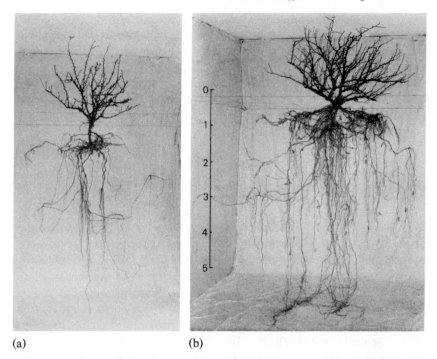

(a)   (b)

Plate 4.5   The effect of nutrients on the root development of eight-year-old gooseberry bushes in a sandy soil, Kent; depth in feet. (a) No manure;
(b) with 110 t ha^{-1} of farmyard manure in the year of planting. The
roots of the manured bush, resting on the ground, should continue
downwards.

ments of basaltic rock or any pockets of farmyard manure, provided there
is no shortage of oxygen.

The root systems of trees often have different types of root systems
depending on soil conditions. Many trees growing in natural forest with
an undisturbed floor of litter tend to have an extensive, much-branched, root
system filling the first few centimetres of the soil and a few deep taproots.
But if these trees are planted in open land where litter cannot accumulate,
and particularly if this is done in rather dry regions, many of these trees will
develop a deeper root system. A good illustration is provided by the cacao
tree.[72] In humid conditions, or badly drained soils, it has a very superficial root system containing many mycorrhizas which, with the layer of leaf
litter that accumulates, forms a mat on the surface. In contrast, in drier areas
on well-drained soils, it has a deep root system, carrying no visible mycorrhiza and forming no surface mat. Similarly the method of soil cultivation

72 Hardy, F., *Trop. Agric. Trin.*, 1944, **21**, 184; *ibid.*, 1943, **20**, 207.

used in orchards and perennial crops affects the root systems of the plants, for the regular hoeing or cultivation of the soil to kill weeds will also kill all the surface-feeding roots. In an orchard undersown with grass where four-year-old apple trees were placed in a strip kept free of weeds using herbicide, it was found that 70 per cent of the new root growth occurred in the strip, and that growth occurred earlier in the year under this strip than under the grass.[73] This example illustrates the effect of competition between root systems for water and nutrients.

Models of root growth

Predicting the growth of root systems might seem to be an impossible task given the many factors shown in this chapter to affect roots. Nevertheless, some general features of predictive use have been elucidated and used as the basis for modelling. The different branching patterns of root systems form the basis for a morphometric analysis of root systems[74] and may provide clear analytical distinction between one species and another. It suffers from the disadvantage, however, that root systems (unlike, for example, branched river systems) branch rapidly with time so that the branch number is continually changing. Several workers have therefore preferred a developmental approach where each member of the root system is numbered from the base of the plant and each class of lateral has distinct properties.[75] Such models arose from the observation that in cereals grown in homogeneous media, the extension and branching of the root members proceeded at constant rates for prolonged periods. Formulae can then be derived to express the frequency of branching and rate of growth of each root member which can, in turn, be used to predict the growth of the whole root system.[76]

For crops growing on a uniform soil profile, the distribution of root length per unit volume generally decreases with depth, z, and may be approximated by:

$$\frac{dL_v}{dz} = -qL_v \qquad [4.2]$$

where L_v is the root length per unit volume of soil and q is a constant independent of depth but is a function of time. Integration with respect to z gives

$$L_v = L_{vo}\, e^{-qz} \qquad [4.3]$$

73 Atkinson, D., *Pl. Soil*, 1977, **49**, 459.
74 Fitter, A. H., *Pl. Cell Environ.*, 1982, **5**, 313.
75 Hackett, C. and Rose, D. A., *Aust. J. biol. Sci.*, 1972, **25**, 669; Rose, D. A., *Pl. Soil*, 1983, **75**, 405.
76 Lungley, D. R., *Pl. Soil*, 1973, **38**, 145.

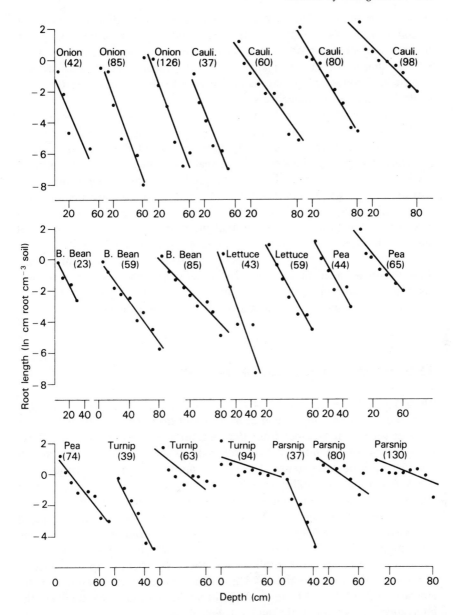

Fig. 4.8 Comparison of predicted and measured distributions of root length with depth for selected vegetable crops. Numbers indicate the number of days of growth. (From Greenwood, D. J. *et al.*, *Pl. Soil*, 1982, **68**, 75.)

where L_{vo} is the root length at $z = 0$. Figure 4.8 shows that for a range of vegetable crops, root length decreases logarithmically with depth[77] except for the growth of storage roots (parsnips) where the total root length in each soil layer is almost constant at many soil depths. Clearly, many soils are not uniform and zones of high mechanical resistance, high nutrient status, etc., would be expected to influence such relations.

Because root length is such an important factor in determining water and nutrient uptake, and because it is very time consuming to measure, further efforts to predict it from more easily observable plant characteristics can be expected. These, combined with numerical solutions to water and nutrient flow equations might then be used to predict optimum root lengths and distributions for crops. For example, calculations using typical uptake rates of nitrogen, phosphorus and water by crops, together with values of maximum inflow into roots, suggest that the root lengths per unit volume of soil (cm cm^{-3}) needed to supply the plant are 0.1 to 1, 1 to 10 and 0.3 to 1 for nitrogen, phosphorus and water respectively.[78] Such calculations also show the importance of knowing the conditions at the soil/root interface, because the root length requirement for water increases from 0.3 to 1 cm^{-2}, to 1 to 5 cm^{-2} when a root/soil contact resistance is included.

The uptake of nutrients by roots

The stages involved in moving a nutrient from a point in the soil into the plant shoot are: first, movement of the nutrient from the bulk soil to the root surface; second, movement from the root surface into the root; and third, translocation of the nutrient from the root to the shoot. This section will consider each of these stages. The present approach to plant nutrition is to consider nutrient availability in relation to mobility of ions and proximity to absorbing surfaces. This concept stresses not only the importance of the nutrient content of soils but also the importance of the size and shape of root systems. In summary, if a nutrient is mobile (nitrate is an example) the total quantity of labile nutrient in the soil largely determines uptake, but if the nutrient is immobile, e.g. phosphate, the amount taken up is a function of the length of root produced.[79] These ideas may now be stated in a mathematical equation[80] and form the basis of studies to show quantitatively the interaction of root properties and nutrient uptake.

77 See also Gerwitz, A. and Page, E. R., *J. appl. Ecol.*, 1974, **11**, 773.
78 van Nordwijk, M. in *Root Ecology and its Practical Application*, p. 207, Int. Symp. Gumpenstein, 1983.
79 Bray, R., *Soil Sci.*, 1954, **78**, 9.
80 See for example Baldwin, J. P., *J. Soil Sci.*, 1975, **26**, 195.

Physiological aspects of uptake

Because roots are selective in their uptake of some ions the relative ionic composition of the shoot system is often very different from that of the soil solution. Table 4.5 shows that plants typically contain more phosphate, potassium and nitrogen relative to calcium than does the soil solution. Barber[81] compared the nutrient content of maize shoots with the concentrations in soil solution and, by assuming that a plant transpired 500 g of water in producing 1 g of dry matter, showed that more phosphate and potassium were present in the plant than mass flow alone could account for, while magnesium and calcium were present in lower amounts. Similarly, it was found that leek plants grown in large tubs in the field accumulated potassium and rejected calcium relative to their concentrations in solution.[82]

The degree of selectivity depends on the plant species: the ionic composition of the growing leaves of a crop differs between species and depends also on the ionic composition of the soil solution. Thus some species, such as lucerne, typically have a higher calcium : potassium ratio in their leaves than do grasses. These differences between plant species become very noticeable if an ion is in excess of its normal value in the soil solution, for then plants can often be classified into accumulators or rejectors of that ion. Some plants (for example, tea) growing in acid soils high in aluminium will accumulate high concentrations of aluminium in their leaves without their growth being affected; some plants growing in salt marshes will accumulate very high concentrations of sodium in their leaves; yet other plant species, growing in these same soils, will have compositions little affected by these high concentrations.

Table 4.5 Comparison of commonly observed concentrations of ions in soil solution with amounts taken up by a crop of winter wheat. (From Gregory, P. J. *et al.*, *J. agric. Sci.*, 1979, **93**, 485)

Nutrient	Concentration in soil solution (M)	Concentration relative to Ca	Uptake by wheat (kg ha^{-1})	Uptake relative to Ca
N	2×10^{-3}	0.4	128	5.6
P	1×10^{-5}	0.002	27	1.2
K	1×10^{-3}	0.2	110	4.8
Ca	5×10^{-3}	1.0	23	1.0
Mg	2×10^{-3}	0.4	11	0.5
S	1×10^{-3}	0.2	9	0.4

81 With Walker, J. M. and Vasey, E. H., *J. Agric. Fd Chem.*, 1963, **11**, 204.
82 Brewster, J. L. and Tinker, P. B., *Soil Sci. Soc. Am. Proc.*, 1970, **34**, 421.

Ionic absorption must maintain overall electric charge neutrality within the plant and this is achieved by an excess of either hydrogen, or bicarbonate or hydroxyl, ions being excreted if there is an excess absorption of cations or of anions, respectively. The source of the hydrogen ions is probably malic or fumaric acids whose rates of synthesis in the absorbing cells are conditioned by the hydrogen ion demand; the bicarbonate anions are a normal by-product of cell metabolism. The pH of the solution outside the root will accordingly rise if more anions than cations (for the same valency) are being absorbed, and fall if more cations are being absorbed. Which of these will happen usually depends on the source of nitrogen the roots are using. If all the nitrogen is being taken up as nitrate, anion uptake is usually in excess, and the pH outside the roots may rise by over one pH unit,[83] while if it is being taken up as ammonium the pH will fall. Changes in the pH of the rhizosphere may also have important consequences for phosphate nutrition as demonstrated for rape plants.[84]

The exact biochemical processes involved in ionic uptake are not yet fully understood but the ability of plants to take up ions selectively, and the observed effects of temperature and metabolic activity on uptake imply an active mechanism. Ion uptake is closely associated with metabolic processes, in particular respiration which produces ATP and hence the energy necessary for ion uptake. During uptake, ions must be transported selectively across plant membranes into the cytoplasm of the cell.

Figure 4.9 illustrates two current theories developed to explain observed features of the uptake process. The carrier-ion hypothesis postulates a phos-phorylated molecule which acts as a carrier and possesses specific binding sites for a particular ion. At the outer boundary of the membrane the carrier and ion are bound together; the complex then diffuses across the membrane to the inner boundary where it is split by a phosphatase enzyme, the ion being released into the adjacent cytoplasm. ATP is required to regenerate the selectivity of the carrier which is brought about by a carrier ATP kinase located at the inner membrane. The phosphorylated carrier then diffuses back to the outer boundary to repeat the cycle.

 A second scheme utilizes the known association of ATPase activity with ion uptake to form an 'ion pump'.[85] The ATPase activity in the membrane results in the production of ADP^- in the cytoplasm and H^+ which is released into the outer medium, thereby establishing pH and electrochemical gradients across the membrane with the inner side being more negatively charged. Cations are thus attracted into the cytoplasm and this depolarizes the

83 Riley, D. and Barber, S. A., *Soil Sci. Soc. Am. Proc.*, 1971, **35**, 301.
84 Hedley, M. J. *et al.*, *New Phytol.*, 1982, **91**, 31, 45; *ibid*; 1983, **95**, 69; with Grinsted, M. J., *op. cit.*, 1982, **91**, 19.
85 Hodges, T. K., *Adv. Agron.*, 1973, **25**, 163; For a more thorough discussion of ion uptake mechanisms see Mengel, K. and Kirkby, E. A., *Principles of Plant Nutrition* (3rd edn), Int. Potash Inst. Berne, 1982.

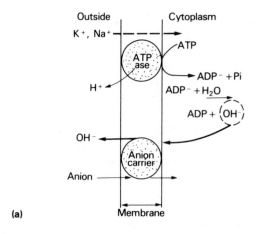

(a)

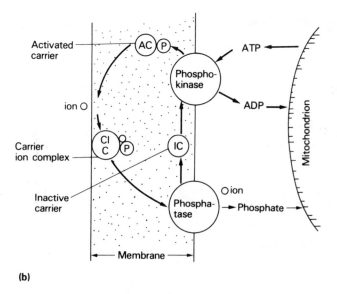

(b)

Fig. 4.9 Diagrammatic representations of the mechanisms of ion uptake; (a) ATPase driven cation pump coupled with an anion carrier, and (b) carrier ion transport involving energy expenditure; P_i inorganic phosphate, other abbreviations in text. (From Mengel, K. and Kirkby, E. A., *Principles of Plant Nutrition*, Int. Pot. Inst. Berne, 1982.)

membrane. Anions are taken up via an anion carrier driven by differences in concentration of cytoplasmic OH^-. This scheme agrees with the typical potential difference across membranes of 60 to 160 mV. Pitman[86] has

86 Pitman, M. G., *Quart. Rev. Biophys.*, 1982, **15**, 481.

combined both hypotheses of uptake to produce a unified scheme that accounts for most of the observed features.

Many statements exist in the literature to the effect that nutrient uptake is restricted to an 'absorbing zone' close to the root apex. This view resulted from anatomical observations suggesting that the endodermis was generally suberized as the root became older and that this could restrict the flow of materials between the cortex and the vascular tissues (see Fig. 4.1). Physiological measurements also showed that the rate of root respiration was faster in the apical regions so providing a direct link between metabolism and ion uptake. These views are now known to be wrong, at least for certain ions. Studies using radioactive tracer, have shown that neither radial movement of phosphorus and potassium into roots, nor transport of these ions to the shoot, is confined to the 'absorbing zone'.[87] Figure 4.10 shows that in barley, uptake and translocation of phosphate from regions 40 cm or more from the apex were as great as close to the apex. Similar results have been determined for the uptake of potassium, ammonium and nitrate ions.[88] However, not all

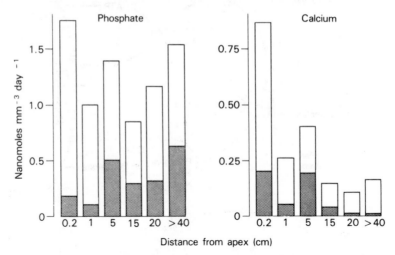

Fig. 4.10   Histograms illustrating absorption (entire column) and translocation (lower section) of phosphate and calcium determined in successive 3.5 mm regions of intact seminal root axes of barley plants about 3 weeks old. Uptake measured over 24 h at 20 °C; external concentrations 3 μM KH_2PO_4, 1.25 μM $CaCl_2$. (Reprinted with permission from Russell, R. S. and Clarkson, D. T. in *Perspectives in Experimental Biology*, Vol. 2, p. 401, Pergamon, 1976.)

87 Clarkson, D. T. and Sanderson, J., *ARC Letcombe Lab. Ann. Rept. for 1970*, p. 16.
88 Clarkson, D. T. in *Physiological Processes Limiting Plant Productivity* (ed. C. B. Johnson), p. 17, Butterworths, 1981.

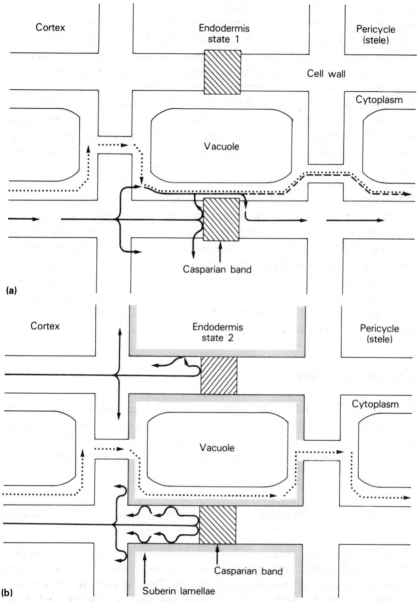

Fig. 4.11 Pathways for radial movement of solutes into the stele. (a) Materials
moving via the apoplast are obliged to negotiate the plasmalemma of
the endodermis by the Casparian band; (b) Deposition of suberin
lamellae prevents direct access to the endodermal plasmalemma from
the apoplast; symplast pathway is undisturbed. . . . . . ., symplast
movement; ———, apoplast movement; – – –, solute from apoplast
moving in symplasm. (Reprinted with permission from Russell, R. S.
and Clarkson, D. T. in *Perspectives in Experimental Biology*, Vol. 2,
p. 401, Pergamon, 1976.)

ions are readily absorbed and translocated by older regions of roots and for some ions the endodermis appears to fulfil its previously described role. In barley the movement of calcium to shoots (Fig. 4.10) was restricted to the younger parts of the root system, and anatomical observations show that there was a close correlation between the decrease in calcium transport and the deposition of suberin lamellae within the endodermis.[89] Similarly, iron uptake in barley plants was also confined to the apical region.[90] The contrasting patterns of uptake of ions may be explained by reference to their movement within the apoplast (mainly between the cell walls) and the symplast (Fig. 4.11). Calcium moves radially through the cortex in the apoplast and, following the development of the Casparian bands in the endodermis, must then traverse the plasmalemma of the endodermal cells to reach the stelar tissues. When suberin lamellae are laid down on the endodermal walls though, the movement of ions is prevented because the plasmalemma is no longer directly accessible from the endodermal cells. In contrast, the symplastic pathway continues to function because numerous plasmodesmata traverse the tangential endodermal walls; radial movement of ions such as phosphate can therefore continue even when the endodermis is suberized. The reason why calcium cannot use the symplasm pathway is not clear; it may be because most of the calcium in the cytoplasm is complexed so that the activity of calcium ions is low.

The soil solution

There are three possible sources from which plants can extract nutrients: the soil solution, the exchangeable ions, and the readily decomposable minerals; and it is very difficult to separate out the relative importance of these sources for any particular plant. This inter-relation of sources may be written:

$$M_{unavailable} \quad \overset{k_1}{\rightleftharpoons} \quad M_{intermediate \atop availability} \quad \overset{k_2}{\rightleftharpoons} \quad M_{exchangeable} \quad \overset{k_3}{\rightleftharpoons} \quad M_{solution}$$

where M represents a nutrient and k_1, k_2 and k_3 are rate constants associated with each step towards the right; $k_1 \ll k_2 \ll k_3$ and appreciable change occurs in months, weeks and minutes, respectively. Thus, if the soil solution is in equilibrium with the exchangeable cations and adsorbed anions, and if any nutrient (except nitrate whose supply is principally from mineralized organic matter) is removed from the solution, at least a part of this loss will be made

89 Robards, A. W. *et al.*, *Protoplasma*, 1973, **77**, 291.
90 Clarkson, D. T. and Sanderson, J., *Pl. Physiol.*, 1978, **61**, 731.

good from the nutrient reserves of the solids. As described in later chapters, the solid materials of the soil keep the soil solution well buffered both for pH and for all nutrients except nitrate.

Plant roots absorb nutrients from the soil solution, and the concentration of nutrients in the soil solution is a factor determining their rate of uptake by roots. Over long periods as, for example, a growing season, the release of nutrients from the more readily decomposable minerals (sources of intermediate availability) in the soil will also be important in sustaining the concentration in solution and the rate of uptake by roots. The importance of this source of nutrients has been demonstrated for potassium.[91] as discussed in Chapter 22.

The soil solution provides the chemical environment of roots and comprises the soil water and its dissolved electrolytes plus small quantities of dissolved gases and water-soluble compounds. Table 4.6 gives ionic concentrations in some American soils and shows that in neutral soils, the concentration of calcium ions is between 3 and 10 mM, and that nitrate is generally present in higher concentration than are the other anions. Typically, the nitrate and bicarbonate concentrations are dependent on microbial and root activity. Concentrations of magnesium, potassium and sodium ions in soils not well supplied with these ions are usually between 10^{-4} and 10^{-3} M, and in the acid soil shown in Table 4.6 both aluminium and manganese concentrations are about 2×10^{-4} M. The table does not show concentrations of phosphate ions. These are frequently difficult to measure because their concentration is much lower than that of other ions, being commonly 10^{-5} to 10^{-6} M.[92]

The composition of the soil solution depends on the moisture content of the soil, the rate of growth of the crop and the activity of the microbial population. These factors are difficult to separate out in the field, though they can be in the laboratory. Table 4.7 gives an example of this separation for an English clay soil that was air dried, then wetted to one of two levels and incubated for periods up to 53 days.[93] The results are consistent with the following generalizations:

1   The concentration of nitrate and chloride varies inversely with the moisture content, since all of these ions are in solution.
2   The concentration of potassium is little affected by change in moisture content, in contrast to the sodium ion concentration which behaves as if the ions are nearly all in solution.
3   The concentration of bicarbonate is relatively unaffected by dilution,

91 Arnold, P. W. and Close, B. M., *J. agric. Sci.*, 1961, **57**, 295.
92 For other examples of soil solution composition see Barber, S. A., *Soil Nutrient Bio-availability: a mechanistic approach*, p. 44, Wiley-Interscience, 1984; Fried, M. and Broeshart, H., *The Soil-Plant System in Relation to Inorganic Nutrition*, p. 19, Acad. Press, 1969.
93 Larsen, S., *J. Sci. Fd Agric.*, 1968, **19**, 693.

Table 4.6 Composition of soil solution in some American soils. Concentrations in meq l^{-1}

	pH	Ca	Mg	K	Na	NH_4	NO_3	Cl	HCO_3	SO_4	cation/anion	Total concentration
Sandy loam*	5.82	8.1	1.7	1.1	0.8	0.4	8.0	0.3	1.2	1.0	1.15	11.3
	6.11	10.6	2.5	2.7	0.6	0.4	9.6	1.6	1.7	3.6	1.02	16.7
	7.19	21.0	1.2	0.7	1.8	0.4	15.6	2.2	1.1	7.0	0.97	25.5
Loam*	7.65	6.4	1.7	0.3	1.4	0.4	4.4	1.9	3.4	1.7	0.90	10.8
Acid loam†	3.9	1.0	1.4	0.4	0.4	0.1	3.8	0.2	n.d.	0.8	0.96	4.8

* Eaton, F. M. et al., *Soil Sci.*, 1960, **90**, 253.
† Vlamis, J., *Soil Sci.*, 1953, **75**, 382. Solution in this acid soil also contained (meq l^{-1}) Al 0.63, Mn 0.58.

Table 4.7 Effect of soil moisture content and period of incubation on composition of the soil solution. Clay soil: pH 5.7, incubated at 20 °C. Concentrations in meq l^{-1}.

	$Ca + Mg$	K	Na	NO_3	Cl	HCO_3	SO_4	$Si(OH)_4$	cation/anion
Moisture content (%)									
After 2 days									
18.3	47.8	1.7	5.6	30.4	3.8	1.8	17.0	0.6	1.04
24.0	37.9	1.5	4.3	22.1	2.6	1.6	14.4	0.6	1.07
Ratio	1.26	1.1	1.30	1.38	1.4	1.1	1.18	1.0	
After 53 days									
18.3	68.9	1.5	6.6	53.1	3.5	1.9	14.3	0.8	1.06
24.0	53.2	1.6	5.2	41.6	2.5	1.9	13.0	0.7	1.02
Ratio	1.29	0.9	1.27	1.28	1.4	1.0	1.10	1.1	

implying that it is controlled by other factors, and the sulphate varies less than expected probably because some is held as an adsorbed phase on soil particles.

The activity of micro-organisms results in an increase in the nitrate content of the soil, with little change in the concentration of the other anions, and in neutral or weakly acid soils the calcium ion concentration increases to balance the nitrate ions produced, a point which is also brought out by Table 4.7.

The soil solution of many leached agricultural soils under normal conditions of moisture content has an osmotic pressure from 20 to 100 kPa, being higher the nearer the moisture content is to the wilting point. Thus, even at the wilting point the soil solution is dilute, and has an osmotic pressure appreciably lower than the 1.0 to 2.0 MPa of the root sap.

It is not feasible at present to measure the chemical components of soil solution *in situ*, so that a major problem in determining its composition is to remove the solution from the soil without altering its composition. The method that achieves this aim most consistently is to displace the soil solution with another liquid (e.g. ethanol or glycerol) from a column of soil. Provided that the column is packed properly, the concentration of ions in successive aliquots of displaced solution remains constant until the displacing liquid appears in the effluent; the displacement may take place over several hours and appears to displace solution from both fine and coarse pores.[94] Suction methods for obtaining solution[95] require careful interpretation because they change the pressure of carbon dioxide and hence the concentration of bicarbonate ions. Moreover, such methods, together with centrifuge techniques, usually employ ceramic materials as filters which can adsorb ions especially phosphate. Solutions obtained in these ways have usually been changed from the true soil solution.

Soil solution extracts are often purported to give estimates of soil solution composition but two significant changes occur in soils as water is added. First, cation exchange occurs between solid and solution because the relative activities of the ions is altered by dilution and second, sparingly soluble salts enter the solution.

The composition of the soil solution has been shown to change over a growing season, becoming less concentrated as the growing season progresses.[96] However, over the winter period, or during the period when there are no crops, the concentrations of ions in solution increase again to their levels prior to cropping (see also p. 711).

---

94 For a general discussion of techniques see Adams, F. in *The Plant Root and Its Environment* (ed. E. W. Carson), p. 441, Univ. Virginia Press, Charlottesville, 1974.

95 For example see Nielsen, N. E., *Pl. Soil*, 1972, **36**, 505.

96 Burd, J. S. and Martin, J. C., *Hilgardia*, 1931, **5**, 455;   Blakemore, M., *J. agric. Sci.*, 1966, **66**, 139.

It is not known at present whether ion concentration or activity in solution determines the rate of uptake of an ion. Uptake usually occurs from dilute solutions where activity is almost the same as concentration but the concentration of ions and the formation of ion pairs may result in an activity lower than the concentration. However, unless the form of the ion when it is taken up is known with certainty, the calculation of activities will not aid the understanding of ion uptake.[97]

Movement of nutrients to roots

As mentioned earlier in this chapter, roots are in direct contact with only a small part of the nutrients in solution or of available nutrients absorbed by the soil solids, so that nutrients must move to the root surface. Depending on the plant species and the size of root system produced, roots may be from 1 mm to several cm apart and the movement of nutrients may have to take place over substantial distances.

There are two different processes whereby nutrients are transferred from the bulk of the soil to the root surface. These processes are mass flow (convection) and diffusion. Mass flow occurs because water is absorbed by roots to meet the loss by transpiration from the shoot; as the water moves to the roots so dissolved ions are also carried to the root surface. Diffusion occurs when ions move along a concentration gradient established between the root surface and the body of the soil; ions diffuse towards the root if they are taken up faster than they are carried to the surface by mass flow and away from the root if the converse pertains.

The rate of nutrient transfer by mass flow (J) is given by the rate of water flux into the root (V) and the concentration of ions in solution (C_l):

$$J = VC_l$$

Typically, in many British soils the concentration of calcium in the soil solution is about 10^{-2} M, of potassium 10^{-3}–10^{-4} M, and of phosphorus 10^{-5}–10^{-6} M. If a crop uses 300 mm of water during growth, this amount of water would carry about 1200 kg ha^{-1} calcium, 12 to 120 kg potassium and 0.9 to 9 kg phosphorus, whereas the crop will take up and translocate to the shoot about 10 to 30 kg ha^{-1} calcium, 50 to 250 kg potassium and about 20 to 30 kg phosphorus (see also Table 4.5). Generally, such calculations indicate that the transfer of calcium, magnesium and sulphate to the root surface by mass flow is faster than the rate of uptake by the crop, that for phosphate

---

97 For relations between concentration and activity see Adams, F., *Soil Sci. Soc. Am. Proc.*, 1971, **35**, 420.

and potassium the rate of uptake exceeds the rate of transfer by mass flow, and that for nitrate the supply and uptake are usually about equal.[98] These calculations are based on averages for the whole growing season and the initial early growth may often occur when temperatures are low and rates of transpiration slow so that mass flow may be overestimated during early growth. Moreover, there are also problems in defining the concentration of ions in the soil solution; most estimates are based on the initial concentration in the bulk soil solution rather than the concentration during growth.

More detailed estimates for the contribution of mass flow to uptake, allowing for changes in rates of transpiration and changes in solution composition, confirm the general conclusions outlined above,[99] although in a three-year study of beech trees growing on an acid brown earth (pH about 4) mass flow was unable to account for the observed uptake of nitrogen, potassium, phosphorus and calcium.[100]

When mass flow is unable to supply sufficient quantities of a nutrient and continued uptake occurs, the concentration of nutrient at the root surface is reduced and a concentration gradient is established. Ions move by thermal motion (Brownian movement) from points of high concentration to points of low concentration when a concentration gradient exists and the diffusion of ions persists until equilibrium is re-established. The distance the concentration gradient extends from the root surface depends upon the rate of diffusion.

The application of diffusion theory to the supply of plant nutrients is now well established and has been thoroughly reviewed by several authors.[101] The basic assumptions are that diffusion only takes place through the water films, that the diffusive flux at any point is proportional to the concentration gradient at that point, that the root is a uniform cylinder and that the flux to the root is entirely radial. The assumption that diffusion only takes place through the water films is justified for, in general, the migration of ions along the clay surfaces to the root is very much slower than the diffusive flux.[102]

The amount of ion diffusing across a unit area in unit time (F) with a concentration gradient dC/dx is given by Fick's first law of diffusion:

$$F = -D\frac{dC}{dx} \qquad [4.4]$$

where D is the diffusion coefficient. The equation applies only to a linear gradient (unlikely to be achieved in many uptake studies) and the minus sign

98 For examples see refs 81 and 82; Wild, A. in *The Chemistry of Soil Processes* (eds D. J. Greenland and M. H. B. Hayes), p. 37, Wiley, 1981.
99 Gregory, P. J. *et al.*, *J. agric. Sci.*, 1979, **93**, 493;   NaNagara, T. *et al.*, *Agron. J.*, 1976, **68**, 67.
100 Prenzel, J., *Pl. Soil*, 1979, **51**, 39.
101 See book by Barber ref. 92; book by Nye and Tinker ref. 7; review by Olsen, S. R. and Kemper, W. D., *Adv. Agron.*, 1968, **20**, 91.
102 Rowell, D. L. *et al.*, *J. Soil Sci.*, 1967, **18**, 204.

indicates that movement is from high to low concentration. For transient state diffusion, which is generally more appropriate to the uptake of ions, Fick's second law can be used to determine the change in concentration with time at a fixed distance x:

$$\frac{dC}{dt} = D \frac{d^2C}{dx^2} \qquad [4.5]$$

This equation applies to linear diffusion, but plant roots are cylindrical and nutrients move radially towards them. Rewriting equation [4.5] for radial diffusion gives:

$$\frac{dC}{dt} = \frac{1}{r} \frac{d}{dr} \left(r D \frac{dC}{dr} \right) \qquad [4.6]$$

where r is the radial distance from the centre of the root cylinder.

The diffusion coefficient, D, differs between media but can be calculated in relation to the diffusion coefficient for the ion in free solution, D_o. If the ion is not absorbed by the soil (e.g. nitrate) then the diffusion coefficient in the soil, D_s, may be written:

$$D_s = D_o \, \theta_v \, f \qquad [4.7]$$

where θ_v is the volumetric water content, expressed as a fraction, and f is the impedance factor, corresponding to the tortuosity factor for gaseous diffusion (see Ch. 9). The impedance factor involves[103]: the true tortuosity of the water films; a term dependent on the viscosity of the water close to the clay surfaces; and a term dependent on the diffuseness of the electrical double layer, which if it occupies a significant proportion of the water film will affect the volume of water through which anions can diffuse. Rowell, working with a Lower Greensand sandy clay loam, found that when the water in the soil was at a suction of 10 kPa, $\theta_v = 0.4$, $f = 0.5$, so $\theta_v f = 0.2$; while at 1.5 MPa suction, $\theta_v = 0.07$, $f = 0.01$, so $\theta_v f = 7 \times 10^{-4}$, or almost 300 times smaller than at field capacity.[104]

If the ion is absorbed by the soil, e.g. phosphate, and the ionic concentration in the soil water is in equilibrium with the amount of that ion absorbed on the soil particles, then, as the root absorbs some of the ions from the solution, the ionic concentration in the solution close to the root will drop, and some of the ions absorbed on the soil surface will go into solution. This reduces the diffusion coefficient such that:

$$D_s = D_o \, \theta_v \, f \, \frac{dC_l}{dC} \qquad [4.8]$$

103 Kemper, W. D. *et al.*, *Soil Sci. Soc. Am. Proc.*, 1964, **28**, 164.
104 See ref. 102.

where C_l is the concentration of ions in the soil solution and C is the concentration of ions per unit volume of soil; dC_l/dC is the reciprocal of the buffer power for that ion and the more strongly buffered the soil, or the more strongly the ion is absorbed by the soil, the smaller is dC_l/dC and the smaller the effective diffusion coefficient.[105] The values of D_o and D_s in Table 4.8 show that the various factors reduce the diffusion coefficient by 10^{-1} to 10^{-5} below its value in water. The diffusion coefficient in moist soil is of the order of 10^{-6} cm^2 s^{-1} for ions that are not adsorbed such as nitrate, 10^{-7} cm^2 s^{-1} for cations such as calcium, magnesium and potassium, and 10^{-9} to 10^{-10} cm^2 s^{-1} for strongly adsorbed ions such as phosphate. Since the average linear distance of diffusive movement of an ion with time is $(2 \, Dt)^{1/2}$, an ion such as NO_3^- in soil (D_s, 1×10^{-6} cm^2 s^{-1}) will move 4 mm in one day whereas K^+ (D_s, 1×10^{-7} cm^2 s^{-1}) and $H_2PO_4^-$ (D_s, 1×10^{-10} cm^2 s^{-1}) will move only 1.3 mm and 0.04 mm respectively.[106] These distances determine the amounts of nutrients that are potentially available to a root by diffusion during the growth of a crop. Such calculations also indicate the very important role that root hairs and mycorrhizas play in the uptake of phosphate by plants.

Table 4.8 Typical diffusion coefficients of ions in solution and in soils

Ion	Medium	Temperature (°C)	$D(cm^2 \, s^{-1})$	Source
K^+	Water	25	1.98×10^{-5}	1
$H_2PO_4^-$	Water	25	0.89×10^{-5}	1
NO_3^-	Water	25	1.90×10^{-5}	1
Ca^{2+}	Water	25	0.78×10^{-5}	1
Mg^{2+}	Water	25	0.70×10^{-5}	1
$H_2PO_4^-$	Sandy clay loam (40% θ_v)	lab	3.3×10^{-9}	2
$H_2PO_4^-$	Sandy clay loam (20% θ_v)	lab	0.3×10^{-9}	2
NO_3^-	Soil	lab	$0.5 - 5.0 \times 10^{-6}$	3
K^+	Soil	lab	2.3×10^{-7}	4
Ca^{2+}	Soil	lab	3.3×10^{-7}	4

1 *Handbook of Chemistry and Physics*, CRC Press.
2 Rowell, D. L. *et al.*, *J. Soil Sci.*, 1967, **18**, 204.
3 Clarke, A. L. and Barley, K. P., *Aust. J. Soil Res.*, 1968, **6**, 75.
4 Vaidyanathan, L. V. and Nye, P. H., *J. Soil Sci.*, 1968, **17**, 175.

105 See for example Vaidyanathan, L. V. *et al.*, *J. Soil Sci.*, 1968, **19**, 94;   Nye, P. H., *Trans. 9th Int. Congr. Soil Sci.*, 1968, **1**, 117.
106 For examples see Farr, E. *et al.*, *Soil Sci.*, 1969, **107**, 385 (onion root, uptake of K and Ca); Drew, M. C. and Nye, P. H., *Pl. Soil*, 1970, **33**, 545 (ryegrass, uptake of P); Olsen, S. R. *et al.*, *Proc. Soil Sci. Soc. Am.*, 1962, **26**, 222 (maize, uptake of P).

Direct measurements of the nutrient concentration in soil close to roots have been made by a few workers and are reported in the books cited earlier. Plate 4.6 shows depletion of phosphate close to the root of rape and the accumulation of phosphate in the root tips. The importance of root hairs in phosphorus nutrition was commented upon earlier in this chapter, and auto-radiographs similar to that shown in Plate 4.6 have been used to quantify their importance. Figure 4.12 shows that root hairs were active in depleting the soil of phosphate in the root hair zone to about 0.4 of the exchangeable phosphate originally present. Their effect, then, is to make available to the plant a much larger volume of soil from which phosphate can be obtained than in the absence of root hairs. For example, if a root of diameter 0.5 mm has 100 root hairs per mm of length, and each root hair is 0.02 mm diameter and 1 mm long, during the first ten days the root hairs will take up four to five times as much phosphate as the root alone would have done.[107] It will take about 16 days before the contribution of the soil from outside the root hair zone will exceed that from the inside of the zone; with potassium this will happen after a few hours.[108]

With the understanding of these basic processes supplying nutrients to plant roots, has come the development of models to predict rates of nutrient uptake by plants.[109] In general, mass flow and diffusion are regarded as additive processes in supplying nutrients rather than interactive processes and this assumption introduces little error.[110]

The flux, F, of nutrients by simultaneous diffusion and mass flow can be written as:

$$F = D_s \frac{dC}{dr} + VC_l \qquad [4.9]$$

where C is the concentration of ions per unit volume of soil and C_l is the concentration of ions in the soil solution. Rewriting equation [4.9] to express the change in concentration with time gives:

$$\frac{dC_l}{dt} = \frac{1}{r}\frac{d}{dr}\left(rD_s\frac{dC_l}{dr} + \frac{r_oVC_l}{b}\right) \qquad [4.10]$$

where r is the radial distance, r_o is the root radius and b is the buffer power

107 Lewis, D. G. and Quirk, J. P., *Pl. Soil*, 1967, **26**, 99, 119, 445, 454 (using wheat); Drew, M. C. and Nye, P. H., *Pl. Soil*, 1970, **33**, 545 found a rather lower value of two- to three-fold for ryegrass.

108 Nye, P. H., *Pl. Soil*, 1966, **25**, 81; with Drew, M. C., *op. cit.*, 1969, **31**, 407.

109 For early examples see Passioura, J. B., *Pl. Soil*, 1963, **18**, 221; Nye, P. H., *Pl. Soil*, 1966, **25**, 81; Brewster, J. L. *et al.*, *Pl. Soil*, 1976, **44**, 295.

110 Marriott, F. H. C. and Nye, P. H., *Trans. 9th Int. Congr. Soil Sci. Adelaide*, 1968, **1**, 127.

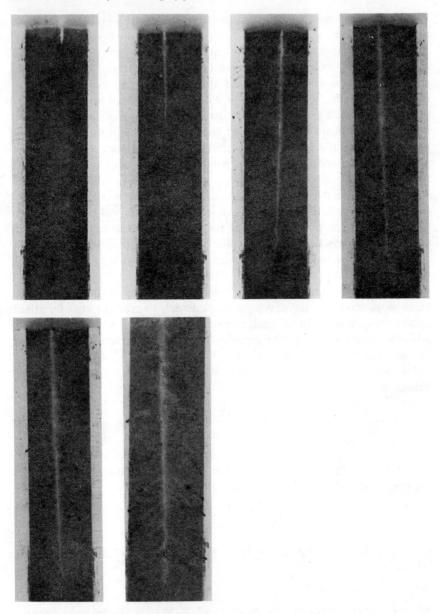

Plate 4.6 Autoradiographs of soil labelled with $H_2^{33}PO_4^-$ and growing a seedling of rape (*Brassica napus*). As the root grows through the soil (1–2 cm day^{-1}) the zone of phosphate removal (light area) becomes wider and the metabolically active root tips show phosphate accumulation. (From Wild, A. in *Chemistry of Soil Processes*, p. 37, Wiley, 1981.)

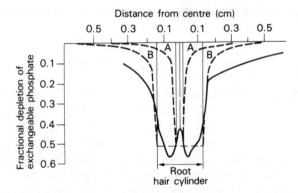

Fig. 4.12 Exchangeable phosphate gradient close to a primary root of rape. Broken curve A calculated assuming root hairs are inactive; B calculated assuming intense root hair activity and uniform depletion within the root hair cylinder; solid curve: observed in the experiment. (From Bhat, K. K. S. and Nye, P. H., *Pl. Soil*, 1973. **38**, 161.)

of the soil. Before this equation can be solved, the initial and boundary conditions must be specified. The initial boundary condition is simply:

$$t = 0, \ r > r_o, \ C_l = C_{li}$$

where C_{li} is the initial nutrient concentration in the soil solution. This condition therefore defines a uniform nutrient distribution around the root. The inner boundary condition (i.e. where $r = r_o$) has been developed assuming that uptake can be described by the Michaelis–Menten equation[111]:

$$I = \frac{I_{max} \ (C_l - C_{min})}{K_m + C_l - C_{min}} \qquad [4.11]$$

where I_{max} is the maximum inflow (uptake/unit length of root), K_m is the concentration at which $I = 0.5 \ I_{max}$ and C_{min} is C_l when net influx is 0. This gives an inner boundary condition:

$$t > 0, \ r = r_o, \ D_s b \frac{dC_l}{dr} + V C_l = \frac{I_{max} \ (C_l - C_{min})}{K_m + C_l - C_{min}}$$

Equation [4.11] was used by Claassen and Barber to predict the potassium uptake by maize growing in four soils, one of which had fertilizer mixed in to give five levels of potassium. The predicted and observed uptake were linearly related ($r^2 = 0.87$) but the predicted uptake was about one-and-one-half times that observed. This overestimation of uptake probably resulted because of the assumption of no competition between roots. Competition

111 Claassen, N. and Barber, S. A., *Agron. J.*, 1976, **68**, 961.

between adjacent roots can be allowed for by expressing a second boundary condition:

$$t > 0, r = r_1, D_s b \frac{dC_l}{dr} + VC_l = 0$$

where r_1 is the mean half distance between root axes. If there is no competition, $C_l = C_{li}$, $r = r_1$ at $t > 0$.[112] This model has been tested under controlled and field conditions and has been shown to give good relations between predicted and observed uptake.[113] Figure 4.13 shows the results of applying such a model to field-grown maize hybrids; it has also been used to predict successfully the effect of root hairs on the uptake of phosphorus under controlled conditions.[114] Sensitivity analysis of such models shows that the rate of root growth and the root radius greatly affected the predicted rates of P and K uptake.[115] Moreover, for both P and K, the parameters describing the soil supply of nutrients were of greater importance than the parameters defining the influx kinetics. Rates of root growth and the shape of the root system can be shown to have a marked effect on the nutrient supply available to a crop. Although this is only one of several factors affecting yield,

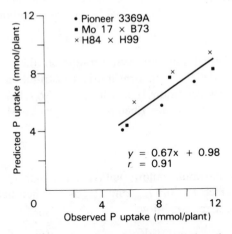

Fig. 4.13 Relation between predicted and observed phosphate uptake by three single-cross hybrids of maize grown in field conditions. (From Schenk, M. K. and Barber, S. A., *Pl. Soil*, 1980, **54**, 65.)

112 Barber, S. A. and Cushman, J. H. in *Modelling Waste Water Renovation Land Treatment* (ed. I. K. Iskander), p. 382, Wiley-Interscience, 1981.
113 For examples see Barber, S. A. and Silberbush, M. in *Roots, Nutrient and Water Influx, and Plant Growth*, p. 65, ASA Spec. Publ., No. 49, 1984.
114 See ref. 15.
115 Silberbush, M. and Barber, S. A., *Agron. J.*, 1983, **75**, 851; *idem*, *Pl. Soil*, 1983, **74**, 93.

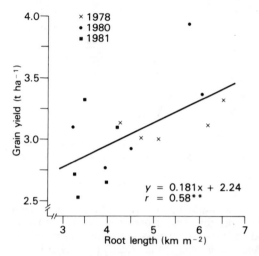

Fig. 4.14 Relation between grain yield of soyabeans and root length at the R6 stage for several varieties of soyabeans grown on Raub silt loam. (From Barber, S. A. and Silberbush, M. in *Roots, Nutrient and Water Influx, and Plant Growth*, Am. Soc. Agron. Spec. Publ., No. 49, 1984.)

Fig. 4.14 shows that yield and root length are related and that improved yields are unlikely to be achieved without also increasing root growth.

Water uptake by roots

As with nutrients, consideration of the anatomy of roots suggests that the primary zone of water absorption will be just behind the root tip where the xylem vessels have differentiated but suberization of the endodermis has not started. Figure 4.15 shows rates of water uptake at various distances behind the root tips of barley and marrow plants. It clearly shows that rates of water uptake are indeed maximal close to the root tip but that water can enter roots even when the endodermis is fully suberized. A study of the uptake of water by various components of the root system of young barley plants showed that about 50 per cent of the water taken up by the main seminal axis came from suberized regions further than 10 cm from the root tip, and together with the associated lateral roots, this region took up 75 per cent of the water transpired.[116] In many perennial plants, it is highly probable that most water uptake is via suberized roots. For example, root growth of apple and plum

116 Sanderson, J., *J. exp. Bot.*, 1983, **34**, 240.

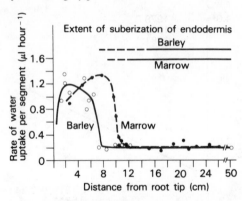

Fig. 4.15   The rate of water uptake at various distances along roots of barley (o) and marrow (●). Note that uptake occurred even where the endodermis was completely suberized. (From Graham, J. *et al.*, *ARC Letcombe Laboratory Ann. Rept.*, p. 9, 1973.)

trees is considerably reduced in mid-summer suggesting that a considerable proportion of the nutrients and water must be absorbed through suberized roots.[117] Similarly, removal of all unsuberized roots from one-year-old Loblolly pine seedlings reduced the total surface area of roots by 42 per cent compared to controls but reduced the rate of water uptake by only 23 per cent.[118]

While it is appreciated that different parts of the root system take up water at different rates, there is currently insufficient evidence to justify considering them separately, and for crops it is usual to calculate the uptake rate assuming the whole of the root length to be equally effective in taking up water. A few workers have done otherwise,[119] but such divisions are no less arbitrary than considering the system as a whole.

Typical values for the inflow of water into root systems are given in Table 4.9. Values of about 10^{-1} to 10^{-3} cm^3 water cm^{-1} root day^{-1} are common in many studies and although they ignore the non-uniform demand for water throughout the day, maximum values are unlikely to be more than around 3–5 times greater than the mean value. Such rates of inflow mean that the resistance offered by the soil to water movement to the root is unlikely to play a major role in limiting water uptake in most situations (see Ch. 11 for more discussion of this point).

During a growing season, water uptake is initially confined to the surface layers of the soil, but as the root system penetrates deeper into the soil and

117 Head, G. C., *J. hort. Sci.*, 1967, **42**, 169.
118 Chung, H. H. and Kramer, P. J., *Canad. J. For. Res.*, 1975, **5**, 229.
119 For example, Greacen, E. L. *et al.*, in *Ecological Studies Analysis and Synthesis*, Vol. 19 (eds. O. L. Large, L. Kappen and E. D. Schulze), p. 86, Springer-Verlag, 1976, calculate water uptake only for roots bearing root hairs.

Table 4.9 Inflow of water ($cm^3 cm^{-1}$ root day^{-1})

Crop	Inflow	Range	Source
Soyabean	3.0×10^{-2}	$0.5-5 \times 10^{-6}$	1
Onion	2.2×10^{-2}		2
Winter wheat	2.0×10^{-3}	$0.7 \times 10^{-3}-2.5 \times 10^{-3}$	3
Ryegrass	7×10^{-4}		4
Cotton		early season 0.1–3.1	5
		late season 0.03–0.86	

1 Allmaras, R. R. *et al.*, *Soil Sci. Soc. Am. Proc.*, 1975, **39**, 771.
2 Dunham, R. J. and Nye, P. H., *J. appl. Ecol.*, 1973, **10**, 585.
3 Gregory, P. J. *et al.*, *J. agric. Sci.*, 1978, **91**, 103.
4 Lawlor, D. W., *J. appl. Ecol.*, 1972, **9**, 79.
5 Taylor, H. M. and Klepper, B., *Aust. J. biol. Sci.*, 1971, **24**, 853.

the upper layers dry, so the zone of maximum root activity moves downward and water uptake from the upper layers becomes less important. If the upper layers are rewetted, they again become the zone of maximum uptake.

Several groups of workers have attempted to predict the amount of water extracted by roots from different layers in soils. This usually involves the numerical solution of soil water flow equations (based on Darcy's law) by dividing the soil into a number of layers and time into a number of short intervals. In essence, water uptake (U, $cm^3 min^{-1}$) from a layer of soil (i) can be determined as a function of the root length in the layer (L_v, cm root cm^{-3} soil), the difference of water potential between the bulk soil (ψ_s, bar) and that in the plant xylem at the soil surface, and the resistances to flow between the bulk soil and the plant xylem. Following Taylor and Klepper,[120] water uptake (U_i) from a soil volume (V_i) can be determined from:

$$U_i = (V_i)(L_{vi})(q_i)(\psi_{si} - \psi_p + \psi_{zi} + \Sigma\psi_{fi}) \qquad [4.12]$$

where q is the average root water uptake rate in V_i (cm^3 water cm^{-1} root $bar^{-1} min^{-1}$), ψ_p is the xylem potential at the soil surface, ψ_{zi} is the water potential loss due to elevation, and ψ_{fi} is the decrease in potential due to frictional forces within the root (dependent on the axial resistance). Figure 4.16 shows that the rate of water uptake by roots of cotton was similar at the same soil water content for roots at all depths. This means that all roots were able to take up water equally effectively irrespective of the age of the root and whether the size of the root system was increasing or decreasing with time. The result also shows that the longitudinal resistance to water flow in the xylem of the root is small compared to other resistances in the system (i.e. the effect of ψ_{zi} can usually be ignored). The observation that roots can take up water when a soil is rewetted after a prolonged drought, without there

120 Taylor, H. M. and Klepper, B., *Adv. Agron.*, 1978, **30**, 99.

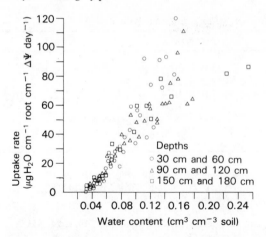

Fig. 4.16 Relation between the rate of water uptake (expressed per cm root per cm potential difference per day) and the soil water content for plants of cotton. (From Taylor, H. M. and Klepper, B., *Soil Sci.*, 1975, **120**, 57, © Williams & Wilkins.)

being a substantial growth of new roots, supports the interpretation of Fig. 4.16; similar conclusions have been reached by other workers.

The size of the resistance to flow axially within the xylem of the root (and hence the size of ψ_f) will have important consequences for the pattern of water uptake. Chapter 11 contains values of axial resistance determined by some workers but many factors affect it. In summary, it may vary between plant species, among soil depths of one species, and between different orders of root; it also will be affected if the xylem is invaded by pathogens. The importance of the axial resistance in determining the pattern of water extraction is also illustrated by Taylor and Klepper. Using experimentally derived values for q (5×10^{-7} cm^3 water cm^{-1} root bar^{-1} min^{-1}), L_v (1.0 cm cm^{-3}), ψ_p (-15 bar), ψ_s (-0.1 bar) and assuming a soil profile comprising four equal volumes of soil (1000 cm^2 × 50 cm deep) they showed that the uptake rate would be almost uniform throughout the profile (0.375 cm^3 min^{-1} in layer 1 to 0.370 cm^3 min^{-1} in layer 4) if $\psi_f = 0$ bar, whereas if ψ_f dropped to 0.07 bar cm^{-1} then water extraction from layer 4 would be only 16 per cent of that in layer 1.

Because of the difficulties in measuring root distribution within a soil profile, some workers have attempted to use in models a simpler volumetric sink term which is added to the continuity equation for soil water flow.[121] In such models, the sink term is considered to be a function of the soil water

121 See for example Feddes, R. A. *et al.*, *J. Hydrol.*, 1976, **31**, 13; Belmans, C. *et al.*, *op. cit.*, 1983, **63**, 271.

content and varies according to the pressure heads known to be critical for water uptake by roots.

These mathematical descriptions of water uptake are useful in highlighting the principal factors determining the amount of water that can be extracted. For a given soil at a uniform water content and with a fixed potential transpiration rate, they indicate that (i) water uptake by crops will depend on the depth of the root system (determining the total quantity of water available for extraction), and (ii) rate of uptake will depend on the root length within each layer (because uptake rate is proportional to root length per unit volume of soil) and on the resistance to flow. Genotypic differences are known to exist in many of these properties and there is scope for modifying the root systems of crop plants. For example, varieties can be bred having xylem vessels with a range of diameters, which would affect the resistance to water flow; its significance for water use in the field has yet to be assessed.[122]

122 Taylor, H. M. in *Adaptation of Plants to Water and High Temperature Stress* (eds N. C. Turner and P. J. Kramer), p. 75, Wiley, 1980; Richards, R. A. and Passioura, J. B., *Crop Sci.*, 1981, **21**, 253.

5

Soil formation and characteristics of soil profiles

Soil properties and processes as described in other chapters are often studied on small samples of soil in the laboratory or at specific field sites. If we are to extrapolate to other locations it is necessary to know about the spatial extent of the property or process and how variable it is. Further, because many measurements are expensive and time consuming we often endeavour to rely on relationships established between field observations and laboratory measurements, so that further investigations might rely more substantially on the field observations. With experience it is possible to recognize regularities in the patterns of soil distribution. For example we might recognize regular patterns of soils if we traverse from hill crest to valley bottom at a number of sites, or as we move from one parent material to another. If we consider the broader scale we can recognize general changes of soil properties when we move from one climatic zone to another. These patterns are useful in helping to make generalizations from a small number of observations or measurements and, further, they can help to set a limit on such generalizations. In considering soil conditions and plant growth, the pedologist can make several useful contributions including:

1   A description of the spatial variability of soil properties and processes.
2   An explanation for the occurrence of particular soil properties at particular locations and their spatial variability.
3   An understanding of the processes of soil formation and the inter-relationships between soil and its environment.

This chapter will deal successively with rock weathering, soil profile formation, the occurrence of organic matter within the soil and the development and nature of major soil types. The interpretation and use of this information through soil survey, soil classification and land evaluation is dealt with in Chapter 24.

Rock weathering

Soils develop during the chemical alteration or decomposition of the mineral particles of the rock or superficial deposits. They may thus be considered a transition phase in the slow changes of unstable minerals to more stable chemical states with concurrent loss of soluble materials. This decomposition alone, however, is not sufficient to distinguish the material as soil rather than simply as decomposed or weathered rock. The material is generally considered to be soil only when plants gain a foothold and the remains of which, together with associated animals, are added to the decomposed rock, thus generating a biological cycle of growth and decay. The important distinguishing factor therefore is the presence of organic matter. The organic residues are usually decomposed continuously and relatively rapidly by the soil organisms. Many of the mineral particles also decompose, but usually at a much slower rate. It is convenient to discuss these processes in two parts: first, in the virtual absence of plant material and the operation of biological processes; and, second, when they are a relevant factor.

The words *soil* and *rock* need definition. Rock is defined as the inorganic mineral material covering the earth's surface, and will therefore commonly include part of the soil. Rock may be hard and solid, such as granite, limestone or other consolidated mineral materials, or it may be loose and unconsolidated, such as gravel, sand or clay, in which case it is often referred to as *regolith*. Many of the solid and hard materials, such as granite, may have a thin layer at the earth's surface which is loose and unconsolidated due to weathering; this is also called the regolith. A soil may be considered to be that part of the regolith that shows vertical differentiation due to biological activity; that is, a soil consists of layers of mineral material whose properties differ from one another, and these differences are due in large degree to biological activity. Simonson[1] identified these two broad processes of soil formation as, first, the accumulation of unconsolidated material at the surface of the earth and, second, the organization of this material into distinctive layers or horizons.

The process of change in rock material, called *weathering*, is a continuous and complex combination of destruction and synthesis. Rock and mineral particles can suffer physical disintegration without change in their chemical composition. The process of comminution of the particles depends on some applied force. These include forces such as overburden pressure, normal load and shearing stress (e.g. the grinding of particles under ice sheets); expansion forces due to temperature change, growth of crystals (e.g. ice) or animal or plant activity; or water pressures which may be important in the processes controlled by wetting and drying cycles. In addition, since many of these forces are applied both at the surface and within a rock mass, it is necessary to state the vertical and lateral confining conditions in order to understand responses such as linear expansion or rock fracture.

In most environments, however, the important changes in the rock and soil minerals are changes in chemical composition brought about by atmospheric and hydrospheric agencies. In addition, these changes in composition may be accompanied by physical disintegration. This group of weathering processes is generally considered under the broad heading of *chemical weathering*, for which water and its dissolved salts and acids are probably the most important agents. The weathered rock material which cannot be considered soil is called *saprolite*. Saprolite is the material resulting from the *in situ* weathering of the original minerals to clays and other secondary minerals, where there is little or no change in volume and the weathered rock material also retains many of the structures found in the original rock. Some of these weathering reactions can occur at depths exceeding 10 m below the surface; saprolite layers of 20 m thickness and depths to fresh rock of 30 m are widespread in parts of Brazil. Reports of weathering to 100 m have been recorded in Nigeria and Uganda, and in parts of Czechoslovakia kaolinite weathering is reported to reach 30 m.[2] The effect of organic acids is generally important only near the surface and hence in the soil forming processes.

The fate of the products of decomposition depends chiefly on the movement of water through the soil. The greater the amount of downward leaching the higher the proportion of the products of weathering that are removed from the zone in which they were formed; they are either redeposited in lower layers of the soil or in the regolith, if the flow is vertical, or are carried into adjacent soil material where lateral flows predominate, or are

---

1 Simonson, R. W. in *Quaternary Soils* (ed W. C. Mahaney), Geo-Abstracts, 1978.
2 For Nigeria see Thomas, M. F., *Trans. Inst. Brit. Geog.*, 1966, **40**, 176. For Uganda see Ollier, C. D., *Zeit. f. Geomorph.*, 1960, **4**, 43. For Czechoslovakia see Demek, J., *Zeit. f. Geomorph.* (Suppl.), 1965, **5**, 82.

carried with the percolating water into the groundwater and then into the rivers.

Plants growing on the soil may counteract some of the downward movement for their roots absorb some of the soluble products of weathering from the lower parts of the soil and transfer them to their leaves and stems, which later fall to the soil surface and decompose, releasing the products of soil weathering on to the soil surface. Burrowing animals such as earthworms, termites and some rodents, also affect the distribution of the products of weathering through the mixing of soil from different layers. Similar effects result from the blowing over of trees by wind or the filling of the large spaces once occupied by their larger roots, and by freezing and thawing at depth in the soil. These processes involving large-scale soil disturbance are known as *pedoturbation*.

The chemical weathering of rocks

Weathering of rocks by water involves two broad but separate sets of processes; the decomposition brought about by the water, and the fate of the products of decomposition. Where water is percolating freely through the medium, the soluble products of the weathering processes will be removed from the weathered mantle and enter the groundwater, and in time may move into rivers. Soil temperature greatly affects the rate at which these weathering processes take place, the warmer the rock in contact with the water, the faster is its decomposition; and the higher the rainfall, provided free drainage of water occurs, the quicker are the soluble products removed. The rate of decomposition of a given mineral particle generally depends on its surface area, so the smaller the particles the faster the rate per unit volume. The rate also depends on the temperature, the pH of the solution, and the chemical composition of the mineral and its crystalline structure.

There are a large number of processes grouped together under the general term chemical weathering. The more important of these are *hydrolysis*, *hydration* and *oxidation*.

Hydrolysis, often considered the most important process of the many which occur in the weathering of silicate and aluminosilicate minerals, involves reactions with water. The cations of a mineral are replaced by H^+ ions formed by dissociation ions of water and the OH^- ions combine with them. The products of hydrolysis can be leached out, remain in solution or become part of the crystal lattice of new secondary silicate minerals.

An example of the general principles involved is given by considering the possible reactions involved in the hydrolysis of microcline, a potassium feldspar:

$$KAlSi_3O_8 + H_2O \rightarrow HAlSi_3O_8 + K^+ + OH^-$$

$$2HAlSi_3O_8 + 14H_2O \rightarrow Al_2O_3.3H_2O + 6H_4SiO_4$$

The products of this reaction are aluminium hydroxide and silicic acid. The silicic acid is soluble and may be removed in drainage waters or recombine with other materials to form secondary silicates such as clay minerals. The potassium that is released may be leached out of the system, taken up by plants or recombined into other minerals. Concomitant with the action of water in weathering is that of oxidation. This is of particular importance in minerals such as biotite which contain Fe^{2+}. Oxidation to Fe^{3+}, which increases the number of positive charges, results in a less stable crystal lattice which is much more susceptible to mechanical and further chemical weathering.

Weathering does not cease once the secondary minerals are formed, because the secondary minerals are stable only between certain concentration limits of soluble silica, alkali and alkaline earth cations and H^+. If these solutes are leached away over time, in strongly weathering environments, the initial secondary minerals weather further to still more stable chemical states.

Susceptibility of rocks to weathering

The rock materials at the earth's surface obviously vary considerably in their mineral composition and consequently in their susceptibility to weathering. A number of attempts have been made to identify sequences of susceptibility of rocks to weathering. For example Parker[3] devised an index of weathering emphasizing the importance of hydrolysis as a weathering process in temperate environments, based on the assumption that alkali and alkali earth elements are the most mobile:

$$\text{WI} = \frac{(Na)a}{(Na\text{-}O)b} + \frac{(Mg)a}{(Mg\text{-}O)b} + \frac{(K)a}{(K\text{-}O)b} + \frac{(Ca)a}{(Ca\text{-}O)b} \times 100$$

where $(M)a$ is the atomic proportion (i.e. (atomic percentage)/(atomic mass)) of element M and $(M\text{-}O)b$ is the strength of the bond with oxygen. Using actual bond strengths, the index may be rewritten[4]:

$$\text{WI} = \frac{(Na)a}{0.35} + \frac{(Mg)a}{0.90} + \frac{(K)a}{0.25} + \frac{(Ca)a}{0.70} \times 100$$

Chesworth[5] gives a procedure for investigating weathering trends, concluding that the general trend in mineral soil horizons is a relative increase in silica, alumina and iron oxides towards the surface. Close to the surface other soil processes such as biotic activity and, in particular, the recycling of elements by plants tend to obscure the weathering trends. In general, however, Ches-

3 Parker, A., *Geol. Mag.*, 1970, **107**, 501.
4 Nicholls, G. D., *Sci. Prog.*, 1963, **51**, 12.
5 Chesworth, W., *Geoderma*, 1973, **10**, 215.

worth concludes that the influence of parent material on the soil tends to diminish towards the surface and through time.

Weathering trends were examined in soils developed in Scotland on a range of non-till-covered rocks (i.e. in the *in situ* weathering residua).[6] The broad conclusions were that in all the soil profiles studied there was a tendency for there to be a high degree of similarity as the soil surface was approached. Under these environmental conditions the intensity of weathering of their parent rocks varied in the following broad sequence:

Serpentine> gabbro> andesite> quartz diorite> granite> feldspathic sandstone> quartzite.

In general the acidic rocks showed much less chemical change than the more basic species.

In terms of the products of the weathering process found in soils as it progresses, the following simplified scheme has been suggested[7]:

Incipient weathering: primary minerals, complex secondary minerals, quartz.
Intermediate weathering: complex secondary minerals, quartz, iron oxides.
Advanced weathering: kaolinite, quartz, iron oxides.
Extreme weathering: oxides of aluminium and iron.

Within this range of minerals quartz has a variable persistence depending upon its particle size. Quartz particles larger than several micrometres in size

Table 5.1 Average composition of igneous rocks and of three surface soils of increasing maturity

	Average of igneous rocks	Barnes loam (South Dakota)	Cecil sandy clay loam (North Carolina)	Columbiana clay (Costa Rica)
SiO_2	60	77	80	26
Al_2O_3	16	13	13	49
Fe_2O_3	7	4	5	20
TiO_2	1	0.6	1	3
MnO	0.1	0.2	0.3	0.4
CaO	5	2	0.2	0.3
MgO	4	1	<0.1	0.7
K_2O	3	2	0.6	0.1
Na_2O	4	1	0.2	0.3
P_2O_5	0.3	0.2	0.2	0.4
SO_3	0.1	0.1	—	0.3
Total	100.5%	101.1%	100.6%	100.5%

6 Smith, B. L. *et al.*, *Trans. R. Soc. Edin: Earth Sci.*, 1982, **73**, 191.
7 Kronberg, B. L. and Nesbitt, H. W., *J. Soil Sci.*, 1981, **32**, 453.

(fine silt) remain in soils for so long that quartz appears to be the most stable state of soil silicon. When finely divided into clay sized particles, however, quartz persists only slightly longer than clay sized feldspars. Even feldspars disappear from the sand and silt fractions relatively rapidly.

Chemical changes on weathering

Table 5.1 illustrates the effect of weathering reactions on the total composition of soils.[8] Here three soils of increasing maturity (South Dakota, North Carolina and Costa Rica), and hence degree of weathering, are compared to the average composition of igneous rocks. The comparisons show general trends only because the parent materials differ from each other and from the average igneous rock. The elemental contents are expressed as weight percentage of oxides, which allows the analysis to sum to 100 per cent.

The first stage of weathering causes considerable loss of Ca^{2+}, Mg^{2+}, Na^+ and K^+ illustrated by comparing the igneous rock with the relatively young Barnes soil. As weathering progresses, further losses of these cations are slower. Most of the alkali and alkali earth cations remaining after early stages of weathering are in larger unweathered mineral grains, and only small proportions of Ca^{2+}, Mg^{2+}, Na^+ and K^+ are retained by cation adsorption on negatively charged secondary mineral particles. These cations are important because, although they are subject to further leaching losses, they contribute to the control of soil pH, and they supply essential elements to plants and soil micro-organisms.

Following the major changes in composition observed between the igneous rocks and the Barnes soil as a result of weathering, the subsequent slower changes are best ascribed to soil development. Consequently the differences in composition between the relatively poorly developed Barnes soil and the well-developed Cecil soil are less than those between the Barnes soil and igneous rock. The differences in elemental composition are small, but there may be substantial differences in mineralogy, and these may lead to substantial differences in availability of ions for plant growth.

Soil development generally involves the steady loss of silicon. This is not shown in the comparison between Barnes and Cecil soils, presumably because of the high silica content of their parent materials. In contrast, the highly weathered Columbiana soil shows evidence of large silica loss. As the SiO_2 content of this soil decreases to less than half the average for igneous rocks, the Al_2O_3 and Fe_2O_3 contents increase three-fold. The solubilities of iron, aluminium and manganese hydroxides are much lower than the solubility of silicon, and the iron and aluminium hydrous oxides are more stable than the silicates which formed under earlier soil conditions.

8 Black, C. A., *Soil Plant Relationships*, Wiley, 1968.

Within the soil, as weathering and soil development progress, so the composition of the soil clay fraction changes. Results presented by Jackson and Sherman[9] illustrate this, identifying 13 stages of soil development as reflected by the prominent minerals in the clay fraction (Table 5.2). However, the degree of development cannot be considered solely in a temporal context. Tardy and co-workers considered the clay minerals formed by weathering of primary minerals under a range of climatic regimes (see Table 5.3).[10]

Table 5.2 Sequence of clay mineral occurrence with increasing soil development

Relative degree of soil development	Prominent minerals in soil clay fraction
1	Gypsum, sulphides and soluble salts
2	Calcite, dolomite and apatite
3	Olivine, amphiboles and pyroxenes
4	Micas and chlorite
5	Feldspars
6	Quartz
7	Muscovite
8	Vermiculite and hydrous micas
9	Montmorillonite
10	Kaolinite and halloysite
11	Gibbsite and allophane
12	Goethite and haematite
13	Titanium oxides, zircon and corundum

Table 5.3 Clay minerals produced from primary minerals in relation to climate

Climatic regime	Primary mineral	Clay mineral formed
Arid	Biotite	Smectite
	Plagioclase	Kaolinite
Temperate	Biotite	Vermiculite
	Plagioclase	Smectite, kaolinite
Humid tropical	Biotite	Kaolinite
	K-feldspar	Kaolinite
	Plagioclase	Kaolinite
Very humid tropical	Biotite	Kaolinite
	K-feldspar	Kaolinite, gibbsite
	Plagioclase	Gibbsite

9 Jackson, M. L. and Sherman, G. D., *Adv. Agron.*, 1953, **5**, 221.
10 Tardy, Y. *et al.*, *Geoderma*, 1973, **10**, 271.

The rate at which soluble products are removed from an area can be estimated from an analysis of river water draining out of the area, provided the river is not receiving effluent from other sources. Few comparable estimates of loss have been made, but many years ago F. W. Clarke collected together the data then available.[11] He showed that European and North American rivers carried away about 40 tonnes of soluble salts per square kilometre of their catchment annually, which corresponds to the solution of 1 cm of soil per 1000 years; while tropical rivers removed only half this amount, due to the soil in the tropical areas being more strongly weathered than in temperate regions. The estimated weights of the various soluble mineral constituents lost per year in kg per hectare are given in Table 5.4.

Table 5.4 Removal of soluble mineral constituents in some rivers and drainage water; amounts in kg ha^{-1} a^{-1}

	CO_3	SO_4	Cl	NO_3	Ca	Mg	Na	K	Al_2O_3 $+ Fe_2O_3$	Si
North American and European rivers	145	55	19	3.9	82	13	23	9	6	17
Amazon	64	17	11	1.2	37	5	10	4	11	18
Drainage water, Broadbalk*	—	63	27	170	178	8	11	3	14	13

* From continuous wheat, unmanured plot[12]

This shows that the Amazon, draining a tropical region, contains much less calcium, magnesium, potassium and sodium than the rivers of temperate regions, but the same amount of silicon and probably more aluminium and iron. The composition of the drainage water from the top 60 cm of the Broadbalk soil is given for comparison.

The formation of clay minerals may involve hydration and some degree of ion exchange, as in the formation of hydrous mica (illite) from muscovite. As stated earlier, oxidation occurs when biotite weathers to form hydrous mica. Essentially these changes are modifications to the structure of the primary mineral, but the changes that result in the formation of the kaolin and montmorillonite groups of minerals are more complex and are discussed in more specialized texts.[13] Broadly, however, kaolinite occurs where leaching removes Ca^{2+}, Mg^{2+}, K^+ and Na^+, and montmorillonite where these ions, and also iron and silica accumulate. In tropical regions, therefore, montmorillonite

11 Clarke, F. W., *Data of Geochemistry*, US Geol. Surv. Bull. No. 70, 1924.
12 Voelcker, A., *J. Chem. Soc.*, 1871, **24**, 276.
13 See for example Dixon, J. B. and Weed, S. B. (eds), *Minerals in Soil Environments*, Soil Sci. Soc. Am., Madison, 1977.

is characteristically present in soils known as Vertisols which form in river basins, whereas kaolinite occurs in soils with free drainage.[14]

The formation of soils

Soil forming processes

Weathering alone produces only the regolith. Soil development begins when plants gain a foothold in the weathered debris. The growing plant and the organisms feeding on it and its litter affect both the nature of rock weathering and the fate of the products of the weathering processes. Plants take up soluble mineral salts from the soil within their root range, transfer them to their leaves and other above-ground parts, and these are then returned to the surface either in leaf drip or in fallen leaves, stems and branches of the vegetation and to the soil in decayed roots. This dead vegetation and the living and dead plant roots are a food source for a wide range of soil organisms, which in turn can affect the soil by their physical and biological activities. Further, some of these organisms are capable of inducing severe reducing conditions in the soil if aeration becomes poor, which introduces a new set of soil processes. In many soils the initial stages of layering within the regolith which Simonson distinguished as soil development consist of the accumulation of organic matter at the surface or intermixed with the surface mineral layers.[1]

The amounts of mineral salts in the biological cycle can be considerably higher than the amounts leaching out of the soil profile into the rivers. In some extreme cases under tropical rainforest conditions, the amounts in outflow in streams and rivers may be almost beyond the limits of detection, but within the vegetation there is a most efficient almost closed nutrient cycle, with the products of litter fall and washing of the canopy taken up directly in a dense root mat close to the surface.[15] Stark described this mechanism as the 'direct nutrient cycling hypothesis', where the nutrients in the Amazonian forest are held mainly in the organic phase in living roots, wood, bark and leaves and in dead organic matter.[16] This hypothesis has been supported by a number of studies in South and Central America.

The amounts of mineral elements returned annually to the soil surface under an oak forest in northern England are given in Table 5.5[17]; the figures should be compared with those above for the amount being removed by

14 For a description of various types of these clays see Dudal, R., *Soil Sci.*, 1963, **95**, 264.
15 See for example Nortcliff, S. and Thornes, J. B., *Acta Amazonica*, 1978, **8**, 245; Jordan, C. F. and Kline, J. R., *Ann. Rev. Ecol. Syst.*, 1972, **3**, 33.
16 Stark, N., *Trop. Ecol.*, 1971, **12**, 24.
17 Carlisle, A. *et al.*, *J. Ecol.*, 1967, **55**, 615.

Table 5.5 The quantities of nutrients washed from a *Quercus petraea* woodland by rainfall and reaching the soil (kg ha^{-1} a^{-1})

	Organic matter	N	P	K	Ca	Mg	Na
Incident rainfall	77	8.7	0.3	2.8	6.7	6.1	50.8
Washed from oak canopy	174	0.9	0.5	16.0	12.3	5.6	32.0
Washed from other vegetation and stem flow	27	−0.3	0.1	10.9	2.0	1.5	6.6
Total in throughfall	277	9.3	0.9	29.7	21.0	13.2	89.4
Total in litter	6670	61.6	3.7	22.2	36.1	6.0	4.0
Total in throughfall as % of litter	4.1	15.1	25.0	133	58.4	222	2210

Table 5.6 Average annual inputs, outputs and net gain or loss of elements from several temperate forest ecosystems

Element	Environment	No.	Input	Output	Net gain (+) or loss (−)	Range of net gain or loss
					(kg ha^{-1} a^{-1})	
Ca	Acid	15	4.4	11.4	−7.0	−0.1 to −20
	Calcareous	4	10.5	102.9	−92.4	−44.8 to −137
Mg	Acid	15	2.0	5.7	−3.7	−1.5 to −7.4
	Calcareous	3	2.2	40.8	−38.6	−7.4 to −75
Na	Inland	13	2.5	10.3	−7.8	−0.6 to −21
	Oceanic	7	23.2	40.0	−16.8	−6 to −27.4
K	Inland	12	1.6	3.5	−1.9	−0.1 to −5
	Oceanic	7	2.6	5.5	−2.9	+0.7 to −8
N	Various	15	7.9	2.0	+5.9	+17.5 to +0.5
P	Various	16	0.26	0.15	+0.11	+0.52 to −0.2
SO$_4$–S	Rural	5	5.6	11.2	−5.6	+1 to −19
	Industrial	6	16.0	21.4	−5.4	+1.2 to −17

rivers. Table 5.6 gives the same kind of information for annual inputs, outputs and net gain or loss of elements from several temperate forest ecosystems.[18]

These figures illustrate the often large amounts of an element in the nutrient cycle relative to that which is leached from the system. For example, amounts of potassium in the cycle are much higher than the annual loss from the range of catchments, more silicon is usually in the cycle than is lost, but the amounts of calcium and magnesium in the cycle may be of the same order as that leached out. Less is known about the amounts of iron and aluminium which cycle since they are rarely measured.

Plants can also directly affect the weathering processes in soils, for rain not only washes salts off the leaves, but also soluble organic compounds, some of which are capable of reducing ferric iron and may form relatively stable ferrous-organic complexes. The roots may also hasten weathering by taking up nutrients from the soil solution, for example the uptake of potassium will lower the potassium ion concentration to a sufficiently low level for the potassium in the mica interlayer to diffuse into solution, so allowing the mica to expand along its broken edges. Rhizosphere bacteria and fungi might also increase the rate of weathering, excreting organic acids capable of chelating di- and trivalent cations, so increasing their rate of dissolution from mineral phases.

Plant litter collects on the soil surface, but its rate and site of decomposition depend on which soil organisms are active, and this, in turn, depends on the kind of litter, on the moisture and temperature regime, and on soil pH. Thus, if conditions favour the larger species of earthworm which form wormcasts, much of the decomposition will take place within the soil under conditions which favour the production of clay–humus complexes, and the earthworms themselves will cause a mixing of the top 10 to 20 cm of soil (they burrow to a depth of 1 m or more). On the other hand, if the soil conditions are unfavourable to all the larger fauna, there will be no mechanism for mixing the litter with the soil, and much of the decomposition will take place on the surface. The amount and distribution of plant roots and their longevity also affect the supply of decomposable organic matter in the soil, and so the distribution and activity of its population.

Finally, if the oxygen supply to micro-organisms is severely restricted e.g. by waterlogging, severe reducing conditions may be created which result in the reduction of ferric to ferrous iron and higher valency manganese to manganous ions. The reduced forms are more soluble, and they may also be present in the soil solution as organic complexes, so resulting in greater loss by leaching. Similar conditions may also result in the reduction of sulphate to sulphide, which on reoxidation to sulphuric acid will increase the soil acidity very

18 Ralston, C. W. in *Forests Soils and Land Use* (ed C. T. Youngberg), Colorado State Univ. Press, 1978.

considerably. These conditions have been widely observed with the drainage of fenland soils and the production of acid sulphate soils.

Soil development

A soil is distinguished from weathered regolith by the vertical differentiation it exhibits due to biological activity, so the properties of soil which are singled out as of importance in most systems of soil classification must be displayed in the soil profile. A soil profile is a vertical face of a soil pit, and typically contains bands or layers or horizons that are visually different from adjacent horizons. The visual differences often merge so that there may be no clear divisions between them. Simonson,[1] in identifying the development of horizons within the weathered rock material as a distinctive feature of soil formation, suggested that these developments could be considered under four broad headings:

1   The addition of soil material.
2   The removal of soil material.
3   The transfer of soil material within the soil system.
4   The transformation of soil material.

Well-drained soils that have suffered relatively little disturbance by man typically possess three or four clearly recognizable horizons: an organic litter layer lying on the soil surface often known as an O; a horizon darkened by the presence of organic matter, the A, which may have lost appreciable amounts of clay or iron and aluminium hydrous oxides (sesquioxides) and be considered an eluvial horizon; a horizon in which soil materials are transformed or which exhibits an illuvial concentration of clay, sesquioxides or other material transferred within the soil, known as the B horizon; and the C horizon which has been little affected by the biological activity in the soil and which is often assumed to be similar to the parent material from which the soil has developed.

Profile development is not confined to well-drained soils, and the characteristics of horizons differ under the influence of impeded drainage or of a fluctuating groundwater table. However, since water moves downwards from the soil surface in most soils for some periods of the year, most soils possess eluvial and illuvial horizons to some degree, though their properties are likely to be strongly affected by the consequences of the water regimes.

Soils are formed under such a variety of conditions that no simple universal specification of horizons can be developed to fit all soils. Under well-drained conditions, rock minerals will weather in the B horizon and the soluble products of weathering are removed in the drainage water, so it is, broadly speaking, both an eluvial and an illuvial horizon. Again, although typically the A horizon contains more visible humus than the B, in some conditions humus can be translocated from the A to the B. It is still possible to retain the ABC nomenclature in these circumstances, but a modifier should be added to the horizon designation to show that it differs from the original simple concept. Thus horizons showing some of the characteristics of the B, but with other characters introduced through impeded drainage or fluctuating groundwater and so showing gley characters, can be designated as a gleyed B or Bg horizon, if enriched with humus as Bh, and with sesquioxides as Bs.

A feature of soils in the field is that in close proximity, they often exhibit considerable differences. Large variations occur in soil properties over short distances in some parent materials such as glacial tills, outwash gravels, loess, colluvial materials and alluvial materials.[19] This variability usually reflects considerable heterogeneity in the top few metres of the land surface. The introduction by the United States Department of Agriculture of the *pedon* as the fundamental unit of soil, arose from a recognition of the occurrence of considerable differences between soils at sites close together.[20] The pedon is thought of as the smallest volume that can represent the soil at a given site. If the variation of soil appears to be random from site to site, it is defined as the soil in an area of 1 m², but if the soil shows any cyclical changes which recur with a linear interval of 2 m to 7 m, the pedon includes one-half the cycle of variation.

Types and forms of humus

The organic compounds washed off the leaves of vegetation by rain, the type of litter falling on the soil, and the processes bringing about its decomposition can have a profound effect on the type of soil being formed. These processes can be seen most clearly in soils that have been uncultivated for long periods of time, such as old forest, old heathland and old grassland. Every undisturbed soil has a characteristic distribution of undecayed and decaying litter and humus on and within its surface, and these distributions are classified under the unfortunate name of humus type; unfortunate because the word humus should be restricted to organic colloids. For instance the humus type under a coniferous forest on an acid sand is different from that under many deciduous forests on loams or clay loams, and these are different from that under grassland on the same site.

Each humus type consists of two or more layers or horizons, each layer having a different form of humus. Typically the top layer is fairly fresh leaf litter and the bottom layer fully humified material, with intervening layers classified by their degree of humification. There is still no generally accepted

19 See for example Nortcliff, S., *J. Soil Sci.*, 1978, **29**, 403.
20 Soil Survey Staff, Soil Classification: a comprehensive system, 7th Approximation, USDA, 1960.

classification or nomenclature for the different humus types[21] and humus forms,[22] and the situation is complicated by many workers using the same name for a humus type as for the dominant humus form constituting that type.[23] Further, there is no recent monograph in which the various types and forms found in either the temperate or the tropical regions have been precisely described.

Three humus types forming the floor of temperate forests[24] have been widely recognized, two of which have the generally accepted names of mull and mor; the third has no generally accepted name, the names used for it including insect mull, mull-like mor, and moder.[25]

Typical mor is a humus type that lies upon and is not mixed into a mineral soil. It is composed of three layers or humus forms: a litter layer, known as the L layer, consisting of recently fallen leaves, which have only just started to decompose, and as a consequence many of the leaves are intact or partially so and the structure of the leaves is clearly visible. The next layer down is a transitional or F layer in which the leaves are undergoing active decomposition but in which numerous leaf fragments are still recognizable. The lowest layer is a humus or H layer, composed of dark structureless humus. This humus type forms a compact blanket, which can often be stripped off the mineral soil, and it can build up to a thickness of 10 or 15 cm in many forests. Typically, the vegetation giving rise to a mor humus roots strongly in the humus horizon, and most or all the roots are infected by mycorrhizas.

Mull humus, on the other hand, is a humus type in which most of the humus is mixed with the mineral soil though a part lies on the soil surface. The part lying on the surface is recognizable plant debris which has suffered little decomposition but which typically disappears within a year, while the part in the soil has lost all trace of the plant remains from which it was derived. Nevertheless it contains many fine dead roots that have not yet been decomposed. This mull, or mould to use the old vernacular name, is crumbly, brown to dark brown in colour, and the organic matter in the top 5 to 10 cm rarely exceeds 10 per cent. In uncultivated soils it has no clearly defined lower boundary. Both the humus type, and the form of humus in the mineral soil are called mull.

Moder humus is principally above the mineral soil surface; it usually has no clear separation from mineral soil and the bottom of the moder layer contains an appreciable amount of mineral grains. It differs from mor in that it is much looser and more porous, and the litter layer is characterized by

---

21 See for example Wilde, S. A., *Soil Sci.*, 1971, **111**, 1.
22 The classical descriptions are given in Kubiena, W., *The Soils of Europe*, Murby, 1953.
23 For temperate soils see Duchaufour, P., *Précis de Pédologie* (3rd edn), Masson, 1970.
24 For a description of humus types in British grassland soils see Barratt, B. C., *J. Soil Sci.*, 1964, **15**, 342; *Geoderma*, 1967, **1**, 209.
25 This classification, but not the actual names, was first put forward by the Danish forester P. E. Müller in *Tidsskr. f. Skovbrug*, 1878, **3**, 1; 1884, **7**, 1.

the leaves being lacerated by the feeding of soil arthropods. A considerable proportion of the layer below the litter can be seen under the microscope to consist of faecal pellets and exoskeletons of the fauna. Moder results from a considerable mixing of plant fragments brought about by animals smaller than the earthworm. Both this humus type and the humus form in this horizon have been called moder. Moders grade into mulls as the earthworm population increases.

The soil, the climate and the vegetation control which of these humus types will be dominant at any one site through their control on the types of soil organisms that can flourish. Mull humus is confined to soils carrying a vegetation of certain species of broad-leaved trees or grasses. Such soils contain a high population of earthworms which feed on decaying leaves on the soil surface and at the same time ingest soil, and excrete the mixture of macerated digested leaf tissue and soil as wormcasts. The depth of the mull horizon is the depth to which earthworms are active. Mulls are therefore restricted to reasonably well-aerated soils that are not more acidic than pH 5, that do not contain a high proportion of coarse sand, and are deep or moist enough to allow the worms to avoid desiccation.

Mor, an extreme type of humus, is formed on soils where the leaf litter is very unpalatable to most members of the soil fauna and where fungi are the principal organisms causing decomposition. The reason for the unpalatability is perhaps that the leaves of the vegetation are high in water-soluble polyphenols which react with proteins to give very stable tanned products.[26] The particular type of polyphenol may be leucoanthocyanins which tend to be high in mor-forming vegetation grown under conditions of low soil fertility.[27] It is still uncertain if this is the sole cause of unpalatability, because leaf litter from the same species of tree can give either a mull or a mor depending upon soil conditions; and it has not been established if the polyphenol content of their leaves differs sufficiently to cause this difference. The F layer is often characterized by a mass of fine white fungal hyphae. The principal arthropods are mites. The soils are acidic and impoverished, often sandy, and the vegetation often coniferous forest.

The moder type is intermediate. The soil or vegetation conditions are unsuitable for earthworms, but the litter is palatable to a number of the larger species of arthropods, such as millipedes, woodlice and beetle larvae, as well as to mites.

26 Handley, W. R. C., *Bull. Forestry Comm.* No. 23, 1954; *Pl. Soil*, 1961, **15**, 37.
27 Davies, R. I. *et al.*, *J. Soil Sci.*, 1964, **14**, 229, 310.

Some well-drained soils

Podzols (spodosols) and podzolization

This soil type, which is usually, but not necessarily, found on well-drained sites, will be described first because, although it is restricted to a narrow range of environmental conditions, it allows a study of some important processes of translocation with the minimum of complicating factors. The most important feature of podzols, or as they are becoming increasingly known, spodosols,[28] is the presence of a horizon with an accumulation of amorphous organic matter and aluminium, with or without iron. As a consequence the criteria used to define a podzol B horizon, or spodic horizon, in most recent soil classifications have been based at least partially on the amounts of organic carbon, aluminium and iron extracted by specific solutions from samples of these horizons.

The historical development of the podzol concept was discussed by Muir.[29] In Western Europe many workers have considered podzols to be restricted to coarse textured parent materials, developed with a grey or bleached horizon underlain by a reddened or darkened illuvial horizon. In Russia, by contrast, the concept of the podzol included only the bleached horizon, the term podzol signifying ash-coloured soil in Russian. More recently the Spodosol order of the United States Department of Agriculture soil classification is based on the Western European concept of podzol, with an illuvial B horizon overlain by a bleached or lighter eluvial albic horizon, which is not necessarily continuous. In the FAO–UNESCO legend to the *Soil Map of the World*, podzols are identified by a spodic B horizon which has an illuvial accumulation of iron or organic matter or both.[30] A continuous white/grey albic horizon is not required above the spodic horizon. A review of the classification of podzols and the podzol process concluded that a bleached horizon was not a necessary feature for a soil to be identified as having been subjected to the podzolization process.[31]

Podzols currently developing are found only in humid regions where precipitation exceeds evaporation. Typically, podzols are found in cool climates, those occurring outside these climates resulting from special local conditions such as the presence of siliceous parent materials. In temperate climates of high latitudes podzols are characteristically associated with a natural vegetation of coniferous forest but are also found under deciduous forests including *Quercus, Fagus, Betula* and *Populus*, heathland and alpine grassland.

28 Soil Survey Staff, *Soil Taxonomy*, USDA Handbook No. 436, 1975.
29 Muir, A., *Adv. Agron.*, 1961, **13**, 1.
30 Dudal, R. *et al.*, *Soil Map of the World*, Vol. 1, *Legend*, FAO–UNESCO, 1974.
31 Bullock, P. and Clayden, B. in *Soils with Variable Charge* (ed. B. K. G. Theng), New Zealand Soc. Soil Sci., 1980.

Characterization

Typically the podzol profile has distinct horizons. Under many types of vegetation, and particularly forests, the upper horizon is a mor humus lying directly on a mineral soil (Colour Plate 1). The mor humus may lie directly on the light-coloured eluvial horizon, the Ea, although in many cases there will be an Ah horizon where organic and mineral matter are intimately mixed in a dark-coloured horizon. Below the eluvial horizon lies the zone of illuvial enrichment, the B horizon. The B horizon may be further subdivided into a zone darkened by organic matter, Bh, and a zone of accumulation with little visual evidence of organic matter, Bs. The Bh/Bs sequence is not universal, and podzols are often found with no visual evidence of organic matter accumulation. The relative thickness of these layers is very variable, with no clear relationship between the thickness of the eluvial and illuvial horizons. In many situations the horizons are not entirely horizontal, but rather the B horizon exhibits considerable downward tonguing into the C horizon below. The mechanisms which produce this tonguing are not fully understood, but it appears in some situations to be related to the former presence of tree roots. The total depth of the podzol profile can vary from a few centimetres to over a metre, the relative depths having no necessary relationship with the age of the soil. An example of a podzol profile is shown in Colour Plate 1.

While it is difficult to generalize about the characteristics of the wide range of soils described as podzols, the following gives a broad characterization of the chemical properties:

1 The pH is usually lowest in the mor humus at the surface and increases down the profile, though in some profiles the Ah may be the most acid.
2 The humus layer and Ah horizon often have a relatively high cation exchange capacity due to their high content of organic matter, but the percentage base saturation is low. The percentage base saturation usually increases down the profile, but because of the low exchange capacity of the Ea and below, the amount of exchangeable calcium and magnesium ions may be highest in the humus layer and Ah horizon. The top of the B horizon may also have a high cation exchange capacity relative to its clay content due to the organic colloids which have accumulated there.
3 The Ea horizon is subjected to intense weathering as can be seen from an examination of the sand grains. These consist of minerals very resistant to weathering or, if weatherable minerals are present, they show signs of being strongly attacked.
4 The illuvial horizon is enriched with material that has been translocated from and through the A horizons, namely organic matter, clay, and iron and aluminium hydroxides. Typically their zone of accumulation occurs at the top of the B horizon, but under some conditions there is a distinct zonation with the peak of organic matter accumulation above the peaks

of accumulation of the other components in a zone that is characterized by a very dark colour.

In essence, therefore, the podzolization process involves the translocation of organic compounds, aluminium and iron. In order to understand the mechanisms of podzolization it is necessary to explain the following:

1 The release of the constituents.
2 The mobilization of the constituents.
3 The processes which bring about the translocation of the constituents.
4 The processes and conditions which bring about the immobilization and deposition of the translocated constituents.

In attempting to summarize the current views on podzolization Mokma and Buurman argue that podzolization follows the sequence[32]:

1 Accumulation of organic matter in and on the topsoil.
2 Leaching and progressive acidification.
3 Weathering.
4 Translocation of organic compounds, aluminium and iron in some form of organo-mineral complex.
5 Immobilization of the organo-mineral complexes.
6 Formation of humus pellets in the B horizon.
7 Cementation of the B horizon.

Formation: influence of organic matter complexing

The formation of organo-mineral complexes has long been a suggested mechanism for the podzolization process. Workers in the 1950s suggested that rain washes soluble organic compounds from either living or recently dead leaves of vegetation.[33] These compounds include polyphenols such as di- or epicatechin, which, under acid conditions, will reduce any ferric ions present, either in weatherable minerals or as films of ferric hydroxide on sand grains, and will form water-soluble complexes with ferrous and possibly aluminium ions.[34] Leaf drip from oak and pine trees in North Carolina was found to contain up to 1 kg ha^{-1} of polyphenols which potentially could mobilize up to 1.5 kg ha^{-1} of iron and half that amount of aluminium per year.[35] Plants growing on acid soils low in nitrogen and phosphate, conditions typical of many of the sandy deposits within which podzols are often found to develop, tend to have a higher content of simple polyphenols than those growing on

32 Mokma, D. L. and Buurman, P., *Podzols and Podzolization in Temperate Regions*, ISM Monog., No. 1, 1982.
33 See for example Bloomfield, C., *Nature, Lond.*, 1952, **170**, 540; *J. Sci. Fd. Agric.*, 1957, **8**, 389.
34 Coulson, G. B. *et al.*, *J. Soil Sci.*, 1960, **11**, 30.
35 Malcolm, R. L. and McCracken, R. J., *Soil Sci. Soc. Am. Proc.*, 1968, **32**, 834.

more fertile soils.[36] The following groups of organic compounds and their complexing power have been presented[37]:

Sugars and amino acids: no complexing of Fe and Al.
Aliphatic acids: varying complexing, partly pH dependent.
Phenolic acids and polyphenols: active complexing but subject to microbial breakdown.
Soluble polymers, fulvic acid precursors, polyphenols: main complexing substances in podzols.

The reaction of humic and fulvic acids with bi- and trivalent cations to form complexes has been discussed by Schnitzer and Kodama,[38] who suggested that the complexes were soluble in water when unsaturated with base cations and insoluble when saturated. There has been some discussion on the relative importance of humic and fulvic acids within the podzol. In a leachate collected between the Ea and Bhf of a humic podzol 87 per cent of the organic matter was fulvic acid.[39] Some workers have suggested a zonation within the zone of illuviation, with the upper, Bh, horizon having a predominance of large humic acid molecules, and the lower, Bs, horizon a predominance of small fulvic acid molecules.[40] The solubility of the two types of complex shows marked contrast; fulvic acid can form water-soluble metal complexes at any pH, humic acid only at pH greater than 6.5. Below pH 6.5 humic acid becomes water insoluble, but exhibits sorption properties that lead to the concentration of components on to the solid phase. Fulvic acid shows a greater affinity for Fe^{3+} and Al^{3+} than for other cations, and this has been used to account for the greater dissolution of iron-rich chlorites and micas than iron-poor minerals.[41] There is some suggestion, however, that the preference of the organic ligands for Fe^{3+} and Al^{3+} may change with pH.[32]

The association of podzols with sandy parent materials may in part be explained in relation to the presence of soluble organic compounds. Often non-sandy soil will have appreciable amounts of calcium and other bases which may form insoluble compounds with the organic materials so preventing the mobilization of iron and aluminium. Similarly, soils with a high clay content are likely to have low amounts of water-soluble organic components because these are adsorbed on to the clay surfaces. Consequently the soils which are most likely to be podzolized are sandy soils containing small amounts of basic cations, aluminium and iron and which easily become acid.

36 Davies, R. L. *et al., J. Soil Sci.*, 1964, **15**, 310.
37 Bruckert, S., *Ann. Agron.*, 1970, **21**, 421.
38 Schnitzer, M. and Kodama, H. in *Minerals in Soil Environments* (eds J. B. Dixon and S. B. Weed), Soil Sci. Soc. Am., Madison, 1977.
39 Schnitzer, M. and Desjardins, J. G., *Canad. J. Soil Sci.*, 1969, **49**, 151.
40 Duchaufour, P., *Pedology*, George Allen and Unwin, 1982.
41 Schnitzer, M. and Kodama, H., *Geoderma*, 1976, **15**, 381;   Adams, W. A. *et al., J. Soil Sci.*, 1980, **31**, 533.

In much of the early work on the podzolization process the investigations concentrated on mechanisms by which the iron and aluminium were mobilized, little attention being paid to the deposition in the B horizon. Bloomfield, investigating the deposition in the B horizon, suggested that Fe^{2+} humates were immobilized as sesquioxides were adsorbed.[33] While there is now little evidence to suggest that transport is in the Fe^{2+} form, the immobilization of Fe^{3+} and Al^{3+} in the metal-organic complexes seems to occur because, as they migrate downwards, they take up further iron and aluminium until a 'saturation value' is reached. When sufficient amounts of aluminium and iron are adsorbed a large immobile polymerized organo-metal compound is formed.[42] Alternative hypotheses accounting for the deposition suggest this takes place as a result of desiccation, as a result of encountering a horizon with a different ionic content in the soil solution, as a result of the oxidation of the organic complex, or as a result of the decomposition of the organic component of the complex by microbial action. Some or all of these may have a role in the deposition, but the deposition as a result of the saturation by iron and aluminium seems most plausible.

Formation: the imogolite hypothesis

Recently a radically different hypothesis of the mechanism of podzolization has been proposed.[43] In this hypothesis the mineral imogolite (a poorly crystalline aluminium silicate) is considered to have a major role in the development of the podzol B horizon. It is suggested that aluminium is transported as a silicate complex (proto-imogolite) and is precipitated in a Bs horizon. while the Bh horizon is formed by adsorption of humus on the hydroxide coated surfaces in the Bs horizon. This proposal may provide an explanation for the formation of podzols with Bs horizons containing imogolite, but it is criticized for disregarding the wealth of experimental research which has suggested the importance of metal–organic complexes in the podzolization process. It does, however, suggest a possible explanation for the conclusions of McKeague and co-workers[44] who suggested it was not possible for fulvic acid complexing to account for all the transport of iron and aluminium from eluvial A horizons to enriched B horizons. The pedogenesis of a gley podzol in New Zealand supports the hypothesis of mobilization and deposition of Al, Si and Fe as positively charged inorganic sols to form a Bs horizon, followed by the formation of a Bh above by migration and precipitation of

42 De Coninck, F., *Coll. Int. du CNRS*, 1981, **303**; Sequi, P., *op. cit.*, 1981, **303**.
43 Farmer, V. C., *Coll. Int. du CNRS*, 1981, **303**, 275;   Anderson, H. A. *et al.*, *J. Soil Sci.*, 1982, **33**, 125;   Farmer, V. C., *Soil Sci. Pl. Nutr.*, 1982, **28**, 571.
44 McKeague, J. A., *Canad. J. Soil Sci.*, 1968, **48**, 27;   McKeague, J. A. *et al.*, *Soil Sci. Soc. Am. Proc.*, 1971, **35**, 33.

organic matter on the already precipitated oxides.[45] The development of a gley podzol in this location is partly explained by the longer development time of the profile when compared to many of the podzols so far investigated in the northern hemisphere.

As mentioned above, the nature of the organic matter in the illuvial horizon has been the source of much discussion. Schnitzer,[46] working on a podzol Bh horizon, and McKeague and co-workers,[47] investigating an iron pan of a humic podzol, found most of the organic matter present in both these soils to be soluble in dilute acid and alkali, indicating that it was predominantly fulvic acid. McKeague,[48] investigating humic acid–fulvic acid ratios in several Canadian podzols, found the molar ratios of pyrophosphate extractable (organically bound) iron and aluminium to fulvic acid to be between 3.5 and 5.0 in the Ea horizons and between 6.0 and 9.0 in the illuvial B horizons. These studies in part support the hypothesis that deposition of the material occurs when the complex becomes saturated with adsorbed aluminium and iron. Ponomareva[49] found that precipitation of iron occurred with less organic matter (fulvic acid) than did aluminium, which might in part be an explanation for the apparent zonation in deposits of iron and aluminium in the illuvial B horizon. However, soils in Wales showed no such pattern with aluminium, although there was the typical pattern of eluviation and illuviation with respect to iron.[50] There are other earlier theories to account for the zonation, but no agreement whether there is a single explanation and, if so, what this is.[51]

Rate of development

The time taken for a podzol to develop its typical profile depends on many conditions. Only a few estimates of the required time have been made because the time of initiation of soil forming processes usually cannot be determined precisely. A time of 1000 to 1500 years has been suggested as sufficient for the development of a mature podzol profile based on comparisons of podzols developed on deposits exposed as the Medenhall Glacier in Alaska retreated. This time span has also been suggested for podzol development in Scandinavia. An examination of podzols developed on beach ridges near Sydney, Australia, suggested that a period of 1000 years is required to develop a distinct B horizon.[52] If, however, the development of

45 Farmer, V. C. *et al., J. Soil Sci.*, 1984, **35**, 99.
46 Schnitzer, M., *Soil Sci. Soc. Am. Proc.*, 1969, **33**, 75.
47 McKeague, J.A. *et al., Canad. J. Soil Sci.*, 1967, **47** 23.
48 McKeague, J. A., *Canad. J. Soil Sci.*, 1968, **48**, 27.
49 Ponomareva, V. V., *Theory of Podzolization*, Israel Program for Sci. Translations, 1964.
50 Adams, W. A. *et al., J. Soil Sci.*, 1980, **31**, 533.
51 See for example Tamm, O., *Medd. Skogsförsöksanst*, 1920, **17**, 49.
52 Burges, A. and Drover, D. P., *Aust. J. Bot.*, 1953, **1**, 83.

an eluviated horizon is taken as evidence of podzolization much shorter time spans have been reported. In Scandinavia a 1 cm eluviated horizon was found to develop in 100 years, and in Scotland an eluviated horizon 0.5 cm thick developed in 20 years.[53] Workers in New South Wales, Australia, investigating restored coastal dunes found clear evidence of a bleached eluviated horizon within nine years of the restoration.[54]

Process of lessivage: brown earths, argillic brown earths

Podzols are typically confined to acid quartz sands under mor humus, yet the same vegetation growing on other parent material types will not give a podzol even though it may give a mor humus. One possible explanation why podzols generally do not develop on non-sandy parent materials is that organic acids are adsorbed on to the clay surfaces and are therefore not available to promote the translocation of iron and aluminium. A second explanation may be that the high concentrations of iron and aluminium make any metal–organic complexes insoluble and immobile. The soil that may develop under these conditions if free drainage prevails is often called a brown earth or simply an earth. In its simplest form the earth consists of a straightforward ABC profile, with no marked eluvial or illuvial horizons. The process of formation of the earth has been investigated only scantily. The term *erdefication* has been coined,[55] which describes the suite of processes that result in the formation of (B) and A horizons (excluding the Ah podzol type), as exemplified by brown earths.

In addition to the normal weathering processes considered above, erdefication includes such processes as wetting and drying of colloidal clay-sized particles; the wedging and burrowing of plants and animals; and other processes that result in the formation of structural peds, pores and cracks. Also included are the humification of organic matter and the physical and biological mixing of organic matter with the mineral matter. In terms of the four processes outlined by Simonson, the processes of transformation are dominant. The (B) horizon is in many respects similar to the cambic B horizon widely used in soil classification. The cambic horizon is an altered horizon: physical alteration is the result of movement of soil particles to such an extent as to destroy most of the original rock structure, or to aggregate the soil into peds or both; chemical alteration is the result of weathering processes and the redistribution or removal of carbonates, iron oxides and other soluble constituents. The cambic horizon has few distinctive features

53 Muir, A., *Forestry*, 1934, **8**, 25.
54 Paton, T. R. *et al., Nature, Lond.*, 1976, **260**, 601.
55 Conacher, A. J. and Dalrymple, J. B., *Geoderma*, 1977, **18**, 1.

and there is a tendency to describe it by referring to the absence of features. Because of its definition pedological features in the cambic horizon are poorly expressed.[56]

In some circumstances the cambic horizon is considered to represent a first step towards more strongly developed horizons, such as the eluviated albic or the clay enriched argillic horizon. Similarly the earth is often considered a precursor to the leached earth and argillic earth. The process of progressive leaching of soluble constituents followed by the dispersion and translocation of clay is widely known by the French term *lessivage* and the resultant soil the *lessivé*. The term *lessivé*, however, has a wide meaning, and may refer to, soils with no clear evidence of clay translocation.

Translocation of clay

The translocation of clay from A horizons to accumulate in the B horizons, which gives rise to a textural B horizon, has long been the focus of pedo- logical investigation. The introduction in recent soil classifications of the argillic horizon and its use as a major diagnostic horizon particularly in the widely distributed Alfisols and Ultisols has further focused attention on the nature of the horizon and the processes which produce it. In many respects the process may be considered in the same broad framework as used for podzolization. The clay must first be mobilized, then translocated from one part of the soil to another and, third, immobilized and deposited. Although clay translocation may take place in sub-optimal conditions, only when optimal conditions prevail for all three sub-processes will clearly iden- tifiable argillic horizons be developed.[57] The general process of clay trans- location seems to occur as follows: when rain falls upon a dry soil, some fine material may be dispersed and it may remain in suspension as the water moves downwards through the major voids and macropores. If the subsoil is dry the downward moving water will be absorbed into the soil mass and the suspended clay will be filtered out and deposited on the walls of the voids.

Many additional factors which favour clay translocation may be considered. These include[58] alternate wetting and drying of the soil, a system of macro voids within the soil, absence of cements such as sesquioxides and carbonates, pH between 4.5 and 6.5 (together with low exchangeable Al and no excess Ca and Mg) or high pH associated with high exchangeable Na levels, and a point of net zero charge different from soil pH (a PNZC 0.5 to 1 unit lower has been suggested). These factors determine the occurrence and intensity of the process of clay translocation. Soluble organic matter may contribute to the

56 Aurousseau, P. in *Soil Micromorphology*, Vol. 2 (eds P. Bullock and C. P. Murphy), A. B. Academic, 1983.
57 McKeague, J. A. in *Soil Micromorphology*, Vol. 2 (eds P. Bullock and C. P. Murphy), A. B. Academic, 1983.
58 Eswaran, H. and Sys, C., *Pedologie*, 1979, **29**, 175.

dispersion of clay, and leaching of soluble salts and carbonates may leave voids that favour local movement of clay and more rapid flow of soil water. The type and size of the clay fraction frequently determine the ease with which the clay is dispersed and the extent of the development of the argillic horizon. It is generally considered that fine clay (<0.2 μm) moves preferentially. Similarly clays with high charge (e.g. montmorillonite) accumulate preferentially in argillic horizons. The medium for translocation of the clay is water, and an ample supply is necessary for the development of argillic horizons. However, an excess of water is not necessarily conducive to the accumulation of clay in the B horizon.

Flocculation has been suggested as a mechanism to account for the accumulation of clay in the B horizon. It is suggested that flocculation may take place if there is a slight increase in the electrolyte content of the soil solution or a contrasting electro-chemical environment (e.g. a calcium rich sub-stratum may promote deposition). As the pH of the B horizon, with a lower amount of organic matter is generally close to the point of zero net charge, this may also induce flocculation.

Once formed, argillic horizons may persist long after the time when the environmental conditions which favoured their development have changed. The soil classification currently in use in England and Wales[59] distinguishes between argillic soils (contemporary or near-contemporary development of the argillic horizon) and palaeo-argillic soils (development of the argillic horizon during the Plio-Pleistocene). The main differences are in the amount and position of the illuvial clay, the palaeo-argillic characteristically has redder colours and a more complex microfabric than does the simple argillic horizon. The palaeo-argillic horizon may have undergone several changes in climate and may be impregnated with carbonates since the original migration of the clay or may be partly destroyed by weathering.

Finally, it is important that a textural contrast within the profile is not taken alone as sufficient evidence of clay translocation. There may be peaks in clay content in the B horizon for reasons other than the translocation of particulate clay, including sedimentary layering in initially non-uniform parent material, formation of clay in place by differential weathering, and the synthesis of clay from components that moved downward in solution. Further, the absence of clay skins may not indicate the absence of clay translocation. Many processes within the soil bring about the disruption of clay skins, including many of the processes considered under the heading erdefication, including wetting and drying of the soil and resulting expansion and contraction.

59 Avery, B. W., *Soil Classification for England and Wales*, Soil Surv. Tech. Monog., No. 14, 1980.

Grassland soils: prairie soils and chernozems

Grasses transport much of the organic matter they synthesize into their roots and hence into the soil. Grasses can add over 2.5 t ha^{-1} dry matter to the soil per year as roots, and natural grassland soils may contain over 12 t ha^{-1} of roots below ground compared with only 2 to 5 t ha^{-1} of above-ground material. Weaver, and his co-workers[60] gave the following weights of organic matter and roots in a Nebraska prairie, in t ha^{-1}:

	Depth in cm				
	0–15	*15–30*	*30–60*	*60–90*	*90–120*
Weight of roots	6.5	1.8	1.57	0.75	0.10
Weight of organic matter	77.4*	65.0	65.0	20.2	10.0

* This figure includes the rhizomes of grasses which occurred in the top 15 cm.

The difference between grassland and forest is also illustrated in Plate 5.1, which shows the profiles, with the distribution of organic carbon down the profile, of a prairie grassland and a deciduous forest soil both on the same parent material, and both from Iowa. Although the prairie soil only contains 50 per cent more organic matter than the forest, it is distributed through a considerably greater depth of soil.

Grass roots are extensive and may have a very great length in the soil; many of the finer roots decompose quickly. The soils typically have a large population of earthworms, and also rodents, which may make extensive burrows and nests in the soil well below the soil surface. Hence little litter accumulates on the surface, most of it being worked into the top layers of the soil, as in the earths.

Grassland vegetation is found in conditions that are unsuited to forests, though many grassland areas of the world may be a consequence of human activity. They are found above the tree line in many mountainous areas, and very extensively in continental areas where summer drought is sufficiently severe for all the available water in the root zone of the soil to be exhausted before the drought breaks. Grasses can withstand these conditions better than deciduous forests in areas of summer drought and winter frosts.

Grasslands occur on many types of soil, but there is one group developed on medium textured, often calcareous, superficial deposits of Pleistocene age widely distributed throughout the continental areas of North America,

60 Weaver, J. E. *et al., Bot. Gaz.*, 1935, **96**, 389; Weaver, J. E. and Zink, E., *Ecology*, 1946, **27**, 115 gives further figures for the annual rate of production of roots by three prairie grasses at different depths in the soil.

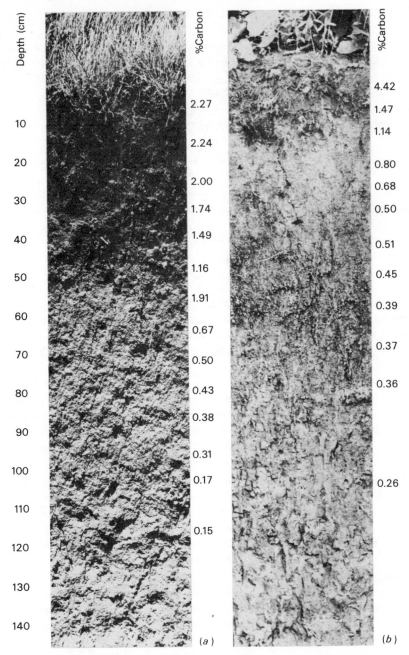

Plate 5.1   The profiles and carbon contents of two soils on a silty loam in Iowa:
(a) Soil under grassland
(b) Soil under hardwood forest. Weight of organic carbon in top 1.2 m
of soil: (a) 170 t ha^{-1}, (b) 105 t ha^{-1}.

Central and Eastern Europe and Central Asia, which have been much studied. The soils under grass in these areas have been divided into prairie, chernozem and chestnut soils.

Typical prairie soils are formed where the summer rainfall is only just too little for deciduous forests, but where there is sufficient rainfall during the rest of the year for considerable leaching to take place. The soils lie in a zone transitional to the leached soils of the cool temperate climates where soils with lessivage and podzols (Alfisols, Inceptisols and Spodosols) predominate. They have been described as leached chernozems.

The chernozem soils are developed under rather more arid conditions where little water leaches out of the root zone, and the chestnut soils under much more arid conditions where the rain water rarely if ever leaches through the profile. Characteristically all these soils have an organic surface horizon with a well-developed fine and friable crumb structure with considerable water stability. In *Soil Taxonomy* it is this property which has been selected as characterizing the diagnostic horizon of the soil order Mollisols, and it is in this order that most of the soils of the grasslands of northern continental land masses of North America and Russia are placed. The mollic epipedon (epipedon is the surface horizon) is defined in terms of its morphology rather than its genesis. It is a mineral horizon, relatively thick, dark-coloured, humus-rich horizon in which divalent cations are dominant on the exchange complex and the structure is moderate to strong. Pedogenetically the mollic epipedon is thought to be formed mainly by subsurface decomposition of organic residues in the presence of divalent cations, particularly calcium. The residues that are decomposed are partly roots and partly organic residues from the soil surface that have been taken underground by animals. Accumulation and turnover of the organic matter in the mollic epipedon are probably rapid.

The prairie soils are in many respects similar to the earths and argillic earths of the more humid areas. The surface layers are often somewhat acid due to leaching, there is usually a washing down of organic matter and clay, and clay formation takes place within the profile as a consequence of weathering. As in the argillic earths in the deciduous forest zone, a pronounced clay pan may be formed both by clay being translocated down the profile and by clay being formed *in situ*. Depending on the degree of development of the argillic horizon and the presence of the mollic surface horizon, they may be classified as Boralfs and Udalfs (argillic horizon present, poorly expressed mollic epipedon) or Mollisols. These soils are well developed in the corn belt of the middle western states of America. They are fertile, or can be made so by adding fertilizers. The rainfall is adequate and if the land is well managed it yields excellent crops of maize and soyabean.

The chernozems, or black earths, to give them the English translation of the Russian word, are so called because of their black colour. The order Mollisols of *Soil Taxonomy*, broadly encompasses most of the soils previously described as chernozem. The typical black layer may exceed 1 m in depth, though the colour lightens with depth. The climate under which these soils develop is moist enough to leach some calcium from the surface layers, so the surface may be slightly acid. Some of the calcium is precipitated lower down as calcium carbonate. If the climate is just sufficiently arid for it to form, the zone of calcium carbonate accumulation begins at about 2 m, and it is present as thin threads or films, which have been called pseudomycelia, because of their resemblance to fungal mycelia. As the climate becomes more arid, the depth at which these begin to form becomes less, and more of the calcium is deposited as soft white concretions. Below the layer of calcium carbonate accumulation under more arid conditions comes a zone in which calcium sulphate is precipitated as gypsum.

The chernozems are well developed on the calcareous loess that extends from Central Europe through Russia into Central Asia, forming the great belt of the Eurasian steppe. They form the short grass prairies of North America which are developed on Pleistocene loess and associated deposits and are well developed in parts of South America. Their main agricultural use is for the cultivation of wheat.

The chestnut soils are developed under still more arid conditions than the chernozems. Their name describes their colour, the colour of the skin of the edible chestnut, *Castanea sativa*. In *Soil Taxonomy* many of the soils previously called chestnut are found in the Mollisol order, chiefly in the suborder Ustolls, characterized by the ustic moisture regime. This moisture

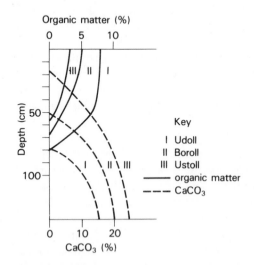

Fig. 5.1   The distribution in depth of organic matter and calcium carbonate in three Mollisols. (Adapted from Duchaufour, P., *Pedology*, George Allen and Unwin, 1982.)

regime produces conditions where the potential transpiration exceeds the rainfall, and at no time of the year does water leach out of the profile. The effect of greater lack of water is to give a grass vegetation with a shorter root system and a smaller annual production of organic matter, so that the depth to which the organic matter reaches, and its content at each depth, decreases. At the same time the layer in which the calcium carbonate and gypsum are deposited comes closer to the surface. Figure 5.1 illustrates the change in organic matter and $CaCO_3$ content and distribution in the profile as the soil moisture regime changes from udic to ustic.

The effect on these grasslands of increasing temperature is three-fold: their organic matter content decreases; their colour reddens; and the calcium carbonate deposits in the subsoil harden. These last two characteristics are also associated with increasing age of the superficial deposits in which they are formed.

Effects of anaerobic conditions

Anaerobic conditions are created when soils become waterlogged. A distinction can be made between soil horizons which are periodically waterlogged and those which are permanently waterlogged. The stress on hydromorphism differs between classification systems, there being three main trends: one approach is exemplified by the classification of the Federal Republic of Germany,[61] which considers hydromorphism as a primary subdivision, distinguishing three main subdivisions: terrestrial, semi-terrestrial (hydromorphic) and sub-aquatic. The second approach is that followed in the classification of England and Wales[59] and FAO/UNESCO[30] where a special class or classes with hydromorphic processes are differentiated. The final approach, exemplified by the USDA *Soil Taxonomy*,[28] considers hydromorphism as a secondary process superimposed on the main pedogenetic process, which are used alone to characterize the primary level of Soil Order. There is no hydromorphic order, but hydromorphic suborders are differentiated within seven of the orders and indicated by the prefix *aqu* (for example aquent and aquod).

Hydromorphic soils

Hydromorphic soils are characterized by an excess of soil water at least for short periods of time. The soil processes which operate under these conditions are called *gleying processes*. Such soils with impeded drainage or with

61 Mückenhausen, E., *Pedology Spec. Ser.*, 1965, **3**, 57.

a seasonal water table within the rooting zone of the vegetation will have zones in which reducing conditions arise. These conditions are induced by water saturation, but only if organic matter contents and soil temperature allow microbial activity. Reduction may take place if the following conditions are met: (i) presence of organic matter; (ii) absence of an oxygen supply; and (iii) presence of anaerobic micro-organisms in an environment suitable for growth. The requirement that an oxygen supply is absent is fulfilled if the soil becomes saturated with water because dissolved oxygen is rapidly consumed by aerobic organisms and the diffusion rate of oxygen in water is very low. The effect of anaerobic conditions on plant growth, and to a large extent on soils, is discussed in Chapter 26, so only their effects on the soil profile and soil forming processes will be summarized here.

The sequence of events that take place with the onset and persistence of anaerobic conditions is likely to be as follows: (i) reduction of remaining oxygen; (ii) reduction of nitrate then manganese (neutral soils) or reduction of manganese then nitrate (acid soils); (iii) reduction of ferric iron to ferrous iron. The reduced manganous and ferrous ions are much more mobile in the soil than their oxidized forms. The result is that soil colours become grey or green–grey instead of brown. The rate of oxidation of organic matter is reduced, so causing an increase in organic matter in the soil and especially in the soil surface.

Under temperate conditions soil horizons having a low permeability are likely to be waterlogged or nearly waterlogged during wet periods, but to be aerated during dry warm periods when the vegetation is transpiring actively. Waterlogging may also occur because the water table is near the surface, as is common in soils of low lying areas.

In acid soils dissolved ferrous iron replaces some of the other exchangeable cations during the waterlogged stage. The displaced cations come into solution and are free to leach out when there is throughflow of water, and the solution will contain aluminium ions, ferrous ions and other cations. As the solution reaches an aerated zone elsewhere, ferric hydroxide will form by oxidation of the ferrous iron. Aluminium ions will displace other cations on the exchange complex.[62]

When considering waterlogged soils a distinction is often made between a groundwater gley soil (GWG) and a surfacewater gley soil (SWG).

Groundwater gleys

Groundwater gley soils occur as a result of a high water table. The onset of anaerobic conditions and the presence of organic matter within the saturated zone leads to the reduction of iron and manganese compounds. These may be lost in drainage waters, or may diffuse upwards to be deposited in aerobic zones. A soil which is more or less permanently waterlogged often has a

62 Brinkman, R., *Geoderma*, 1970, **3**, 199.

grey or grey–green colour due to the presence of ferroso-ferric hydroxides, or ferrous sulphate. The presence of large amounts of ferrous sulphide may result in the horizon having a black colour. The chemistry and alteration of manganese are not so clearly established and the importance of manganese in determining the characteristic gley colours and morphology is not fully known.

Above the water table it is likely that the soil is still partially anaerobic. In particular it seems likely that ped interiors will have anaerobic conditions and high concentrations of manganous and ferrous ions. In contrast the ped surfaces in this zone will probably be aerobic and here the ferrous and manganous ions will be oxidized, removed from solution and deposited. At the ped surface, hydrated ferric oxides such as lepidocrocite or goethite are formed. This gives rise to the mottled appearance characteristic of many horizons which are intermittently waterlogged. The contrast between ped interiors and exteriors is shown by the results from analysis of samples of a groundwater gley soil near Reading, England.[63] The information below is for total and free (mobile) iron from peds approximately 2 cm in diameter.

	Total iron	Free iron
	$g\ Fe\ g^{-1}\ Soil$	
Grey ped centres	0.027	0.0085
Brown ped coatings	0.15	0.13

Most of the free iron had apparently migrated to the ped surfaces by diffusion.

The water table is rarely static within the profile; there may be fluctuations over a few hours, but more significant are seasonal variations which result in part of the soil being aerobic during the drier part of the year and anaerobic during the wetter parts. As a consequence there is likely to be a marked seasonality in the movement of iron, with 'flushes' during the wetter periods. The development of the characteristic morphology of grey ped interiors and brown or ochreous exteriors will occur when the soil is saturated and probably during the early stages of drainage. If, however, the soil is dry and is then 'wetted up', the tendency will be for the soil adjacent to the pores on ped exteriors to be subject to anaerobic conditions and the soil in the ped interiors to remain aerobic until saturation is approached, assuming that aerobic conditions were achieved during the drying process. Root channels may contain air when the bulk of the soil is waterlogged and the walls may be stained brown, but if the channels are filled with water the organic matter from the roots causes rapid reduction in the zone around the channels.

63 Rowell, D. L. in *The Chemistry of Soil Processes* (eds D. J. Greenland and M. H. B. Hayes), Wiley, 1981.

The actual pattern of oxidizing and reducing conditions in the soil is exceedingly complex, varying not just with the position of the water table but also with the size of peds, presence of pores, both macro and micro, and the distribution of organic matter and organic products. As a consequence the morphology of gleyed soils is often complex with the matrix colours and mottle pattern and colour showing considerable variability.

Surfacewater gleys

Surfacewater gley soils occur where drainage is impeded within the soil above an impervious or slowly permeable horizon. Gleying again results from the onset of anaerobic conditions within the soil. Where waterlogging results from slowly draining pores, the ped surfaces may become anaerobic while the ped interiors remain aerobic, particularly if the water contains reductants from the surface organic layers of the soil or from the litter or living vegetation. The resulting soil morphology reflects the reducing conditions which prevail at the pore/ped surfaces and the oxidizing conditions of ped interiors, giving grey ped exteriors and brown ped interiors. In the A horizons of both groundwater and surfacewater gley soils, grass roots are frequently found surrounded with iron in the form of rusty sheaths. Ferrous-rich waters move to the roots during wet periods to provide the transpiration needs of plants. As the soil dries the ferrous iron is oxidized and deposited. In surfacewater gley soils, where the gleyed horizon is underlain by an aerobic region, the iron and manganese may be oxidized and precipitated. In general the iron tends to be precipitated out first with the manganese below.

The distinction between surfacewater gley and groundwater gley is often not clear cut. Many soils are influenced both by a high groundwater table producing a waterlogged lower profile, and if the upper horizons are of restricted or only slow permeability waterlogging may develop in the upper horizons also. The balance between the degree of waterlogging, the soil's permeability and structure, and the nature of the organic matter present will influence the morphology of the soil.

A feature of some gleyed soils is the occurrence of nodules of iron and manganese, often in the form of mixed iron–manganese oxide. These may be formed around the nucleus of another mineral particle such as a quartz grain and are built in concentric layers.[64] The mechanism of movement of iron and manganese to the nodule may be the same as that suggested for the migration from ped interior to ped surface, but why the nodules occur in the particular location is not clear. It has been suggested that there may be areas which are locally well aerated while the surrounding soil is saturated.

The soil mottling often found in gley soils has been widely used to indicate

64 Furukawa, H. *et al., South East Asian Studies,* 1976, **14**, 365.

soil moisture regimes. For example the Soil Survey of England and Wales gave a division into five drainage classes with associated profile morphologies.[65] This scheme has been superseded but it still gives a general indication of the manner in which the relationship has been used. A study in the Thames Valley, England, endeavoured to establish similar relationships.[66] These relationships are in some cases strong, but in other cases it is apparent that the mottling reflects not a current but a relict moisture regime, which may have been associated with a former wetter climate, or conditions prior to artificial drainage. Furthermore, mottling does not always form in saturated soils. There may be a lack of bacterial substrate, the original parent material may have been low in iron, or iron may have been completely leached from the profile. Moisture regimes and mottles are not directly related, but they are coupled by means of a complex, interacting soil forming process. Furthermore, mottling alone is not sufficient to identify the differences between the conditions which bring about waterlogging, for example differences between groundwater gley, surfacewater gley and pseudogley.

Leached tropical soils

For many years it was considered that soil development within the tropical lands produced a suite of soils exclusive to the area. As more knowledge is gained about the tropical zone it becomes apparent that within it are probably found representatives of all soils normally associated with regions outside the tropics, but in addition there are groups of soils which appear to have no extra-tropical representatives. Further, it is no longer thought that the processes which prevail in the tropics are substantially different from those in temperate zones. The differences that do exist are in rates of operation of these processes and their duration.

Formation

The rate or intensity of weathering is often much greater in the tropics because the soil temperatures are higher than in higher latitudes, and furthermore there is generally far less seasonal variation in temperature. The intensity of the weathering is therefore dependent upon the length of time the regolith volume is moist; and the fate of the products of the weathering processes depends upon the intensity of leaching. Some French workers have distinguished 'total hydrolysis' as weathering specific to tropical soils in the absence of organic acid anions.[40] This, they argue, produces a distinct tropical

65 Corbett, W. M. and Tatler, W., *Soils of the Beccles District*, Soil Survey of England and Wales, 1970.
66 Moore, T. R., *Geoderma*, 1974, **11**, 297.

weathering process in which the end products of the process are similar, no matter what the primary minerals were. This 'total hydrolysis' favours the elimination not only of the basic cations but also silica which is considered to be almost as mobile. On the other hand the oxides of iron and aluminium are only slightly mobile and accumulate. Provided the water table is deep, weathered regoliths on old land surfaces may be devoid of minerals to a considerable depth. Results from Kenya[67] on the weathering of basic igneous rocks, suggested that in the early stages of weathering, smectites were produced. As weathering proceeded the smectites were decomposed due to the lowering of silica and magnesium concentrations in solution because of the free drainage and rapid leaching favoured by an extremely well-aggregated and permeable subsoil. The result was to give an acid soil with a clay fraction dominated by kaolinite and some gibbsite.

The nature of the original material will obviously greatly influence the nature of these weathering processes. In general terms basic rocks weather to kaolinite and gibbsite via montmorillonite, acidic rocks weather directly to kaolinite and eventually gibbsite. The drainage within the regolith is also likely to influence the weathering products: under poor drainage the loss of silica is slowed down and clays are synthesized. Under free drainage silica loss is unimpeded, bases are readily leached giving rise to a low pH and kaolinite is the dominant clay mineral.

An important difference between soils in temperate regions and in many tropical regions is that soils on the tropical land surfaces were not disturbed by ice sheets or periglacial action in the last Ice Age. In many parts of the temperate zone, weathering and soil formation recommenced at the conclusion of the last glacial stage, 10 000 to 20 000 years ago, on land surfaces glaciated or modified by periglacial processes. While there is evidence of fluctuations of climatic belts within the tropics during the Pleistocene, renovation of parent materials did not take place widely within the tropics. It is possible that some of the existing surfaces in Tanzania are of Jurassic age, but many are certainly Pliocene and some probably early Miocene.[68] The result is that weathering may have been in progress, more or less uninterrupted, for many millions of years, under a variety of climates so that not only may the weathered layers be very deep, but the relative levels may have been subjected to substantial changes due to tectonic activity and partly due to changes in the base level to which rivers are grading. As a consequence, the weathered mantle lying on a hilltop at the present time may have been formed in a valley bottom; and land that was a plain in which deep weathering was taking place may now have a variety of levels due to rifting, faulting and mountain building on the one hand and river terracing on the other.

67 Kantor, W. and Schwertmann, U., *J. Soil Sci.*, 1974, **25**, 67.
68 Stephens, C. G., *Geoderma*, 1971, **5**, 5.

Consequently it is only possible to interpret the conditions which led to the present soil and regolith cover correctly if the geomorphological history of the area is known. In some cases it may be that the soil features which are seen do not relate to the present conditions, but are relict features related to different combinations of soil forming factors from the past; for example soils are often related to substantially different climatic conditions which prevailed in some regions during the Quaternary.[69] In contrast there are situations within the tropics where soil development has been of particularly short duration, for example where volcanic materials have been recently deposited. Often however, the rate of soil development upon these recently deposited materials is extremely rapid.[70]

Classification

While it is probably true that there are representatives of all soil groups within the tropics, there are broad groups of soils which seem to be restricted to the tropical zone, or have developed under what was previously a tropical environment. This broad group of soils was described by Kellogg as Latosols.[71] According to Kellogg these zonal soils of the tropical and equatorial region have as their dominant characteristics: (i) a low silica : sesquioxide ratio in the clay fraction; (ii) low base status; (iii) low activities of clay; (iv) low content of most primary minerals; (v) low content of soluble constituents; (vi) a high degree of aggregate stability, and (vii) probably some red colour. In this broad framework it is possible to include most of the red and yellow soils of the tropics.

Soil classification in the tropics is at a relatively early stage of development and there is no consensus on the relative importance of soil properties for class separation. Many of the soils broadly described as Latosols would be allocated to the soil orders Oxisols, Ultisols and to a degree Alfisols in *Soil Taxonomy*, but many problems have arisen in attempting to allocate soils to these classes. The system of classification produced to accompany the *Soil Map of the World* by FAO–Unesco is widely used internationally. A third scheme was produced in 1964 by the Commission for Technical Co-operation in Africa (CCTA)[72] to accompany the *Soil Map of Africa*. This third scheme, using terminology widely used by workers from a number of countries, identifies two broad groups of soils: ferruginous soils and ferralitic soils, and associated with these two soil groupings it is convenient to discuss the broad features of the soil forming processes.[73]

69 See for example Isbell, R. F. and Field, J. B. F., *Geoderma*, 1977, **18**, 155.
70 Colmet-Daage, F., *Cah. ORSTOM Ser. Ped.*, 1967, **5**, 353.
71 Kellogg, C. E., *Commonw. Bur. Soil Sci. Tech. Comm.* No., 46, p. 76, 1949.
72 D'Hoore, J., *CCTA Pub.*, 1964, 93.
73 Young, A., *Tropical Soils and Soil Survey*, Cambridge Univ. Press, 1976.

Processes

The first soil forming process, *ferrugination*, consists of a suite of individual processes: chemical weathering of moderate to high intensity, complete leaching of soluble salts, moderate leaching of bases, partial leaching of silica, and release and dehydration of hydrated iron oxides and their deposition as coatings on other soil constituents. Occasionally iron oxides are deposited and may harden as concretions. Kaolinite forms with some goethite and haematite, small amounts of illite and montmorillonite, but not gibbsite. There is generally some clay translocation and the formation of a blocky structure within the B horizon, which may contain clay skins together with iron oxide (ferriargillans).

Soils subject to this soil forming process are often situated within the region of the tropics subject to marked wet and dry seasons, broadly referred to as savanna. Depending upon the intensity of the leaching, the nature of the original parent material and degree of weathering, the base status of these soils is variable. The mineralogy of the soils reflects the loss of silica from 2 : 1 clay minerals and the possible synthesis of kaolinite in the B horizon.

The second process, *ferralitization*, is the product of intense weathering with the almost complete breakdown of all minerals except quartz, together with complete leaching of soluble salts, strong leaching of exchangeable bases, partial leaching of silica and to a lesser degree iron and aluminium. Kaolinite is formed with some goethite and gibbsite, but no, or very little, montmorillonite is present. Clay translocation is rare. A fine microstructure stabilized by iron often develops, referred to by some workers as pseudosands.[74]

It is often suggested that these soils form the zonal soil of the tropics, being associated with the moist tropical rainforest. Soils with similar properties have been identified beyond the boundaries of the humid tropical zone, distinguished by Young[73] as weathered ferralitic soils, and are considered to be extremely old soils. Young describes the zonal soil of the humid tropics as leached ferralitic soil.

A fundamental difference between the ferruginous and ferralitic groups of soils is the frequent presence of a zone of clay translocation in the former and its general absence in the latter. This contrast has been variously accounted for by the stabilizing effects of the iron and aluminium and the low pH of the ferralitic soil being close to point of zero charge and hence not favouring dispersion of the clay. The contrasting climatic conditions under which the two soils are found, in particular the absence of a sharp seasonal contrast between wet and dry seasons in the case of the ferralitic soils, have also been suggested as possible explanations for the absence of a horizon of clay accumulation in these soils. A number of workers have suggested a

74 Gallez, A. *et al.*, *Soil Sci. Soc. Am. Proc.*, 1975, **39**, 577.

chronosequence of soil development in which soils would develop through ferruginous to ferralitic soil.[75] Such a development sequence necessitates the development of an argillic B horizon during ferrugination and its destruction and disappearance during ferralitization.

A similar degradation sequence is discussed by Eswaran and Sys,[76] that of the change from an Ultisol (ferruginous soil) to Oxisol (ferralitic soil). They hypothesized that the horizon of accumulation is progressively moving downwards because: (i) clay translocation features are being removed from the upper part of the profile; (ii) deposition may be occurring deeper within the profile (this may vary considerably within the landscape and as a consequence Ultisols and Oxisols may occur in close association).

One feature often widely associated with the leached tropical soils is the red or reddish colours. In general the colour of soils in hot climates is much more intense than that found in temperate climates, soils often being either strongly ochreous or red in colour. The oxides of iron and aluminium are among the major components of soils in the tropics. These oxides exist chiefly as amorphous and crystalline inorganic forms. A small proportion of the iron and aluminium is also present in the soils in the form of organic complexes. The large amount of iron present in tropical soils is in part explained by the intense weathering and the fact that the released iron and aluminium oxides are not removed by organic complexes in drainage waters. It has been suggested that the colour is intense because the processes of crystallization of the free iron oxides are not prevented by the presence of free organic matter as in many temperate soils.[77] In soils of the tropical zone iron oxide occurs in crystalline forms, with goethite giving the ochreous colour and haematite the red colour. Further, goethite is associated with gradual crystallization under a humid climate, and haematite is associated with rapid crystallization, most probably under seasonally arid conditions.

Plinthite (laterite)

Laterite is a hard material, rich in iron oxides. This hardness is retained even when the material is immersed in water. The iron occurs mainly as goethite, haematite and amorphous iron oxides. The material is usually coloured reddish brown with a moderately high density (2.5 to 3.6) and usually contains secondary aluminium. The silica content is generally low, but some quartz and sometimes kaolinite is present. Laterite often occurs on remnants of old land surfaces. A soft but similar material is less widespread and called soft laterite. The word laterite was first coined by the geologist Buchanan, in 1807,[78] for a particular type of iron-rich material which was quarried and

75 Sys, C., *Pedologie*, 1967, **17**, 284.
76 Eswaran, H. and Sys, C., *Pedologie*, 1979, **29**, 175.
77 Schwertmann, U. *et al., Trans. 10th Int. Congr. Soil Sci.*, 1974, **6**, 232.
78 Buchanan, F., *A Journey from Madras through the countries of Mysore, Canara and Malabar*, East India Co., 1807.

cut up in the shape of bricks; these were exposed to the weather for some time which caused them to harden, and then used for buildings. He derived the term laterite from the latin word *later* (a brick). The soft laterite described by Buchanan has proved to be far less common than the hardened material, although later descriptions of the type site suggested that 'soft' was probably a misleading description as pick axes were necessary to excavate the material.[79] Regrettably the word laterite has been used to describe a wide range of materials.[80] The term *plinthite* was introduced in 1960 by USDA in its soil classification, and is now described as follows: 'Plinthite is an iron-rich, humus-poor mixture of clay with quartz and other diluents. It commonly occurs as dark red mottles, which usually are in platy, polygonal or reticulate patterns. Plinthite changes irreversibly to an ironstone hardpan or to irregular aggregates on exposure to repeated wetting and drying, especially if it is exposed also to heat from the sun.'

The origin of the iron that has accumulated in laterites is not certain. It could all be residual, left after weathering has removed most of the silica, and perhaps some of the alumina, from the parent material (intense ferralitization), or it could have been brought into this horizon by groundwater either vertically from below or laterally from adjacent soils. The first of these is described as *relative accumulation*, the second *absolute accumulation*. Of the two the first seems unlikely because ferralitic soils of similar composition show no hardening properties; and second, micromorphological evidence of lateritic materials indicates additions of iron.

The most widespread mechanism of plinthite formation is probably precipitation of iron within a zone of groundwater fluctuation. Such fluctuations are produced in the seasonally arid climates of the tropics. This process has been suggested as being broadly analogous with groundwater gleying of temperate lands. When the water levels are high, the iron is held within the waterlogged zone in ferrous forms. When the water level falls iron may be moved vertically downwards, but then oxidized and precipitated. The resulting ferric compounds are relatively insoluble and remain in this form during later periods of waterlogging. After exposure to air it becomes very hard.

A wide variety of lateritic materials has been observed. They are classified on the basis of morphology, composition and position of occurrence in relation to landforms and soil profiles.[73]

Saline soils

Under hot, arid conditions soluble salts accumulate in the surface of soils whenever the groundwater comes within a few metres of the surface. This

79 Stephens, C. G., *J. Soil Sci.*, 1961, **12**, 214.
80 Maignien, R., *Review of Research on Laterites*, Vol. IV, UNESCO Nat. Res. Ser., 1966.

may happen, under natural conditions, in the flood plains of rivers, the low-lying shores of lakes, and in depressions in which drainage water accumulates – in fact, in any region where marsh, swamp or other ill-drained soil would be found in humid regions. The amount of salts that accumulate depends on the salt content of the groundwater and the length of time salts have been entering the region. During dry periods the surface of these soils is covered with an efflorescence, or salt crust, which is dissolved in the soil water each time the soil is wetted.

The soils typically have an uneven surface, being covered with small puffed up spots a few centimetres high that are enriched in salts, for as explained on p. 336 and as illustrated in Plate 5.2, salts concentrate in the most salty areas because these areas remain moist longest after the onset of drought.

The soils normally show no change of structure down the profile, implying that the soil is barely affected by soil weathering and soil forming processes, except that they may show signs of gleying in the subsoil. Such soils are often called solonchaks, a word adopted in Russia for soils on recently deposited calcareous river terraces; Hilgard[81] called them white alkali, but they are now called saline or halomorphic. Usually they are low in humus, because the natural vegetation cannot make much annual growth on them. The salts usually present in the soil are the sulphates and chlorides of sodium and calcium, though nitrates occur in a few places, and magnesium sometimes constitutes an appreciable proportion of the cations. Under these conditions, the pH of the soil is below 8.5 and the soil surface is light coloured. However, under some conditions an appreciable proportion of the salts present may be sodium carbonate which will raise the pH of the soil to 9 or even up to 10. If other salts are only present in small concentrations, this sodium carbonate may cause the humic matter in the soil to disperse and take on a black colour giving the black alkali soil of Hilgard, which will be discussed in the next section.

Saline soils may contain over 250 t ha^{-1} of salt in the top 120 cm of soil, that is, the salts may constitute over 1 per cent by weight of the soil, though many saline soils contain less than this. The natural vegetation on such soils has a very high ash content, up to one-quarter of the air-dry plant may be ash, and the greater proportion of the ash may be soluble salts, typically sodium chloride. Hence, the vegetation will also bring salts to the soil surface, but its effect is probably small, amounting to under 200 kg ha^{-1} annually,[82] owing to the small amount of total growth made per year.

The source of the salts in naturally saline soils is usually the groundwater, which is enriched with salts from two sources. Part, sometimes all, is derived from the weathering of rocks in the upper reaches of the river, and part is sometimes derived from salt deposits laid down in early geological periods

81 Hilgard, E. W., *Soils*, New York, 1906.
82 Kovda, V. A., *Pedology*, 1944, Nos 4–5, 144.

(a)

(b)

Plate 5.2   (a) and (b) Salt soils in the Karun delta, Iran. Note the way the salt efflorescences occur in separate patches.

in strata through which the groundwater moves. But soils in continental areas can also receive soluble salts in the rain, and much of these salts may come from the salt crusts formed elsewhere, which are picked up by the wind after they have become dry and loose on the soil surface.[83] Saline soils have also been produced artificially by faulty irrigation, for irrigation always involves putting salts on the land as well as water. Hence, salt control is a fundamental part of irrigation and will be discussed in detail in Chapter 27.

Sodic soils

When the water table in a natural saline soil falls, so that salts no longer accumulate in its surface, the rain water washes the salts down the profile, and this process sometimes causes considerable chemical changes to take place in the profile. If the salts are predominantly calcium or if, during the process of washing out, over 90 per cent of the exchangeable ions remain calcium, then the saline soil is converted into the steppe or semi-desert soil appropriate to its region.

Many more radical changes in the surface soil take place if the calcium reserves in the soil are so low that, during the washing out of the salts, an appreciable proportion of the exchangeable calcium ions are replaced by sodium. Sodium ions only need to constitute 12 to 15 per cent of the exchangeable ions to reduce the water stability of the soil structure sufficiently for the clay and humic particles to disperse. This harmful effect is accentuated by sodium carbonate being formed in the soil solution during the final stages of the washing out of the salts, causing the pH of the soil solution to rise, often above 9, and consequently increasing the ease of dispersion of the fine particles.

For a long time the source of the sodium carbonate produced during this washing out was not understood, although P. de Mondésir[84] in 1888, and K. K. Gedroiz[85] in 1912, gave essentially the correct explanation. Carbonate and bicarbonate anions are being continually produced in the soil by the carbon dioxide given off by the plant roots and soil organisms, and these anions must be neutralized by cations or hydrogen ions. These cations will be obtained from the exchangeable cations in the exchange complex unless there are reserves of calcium carbonate in the soil. Hence, if there is an appreciable proportion of exchangeable sodium ions in the soil, enough will come into the soil solution to give what is in effect a solution of sodium carbonate strong enough to raise the pH of the soil to 9 or over. There is

---

83 For an example from Australia see Hutton, J. T., *Trans. 9th Int. Congr. Soil Sci.*, 1968, **4**, 313.
84 De Mondésir, P., *Comp. Rend.*, 1888, **106**, 459.
85 Gedroiz, K. K., *J. Exp. Agron.* (Russian), 1912, **13**, 363.

also a second process which produces sodium carbonate in soils. If the soil contains sulphates, is low in accessible ferric iron, and is subjected to reducing conditions, the sulphate will be reduced to sulphide which, in the absence of ferrous ions or ions of other metals having insoluble sulphides, will be lost as hydrogen sulphide; and electrical neutrality of the solution will be maintained by the dissociation of carbonic acid to give carbonate anions (see Ch. 27). Reducing conditions will commonly occur in these soils as soon as their permeability drops due to the salt content dropping.

These conditions of high alkalinity and low salt content lead to the clay and organic matter particles becoming deflocculated and the soil structure water-unstable. The soil surface becomes dark-coloured, often black, due to the dispersed humic particles; the surface typically dries into large, very hard, prismatic units having well-defined edges and smooth surfaces; and clay particles tend to wash down the profile, giving an incipient clay pan. *Soil Taxonomy* recognizes this horizon as a natric diagnostic horizon, a special form of the argillic horizon in which more than 15 per cent of the exchange complex is saturated with sodium ions. Deflocculated soils are extremely difficult to handle, for they are very plastic and sticky when wet and form hard compact clods when dry.

The second stage in the washing out of salts, when there is an appreciable proportion of exchangeable sodium in the exchange complex, is for clay and organic matter to move down the profile into the developing clay pan, with the consequence that the profile becomes banded rather like a podzol. The details of the type of soil developed as leaching proceeds depends on local circumstances, particularly on the soil texture and type of clay present. On the Russian steppes, where these were first studied, the surface soil is dark grey, owing to the deflocculated humus, then comes a pale layer, and then another dark, very compact layer having a very sharply defined upper surface and merging gradually into the subsoil with increasing depth. The darker colour of the compact layer compared with the layer above it may be due to its higher clay content, for it does not always have a higher content of organic matter.[86] The top two layers have lost much of their clay, and have a loose, porous, laminar structure, whose upper surfaces may be paler than their lower, possibly because of silica being deposited on them. The clay pan cracks on drying into well-defined vertical columns having a rounded top and smooth, shiny, well-defined sides as shown in Plate 12.1. These can be broken into units about 10 cm high and 5 cm across with a flat base. Below this the columns break into rather smaller units with a flat top and bottom which on light crushing break up into angular fragments.

As the leaching of these desalinized soils proceeds, the upper two horizons deepen, and often become slightly acid in reaction. Gedroiz, who was one

86 For illustrations of this for well-leached Minnesota soils see Brown, A. L. and Caldwell, A. C., *Soil Sci.*, 1947, **63**, 183.

of the first to give a plausible account of the chemistry of the process, noticed that the content of amorphous silica in these horizons, and particularly in the top darker horizon, increased. Historically these soils in Hungary have been called solonetz in the early stages and solod in the later stages of their development, and they have been extensively studied on the river terraces of southern Russia and Central Asia.[87] Gedroiz assumed that they formed under the influence of exchangeable sodium, as given in the account above, and this assumption has been accepted by many other workers. But there are large areas in western Canada and the United States[88] and in Australia[89] where soils having the morphology of these solonetz and solods are found,

Table 5.7   USDA classification of salt-affected soils. (From Landon, J. R. (ed.), *Booker Soil Manual*, Longman, 1984)

Soil	EC_e $(dS\ m^{-1})$	ESP	pH	Description
Saline soils	> 4	< 15	Usually <8.5	Non-sodic soils containing sufficient soluble salts to interfere with growth of most crops
Saline–sodic soils	> 4	> 15	Usually <8.5	Soils with sufficient exchangeable sodium to interfere with growth of most plants, and containing appreciable quantities of soluble salts
Sodic soils	< 4	> 15	Usually >8.5	Soils with sufficient exchangeable sodium to interfere with growth of most plants, but without appreciable quantities of soluble salts

Note   Although fairly widely accepted, the values for EC_e, ESP and pH should be regarded as indicative rather than as fixed critical values. The effects of increasing ESP, for example, gradually worsen rather than rapidly change soil conditions as a value of 15 is reached. Local experience should be compared with measured values wherever this is possible. The presence of gypsum, in particular, in a soil can mitigate the effects of high ESP values.

87 For some detailed work, see Gedroiz, K. K., *Nosovka Agric. Exp. Stn. Bull.*, **38**, 1925; **44**, 1926; **46**, 1928; Vilensky, D. G., *Salinised Soils*, Moscow, 1924; Kovda, V. A., *Solonchaks and Solonetz*, Moscow, 1937; Joffe, J. S. in his book *Pedology*, New Brunswick, 1938, has given a long summary of Gedroiz's and Vilenksy's work.
88 See for example Kelley, W. P., *Soil Survey Assoc.*, 1934, **15**, 45 (California); Kellogg, C. E., *Soil Sci.*, 1934, **38**, 483 (North Dakota); Rost, C. O. and Maehl, K. A., *Soil Sci.*, 1943, **55**, 301 (Minnesota); MacGregor, J. M. and Wyatt, F. A., *Soil Sci.*, 1945, **59**, 419 (Alberta); Bentley, C. F. and Rost, C. O., *Sci. Agric.*, 1947, **27**, 293 (Saskatchewan).
89 Stace, H. C. T. *et al.* (eds), *Manual of Australian Soils*, Rellim, 1968.

yet where sodium forms a minor proportion of the exchangeable ions. It is possible that originally they contained enough exchangeable sodium for the solonetz–solod morphology to develop in the profile, but that most of this sodium has been lost by leaching. However, most of these soils that are now low in exchangeable sodium contain over 40 per cent, or even over 50 per cent, of exchangeable magnesium. The effect of magnesium in causing dispersion is discussed in Chapter 27.

USDA has classified saline and sodic soils by means of the electrical conductivity and exchangeable sodium percentage of the soil extract (see Ch. 27) and pH. Table 5.7 summarizes the classification.

6

The inorganic components
of the soil

The weathering processes described in the previous chapter produce the mineral fabric of soil. The components of the fabric include mineral grains inherited from the parent material and clays which are derived from primary minerals as weathering products. The proportion of primary minerals, and their size, and the proportion of clays and their mineralogy, play a very large part in determining the physical and chemical properties of a soil, properties that are discussed in later chapters. This chapter gives a description of the size distribution of the particles and the basic structure of the clay minerals and iron and aluminium hydrous oxides.

Texture and particle size distribution

The earliest classifications of soils were based on an assessment of their ease of cultivation and management. Soils containing a high proportion of coarse particles are easy or 'light' to cultivate, usually free draining and never sticky. Soils high in clay on the other hand do not drain well, are sticky and plastic when wet and usually become very hard on drying. The combined effect of these properties, which affect ease of cultivation, is referred to as *texture*, soils being classified in terms like heavy, medium and light. But since these properties are associated with different particle size compositions, they are commonly related to the preponderant particle sizes: clayey, loamy or sandy; or as fine, medium or coarse textured, again referring to particle size.

However, the mechanical and physical properties of soil depend also on properties other than particle size composition. The exact nature of the clays present is important as is the presence and distribution of other inorganic and organic coatings and accretions, so that there is not a very close correlation between 'texture' and particle size composition. Soil texture then has no universally accepted definition. Sometimes it is defined with reference solely to the size distribution of the particles forming the soil, though keeping in mind the farmer's concept of workability and attempting to be as consistent as possible with the farmer's description.[1] This type of classification is illus-

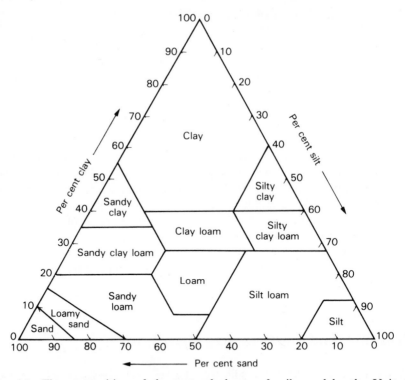

Fig. 6.1   The composition of the textural classes of soils used by the United States Soil Survey. (Sand, 2–0.05 mm; silt, 0.05–0.002 mm; clay, below 0.002 mm).

1 Childs, E. C., *J. Soil Sci.*, 1974, **25**, 408.

trated in Fig. 6.1 in which the textural class of a soil is determined from its particle size distribution, but the boundaries between the classes are chosen so that classes fit as well as possible to the farmer's concept of what properties these classes should have. This particular classification is used by American and British Soil Surveys.

This use of the concept of soil texture as a measure of the particle size distribution in a soil is thus dependent on there being a correlation between textural class as so defined and the farmer's appreciation of the workability of the soil. In general this correlation is good for most of the soils of Western Europe and the northeastern parts of the USA, although it breaks down very badly for many tropical and subtropical soils. It was to improve this correlation that the American soil surveyors chose 50 μm instead of 20 μm for the upper limit of the silt.

An experienced soil surveyor can distinguish about 15 to 20 textural classes from the feel and handling of soils over a range of water contents. Attempts to find objective methods of assessing texture in terms of mechanical properties and behaviour[2] have met with no more success than has particle size analysis, but this is inevitable without a more objective definition of it.

Interest in methods of particle size analysis of soils, and of other materials, has existed for many centuries,[3] but an exact analysis of the finer fractions was not possible until the principles of sedimentation analysis were discovered in the middle of the nineteenth century, after which development was rapid.

It is convenient to consider the procedures used for size distribution analysis (or 'mechanical' analysis as it was formerly called) in three steps. First the initial preparation of the sample, involving gentle grinding and the rejection of stones and debris retained by a 2 mm sieve. With some soils containing partially weathered mineral matter, concretions, chalk and weakly bound aggregates the vigour of grinding and sieving can considerably affect results. The next step, the removal of cementing agents and the deflocculation of colloids, also introduces some arbitrariness into the procedure. Depending on the exact treatment and the soil composition a variable amount of cementing material may be removed, and a varying proportion of the actual clay content may be dispersed. The standard International A method consists of dissolving the more effective part of the organic matter by boiling with hydrogen peroxide, removing calcium carbonate and sesquioxides with dilute hydrochloric acid, washing out of excess acid, and finally addition of dilute sodium hydroxide and shaking for 6 hours to deflocculate the clay. More rapid alternatives include dispersing the soil in 0.5 per cent sodium hexametaphosphate solution using high speed stirring for 5 minutes in a suit-

2 Towner, G. D., *J. Soil Sci.*, 1974, **25**, 298.
3 Keen, B. A., *The Physical Properties of Soils*, 1931;   Krumbein, W. C. and Pettijohn, F. J., *Manual of Sedimentary Petrology*, 1938.

able macerator, when the hexametaphosphate dissolves and complexes some of the di- and trivalent cations present.

For soils containing large amounts of sesquioxide, reducing and complexing reagents are often used. Ultrasonics, with or without other dispersing agents, are also frequently used to disperse soils. Depending on the nature of the soil considerable differences in results can arise following different methods of dispersion.

The final stage, the analysis of the particles into different size classes, depends on the separation of the coarse sand (2 mm–200 μm) particles by sieving, followed by sedimentation of the remainder. Sedimentation analysis is based on Stokes' law (1851) which relates the size of a sphere to the resistance to its motion in a fluid under non-turbulent conditions:

$R = 6\pi r\eta v$, where R is the resistance force, r the radius of sphere, η the dynamic viscosity of fluid and v the relative velocity.

On equating this with the net downward force the sedimentation velocity v becomes

$$\frac{2gr^2}{9\eta}(d_s - d_l)$$

$d_s - d_l$ being the difference in density between the sphere and the fluid. Fortunately most inorganic constituents of soils have rather similar densities, so that an average value of about 2.6 Mg m^{-3} can be used without serious error. Soil particles are not spherical, and some are very far from spherical, but again this is not a significant drawback since the effect of particle characteristics on soil properties is at least as well reflected by their settling velocities or equivalent spherical diameters as by their exact dimensions.

There is an upper size limit of application of Stokes's law at about 80 μm when the velocity of sedimentation of soil is high enough to cause turbulence and the resistance to motion increases more rapidly. Fortunately, separation by sieving is feasible above about 40 μm, though sieves of this size are somewhat fragile and prone to rapid wear. There is also a lower size limit to sedimentation because very fine particles are subject to Brownian movement causing them to diffuse upwards against the direction of settling. This effect becomes serious for particles of about 50 nm diameter, but can be overcome by increasing the settling force using a centrifuge.[4]

For agricultural purposes accurate mechanical analysis has not proved of great value. In pedology it provides important evidence of origins and movement of soil material within the soil profile.[5] The actual size limits chosen for the different fractions and the names given to them have varied from time to time and from country to country. Atterberg[6] originally proposed four

4 Dallavalle, J. M., *Micromeritics: the technology of fine particles*, Pitman, 1948.
5 Robinson, G. W., *Soils: their origin, constitution and classification*, Murby, 1949.
6 Atterberg, A., *Int. Mitt. Bodenk.*, 1912, **2**, 312.

groups: coarse sand, 2 to 0.2 mm; fine sand, 0.2 to 0.02 mm; silt, 0.02 to 0.002 mm; and clay, finer than 2 μm. The division between silt and sand is now more commonly put at 50 μm, and some subdivisions may be made within any of the fractions.

Practical details of methods and precautions are given in many standard texts.[7] The most commonly used sedimentation methods are pipette sampling of the suspension to determine concentrations at chosen depths and times, and the measurement of suspension density at different times with a hydrometer, but many other methods have been described.

Results are conveniently expressed as cumulative percentage curves, by plotting the percentage of particles smaller than a certain size against the size, or more commonly the logarithm of size or of settling velocity. In this way the actual size limits used are not important, having no effect on the curve.

When the main interest is in the clay content or properties associated with clay content, then it may be more directly useful to measure the surface area of the soil constituents by sorption methods rather than measure the particle size distribution. If the interest is in properties relating to cultivation, then measurements of mechanical properties at different water contents may be preferred. In general, particle size distribution analysis is inevitably a somewhat arbitrary exercise, formerly used as an index of many other properties, most of which can and should now be measured directly rather than inferred indirectly.

Mineralogical composition of soil particles

The inorganic constituents in sand, silt and clay are usually crystalline, that is, the particles have a regularly ordered and repeating arrangement of atoms governed by lattice coordination properties. Thus the methods used to investigate structure at this level are most often based on the crystalline properties of the mineral rather than, for example, their wet chemistry. These methods differ to some extent depending on particle size. It is usual to examine sand and to some extent silt size minerals using the optical microscope to view the crystals, but clay and fine silt cannot be seen directly. An electron microscope, used in either the scanning or transmission mode, can give valuable visual detail about the < 2 μm fraction, but generally X-ray diffraction methods yield more information. Amorphous inorganic materials also exist in some soils, and to a considerable extent in Andosols in which a clay size fraction consisting of poorly defined silica–alumina gels can be an indication of the volcanic origin of the soil. The definition of 'amorphous' compounds

7 Soil Survey of England and Wales, *Soil Survey Laboratory Methods*, 1974; Gee, G. W. and Bauder, J. W., in *Methods of Soil Analysis*, Part I (ed. A. Klute), p. 383, Am. Soc. Agron., Madison, 1986.

is to some extent operational, in the sense that the usual X-ray techniques are not sensitive enough to respond to the degree of order which exists in the (sometimes very small) particles in the clay fraction of such soils. Characterization may be partially achieved physically by their thermal properties and chemically by their surface charge.

Sand and silt fractions

These particles can be divided into two main groups: crystalline mineral particles derived from primary rock and rock fragments; and microcrystalline aggregates composed, for example, of calcium carbonate, ferric or aluminium hydroxides, or silica, which have been formed either from products of weathering or from residues of plant and animal life. There may also be present crystals formed in the soil such as calcite and possibly quartz, as well as minerals formed during the weathering process, and perhaps non-crystalline or poorly crystalline residues of the weathering of rock minerals.

The principal minerals found in the silt and sand fraction of the soil are:

1   Quartz.
2   Feldspars.
   (a) Microcline and orthoclase,[8] $KAlSi_3O_8$, which are potash feldspars. Both are resistant to weathering, microcline being the more resistant.
   (b) Plagioclase, a series of mixed crystals having albite or sodium feldspar, $NaAlSi_3O_8$, and anorthite or calcium feldspar, $CaAl_2Si_2O_8$, as end members. The sodium-rich members are about as resistant to weathering as orthoclase, while the calcium-rich members weather more easily.
3   Micas.
   (a) Dioctahedral (a term defined below) micas usually containing no divalent metals: muscovite, a potassium aluminium silicate, $H_2KAl_3(SiO_4)_3$, which is fairly stable.
   (b) Trioctahedral micas, usually containing divalent metals: biotite, a potassium magnesium iron aluminium silicate not very resistant to decomposition. No true micas containing calcium are known, and sodium mica is very rare.
   (c) Glauconite: a potassium mica, relatively low in aluminium but high in ferric iron and containing some magnesium and ferrous iron.
4   The ferromagnesian minerals, which are low in aluminium and are divided into the pyroxenes $(MgFe)SiO_3$, the amphiboles $(MgFe)_7(Si_4O_{11})_2(OH)_2$

---

8 The following formulae are given only to show the type of constitution. The composition of actual specimens may differ considerably from the type formula.

and the olivines $(MgFe)_2SiO_4$, where $(MgFe)$ refers to one ion only. Magnesium and ferrous iron are completely interchangeable in these minerals, and in general there is a certain amount of replacement of either by calcium; while in the amphiboles, and particularly in hornblende, sodium, potassium, calcium and aluminium may replace part of the silicon. These minerals are usually not very resistant to decomposition.

5   Various minerals, such as zircon, garnet, apatite, ilmenite $FeTiO_3$, the iron oxides haematite Fe_2O_3, and magnetite Fe_3O_4.

Igneous or primary rocks are composed of these minerals, and Table 6.1 gives a simple classification of the principal rocks and their mineral constitution.

6   Certain clay minerals which may be present in large particles, such as vermiculites and chlorites, occur in the sand fractions of some soils derived from certain basic igneous rocks. Kaolins are also often present in the silt fraction of soils, particularly if derived from granites. These clay minerals may be cemented by hydrous iron or aluminium oxides or dehydrated iron oxides.

Quartz is by far the commonest mineral in most temperate soils, and also the most resistant to decomposition. In soils derived from sedimentary deposits it often makes up 90 to 95 per cent of all the sand and silt particles. Soils directly derived from primary rock contain much less, the actual quantity depending on the quartz content of the rock itself and the amount of weathering which has occurred. The sand and silt particles in soils derived from basic igneous rocks can also differ from those in nearly all other types of soil in that they can contain large particles of silicate minerals classed as clay minerals, such as vermiculite and chlorite, more usually found in the clay fraction.

As mentioned above, most of these minerals can be identified by the optical microscope, either as separated crystals or when the soil sample has been prepared as a thin section on a microscope slide.[9] In this technique a soil aggregate is impregnated with a setting resin (such as 'Araldite'), fixed to a microscope slide, then the hardened block ground finely parallel to the slide. The section is viewed using light transmitted through the crystals, and has the advantage that the sand and silt sized particles are observed in their natural arrangement to other soil components. The microscope itself is fitted with crossed polarizers which split the light into two opposed planes of vibration. If this light has also been polarized because of the nature of the crystal through which it has passed then, at characteristic angles of the crystal to the

---

9   See for reference, Bullock P. *et al.*, *Handbook for Soil Thin Section Description*, Int. Soc. Soil Sci., Waine Res. Publications, Albrighton, UK, 1985. For applications in soil recognition, *Soil Micromorphology and Soil Classification* (eds L. A. Douglas and M. L. Thompson), Soil Sci. Soc. Am. Spec. Publ. No. 15, Madison, 1985.

Table 6.1 The mineral constitution of the principal igneous rocks

| Size of mineral crystals | | | Constitution | | | | | |
Coarse	Medium	Fine	Quartz	Alkali feldspars	Plagioclases	Micas	Pyroxenes amphiboles	Olivines
Granite	Quartz porphyry / Felsite	Rhyolite	×	×	+	+	o	−
Syenite	Microsyenite	Trachyte	o	×	o	+	o	−
Diorite	Microdiorite	Andesite	o	+	×	+	×	−
Gabbro	Dolerite (diabase)	Basalt	−	−	×	−	×	+
Peridotite } Serpentine }		Picrite-basalt	−	−	o	−	+	×

× plentiful   + less plentiful   o rare   − absent

beam, light extinction will occur (the crystal will appear dark). Quartz has this property most strongly and shows sharp extinctions. Other minerals may be distinguished by their refractive indices, birefringence colours or interference in convergent light.[10]

A knowledge of the amounts of these minerals in a soil is of value for both physical and chemical reasons. The large particles are important structurally and their presence affects permeability. Their size has the corollary that the surface area of sand and silt particles is low, so their value as a chemical adsorbing surface is not generally significant. However, both micas and feldspars have a vital role in the fixation and slow release of K^+, and this is dealt with more fully in Chapter 22. Any of the heavy minerals may be valuable pedogenic indicators especially with respect to parent material, the degree of weathering and in some instances the process of translocation both at the pedon and landscape level.

The clay fraction

The clay fraction is typically differentiated mineralogically from the silt fraction by being composed predominantly of minerals which are formed as products of weathering and which are not found in unweathered rocks. These minerals may occur in particles larger than 2 μm, but are usually present as particles smaller than this. They are much more resistant to weathering in the soil than are rock minerals ground to a comparable particle size, and they comprise the particles that carry the physical and chemical properties characteristic of clays.

The coarser clay fractions, particularly those larger than 0.5 μm in diameter, may contain appreciable proportions of quartz and sometimes of mica, but the fractions finer than 0.1 μm are almost entirely clay minerals or other products of weathering, such as hydrated ferric, aluminium, titanium, and manganese oxides.

At one level, the arrangement of atoms in a clay mineral is fairly easy to describe, and only the simplest outline will be given in this book.[11] This does not mean that a detailed structural understanding is not important, and for some aspects of soil chemistry, for example the thermodynamic stability of

10 For a student text, Gribble, C. D. and Hall, C. D., *A Practical Introduction to Optical Mineralogy*, Allen and Unwin, 1985.
11 More complete structural accounts are given by Brown, G. *et al.*, *The Chemistry of Soil Constituents* (eds D. J. Greenland and M. H. B. Hayes), Ch. 2, Wiley, 1978; Dixon, J. B. and Weed, S. B. (eds), *Minerals in Soil Environments*, Soil Sci. Soc. Am., Madison, 1977; Weaver, C. E. and Pollard, L. D., *The Chemistry of Clay Minerals, Developments in Sedimentology*, No. 15, Elsevier, 1973; Gieseking, J. E. (ed), *Soil Components*, Vol. 2, *Inorganic Components*, Springer-Verlag, 1975.

the crystal, it is vital. What is certainly of general value so far as the soil in relation to plant growth is concerned is to gain an appreciation of surface, as opposed to bulk, features. Of course the appearance of the surface at the atomic level, the nature of the chemical groups and the overall electrical charge, if any, will all be influenced by the mineralogy of the solid, but important generalizations with wide applicability can be made on the basis of a simple model. It will be part of the purpose of this chapter to show how the surface chemistry of soil particles is affected by the solid itself, whilst Chapter 7 deals mainly with the influence of the solution phase on surface properties. Much of soil chemistry is a result of the balance between these two contributors to the interface.

Clay minerals are aluminosilicates, most of which are in the form of platey sheets. Their dimensions in terms of length and breadth may be very large compared to the thickness of the sheets, rather like a piece of paper. The term aluminosilicate is only a general one but correctly gives the information that the common elements in the crystals are aluminium, silicon and oxygen. The latter is the dominant structural element. The O^{2-} ion is relatively large (0.14 nm in radius) so solid oxides are arrangements of more or less close-packed oxygen ions with much smaller cations filling the spaces between the O^{2-} spheres in order to balance the electric charge. The Si^{4+} ion is very small (0.026 nm radius) due to its high positive charge, and can occupy the space between 4 oxygens arranged as a tetrahedron (pyramidally). This provides the extremely stable SiO_4 structural unit which is dominant in rock-forming silicate minerals. The silica tetrahedra are able to polymerize one-dimensionally (to form chains and ribbons – inosilicates), two-dimensionally (to form sheets – phyllosilicates) and three-dimensionally (to form crystals such as quartz – tectosilicates).

So far as clay minerals are concerned, it is the two-dimensional condensation which is of interest. The tetrahedral units polymerize in a plane into an hexagonal arrangement and this, like carbon rings in graphite, can extend in the plane to include hundreds of hexagons and form the so-called 'tetrahedral sheet' (Fig. 6.2).

Each Si also has an O linked to it at right angles to the plane of the page, and this is the O through which the tetrahedral sheet is linked to other structural elements. The Si–O–Si bonds are referred to as siloxane, and a surface featuring this arrangement is known as a siloxane surface.

Diagrams such as Fig. 6.2a mislead in one important respect. In fact the Si ions fit into the small tetrahedral spaces between touching O^{2-} ions, so the space at the centre of the hexagon is much smaller than indicated: it is about the size of an oxygen ion (Fig. 6.2b). This space will be seen to be of significance in the chemistry of siloxane surfaces.

The Al^{3+} ion does not form tetrahedra. The ion is twice the size (0.053 nm) of Si^{4+} and cannot fit into the tetrahedral 'hole' between O^{2-} ions arranged in this way. A larger space is formed when they are arranged octahedrally,

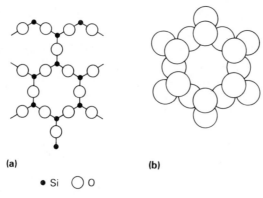

(a)   (b)

• Si   ◯ O

Fig. 6.2 Diagrammatic representation of silica tetrahedra.

that is, six ions at right angles to each other around the cation. The term octahedral is used because an eight-sided solid would be produced by joining the vertices produced by the six ions so arranged. This unit too can polymerize in two dimensions and form an octahedral sheet. So far as Al^{3+} is concerned, only two-thirds of the octahedral spaces need to be filled in order to achieve charge balance (forming a dioctahedral, gibbsite $Al(OH)_3$, layer); however divalent ions such as Mg^{2+} (0.072 nm radius) can also form octahedral layers, and in such a case all the spaces must be filled for charge balance (giving a trioctahedral, brucite $Mg(OH)_2$, layer). Looking edge on (rather than perpendicular as in the Si case above), the polymerization may be represented as shown in Fig. 6.3.

The tetrahedral sheet of polymerized SiO_4 units and the octahedral sheet (whether di- or trioctahedral) are the basic building blocks of most clay minerals. It turns out that it is geometrically possible, though with some distortion and strain in the sheets, for oxygens to be shared between the two types of sheet lying one on top of the other. The oxygens involved are the ones at right angles to the Si sheet and the oxygens linked to the Al and

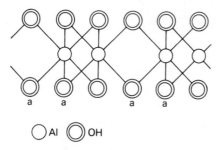

◯ Al   ◎ OH

Fig. 6.3 Diagrammatic representation of alumina octahedra.

marked 'a' in Fig. 6.3. The result is that the O of Si–O becomes the O of Al–OH, producing the link Si–O–Al (Fig. 6.4).

Only three further key ideas are required to gain a general understanding of the most important layer clay minerals in soils. It will be seen from the side view of the octahedral layer that each metal ion has a spare O (not coordinated to the rest of the sheet) both above and below. The tetrahedral Si, though, has only one spare O which is not coordinated fully. This is a vital difference. It means that in addition to the simple linking of a tetrahedral to an octahedral layer (a 1 : 1 clay mineral, a group which includes kaolinite), the latter may also be sandwiched between two tetrahedral layers by the same type of sharing (to form a 2 : 1 clay mineral, a group which includes mica). However, this cannot be vice versa: a tetrahedral sheet cannot be sandwiched between two octahedral layers as the former does not have sufficient spare oxygens to share. This eliminates the possibility of an endless series of alternate octahedral and tetrahedral layers building up thick crystals. The unit layers, often called lamellae, of clay minerals therefore remain thin compared to their areal extent. The thickness of a 1 : 1 lamella is 0.7 nm and that of a 2 : 1 lamella close to 1.0 nm. This limitation on the bonding and hence arrangement of the tetrahedral and octahedral layers is the first key idea.

The second, however, adds a simple refinement by recognizing the importance of hydrogen bonding as a factor affecting structure in lamellar compounds, in addition to full chemical linkage. In the case of the 1 : 1 kaolinite type mineral, for example, the two faces are chemically quite different: one is a siloxane surface and the other is the octahedral sheet surface consisting (for an Al-dioctahedral structure) of aluminol, Al–OH, groups. When two of these lamellae lie adjacent and parallel to one another, the H of the Al–OH in one lamella can to an extent be shared with the O of the Si–O in the other. This is favoured by the geometric similarity of the O arrangement in the two layers (see also above). Although an H bond is

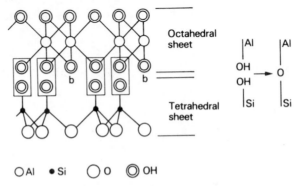

Fig. 6.4 Diagrammatic representation of condensation of octahedral and tetrahedral sheets.

weak compared with most ionic or covalent bonds, a large number will be formed when two extensive (perhaps over an area of several hundred nm^2) lamellae are in contact. So it is that 1 : 1 lamellae can be joined to form stable crystals; indeed it is often observed that the edge (as opposed to the face) area of well-formed kaolinite crystals is around 15–20 per cent of the total. Only very vigorous chemical treatment can disrupt the hydrogen bonds at all rapidly (for example, intercalation can be achieved using very strong salt solutions); 2 : 1 minerals cannot be hydrogen bonded in this manner as the octahedral layer is already sandwiched chemically between two siloxane layers and there are no further OH groups available. Thus the physical and chemical properties of, for example, the vermiculite or smectite group of minerals (see below) are much more dominated by the behaviour of the lamellae than is the case for kaolinites.

There is in fact one way in which H bonds can be involved with 2 : 1 lamellae, though the mineral product is not very common in soils. If an extra layer of gibbsite or brucite is formed (by, for example, precipitation from solution) between adjacent 2 : 1 lamellae then the possibility exists for the octahedral intercalated layer to hydrogen bond firmly into the structure. This produces a so-called 2 : 2 mineral – the chlorite group, but this is best thought of as a 2 : 1 clay with an hydroxide interlayer. Chlorites are still under-investigated as components of soil mineralogy.

One final idea remains to give a general picture of the differences between the major groups of clay minerals. This is the concept of isomorphous (or isomorphic) substitution. Most soil minerals are formed as part of a weathering sequence by the action of aqueous acidic dissolution of primary minerals. The aqueous solution is likely to contain a variety of metal ions, depending on the geochemistry of the environment, and not just the ions (such as Si, Al and Mg) of the idealized structures described above. It has been stated that the metal ions are in the spaces between oxygen ions regarded as simple packed spheres, in other words the structure has been depicted in terms of coordinating properties of ions rather than their valency. During the process of formation of a clay mineral, ions of different (generally lower) valency become substituted into metal positions in the oxygen structure, and this happens because the substituting ion is of a similar size to the 'hole' available, not because it happens to be of the correct valency. It is the geometry of the structure, not its final electric charge (which has no need to be zero) which is the controlling factor. Hence Al^{3+} may substitute for Si^{4+} in the tetrahedral layer (and when it does the structure becomes even more distorted from the hexagonal); octahedral layer Al^{3+} may be substituted principally by Mg^{2+}, Fe^{2+} and Fe^{3+} (of which the last named does not of course change the charge). Numerous other ions, e.g. Mn, Ti, V, Zn, Ni and Cu may also be found, but generally in much smaller proportions. It should be noted that Ca^{2+} (a large cation, radius: 0.10 nm) is not found as a substituting ion.

The fact of isomorphous substitution thus gives clay minerals an overall electric charge, almost always negative, and permanent in the sense that it is a function of crystal composition. The charge on the solid is balanced by exchangeable cations attracted electrostatically from solution to the surface of the clay. It is the 2 : 1 and 2 : 2 minerals for which the permanent charge is marked and 2 : 1 clays are classified by the extent of it. Micas are clays of high surface density of charge, the vermiculite group has intermediate values, while the smectite (montmorillonite) group has the lowest surface charge. Isomorphous substitution seems to be rather variable in chlorites, and of very little importance for kaolinite minerals.

Two final points should be made in conclusion to this section on clay mineral layer structures. The first is that although the description has been in terms of apparently wholly ionic crystals there will be a degree of co-valency in the bonding. This will be most likely in an Si–O link in which, according to Pauling's rules,[12] some 50 per cent covalent character would be expected; Mg–O bonds should on this basis be mostly ionic. Second, it should not be thought that the process of formation of secondary clay minerals involves the simple coming together of preformed layers. The outline given above in terms of tetrahedral octahedral layers and their joining was to give a simple picture of the result, not to describe a process which in reality will involve a complex series of dissolution and re-formation events under close thermodynamic control.

Recognition of clay minerals

X-ray diffraction techniques can be used to determine the repeating distance in a crystal. All the clay minerals are rather similar in their atomic arrange-ment so far as the two dimensions of their area – the a and b dimensions of their unit cells – are concerned, but the main groups have different c axis thicknesses and so it is to be expected that measurement of this dimension (1 : 1, 0.7 nm; 2 : 1, 1.0 nm; 2 : 2, 1.4 nm) should easily distinguish between them. However, the method is much more powerful than this. The repeating distance includes the region between the lamellae and this, for 2 : 1 minerals and swelling chlorites, must include the exchangeable cations held on inter-lamellar surfaces to balance the permanent negative charge of the solid. Experimentally, the amount of swelling of this space may be manipulated in order to distinguish the five main clay mineral groups.[13]

12 Pauling, L., *The Nature of the Chemical Bond* (3rd edn), Cornell Univ. Press, 1960.
13 Brindley, G. W. and Brown, G. (eds), *Crystal Structures of Clay Minerals and their X-Ray Identification*, Min. Soc. (Lond.), Monog., No. 5, 1980: this is a definitive reference book.

Kaolinite group

Kaolinite often forms large, characteristically hexagonal crystals, and an example is shown in Plate 6.1a. Chemically, little or no isomorphous substitution occurs, so there is a silicon in every tetrahedral space and an aluminium in two-thirds of the octahedral spaces. The unit repeating cell is:

$$6O \quad 4Si \quad 4O \quad 2OH \quad 4Al \quad 6OH \qquad \text{as ions in layers, or}$$
$$(Si_4)^{IV}(Al_4)^{VI}O_{10}(OH)_8$$

where ^{IV} indicates the tetrahedral layer and ^{VI} the octahedral layer. It should be noted that because of the space at the centre of the SiO_4 hexagons there is an unreacted octahedral OH group just underneath, and this accounts for the 2OH in the centre of the layer structure above: the unit cell covers two hexagons. Because of the distortion involved in the joining together of the idealized tetrahedral and octahedral layers the hexagonal hole in the siloxane

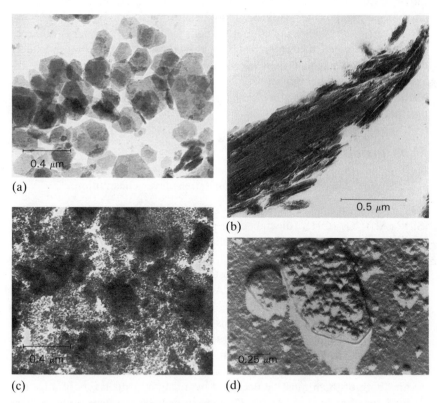

(a)

(b)

(c)   (d)

Plate 6.1 (a) Well-crystallized kaolinite showing characteristic hexagonal form.
(b) Clay-sized soil particle with a pronounced micaceous core.
(c) Freshly precipitated ferrihydrite.
(d) Kaolinite crystals coated with ferrihydrite.
(b) by courtesy of P. Kane; (c) and (d) by courtesy of A. Saleh.

surface has become, in a strict crystallographic sense, ditrigonal. It is thus referred to as the ditrigonal cavity. It is a feature of all layer mineral siloxane surfaces and, as mentioned above, is about the diameter of an O^{2-} ion in size.

When the c axis spacing is measured by X-rays for an oriented specimen (that is, with the majority of the crystals lying in the same plane) there is the expected strong reflection at 0.71–0.73 nm. There is also a strong line at the half distance of 0.355 nm because some submultiples are a feature of X-ray patterns. Both of these lines are unaffected by chemical treatment such as ion exchange, and this would be expected since the kaolinite lamellae, which are held together strongly by hydrogen bonds, are non-swelling. When the sample is heated to 550 °C dehydroxylation occurs, the structure collapses and no X-ray peaks are observed. Neither the 0.7 nm peak nor the high temperature collapse is a property of any of the other main groups.

An important 1 : 1 mineral in addition to kaolinite is halloysite, which is an hydrated form. The presence of internal layers of water sometimes causes the layers to curl and form tubes, and these are observable in the electron microscope. Halloysite gives a broad peak somewhere between 0.7 and 1.0 nm at room temperature, but it collapses to the standard value of 0.71–0.73 nm when the water has been driven off (the temperature usually chosen is 335 °C, before dehydroxylation occurs). Like kaolinite, no peaks are observed after heating to 550 °C.

The 1 : 1 minerals are also characteristically observed by thermal methods (see below).

Mica group

A 2 : 1 mineral has the ideal (no substitution) layer constitution:

6O   4Si   4O   2OH   4Al   2OH   4O   4Si   6O   or
$(Si_8)^{IV}(Al_4)^{VI}O_{20}(OH)_4$ (dioctahedral: pyrophyllite)
$(Si_8)^{IV}(Mg_6)^{VI}O_{20}(OH)_4$ (trioctahedral: talc)   :

Micas may have isomorphous substitution in both layers, but predominantly it is in the tetrahedral layer (Al for Si). The total layer charge may be up to 2^- per unit cell, requiring two exchangeable cation charges to balance the deficit at the surface. This cation is always K^+. The unhydrated ion is about the same size (0.0138 nm) as an O^{2-} so 'fits' almost exactly in the ditrigonal cavity. For this reason, the K^+ is specifically adsorbed (see Ch. 7) at the siloxane surface. Also, since each unit cell has two ditrigonal cavities both of them must be occupied by a cation. Thus layers of mica can be held together by unhydrated K^+ ions in every cavity which allows very close contact between the lamellae. So strong is the mutual attraction that the K^+ is often thought of as being part of the crystal rather than strictly an exchange ion. However, the K^+ is exchangeable as it can, at extremely slow rates, diffuse out of the crystal as it becomes weathered.[14] Plate 6.1b shows a particle of weathering clay with a strongly micaceous core. Its parallel, laminar structure is obvious, though it is not so clear in this case that all the less well-defined particles adhering to it are in fact derived from the core.

Both dioctahedral and trioctahedral micas occur. Muscovite, in which there is little substitution of the Al in the octahedral layer, is the typical diocta-hedral and biotite, in which the Mg ions are often replaced by Fe^{2+}, Mn^{2+} and other divalent ions, is a common trioctahedral. The potassium ions, however, bind adjacent sheets of dioctahedral micas together more strongly than they do those of the trioctahedral, possibly because the direction of the dipole of the hydroxyl ion below the hexagonal hole is vertically upwards in the trioctahedral, but twisted from the vertical in the dioctahedral. This may be the reason biotite micas lose their potassium more easily than muscovite micas. An important K-releasing mica species is illite. The name has been used rather loosely in the literature to describe a variety of micas with struc-tures to an extent weakened by incorporation of water molecules between some lamellae. Illites are important agriculturally as a source of potassium, available, not immediately, but over the course of a growing season.

The X-ray diffraction pattern observed for micas is relatively straightfor-ward. There is a very sharp peak at 1.01 nm, which is almost the thickness of the elementary 2 : 1 lamella (there is no water between them) together with submultiples at 0.5 and 0.34 nm. This observation is still made after heating to 550 °C since in the case of 2 : 1 minerals no significant dehydroxy-lation occurs.

Vermiculite group

Vermiculite, like micas, may have substitution in both layers, but the overall charge per unit cell is lower (1.2^- to 1.8^-) and often distributed in a more complicated way. In some samples the net negative charge is composed of a high tetrahedral substitution combined with a small positive charge in the octahedral layer in which the amount of Mg is higher than expected. The lower overall charge has the consequence that the lamellae are not held together as strongly as for mica, and so are able to move apart sufficiently to accommodate hydrated exchange cations. These are most usually Mg^{2+} which, when adsorbed between adjacent layers, have a shell of nearest neigh-bour water molecules about 0.4 nm in diameter. The arrangement is that the Mg ions are midway between the sheets (and so are not associated with the ditrigonal cavities) with the water molecules propping the sheets apart to give an X-ray repeating distance of 1.40–1.42 nm. The water of hydration, like

14 Chute, J. H. and Quirk, J. P. have measured diffusion coefficients of around 10^{-23} to 10^{-24} m^2 s^{-1} for K^+ out of clay-sized particles of (micaceous) Fithian illite. *Nature, Lond.*, 1967, **213**, 1156.

the chemically bound water in halloysite, is lost by 335 °C, at which stage the spacing has collapsed to the mica-like 1.0 nm.

Smectite group

The unit cell charge is now down to 1^- or below, and usually the substitution is in the octahedral layer well away from the surface. The attraction between adjacent lamellae is consequently weak and almost any hydrated cation can push them apart and act as an exchanger at the interlayer surfaces. This means that it is important in X-ray work to specify the exchangeable cation so that hydration properties are also known. Mg^{2+} is usually chosen. The equilibrium spacing is a little larger than for vermiculite: around 1.5 nm, and this collapses to 1.0 nm on heating. The smectite minerals (usually montmorillonite) may be distinguished from those in the vermiculite group by attempting to exchange the water molecules of hydration by ethylene glycol. This is possible for smectite and, because the glycol molecule is larger than water, the c-axis spacing rises to around 1.7 nm. The vermiculite space, however, does not expand when glycol is added because the layers are more strongly attracted, and so the 1.7 nm peak is not seen after glycol treatment.

The most important property of members of the smectite group of clay minerals is their ability to swell and finally to disperse into particles only one lamella thick. Because the swelling property can be so marked in the behaviour of Vertisols, which contain mainly montmorillonite in the clay fraction, and dispersion is such a dramatic event in a poorly managed sodic soil, it has sometimes been thought that montmorillonite is much the most surface active of the clay minerals. In fact, it has the lowest charge density of any 2 : 1 minerals: the important point is that all its internal surface (some 800 m^2 g^{-1}: see below) is available for chemical reaction and hydration. Although in most circumstances divalent exchangeable cations like calcium hold the lamellae together such that swelling ceases at only three water layers (1.7–1.8 nm) separation, monovalent sodium on the surfaces can lead to extreme structural instability. It will be shown in the next chapter that some monovalent cations are associated with thick diffuse layers, especially at low electrolyte concentration. In the case of smectites, the attractive forces between the lamellae under sodic conditions are not sufficient to prevent the formation of extensive diffuse layers in solution and the mineral swells to accommodate them. The practical consequences of this behaviour are examined in Chapter 27.

Chlorite group

It is often found that there is very considerable tetrahedral substitution in chlorites. Rather as with some vermiculites this may be partially compensated internally by a positive charge on one or both of the octahedral layers; this is often brought about by excess Al^{3+}. The range of composition for chlorites, both for dioctahedral gibbsite and trioctahedral brucite intercalated hydroxy interlayer clays, is reflected in the fact that both swelling and non-swelling examples are known. The latter show a c-axis spacing of 1.40–1.43 nm irrespective of heat or other treatments (i.e. this is the thickness of the 2 : 2 lamella), but this moves to around 1.70 for a glycol-treated swelling chlorite. The distinction from 2 : 1 (swelling) smectites is of course the reduction to 1.4 nm, not 1.0 nm, with heating to remove interlayer water.

Clay minerals in soils

Although at a simple level clay minerals fall into the five classes described above, the result of chemical weathering and pedological movement nearly always produces a complex situation in practice. The X-ray trace of a separated soil clay (from which organic matter has been removed) may well show several of the peaks expected for individual components, but often against a background of further information which it is not possible to resolve easily. Part of the reason for this is that the clay mineral particles will not be arranged perfectly parallel to one another in the diffractometer sample, so reflections other than those due to the lamellar thickness are being observed. The crystallinity of the clay fraction may also be poor and this may give rise to peak broadening. Other peaks, and a large number may be due to hydrous oxide particles as well as clay minerals, may then be overlapped and become hidden.

Another reason for difficulty in interpretation is more fundamentally concerned with the constitution of soil clay minerals themselves: it is that the lamellae may not have crystallized to give a single type of mineral. For example, smectite and mica can form a crystal unit in which they are regularly alternated. This will provide a repeating distance which is characteristic of the alternation or interstratification of the layers. Spacings diagnostic of the 'pure' components will not be seen clearly unless there is also some of that clay present as well. Unfortunately, there is apparently no requirement for regular interstratification, and such minerals pose special problems. Smectite/mica interstratifications are common in soils derived from sedimentary materials; smectite/kaolinite is also found, as are virtually all combinations of the main groups.[15]

Other clay minerals

There are other non-platey clay minerals which are less common in soils.

15 Brown G. *et al.* (ref. 11); MacEwan, D. M. C. and Ruiz Amil, A., Ch. 8 in Soil Components, Vol. 2 (ref. 11); Sawney, B. L., Ch. 12 in *Minerals in Soil Environments* (ref. 11).

Examples are palygorskite, sepiolite and zeolites. Their structure and chemistry are described in more specialized texts.[16]

Supplementary techniques of analysis

Although semi-quantitative X-ray diffraction analysis of clay minerals is the primary tool, the complexity of most soil clays means that a wide range of other methods is useful, each of which might be expected to add something to understanding. The following[17] are usually of value.

X-ray fluorescence analysis (XRF): total elemental composition

The chemical composition of minerals used to be obtained by destructive dissolution in either strong acid or base flux (Na_2CO_3) and then determining the elements in solution by conventional chemical means. This still has some use, but the process most frequently used now, for all but the lightest elements, is XRF. Each element has a characteristic set of X-ray wavelengths which are produced as secondary (fluorescent) radiation on irradiation with a primary beam of X-rays. The secondary X-rays are then measured for intensity. This property can be used to estimate the absolute amount of the element concerned by comparing the intensity with that provided by standards made up in a similar matrix. Clay minerals are usually pressed into the form of a solid disc and this is placed in the path of the X-rays. The technique is therefore non-destructive, and is potentially able to measure the amounts of the vast majority of elements likely to be present in a sample. The elemental composition so derived can, in some instances, be back-calculated to a clay mineral unit cell formula. Any contamination (say with oxides) or mixing with other clay minerals must either be known accurately or experimentally eliminated from the sample. In general, too, a good deal of other information about the chemistry and behaviour of the material will need to be known. Examples might be its charge density (see below) and total potassium content; the latter is often assumed to be directly related to mica content. For further details of XRF techniques, both in theory and practice, see for example Jones.[18]

---

16 Ovcharenko, F. D. *et al.*, *The Colloid Chemistry of Palygorskite*, Israel Program for Scientific Translation, Jerusalem, 1964;   Barrer, R. M., *Zeolites and Clay Minerals as Sorbents and Molecular Sieves*, Acad. Press, 1978;   Zelazny, L. W. and Calhoun, F. G., Ch. 13 in *Minerals in Soil Environments* (ref. 11).

17 In addition, the (so far) lesser-used methods – Mössbauer spectra, neutron scattering, X-ray photoelectron spectroscopy, NMR, ESR and photoacoustic spectroscopy – are described in Stucki, J. W. and Banwart, W. L. (eds), *Advanced Chemical Methods for Soil and Clay Mineral Research*, D. Reidel, Dordrecht, 1980.

18 Jones, A. A. in *Methods of Soil Analysis* (2nd edn) (eds A. L. Page, R. H. Miller and D. R. Keeney) Part 2, Ch. 5, Am. Soc. Agron., Madison, 1982.

Electron microscopy

Plate 6.1 shows electron micrographs[19] of some components of the clay fraction. Mica and kaolinite are often large enough and distinctive enough to be unambiguous, though smectite particles are usually too small to be seen individually. Further refinements such as electron diffraction and electron probe can give crystalline and elemental information from selected areas of the picture.

Thermal analysis

This is conventionally divided into thermogravimetric analysis (TGA) in which the weight lost – in the case of clay minerals, due to water – is measured over a programmed temperature rise; and differential thermal analysis (DTA). In the latter technique the difference in temperature between the sample and an equal weight of calcined alumina (held in an identical crucible) is monitored as the amount of heat supplied to both is increased. The sample may at some definite temperature go through, for example, an endothermic reaction such as the loss of water due to dehydroxylation. Around this temperature the sample will be cooler, to an extent dependent on the amount of substance present undergoing the reaction, than the reference alumina which is assumed not to undergo any thermal reactions on heating. If the sample were to react exothermically then it would at that stage become hotter than the reference. Both TGA and DTA curves may be characteristic for clay minerals due to differences in water content and structure. The methods work as satisfactorily for very fine grained clays as for well-crystallized materials, and a particular advantage of DTA is that it is sometimes very sensitive to small changes in lattice substitutions. The disadvantages of the technique are peak overlaps and the difficulties of quantification.[20]

Infrared analysis

This technique determines the vibration frequencies found for atoms in chemical groups mostly in the surface region of the mineral. The assignment of frequencies to individual species has been considerably researched[21] and extensive tables exist to help interpretation for a new sample. As with several

19 For a comprehensive collection of soil and clay micrographs see Smart, P. and Tovey, N. K., *Electron Microscopy of Soils and Sediments*: Examples (1981), Techniques (1982), Clarendon Press, Oxford. Also useful is Gard, J. A. (ed), *The Electron-Optical Investigation of Clays*, Min. Soc. (Lond.), Monog., No. 3, 1971.
20 For further details, see Mackenzie, R. C. and Caillère, S. in Ch. 16 of *Soil Components* (ref. 11) or at greater length, Mackenzie, R. C. (ed.), *Differential Thermal Analysis*, Vol. 1 (1970), Vol. 2 (1972), Acad. Press, London.
21 Farmer, V. C. and Palmieri, F. in Ch. 17 of *Soil Components* (ref. 11) or, at greater length, *The Infrared Spectra of Minerals* (ed V. C. Farmer), Min. Soc. (Lond.) Monog., No. 4, 1974.

of the methods used to help elucidate clay mineral structure, reasonably clear interpretations can be made for individual minerals of known composition; however, IR gives much more ambiguous results for mixed mineral systems, and generally has limited use as a routine tool in a soil mineralogy laboratory.

Surface charge density

It has already been mentioned that one of the key distinguishing features of clay minerals is the charge per unit cell. It will therefore be the case that the actual measurement of the average specific charge density will be a useful confirmatory technique for clay minerals. It requires two measurements: the specific surface area and the total negative charge, usually assumed to be equal (and opposite) to the cation exchange capacity (CEC – but see Ch. 7).

Surface area   The theoretical surface area of lamellae surfaces in a $2 : 1$ mineral can be shown to be around $750–800$ m^2 g^{-1}. This may be worked out from the area of the unit cell (0.46 nm$^2 \times 2$: top and bottom) of which there are 6×10^{23} per unit cell atomic weight, which say, for, montmorillonite, is around 720 g. This result is only true for the clay mineral in practice if all the faces of the lamellae are accessible for measurement of area. The laboratory methods for determining area may be divided into those in which some adsorbate penetrates between the lamellae to measure both internal as well as external surface, and those in which adsorbate cannot do so, and so only measures external area. An example of the latter is dinitrogen gas which may be sorbed at low temperature (usually the boiling point of liquid nitrogen, 77 K) on to the sample. The N$_2$ adsorption isotherm can yield the specific surface area by use of the Brunauer–Emmett–Teller (BET) equation.[22] Internal area may be measured by analysis of the adsorption behaviour of cetylpyridinium bromide,[23] water[24] or ethylene glycol.[25]

Surface charge   It is again important to distinguish between the charge over all surfaces, internal and external to the lamellae stacks, and that on the external area only. The lamellae of mica, for example, are so strongly held together that any cation exchange measurement only determines the charge on the external surfaces, but for smectites all charge balanced by exchangeable cations would be included. Methods for charge determination are considered in more detail in the next chapter.

22 Gregg, S. J. and Sing, K. S. W., *Adsorption, Surface Area and Porosity* (2nd edn), Acad. Press, 1982. Greenland, D. J. and Mott, C. J. B., Ch. 4 in *The Chemistry of Soil Constituents* (ref. 11).

23 Greenland, D. J. and Quirk, J. P., *J. phys. Chem.*, 1963, **67**, 2886; *idem, J. Soil Sci.*, 1964, **15**, 178.

24 Ormerod, E. C. and Newman, A. C. D., *Clay Min.*, 1983, **18**, 289. The authors distinguish between external and internal areas of Ca-montmorillonite using water adsorption.

25 Eltantawy, I. M. and Arnold, P. W., *J. Soil Sci.*, 1974, **25**, 99.

The surface density of charge is simply the charge per unit area, but this is only valid if both quantities discussed above have been determined with respect to the same surfaces. For 2 : 1 minerals the charge density ranges from one charge per $0.46\ nm^2$ in micas to about $1.30\ nm^2$ for smectites. Kaolinites have similar charge densities to smectites, though their exchange capacities are in absolute terms much smaller. This is because kaolinite lamellae cannot expand and so the mineral exhibits charge only on its external crystal surfaces, which is not the case for the swelling clay. Kaolinite surface charge varies with solution concentration and pH and in this respect behaves like hydrous oxides. It will be discussed in Chapter 7.

Summary for clay minerals

A knowledge of the clay mineralogy of a soil provides information pertinent to both its physical and chemical properties. In chemical terms it leads to an understanding of the permanent electrical charge of the mineral particles and to details of the surface structure – such as the ditrigonal cavity – which will influence its interaction with the solution phase. The mineral structure also leads to an expectation that there will also be surface chemical groups such as Al–OH and Mg–OH which may have important properties. These will exist not only on any octahedral face not hydrogen bonded to another lamella, but also at the edges of the crystal where the repeating areal pattern breaks.

Other inorganic components

Many soils contain amounts of hydrated oxides, for example of iron, aluminium, manganese and (usually to a lesser extent) titanium oxide. Because of the variety of structures and forms encountered, terminology with respect to these minerals is sometimes loose: descriptions such as hydrous oxides and sesquioxides are used to cover the group generally, as well as the more exact oxide (such as haematite, Fe_2O_3) and oxy-hydroxide (such as goethite, α-FeOOH). In temperate soils these hydrous oxides may have a highly disordered structure and be present in only small proportion, but in tropical soils in which more intense weathering has taken place, the minerals may be both well crystalline and constitute a large proportion of the $<2\ \mu m$ fraction. The most widespread of these hydrous oxides are the ferric, which impart the yellow, brown and red colours to soils, and they may appear as uniformly dispersed over the soil particles, or present as localized streaks or stains. In some soils they are present as discrete particles, whose size varies from submicroscopic up to the size of gravels, often with a smooth rounded surface, and which are usually very hard. But they may be present as cements which bond soil particles together into concretions of indefinite shape, or into a pan; and these vary in strength from being soft enough to be crushed between the fingers up to massive ironstone.

Recent work has done much to clarify the structural types of the iron minerals,[26] of which three are discussed below.

1 *Ferrihydrite*   This has the approximate formula $Fe_5(O_4H_3)_3$ and is the initial product of the hydrolysis of Fe^{3+} in water. Ferrihydrite is found both in soils, for example in the B_{fe} horizons of podzols, and in surface waters. The mineral characteristically forms very small (5–7 nm diameter) spherical particles with a surface area of 200–300 $m^2 g^{-1}$, and is illustrated in Plate 6.1c. Such a large surface area can develop a considerable positive charge and can as a result associate with the surfaces of clay minerals. Plate 6.1d shows the coating of kaolinite crystals with ferrihydrite, and it is apparent that the oxide remains as discrete particles rather than forming a smooth 'coat'. The surface electrical charge of both components will be modified by this association, which varies in effectiveness with pH. Ferrihydrite is thermodynamically unstable and under natural conditions slowly converts to goethite or by internal dehydration to haematite.

2 *Haematite*   This mineral forms hexagonal plates and is given the formula of $\alpha\text{-}Fe_2O_3$. It is found mainly in tropical soils where average temperatures are higher than in temperate areas. Ferrihydrite is necessary initially for haematite to crystallize. The oxide is bright red and is an important source of this colour in soils.

3 *Goethite*   Although goethite, $\alpha\text{-}FeOOH$, is generally found in cooler, wetter environments than haematite, it can be expected in almost any soil type or climatic zone, because of its thermodynamic stability. Its colour is yellowish brown and forms, like haematite, from a ferrihydrite precursor, often in the presence of organic matter. Goethite is recognized by X-ray diffraction through a strong reflection at 0.418 nm. The mineral forms characteristic acicular crystals and these are often recognizable in the electron microscope.

One of the chief interests so far as goethite is concerned is that its surface has been precisely characterized. Parfitt and his colleagues,[27] using infrared spectroscopy, have shown that the surface contains OH groups linked to a single Fe^{3+}, linked to 2 and linked to 3 Fe^{3+} ions. There are also O^{2-} ions in the surface layer. They have also shown that only the singly linked Fe–OH group is chemically active, in the sense that the OH ligand can be replaced by some other. From consideration of the position of these groups on the predominant (80 per cent) 100 plane of the crystal,

26 For a review see Schwertmann, U. and Taylor, R. M. in Ch. 5 of *Minerals in Soil Environments* (ref. 11).
27 Parfitt, R. L. *et al.*, *J. Chem. Soc. Faraday. I*, 1976, **72**, 1082. See also the discussion in Ch. 7.

it can be calculated that the maximum density of reactive OH sites is around 4.4 μmol m^{-2}. Thus, a combination of a measurement of the surface area using nitrogen (there is no internal surface for an oxide as there can be for some clay minerals) with the charge density information gives a figure for the maximum ligand reactivity of a sample of goethite. This is considered with respect to some anionic ligands in Chapter 7. Calculations such as this can only be done if the hydrous oxide crystal forms in a reproducible fashion. Goethite is much the best example, though attempts to calculate theoretical charge density have been made for several other minerals, including gibbsite and kaolinite.[28].

Other iron oxides

Lepidocrocite (γ-FeOOH) forms in hydromorphic soils and its presence is indicative of gleying conditions; maghemite (γ-Fe$_2$O$_3$), a ferromagnetic mineral, is found in highly weathered soils of the tropics where it is frequently associated with haematite.

  All iron oxides may have 'foreign' ions in their structures instead of Fe. Ferrihydrite is able to accept both Si and Al in its lattice and these replacements persist in the structures of the pedogenically derived goethite or haematite. The former has been shown to have up to 30 mole per cent Al under natural conditions. The lack of perfection in the crystal provided by isomorphous replacement gives rise both to changes in surface properties, especially the zero point of charge (Ch. 7), and also in some cases to permanent charge. Thus, although 'type' iron oxides are well understood in terms of bulk and surface properties, a natural example derived from soil may present considerable problems in terms of its characterization.

Other hydrous oxides

Gibbsite is the only form of aluminium hydroxide at all common in soils. It is not usually found in large amounts except in some highly weathered tropical soils: the USDA classification has a 'Gibbs-' Great Group formative prefix to indicate soils rich in gibbsite (e.g. a Gibbsihumox). Synthetic crystals have a regular hexagonal shape in the electron microscope, and it is thought that the chemical activity of the crystal is restricted to the Al–OH aluminol groups on the edges of the hexagons. Unfortunately, the chemical environment of these groups is not as well characterized as in the case of goethite, but it is certain that surface ligands on gibbsite can be exchanged by ligands from the soil solution.[29] The chemistry of Al itself in solution and the hydro-

28 Sposito, G., *The Surface Chemistry of Soils*, Ch. 1, Oxford, 1984.
29 Hingston, F. J. *et al.*, *J. Soil Sci.*, 1972, **23**, 177; ibid., 1974, **25**, 16. The authors compare the behaviour of goethite and gibbsite.

lysis of Al^{3+} to form gibbsite are not entirely understood; the behaviour of aluminium hydroxy species as a factor in the generation of soil acidity is reviewed in Chapter 25.

Manganese oxides[30] are difficult to characterize due both to the number of individual and (apparently) mixed forms occurring and also the sometimes uncertain valence state of the Mn in the lattice. However, Mn is a plant micronutrient (see Ch. 23) and oxides form a source for the low concentration required in the soil solution. The most common form is birnessite, but the structure is not known for certain and it shows variable composition. For this reason, the surface has not been closely described at the atomic level, and this means that exact mechanisms of surface ion exchange are not known either. The chemically active groups are likely to be Mn–OH.

Although quartz is usually most important in the sand and silt fractions of soils, it can also be found in the clay fraction. The mineral has numerous diagnostic lines in the X-ray diffraction scan. It also has a particularly distinctive reversible transition at 573 °C in the DTA trace: so unvarying is this temperature that it is sometimes used to calibrate the DTA sample thermocouple temperature.

General summary

It was stated earlier in this chapter that a major objective of the determinative mineralogy of soils is to gain some appreciation of surfaces, rather than only the solid structure. The simplest summary that might be made of the various surfaces encountered would be that mineral surfaces can be divided into only two types: the permanent charge type, exemplified by the siloxane surface, and the variable charge type, as shown by oxidic surfaces. The former is charged due to the properties of the *solid* (isomorphous substitution) and the latter by the properties of the *solution*, specifically the pH which controls the extent to which the surface OH groups are either protonated or ionized (hence, variable charge). It is, however, generally thought useful to make a further category of surface type – though it is really only a combination of the two already described – and that consists of those soils in which the permanent charge particles have oxidic coats. Sanchez,[31] for example, suggests that the mixed category describes the ion exchange properties of most 'red' tropical soils (except perhaps the oxidic, gibbsitic and ferritic families of Oxisols, Ultisols and Alfisols) more adequately than either the layer silicate or the pure oxide system. Modern attempts at mathematical modelling of surface conditions may, however, ultimately make the mixed surface category superfluous.

30 McKenzie, R. M., Ch. 6 of *Minerals in Soil Environments* (ref. 11).
31 Sanchez, P. A., *Properties and Management of Soils in the Tropics*, Ch. 4, Wiley, 1976.

Surface chemistry of soil particles

The previous chapter has described the main inorganic components in soil, and the main features of the constitution of the organic fraction are outlined in Chapter 18. It will be obvious from these accounts that, with only rare exceptions, the compounds which together make up the solid phase of a soil are either permanently electrically charged or may become so in contact with pore solution. The electric charge, which has its main chemical and physical effects at the particle surface (the interface with the solution), is perhaps the key feature of the chemistry of soils. The effect of charge, itself the product of weathering and soil development, determines much of the behaviour of plant nutrients, the reaction of added fertilizers and the capability of the soil profile to adsorb potential pollutants.

This chapter is therefore predominantly concerned with the effects of surface charge and charge density on the equilibrium properties of anions and cations in contact with the solid phase. This is preceded by a brief examin-ation of the solution phase, which has its own properties irrespective of the solid. Together the solution and the interface tell us a significant amount about the environment from which plants absorb their mineral nutrients. The emphasis will be on equilibrium properties, but a word of caution is necessary. It is likely that soils containing growing roots, for example, or an active microbial population, or a rapidly fluctuating water content are never at anything like true equilibrium, and this must be borne in mind as a limitation on what follows. It is also still very difficult to describe the heterogeneity of most soil conditions and this means that generalized pictures or simple equations will not provide a *full* picture of a whole soil body – even at the pedon level.

The soil solution

The chemical composition of the pore solution in a soil is of immediate interest in terms of plant nutrition (see p. 150) and modern methods (especially inductively coupled plasma (ICP) emission spectrometry and ion chromatography) have greatly eased the problem of accurate analysis. Some examples of soil solution compositions are given in Chapter 4, but it is useful to list here the main elements likely to be found: cations, Na, K, $N(NH_4^+)$, Mg, Ca, Al, Mn, Fe, Cu, Zn, Sr, Ba; anions (or, in some cases, uncharged species), $N(NO_3^-)$, Cl, $C(HCO_3^-)$, F, P, S, B, Si, Sc, As. There will also be a significant amount of dissolved organic carbon (i.e. other than as bicarbonate/carbonate) in compounds of largely uncharacterized structure. The list of elements is not exhaustive. Kinniburgh and Miles,[1] who used a method of extraction of soil solution through its immiscible displacement by a dense fluorocarbon in a high speed centrifuge, calibrated their ICP to determine also Li, V, Co, Ni, Y, Mo, Cd and Pb. Although they did not find significant amounts of these elements in solutions obtained from 10 (mainly calcareous) soils from the South of England, their presence in soils formed from other parent materials or through contamination might certainly be expected.

Unfortunately, a simple listing of components in the solution is only the start of a full description. Flame methods of analysis (ICP, atomic absorption spectrometry (AAS) or flame photometry) measure the concentration of the total quantity of the element present, irrespective of what chemical species it is in. For example, the median value for the concentration of Cu in Kinniburgh and Miles's soils was $10^{-6.1}$ M, but that did not mean that the concentration of the copper ion, Cu^{2+}, in the soil solution had that value. In the case of Cu it would have been possible to have determined the true Cu^{2+} activity directly using a reversible (specific ion) electrode for copper which, like the glass electrode for H^+, shows a potential proportional only to the simple, chemically unbound ion to which it is sensitive, and not to complexed

1 Kinniburgh, D. G. and Miles, D. L., *Environ. Sci. Tech.*, 1983, **17**, 362.

species.[2] However, in a soil solution copper is subject to hydrolysis at normal soil pH values and it is also complexed by other components of the soil solution, especially the soluble organic matter. The degree to which this happens depends both upon pH and the thermodynamic activity of these other species, and this is a function of the total ionic strength of the solution. Trace metals like copper, which take part in numerous chemical associations, are among the most difficult to speciate, i.e. to list the amounts of all possible species that contain the element of interest. Even what might be thought of as 'simple' ions need to be speciated in order that the properties of the soil solution can be fully understood.

The case of a dilute solution of gypsum illustrates the principles of speciation and the calculation of the ionic activities involved. When calcium sulphate is dissolved in water a proportion of the calcium and sulphate ions associate to form the uncharged solution species $CaSO_4^0$ (where the superscript ^{0} indicates the zero charge). This entity is in dynamic equilibrium with its charged components according to the association reaction

$$Ca^{2+} + SO_4^{2-} \rightleftharpoons CaSO_4^0$$

$$\text{and } K° = \frac{(CaSO_4^0)}{(Ca^{2+})(SO_4^{2-})}$$

where parentheses indicate activities and $K°$ is the association constant. For $CaSO_4^0$ log $K°$ is positive and has the value 2.31, i.e. a constant by no means unfavourable to the formation of the complex. It must be emphasized that the above equation does not refer to the precipitation of solid gypsum. This process would be:

$$Ca^{2+} + SO_4^{2-} + 2H_2O \rightleftharpoons CaSO_4.2H_2O(s)$$

and here the solubility product, log K_{sp}, is -4.64.

The degree of complex formation can only be calculated with a knowledge of the individual ion activities, so long as $K°$ is known, and these are derived by a version of the Debye–Hückel equation[3] or the Davies equation[4]:

$$\log \gamma_i = -Az_i^2 \left(\frac{\mu^{1/2}}{(1 + \mu^{1/2})} - 0.3 \, \mu \right)$$

where γ_i is the activity coefficient of an ion, i, of valency z_i; A is a constant and μ is the ionic strength of the solution. The latter term

$$\mu = 1/2 \, \Sigma \, c_i z_i^2$$

2 But the method is limited in terms of direct measurement to the specific ion electrodes available: Na, NH₄, Ag, Cu, Cd, Pb, F, Cl, Br and I.
3 See for example Ch. 2 of W. L. Lindsay's *Chemical Equilibria in Soils*, Wiley, 1979, for a discussion of the various forms of this equation.
4 Davies, C. W., *Ion Association*, Butterworth, 1962.

cannot be calculated unless the concentration, c_i, of each individual ion is known, and this is just what is *not* known given that the original question was to determine the degree to which the initial concentration had been changed by complexation! This summarizes the circle of unknowns which requires that a guess is made at one of the quantities, followed by an iterative procedure which rapidly refines the values for each species until all equations are obeyed with trivial error. After doing this for the $CaSO_4$ case,[5] it turns out that for, as an example, a 10 mM solution, the degree of complexation is 28.9 per cent, and so the original amount dissolved gives a concentration for Ca^{2+} and SO_4^{2-} of 7.11 mM rather than 10 mM. The ionic strength is 0.028, not 0.04 which would have been calculated on the basis of no complexation, and this gives an activity of 3.762 mM for the two ions. Even this is not the end of the story. If it is assumed that the solution is at pH 7 then to a small extent the calcium ion will be hydrolysed to $CaOH^+$ (log $K° = -12.7$) and the sulphate will associate with the proton to form HSO_4^- (log $K° = -1.98$); the existence of these other ions illustrates the direct importance of the chemical properties of the solvent. Table 7.1 summarizes the situation for the 10 mM solution of calcium sulphate at pH 7. This table ignores the formation of even less favoured species, such as $H_2SO_4^0$ and $Ca(OH)_2^0$, but these will be present in very minute quantities.

Table 7.1   Values for solution of 10 mM $CaSO_4$ at pH 7

	Ca^{2+}	SO_4^{2-}	$CaSO_4^0$	$CaOH^+$	HSO_4^-
Analytical concentration, mM	10	10	–	–	–
Equilibrium concentration, mM	7.11	7.11	2.89	7.4×10^{-6}	3.6×10^{-5}
Activity coefficient	0.529	0.529	1	0.853	0.853
Equilibrium activity, mM	3.762	3.762	2.89	6.4×10^{-6}	3.1×10^{-5}
−log (molar activity)	2.425	2.425	2.539	8.194	7.517

The calculation required to produce the figures in Table 7.1 is sufficiently tedious to make a computer solution attractive. For real soil solutions with a minimum of perhaps 15–20 components it becomes essential, as these components will generate tens, or if organic species are taken into some kind of account, hundreds of possible complexes. A number of computer programs now exist to calculate the amounts of these species. They differ both in the number of components they take into account, the manner of treating (if at all) dissolved organic matter, and the values for association constants used to solve the equations. It may be hoped that a self-consistent set of $K°$ values will eventually be accepted by all workers, but the organic matter problem is impossible to deal with save by using simplifying assumptions of arguable

5 For details of the calculation, see Adams, F., *Soil Sci. Soc. Am. Proc.*, 1971, **35**, 420, or Bolt, G. H. and Bruggenwert, M. G. M. (eds), *Soil Chemistry, A. Basic elements*, Ch. 2, Elsevier, 1978.

validity. The two main lines of approach are first, to treat the soluble fulvic material as a mixture of mainly simple organic acids with functional groups whose formation constants for interaction with metals are known, and thus programmable. A possible set[6] might be acetic, glutamic, phthalic, salicylic and tartaric acids and glycine, all in some arbitrary proportion. The second approach is to regard the dissolved carbon as consisting of so-called quasi-particles, hypothetical entities which each have the functional group activity of the organic fraction as a whole. The quasiparticles are 'mean fulvic acid units', the properties of which may be determined by potentiometric titration curves. Complete tests of these two approaches have not yet been made, but the models have been compared by Sposito and colleagues[7] for the speciation of Cu and Cd. The result seemed to show some advantage to the quasiparticle model, though it is not easy to incorporate it within a computer algorithm.

GEOCHEM

The best known speciation package in use in soil chemistry is GEOCHEM, developed in 1979 by Sposito and Mattigood.[8] The data included in the program is extensive: more than 2000 aqueous species can be considered. Nearly 900 of them contain organic ligands so the validity of the model used is crucial to the success of speciation in many instances. GEOCHEM is also useful for a wider understanding of the soil system from which the solution has been extracted. Because the speciation procedure calculates single ion activities under known conditions of oxidation potential and pH, it becomes possible to compare suitable ion activity products with the known solubility of any solid mineral or mineral mixture (so long as it is thermodynamically characterized) likely to be present in a soil system. As a simple example,[9] if the true ion activities of Al^{3+} and SO_4^{2-} have been calculated at the known pH of the solution, then it is easy to determine its degree of saturation with respect to basic aluminium sulphate, basaluminite, $Al_4(SO_4)(OH)_{10}$, through the product $(Al^{3+})^4 \times (SO_4^{2-})/(H^+)^{10}$. This principle can be extended to a consideration of the stability of aluminosilicates, including clays, by using the figure determined for $(H_4SiO_4^0)$, and this example is considered later in the chapter. Precise information on ion activities is a valuable aid in predicting the behaviour of soil solution composition when changes, perhaps in water content or pH, occur. GEOCHEM also has a built-in subroutine to calculate cation exchange behaviour on constant charge surfaces. This is possible because cation exchange, like solubility, can be described on a thermody-

6 Morel, F. M. *et al.*, *Environ. Sci. Tech.*, 1975, **9**, 756.
7 Sposito, G. *et al.*, *Soil Sci. Soc. Am. J.*, 1982, **46**, 51.
8 Sposito, G. and Mattigood, S. V., GEOCHEM: a complete program for the calculation of chemical equilibria in soil solution and other natural water systems. Kearney Foundation of Soil Sci., Univ. Calif., Riverside, 1979.
9 For a discussion of the solubility of basaluminite see Adams, F. and Rawajfih, Z., *Soil Sci. Soc. Am. J.*, 1977, **41**, 686.

namic basis and so observed solution activity of a suite of cations has a direct relationship to composition in the surface adsorbed layer. As with organic complexes in solution, the only limitation to success is the validity of the model used. What is required for all computer analyses is an accurately predictive model valid over wide conditions of concentration and pH, and this emphasizes the reason why much recent work in soil chemistry has had this as a primary aim, rather than necessarily trying to understand detailed mechanisms or structures within the system.

Summary of main components of the soil solution

Many of the complexes present in soil solution will be present in very small amounts and have no influence on the properties significant to plant growth: they are only worth calculating in order to refine activity values for the ions of real interest. Some limited generalizations can be made about the complexing abilities of these.

Na^+, K^+, NH_4^+

These cations have trivial complexing ability and to good approximation their activities can be calculated without reference to other species. However, Cl^- and, more significantly, SO_4^{2-} can form complex species with them, e.g. $NaCl^\circ$, $NH_4SO_4^-$ though there are not likely to be circumstances in soil solutions where this has much significance.

Ca^{2+}, Mg^{2+}

Both form complexes with SO_4^{2-}, HCO_3^- and $H_2PO_4^-$, among other anions, and so accurate cation activities depend on calculating extent of complexation. Hydrolysis is usually insignificant.

Al^{3+}

This complexes to a greater extent than Ca^{2+} and Mg^{2+} with SO_4^{2-}, and extremely readily with F^-. Hydrolysis of Al^{3+} at pH values above about 4 is usually of major importance (see Ch. 25) and its effect must be calculated in addition to complex formation. The combination of effects serves to reduce (Al^{3+}) to a very low level in anything but very acid soils.

Trace metals

Complexing is often considerable with organic fractions, but there is no general pattern of behaviour. In addition, some association constants, even for inorganic complexes with, e.g. HCO_3^-, and $H_2PO_4^-$, are not always known with certainty, and have to be estimated from ionic properties. Hydrolysed species are usually important.

Anions

HCO_3^- (and CO_3^{2-}), $H_2PO_4^-$ (and HPO_4^{2-}, PO_4^{3-}) and SO_4^{2-} may all form complexes to some extent, and OH^- activity will determine hydrolysis. Cl^- is not a strong complexer but $CaCl^+$ and $MgCl^+$ have been suggested as important in some cation exchange experiments.[10] NO_3^- behaves in a similar way, and so, for example, $CaNO_3^+$ is formed (log $K° = -4.80$). Cl^- and NO_3^- are often regarded as 'indifferent' anions with respect to their adsorption on soil particle surfaces, i.e. they have no specific interactions with the surface in addition to coulombic forces, and this to a very large extent also characterizes their behaviour with respect to othci species in typical soil solutions.

It will be evident that accurate characterization of the soil solution will be of importance in understanding many chemical reactions in the soil. An example, referred to again on p. 870 in the context of acidity, is the toxicity of soil solutions containing aluminium to coffee seedlings. Pavan and Bingham[11] obtained saturation extracts from a set of Oxisols and Ultisols in which the coffee had been grown, measured and speciated the dissolved Al, then separately correlated the observed toxicity (in terms of the reduction of root and shoot weight) against (Al^{3+}) and total Al in solution. It may just be noted here that the better fit to their observations for both root and shoot was when Al^{3+} alone was considered, and the implication is that this, and not total soluble Al, is the toxic factor to coffee plants.

Interaction of solution ions with the surface

It has been shown in Chapter 6 that soil particle surfaces are essentially either of the layer silicate type, with which on the whole only electrostatic interactions with solution ions occur, or of the oxide type (which includes organic matter surfaces), at which chemical adsorption to form surface compounds or complexes can take place. So far as the former is concerned, cations from solution are attracted and anions repelled to an extent partly dictated by the surface density of negative charge on the surface. It might be expected, since there is no chemical reaction of the ions with the surface, that all cations of the same charge in the soil solution would behave similarly at such surfaces. While it is true that almost any small cation might be used to estimate the charge on the particles by exchange (see section on cation exchange capacity,

10 Sposito, G. *et al.*, *Soil Sci. Soc. Am. J.*, 1983, **47**, 51. The authors demonstrate the difference between using a non-complexing perchlorate and a chloride ionic medium for exchange of Ca and Mg with Na on Wyoming Bentonite.

11 Pavan, M. A. and Bingham, F. T., *Soil Sci. Soc. Am. J.*, 1982, **46**, 993; with Pratt, P. F., *op. cit.*, 1201.

below), the actual distribution of cations in the solution close to the surface varies markedly with species.

A key difference between one cation and another lies in the strength with which water molecules of hydration are held by the cation. Not surprisingly, di- and trivalent cations hold water much more tenaciously than monovalent ones. It has been estimated that some 6–8 water molecules form an effective primary hydration shell round a divalent alkaline earth cation like Mg or Ca, and that both shell and ion move together as a unit in the solution. There is also a much less strongly held secondary shell of up to 15 water molecules in number, though only in dilute solution. Monovalent cations also form primary shells, though with fewer, less strongly held water molecules as compared to the divalent ion. The attraction of a water molecule to an ion becomes less the larger the ion, and so in the monovalent series the strength with which the solvation shell is held is at its strongest for Li^+ (which probably has six nearest neighbours), is weaker for K^+ and very weak for Cs^+. It is presumably also very weak for complexes such as $CaCl^+$. None the less, even a single hydration shell, such as certainly exists for Li^+ and Na^+, forms an effective dielectric screen round the ion so that it can no longer be considered as a 'point' charge. The physical effect of a strongly held set of solvent molecules is of course also to limit the closeness of approach of the ion to the surface itself.

The diffuse electrical double layer

The classical approach to understanding the distribution of cations in the solution close to a negatively charged surface, especially those of the layer silicate type, is the application of diffuse layer theory. This provides a mathematical description of the decrease in electric potential with distance away from a charged plane surface. The distribution of ions from a salt solution can also be calculated, under certain assumptions, from well-established physical principles,[12] and Fig. 7.1 illustrates a typical distribution curve for cations and anions away from the surface. The reason for the existence of such distributions, rather than the counter ions simply being attracted in a single layer up against the solid, can be pictured as a balance between electrical attraction for the cations and a diffusion away from the surface due to the concentration gradient (established as a result of the attraction). However, the relevance of diffuse layer calculations to clay and soil systems depends on the appropriateness of the assumptions made in developing the simple theory. The most important of these is the requirement that the negative charge is uniformly spread out on the surface rather than being effective

12 For a concise account of diffuse layer derivations and computations, see Appendix III of H. van Olphen, *An Introduction to Clay Colloid Chemistry* (2nd edn), Wiley, 1977.

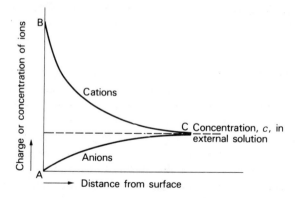

Fig. 7.1 The ionic concentrations in an idealized diffuse layer. AB: negative charge on the surface (balanced by an excess of cations over anions). At C the exponential cation and anion curves have effectively reached the equilibrium values of the pore solution.

only at specific spots (chemical groups), and this is why the theory is usually described in terms of permanent rather than variable charge minerals. This condition of uniformity of charge distribution will be most nearly met if the charges in the solid and solution are at some distance apart and separated by a dielectric which screens their point effect. Thus clays for which isomorphous substitution is in the octahedral layer will be expected to be in equilibrium with more diffuse-like layers than clays of similar charge density but with substitution in the tetrahedral layer. More important is the effect of the solvation sheath round the solution cation: those which retain their nearest neighbour waters cannot approach closer than about 0.28 nm to the surface, and this, combined with the dielectric properties of the shell, means that the cation cannot electrically neutralize any specific charge at the surface. Thus on these grounds alone, it would be predicted that monovalent ions like Li^+ and Na^+ and divalent ions such as Mg^{2+} and Ca^{2+} would obey a simple diffuse layer picture more closely than, for example, K^+, NH_4^+ or complexes such as $CaNO_3^+$, and this appears to be so. The case of K^+ (and for similar reasons, NH_4^+) is further distinguished because its unhydrated size, being very close to that of O^{2-}, allows it to fit snugly in the ditrigonal siloxane cavity on 2 : 1 mineral surfaces. This is energetically favourable and makes K^+ less of a diffuse layer ion when balancing charge at a siloxane surface rather than at an oxide type. Experimentally, it has been established[13] that the strength of binding of K^+ to montmorillonite is greater than that of Ca^{2+}, largely because the closer distance of approach of the monovalent ion is a more important factor than its lower charge. The diffuse layer theory does

13 Hutcheon, A. T., *J. Soil Sci.*, 1966, **17**, 339.

not take such differences into account – only charge is important. Cations like K^+ which are able to shed hydration water and move close or even right on to the surface are regarded as being, to a greater or lesser extent, *specifically* adsorbed.

Although the diffuse layer approach is only of value in its full or unmodified form for a few ions, it is important to understand its consequences because most cations will exhibit some degree of diffuse-layer behaviour at any charged surface. There are four properties of the layer of importance. First, its thickness decreases as the electrolyte concentration in the solution increases, and at equal equivalent (moles of charge) concentrations of electrolyte, divalent cations decrease the thickness of the diffuse layer more than monovalent, and trivalent more than divalent. As a corollary to this, if clay particles are suspended in electrolyte, the diffuse layer is approximately twice as thick if the exchangeable ion is sodium than if it is calcium. At concentrations as high as 10^{-1} M, the thickness with both types of ion is little more than that required for a single layer of hydrated cations, so the adsorbed counter ion layer is both well defined and has little diffuse character; at more typical soil solution concentrations the usually predominant Ca^{2+} has a layer thickness of only between about 5 and 10 nm.

Secondly, the higher the solution concentration the lower the potential on the surface of a permanent charge mineral. In this respect, the surface–solution interface is acting in the same way as a capacitor in an electrical circuit, and this is a useful analogy so long as its limitations are recognized. Instead of two fixed plates on which charge is accumulated at some given applied voltage, the negative plate (the clay surface) is fixed both geometrically and in its charge, but the positive plate consists of the mobile ions from the electrolyte. Just as the value of the capacitance is determined by the distance apart of the plates and the nature of the dielectric, so is that for the clay–electrolyte system determined by their distance apart (the thickness of the diffuse layer) and the value of the dielectric constant of the solvent. It is therefore not surprising that higher concentrations (compressed diffuse layers) are in equilibrium with the fixed charge of the surface at lower potentials relative to the bulk solution or that this potential is lower for divalent than monovalent ions. Thus soils will tend to be well flocculated at most common soil solution concentrations because particles can approach to close distances without either repulsion due to large negative potentials or overlap of diffuse layers. If overlap were to occur then the concentration of electrolyte between the particles would necessarily rise relative to the rest of the soil solution, and this would produce a difference of osmotic pressure tending to push the particles apart. Of course the particles in most soils are bonded by molecular links (e.g. organic matter) as well as electrostatically, but it is well known that the forces of attraction can be overcome in sodic soils in which the electrolyte concentration becomes significantly reduced. The consequences of swelling pressure and eventual dispersion of salt affected soils are dealt with in Chapter 27.

The third point of interest is the obvious one that the properties of the diffuse layer are determined by the intensive and not extensive properties of the equilibrium solution. This means that, for example, the ratio of H^+ to Ca^{2+} in the adsorbed layer has some equilibrium value irrespective of the amount of solution with which it is in equilibrium. It is for this reason that pH measurements on temperate soils are often made by determination on a suspension in 0.01 M $CaCl_2$ rather than water, as the former will alter less the diffuse layer conditions – and hence the pH of the solution in the soil pores. Such measurements are discussed in more detail in Chapter 25.

Fourthly, the concentration of anions in the solution is lower in the diffuse layer than in the bulk of the solution. Hence, if a dilute sodium chloride solution is added to a dry sodium clay, and if the clay does not adsorb any chloride, the concentration of chloride in the solution will rise because the anion will be partially expelled from the adsorbed layer. This phenomenon is known as negative adsorption (a possibly confusing term) or better, anion repulsion, and it needs to be taken into account when anion equilibrium in the soil solution is being measured: the more dilute the solution, the more important is the exclusion from the surface region.

Measurements of the anion repulsion of chloride allow the calculation of the effective volume of solution from which the chloride has been excluded and this can be used to calculate the area of the surfaces on which the diffuse layer has developed.[14] It has been shown for 2 : 1 minerals[15] that for electrolytes which contain a cation which is to an extent specifically adsorbed, and so does not form a fully developed diffuse layer, then in consequence the anion is not repelled as much as expected. The calculation thus gives a falsely low apparent surface area. There is also uncertainty for clays such as montmorillonite as to the effect the cation has on the size (and so surface area) of the clay crystallite in suspension. These two points have rather limited the analytical use of the method with 2 : 1 minerals to sodium saturated systems. For tropical soils, usually dominated by kaolinite having a fixed particle size, Uehara and Gillman[16] recommend the use of KCl for area determination by anion repulsion. The fact that this appears satisfactory is interesting confirmation that K^+ exhibits less specific behaviour at variable charge sites than at siloxane surfaces.

Anion repulsion is also important in understanding solute movement in soils. Because anions are excluded from the surface region, the actual volume available for salt movement is less than might be calculated by water-filled porosity alone. Such exclusion will also lead to regions in small (and sometimes 'dead-end') pores becoming incapable of contributing to movement at

---

14 This was first described by R. K. Schofield in *Nature, Lond.*, 1947, **160**, 408 and experimental results given in Schofield, R. K. and Talibudeen, O., *Disc. Faraday Soc.*, 1948, **3**, 51.

15 Edwards, D. G. *et al*, *Trans. Faraday Soc.*, 1965, **61**, 2808 (three papers).

16 Uehara, G. and Gillman, G., *The Mineralogy, Chemistry and Physics of Tropical Soils with Variable Charge Clays*, Ch. 6, Westview Press, Boulder, Colorado, 1981.

all. An example of a model which takes anion exclusion into account in this way has been given by Thomas and Swoboda.[17]

It will be apparent that the main use of the diffuse layer approach is in describing effects, many of which have practical significance, at some distance from the surface in a region beyond that in which surface complexes are formed. It was very early seen as a limitation of the theory that, because it took no account of ion size – let alone the effects of hydration, its prediction of, for example, the concentration of ions close to the surface was bound to be physically nonsensical. An attempt was made by Stern to distinguish the region about two water molecules thick (0.5 nm) as having different properties to the truly diffuse layer beyond it. The Stern layer was itself divided later into a part immediately close to the surface – the inner Helmholtz plane – and that at least one water molecule distance away from it, the outer Helmholtz plane. Used as a mathematical model this generally improves the application of diffuse layer theory but still sheds little light on the true situation close to the surface. The importance of the Stern approach is that it was the first real attempt to model the interface region by dividing it by notional planes parallel to the solid surface, each plane at a different (and perhaps calculable) electric potential. This idea has been much expanded, as will be indicated below, in modelling the more complicated situation at variable charge surfaces.

In summary, a diffuse layer of ions may extend from a charged surface, but only after the ions in solution have formed any specific interaction with the surface. The ions which typically exhibit diffuse layer behaviour are those which do not on the whole form such complexes. These include the cations Li^+, Na^+, Mg^{2+} and Ca^{2+} at charged siloxane surfaces; whereas at hydroxylated sites Ca^{2+} and perhaps Mg^{2+} are more likely to be specifically adsorbed; K^+ may be specifically adsorbed or be present as a diffuse layer ion. Diffuse layer anions include $HClO_4^-$, NO_3^-, Cl^-, Br^-.

Surface complexes

It is now necessary to examine the interface region, in which surface complexes with solution ions are formed. The modern description, following on from the discussion above on solvation properties of ions, makes a distinction between those ions which can balance charge at the mineral surface itself and form so-called *inner sphere complexes* and those which are kept apart from it by a single water molecule to form *outer sphere complexes*. This is a similar, though not quite identical, idea to the *inner and outer Helmholtz planes*. The reason why complexes occur will of course be due to the properties of both the ions in solution and the surface itself, as the specificity of

17 Thomas, G. W. and Swoboda, A. R., *Soil Sci.*, 1970, **110**, 163.

K^+ at the ditrigonal cavity demonstrates. Chemical reactions though, are far more likely when the surface itself is populated with discrete chemical groups able to ionize a proton into solution. It has been seen in Chapter 6 that the hydroxylated variable charge surface is common in most soils, especially weathered tropical ones containing $1:1$ minerals and hydrous oxides. Because the electric charge in such materials is right at the mineral surface, their ability to form both inner and outer sphere complexes will be enhanced, but the fact that the charge is also part of a reactive chemical group provides its more significant property. Virtually all anions of interest to plant growth, other than the diffuse layer pair Cl^- and NO_3^-, form firmly bound surface species by making partly covalent bonds at hydroxyl sites. The anions do this not by addition, but by complete replacement of the charged group at the surface. At its simplest, the process can be written for a positively charged surface site as:

$$S-OH_2^+ + A^{x-} \rightleftharpoons S-A^{(1-x)} + H_2O$$

where S is the solid and A^{x-} an anion. This process is referred to as *ligand exchange* because both OH_2 and A form part of the coordination shell of the metal ions at the surface of the solid. In the most obvious sense, the new ligand A is forming an inner sphere complex (there is no water molecule between it and the surface), but it is worth distinguishing the ligand mechanism separately because such surface complexes are much less easily replaceable than a species attracted (only) electrostatically.

The above mechanism is only part of the story. Many of the ligands at the particle surface are hydroxyl, rather than water, and the more so as pH rises. Surface hydroxyl ligands would be expected to be more strongly held to the metal cation of the solid and thus less easily replaceable than water ligands by some other ligand species from solution. It is therefore assumed that a probable first step is the protonation of the hydroxyl ligand

$$S-OH + H^+ \rightarrow S-OH_2^+$$

Although S–OH is shown for simplicity as an uncharged group, it may in fact carry a part negative charge, but the charge will depend on the composition of S. The step will be encouraged if the anion itself has a protonated form in solution: an example of this is provided by H_4SiO_4, silicic acid, which is the conjugate acid of the silicate anion $H_3SiO_4^-$. The surface hydroxyl, being usually negatively charged, encourages the ionization of the proton from the acid, and then the proton is added to the surface as above. It has been shown experimentally[18] that the surface does indeed become more negatively charged as anion sorption proceeds, and this is certainly in accordance with the displacement of the water ligand. This evidence also illustrates a vital distinction between the ions which take part in surface complex formation

18 Hingston, F. J. *et al.*, *Nature, Lond.*, 1967, **215**, 459.

and those which are part of the diffuse ion swarm: the interface ions are charging ions. Their adsorption changes the net surface charge density and so also the resultant electric potential at points away from the surface. Another way of putting this is to say that the new ligand itself becomes a part of the surface, the properties of which are now influenced by what has been adsorbed. It is this fact which makes the charge characteristics of a soil particle difficult to measure unambiguously and which has led to confusion in, for example, the definition of zero points of charge (see below).

The main characteristics of anion adsorption at ligand sites will now be described. First, the anions which behave in this way are usually oxyanions derived from weak acids. This ensures that there is (at least) one protonated form possible. Examples are the orthophosphate anions ($H_3PO_4/H_2PO_4^-$, HPO_4^{2-}, PO_4^{3-}), silicate ($H_4SiO_4/H_3SiO_4^-$), borate ($H_3BO_3/H_4BO_4^-$), molybdate ($H_2MoO_4/HMoO_4^-$) and selenite ($H_2SeO_3/HSeO_3^-$). When such anions act as replacing ligands according to the scheme outlined above, surface species of the following type are obtained:

(a) $Al-OH^{\delta-} + H^+ \rightarrow Al-OH_2^{(1-\delta)+}$

$Al-OH_2^{(1-\delta)+} + H_3SiO_4^- \rightarrow Al-OSi(OH)_3^{\delta-} + H_2O$

(b) $Fe-OH_2^{(1-\delta)+} + HPO_4^{2-} \rightarrow Fe-O-PO_2OH^{(1+\delta)-} + H_2O$

In these examples the hydroxyl ligand has been given the charge $\delta-$; the Al and Fe cations are part of hydrous oxide solids or occur at the edge of clay crystals. It will be seen that oxyanion ligands are well suited to this replacement since, so far as the metal cation of the solid is concerned, its surrounding coordination shell of oxygen atoms is maintained despite the change of ligand. So long as the replacing species can be accommodated on the atomically non-smooth surface either in terms of its bulk or interactions with other surface groups, then the process is energetically favoured. Sometimes the surface topography and disposition of other surface groups actually encourages the formation of strong surface complexes. Of the anions mentioned, phosphate is by far the most important agriculturally, and it is well known that it often becomes extremely strongly adsorbed on a wide variety of soils. This is because it is thought that phosphate is frequently able to form a bridging ligand by making a second replacement of a surface hydroxyl when this is geometrically feasible:

Much of the work to demonstrate this has been carried out using the iron oxide, goethite. The surface arrangement of hydroxyl ligands on its predominant 100 crystal surface has been shown to be exactly compatible with the size of the phosphate bridge and binuclear sorption has been unambiguously demonstrated.[19] However, a similar adsorption is also possible at pairs of Al–OH sites. It will be seen that adsorption of phosphate on, say, goethite has become tantamount to forming a partial surface skin of iron phosphate on the iron oxide particle, and this has consequences for both the oxide as an adsorbing surface and the subsequent solubility relationships of Fe and P in the soil solution. It provides some explanation for the extremely low desorption of phosphate by water alone: the phosphate may only subsequently be released by a favourable pH change – the addition of OH – which, in the field, though only rarely encouraged in laboratory experiments, may happen microbially.

pH is an important determinant of the ligand reaction, as would be expected in a mechanism involving hydroxyl surface species. The most

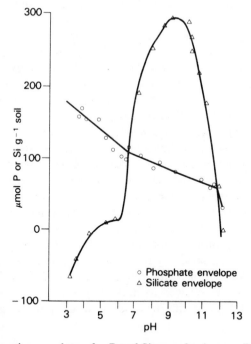

Fig. 7.2   Adsorption envelopes for P and Si on a clay loam soil formed on olivine basalt from Northern Ireland. (From Obihara, C. H. and Russell, E. W., *J. Soil Sci.*, 1972, **23**, 105.)

19 Parfitt, R. L. *et al.*, *J. Chem. Soc. Faraday I*, 1976, **72**, 1082.

common observation is that the greatest amount of anion adsorption occurs at the pH equal to the first pK of the anion irrespective of the charge on the solid. The graph of maximum adsorption against pH is referred to as an adsorption envelope, and Fig. 7.2 provides an illustration. For silicate there exists only one conjugate acid under soil conditions, and the envelope for Si is a broad peak at a pH of about the pK value of 9.6. Orthophosphoric acid ionizes to give three anions; as pH is increased the valence of the P species also increases and adsorption, as indicated by the envelope, declines steeply. In this case, a maximum at the pH equal to pK_1 is not seen as it is below the experimental range.

The correlation between the pH and the pK values, and features of the adsorption envelope provide the simplest summary of the behaviour of a wide range of weak acid anions, but it has not been possible to explain all characteristics of the adsorption in terms of the properties of the solution ion. As adsorption proceeds the surface itself becomes more and more negatively charged, and this discourages further anion bonding. Surfaces themselves differ not only in the density of ligand sites and point of zero charge (see below), but also in the chemical reactivity of the sites to exchange. All of these features have been shown to affect the amount of uptake under given conditions.[20]

Four further anions of importance in soil solution need some mention in this context. Fluoride, though not an oxyanion, turns out to behave in a very similar fashion to the group already discussed. This is because F$^-$ has similar ionic characteristics to OH$^-$ and often substitutes for it in mineral structures. Thus surface species such as Al–F$^{\delta-}$, corresponding to Al–OH$^{\delta-}$, are easily formed (just as in solution, fluoride complexes with aluminium at least as strongly as hydroxyl). Indeed, the amount of chemically active surface OH groups, especially in soil containing allophane, can be estimated directly by the release of alkalinity following the addition of NaF, usually at a constant pH of 6.8.[21] Secondly, both sulphate and carbonate show some evidence that they can form outer sphere complexes at low solution activity as well as forming ligand bonds. It has been shown that sulphate can form bridging ligands in a similar way to phosphate, but the evidence for this is mainly through the interpretation of infrared spectra.[22] Comparisons are made between the absorption bands of the solid before and after sorption, and differences are postulated as being due to the formation of the new bonds of the surface complex. As well as helping to confirm the bridging ligand mechanism for sulphate in this way, infrared analysis has also given the

20 For an authoritative survey of the mechanisms involved, see R. L. Parfitt's review on anion adsorption by soils and soil materials in *Adv. Agron.*, 1978, **30**, 1.
21 Perrott, K. W. *et al.*, *J. Soil Sci.*, 1976, **27**, 58.
22 Parfitt, R. L. and Smart, R. St. C., *J. Chem. Soc. Faraday I*, 1977, **73**, 796.

extremely valuable information that ligand anions only exchange with surface hydroxyls which themselves have only a single coordinate link with the cation in the solid. Thus, a group such as Fe(OH)Fe will not take part in exchange reactions. Sample preparation for infrared work involves the drying of specimens and this may encourage the formation of binuclear ligands in cases such as sulphate; even NO_3^- has been shown to be ligand bonded after intense drying,[23] something which certainly does not happen in a normal moist soil. Such observations carry the interesting implication that wetting and drying sequences, infrequently examined experimentally, may be important in surface compound formation in soils. Certainly in a continuously moist soil sulphate is more weakly held and its adsorption isotherm more reversible to concentration than for phosphate. This has been extensively tested in both the laboratory and the field in New Zealand,[24] where the ability of phosphate to displace sulphate, but not vice versa, has been clearly shown.

Finally, there is the question of organic acid anions. This is a matter of key interest because soluble fulvic material is often described as a polynuclear weak (oxy)acid, and so should behave as a multi-ligand. Careful experiments have been done using low molecular weight organic acids and the evidence is that, with the probable exception of the simplest anion acetate, all others (for example, oxalate, citrate, benzoate) adsorb by a ligand exchange mechanism. Work using humic and fulvic extracts from soil[25] show a similar picture, but the evidence is impossible to make properly quantitative. It is very likely that an organic molecule with many reactive surface groups can make numerous ligand bonds due to the flexible three-dimensional shape apparently characteristic of organic colloids, and that very stable mineral/organic complexes are formed as a result. This is one of the reasons why it is often extremely difficult to free a humic suspension from inorganic 'ash', even by repeated solubilization and reprecipitation cycles; the components remain linked to each other, even in the colloid phase.

The position with respect to the adsorption of anions can now be summarized. At a constant negative charge surface, where the only mechanism for sorption would be specific at the siloxane ditrigonal cavity, only F^- – and then only weakly – has the possibility of exchange with OH. All other anions will be repelled. However, at variable charge surfaces, and these can include 2 : 1 clay mineral edges as well as 1 : 1 clays and oxyhydroxides, the picture changes radically. Anions of interest to plant growth can be divided into the following groups:

23 Parfitt, R. L. and Russell, J. D., *J. Soil Sci.*, 1977, **28**, 297. This important paper summarizes IR evidence for a wide range of anions.
24 Gregg, P. E. H. and Goh, K. M. have reported the field behaviour of sulphate following treatment with ^{35}S-labelled gypsum. *New Zealand J. agric. Res.*, 1978, **21**, 593; *ibid.*, 1979, **22**, 425 (also further papers in series).
25 Parfitt, R. L. *et al.*, *J. Soil Sci.*, 1977, **28**, 289.

1 Ligand bonded
   (a) Form two or more ligand bonds and so are poorly desorbable, e.g. organic acids (not acetic), fulvic and humic material, orthophosphate, selenite.
   (b) Form a single ligand bond, and so are more easily desorbable than 1(a), e.g. silicate, borate, fluoride.
   (c) May form ligand bonds, but may alternatively form outer sphere complexes, e.g. carbonate, sulphate.
2 Non-ligand bonded
   These anions are usually to be found as outer sphere complexes or in the dissociated diffuse layer, e.g. chloride, nitrate. These two important anions, because they have a small degree of complexing ability, may show a little specificity in their adsorption on to soil components. Only per-chlorate is regarded as showing completely non-specific behaviour, and because of this has been recommended for laboratory experiments in which an indifferent anion is required.[26]

It should be noted finally that the sorption of anions has been discussed almost wholly in terms of the simple species. There is so far little quantitative information concerning the behaviour at surfaces of the charged complexes that are formed in soil solution. This can be seen as part of the general ignorance which exists of the formation and stability of complexes with three or more components (of which the surface site may be one).

Sorption of metals

The elements of the periodic table classified as metals can, for the purpose of discussion of sorption properties in soil, be divided into two broad groups. First there is the A Group: alkali and alkaline earth metals, principally Na, K, Mg and Ca. Although these metals do form solution complexes (especially Ca which is to an extent intermediate between the A and B groups being described) the property does not dominate their chemistry. This means that complex formation can often be disregarded and, for example, cation exchange of these ions described by simple equations (see p. 277). Most other metals – the B Group, such as Cu, Zn, Cr and Mn, and the transition series Fe, Co and Ni – form solution complexes strongly, and this fact must be taken into account in understanding their sorption on to variable charge surfaces. There are two considerations which help in giving a qualitative understanding of the behaviour of these metals. In the first place they often form complexes with OH^-, i.e. they become oxycations (for example, $ZnOH^+$). By analogy with the behaviour described above for oxyanions, it

26 See ref. 10.

will be expected that oxycations will ligand exchange at suitable sites on a surface, probably releasing protons. The characteristics of the envelope will also be the opposite of what is shown in Fig. 7.2. In this the high pH limb of the curve is steeply descending (negative ions being repelled from an increasingly negative surface), but for metals it is the lower pH (upward) limb that is steep (positive species being attracted to an increasingly negative surface). Figure 7.3, taken from the work of Kinniburgh and Jackson, shows the adsorption (plotted as per cent) of some metals on amorphous alumina and ferric hydroxide gels over the pH range 4 to 10. The rather different behaviour of Mg and Sr from that of the other metals should be noted. There is no downward limb to the adsorption curves at high pH because the metals are precipitated from solution.

The second point to note about metal adsorption is that as a consequence of complex formation in solution, there must be competition between the free metal ions, the charged complexes containing the metal, and the complex forming species itself for sites on the surface. There can therefore be no quantitative prediction possible about the extent of metal sorption unless the equilibrium constants are known for all possible interactions. Progress has been made in understanding the adsorption of Cu on simple minerals from

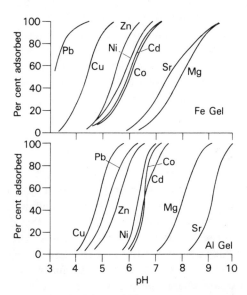

Fig. 7.3 The per cent adsorbed from an initial solution concentration of 1.25×10^{-2} M (in 1 M $NaNO_3$ background electrolyte) for eight metals on to freshly precipitated iron hydroxide and aluminium hydroxide gels. (From Kinniburgh, D. G. *et al.*, *Soil Sci. Soc. Am. J.*, 1976, **40**, 796.)

solutions containing only one ligand other than OH by Farrah and Pickering.[27] Soil solutions will of course contain dissolved organic carbon with a variety of complex-forming ligands, so an authentic description will require considerable computing power. It will also be dependent on a reasonable model for the organic material itself, a problem referred to at the beginning of this chapter.

Summary

In the previous sections a major distinction has been drawn between diffuse layer behaviour, which to some extent will always be a feature of an equilibrium between mobile ions and a charged particle, and complex formation at the surface due to inner or outer sphere bonding. The latter is a charging process and changes both the electrical and chemical features of the surface. Complexes occur at both constant charge and variable charge surfaces. At the former, specific adsorption can occur, especially at the ditrigonal cavity where some monovalent ions can lose a water molecule of hydration and form inner sphere complexes; ligand bonding is not important. In contrast, the latter is often the key mechanism for adsorption on to variable charge oxy-sites whether they exist as part of a mineral surface, as groups on a ligand sorbed on to a mineral or as organic matter. Variable charge minerals in soil will very often have some permanent charge also and Fig. 7.4a, which is a schematic picture of the surface and its associated planes in the solution, incorporates this. The potential drop across the planes is linear in the surface region, but of course exponential across the diffuse layer. Figure 7.4b is also schematic, and attempts to show where adsorption of oxyanions, oxycations and simple cations may occur. Such diagrams – the planes may be defined in a number of ways – are the basis for model calculations.

Surface charge

Given a model of the surface region and the way it becomes modified by complexing species, it is possible to understand something of both how the overall charge varies with soil conditions and why it is that different methods of measuring it give rise to different numerical results. The negative charge on a soil has often been identified with the CEC, and this important variable will be considered below. However, it is first useful to consider under what circumstances a variable charge surface may show no net charge – that is, when negative and positive charges are equal in number.

27 Farrah, H. and Pickering, W., *Aust. J. Chem.*, 1976, **29**, 1167, 1177.

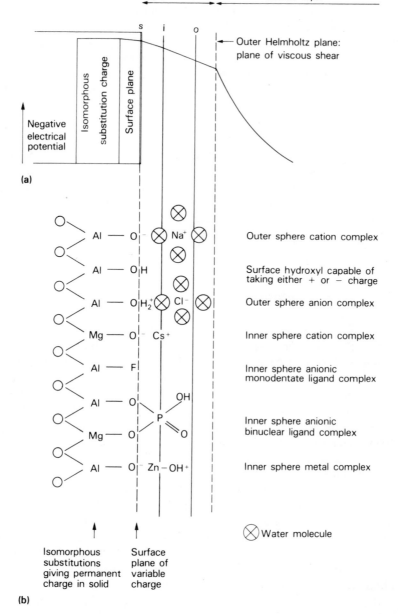

Fig. 7.4 Schematic representation of complex formation in the adsorption region of a variable charge surface of a mineral carrying some permanent charge. The s plane is that of the surface hydroxyls, and the i and o planes are associated with inner sphere and outer sphere complexes, respectively. (a) Based on Fig. 1 of Kleijn, W. B. and Oster, J. D., *Soil Sci. Soc. Am. J.*, 1983, **47**, 821; (b) based on Sposito, G., *Surface Chemistry of Soils*, p. 178, Oxford, 1984.)

The charge on a surface can be expressed in two ways. First it is the sum of the permanent isomorphous substitution charge density (σ_o) and the charge due to the ionization of protons from the surface – the pH-dependent variable charge density (σ_H). These will be balanced by an equal but opposite counter charge – and this is the second way of expressing the particle charge – found in the surface complexed layers. Using the terminology of Sposito, this balance may be expressed as

$$\sigma_o + \sigma_H = - (\sigma_{is} + \sigma_{os} + \sigma_D)$$

where σ_{is} and σ_{os} are the surface charge densities of inner sphere and outer sphere complexes, respectively, and σ_D is the dissociated charge density in the diffuse layer. Of the five quantities, only σ_o can be expected to be a constant, irrespective of experimental procedures. Measurements of the variable charge component, σ_H, are therefore seen to depend crucially on the balance of ions complexed or dissociated from the surface during the measurement itself.

Points of zero charge

A point of zero charge is associated with variations in σ_H, so it is understandable that the definition of this term is the pH value at which the zero charge is considered to occur. For many years it was thought that variable charge soils and minerals were characterized by a single point of zero charge (PZC) and that this was only a function of the ionization equilibrium of protons from water ligands on the surface. Strictly though, if the zero point does only refer to σ_H ($\sigma_H = 0$) then what is being defined is a point of zero net proton charge, PZNPC. This should be measurable by acid/base (potentiometric) titration, in which the difference between the amounts of H^+ and OH^- adsorbed by the sites can be estimated by the proton consumption of the suspension compared to that of the suspending electrolyte solution alone. In fact, it is difficult, if not impossible, to be sure that proton consuming reactions other than the simple equilibrium with oxy-sites are taking place: cation exchange of H^+, complex formation and solid dissolution are other (simultaneous) possibilities. It is consequently unlikely that the PZNPC can unambiguously be measured except for well-characterized hydrous oxides. In 1979, Parker and his colleagues in Virginia recommended[28] two further zero points, both based on methods of charge measurement. The first is related to PZNPC (and in consequence has the same problem with respect to proton consumption) in that it is also usually determined by acid/base titration: several are carried out, but over a range of suspending electrolyte concentration (strictly,

28 Parker, J. C. *et al.*, *Soil Sci. Soc. Am. J.*, 1979, **43**, 668: an important paper in the development of ideas on surface charge.

ionic strength). The pH at which the titration curves intersect is the point of zero salt effect (PZSE), and some examples are shown in Fig. 7.5. The reason why the crossover point has been regarded as a legitimate estimate of PZC can be understood by going back to the electrical condenser model of the electrified interface, mentioned above in the account of the diffuse layer. The mathematical derivation indicates that the surface charge is related to the potential there by a capacitance term directly including the (square root of the) solution concentration. Thus when the charge is zero, so must the potential, irrespective of the value of the concentration: hence all poten-

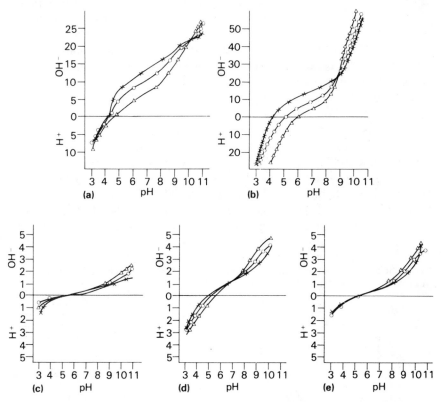

Fig. 7.5 Points of zero salt effect (PZSE) for (a) aluminium hydrous oxide, (b) ferric hydrous oxide, (c) kaolinite, (d) kaolinite coated with ferric hydrous oxide and (e) kaolinite coated with aluminium hydrous oxide measured in 10^{-3} M (x), 10^{-2} M(o) and 10^{-1} M($\triangle$) NaCl. The units of the ordinate are cmol kg^{-1}. (From Hendershot, W. H. and Lavkulich, L. M., *Soil Sci. Soc. Am. J.*, 1983, **47**, 1252.)

The Fe curves (b) and (d) cross clearly at the PZSE and the coat increases the value for kaolinite; (c) and (e) are similar but the curves remain separate and have a PZSE below pH 3. The Al hydrous oxide alone shows curves which cross at a pH above 9 but shows complex behaviour at low pH.

tiometric titration curves should cross at a single pH which defines that zero point. The methodology means that the PZSE actually occurs when $\partial\sigma_H/\partial\mu = 0$, and this is not the same conditions as $\sigma_H = 0$. (μ is the ionic strength.)

The second term introduced by Parker is the point of zero net charge, PZNC. This is determined by measuring the amounts of both cations and anions of a swamping electrolyte which are adsorbed by the soil over a suitable pH range, a method first used by Schofield in 1949. In one of the classic papers in soil science (and with one of its most memorable literature references!),[29] he estimated the surface adsorption of NH_4^+ and Cl^- on to the surfaces of an illitic subsoil following equilibration with NH_4Cl over a range of pH values. The results are given in Table 7.2. Although Schofield corrected for anion repulsion, this has been omitted from the table for simplicity. The Rothamsted soil used showed no zero point within the pH range 2.5 to 7.5 because its charge characteristics were dominated by the permanent negative charge on the illite; but the table shows how the net charge – the difference between the negative and positive surface charge – varies with pH. For a zero point to be shown, σ_H must of course be sufficiently important that enough positive charge is generated at low pH.

Figure 7.6 shows results obtained for two New Zealand soil clays derived from volcanic soils compared with the oxides, silica and alumina, of which the clays are principally composed.[30] The charge values were obtained by the Schofield method, but using NaCl, and show PZNC values at pH values between 5 and 6. It will be noticed that in this case the electrolyte used consisted of ions unable to form inner sphere complexes (an 'indifferent' elec-

Table 7.2   Positive and negative charges on a Rothamsted subsoil determined by the Schofield method from 0.2 M NH_4Cl (see text). The charge is expressed in cmol kg^{-1} oven-dry soil

pH	Excess of NH4 over Cl adsorbed	Positive charge	Negative charge
2.05	19.5	2.0	21.5
2.3	21.3	1.8	23.1
2.6	21.7	1.7	23.4
3.1	21.6	1.7	23.3
3.3	21.7	1.6	23.3
3.8	22.3	1.1	23.4
5.5	24.0	0.5	24.5
6.2	25.7	0.1	25.8
7.15	27.0	−0.1	26.9
7.4	28.2	0.0	28.2

29 Schofield, R. K., *J. Soil Sci.*, 1949, **1**, 1.
30 Perrott, K. W., *Clays Clay Miner.*, 1977, **25**, 417.

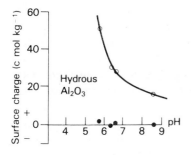

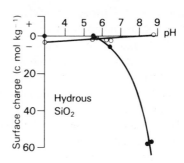

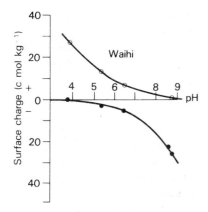

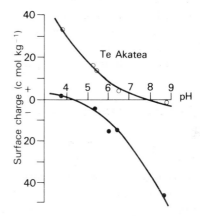

Fig. 7.6 Surface charge variation with pH for two hydrous oxides and a pair of allophanic soils (vitric andepts) from New Zealand.
o – positive charge
● – negative charge
For reference see text.

trolyte), and thus not able to displace any already present. The condition for the PZNC thus determined must therefore be that $\sigma_{os} + \sigma_D = 0$. Had the electrolyte been able to displace all inner sphere complexes then $\sigma_{is} + \sigma_{os} + \sigma_D = 0$. The exact numerical value of the PZNC is therefore critically dependent on the ability of the electrolyte to form surface complexes. In retrospect, Schofield's results are, in detail, difficult to interpret because NH_4^+ acts to some extent like K^+ and may well have formed – to an unknown extent – some inner sphere complexes on the siloxane surface of the illite; i.e. the net charge measured was method dependent.

Having now defined PZNPC, PZSE and PZNC it may be wondered what is left of the old idea of the PZC. In fact, this too is definable, and is different again from the other three. The condition of there being no charge on the particle must mean from the charge balance equation above, that the sum of all charges on the surface is zero, and so σ_D (the diffuse layer charge) is zero. This *defines* the surface as consisting of surface complexes as well as (mineral) surface groups. The PZC is the most difficult to measure with certainty, but it can be attempted by procedures which measure the electrophoretic mobility of suspended particles.[31]

The conditions under which these four points of zero charge operate have been discussed by Sposito.[32] It is sufficient to emphasize here that the differences between them can be understood through appreciation of the methods used to determine them, and by always asking the question about the possible formation of surface complexes which themselves charge the surfaces. It is also useful to make the generalization that if inner sphere complexes are not formed (for example by using 1 : 1 indifferent electrolytes for swamping or displacement steps) then the PZC, PZNC and PZSE are the same. A further consequence is that, under the same condition, the permanent charge density δ_O can be determined precisely by the Schofield method at the PZNPC.

Cation exchange capacity (CEC)

It is helpful to begin with to observe that the term CEC is concerned with the capacity of the soil to exchange cations, not simply its capacity for exchangeable cations. The term is often used as though it were equivalent to cation capacity – with the corollary that such a capacity is equal and opposite to some fixed negative charge on the particles. CEC has the virtue that it describes a rather abstract idea – the negative charge density of a soil – by a description of its effect on a property of interest in terms of nutrient holding power. In fact, in the light of the discussion on zero points, the value of the CEC will be dependent on any surface complexes formed by the electrolyte used to measure it, and of course it will depend on the pH of the electrolyte through its influence on σ_H.

In essence, all methods of determining CEC are variations of the Schofield method, though only that part of it which measures cation holding. Either, adsorbed cations are replaced by others from an electrolyte such as KCl, then measured and summed, or the added (probe) electrolyte cation itself is replaced by another and determined. In the former case, the cation sum (usually Ca, Mg, Na, and to an extent Al) is referred to as the effective cation exchange capacity, ECEC. It has the advantage that, because the electrolyte is unbuffered, the cation exchange occurs at the pH of the soil, and this may

31 James, A. M., *Surface Colloid Sci.*, 1979, **11**, 121.
32 Sposito, G., *Soil Sci. Soc. Am. J.*, 1981, **45**, 292. This important paper should be obligatory reading for any student of soil chemistry.

better reflect field behaviour. A probe electrolyte, on the other hand, is often buffered to a known pH: examples are ammonium acetate at pH 7 or barium chloride–triethanolamine (Ba–TEA) at pH 8.2. In any case the probe cation should always be specified in the quoted CEC by expressing the units as, for example: c mol (NH_4^+) kg^{-1}. The CEC is thus identified as a component of PZNC, with the qualifications on this already noted. The same can be said of measurements of anion exchange capacity (AEC) by determination of the amount of anion adsorbed.

CEC of variable charge soils

It is now recognized that (both cation and anion) exchange capacities of variable charge soils – mostly of the tropics – should not be measured by the methods long used for the largely permanent charge soils of temperate regions where such techniques were first developed. Table 7.3, taken from the work of Sanchez,[33] gives a comparison between the ECEC of several Colombian soils and estimates of their CEC buffered at pH values of 7 and 8.2. As would be expected from their predominantly kaolinitic mineralogy, the contribution of σ_H is developed strongly as pH rises. It should be noted that because the base saturation, which is the percentage of the exchange capacity taken up by cations other than the acid ions H$^+$ and Al^{3+} (see Ch. 25) will be dependent on the CEC figure, systems of soil classification such as *Soil Taxonomy* (see Ch. 24), which depend on base saturation to differentiate classes, have carefully to specify laboratory methodology for exchange capacity. In its absence there are practical as well as logical difficulties of definition.

The effect of variation in pH on the CEC of soils containing both variable and permanent charge components has been discussed by Uehara and Gillman.[34] The theory combines diffuse layer theory and estimates of the reversible surface charge. The diffuse layer approach expresses the variable surface charge density (including complexes) as:

$$\sigma = \left(\frac{2c\epsilon kT}{\pi}\right)^{\frac{1}{2}} \sinh \frac{ze}{2kT} \psi_o$$

where ψ_o is the surface potential
    c is the concentration of counter ions, valency z, in the equilibrium solution
    ϵ is the dielectric constant of the solution
    e is the electronic charge
and k, T and π have their conventional meanings

33 Sanchez, P. A., *Properties and Management of Soils in the Tropics*, Ch. 4, Wiley, 1976.
34 See ref. 16. This book together with Theng, B. K. G. (ed), *Soils with Variable Charge*, New Zealand Soc. Soil Sci., 1980, are valuable texts for this topic.

Table 7.3 Comparison of the ECEC (M KCl) with the CEC determined at pH values 7 and 8.2 for some Colombian soils; also tabulated is the consequent apparent difference in base saturation. (From Sanchez, P. A. quoting Leon, L. A., unpublished PhD thesis, University of California, Riverside, 1967)

Soil	Mineralogy	pH	Organic matter (%)	Cation exchange capacity cmol kg^{-1}			Base saturation	
				M KCl (ECEC)	CaOAc pH 7	Ba-TEA pH 8.2	M KCl (ECEC) (%)	Ba-TEA pH 8.2 (%)
Andept	Allophanic	6.0	6	3	19	23	88	16
Andept	Kaolinite-intergrade	5.2	36	8	65	110	4	1
Tropept	Kaolinite-intergrade	4.4	4	5	13	18	6	2
Aquept	Kaolinite-intergrade	4.5	18	22	42	62	81	29
Humult	Kaolinite-intergrade	4.5	15	6	15	21	50	16
Ustox	Kaolinite-intergrade	4.9	2	1	6	7	23	14
Orthox	Goethite	5.1	10	3	17	24	74	10

For a variable charge surface, however, there is a further constraint: the potential, ψ_o, is controlled by the activity of the potential determining H^+ ions. This dependency can be expressed by the Nernst equation

$$\psi_o = \frac{2.303kT}{e}(pH_o - pH)$$

where pH_o is the point of zero charge.

Bringing these two equations together results in a relationship between surface charge and pH

$$\sigma = \left(\frac{2c\epsilon kT}{\pi}\right)^{\frac{1}{2}} \sinh 1.15z(pH_o - pH)$$

and indicates that the negative charge on such a soil increases with electrolyte concentration (c), counter ion valence (z) and the difference ($pH_o - pH$). Thus to measure the exchange capacity of an oxidic soil which in the field has, say, a pH of 4 and a soil solution concentration of 1 mmol charge per litre by using a molar salt at pH 8.2 is obviously incorrect. In cases like this it is more realistic to measure unbuffered ECEC. An even more useful method is based on that of Bascomb.[35] In this the soil is saturated with Ba^{2+} by washing several times with $BaCl_2$. The supernatant is discarded and dilute $MgSO_4$ solution added. This displaces all the retained Ba by compulsive exchange because the highly insoluble $BaSO_4$ is precipitated within the suspension and an equivalent quantity of Mg taken up by the soil. This amount is easily estimated by its drop in solution concentration. The advantage of this procedure is that it can be carried out at the pH and electrolyte concentration of the original soil solution. Uehara and Gillman suggest that the ionic strength (μ) of the solution be adjusted by using the electrical conductivity (EC in mS cm^{-1}) of the 1:5 soil:water suspension and applying the formula

$$\mu = 0.0446 \, EC - 0.000 \, 173$$

This step may be avoided if the value of μ of 0.006 (0.015 M $MgSO_4$) is taken as being broadly representative of the soil solution in tropical soils. Figure 7.7 gives an estimate of the $CEC/\mu^{\frac{1}{2}}$ relationship (σ is proportional to $c^{\frac{1}{2}}$ above) for some Australian soils.

Although the equation relating charge to pH predicts the general features of the dependency of CEC on other variables, it is in detail incorrect. It does not take surface complexes into account so far as the diffuse layer expression is concerned and, unfortunately, it seems certain that even 'pure' oxides do not obey the Nernst equation exactly. One reason for this is that uptake of

35 Bascomb, C. L., *J. Sci. Fd Agric.*, 1964, **15**, 821.

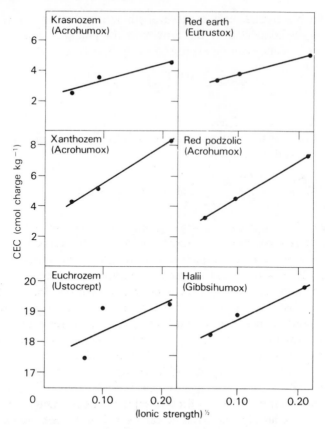

Fig. 7.7   Relationship between CEC and $\sqrt{\mu}$ at pH 5 for six soils. All from north Queensland, Australia, except the Halii (Hawaii). (From Gillman, G. P., *Aust. J. Soil Res.*, 1981, **19**, 93.)

H^+ on the surface inevitably changes the composition of the soil solution, and this also is not taken into account. However, the Uehara and Gillman approach has proved useful in understanding the principles of management for variable charge soils. These frequently have a low net negative charge and the aim of management is to increase nutrient holding power by raising the value of σ. The relationship suggests that this can be done by: (a) increasing pH (liming); (b) decreasing pH_o by, for example, adding organic matter or ions like phosphate or silicate which will increase negative charge by forming inner sphere ligand complexes; or (c) increasing soil solution concentration, especially with divalent ions (and this will be a consequence of liming or the addition of fertilizer).

The foregoing account has dealt essentially with a chemical description of the electrified surface region. It is now necessary to examine the reactions which occur when changes are made in the soil solution phase. The vast majority of agricultural management practices will have their first chemical effect on the soil solution: fertilizer will dissolve and may acidify, irrigation will add or dilute salts, and crop growth itself will redistribute and extract nutrients. Many of these processes are of special interest and have chapters devoted to them, for example, soil acidity (Ch. 25), salinity (Ch. 27) and fertilizer behaviour (Ch. 20, 21, 22). The rest of this chapter will therefore be concerned with the theoretical background which is of common interest to a number of these themes, concentrating especially on adsorption properties and cation exchange.

Adsorption isotherms

It has been amply indicated in previous sections that soil particle surfaces, whether they be of the constant or variable charge type, are able to hold ions derived from the soil solution in close proximity to the solid. In this region there will be built up a surface excess (over what would be expected without an adsorption process) and this acts as a potential store of nutrients able to replenish the soil solution when they are removed by root or microbial uptake. The literature of soil science contains a very large number of papers describing adsorption quantitatively and then attempting to understand the phenomenon in terms of processes at the surface. In fact it turns out that mathematical descriptions of adsorption are never specific to a single process, and other evidence is always required.

The most usual type of experiment carried out is simply to equilibrate subsamples of a soil or soil clay at constant temperature (*isotherm*) with a number of aliquots of solutions containing different concentrations of the adsorbate of interest, and then determine the amount remaining after adsorption. The time to equilibrium varies widely with the system being studied: a simple cation exchange may be essentially completed within minutes, whereas the adsorption of orthophosphate (which may be an acid/base catalysed formation of inner sphere bonds) can continue increasing for two days – and even after this proceed at a very slow rate for some months: in general, the time to equilibrium must be determined in each experimental case. Care must be taken to avoid physical comminution of the soil particles themselves during the equilibration, and it has been recommended that only very gentle agitation of the tubes containing the suspension (on slowly revolving rollers) be permitted in cases when equilibrium times are long.[36]

36 Barrow, N. J. and Shaw, T. C. tested experimental methods for adsorption measurements. See their series on the slow reaction between soil and anions; for P: *Soil Sci.*, 1974, **118**, 380; ibid., 1975, **119**, 167, 190, 311; for Mo: *ibid.*, **119**, 301; for F: *ibid.*, 1977, **124**, 265; for SO$_4$: *ibid.*, **124**, 347.

It may also be necessary to allow for change in electrolyte concentration because of anion repulsion.

The graphical plot of the amount adsorbed by the soil against the equilibrium solution concentration is termed the adsorption isotherm. These almost invariably have a steep portion at relatively low solution concentrations followed by a less steeply rising curve as the concentration becomes higher (see Fig. 7.8a). Curves such as these have been described by one of two well-known isotherm equations, the Langmuir and the Freundlich (or, van Bemmelen–Freundlich), or to a lesser extent by the Temkin equation. The first of these has probably been the most used, and expresses the amount adsorbed, x, at a concentration, c, as:

$$x = \frac{kKc}{1 + Kc}$$

in which both k and K are constants.

A linear plot is given by either c/x against c (which is more common in the literature) or by x/c against x; from either of these graphs values of K and k can be extracted from the linear slope and intercept. It is sometimes found that this linearization process produces two intersecting straight lines, and so an extended Langmuir equation containing two sets of constants, but otherwise exactly as above, is written

$$x = \frac{k_1 K_1 c}{1 + K_1 c} + \frac{k_2 K_2 c}{1 + K_2 c}$$

Because it has been thought that this equation applies because the adsorbate is held by two sets of sites each having a different energy of adsorption, the relationship is sometimes called a two-surface Langmuir equation.[37]

The Freundlich equation is

$$x = ac^b$$

in which both a and b are constants; b is fractional and is sometimes written as $1/n$. A linear plot is obtained for $\log x$ against $\log c$ when this relationship is obeyed. The Temkin equation gives a linear plot for x against $\log c$, as its equation is

$$x = \alpha + \beta \log c$$

with α and β as constants.

In practice all of these equations may be tried in a particular case to produce 'best fit' of data. Figure 7.8b to d shows the linearization of the points shown in Fig. 7.8a for the adsorption of phosphate on two soils from Australia. It may be concluded that in this case the two-surface equation best

37 Holford, I. C. R. *et al.*, *J. Soil Sci.*, 1974, **25**, 242.

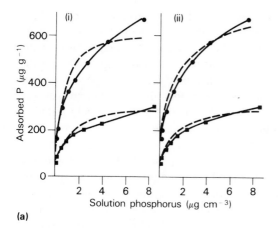

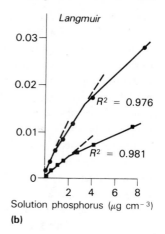

(a)

(b)

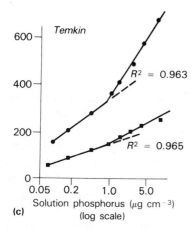

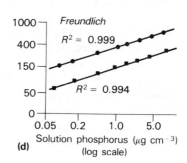

(c)

(d)

Fig. 7.8 (a) Comparison of observed and predicted adsorptions using four
equations.
Observed ● Euchrozem
■ Black earth
Predicted (i) —— two-surface Langmuir
——— one-surface Langmuir
(ii) —— Freundlich
——— Temkin
(b) to (d) Effect of transformation of phosphate adsorption data on to
linear plots of isotherms
(b) Langmuir (c/x against c), (c) Temkin (x against $\ln c$),
(d) Freundlich ($\ln x$ against $\ln c$).
(From Mead, J. A., *Aust. J. Soil Res.*, 1981, **19**, 333.)

describes the experimental observations – and this seems often to be the case for phosphate on a wide range of soils – closely followed by the Freundlich.

Unfortunately, there is no general theory which will predict which of the isotherm equations is most suitable. It has been shown by Sposito[38] that the Langmuir equation is a quite general expression with no single mechanistic model required to provide it, and similarly that the so-called two-surface equation can always be 'obeyed' by isotherms of the type usually found in soil systems. In the absence of independent evidence, the identification of the constants in the Langmuir equations with characteristics of the adsorption (such as K with energy of adsorption) is not justified. The Freundlich expression can be regarded as being produced by a set of Langmuir sites with a log normal spread of energy levels, so similar problems of understanding occur with this also. All this does not mean that the use of adsorption equations is unjustified: for example, they provide a useful method for predicting adsorption at values of solution concentration not used experimentally. However, the equations have no theoretical status for ion adsorption.

Adsorption/precipitation

So general is the Langmuir expression that it can in some circumstances even describe the process of precipitation of a sparingly soluble compound. Confusion as to whether precipitation or adsorption is the dominant process leading to a drop in solution concentration is caused for two reasons. First, many investigators only measure the concentration of one ion of interest as it becomes removed from solution; secondly, if surface species do build up beyond the amount in the monolayer, then a true new compound may indeed be formed on the original surface.

These points may be illustrated by sorption (a word implying neither adsorption or precipitation) on to well-characterized calcium carbonate surfaces. Figure 7.9 gives results for the sorption of Cu, plotted on a log/log scale. In this instance, the complex shape of the relationship clearly indicates that more than a single process operates over the concentration range used. At low activities of Cu it appears that true adsorption does occur, and while no pH change is associated with it there is a small decrease in solution calcium. At very much higher Cu activities (greater than about -8 in the figure) sorption is accompanied by a pH decrease but an increasing Ca activity in solution. These additional observations, combined with careful control of CO_2 partial pressure in the reaction vessel, made it possible through solubility considerations to conclude that $Cu(OH)_2$ was being precipitated at high Cu levels. However, there are a number of changes of slope

38 Veith, J. A. and Sposito, G., *Soil Sci. Soc. Am. J.*, 1977, **41**, 697; Sposito, G., *op. cit.*, 1982, **46**, 1147.

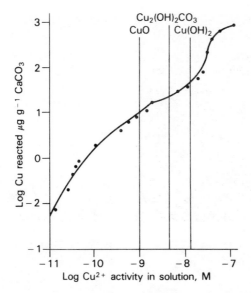

Fig. 7.9 Cu adsorption and precipitation reaction on $CaCO_3$ (log/log plot). Vertical lines represent Cu^{2+} activity for CuO, $Cu_2(OH)_2CO_3$ and $Cu(OH)_2$ at pH = 8.30 and P_{CO_2} = $10^{-3.49}$ atm. (From Papadopoulos, P.I., unpublished PhD thesis, Univ. Reading, 1985.)

in the curve between the initial (adsorption) linearity and the $Cu(OH)_2$ region, and the surface species which form prior to the regular hydroxide cannot be determined with certainty by sorption or solubility data. Electron probe analysis of the surface would be required.

A second example shows that precipitation can be unambiguously demonstrated if enough forms for direct observation. Plate 7.1 shows the build-up of precipitated calcium phosphate on calcite by scanning electron micrographs. The precipitated compound is observed as small crystals on the calcite surface rather than as a surface coat. The form of the precipitate was confirmed to be mainly dicalcium phosphate by Freeman and Rowell[39] using electron probe, X-ray diffraction and isotopic exchange – as well as simple solution measurements. It was this combination of evidence which made the mechanism clear.

Solubility

The reverse process to precipitation is solubility, and a considerable literature exists on the possible dissolution of sparingly soluble compounds into the soil

39 Freeman, J. S. and Rowell, D. L., *J. Soil Sci.*, 1981, **32**, 75.

Plate 7.1   Scanning electron micrographs of calcite reacted with phosphate solutions for 10 days. (a) 0 μg P ml^{-1} in the initial reacting solution, (b) 10 μg ml^{-1}, (c) 25 μg ml^{-1}, (d) 50 μg ml^{-1}, (e) 100 μg ml^{-1} (f) 500 μg ml^{-1}. Scale in μm. Solid : solution ratio = 2.25.

solution. It was shown in the first part of this chapter that, because the calculation of solution activities can now be more reliably achieved, it has become easier to answer the question as to whether some known solid 'controls' the solution concentration with respect to those ions produced by dissolving the solid. So far as soil solutions are concerned, the calculation of the activities of just a small number of ions of interest will probably be insufficient: as complete an analysis as possible of the solution should be made. Solubility considerations have often been used to understand the behaviour of added phosphate in soils, and this example is given without theoretical background in Chapter 21.

The method will be illustrated here by a calculation of the mineral equilibrium between gibbsite and kaolinite. It has sometimes been thought that soil minerals are of such limited solubilities that dissolution reactions can have virtually no significance to solution composition. In fact, not only does mineral solubility have a crucial role in the level of, e.g. Al species in solution, but also in the long term, the production of new colloidal species. Further details of the weathering sequences in which mineral stability plays a part will be found in Chapter 5.

Consider first the solubility of gibbsite, $Al(OH)_3$

$$Al(OH)_3 + 3H^+ \rightarrow Al^{3+} + 3H_2O$$

Assuming the activity of gibbsite (solid) and water each to be unity, then the pK of this reaction is $pAl^{3+} - 3pH$ (p $= -\log$, as for pH). The equilibrium constant, K, however, is also related to the standard free energy change of the reaction:

$$\triangle G_r = -RT \ln K$$

Assuming that accurate thermochemical data exist for free energies of formation ($\triangle G_f$) of the components of a reaction, then $\triangle G_r$ may be calculated simply as the difference between $\triangle G_f$ (products) and $\triangle G_r$ (reactants). For gibbsite solubility this approach gives $\triangle G_r = (\triangle G_{Al_3} + 3\triangle G_{H_2O} - \triangle G_{Al(OH)_3}) = -45\ 934$ J and hence p$K = -8.04$ (at 25 °C).

The same procedure gives for kaolinite solubility:

$$Al_2Si_2O_5(OH)_4 + 6H^+ \rightleftharpoons 2Al^{3+} + 2H_4SiO_4 \text{ (aq)} + H_2O$$
$$pK = 2pAl^{3+} + 2pH_4SiO_4 - 6pH = -8.02$$

The free energy of formation of kaolinite is less certain than that of gibbsite because of possible lattice substitutions and imperfections. The pK may alternatively be determined by the measurement of the experimental values of Al^{3+}, H_4SiO_4 and pH in an aqueous solution in long-term equilibrium with the mineral. The figure of -8.02 above was determined in this way by Kittrick.[40] It will be noticed that both the pK relationship for gibbsite and

40 Kittrick, J. A., *Soil Sci. Soc. Am. J.*, 1980, **44**, 139.

for kaolinite can be written in terms of a compound variable: $pH - 1/3p\ Al^{3+}$

$$\text{gibbsite} \quad pH - 1/3p\ Al^{3+} = 2.68$$
$$\text{kaolinite} \quad pH - 1/3p\ Al^{3+} = 1/3p\ H_4SiO_4 + 1.34$$

These may be plotted as in Fig. 7.10, and the point of intersection of the lines for the two minerals indicates the solution concentration of H_4SiO_4 ($\approx 10^{-4.04}$ M) for which gibbsite and kaolinite are in stable equilibrium. The region above a stability line indicates supersaturation and the region below undersaturation with respect to that component, and either condition has an implication for the stability of the mineral.

It is not always straightforward to interpret solubility diagrams when actual soil solutions are considered. Karathanasis and his colleagues, investigating the equilibrium between the two minerals considered above and hydroxy-interlayered vermiculite, found that in terms of solution activities the kaolinite in the B horizons of the soils they investigated appeared to be more soluble than calculated from the simple formula. The apparent solubility is also indicated in Fig. 7.10. It also appeared that the degree of supersaturation in soil solutions derived by centrifugation from Ap horizons was even more

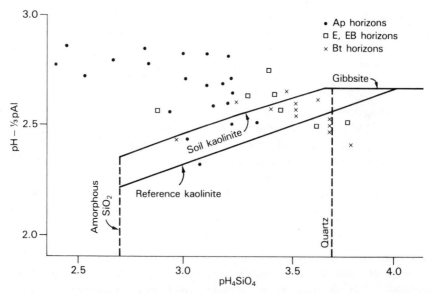

Fig. 7.10   Stability diagram for reference kaolinite, gibbsite and 'soil kaolinite' in the Al_2O_3–SiO_2–H_2O system between the limiting solubilities of amorphous silica and quartz. The plotted points are observed soil solution compositions for selected topsoil and subsoil horizons of 14 Ultisols from southern USA. (From Karathanasis, A. D. *et al.*, *Soil Sci. Soc. Am. J.*, 1983, **47**, 1247.)

pronounced. The reason for this is presumably connected with both the difficulty of accounting quantitatively for the complexing effects of dissolved organic carbon as well as the thermochemically undefined nature of many minerals (in this case the hydroxy-interlayered vermiculite) in weathered soils.

Cation exchange

The ability of soils to exchange one cation for another on leaching was one of the earliest observations in soil studies. The importance of the phenomenon to many processes can hardly be exaggerated: just as adsorption and precipitation theory are widely used to interpret the chemistry of phosphorus in soil, so cation exchange considerations govern much of the behaviour of potassium, another major plant nutrient. Much of the same theory is also involved with one of the most important management problems in modern irrigated agriculture, namely sodicity and salinity. Both of these topics are dealt with elsewhere in this book. The background theory is now highly developed and no attempt will be made to cover it in detail here. It is however important to realize, especially in a simple presentation, that as with adsorption isotherms, cation exchange models say nothing about surface mechanisms: the equations are derived from a consideration of thermodynamic properties of ions, and these are independent of process.

Both of the examples mentioned in the previous paragraph are essentially concerned with the exchange of monovalent for divalent ions on the solid surface. Using K^+ as an example, the equation is

$$CaX_2 + 2K^+ \text{ (aq)} \rightleftharpoons 2KX + Ca^{2+} \text{ (aq)}$$

where X represents the negatively charged surface. It should be noted that the equation assumes that there are enough sites on the surface for the reaction to proceed. Using a traditional mass action approach, the exchange equilibrium constant K_{ex} is:

$$K_{ex} = \frac{(KX)^2(Ca^{2+})}{(CaX_2)(K^+)^2}$$

in which the parentheses ( ) are activities. This is easy to understand in the case of the solution ions but is a more obscure concept for the adsorbed species. In fact, the main difficulty in devising ion exchange models is to provide sound estimates for the activity coefficients of ions held on the exchange complex. The main differences between the several equations found in the literature is in the convention used to interpret expressions like (KX). For example, the procedure used by Vanselow, as early as 1932, was to identify the surface activity with the mole fractions of the two components of the surface phase

$$(KX) = \frac{[KX]}{[KX + CaX_2]} \quad \text{and} \quad (CaX) = \frac{[CaX]}{[KX + CaX_2]}$$

where square brackets indicate the concentration (for example, in mol kg^{-1}) so that

$$K^V = \frac{[KX]^2}{[CaX_2][KX + CaX_2]} \cdot \frac{(Ca^{2+})}{(K^+)^2}$$

Inherent in this approach is that the activity coefficients of the ions on the surface are regarded as unity. Despite this, the Vanselow approach has proved useful, though experience shows that it does not always work well if used over a wide range of surface compositions, or even generally for soils as compared with single minerals.

An alternative idea is to use the concept of charge balance rather than expressing the exchange equation in terms of mole amounts. This emphasizes that only half as much Ca^{2+} as K^+ can be adsorbed on to the surface in order to balance the (assumed) fixed charge

$$2Ca_{1/2}X + 2K^+(aq) \rightleftharpoons 2KX + Ca^{2+}(aq)$$

In this, the number of equivalent charged sites (moles of charge on the surface) is fixed, but to compensate the solution concentrations themselves must now be expressed in moles of charge (equivalents) rather than moles.

$$K_{ex} = \frac{(KX)^2(Ca^{2+})}{(Ca_{1/2}X)^2(K^+)^2}$$

The similarity of this to the Vanselow approach is completed by the assumption that the activity of each ion on the surface is equal to its *equivalent* fraction there. The K_{ex} so derived is the basis of the Gapon approach to cation equilibria. The equation is written as above, but using half quantities

$$Ca_{1/2}X + K^+(aq) \rightleftharpoons KX + \tfrac{1}{2}Ca^{2+}(aq)$$

$$K'_{ex} = \frac{(KX)\ (Ca^{2+})^{1/2}}{(Ca_{1/2}X)\ (K^+)}$$

The Gapon constant, K^G, is then defined using solution concentrations rather than activities:

$$K^G = \frac{\sqrt{[Ca^{2+}]}}{[K^+]} \cdot \frac{[KX]}{[Ca_{1/2}X]}$$

The introduction of concentration may seem inappropriate in such an equation. However, the Gapon equation has been most widely used in salinity studies,[41] and it turns out that the ratio of equivalent concentrations of sodium and (square root) calcium is very similar to the ratio of the corresponding activities over the range usually encountered in practice. The use of the equation in salinity studies is briefly considered below.

Potassium exchange

The ratio of the solution activities, not the Gapon concentration, has been used to provide a measure of the intensity (I) of potassium in soil solution. The intensity, $(K^+)/\sqrt{(Ca^{2+})}$, is sometimes expressed as $(pK - 1/2pCa)$ which in this form has been called the potassium potential in solution. The reason for the inclusion of Ca in this definition is that Ca^{2+} is very often the dominant cation in soil solution, so the exchange properties of K^+ cannot be considered in isolation from the cation with which it is most likely to exchange. If the amount of exchange is limited, for example when potassium fertilizer is added, or potassium taken up by crops, then $[Ca_{1/2}X]$ remains approximately constant. In such a case it can be included as part of a new equilibrium constant:

$$[KX] = K' \frac{(K^+)}{\sqrt{(Ca^{2+})}}$$

Commonly, $[KX]$ is approximated by the amount or quantity (Q) of K^+ adsorbed or desorbed by the soil when it is in contact with a solution of a particular intensity value. A plot of this amount (often written as $\triangle K$) against $(K^+)/\sqrt{(Ca^{2+})}$ is a quantity/intensity (Q/I) relationship. A typical curve (Fig. 7.11) shows a characteristic shape. It often has an approximately

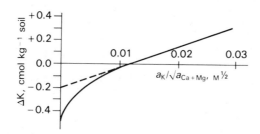

Fig. 7.11 The exchangeable potassium (Q) against activity ratio (I) (Q/I) curve for a soil formed on Lower Greensand in southern England. Mg is assumed to behave in the same way as Ca and is usually incorporated into the monovalent/divalent activity ratio. Note, $a_K = (K^+)$.

---

41 See for example Bresler, E. *et al.*, *Saline and Sodic Soils*, Advanced Series in Agricultural Sciences, Vol. 10, Springer-Verlag, 1982.

linear portion above (K^+) of around 0.01, with a much steeper portion below this. The linear part is thought to represent K^+ adsorbed on planar, relatively low energy sites, and it has been suggested that the curved position represents potassium fixed in geometrically selective adsorption sites in holes, cracks or wedge-shaped spaces at the broken or weathered edges of mica clays. Some evidence for this and further discussion of the use of Q/I curves for characterizing the behaviour of potassium in soils is given in Chapter 22.

Salinity/sodicity

As mentioned above, the Gapon equation itself (using concentration terms) is widely used in understanding the exchange of sodium for calcium in saline soils. In practice, such soils often contain significant amounts of exchangeable magnesium, and this is assumed to behave as an exchange ion in just the same way as calcium. The Gapon relationship is thus

$$\frac{[NaX]}{[(Ca + Mg)_{1/2}X]} = K^G \frac{[Na^+]}{\sqrt{[Ca^{2+} + Mg^{2+}]}}$$

or in charge (equivalent units)

$$\frac{[NaX]}{[(Ca + Mg)X]} = K^G \frac{[Na^+]}{\sqrt{[Ca^{2+} + Mg^{2+}]/2}}$$

This equation leads directly to the definition of the exchangeable sodium percentage (ESP) – a characteristic of the ions adsorbed on the solid, and the sodium adsorption ratio (SAR) – a characteristic of the ions dissolved in soil solution. Assuming Na^+, Ca^{2+} and Mg^{2+} to be the only cations of significance, then

$$ESP = \frac{100[NaX]}{[(Ca + Mg + Na)X]}$$

and $(Ca + Mg + Na)X = CEC$ (all in units of mol charge)

$$SAR = \frac{[Na^+]}{\sqrt{(Ca^{2+} + Mg^{2+})/2}}$$

and so finally

$$ESP = \frac{100SAR/K^G}{1 + (SAR/K^G)}$$

The consequences of soil salinity, the use of the above definitions and the ameliorative management required on such soils are further discussed in Chapter 27.

Cation exchange isotherms

Use has been made of complete binary exchange isotherms in order to answer questions about the free energy of exchange of one ion for another. The thermodynamics which lie behind such a procedure are involved, and the interested reader may be referred to Talibudeen,[42] Bolt[43] and Sposito[44] for up to date accounts of this specialized area.

42 Talibudeen, O. in *The Chemistry of Soil Processes* (eds D. J. Greenland and M. H. B. Hayes), Ch. 4, Wiley, 1981.
43 Bolt, G. H. in *Soil Chemistry, B. Physico-chemical models*, Ch. 2, Elsevier, 1979.
44 Sposito, G. in *The Thermodynamics of Soil Solutions*, Ch. 5, Oxford, 1981. This book, together with its companion volume *The Surface Chemistry of Soils*, 1984, is authoritative.

8

The temperature of the soil

Source and transfer of heat

The soil derives its heat almost entirely directly from the sun and loses much
of it by radiation back into the sky. The opacity of the atmosphere to solar
radiation and to longwave back radiation is therefore an important factor in
quantitative studies on soil temperature. The sun has an effective radiation
surface temperature of 6 000 K, so the incoming solar radiation has a fairly
broad wavelength band centred at about 0.5 μm (shortwave radiation). As
the effective radiation surface temperature of the earth is of the order of 280
to 300 K, it radiates energy in the broad wavelength band 3 to 100 μm but
centred around 10 μm (longwave radiation). The clear atmosphere is fairly
transparent to incoming solar radiation and it has a transparency window
around 10 μm which is bounded on each side by water vapour and carbon
dioxide absorption bands centred on 6 μm and 20 μm, respectively.

Water droplets absorb and scatter both the short and the long wavelength
radiant energy, and quite small amounts of cloud will reduce the input of
solar energy to the soil, but reduce the loss of heat from the soil by absorbing
and re-radiating back to the soil surface longwave energy. The mean fluxes
of solar energy for each month at Rothamsted are given in Table 8.1. Figure
8.1 shows the difference between the actual incident flux of energy and that
if there were no cloud.

In Chapter 11 the fate of the incoming solar radiation that falls on the soil
will be discussed in some detail, and only a brief outline of the heat budget

Table 8.1 Mean value of the daily and monthly radiation at Rothamsted. Period 1956–69

	$kJ\ cm^{-2}\ day^{-1}$	$kJ\ cm^{-2}\ month^{-1}$	
January	0.23	7.1	
February	0.40	11.1	
March	0.77	24.0	
			42.2
April	1.13	33.9	
May	1.55	48.0	
June	1.76	52.7	
			134.6
July	1.53	47.4	
August	1.25	38.9	
September	0.96	28.8	
			115.1
October	0.54	16.8	
November	0.25	7.5	
December	0.15	4.6	28.9
Total for year			320.8

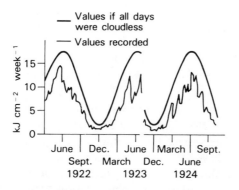

Fig. 8.1 Solar radiation received at Rothamsted Experimental Station, Southern England.

of a soil will be given here. Only a part of the incoming radiation R received by the soil is absorbed, for a fraction s, the albedo of the surface, is reflected or scattered, so the amount absorbed is $R(1 - \alpha)$. The absorbed radiation is dissipated in four different ways: part is re-radiated back to the sky as long-wave radiation, L_o part is used for evaporating water from the soil and is dissipated as latent heat; part may raise the temperature of the surface soil and be dissipated as sensible heat to the air, since it will raise the temperature of the air in contact with it; and part will be conducted into the body of the soil.

When the soil surface is moist, most of the net radiation absorbed, $R(1 - \alpha) - L_o$, is used to evaporate water, but as the soil becomes drier, an increasing quantity is dissipated either as sensible heat to the air or as a heat flux into the soil. Table 8.2 gives an example of this distribution of sinks for the absorbed radiation in a wet and a moist soil in Arizona.[1] In this example the wet soil is dissipating rather more energy as latent heat than it is absorbing as net radiation, so it is drawing heat from the soil and cooling its surface; whereas the moist soil is dissipating only about 70 per cent of the net radiation it is absorbing as latent heat, and 10 per cent is entering the soil and warming its surface. The reason less heat is absorbed by the drier soil is partly because its albedo is greater, so more of the incoming radiation is reflected, and partly that its surface temperature is higher so it is losing more heat by back radiation. This transference of a part of the absorbed heat to sensible heat is the cause of the air temperature following the soil temperature, as shown in Fig. 8.2.

Table 8.2 Dissipation of absorbed energy by a bare soil. Fluxes in J cm^{-2} day^{-1}

Soil	Net radiation	Latent heat term	Sensible heat to air	Heat to soil
Wet	1690	1730	−4	−33
Moist	1370	940	289	142
		incoming radiation 3060		

Figure 8.2 also shows that the temperature of the surface layer for a bare Rothamsted soil on a clear sunny day in July is controlled by the rate it is absorbing energy. The surface temperature is seen to vary in phase with the incoming radiation during the day; but during the night it continues to fall, though much more slowly than during the day. The air temperature curve at 1.2 m above the soil surface lags behind the soil temperature curve during the day and reaches a lower maximum value, but during the evening and night it is almost the same as the soil surface. The soil temperature below the soil surface follows the changes in the surface temperature though it lags behind the surface and the diurnal variation is reduced, and both these changes become more marked with depth. This behaviour is shown in Fig. 8.3 for the bare Rothamsted soil in early August. It shows that at 20 cm depth, the maximum temperature occurs at about 19.00 hours and the diurnal

1 Van Bavel, C. H. M. and Fritschen, L. J., *UNESCO Arid Zone Res.*, 1965, **25**, 99.

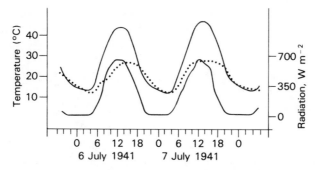

Fig. 8.2 Relation between incoming solar radiation and the air and soil surface temperatures on two cloud-free days at Rothamsted. Upper full curve: Temperature of the surface of the soil. Lower full curve: Solar radiation received by soil surface. Middle dotted curve: Air temperature 1.2 m above the ground (in the screen).

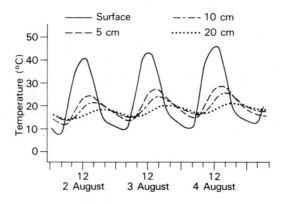

Fig. 8.3 The damping of the daily temperature wave with depth in a bare Rothamsted soil.

variation is already small, and it becomes negligible at about 40 cm: a result found in many other places for clear warm or hot days.[2]

The mean monthly soil temperatures also show a clear seasonal trend, that for the surface soil being approximately in phase with the incoming solar radiation, and the seasonal change becoming smaller, and the time at which the maximum temperature is reached becoming later in the year with depth. Figure 8.4 shows the seasonal change of temperature at three depths for an

2 For Muguga, Kenya, see McCulloch, J. S. G., *Quart. J. R. Meteorol Soc.*, 1959, **85**, 51. For Illinois, Carson, J. E. and Moses, H., *J. appl. Meteorol.*, 1963, **2**, 397. For Arizona desert, Guild, W. R., *J. Meteorol.*, 1950, **7**, 140.

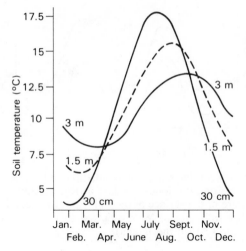

Fig. 8.4   Annual variation of soil temperature with depth under short grass; Oxford, Radcliffe Observatory.

Oxford soil,[3] in this example under short grass. Whereas at a depth of 30 cm the maximum temperature is reached in mid-July and the range between mid-winter and mid-summer is 14 °C, at 3 m it is not reached till September, and the range is only 5.5 °C. In temperate regions the seasonal variation extends to several metres depth, but in equatorial regions, as for example at Muguga, Kenya (latitude 1 ° 13′ S) it is small at 1 m.

Heat flux

Change in temperature of any given volume of soil depends on the difference between the rates of flow of heat, the heat fluxes, into and out of that volume. The heat flux dQ/dt depends on the temperature gradient dT/dz in the soil and on the thermal conductivity k. In any given soil volume the heat flux is given by

$$dQ/dt = -k \; dT/dz$$

where z is measured along the gradient. The thermal conductivity of the soil is very dependent on its air and water content, for the value of k for air is $0.025 \text{ W m}^{-1} \text{ K}^{-1}$ while that for water is $0.6 \text{ W m}^{-1} \text{ K}^{-1}$ and for soil solids about $9 \text{ W m}^{-1} \text{ K}^{-1}$. The more compact the soil, and the larger the water content, the greater its thermal conductivity.

3 Original data by Rambaut, A. A., Radcliffe Observations 1898–1910, and the data for Fig. 8.4 computed from Keen, B. A., *The Physical Properties of Soil*, Longman, 1931.

Since the rate of change of temperature at a given depth depends on the difference between the heat flux into and out of the soil at that depth, and on its thermal capacity, it is given by:

$$\frac{dT}{dt} = \frac{k}{\rho c}\frac{d^2 T}{dz^2}$$

where ρ is the density of the soil and c its specific heat capacity. The quantity $k/\rho c$ is called the thermal diffusivity of the soil, or sometimes its temperature conductivity. The diffusivity (D) initially increases with increasing moisture content since k increases more rapidly than ρc in moderately dry soil, but at higher moisture content k increase less rapidly and so diffusivity reaches a flat maximum and finally decreases again. Figure 8.5 shows the relation between the thermal conductivity, and the diffusivity of a sandy loam soil, where the moisture content has been expressed on a volume basis.[4]

Thermal diffusivity is the soil factor involved in determining the attenuation with depth and the velocity of the daily and annual wave of temperature in soil. The fractional decrease of amplitude with depth is given by $e^{-z\sqrt{\omega/2D}}$ where z is depth, ω is the angular velocity of the daily or annual cycle and equals $2\pi/t$ radians per seconds where t is the number of seconds in a day or a year, and D is the thermal diffusivity of the soil. If, for example, D is 5×10^{-3} cm^2 s^{-1} then this formula shows that the amplitude decreases to half that at the surface at 8 cm depth for the daily wave and 1.6 m for the annual one. At a depth $z = \sqrt{2D/\omega}$ the amplitude of the temperature wave is e^{-1} or 0.37 times the

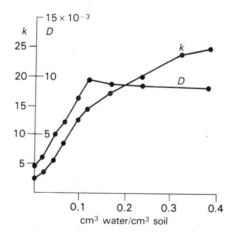

Fig. 8.5   Variation of thermal conductivity, k (W cm^{-1} K^{-1}) and thermal diffusivity, D (cm^2 s^{-1}), of a sandy loam soil with moisture content. The saturated soil held 0.38 cm^3 water cm^{-3} soil.

4 Moench, A. F. and Evans, D. D., *Soil Sci. Soc. Am. Proc.*, 1970, **34**, 377.

amplitude at the surface, and this depth is called the damping depth. For the example given above this is 12 cm for the daily and 2.2 m for the annual wave.

A temperature gradient in an unsaturated soil, that is, a soil that contains air, causes water to move from warmer to cooler parts of the soil, for water will evaporate in the warmer region, diffuse as vapour through the air spaces and condense in the cooler region, so this transfer of vapour also transfers heat which increases the apparent thermal conductivity of the soil.[5] In soils in which the soil water tension does not exceed 1.5 MPa the vapour transfer contribution to the conductivity is not usually an important factor, for it is unlikely to increase the apparent conductivity by more than 10 or 20 per cent.

The temperature of bare soil

The factors which affect the temperature of a given soil are the amount of radiation it receives, its albedo, and its moisture and air contents, and any factor that affects the rate of evaporation of water from the moist soil. The amount of radiation a soil receives depends on its aspect, and in the northern hemisphere a south-facing slope is warmer than a horizontal surface, which, in turn, is warmer than a north-facing slope. Correspondingly a west-facing slope is normally somewhat warmer than an east-facing slope. This effect of aspect can be very important in the spring when soils are warming up. Thus Ludwig[6] at Oxford showed that maize germinated appreciably quicker if sown on a south than on a north slope if sown in late March or April, but if sown in May there was little difference because the soil temperature on the north slope was by that time high enough for good germination.

Moist soils have albedos of about 0.10 to 0.15, and this increases as the soil dries, and light-coloured soils have higher albedos than dark. The greater the albedo of a soil the smaller the temperature fluctuations tend to be. Thus, covering the soil in summer with a thin layer of white powder will reduce the surface temperature, and a layer of black powder will increase it. For example it was found[7] that dusting the surface of soil with either chalk or a light-coloured soil reduced the maximum daily soil temperature at a depth of 5 cm by 5 °C compared with dusting it with a black or grey soil, although the daily minimum temperature was only reduced by 1 °C. These figures refer to the mean temperatures during three weeks in April, and maize sown at the beginning of April took 20 days for 50 per cent of the seeds that finally germinated to germinate under the dark-coloured film compared with 32 days under the light, and the final germinations were 70 and 44 per cent of the seed sown. Similarly, Stanhill[8] working in Israel showed that under hot desert

5 de Vries, D. A., *Soil Sci.*, 1952, **73**, 83; *Neth. J. agric. Sci.*, 1953, **1**, 115.
6 Ludwig, J. W. *et al.*, *J. Ecol.*, 1957, **45**, 205.
7 Ludwig, J. W. and Harper, J. L., *J. Ecol.*, 1958, **46**, 381.
8 Stanhill, G., *Agric. Meteorol.*, 1965, **2**, 197.

conditions a film of white magnesium carbonate 0.05 mm thick increased the albedo from 0.31 to 0.64 and reduced the maximum daily temperature at 2 cm depth by about 7 to 10 °C over a period of a month, though it increased the minimum temperature by about 2 °C.

Mulches applied to the surface of a soil affect the amount of heat received and the way it is dissipated. Mulches of dead vegetation, of straw, stover or dead grass, for example, immobilize to some extent the air within the mulch; and because still air has a very low thermal conductivity, heat is only slowly transmitted from the surface of the mulch to the soil surface. Thus soils under these mulches remain cold in spring in regions which have a cold winter and germination of seed sown under such mulches is delayed and usually poor. This can be very damaging in regions with cold winters and hot summers where crops requiring warm soils in which to germinate, such as maize or sorghum, are grown; for these mulches will delay sowing dates and so shorten the growing season.[9] As will be discussed in Chapter 13, this type of mulch can be very valuable for soil conservation if this effect on shortening the growing season can be overcome. In addition the soil remains moist and often rather poorly aerated. Mulches are also useful in hot climates to prevent excessively high soil temperature which can damage roots.[10]

A film of transparent polythene laid on the soil surface will increase the soil temperature. This is because, although it is transparent both to solar radiation and to back radiation during clear nights, its under-surface becomes covered with a film of water droplets which act as a barrier to radiation losses from the soil surface into outer space. Table 8.3 gives the effect of different types of mulch or surface covering on the soil temperature at Manhattan, Kansas, for the summer months[11] and it shows very clearly the higher mean

Table 8.3 Effect of different mulches on the soil temperature (°C); Manhattan, Kansas

Depth (cm)	Polythene film	Bare soil	2.5 cm layer gravel painted		Wheatstraw (10 t ha^{-1})
			Black	*Aluminium*	
1	34.4	29.1	27.4	25.9	23.8
4	31.8	28.0	27.1	25.4	23.6
16	29.0	26.8	26.0	24.2	23.0

9 For an example with maize in the US see van Wijk, W. R. *et al.*, *Soil Sci. Soc. Am. Proc.*, 1959, **23**, 428; Burrows, W. C. and Larson, W. E., *Agron. J.*, 1962, **54**, 19.
10 Harrison-Murray, R. S. and Lal, R. in *Soil Physical Properties and Crop Production in the Tropics* (eds R. Lal and D. J. Greenland), p. 285, Wiley, 1979.
11 Taken from Hanks, R. J. *et al.*, *Soil Sci.*, 1961, **91**, 233. For other examples see Waggoner, P. E., *Trans. 7th Int. Congr. Soil Sci.*, 1960, **1**, 164; Adams, J. E., *Agron. J.*, 1962, **54**, 257; Miller, D. E. and Bunger, W. C., *op. cit.*, 1963, **55**, 417.

soil temperatures under transparent polythene, and the lower under mulches. This additional warmth in the soil under polythene is of increasing commercial importance for the early production of high-priced crops such as strawberries and cantaloupe melons.

The effect of the moisture content of a soil on its temperature is complex. On the one hand, a moist soil conducts heat upwards or downwards much better than a dry soil, but during a sunny day the surface of a dry soil warms up much quicker, and during a clear night cools much quicker than a wet surface. This greater temperature fluctuation at the dry surface is rapidly damped with depth, and at 7 to 10 cm there need be little difference in temperature due to the soil's moisture content. In the same way a sandy soil and a clay soil in the same region will have very different surface temperatures, that for the sand having a much greater diurnal variation, but at depths of 5 to 10 cm the temperature difference between the soils can be small. Sandy soils are, however, commonly said to warm up quicker in the spring than clay soils, and in fact sands are often referred to as warm soils and clays as cold soils. The reasons for the differences in the rate of germination and early growth, however, may not be due to differences in soil temperature but to differences in soil aeration and pore space distribution. Heavy-textured soils in spring are typically wet after the winter and possess few air-filled pores; seeds tend to germinate poorly in them and they contain few pores into which the young roots can grow. On the other hand, in spring, lighter-textured soils contain considerably more air-filled pores and it may be that fact that accounts for the earlier spring crops on these soils. This will result in the surface of the sandier soils warming up much quicker in the middle of the day than the clay soils, but because their thermal conductivity will be lower, due to their higher air content their temperature at, say, 5 to 10 cm deep need not be very different. This appears to be what is found in practice; for instance, germination in the sandy loam soil at Woburn is earlier than in the clay loam at Rothamsted, yet in spring its temperature is only about 0.5 °C higher at a depth of 10 cm, while in autumn its temperature is about the same and in winter it may freeze earlier. There is still very little published data on the temperature difference between soils of different texture in the same locality.

Irrigating a soil in summer will always lower its surface temperature, but this cooling effect need not go very deep. Smith[12] in Arizona has given an example where irrigation cooled the soil by 5 °C at 2.5 cm, 2 °C at 5 cm and 1 °C at 7.5 cm depth. Conversely, if the soil is loose before irrigation, irrigation may raise the temperature at 5 cm, because of its higher thermal conductivity. Thus Brooks[13] found that the soil temperature at 10 cm was between 3 and 4 °C warmer when irrigated than on a neighbouring plot which

12 Smith, G. E. P. *et al.*, *Ariz. Agric. Exp. Stn. Tech. Bull.*, No. 37, 1931.
13 Brooks, F. A., quoted by Richards, S. J. *et al.* in *Soil Physical Conditions and Crop Growth*, Agron. Monog. No. 2, p. 303, Acad. Press, 1952.

had a loose dry tilth, although the maximum surface temperature of the dry soil was appreciably higher.

This effect of irrigation in increasing the thermal conductivity of a soil can be important in practice at night time during periods of clear skies and strong temperature inversion when radiation frosts are liable to occur, particularly if the surface soil is bare and dry because of its low thermal conductivity. If the soil is made moist, or if it is rolled to be made more compact[14] its conductivity is increased, more heat can be conducted from the body of the soil to its surface, and its surface temperature will drop much less; for although the total radiation loss is the same, a greater proportion of the heat radiated will come from the soil and a smaller proportion from the air. These effects are large enough to prevent a light frost from harming crops at night.

The influence of vegetation on soil temperature

Vegetation has the same general effect as a mulch reducing both the diurnal and seasonal fluctuations, because it intercepts a part or all of the incoming radiation and of the back-radiation from the soil. Its effect depends on the degree with which its leaves shade the soil surface from the sky, and in general its effect is roughly proportional to the proportion of soil shaded. Under a complete canopy, the leaves will absorb all the incoming solar radiation and will be the direct source of all back-radiation from the surface into space. The soil surface will only receive longwave radiation from the canopy and its back-radiation will all be absorbed and scattered by the canopy. In general the air temperature above a crop will be lower than the surface soil temperature on a clear night, whereas it will be the same as the surface temperature of a bare soil. The difference between the air temperature above a crop and the surface soil temperature on a clear day depends on the rate of transpiration by the crop, but if the crop dissipates any of its net radiation as sensible heat the air temperature will typically be higher than the surface soil temperature.

Vegetation affects the seasonal changes in the soil surface temperature. Soil under vegetation warms up more slowly in the spring and cools down more slowly in the autumn than bare soil. But little is known about the magnitude of this effect or how it varies from year to year. At Rothamsted at 10 cm below the surface, bare soil is always cooler than turf, except in June and July, and the difference between these two is largest in autumn. On a ten-year average (1930–39) soil under turf is 1.2 °C warmer in October and November than bare, but is only 0.6 °C warmer during the winter and spring, while from May to August it is within 0.3 °C of the bare soil temperature.

14 For an example from Victoria, Australia, see Bridley, S. F. *et al.*, *Agric. Meteorol.*, 1965, **2**, 373.

The effect of vegetation on the heat flux into and out of a soil depends on the thickness and height of its canopy. A complete canopy of thick grass differs from that of a forest in that the thick grass maintains an almost stagnant body of air among its leaf blades, which insulates the soil surface against appreciable heat fluxes, whereas in a forest there is a deep column of fairly still air in which convection can take place easily. This allows heat to be transferred from the soil to the air whenever the air becomes cooler than the soil surface, although in many forests the flux from the soil to the air may be low because of the large proportion of air pores in the forest floor.

This difference in type of canopy can be of great importance in affecting the incidence of frost during clear nights in spring. Thus Cornford[15] found the minimum air temperatures 90 cm above the ground in some flat land in Kent after a clear still night at the end of May were: above bare soil 9.7 °C, in a wood 9.4 °C, above a short grass meadow 7.6 °C and above a long grass meadow 6.1 °C. The small difference between the wood and the bare soil is due to the circulation of the air mass in the wood brought about by its being cooled at canopy level and warmed at the soil surface. This cooling effect of vegetation only affects the air near the canopy, for the air a few metres above the canopy will be appreciably warmer. As already noted, straw mulches have the same effect as grass and as another example from Kent, Rogers[16] found that in some strawberry beds there could be 7.5 °C temperature difference on either side of the straw mulch on a clear spring night, and the minimum air temperature might be 4 °C lower over a straw mulch than a bare soil.

Since frost damage in spring can be serious on fruit and horticultural crops, it is worth while summarizing the methods the grower has at his disposal for minimizing this damage. These fall into two groups – ensuring the maximum transfer of heat from the soil to the air around the frost sensitive parts of the crop, and preventing cold air reaching these sensitive parts. The former is achieved by keeping the soil surface as free from weeds and mulch as possible, as well as compact and damp. This not only ensures the maximum transfer of heat from the subsoil to the air, but also the maximum transfer of water from the soil to condense as dew on the plant. The latter can be achieved either by ensuring that air draining into the area is as warm as possible, which can be achieved by treating the higher lying land in this way, or by managing the crop so its sensitive parts are as high as possible above the ground. Careful management can often raise the temperature by 5 °C, or even more, compared with poor management.

This blanketing effect of vegetation or mulches can also be important in winter, for it reduces the penetration of frost into the soil. Thus, Salisbury[17] quotes an example where frost penetrated a sandy loam to a depth of 5.5

15 Cornford, C. E., *Quart. J. R. Meteorol. Soc.*, 1938, **64**, 553.
16 Rogers, W. S., *Imp. Bur. Hortic., Tech. Comm.*, 15, 1945.
17 Salisbury, E. J., *Quart. J. R. Meteorol. Soc.*, 1939, **65**, 337.

to 8.5 cm if bare, to a depth of 2.5 to 3.5 cm under rough grass, to less than 2 cm under some bushes, to 1.5 cm in an open hazel copse where there was no litter and did not enter at all where there was litter. Schofield[18] found at Rothamsted and Woburn that a grass cover was even more effective, being equivalent to 10 cm of soil as is shown in Fig. 8.6; that is, if frost penetrated 20 cm in bare soil, it only penetrated 10 cm under short grass.

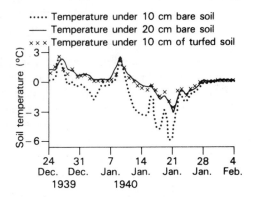

Fig. 8.6 The effect of a turf covering on the subsurface temperature of the Rothamsted soil during a hard frost.

Post and Dreibelbis[19] in Ohio, have given another example; they found that soil became frozen to a depth of 5 cm or more many times during the winter when under wheat, only a few times under grass and not at all under forests. Snow also protects the soil against penetration of frost; in an intensely cold winter, during which there is a great deal of snow, the soil may be warmer and frozen to a lesser depth than in a milder winter without snow.

This effect of vegetation in reducing fluctuations in the soil surface temperature, and in particular the maximum temperature in the middle of the day, is of particular importance in the tropics. Tree crops such as coffee and cocoa, which do not give complete canopy cover of the soil, either need to be planted under shade trees or to have the soil surface covered with a mulch, for only in this way can they maintain an active root system in the surface soil, which is required for the uptake of nutrients.

The effect of soil temperature on crop growth

The growth of plants is profoundly affected by temperature. From sowing onwards, the soil temperature plays an important role in the rate and dur-

18 Schofield, R. K., *Quart. J. R. Meteorol. Soc.*, 1940, **66**, 167.
19 Post, F. A. and Dreibelbis, F. R., *Soil Sci. Soc. Am. Proc.*, 1942, **7**, 95.

ation of crop growth. When soils are moist, temperature is usually the dominant environmental factor determining the rate of germination, and soil temperature frequently remains the dominating influence during seedling growth and for root growth, provided the structure and nutrient status of soils are adequate. The literature on the responses of plants to temperature is diverse and general principles are emerging only slowly.[20] Evidence shows that soil temperature affects many aspects of growth including the development and growth of root systems and the rates at which they take up water and nutrients; the development and expansion of leaves; dry matter production and root : shoot ratios; and flowering and harvestable yield.

Generally, plants start to grow once some minimum (or base) temperature is reached, and the rate of growth increases up to the optimum temperature. Thereafter, the rate decreases to zero at the maximum temperature for growth. Considerable effort has been expended in the past to define the optimum temperature for growth.[21] The base, optimum and maximum temperatures differ between crop species (see Table 2.2 for base temperatures) and may also differ between growth stages in the same crop. For example, in sugar beet the optimum temperature for germination was higher than the optimum for growth later in the season[22] although this effect is not always found[23] and may result from the greater influence of some other environmental factor.

The importance of soil temperature for crop growth lies in the fact that for many crops, particularly cereals, the shoot meristem, which is the site of temperature perception, lies below the soil surface for appreciable periods. Thus Watts[24] found that the expansion of leaf area in maize crops, grown in the field with different mulches to induce different soil temperatures, was closely correlated with mean daily soil temperatures at 5 cm. Similarly, the time of leaf appearance and the rate of extension in field-grown winter wheat were also controlled by soil temperature.[25] In plants where the shoot meristem is not located below ground for substantial periods then the controlling temperature is that of the air.[26] However, even when the shoot meristem moves about the soil surface its temperature may still be influenced by the soil if the transpiration rate is high and the temperature of the water is

20 See general reviews by Cooper, A. J., *Res. Review*, No. 4, Commonw. Bur. Hort. and Plant. Crops, East Malling, Kent, 1973; Nielsen, K. F. in *The Plant Root and Its Environment* (ed E. W. Carson), Univ. Virginia Press, 1974. A general consideration of quantitative effects of temperature on plant growth can be found in Johnson, I. R. and Thornley, J. H. M., *Ann. Bot.*, 1985, **55**, 1.
21 See for example Brouwer, R., *Jaarb. Inst. biol. scheik. Onderz. Landb. Gewass.*, 1962, 11.
22 Radke, J. and Bauer, R. E., *Agron. J.*, 1969, **61**, 860.
23 See Ch. 2 for an example with pearl millet where the base temperature was constant for many developmental stages.
24 Watts, W. R., *Expl Agric.*, 1973, **9**, 1.
25 Hay, R. K. M. and Wilson, G. T., *J. agric. Sci.*, 1982, **99**, 403.
26 Bull, T. A., *J. appl. Ecol.*, 1968, **5**, 61.

different from the air temperature.[27] The effects of root temperature, then, may be transmitted to other parts of the plant either directly as a consequence of the temperature of the transpiration stream or indirectly via the production and transport of hormones originating in the root.

When considering the effects of temperature on growth, it is useful to distinguish between the effects on the rate of a process and on the duration of a process. Generally, the rate of a process increases as temperature increases from the base temperature to the optimum, and the duration is inversely related to temperature in this same range. This means that the optimum temperature for achieving the maximum size of a plant or plant organ will be less than the optimum for maximum rate of growth.[28] For example, Fig. 8.7 shows that while the maximum rate of extension of maize leaves occurred at root temperatures of 30 °C, the maximum final length was achieved at 20 °C. Similar examples are provided by the growth of grain of wheat crops (see Fig. 2.12), the growth of spikelets in the ears of wheat,[29] and the dry matter production and leaf growth of pearl millet.[30] While exam-

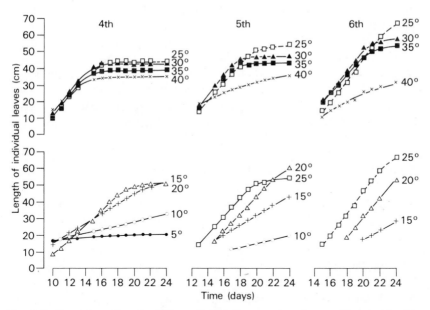

Fig. 8.7 Length of fourth, fifth and sixth leaves of maize seedlings with roots at temperatures shown and tops at 20 °C. (From Grobellaar, W. P., *Meded. Landbou. Wageningen*, 1963, **63**(5), 1.)

27 Watts, W. R., *J. exp. Bot.*, 1972, **23**, 713.
28 Monteith, J. L. in *Soil Physical Properties and Crop Production in the Tropics* (eds R. Lal and D. J. Greenland), p. 249, Wiley, 1979.
29 Sofield, I. *et al.*, *Aust. J. Pl. Physiol.*, 1977, **4**, 785.
30 Squire, G. R. *et al.*, *J. exp. Bot.*, 1984, **35**, 599.

ples of this general phenomenon exist for shoot growth, there are no comparable studies of the response of root systems to temperature. Most studies with roots have been with excised roots or with young radicles and indicate an increasing rate of extension as temperature increases to the optimum.[31] There are, though, no studies of the duration of growth, and while the root system as a whole may reach a clearly defined maximum size (see Ch. 4), there is currently no comparable description of the effect of temperature on an individual member of the root system as there is for leaves. Thus, it has still to be shown that effects of temperature on root growth can be described in the same way as for leaves.

The practical consequences of low soil temperatures have tended to dominate the scientific literature because the length of growing seasons in Northern Europe and North America are directly related to temperature and hence to the nature of crops that can be grown and to the levels of dry matter that can be produced.[32]

In the tropics, high temperatures may also exert a deleterious effect by reducing germination and crop growth. For maize at Ibadan, Lal[33] found that temperature at the soil surface could exceed 60 °C during the central 3–4 hours of the day and therefore inhibit germination. Similarly, the emergence of sorghum grown in an Alfisol, where the temperature at 50 mm (the depth of sowing) always exceeded 40 °C at 14.00 hours, was lower (45.6 per cent at 14 days after sowing) than that where 500 g m^{-2} of kaolin had been applied to the soil surface (80.9 per cent) and higher than that where 500 g m^{-2} of charcoal had been applied (10.5 per cent). There were differences, however, between genotypes in their ability to emerge at high soil temperature.[34]

The early effects of soil temperature on plant growth may persist throughout the life of a crop. Thus at Kitale in Kenya, it was found[35] that the substantial reduction in yields of maize of about 0.6 t ha^{-1} week^{-1} which occurred when planting was delayed, was a consequence of decreased soil temperature following the start of the rains. Final yield was highly correlated with the total weight of the plants five weeks after emergence which was correlated, in turn, with the mean temperature experienced by the shoot meristem during these weeks.

The rate at which roots can take up water and nutrients from the soil is affected by temperature. The effect of low soil temperature in reducing the rate of water uptake may be important in springtime in causing water stress in crops because the soil can be cool but the sun bright and the wind dry, giving conditions encouraging fast rates of transpiration. The effects of

---

31 See for example Blacklow, W. M., *Crop Sci.*, 1972, **12**, 647.
32 For maize in England see Bunting, E. S., *J. agric. Sci.*, 1976, **87**, 577.   For tomatoes in England see Gray, D. *et al.*, *op. cit.*, 1980, **95**, 285.
33 Lal, R., *Pl. Soil*, 1974, **40**, 129.
34 Wilson, G. L. *et al.*, *Indian J. agric. Sci.*, 1982, **52**, 848.
35 Cooper, P. J. M. and Law, R., *J. agric. Sci.*, 1977, **89**, 355; *ibid.*, 1978, **91**, 569.

temperature on nutrient uptake, particularly phosphate, have been studied by several workers (see later chapters) and generally the rate of phosphate uptake increases with increasing temperature. However, as Moorby and Nye[36] have shown, this may be simply an effect on root growth since rape grown at temperatures of 10 to 23 °C had constant phosphate inflow; only when the whole root system was at 5 °C was inflow reduced. .

36 Moorby, H. and Nye, P. H., *Pl. Soil*, 1984, **78**, 283.

9

The soil atmosphere

The soil pores that are not filled with water contain gases, and these gases constitute the soil atmosphere. Its composition differs from that of the free atmosphere because the plant roots and organisms living in the soil remove oxygen from it and respire carbon dioxide into it, so that it is richer in carbon dioxide and poorer in oxygen than the free atmosphere. Since the roots of most crops can only function actively if they have an adequate oxygen supply, there must be present in soils mechanisms or processes which allow the transfer of oxygen from the atmosphere to the soil organisms and plant roots, and of carbon dioxide from soil to the atmosphere at rates adequate to meet the needs of the crop. The two factors affecting the magnitude of these transfers are the rate at which the soil organisms and the plant roots are converting oxygen into carbon dioxide, and the rate that oxygen can move from the atmosphere to the sites of active oxygen demand.

The rates of oxygen consumption and carbon dioxide production by microorganisms in most agricultural soils adequately supplied with oxygen, depend on the soil moisture content and temperature, on the ease of decomposition of the organic matter in the soil, and also on the activity of the crop's roots and their associated rhizosphere organisms. The respiratory quotient, RQ, of a well-aerated soil, that is the ratio of the volume of carbon dioxide produced to the volume of oxygen consumed, is close to unity[1]; it only rises above unity when there are anaerobic pockets present in the soil. A determination of this

---

1 For an example see Rixon, A. J. and Bridge, B. J., *Nature*, 1968, **218**, 961.

quotient in a soil is therefore a sensitive method for checking if the whole body of the soil is aerobic.

There is a very wide range in the published figures for oxygen consumption and carbon dioxide production in soils, because this depends on so many factors. Monteith[2] found, for a bare Rothamsted clay loam soil, a carbon dioxide flux of 1.5 g m^{-2} day^{-1} in winter and 6.7 g m^{-2} day^{-1} in summer, and the temperature dependence of the flux followed approximately:

$$R = R_0 Q^{T/10}$$

where R_0 is the flux at 0 °C and R at T °C. He found Q had a value of about 3, and the results of these determinations are given in Fig. 9.1. The mean daily flux was 4 g m^{-2} and the annual flux 1.44 kg m^{-2} corresponding to a loss of 0.4 kg m^{-2} of carbon. In this example the carbon content of the top 46 cm of soil was 12.2 kg m^{-2}, so this annual loss represents 3.1 per cent of the soil carbon. He also found that beans, kale and short grass increased the summer flux by 2 to 3 g m^{-2} day^{-1}, but barley gave fluxes up to 10 g m^{-2} day^{-1} in excess of the bare soil. These figures should be compared with the rates of carbon dioxide assimilation of the crops of 14 to 22 g m^{-2} day^{-1}.

These figures of Monteith may be rather low, because Currie,[3] also at Rothamsted, but using a more direct method of measuring fluxes, obtained much higher summer figures than Monteith. His results are given in Table 9.1, and since he measured both the oxygen consumed and carbon dioxide

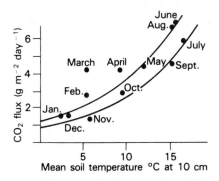

Fig. 9.1 The relation between the daily soil respiration and mean soil temperature at Rothamsted for a bare soil (October 1960–September 1961). The two curves are the plot of $R_0 Q^{T/10}$ for $Q = 3$ and $R_0 = 1.2$ upper curve and $R_0 = 0.9$ lower curve.

2 Monteith, J. L. *et al.*, *J. appl. Ecol.*, 1964, **1**, 321. For similar results on a sandy clay loam under potatoes and kale, see Brown, N. J. *et al.*, *J. agric. Sci.*, 1965, **64**, 195.
3 Currie, J. A. in *Sorption and Transport Processes in Soils*, Soc. Chem. Ind. Monog., No. 37, 152, 1970.

Table 9.1   Oxygen consumption and carbon dioxide production from a bare soil and a soil under kale at Rothamsted (g m^{-2} day^{-1})

	July		January	
Soil	*cropped*	*bare*	*cropped*	*bare*
Oxygen consumption	24	12	2.0	0.7
Carbon dioxide production	35	16	3.0	1.2
Soil temperature at 10 cm	17 °C		3 °C	

produced, was able to demonstrate that the respiratory quotient was almost exactly 1.00, except for the land under kale in July when it was 1.05. In another experiment he showed that the soil under a kale crop, which gave a dry matter yield of about 12 t ha^{-1}, absorbed about 17.5 t ha^{-1} of oxygen during the 18-week period from July to November, when it was making most of its growth, of which probably about half was due to the oxygen absorbed by the roots and their associated organisms.[4] The results of many other workers tend to fall within the ranges given above. Clark and Kemper[5] quote mean figures for the oxygen consumption of bare soil as between 2.5 and 5 g m^{-2} day^{-1}, which are doubled if the soil is cropped. Romell[6] working in Sweden, found consumptions of between 13 and 20 g m^{-2} day^{-1} for forest soils and an *RQ* of about unity. Hilger[7] working in Zaïre (Belgian Congo) found respirations of 12 to 15 g m^{-2} day^{-1} of carbon dioxide under forest, but if this was cleared and cultivated it rose up to 35 g, and Schulze[8] working in Puerto Rico found respirations up to 10 g under deciduous forest, which would be a fairly open type of forest, and 25 to 50 g m^{-2} day^{-1} under gallery or high-rainfall forest.

The rate of oxygen uptake by soils is usually large compared with the amount of oxygen present in a soil. If a soil is using 7 g m^{-2} day^{-1} of oxygen, and if this use is all assumed to take place in the top 25 cm of soil, this amounts to 2.8×10^{-5} g cm^{-3} soil day^{-1} or 3.25×10^{-10} g cm^{-3} s^{-1}. If the soil contains 20 per cent by volume of air, and if this air contains 20 per cent of oxygen, there will be 5.8×10^{-5} g cm^{-3} soil of oxygen, which is only about twice the daily use. Thus, if the surface of the soil was sealed completely against the entry of oxygen into the soil, the soil's oxygen supply would only last for about two days.

4 Currie, J. A. in *Rothamsted Ann. Rept. for 1967*, p. 33.
5 Clark, F. E. and Kemper, W. D. in *Irrigation of Agricultural Lands*, Agron. Monog., No. 11, 472, 1967.
6 Romell, L. G., *Medd. f. Stat. Skogsförsöks-anstalt*, 1922, **19**, 125.
7 Hilger, F., *Bull. Inst. Agron. Gembloux*, 1963, **31**, 154.
8 Schulze, E. D., *Ecology*, 1967, **48**, 652.

Mechanisms of gaseous transfer

The principal paths along which oxygen and carbon dioxide move in the soil are those pores containing air which form a continuous system stretching from the surface into the deeper layers of the soil. Problems of soil aeration are normally only of importance when this system of pores is absent, either due to the soil being waterlogged or to the coarsest pores being sufficiently fine that they are not emptied of water by drainage. There are a number of processes operative which help in these gaseous transfers. Changes in soil temperature and in atmospheric pressure cause the soil gases either to expand or contract; rain carries dissolved oxygen into the soil as well as pushing a body of air in front of its wetting front as it penetrates into the soil; and even gusts of wind blowing over the soil surface will cause some atmospheric air to be sucked into the top 1 or 2 cm of a loose soil.[9] But Romell[6] in a classic paper showed that the only process which is normally of any significance is gaseous diffusion along the concentration gradients in the air-filled pores. This allows oxygen to diffuse to the root from the atmosphere and carbon dioxide from the root to the atmosphere.

The rate of gaseous diffusion from, say, the neighbourhood of the plant root to the atmosphere depends on the difference in gas concentration between these two regions, the length of the diffusion path and the diffusion coefficient of the gas; and it is much less dependent on the shape of the air-filled pores than is the hydraulic conductivity of the soil. For many soils under normal field conditions the rate of diffusion is approximately proportional to the proportion of the soil volume that contains air.

Early work by Penman[10] and van Bavel[11] suggested that, if D is the diffusion coefficient of a gas in the soil and D_0 that in free air, $D/D_0 = 0.6\,S$ to a first approximation, where S is the proportion of the soil volume occupied by air. Later work by Currie[12] showed that if this was written $D/D_0 = \alpha S$, the value of α depends both on the soil structure and on the moisture content. Some of his results are illustrated in Fig. 9.2.[3] The relation between α and S for a dry sand and a wet sand is quite different because compaction of a dry sand does not fundamentally alter the shape of the pore space, while wetting a sand introduces water wedges around all the points of contact of the sand particles. This alters the shape of the air pores very considerably and so has a much greater effect on the diffusion coefficient. Columns of soil crumbs behave like sand grains when wet, but once a certain moisture content, or air space, has been reached, the value for α becomes about 0.6. The point at which this happens is when all the pore spaces between the

9 Kimball, B. A. and Lemon, E. R., *Soil Sci. Soc. Am. Proc.*, 1971, **35**, 16.
10 Penman, H. L., *J. agric. Sci.*, 1940, **30**, 437.
11 van Bavel, C. H. M., *Soil Sci.*, 1952, **73**, 91.
12 Currie, J. A., *Brit. J. appl. Physics*, 1960, **11**, 314; ibid., 1961, **12**, 275.

crumbs contain air, but the crumbs themselves contain little if any. This shows that in structured soil, the diffusion coefficient is primarily determined by the intercrumb air spaces, and the air spaces within the crumbs only make a minor contribution. The shape of the curve for any particular soil depends on the soil structure, and the actual value of α at the wet end tends to be higher for a well-structured than a puddled soil.[13]

Factor α is composed of two parts, one part being the true tortuosity and the other a shape factor. Since the air pores are not straight tubes, but follow tortuous paths around the soil particles, the actual distance between two points in the path, l_e, is greater than the straight line distance, l, between them. The ratio l_e/l is called the tortuosity of the path, and is greater the more tortuous the path. The effect is to decrease the apparent diffusion coefficient in the soil by a factor $(l/l_e)^2$, which is known as the tortuosity factor, which becomes smaller as the path becomes more tortuous. Further, the pore is not a cylindrical tube but has a very varying cross-sectional area, and this still further reduces the coefficient by a shape factor f, so that $\alpha = f(l/l_e)^2$. Figure 9.2 shows that at high moisture content when there are only few air-filled pores, α can be as low as 0.1 or 0.2, while it is 0.6 for a dry soil. It is not possible in practice to determine f and l/l_e separately, and α is often taken as equal to the tortuosity factor. Ball[14] was able to determine l/l_e and an effective pore radius and number from measurements at a number of water contents of D/D_o, gas permeability and air-filled porosity. In a silt loam subsoil the pores appeared to be from two to seven times the length of the sample, the drier the soil the more direct the pores.

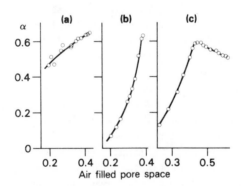

Fig. 9.2 Plot of α against the air content of sand and soil: (a) dry sand; (b) wet sand; (c) wet soil crumbs.

13 Bakker, J. W. and Hidding, A. P., *Neth. J. agric. Sci.*, 1970, **18**, 37.
14 Ball, B. C., *J. Soil Sci.*, 1981, **32**, 483.

Currie[15] compared the diffusion coefficient of a gas in a dry soil in its natural structure with the diffusion coefficient within the individual soil crumbs, and found that the within-crumb diffusion coefficient was only about one-fifth that of the between-crumb coefficient for a given air space. He used a theoretical equation developed by Burger[16] which can be written

$$D_c/D_o = S_c/[1 + (k_c - 1)(1 - S_c)]$$

where D_c and S_c are the diffusion coefficient and volumetric air space within the crumb and k_c is a constant. He showed that k_c was a useful measure of soil structure, being high for poorly structured soils, and is, in fact, a measure of the tortuosity of the larger pores within the crumb. The value of k_c varied from 4.2 for crumbs from an old Rothamsted pasture soil to 7.9 for a Rothamsted arable soil that had received no farmyard manure for over 100 years and was in poor physical condition, to 11.0 for crumbs of a silty clay soil from a restored open cast coal site, in which the soil had completely lost its structure. This was not connected with the air space, for the air spaces of the three groups of crumbs were 0.36, 0.25 and 0.30.

Diffusion of oxygen through a water film is about 10^{-4} times that in air, the diffusion coefficients being approximately 2.4×10^{-5} compared with 2.1×10^{-1} cm^2 s^{-1}. It will therefore diffuse quicker through a column of air 1 m deep than through a water film 1 mm thick. But transport of oxygen through a water film in the soil is affected by the presence of soil organisms, for they live in the water and will themselves be taking up oxygen from it. This means that oxygen can only diffuse at a significant rate through water films if they are very thin, the critical thickness depending on the level of microbial demand for oxygen. Currie[17] has shown that if oxygen is diffusing into a spherical waterlogged crumb, the maximum radius r of the crumb for which oxygen can just reach the centre, so the whole of the crumb contains some oxygen, is given by

$$r^2 = 6DC/M$$

where D is the diffusion coefficient of oxygen in the crumb, C is the concentration of oxygen in the water at the outer surface of the crumb, and M is the rate at which the micro-organisms and plant roots in the crumb are using up oxygen. The assumption that must be made for this simple equation to be valid is that M is a constant independent of the oxygen concentration around the organisms, an assumption which Greenwood[18] has shown to be reasonable. The same result was obtained by Greenwood and Berry,[19] but they gave a more general solution of the differential equation by allowing M

15 Currie, J. A., *J. Soil Sci.*, 1965, **16**, 279.
16 Burger, H. C., *Phys. Ztschr.*, 1919, **20**, 73.
17 Currie, J. A., *Soil Sci.*, 1961, **92**, 40.
18 Greenwood, D. J. and Goodman, D., *J. Sci. Fd Agric.*, 1965, **16**, 152.
19 Greenwood, D. J. and Berry, G., *Nature, Lond.*, 1962, **195**, 161.

to depend on the oxygen concentration. Experimental confirmation was achieved by Greenwood and Goodman.[20]

The value of D depends on the size and tortuosity of the pore space within the crumb. With $D = aD_o$ where D_o is the diffusion coefficient of oxygen in air, Currie[17] gave values of a for dry soil crumbs of between 0.08 and 0.13 for an old Rothamsted pasture soil whose crumbs had a porosity of 0.34 cm^3 cm^{-3}; 0.034 to 0.063 for an arable soil which had received regular annual dressings of farmyard manure with a crumb porosity of 0.27 cm^3 cm^{-3}; and of 0.015 to 0.044 for an arable soil which had received no manure or fertilizer for over a century and whose crumb porosity was 0.21 cm^3 cm^{-3}. Thus, D for wet crumbs probably falls within the range 0.5 to 2.0 $\times$ 10^{-6} cm^2 s^{-1}. The value of M depends on the soil temperature, the amount of decomposable organic matter and the activity of the plant roots. At Rothamsted for a bare soil in winter it is about 50 $\times$ 10^{-12} g cm^{-3} s^{-1} of oxygen corresponding to about 1 g m^{-2} day^{-1}, going up to over 500 $\times$ 10^{-12} for cropped soil in summer. C is the concentration of oxygen in the soil water at the air–water interface, and if the oxygen concentration in the air just outside the wet crumb is assumed to be the same as in the free atmosphere, it is equal to about 10^{-5} g oxygen per cm^3 water, a figure somewhat dependent on the temperature, as the solubility of oxygen in water is temperature dependent. Putting these figures into the equation gives values of r ranging from about 0.4 to about 1 cm, so that, under summer conditions, waterlogged crumbs larger than about 1 cm in size are likely to have an anaerobic core, whereas in winter the maximum size of crumb that is aerated throughout is over 2 cm.

Wet well-drained soils of rather coarse structure can therefore consist of a mosaic of anaerobic volumes embedded in an aerobic matrix often revealed by mottling. This result explains an old result of Russell and Appleyard[21] that the composition of the gases dissolved in the soil water is quite different from that in the soil pores in that it is almost oxygen-free and consists primarily of nitrogen and carbon dioxide. It also explains why denitrification can proceed in soils in which the oxygen content of the soil air has not been appreciably reduced below its value in the atmosphere. Denitrification is discussed in Chapter 19.

The application of this diffusion exchange process to the transfer of carbon dioxide predicts that the carbon dioxide content of the soil air should not rise above 1 per cent if there are no anaerobic pockets.[22] This is primarily due to the solubility of carbon dioxide in water being 31 times greater than that of oxygen, for the ratio of its diffusion coefficient in water to that of oxygen is 0.74. This has the consequence that if the oxygen and carbon dioxide fluxes

20 Greenwood, D. J. and Goodman, D., *J. Sci. Fd Agric.*, 1964, **15**, 579.
21 Russell, E. J. and Appleyard, A., *J. agric. Sci.*, 1915, **7**, 1.
22 Greenwood, D. J., *J. Soil Sci.*, 1970, **21**, 314.

in a waterlogged crumb are equal[23] as they usually are in the absence of anaerobic pockets, for each 1 per cent reduction in the oxygen gradient between the inside and the outside of the crumb, there is only a 0.05 per cent increase in the carbon dioxide concentration, so that if the centre of the crumb is just free from oxygen, so the difference in oxygen concentration is 20 per cent, the difference in carbon dioxide concentration is 1 per cent.

The theory that the principal mechanism of gas exchange between the soil and the atmosphere is by diffusion through air pores would predict that a soil in which the surface structure has collapsed into a thin crust should rapidly become anaerobic whenever the crust becomes wet, for its pores are too fine to be emptied of water by drainage. Surface crusts, as distinct from the waterlogging of the soil surface, are not found to be as damaging to the aeration as would be expected.[24] This is due to the effect of wind[25] for provided there are a certain number of cracks and holes through the crust, as there usually are, wind blowing across the surface of the soil will suck air through these holes, and through the air spaces just under the crust, and so maintain an oxygen concentration there not very different from that of the atmosphere.

A further consequence of the diffusion mechanism for gas exchange is that if the soil layers are absorbing oxygen at a uniform rate down the profile, a depth will be reached at which the atmosphere will become oxygen-free. This depth need not be very great if the air porosity of the soil is low and the biological activity of the subsoil is high, but in general the principal zone of biological activity is the surface soil. However, if crops are taking their water from depth, the roots at depth will have an appreciable oxygen demand so the oxygen diffusion path will be appreciable. As an example, Lipps and Fox[26] found that when lucerne was growing actively and taking water from a depth of 2 m, the oxygen content at that depth was 16 per cent, but in the autumn when active growth ceased and the oxygen demand of the roots fell, it rose to 20 per cent, and it maintained this concentration down to a depth of 3.6 m.

Measurement of aeration

The concentration of oxygen in soil air appears to be an important factor for plant roots and many attempts have been made to extract samples of soil air and analyse it, originally by classical gas analysis techniques and more recently by sensitive physical methods. However, gas sampling extracts air mainly from the larger, better aerated pores and samples are also likely to

23 Bunt, J. S. and Rovira, A. D., *J. Soil Sci.*, 1955, **6**, 119, 129.
24 Brown, N. J., Fountaine, E. R. and Holden, M. R., *J. agric. Sci.*, 1965, **64**, 195.
25 Farrell, D. A. *et al.*, *Soil Sci.*, 1966, **102**, 305.
26 Lipps, R. C. and Fox, R. L., *Soil Sci.*, 1964, **97**, 4.

contain some atmospheric air, either from dead space in the sampling equipment or by leakage down the outside of the sampling tubes. Oxygen concentrations were often found to be lower than those in the air above the soil but rarely below 10 per cent by volume, with carbon dioxide concentrations rising to a few per cent.

More recent studies using very small samples show much lower oxygen concentrations and in waterlogged soils the concentration can fall to virtually zero.[27] It is the oxygen concentration at the respiring surface that presumably is important, and since these surfaces are always covered by water films of varying thickness offering considerable resistance to the passage of oxygen, it is unlikely that the composition of soil air as found by sampling can be a useful measure of aeration status. We are also aware of the common occurrence of anaerobic zones in an otherwise aerobic soil, and the dependence of oxygen demand on soil temperatures, so complicating the problem still further. Soil air composition can provide useful evidence of the overall effects of such factors as temperature, pH, and the supply of respirable substrate on oxygen demand but cannot be expected to correlate directly with root growth and activity.

Concentrations at points in the soil have been measured using coated platinum electrodes[28] and attention has also been given to the measurement of oxygen diffusion rates. Stolzy and Letey[29] used a platinum wire cathode and calomel half-cell anode, and by applying a potential difference of about 0.6 volts, sufficient to reduce elementary oxygen at the cathode surface to water, were able to measure the final steady current flowing. This depended on the rate of supply of oxygen to the cathode surface where the concentration was effectively zero, as is likely at the root surface when supply only just balances demand. The measurement of current can readily be converted into the corresponding oxygen diffusion rate, or ODR, and this appears to become critical if it falls below 3–4 ng oxygen cm^{-2} s^{-1}. This method, which is often referred to as the polarographic method, has been criticized by McIntyre[30] who showed that it is subject to a number of errors which seriously affect interpretation of results. More recent work[31] avoids the principal errors by carrying out a preliminary determination of the appropriate potential difference to be applied to avoid reactions other than the reduction of oxygen. This shows significantly lower values of oxygen flux density which indicate the critical value for root extension at 10 °C to be about 1 ng cm^{-2} s^{-1}. Results correlated more closely with root growth, over a wide range of moisture conditions, than did either ODR or oxygen concentration in solution as found

27 Blackwell, P. S., *J. Soil Sci.*, 1983, **34**, 271.
28 Greenwood, D. J. in *Soil Physical Conditions and Crop Production, MAFF Tech. Bull.* No 29, 261, 1975. See also ref. 27.
29 Stolzy, L. H. and Letey, J., *Adv. Agron.*, 1964, **16**, 249.
30 McIntyre, D. S., *Adv. Agron.*, 1970, **22**, 235.
31 Blackwell, P. S. and Wells, E. A., *Pl. Soil*, 1983, **73**, 129.

by the coated electrode method. The calculated respiration rate of the root tissues agreed closely with measurements made earlier.[32]

Soil aeration and microbial activity

Most aerobic soil micro-organisms have a very efficient enzyme system for absorbing free oxygen based on cytochrome oxidase as their terminal oxidase system, which only falls to half the rate for a fully aerobic system when the oxygen concentration in the solution at the site of the enzyme has fallen to about 2.5×10^{-10} g cm^{-3}.[33] For most organisms, however, the site of enzyme activity is not at the outer surface of the organisms in contact with the solution, but within the bacterial cytoplasm which is separated from the solution by a cell wall, and sometimes by extracellular polysaccharide gums. Thus, the actual oxygen concentration in the bathing solution necessary to reduce the activity of the terminal oxidase by half is considerably greater than this, and in the complex environment of the soil, lies between 1 and 2×10^{-7} g cm^{-3}.[34] This should be compared with the concentration of oxygen in solution when in equilibrium with air at 20 to 25 °C of about 10^{-5} g cm^{-3}, which is between 50- and a 100-fold larger. There are, however, a number of organisms incapable of anaerobic metabolism which have their growth affected by oxygen concentrations considerably higher than this, presumably due either to the organisms not using a cytochrome-oxidase system, or using this system but having a high resistance to oxygen diffusion to the sites of enzyme activity.

This result has an important consequence for it implies that a soil is fully aerobic as long as the oxygen concentration in the solution does not fall below the figure corresponding to a partial pressure of 1 per cent of the pressure of oxygen in the free atmosphere. Until the oxygen concentration has fallen to about this, the microbial population will use a fully aerobic metabolism, but, once it has fallen below, the metabolism will rapidly become characteristic of an anaerobic system. Products of reduction such as nitrous oxide from nitrate, or hydrocarbons such as methane and ethylene from organic matter decomposition, will occur only if there are soil volumes containing active micro-organisms and an oxygen concentration of less than between 0.6 and 1.2×10^{-8} g cm^{-3}. Now that the presence of these gases or vapours can be easily detected and their concentration measured by the use

32 Willis, A. J. and Yemm, E. W., *New Phytol.*, 1955, **54**, 163; Clarkson, D. T. *et al.*, *Planta*, 1974, **121**, 81; Barber, D. A., *New Phytol.*, 1974, **73**, 91.

33 Griffin, D. M., *New Phytol.*, 1968, **67**, 561.

34 Greenwood, D. J., *Pl. Soil*, 1961, **14**, 360; see also ref. 20.

of gas chromatography, this technique offers one of the most reliable methods available for determining if the whole body of the soil is aerobic or if there are anaerobic pockets within the soil crumbs or clods.

The effect of carbon dioxide concentration in the soil air, as distinct from the oxygen concentration, on microbial activity has been little studied. Macfadyan[35] found that, on an acid sandy heath, a concentration of 0.25 per cent carbon dioxide markedly reduced its oxygen consumption, and a concentration of 0.8 per cent suppressed it. He also found that, for the soils he worked with, the finer the texture the higher the carbon dioxide concentration needed to reduce the oxygen consumption of the soil by a given proportion; and a grass pasture on clay was the most tolerant of his soils. This implies that the sensitivity of the soil organisms to carbon dioxide increases the better aerated the soil in its normal range of moisture conditions.

Poor aeration affects root growth and functioning principally through the products of reduction metabolized by bacteria when using an anaerobic metabolism, as will be discussed in more detail in Chapter 26; for the root itself contains efficient cytochrome-oxidase systems. It may be necessary to maintain a rather higher oxygen concentration just outside the root epidermis than outside a bacterial cell, for oxygen may have to diffuse through the living protoplasm of several cells, or in the air spaces around several cells if the whole root is to be adequately aerated.

Soil aeration and crop growth

Poorly drained and badly aerated soils occur in many parts of the world and it has been estimated that about 12 per cent of the world's soils have excess water.[36] Because of the variety of soils and climates where anaerobicity occurs, it is not surprising that plants exhibit a variety of responses to poor aeration.

Plant roots need oxygen for respiration and for most agricultural crops the source of oxygen is the soil air. As seen earlier in this chapter, the flux of oxygen into soils in response to respiration by roots and micro-organisms is about 7 to 35 g m^{-2} day^{-1} during the summer in temperate climates and about one-tenth of this during the winter. This compares with a volume of oxygen contained in the soil profile to 1 m depth of about 57 g m^{-2} (assuming an air-filled porosity of 0.2) which, as discussed earlier in this chapter, is sufficient for about 2 days' respiration in hot weather but for much longer in cooler

35 Macfadyan, A., *Soil Biol. Biochem.*, 1970, **2**, 9.
36 Dudal, R. in *Plant Adaptation to Mineral Stress in Problem Soils* (ed. M. J. Wright), Spec. Publ. Cornell Univ. Agric. Exp. Stn., p. 3, 1976.

weather when respiration rates are slower. The amount of oxygen dissolved in soil water is typically no more than about 3 g m^{-3} soil.

When soils are inundated with water, the growth and development of roots may be affected before total depletion of oxygen in the tissues occurs (anoxia). The principal effects are usually to increase the ethylene content and to decrease the internal concentration of oxygen (hypoxia). Plant species differ in their rates of production of ethylene; for example, while rice produces ethylene slowly, white mustard has much faster rates of synthesis. This difference in ethylene production leads to differences in the root growth of the two species, elongation of rice roots being stimulated while that of white mustard is inhibited.[37]

Deoxygenation of the soil results in a series of chemical and biochemical reductions in the soil[38] (see Ch. 26) which contribute to root injury. Toxic substances such as NO_2^-, Mn^{2+}, Fe^{2+} and S^- together with microbial metabolites may accumulate in concentrations injurious to root metabolism. Anaerobic respiration of roots may form products (e.g. ethanol, acetaldehyde) that are harmful when accumulated in high concentrations. The notion that ethanol production is primarily the cause of cell death in flood-sensitive species in anaerobic conditions[39] now seems doubtful. It is more likely that the main effect of anaerobic respiration is to produce insufficient ATP for growth and maintenance because supplies of easily respired substrates are soon used and translocation of substrates from the shoot to the root may be curtailed.[40] This latter hypothesis is supported by the observation that addition of carbohydrate may extend root survival. Moreover, several workers have shown that the adenylate energy charge (AEC given by $([ATP]+0.5\,[ADP])/([ATP]+[ADP]+[AMP])$, representing the energy status of cells and the effectiveness of respiration in maintaining them, is markedly reduced by anoxia. For example, the AEC of freshly excised root tips of maize was 0.9 in air but decreased to 0.6 after 30 minutes in nitrogen and to 0.15 after ageing in air to deplete soluble sugars by respiration followed by 30 minutes of anoxia; addition of glucose increased AEC again to 0.6.

Plants growing in anaerobic soils differ in susceptibility to anoxic conditions. For example, in oxygen-free environments cotton roots survived only 0.5 to 3 hours (at 30 °C) while rice seminal roots survived 4 to 5 days (at 25 °C) and wetland species such as yellow flag (*Iris pseudacorus*) and reed

37 Konings, H. and Jackson, M. B., *Z. Pflanzenphysiol.*, 1979, **92**, 385; for further discussion see Jackson, M. B. and Drew, M. C. in *Flooding and Plant Growth* (ed. T. T. Kozlowski), p. 47, Acad. Press, 1984.

38 For a review see Ponnamperuma, F. N. in *Flooding and Plant Growth* (ed. T. T. Kozlowski), p. 9, Acad. Press, 1984.

39 Crawford, R. M. M. in *Plant Life in Anaerobic Environments* (eds D. D. Hook and R. M. M. Crawford), p. 119, Ann Arbor Sci., 1978.

40 Saglio, P. H. *et al.*, *Plant Physiol.*, 1980, **66**, 1053; see also Drew, M. C., *Pl. Soil*, 1983, **75**, 179 for a review.

(*Phragmites australis*) survived up to 1 month (at 22 °C).[41] Such differences between species have fuelled speculation that metabolic or structural adaptations or both might allow some species to survive anoxic conditions. There is, however, no generally accepted explanation. A general metabolic theory was developed by Crawford which essentially proposed that tolerance to flooding was related to the capacity to avoid the generation of potentially toxic concentrations of ethanol during the regeneration of NAD from NADH in the glycolytic pathway.[42] In flood-susceptible species, ethanol derived from glycolysis via phosphoenolpyruvate (PEP) was assumed to cause cell death in roots whereas in flood-tolerant species the PEP was converted to oxalo-acetate and thence to malate allowing regeneration of NAD; malate was assumed to be less toxic than ethanol. However, the evidence for this scheme is incomplete and some of the evidence contradicts it.[43]

In many herbaceous, wetland species, the roots develop gas-filled cavities (aerenchyma) thereby increasing the cross-sectional area for gaseous diffusion (Plate 9.1). The formation of aerenchyma seems to contribute to the survival of many species and its formation even in aerobic conditions in species such as rice suggests that it is genetically controlled. Plant species not normally grown in wetland conditions can also develop aerenchymatous roots when waterlogging occurs. Adventitious roots produced from the base of the shoot have a partially broken-down cortex allowing movement of air within the root although the fraction of root volume filled with air is smaller than in wetland species. The production of such roots in, for example, maize and wheat[44] partially offsets the death of roots caused by flooding and their structure allows the passage of ions to the xylem and hence their translocation to the shoot.[45]

The mechanism controlling aerenchyma formation has not been much studied but is thought to involve endogenous ethylene production, at least in maize.[46] Cell lysis of the cortex of maize gave rise to aerenchyma in response to oxygen shortage (hypoxia) and was particularly well developed 3–4 cm behind the root tip.[44] Low oxygen concentrations in flooded soils stimulate ethylene production endogenously and this is thought to increase lysis of the cells. This view is supported by the observation that roots growing in well-aerated solutions supplied with ethylene (as low as $0.1\mu l\ l^{-1}$) will

41 For cotton see Huck, M. G., *Agron. J.*, 1970, **62**, 815.   For rice see Bertani, A., *et al. J. exp. Bot.*, 1980, **31**, 325.   For wetland species see Crawford, R. M. M., *Trans. Bot. Soc. Edinburgh*, 1982, **44**, 57.
42 Crawford, R. M. M., *J. exp. Bot.*, 1967, **18**, 458; see also ref. 4.
43 For a discussion of the metabolic theory and the evidence, see Jackson, M. B. and Drew, M. C. in *Flooding and Plant Growth* (ed T. T. Kozlowski), p. 47, Acad. Press, 1984.
44 For maize see Drew, M. C. *et al.*, *Planta*, 1979, **147**, 83.   For wheat see Trought, M. C. T. and Drew, M. C., *J. exp. Bot.*, 1980, **31**, 1573.
45 For a review see Drew, M. C. and Lynch, J. M., *Ann. Rev. Phytopath.*, 1980, **18**, 37.
46 Konings, H., *Physiol. Plant.*, 1982, **54**, 119.

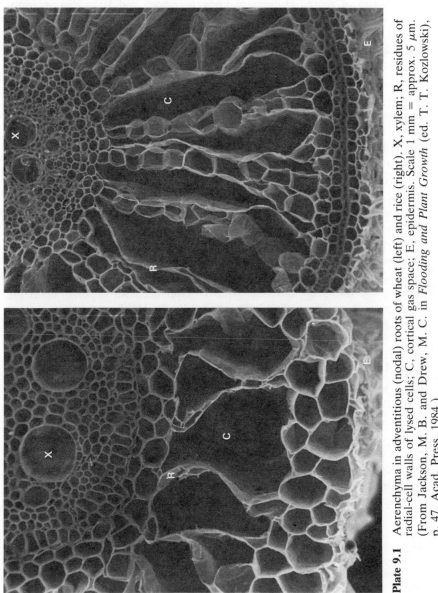

Plate 9.1 Aerenchyma in adventitious (nodal) roots of wheat (left) and rice (right). X, xylem; R, residues of radial-cell walls of lysed cells; C, cortical gas space; E, epidermis. Scale 1 mm = approx. 5 μm. (From Jackson, M. B. and Drew, M. C. in *Flooding and Plant Growth* (ed. T. T. Kozlowski), p. 47, Acad. Press, 1984.)

develop aerenchyma identical in appearance to that induced by oxygen deficiency.[47]

Structural adaptations to waterlogging also occur in tree species. Comparison of rooted cuttings of Lodgepole pine and Sitka spruce grown under waterlogged conditions showed that although root elongation of both species stopped within a few days of flooding, more root tips and more of the basal region of the roots survived in pine than spruce.[48] The greater tolerance of pine to waterlogging may have resulted from the production of large gas-filled cavities in the stele; these were absent in the spruce.

In leguminous plants, waterlogging may induce structural changes in the nodules.[49] For example, in a study of cowpeas (*Vigna unguiculata*) water-logged for either 0, 4, 8, 16 or 32 days the cross-sectional area of cortex in the nodules was increased from a mean value of about 40 per cent of the total area to 50 per cent after 4 and 8 days of waterlogging and to 60 per cent after 32 days of waterlogging.[50] This increased cortication was due to the production of additional outer cortical material and to the formation of lenticels within the cortex. Lenticels are believed to improve gas exchange and hence allow continued symbiotic nitrogen fixation.

Many crops may be subjected to transient periods of waterlogging because of heavy rainfall exceeding the ability of the upper soil layers to drain or because of water tables rising into the root zone. The optimum depth of the water table varies for different types of crop and also depends upon the experimental conditions (soil type and weather during the growing season).[51] Table 9.2 shows the influence of season on the effects of the depth of water tables on yields and shows also that transient flooding of the surface layers of soil for even a few days reduces yields more than shallow but constant water tables. The effects of fluctuating water tables have not been investi-gated widely but lysimeters at the ex-ARC Letcombe Laboratory have allowed controlled studies of their effects.[52] Waterlogging in England usually occurs during the winter period when temperatures are low so that winter barley and winter wheat both survived water tables at 10 cm for a four-month period during the winter (December until March inclusive). However, winter waterlogging restricted tillering and decreased the number of ears, thereby reducing yields of winter barley grown on a clay by 30 per cent and yields of winter wheat grown on clay and sandy loam soils by 24 and 21 per cent,

---

47 Drew, M. C. *et al.*, *Planta*, 1981, **153**, 217;   Campbell, R. and Drew, M. C. *Planta*, 1983, **157**, 350.
48 Coutts, M. P. and Philipson, J. J., *New Phytol.*, 1978, **80**, 63 and 71.
49 Pankhurst, C. E. and Sprent, J. I., *Protoplasma*, 1975, **85**, 85.
50 Minchin, F. R. and Summerfield, R. J., *Pl. Soil*, 1976, **45**, 113.
51 For a review see Williamson, R. E. and Kriz, G. J., *Trans. Am. Soc. agric. Eng.*, 1970, **13**, 216.
52 Cannell, R. Q., *et al.*, *J. Sci. Fd. Agric.*, 1980, **31**, 105.

Table 9.2 Effects of static and transient high water tables on yields of crops. Yields are expressed as percentages of the best treatment in each experiment. (Adapted from Cannell, R. Q. and Jackson, M. B. in *Modifying the Root Environment to Reduce Crop Stress* (eds Arkin, G. F. and Taylor, H. M.), 21, Agron Monog. No. 4, *Am. Soc. agric. Engng.*, 1981)

	Crop			
Water table regime	Maize[1]	Maize[2] Wet year / Dry year	Wheat[2] Means of two yrs	Sugar beet[3]
Free-drained	100			95
Constant at 15 cm	75			87
30 cm	81			92
55 or 60 cm		59 100	91	100
90 cm		87 96	100	
120 cm		100 85	86	
150 cm		78 72	74	
Surface flooding[†] for 1 day		100 100	100	
2 days	63	70 84		
3 days		56 61	91	
4 days	63	44 42		
5 days		28	79	
7 days			51*	
11 days			26*	

* One year only.
[†] Flooding in three successive weeks for maize in first column and every 14 or 18 days for other maize and wheat.
[1] Lal, R. and Taylor, G. S., *Soil Sci. Soc. Am. Proc.*, 1969, **33**, 937.
[2] Chaudhary, T. N. *et al.*, *Agron. J.*, 1974, **66**, 32; ibid., 1975, **67**, 745.
[3] Shalhevet, J. *et al.*, *Israel J. agric. Res.*, 1969, **19**, 161.

respectively, when compared with crops where the winter water table was maintained at 90 cm.[53]

The effects of poor aeration in soils may be alleviated in several ways by soil and crop improvement and management.[54] Installation of drainage is widely practised and is discussed in Chapter 13. Application of chemicals such as calcium peroxide to seeds is another remedial measure that has been attempted. The peroxide provides oxygen to the zone of soil immediately surrounding the seed and may allow faster germination and more rapid early

53 Cannell, R. Q. *et al.*, *Pl. Soil*, 1984, **80**, 53.
54 For discussion see Cannell, R. Q. and Jackson, M. B. in *Modifying the Root Environment to Reduce Crop Stress* (eds G. F. Arkin and H. M. Taylor), p. 141, Am. Soc. agric. Eng., Monog. No. 4, 1981.

growth. There has been some limited success with this technique in promoting early growth in rice.[55]

Differences between cultivars in their response to waterlogging have been found in several crop species and this opens the possibility of selecting varieties adapted to locations where waterlogging is common. For example, five cultivars of *Phaseolus vulgaris* (field beans) had very similar root growth in conditions of optimal solution culture but diverse patterns of growth when exposed to localized anoxia.[56]

55 For a review, see Leaver, J. P. and Roberts, E. H., *Outlook on Agric.*, 1984, **13**, 147.
56 Schumacher, T. E. and Smucker, A. J. M. , *J. exp. Bot.*, 1984, **35**, 1039.

10

The behaviour of water in soil

Not only is water of direct importance to plants but it plays many parts in soil, acting as a solvent, hydrolysing reagent, temperature buffer, swelling agent and weakener of the soil fabric. High water content facilitates water and solute movement; it also reduces the amount and rate of movement of oxygen in soil, sometimes seriously.

The behaviour of water in soil depends both on its own somewhat unusual properties and on the forces acting on it, including those forces resisting movement, which themselves depend on the geometry of the pores and on the interaction between water and the solid surfaces with which it is in contact.

The pore space of soil may be regarded as a system of interconnecting irregular tubes within which water can be held by capillary forces (surface tension and adhesion) and into which it may be drawn just as it is into a capillary tube. In a vertical cylindrical capillary the height of rise of water can be found by equating the upward and downward forces, giving the formula for capillary rise

$$h = \frac{4T}{g\rho d} \cos \alpha$$

where T is surface tension, g the gravitational force per unit mass, ρ the density of water, d the capillary diameter and α the angle of contact between water and soil. This last is usually zero, so that its cosine is unity, and is hence often ignored. Introducing actual values the formula reduces to

$$h \text{ approximately equal to } 3 \times 10^{-5}/d$$

so that, for example, in a pore of 3 μm equivalent cylindrical diameter capillary rise would be 10 m, which is the height of water column giving 1 atmosphere or 1 bar pressure at its base. In other words a suction just exceeding 1 bar is needed to empty such a pore. An alternative statement is that negative pressures of up to 1 bar can exist within the water held in the pore, tending to draw the pore walls together and sometimes causing contraction, as occurs in clays during drying.

Comparing soils of different particle size composition, the higher the clay content the larger the total volume of pores, but the pores within clay aggregates are very narrow and hence the capillary force can be very large. In order to extract water from fine pores not only is a large suction necessary but the resistance to movement is high, the rate of movement in a cylindrical tube being proportional to the fourth power of the diameter (Poiseuille's law). So as soil dries, not only is the plant less able to extract water but the rate of movement to the root surface falls rapidly as progressively finer pores empty, and when the rate of supply falls below the needs of the plant then wilting may occur. The water content of a soil at which permanent, irreversible, wilting occurs is called the wilting point of that soil and usually corresponds to a suction of about 15 bar (1500 kPa) and hence to the emptying of pores down to 0.2 μm diameter. At these higher suctions water content does not change much with suction, nor is equilibrium attained in less than a few days. So although for practical purposes a 15 bar percentage water content, as found using a pressure plate, is a useful approximation to the permanent wilting point it is not possible to be very definite about the exact suction, nor can one dismiss the slow rate of water movement as a contributing factor in water availability. There may be considerable residual water held at greater suctions than 15 bar, especially in clay soils, and this is essentially unavailable to plants. In most clay soils the pores are not rigid and so there may be release of water from pores by their collapse rather than by capillary emptying, and this is accompanied by shrinkage. The volumetric shrinkage is often equal to the volume of water lost and is then called normal shrinkage[1] and points to pore collapse rather than pore emptying. Other evidence[2] also suggests that much of the plant available water in clay soils is held in pores narrower than 0.1 μm which contract on drying, and in which suctions in excess of 30 bar would be needed to remove water if the pores were rigid. Hence plant available water in heavy soils cannot be equated with that held in pores between 50 and 0.2 μm, as it can in non-swelling soils.

At high water contents the main forces acting on soil water are gravity and surface tension, but as soils dry other forces become more important such

1 Haines, W. B., *J. agric. Sci.*, 1923, **13**, 296.
2 Lawrence, G. P. *et al.*, *J. Soil Sci.*, 1979, **30**, 499.

as the adhesion forces between water and soil surfaces, and osmotic pressure and overburden pressures have also to be taken into account.

The intensity factor of soil water retention

It is usually difficult to analyse exactly the contribution of each of the individual forces acting on soil water, especially as they vary from point to point. However, the result is a pressure deficit within water in soil. This is commonly referred to as soil water suction or tension and it can be measured using a suitable manometer or pressure gauge and can be related, as we have seen, to an effective pore diameter using the formula for capillary rise. The pressure within soil water is generally negative with respect to that of the atmosphere, and may be expressed in conventional pressure units, pascals or bars, or as the height of an equivalent column of water ('water head'). Since this latter expression gives an inconveniently long scale of 0 to 10^5 metres from saturated to oven dry and changes in water content are very small at the drier end, Schofield[3] suggested taking the decadic logarithm of the height of water head, expressed in centimetres, and called this pF, the p by analogy with Sorensen's logarithmic scale of hydrogen ion concentration, pH, and the F as a symbol for thermodynamic free energy.

A pressure difference between free water and water in soil implies a difference in energy of the water, and this energy difference leads to other ways of expressing the intensity factor of soil water. Buckingham[4] recommended the use of the concept of potential, or more strictly the potential difference between free water and water in soil. Potential is defined as the work done per unit quantity of water in transferring an infinitesimal quantity, so as not to disturb conditions, reversibly and isothermally from pure free water at a given level to the point in soil under consideration. The given or chosen reference level may be the same as that of the point in soil in question, but in general we are dealing with a vertically extended system in which there will be differences in level and hence differences in the gravitational component of potential, the reference level being necessarily fixed for any one system. It should be noted that we are only interested in differences in potential between two points or systems, we cannot ascribe absolute values to potential.

Each force acting gives rise to its own component of potential, for example gravity gives rise to gravitational potential (ψ_g), osmotic forces to osmotic potentials (ψ_o), capillary forces to capillary or matric potentials (ψ_c), and the components are summed algebraically to give the total potential (ψ_T) so that

3 Schofield, R. K., *Trans. 3rd Int. Congr. Soil Sci.*, 1935, **2**, 37.
4 Buckingham, E., *USDA Bur. Soils Bull.*, No. 38, 1907.

$\psi_T = \psi_g + \psi_o + \psi_c$, etc. But, as Buckingham pointed out, we do not need to consider the individual components, nor do we need to understand all the individual forces acting in order to use potential, but only those which are not uniform or whose effect depends upon position like gravity. Since most of the forces between soil and water are attractive, water potential is normally negative and care must be taken in nomenclature. A decreasing potential is one that is more negative, and corresponds to a larger suction. The principal advantage of potential as a concept applied to soil water is that its uniformity in a system is a criterion of mechanical equilibrium. Conversely if it is not uniform, movement will occur to equalize potential throughout a system, the rate of movement being proportional to the gradient of potential and in the direction of the maximum gradient.

In a system at equilibrium the individual components of potential may vary from place to place, but the sum of all components must be everywhere the same. In a bucket of water at rest the higher gravitational component of potential at the surface is exactly equalled by the higher hydrostatic pressure potential at the bottom of the bucket. Similarly in a vertical column of moist soil at equilibrium, the lower, more negative, capillary or matric potential at the surface in the drier soil is exactly balanced by the higher, more positive, gravitational potential of the water there. If this were not the case then the water would not be at equilibrium and redistribution would occur until uniformity of potential was established. Potential may be expressed as energy or work per unit volume, or per unit mass or per unit weight. Energy per unit volume has the same dimensions as force per unit area and so is numerically equal to suction and may be expressed in the same pressure units, but they are quite distinct concepts and each has its own particular application. Units are given in Table 10.1.

A third way of expressing the intensity factor of soil water retention is in terms of thermodynamic free energy, or more strictly the partial Gibbs free energy or chemical potential. This concept is similar to that of mechanical potential but thermodynamics takes account of heat changes and so distinguishes between total energy and that part of the total energy that can be made to do work and which is therefore regarded as available or free, as distinct from the unavailable energy which is dissipated, for example in a change of state. This is particularly important when considering a system containing water in liquid and vapour, or liquid and solid, states. Similarly during removal or addition of water to a soil there are heat changes in the extension or diminution in air–water interfaces which can only readily be taken into account in thermodynamic terms.[5] Thermodynamics thus facilitates the calculation of relationships between the free energy of soil water and its

5 Edlefsen, N. E. and Anderson, A. B. C., *Hilgardia*, 1943, **15**, 31;   Bolt, G. H. and Frissel, M. J., *Neth. J. agric. Sci.*, 1960, **8**, 57;   Spanner, D. C., *Introduction to Thermodynamics*, Acad. Press, London, 1964.

Table 10.1 Units of soil water suction, and its equivalents as matric potential, cylindrical pore diameter, relative humidity and freezing point depression

Suction			Equivalent as			
			matric potential	*cylindrical pore diameter*	*relative humidity*	*freezing point depression*
(m)	*(bar)*	*(pF)*	*(J kg^{-1} or kPa)**		*(%)*	*(K)*
10^{-2}	10^{-3}	0	-10^{-1}	3 mm	100	
10^{-1}	10^{-2}	1	-1	300 μm	100	
1	10^{-1}	2	-10	30 μm	99.99	
10	1	3	-10^2	3 μm	99.93	0.08
10^2	10	4	-10^3	300 nm	99.28	0.8
10^3	10^2	5	-10^4	30 nm	93.0	8.0
10^4	10^3	6	-10^5		48.4	
10^5	10^4	7	-10^6		0.07	

* kPa is strictly a unit of pressure although commonly used for potential.

freezing point and its vapour pressure, which is both practically useful and theoretically revealing. The term pF was originally intended to draw attention to this relationship with thermodynamic free energy at a time when F was commonly used to denote free energy at constant pressure, Gibbs free energy.

It is necessary to recognize the distinctions between these three concepts of the intensity factors of soil water retention (i.e. suction, potential and free energy) and to apply the one appropriate to the particular problem under consideration. Soil water suction or tension is an internal pressure within the soil that can be readily measured and has a quantitative effect on the mechanical properties of the soil. For water movement, gradients of potential are important and at high water contents this is largely gravitational. Chemical potential or free energy may be more closely related to the availability of water to plant roots, includes osmotic effects, and takes account of thermal changes and hence of change of state.

It is of course equally important to distinguish between the intensity factors and the capacity factors of soil water, the water 'potential' and the amount present, just as it is necessary in thermal problems to distinguish between the thermal intensity, i.e. temperature, and the amount of heat present. But additional difficulties arise with water in soil. As discussed above there is no single unambiguous intensity factor corresponding to temperature. Suction and matric potential can be related to water content but unlike thermal capacity the relationship is neither linear nor unique. Total potential and chemical potential do not relate to water content. Neither are there any absolute values of the intensity factors except for suction, and in practice this is never referred to a zero pressure.

For most practical purposes these difficulties are not serious, and cause no problems but, for example, in dealing with vertical infiltration it is the gradients of the sum of the matric and gravitational potential that are important. The sum of the matric and gravitational potential is sometimes called the hydraulic potential. At low moisture contents and for horizontal movement then only matric potentials need be considered. It is common and convenient to attribute to a particular soil–water system a particular potential, but this can be misleading unless it is clear to what arbitrary zero it is being referred, usually saturated soil at the same level, in which case it is probably the capillary or matric component of potential that is being considered. It is this matric potential to which pF usually refers, though it may include osmotic potential, but only if measurement is through a semi-permeable membrane, that is, a barrier which allows the passage of water but not of solutes, such as air or an ice–water interface as in psychrometry or freezing point determination. The osmotic potential component of humid soils rarely exceeds the equivalent of 200 kPa (2 bar) osmotic pressure when the soil is at the permanent wilting point. Some components of potential are shown in Fig. 10.1.

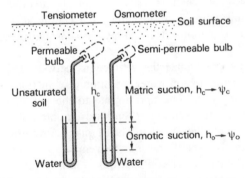

Fig. 10.1   Matric and osmotic components of soil water suction as shown by a tensiometer and an osmometer.

The moisture characteristic curve

The relationship between water content and water suction is a very important property of a soil and was called the moisture characteristic curve by Childs,[6] now more usually called simply the moisture characteristic. Since each suction corresponds to the emptying of pores of a certain diameter, then a range of pore diameters can be attributed to each increment of water withdrawn. The

6 Childs, E. C., *Soil Sci.*, 1940, **50**, 239.

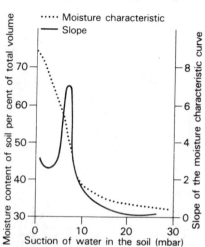

Fig. 10.2 The moisture content–suction curve and its differential for a Gault clay. (Soil initially in crumbs 1–2 mm size.)

slope of the moisture characteristic thus represents a pore size distribution curve, the steeper the slope the greater the volume of pores of size corresponding to the suction or negative pressure needed to empty them (Fig. 10.2). This applies only to soil (and other materials) with rigid pores. Pore size here really refers to the narrowest parts of irregular pores since it is those that control water retention, the suction to withdraw water through them being greater than for the wider parts of the pore. Once past the narrowest point of a pore, the water level falls rapidly until it reaches the next 'neck' or constriction. To refill a pore the suction has to be reduced to a value corresponding to the widest part of that pore, and from there it will then rapidly ascend to the next higher 'neck'. It follows that at a given suction the soil holds more water during drying than it takes up on wetting, and the wetting and drying curves do not coincide, the soil behaving as if it were reluctant both to drain and to refill. This phenomenon is called hysteresis. Figure 10.3 shows the resulting hysteresis loop. An analysis of this hysteresis in water relations is given by Poulovassilis[7] and is discussed by Childs.[8] One consequence of hysteresis is the uncertainty introduced in predicting water content from a knowledge of tension, and vice versa, unless the wetting history of the soil is known. The moisture characteristic is always measured on drying soil so that the water content or volume of any particular group of pores is associated with the diameter of the corresponding pore necks.

7 Poulovassilis, A., *Soil Sci.*, 1962, **93**, 405.
8 Childs, E. C., *An Introduction to the Physical Basis of Soil Water Phenomena*, Wiley, London, 1969.

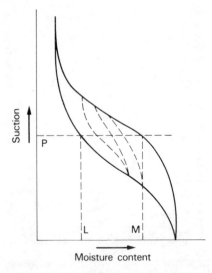

Fig. 10.3 Diagrammatic suction–moisture content curve showing hysteresis loops. At a suction of *P*, the soil may have any moisture content between L and M.

Hysteresis also plays a part in the persistence of wetting fronts in soil. After water has been applied to a dry soil and has wetted the upper layers, movement into the dry lower layers soon ceases, with a sharp boundary between the two. The upper layers are tending to dry and the lower layers to become wetter, and for the same soil water suction the upper drying layer then has a higher water content than the lower wetting soil. Wetting fronts can often be seen on the sides of earth cuttings after summer showers.

The moisture characteristic of a soil thus reveals not only the amount of water held in a soil at different suctions, but also shows the availability of water in that soil, and the total capacity of that soil for holding plant available water. It can be interpreted as a pore size distribution curve, and leads to useful estimates of conductivity at all water contents. It gives the equilibrium water content profile of a soil above a water table, including the thickness of the capillary fringe. This is the depth of soil above the water table which is saturated but under suction, and in which drains cannot reduce the water content since they can accept water only under zero or positive external pressure, not under negative. The moisture characteristic can also be used to infer suction from measured moisture content and vice versa, but of course only if the wetting history of the soil is known, because of hysteresis.

Measurement of soil water suction

The negative pressure in soil water may be measured by allowing equilibrium to be attained between soil water and a manometer separated from it by a suitable porous membrane which allows the passage of water but excludes air. A commonly used form of apparatus, called a tensiometer, consists of a water-filled hollow porous ceramic probe, which is buried in the soil and is connected by narrow flexible tubing to a vacuum gauge or mercury manometer. Its response depends on good contact with the soil, and it fails when the soil water tension exceeds about 0.8 bar, beyond which bubbles of air accumulate inside the ceramic probe, expanding as the pressure falls and cutting off hydraulic contact. For measuring changes in water content with tensions up to this same limit in the laboratory the soil is placed on a suitably porous support in a Buchner funnel, the stem of which is connected by a U-tube to the lower end of a burette.[9] The difference in water levels in the two arms gives the suction applied and the differences in burette readings give the volumes of water withdrawn at successive increments of suction. Suctions higher than 0.8 bar can be achieved by using gas pressure above the porous membrane, using a strong enclosing chamber, and so bringing the membrane to a suction equal to the applied pressure, when water moves out of the soil to come to equilibrium with the membrane. The apparatus is called the pressure plate or pressure membrane, and is used up to about 20 bar pressure. Factors limiting the useful range of pressures include the ever slower rate of equilibration at lower moisture content, and the slow but continuous passage of gas through the membrane due to its change in solubility with fall in pressure across the membrane. This solubility 'leak' carries away water vapour and so interferes with the attainment of equilibrium. It is possible to use gas pressure up to 2000 bar but the method is rarely used beyond 20 bar because of the length of time needed. Larger suctions may be investigated from their effect in reducing the vapour pressure of the water in the soil. The relationship between suction h and relative humidity H is

given by $h = \dfrac{RT}{mg} \ln \dfrac{100}{H}$, and between pF and relative humidity is

p$F = 2 + \log 2.303 \dfrac{RT}{mg} + \log (2 - \log H)$ where R is the gas constant, T

is temperature, m the molecular mass of water, g the gravitational force on unit mass, and H the relative humidity expressed as a percentage.

Inserting appropriate values for the constants gives p$F = 6.502 + \log (2 - \log H)$ from which it can be shown that in soil drier than pF 4.5 (3 MPa or 30 bar suction, $H = 97.5$ per cent) the relative humidity falls rapidly with increasing water suction. In this range the humidity is not very sensitive to temperature change, unlike in wetter soil, and very simple methods can be used to measure it. Equilibration of soil to known equivalent suctions can be used to find the relation between suction and moisture content beyond 3 MPa suction by standing over saturated solutions of appropriate salts.

9 Haines, W. B., *J. agric. Sci.*, 1923, **20**, 97.

At humidities above 97.5 per cent temperature control becomes critical and it is also necessary to use a very sensitive method of measurement of humidity. Thermocouple psychrometry[10] shows promise but demands great attention to detail, including control of temperature.

Other methods used for finding the chemical potential of soil water include freezing point depression[11] and equilibration through a semi-permeable membrane with the osmotic pressure of solutions of polyethylene glycol.[12] A major difficulty with the freezing point method arises from supercooling which obscures the initial freezing temperature. Osmotic equilibration is slow and the variability in composition of polyethylene glycols makes calibration necessary for each batch, which can raise other problems depending on facilities available.

Unlike tensiometry and pressure plate methods these latter indirect techniques respond to the effect of dissolved salts. None includes the gravitational component of potential but this is very easily taken into account if necessary.

In the field one of the most popular methods of finding matric suction or potential beyond the range covered by tensiometers is by porous absorbers whose moisture content comes to equilibrium with the soil water suction, and by measuring their electrical conductivity these changes can be followed. Calibration is needed before the absorbers are buried. Soluble salts in the soil interfere, though with plaster-of-paris blocks the slight solubility of the material itself reduces the effect of salt.[13]

The water content of soil

Water content or concentration is expressed in many ways including mass or volume fraction, as a depth of water equivalent to that contained within a particular soil layer or the soil profile, or as a soil water deficit, which is the depth of water needed to return the water content to its equilibrium value after saturation and subsequent natural drainage, this latter as a moisture content being called field capacity (see p. 353).

Unlike the intensity factor of soil water retention the water content is relatively unambiguous and different ways of expressing it are all readily interconvertible. The dry state from which it is measured is somewhat arbitrary since there is no obvious distinction between free water and water of consti-

10 Monteith, J. L. and Owen, P. C., *J. Scientific Inst.*, 1958, **35**, 443;   Brown, R. W. and van Haveren, B. P. (eds), *Psychrometry in Water Relations Research*, Utah Agric. Expt. Stn., Utah State Univ., Logan, 1972.
11 Schofield, R. K. and Da Costa, J. V. B., *J. agric. Sci.*, 1938, **28**, 644;   Richards, L. A. and Campbell, R. B., *Soil Sci. Soc. Am. Proc.*, 1949, **13**, 70.
12 Zur, B., *Soil Sci.*, 1966, **102**, 394;   Waldron, L. J. and Manbeian, T., *Soil Sci.*, 1970, **110**, 401.
13 Marshall, T. J. and Holmes, J. W., *Soil Physics*, Cambridge Univ. Press, 1979.

tution, but this normally causes no difficulty in practice. The standard method for measuring water content is to find the loss in mass on drying at 105 °C, which leads to mass fraction or percentage of water in unit mass of 'oven dry soil'. Volume fraction can be derived from mass fraction by multiplying by the ratio of soil bulk density to water density, and from this it is a simple step to find the equivalent depth of water in any given depth of soil. If the field capacity of the soil is known, soil water deficit can be derived, though it is more usual to calculate it from estimated losses by evaporation and transpiration, used in a notional water balance.

Difficulties arise in swelling soils where the volume of soil changes with moisture content, so changing the frame of reference for the volume fraction of water. Also there is the problem of comparing changes with depth when the depths too have changed with changing water content, and in a cracking soil there is the additional difficulty of including in a volumetric sample an appropriate proportion of crack space.

A commonly used field method for measuring water content is based on the slowing of fast neutrons by the nuclei of hydrogen atoms, these occurring principally in water, and counting the resultant cloud of slow ('thermalized') neutrons using a boron trifluoride counter. With a probe containing a source of fast (4 MeV) neutrons and a counter of slow (1/40 eV) neutrons the count rate is proportional to the concentration of hydrogen atoms around the probe, and therefore to the concentration of water. This volume varies with the water content of soil, but is typically a sphere of about 30–40 cm diameter. Various factors affect the count rate, and for accurate results the instrument must be calibrated in each soil in which it is used. Because it responds to water in a volume of soil it cannot identify sharp changes in water content as occur across a wetting front. Nor is it entirely reliable within about 20 cm of the surface, where not only is there escape of neutrons into the air, but often a higher and more variable content of organically bound hydrogen in plant roots and other organic materials which are concentrated in the surface layers. The probe is normally used with semi-permanently installed access tubes, and great care must be taken during their installation to avoid compression or air gaps around the access tubes. In swelling soils this is not always possible.[14]

Other methods of estimating moisture content in the field include the measurement and calibration in terms of water content of electrical conductivity or capacity, or thermal conductivity, of soil itself or of buried porous absorbers, mechanical resistance to probe penetration and gamma ray attenuation.[15]

14 Eeles, C. W. O., *Inst. Hydrol. Rept.*, No. 7, Wallingford, England, 1969.
15 See ref. 13.

The movement of water in soil

Where there is a difference of water potential between two points, the water is not in mechanical equilibrium and there will be a tendency for water to move to equalize the potential, the rate of movement being proportional to the gradient of potential and to the conductivity of the soil (Darcy's law 1856)

$$\frac{dQ}{dt} = -K\frac{d\psi}{dl}$$

where dQ/dt is the volume rate of flow across a unit area perpendicular to the direction of flow, $d\psi/dl$ is the gradient of potential expressed as a water-head and K is the hydraulic conductivity of the soil. The minus sign appears because the direction of flow is opposite in sign to that of the gradient.

Hydraulic conductivity depends on the size and continuity of the conducting pores and on the viscosity of water. Under a given potential gradient movement is fastest when the soil is saturated and all pores are conducting water. As soil loses water by drainage or in other ways the larger pores empty first, and because rates of flow are proportional to the fourth power of pore diameters the larger pores contribute most to conduction, and so conductivity falls very rapidly with falling moisture content. This effect is more pronounced in coarse textured soils since they generally have rather few small pores and flow in thin surface films is negligible. The hydraulic conductivity depends not only on the soil pore characteristics but also on the viscosity of water. In order to avoid this dependence of conductivity on the properties of the fluid the conduction coefficient is sometimes expressed as intrinsic permeability, K', the flow equation becoming

$$\frac{dQ}{dt} = -\frac{K'}{\eta} \cdot \frac{g\rho \, dh}{dl}$$

where η is the viscosity of the fluid, g the gravitational force per unit mass, ρ the fluid density and dh/dl the gradient of hydraulic head or suction.

In many cases of water movement in unsaturated soil it is often convenient, though physically incorrect, to treat it as a diffusion process down a concentration gradient rather than as mass flow down a potential gradient. By defining a water diffusivity as the product of conductivity and the rate of change of suction with water content (c), $D = K \, dh/dc$, then the flow equation becomes

$$\frac{dQ}{dt} = -D\frac{dc}{dl}$$

expressing movement as the product of a diffusion coefficient, called the diffusivity of soil water, and the gradient of water concentration. This facilitates mathematical analysis and the application of existing solutions for diffusion and heat flow.

Hydraulic conductivity has the dimensions of a velocity, and is usually expressed in metres per day. Permeability dimensions are those of an area, m² or cm², and diffusivity an area per unit time, m² day⁻¹ or cm² second⁻¹. Figure 10.4[16] shows the relationship between conductivity and suction, between diffusivity and moisture content and between suction and moisture

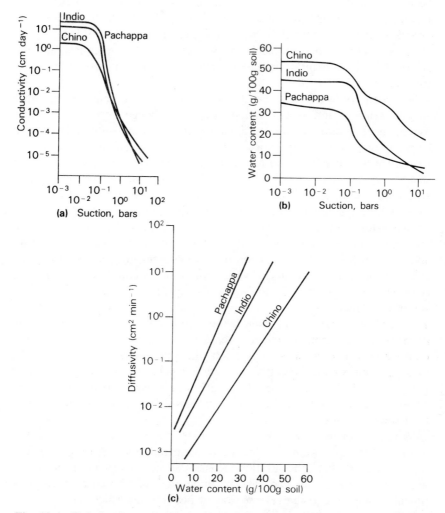

Fig. 10.4   Relation between the conductivity and suction, and between diffusivity and moisture content for Pachappa sandy loam, Indio loam and Chino clay. (1 bar = 100 kPa).

16 Gardner, W. R., *Soil Sci.*, 1960, **89**, 63;   Gardner, W. R., *Arid Zone Res. UNESCO*, 1961, Vol. 15, 37.

content for a Californian clay soil, and a sandy loam, showing how rapidly the conductivity and diffusivity can fall with increasing suction and decreasing moisture content.

The potential gradients involved in water movement include components due to gravity, to hydrostatic pressure and to capillarity or surface tension. As mentioned earlier the sum of these is called the hydraulic potential. It may be expressed in various ways; as an equivalent hydraulic head for use in Darcy's law when the transmission coefficient is called hydraulic conductivity, or as a volumetric potential in the intrinsic permeability form of Darcy's law. Since soil pores are permeable to solutes, osmotic potentials are not concerned in movement of liquid water, but they can be important in vapour movement.

Movement of water also occurs along gradients of temperature, since surface tension and water vapour pressure change with temperature. Temperature induced movement is mainly as vapour diffusion, and this can be many times faster than expected from Fick's law. The increased rate of movement is probably due to temperature gradients being steeper across air spaces, whose thermal conductivity is low, than across the water films separating the air spaces. Diffusion of water vapour is then rapid across air gaps under steep temperature gradients, with rapid liquid phase movement across water films at relatively uniform temperature. The overall rate of movement thus depends on the distribution and relative proportions of air and water in the soil, as well as on the overall gradient of temperature, and can be much faster than simple vapour diffusion would allow. Under some semi-arid conditions with large diurnal fluctuations of temperature and low soil water contents there can be significant amounts of water vapour moved up at night into the rooting zone by temperature gradients.[17] Another situation where temperature gradients can lead to significant amounts of water movement is in freezing soils when liquid water moves up to the colder surface region where the surface tension is higher and therefore the matric potential lower, and because of the effective removal of water from the system on conversion into ice.[18] This process is involved in some forms of frost heave of soils.

Movement of water vapour does not carry dissolved salts with it, unlike movement of liquid water, so that where temperature induced vapour movement occurs the salt concentration increases at the hotter end of the system and water at the cooler end. The resulting gradient of matric or hydraulic potential may lead to the reverse flow of liquid water which reinforces this separation of salt to the hotter areas.[19]

There are many excellent accounts of the behaviour of water in soil.[20]

17 Rose, C. W., *Aust. J. Soil Res.*, 1968, **6**, 31.
18 Dirksen, C. and Miller, R. D., *Soil Sci. Soc. Am. Proc.*, 1966, **30**, 168.
19 Marshall, T. J. and Gurr, C. G., *Soil Sci.*, 1954, **77**, 147.
20 Hillel, D., *Fundamentals of Soil Physics*, Acad. Press, London, 1980; also refs 8 and 13.

Unfortunately there is still some confusion of terminology and of viewpoint. The recommendations of the sub-committee on soil physics terminology of the International Society of Soil Science[21] are discussed and amplified by Rose.[22]

Entry of water into soil

The infiltration rate of most medium- or fine-textured soils kept flooded becomes less with time for a number of reasons. If the soil is initially dry, wetting the soil will entrap air in the coarser pores; and even if the soil is initially moist some of the air present in the wider pores may become entrapped. These air bubbles will block pores and slow up the passage of water through them, although with time the air in these bubbles may dissolve in the water and diffuse into the atmosphere. Again most soil crumbs, if they contain much clay or organic matter, swell on wetting, and this swelling will also reduce pore diameters. This swelling may be a slow process for the heavier-textured soils, such as clay loams and clays,[23] and cracks produced by the crop drying the soil in summer often persist through the autumn and early winter, allowing water to drain through the soil. These cracks may become almost sealed by late winter and spring so the soil becomes almost impermeable.

The infiltration rate of a soil can only be maintained if the system of coarser pores is maintained. The zone where this system is most likely to collapse is in the surface of the soil, for wet soil crumbs are weak and can easily be broken by the impact of raindrops, which cause soil particles to become detached from the crumbs and block the coarser surface pores. The crumbs and the walls of the coarser pores in some soils may collapse spontaneously on wetting, and this is particularly liable to happen if a dry soil is suddenly flooded; or they may collapse slowly if the soil becomes waterlogged for any reason. The maintenance of permeability in the surface layer of the soil is one of the major problems of good soil management in soils of rather low structural stability. Failure to maintain permeability can lead to loss of crop through poor aeration, and to loss of water or soil by run-off and soil erosion.

Infiltration of water into initially dry soil is at first largely controlled by the gradient of matric potential and the nature of the surface pores, but as wetting proceeds and the matric potential gradient falls then gravity becomes relatively more important. The rate decreases with time, and the depth to the wetting front, above which the moisture content is usually high and fairly constant, is initially approximately proportional to the square root of elapsed

21 *Int. Soc. Soil Sci. Bull.*, No. 49, 1976, p. 76.
22 Rose, D. A., *J. Soil Sci.*, 1979, **30**, 1.
23 Emerson, W. W., *J. Soil Sci.*, 1955, **6**, 147.

time. The rate of flow approaches the hydraulic conductivity after a long time when the soil has wetted to depth and the hydraulic gradient is simply gravitational, unity when expressed as a hydraulic head per unit height change. If i is the cumulative infiltration and t elapsed time, Philip[24] suggested using $i = St^{0.5} + At$ where S is called the sorptivity of the soil. At large values of t, di/dt (the flux v) becomes equal to K, the hydraulic conductivity, and therefore to A since $v = 0.5St^{-0.5} + A$.

Very permeable soils have infiltration rates as high as 10^{-1} mm s^{-1} or about 10 m day^{-1}, while soils of low permeability have rates of 10^{-4} mm s^{-1} or 10 mm day^{-1}, or less.

When a limited amount of water is added to a deep dry porous soil the water tends to move down with a fairly sharply defined wetting front, but once the free water on the soil surface has all percolated into the soil, the wetting front soon becomes almost stationary. This line of demarcation between the wet and the dry soil may remain fairly sharp for a matter of days or sometimes weeks; and it is likely to remain sharper the steeper the slope of the moisture-characteristic curve at the suction of the water in the front, and to become diffused quicker the more gentle the slope of this curve at this suction.[25] Figure 10.5 illustrates this difference of behaviour between soils[26];

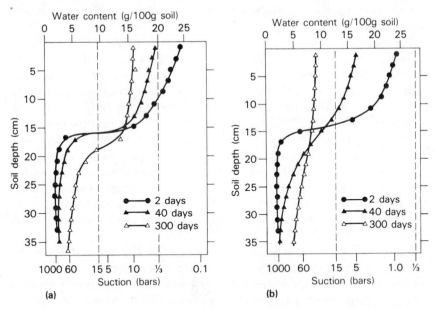

Fig. 10.5 Rate of movement of a limited volume of water into a dry soil column. (a) Clarion loam. (b) Mumford silty clay loam. (1 bar = 100 kPa).

24 Philip, J. R., *Soil Sci.*, 1957, **84**, 257.
25 Whisler, F. D. and Klute, A., *Soil Sci. Soc. Am. Proc.*, 1965, **29**, 489.
26 Sykes, D. J. and Loomis, W. E., *Soil Sci.*, 1967, **104**, 163.

it shows how the boundary between the wet and the dry soil remains fairly sharp for the Clarion loam, but disappears after a time for the Mumford silty clay loam. The persistence of the wetting front is partly due to hysteresis as mentioned earlier, and partly to the resistance to flow of water in the thin films across the wetting front and into the dry soil. With time the front will always become diffuse because of vapour transfer, particularly if there is a temperature gradient from the warmer moist soil to the cooler drier soil.

Richards[27] has given a good example of the slowness of water movement into some dry soils. He found it took over three months for equilibrium to be reached when 7.5 cm of air-dry soil was wetted with water under a suction of 1.5 bar, though if the soil was previously damp, equilibrium was attained in two to three days.

The same effect is seen if a water table rises in a dry soil and then stays at a constant level.[28] Water will be sucked up by capillarity into the dry soil giving a fairly uniform wetting front, but after a time the front will cease to rise and the relatively sharp boundary will only slowly become diffuse.

In the field, however, temperature gradients can be an important cause of water movement in a soil profile. An experiment by Edlefsen and Bodman[29] at Davis, California, illustrates this point very clearly. They irrigated heavily a deep silt loam, and then covered its surface to prevent any evaporation of water. They then determined the amount of water in each 15 cm layer down to 2.7 m at increasing intervals of time after the soil surface had ceased to be waterlogged, which was on 3 October 1934. Some of their results are given in Table 10.2 which shows that from the fourteenth to the fifty-eighth day after the commencement of the experiment 10 cm of water had drained out of the profile, of which 4 cm had come from the first 1.2 m. Drainage almost ceased throughout the winter and spring, but as the soil warmed up in the summer another 13.4 cm left the profile, while during the next summer a further 3.3 cm left. This soil was still losing water during summertime by drainage two and a half years after this irrigation. Water movements in this

Table 10.2 Water draining out of a silt loam profile (Davis, Calif.)

Sampling dates	3/10/34– 6/10	6/10– 17/10	17/10– 30/11	30/11/34– 8/5/35	8/5– 31/8	31/8/35– 22/1/37
No. of days of drainage	3	14	58	217	332	842
Cm water draining out of						
top 1.2 m soil	24.2	3.8	4.0	0.5	6.8	1.0
top 2.7 m soil	44.5	10.4	10.1	2.0	13.4	3.3

27 Richards, L. A., *J. Am. Soc. Agron.*, 1941, **33**, 778.
28 Moore, R. E., *Hilgardia*, 1939, **12**, 383.
29 Edlefsen, N. F. and Bodman, G. B., *J. Am. Soc. Agron.*, 1941, **33**, 713.

profile are complicated by temperature gradients, for during the spring of 1936 the top 1.2 m and bottom 45 cm were both losing water, but the middle section from 1.2 to 2.2 m below the surface was gaining water.

Evaporation of water from bare soil

The rate of loss of water from a damp soil is controlled by the evaporating power of the air and the heat energy falling on it, and so long as the soil can supply water to the surface fast enough to keep it moist, this rate of loss persists. The rate falls sharply, however, as soon as the rate of loss exceeds the rate of supply sufficiently for the surface to become dry; for then instead of the water vapour being produced at the soil surface, and diffusing immediately into the air to be readily removed by the air currents, it is produced below the soil surface and must diffuse through the soil pore space under a small concentration gradient before it reaches the atmosphere. Even a dry layer 1 to 2 mm thick can appreciably reduce the rate of evaporation.[30]

The magnitude and rate of evaporation losses from deep soils without a water table can be illustrated with results obtained from the fallow drain gauges at Rothamsted. The rate of loss of the first 12 mm of water is controlled by the evaporating power of the air, and in summer this lasts for about five days. The rate then drops rapidly from 2.5 mm per day to about 1 to 2 mm per week, so that it is only after a six-week drought that 23 mm have been lost by evaporation from the bare soil. The three-month drought of 1921, which lasted from June to August, only caused a total loss of 32 mm, and it is unlikely that more than 6 mm of water moved up from the soil layer below 22 cm in depth into the layer 0 to 22 cm deep. Figure 10.6 illustrates this result for the moisture content of a bare soil at Rothamsted on 27 June 1870, after a prolonged drought. Within the accuracy of the determination no water had moved up from below 45 cm and only about 6 mm from below 22 cm. The figure also gives the moisture distribution in a neighbouring plot carrying barley, and shows by way of contrast how the crop reduced the water content of the layer 110 to 135 cm deep. The figure further illustrates the importance of specifying the moisture contents on a volume and not a weight basis, for the fallow plot has a larger average pore space than the cropped, and hence can hold more water. If the comparison were made between moisture contents calculated on a weight basis, the barley would be judged to have removed much more water from the subsoil than it in fact had done.

This result, that a bare Rothamsted soil can only lose about 12 mm of water rapidly by evaporation, is not necessarily true in the laboratory, for there Penman[31] showed that it could lose about 30 mm at the rate of 2.5 mm

30 Penman, H. L., *J. agric. Sci.*, 1941, **31**, 454.
31 See ref. 30.

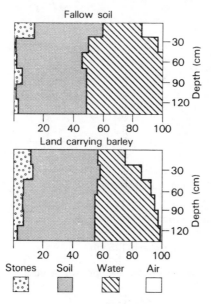

Fallow soil

Land carrying barley

Stones   Soil   Water   Air

Fig. 10.6   Volumes of soil, water and air at different depths of a fallow and of a cropped soil at Rothamsted on 27 June 1870, after a prolonged drought (R. K. Schofield).

per day, that is, at the same initial rate as the drain gauge on typical June days but for over twice as long. But after it has lost about 30 mm its rate of loss falls very rapidly to the 2 mm or less per week, just as with the drain gauge. The difference in behaviour between the drain gauge and the laboratory soil is due to the high evaporation rate under field conditions which dries the soil surface and so prevents water moving to the surface at an appreciable rate. When the surface of the laboratory column was heated by shining an electric fire over it for 12 hours a day, to simulate the effect of sunny June weather, the soil then behaved like the drain gauge soil, losing about 12 mm of water at the rate of 2.5 mm a day before the rate of loss dropped rapidly to the 2 mm or less per week. Figure 10.7 illustrates this effect of potential evaporative power of the air on the rate of loss of water from a cylinder of Indio loam soil when the temperature is kept constant.[32] It shows as Penman found, that initially the rate of loss is equal to the evaporative demand, but that once a certain amount of water has been lost, which is as soon as the rate of upward movement of water to the surface is less than the potential demand, the rate drops to that controlled by diffusion of water vapour to the surface of the soil. The dotted curve in the graph gives the locus of points at which the rate of liquid supply becomes limiting; and all

32 Gardner, W. R. and Hillel, D., *J. Geophys. Res.*, 1962, **67**, 4319.

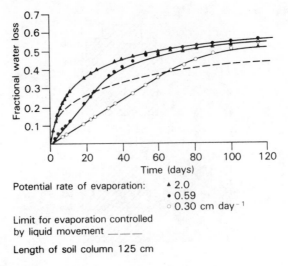

Potential rate of evaporation: ▲ 2.0
                                 ● 0.59
                                 ○ 0.30 cm day⁻¹

Limit for evaporation controlled
by liquid movement _ _ _ _

Length of soil column 125 cm

Fig. 10.7   Cumulative water loss by evaporation from a wet column of an Indio loam soil.

the drying curves above this line are parts of the same diffusion-controlled curve. Once diffusion controls the rate of loss from the column, the total amount of water lost after t days is proportional to $\sqrt{t}$, where t is the number of days after the start of diffusion control. This result would be expected on theoretical grounds if the water loss is controlled solely by diffusion.[33]

A further consequence of these results is that during the winter half-year at Rothamsted, from October to April, the evaporation from the soil is controlled by the evaporative power of the air, or by the amount of radiation received from the sun; for this is sufficiently low, and the rainfall sufficiently frequent, for the soil surface never to be dried out. But during the warmer months, from May to September, and even in some years in March, April and October, the evaporation is normally controlled by the number of days the soil surface remains wet, which in turn, depends on the frequency of the rain showers and on their intensity.[34]

This effect of rainfall distribution is of great practical importance in arid countries having summer rainfall, as in the dry farming wheat belt of the Canadian prairies where summer fallows are taken to store water for a succeeding crop. Frequent light falls of rain on such fallows are evaporated soon after they fall and so cannot contribute to the soil moisture supply. Thus Hopkins[35] has calculated that at Swift Current, Saskatchewan, if 25 mm of rain fell on one day, 16 mm remained in the soil ten days afterwards, whereas

33 Rose, D. A., *Brit. J. appl. Phys.*, 1968, Series 2, **1**, 1779.
34 Schofield, R. K. and Penman, H. L., *J. agric. Sci.*, 1941, **31**, 74.
35 Hopkins, J. W., 1940, *Canad. J. Res.*, 1940, **18C**, 388.

if rain fell as five separate showers, each of 5 mm on five consecutive days, only 8 mm remained ten days afterwards.

The depth from which a bare soil loses water in the tropics depends on a number of factors. One important factor is the amount of soil cracking that takes place during drying, for water vapour diffuses very much faster from the surface of even deep cracks into the atmosphere than through the equivalent depth of soil. The loss is therefore greater from deep-cracking montmorillonitic clays than from kaolinitic soils. Further, the greater the average windspeed over the soil surface, the more rapid the transport of water vapour from the sides of the cracks into the atmosphere.[36] Gusts of wind can cause air currents in the surface layers of loose or open soils and also increase considerably the water vapour flux through a dry surface crust.[37] A second factor is the magnitude of the temperature gradient in the soil profile and its diurnal variation, for this affects the rate and direction of diffusion of water vapour in the profile. This is usually more important in the surface layers of the soil than in the deep subsoil, both because the temperature and suction gradients are greater there and also because the air space is greater. Thus, Rose,[38] working at Alice Springs, central Australia, showed that, under the very strongly drying conditions prevailing there, the downward flux of water vapour during the day in the surface soil was of the same order of magnitude as the upward flux of liquid water once the soil suction exceeded 0.3 bar.

There is still little direct field evidence on the depth from which soils lose water in subtropical and tropical regions having a long hot dry season. Part of the difficulty of interpreting some of the earlier work is due to the neglect of loss of water from the subsoil by slow drainage, as illustrated in Table 10.2. Pereira[39] studied both the loss of water and the rise in suction of the soil water of a bare soil at Kongwa in central Tanzania after the rainy season in two successive years, and in each it took about one month for the soil at 15 cm and three or four months for the soil at 30 cm to be dried to a suction of 15 bar. At the end of a seven-month dry season the soil had been dried to a suction of about 7 bar at a depth of 45 cm, and to about 1 bar at a depth of 60 cm. The mean monthly soil temperature at 7.5 cm depth towards the end of the dry season exceeded 33 °C.

The amount of water a soil can lose by evaporation after it has been wetted depends not only on the evaporative power of the air, but also on the rate of downward movement of the water in the subsoil, and on the amount of water the surface soil can hold against drainage. On this basis a dry soil flooded by irrigation may hold more water in its surface than if wetted slowly,

36 Adams, J. E. *et al.*, *Soil Sci. Soc. Am. Proc.*, 1969, **33**, 609. Adams, J. E. and Hanks, R. J., *Soil Sci. Soc. Am. Proc.*, 1964, **28**, 281.
37 Farrell, D. A. *et al.*, *Soil Sci.*, 1966, **102**, 305.
38 Rose, C. W., *Aust. J. Soil Res.*, 1968, **6**, 31, 45;   Fritton, D. D. *et al.*, *Soil Sci. Soc. Am. Proc.*, 1967, **31**, 599.
39 Pereira, H. C. and Wood, R. A., *Emp. J. expl. Agric.*, 1958, **26**, 213.

due to the trapped air reducing the rate of infiltration into the subsoil, with the consequence that an irrigated soil is likely to lose more water by evaporation if it is irrigated by flooding than by sprinkler.[40]

Soils with a water table near the surface

Evaporation from a bare soil with a water table fairly close to the surface, and by this is usually meant within 2 m of the surface, is controlled both by the evaporative power of the air and by the rate at which water will move up from the water table to the soil surface by liquid flow through the capillary pores and water films in the soil under the suction gradient caused by the soil surface drying out. This, in turn, depends on the relation between the capillary conductivity and the suction in the soil profile. Soils whose capillary conductivity falls relatively uniformly with increase in suction such as many alluvial fine sandy or silty loams, will lose water by evaporation at a greater rate from a deeper water table compared with a soil possessing a well-marked field capacity at a fairly low suction. It is possible that losses of the order of 0.5 mm day^{-1} can take place under fairly strongly drying conditions from a water table 2 m below the surface in certain soils.[41]

Under dry conditions the rate of loss of water by evaporation can be very considerably reduced in soils with a fairly high water table either by applying a mulch of dry sand or stones, or by hoeing the surface to let it dry out, so ensuring that water transfer through the mulch to the atmosphere only takes place through the vapour phase. This reduction in evaporation can be very marked for soils with a water table between 1 and 2 m below the soil surface.

Evaporation of water from saline soils

Evaporation of water from a saline soil differs from that of a non-saline soil in that, under comparable conditions, the saline soil is likely to lose more water by evaporation during a drought than the non-saline, because it will maintain a moist surface for a longer time. There are two principal reasons for this. In the first place, as water evaporates from the surface of the saline soil, the salt concentration in the soil surface increases, which lowers the vapour pressure of the solution and increases its osmotic pressure. This reduces its rate of evaporation and so allows solution to move up from the subsoil for a longer time than if no salts were present. Once the soil surface becomes dry and covered with salt crust or efflorescence, the rate of evaporation drops, just as in a salt-free soil. In consequence of these effects, salts may increase the depth from which water can move up from a water table to be evaporated at the surface, though the effect is most noticeable at

40 Bresler, E. *et al., Soil Sci. Soc. Am. Proc.*, 1969, **33**, 832.
41 Richards, L. A. *et al., Soil Sci. Soc. Am. Proc.*, 1956, **20**, 310;   Gardner, W. R. and Fireman, M., *Soil Sci.*, 1958, **85**, 244; see also ref. 28.

moderate concentrations, for at high concentrations a salt crust soon forms.[42]

Vapour pressure gradients will also move salts to the surface indirectly, for during daytime when evaporative conditions are high, there is a marked downward temperature gradient, which will give an appreciable downward vapour flux. The resulting accumulation of water at a shallow depth changes the hydraulic potential gradient leading to an increased upward flux of water carrying salts with it. The daytime temperature gradient acts as a pump, pumping salts from the subsoil into the surface as long as the soil remains moist enough for the upward movement of liquid water to remain appreciable.[43]

These effects have the consequence that once a salt patch starts to form on the soil surface, it grows at the expense of the neighbouring soil; so instead of the soil being covered with a uniform thin film of salt, it tends to have a large number of salt patches. This is illustrated in Plate 5.2. It is seen to a much less extent if water is moving up from a water table, or if the soil surface is almost dead level as happens sometimes in an old lake bed. It can also have very important results in irrigated fields where much salt is present. The salt patches can be prevented only by maintaining a level surface and a uniform permeability in the surface soil, and by avoiding water standing in pools. It is also essential for the water table to be sufficiently deep so that it cannot influence the moisture regime of the surface soil.

42 Quayyum, M. A. and Kemper, W. D., *Soil Sci.*, 1962, **93**, 333.
43 Jackson, R. D. *et al.*, *Nature, Lond.*, 1965, **205**, 314.

11

Water and crop growth

The water requirements of crops

Water is essential for plant survival and growth. In contrast to most nutrients, which are retained in the plant, the characteristic of water is its flow from the soil through the roots and stems to the leaves and then into the air. Water is used in photosynthesis to form sugars, as a solvent for biochemical reactions and for the translocation of cell constituents and for the support of the plant by turgor, but only about 1 per cent of the water taken up by plants is used for these metabolic activities. Most of the water taken up by plants is vaporized (evaporated) into the surrounding air predominantly via the stomata. Evaporation of water that has passed through the plant is called transpiration, and if the plant is growing in a moist soil the rate is controlled by the microclimate outside the leaves.

Plants use large quantities of water during growth and every day an actively growing crop may use several times its own mass of water. For example, an average crop of wheat in the UK weighing 5 t ha^{-1} (dry matter) during mid-

June will be evaporating about six times its own mass daily. Similarly, it is estimated that a maize plant in Kansas will lose over 200 litres of water during its life, equivalent to about 100 times its own fresh weight.[1] Transpiration often occurs simultaneously with evaporation from the soil surface and there is often no easy way of separating these two sources of water vapour. Both pathways of loss are evaporative, involving a change of phase that requires energy known as the latent heat of vaporization. In recognition of the difficulty of distinguishing the sources of water vapour, the term evapotranspiration is frequently used to denote the overall process of water transfer from both soil and vegetation to the atmosphere.

When there is a freely available supply of water, the rate at which water vapour diffuses into the atmosphere is controlled by two factors. First the energy arriving at the evaporating surface determines the amount of water that can be evaporated, and second, the dryness and speed of the air together determine the rate at which water vapour can be dispersed into the free atmosphere. The relative importance of these two factors has long been the subject of debate: over 2000 years ago Aristotle concluded that 'wind is more influential in evaporation than the sun'.

The response to this atmospheric demand for water is the movement of water from the soil through the plant and into the atmosphere. In most crops, loss is mainly through the leaves with about 90 per cent of the water lost through the stomata[2] and only a small amount lost through the cuticle. However, there is variation between crops in the resistance to flow through the cuticle associated with the thickness of epicuticular wax, which may be physiologically significant in conferring drought resistance. For example, on average, rice varieties have from one-quarter to one-half the cuticular resistance of sorghum and maize, but there are substantial differences between varieties, especially of rice.[3]

Radiation balance

Energy arrives at and departs from crop surfaces in a number of forms and by a number of processes. The energy of solar radiation is concentrated in that part of the spectrum with wavelengths of 0.3 to 4 μm (referred to here as shortwave), in contrast to the longwave radiation (4–80 μm) that origi-

1 Miller, E. C., *Plant Physiology* (2nd edn), McGraw-Hill, 1938.
2 Larcher, W. in *Physiological Plant Ecology*, Springer-Verlag, 1975, states that cuticular water loss varied from 0.05 per cent of total water loss in xerophytic cacti to 32 per cent in an herbaceous plant.
3 Yoshida, S. and de los Reyes, E., *Soil Sci. Pl. Nutr.*, 1976, **22**, 95; see also O'Toole, J. C. *et al.*, *Physiol. Plant.*, 1979, **47**, 239 for a detailed comparison of an upland rice from West Africa and a paddy rice from the Philippines.

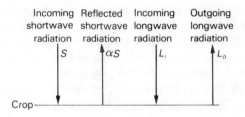

Fig. 11.1   Components of the radiation balance at a crop surface.

nates from surfaces of lower temperature such as clouds and the earth. Different proportions of these shortwave and longwave radiations are absorbed, reflected or emitted depending upon the nature of the surface. Net radiation, R_n, is the algebraic sum of these various processes and is the difference between the total downward and upward radiation fluxes for a unit area of surface (Fig. 11.1). Thus

$$R_n = S(1 - \alpha) + L_n \qquad [11.1]$$

where S is the total incident shortwave radiation, α is the reflection coefficient or albedo of the absorbing surface (*i.e.* $S(1 - \alpha)$ is the amount of shortwave energy absorbed) and L_n is the net longwave radiation ($L_i - L_o$).

During the summer at Rothamsted about 80 per cent of the incoming radiation is absorbed, of which 30 per cent is re-radiated as longwave radiation.[4] When crops are subjected to drought, the proportion of incoming radiation retained as net radiation is decreased because canopy temperature increases and more radiation is re-radiated as longwave radiation. As will be shown later in this chapter, net radiation is an important quantity because it is the energy available at the ground surface for the processes of evaporation, air and soil heating, and photosynthesis.

During the day, net radiation is usually positive because there is normally more radiant energy arriving at a surface than leaving it, whereas at night the reverse is usually the case. Crops receiving similar amounts of incoming radiation may absorb different amounts of net radiation because of differences in reflectivity or surface temperature or both. The greatest absorption of radiation usually occurs in the upper parts of the canopy. It is possible to manipulate the interception of radiation by changing plant spacing and population, and also by changing the architecture of the canopy. However, the greater interception of energy to promote growth may also lead to increased transpiration which may in turn result in water shortage, a smaller leaf area, and a reduced amount of R_n intercepted. When sweet corn was water-stressed early in growth, Campbell[5] found that R_n measured below the canopy

4 Penman, H. L., *Brit. J. appl. Phys.*, 1951, **2**, 145;   Monteith, J. L. and Szeicz, G., *Quart. J. R. Meteorol. Soc.*, 1961, **87**, 159.
5 Campbell, R. B. *et al., Agric. Meteorol.*, 1981, **23**, 143.

increased from 32 to 46 per cent of the net radiation measured above the canopy, and when stressed later it increased to 51 per cent; yields of fresh cobs were correspondingly reduced by 25 and 45 per cent of the non-stressed yield.

The value of albedo (α) varies according to the nature of the surface and the angle of incidence; the higher the sun is in the sky, the lower is the albedo. It also depends on the wavelength of the incident radiation, being small for visible light (0.3–0.7 μm) but much larger in the near infrared (0.7–4.0 μm).[6] Pale green and dark green leaves have much the same albedo but the structure of the canopy is important because it affects both the angle the leaves make with the incoming light, and the proportion of energy reflected by leaves low down in the canopy that may be absorbed by leaves higher up. Most rock, sand, and soil surfaces reflect 10–30 per cent of the incident solar radiation and most crops around 15–30 per cent. Moist soils have albedos of about 0.10–0.15 and this increases as the soil dries to around 0.25–0.45 for dry sandy soils. Light-coloured soils have higher albedos than dark-coloured soils; peat typically lies in the range 0.05–0.15. The albedo for clear water is about 0.05 and for fresh snow 0.80–0.95. Short annual crops have values of 0.10–0.20 depending upon the leaf area and the type of crop. Some parts of crops, particularly flowers, may reflect more than other parts as, for example, the flowers of rape which have an albedo of 0.4–0.5.

Net radiation can be measured using net radiometers but these are frequently only available for research purposes.[7] More commonly, R_n is calculated from meteorological measurements by assuming a value for α. The incident shortwave radiation, S, can be measured using a pyranometer or can be calculated from measurements of the number of sunshine hours. The amount of longwave radiation re-radiated may also be calculated from meteorological data. Penman[8] showed that, as a basis for estimating potential evaporation, $R(= S + L_i)$ and L_n could be calculated accurately enough from standard meteorological measurements using the relations:

$$R = R_T \, (a_1 + 0.62n/N) \qquad [11.2]$$

$$L_n = \sigma T^4 \, (0.47 - 0.065 \, \sqrt{e_d})(0.17 + 0.83n/N) \qquad [11.3]$$

where R_T is the incoming solar radiation if the atmosphere were completely transparent (calculated from the solar constant, the latitude, and time of year and is given in suitable tables), n/N is the hours of sunshine as a proportion of the maximum possible hours (i.e. as if there were no cloud), σ is the Stefan–Boltzmann constant (5.67×10^{-8} W m^{-2} K^{-4}), T is the temperature of the absorbing surface, e_d is the vapour pressure of water in the air (here

6 For some values see Yocum, C. S. *et al.*, *Agron. J.*, 1964, **56**, 249.
7 For an example of their use see Biscoe, P. V. *et al.*, *J. appl. Ecol.*, 1975, **12**, 227.
8 Penman, H. L., *Proc. R. Soc. Lond.*, 1948, **A193**, 120. The numerical values given here are the revised figures given in *J. agric. Sci.*, 1970, **75**, 69.

expressed in mbar), and a_1 is a constant depending somewhat on latitude, being about 0.16 at Rothamsted and 0.25 in tropical regions.[9] The ratio of R_n to R therefore varies with the temperature and reflectivity of the surface. For example, the ratio varied from 0.58 for open water (where the upward reflection of longwave radiation is low, the surface is cool, and shortwave reflectivity is low) to 0.25 for a desert (where surface temperatures are high and shortwave reflectivity is high).[10]

Estimation of evaporation

The energy available for evaporation can be derived from the general form of the energy balance equation, which can be written as:

$$R_n = C + G + J + M + \lambda E \qquad [11.4]$$

where R_n is the net radiation, C is the energy used to warm up the air in contact with the evaporating surface, G is the heat energy conducted to warm the materials below the evaporating surface, J is the energy stored or retained by the surface, M is the net storage of energy during photosynthesis, and λE is the energy used to evaporate water where E is evaporation and λ is the latent heat of vaporization. For evaporation from open water and a bare soil surface J and M can be omitted.

Examination of the terms of the energy balance for crops allows a number of simplifications to be made. In bright sunlight R_n may be 300 to 500 W m^{-2} (up to 800 W m^{-2} in the tropics) and M represents the net storage of energy in photosynthesis. Using values given by Monteith[11] it can be shown that M is usually small in comparison with R_n and can therefore be ignored when the balance is calculated. Typically, the maximum rate of photosynthesis is 2 to 5 g CO_2 m^{-2} h^{-1}, producing 1.4 to 3.5 g carbohydrate m^{-2} h^{-1}. The heat of combustion of carbohydrate is about 17 kJ g^{-1} giving values of M ranging from 7 to 17 W m^{-2}.

Similarly, the heat content of the standing crop, J, changes very little during the day. Again, using values given by Monteith[12] and assuming that leaves have the same specific heat as water, a leaf 1 mm thick will have a heat capacity of 4.2 kJ m^{-2} °C^{-1}. This means that if leaf temperature increased by 5 °C per hour (as might happen on a clear sunny day) then J would be 5.8 W m^{-2}, about 1 per cent of R_n. Over whole days, J is always a small term in the energy balance but may be appreciable under certain circumstances in the short term.

9 Glover, J. and McCulloch, J. S. G., *Quart. J. R. Meteorol. Soc.*, 1958, **84**, 56.
10 Stanhill, G. *et al.*, *Quart. J. R. Meteorol. Soc.*, 1966, **92**, 128.
11 Monteith, J. L., *Principles of Environmental Physics*, p. 152, Arnold, 1975.
12 See ref. 11, p. 158.

As discussed in Chapter 8, on clear sunny days there may be an appreciable heat flux, G, into the soil from sunrise until about three hours after midday, and a corresponding flux out from then until sunrise. The exact relation between the amount of heat stored in the soil and the heat balance of the surface depends on many physical factors, the most important being the heat capacity and thermal conductivity of the soil, the diurnal change of air temperature, the turbulent diffusion coefficient for heat transfer between the surface and the air, the windspeed and the surface roughness. In dry, bare soil G may be 0.3 to 0.6 of R_n but when soil is covered by vegetation, it is usually about 0.05 to 0.1 of R_n.

The small changes in the terms M, J and G relative to R_n means that over periods of several days, no great error is made in assuming that net radiation is converted into either latent heat by evaporating water, or sensible heat by warming the air.

Sensible and latent heat fluxes from the absorbing surface involve the transfer of warm and humid air from the surface into the atmosphere, as a result of gradients of temperature, T, and vapour pressure, e, above it. Although these gradients are constantly varying at any one location because of the turbulence of the atmosphere, if the gradients are averaged over appreciable areas of land surface and over appreciable intervals of time (usually greater than one hour), then the fluxes can be related to the gradients by the relations

$$C = -\rho c_p \, K_h \, \partial T / \partial z \qquad\qquad [11.5]$$

$$\lambda E = \frac{-\rho c_p K_v \partial e / \partial z}{\gamma} \qquad\qquad [11.6]$$

where ρ is the density of air, c_p is the specific heat capacity of air at constant pressure and λ is the heat of vaporization of water at the prevailing temperature (2.45 kJ g^{-1} at 20 °C). γ is the psychrometric constant, which is the ratio $p c_p / 0.622\lambda$ (where p is the total air pressure), and has a value of 0.66 mbar K^{-1} in air at atmospheric pressure and at 20 °C. The ratio of the sensible to the latent heat transfer is known as the Bowen ratio (β), and may be written as:

$$\beta = \frac{C}{\lambda E} = \frac{\rho c_p \, K_h}{\dfrac{\rho c_p \, K_v}{\gamma}} \frac{\partial T / \partial z}{\partial e / \partial z} \qquad\qquad [11.7]$$

The quantities K_h and K_v are the diffusion coefficients for turbulent exchange (sometimes called eddy coefficients) for heat and vapour transfer and they are often assumed to be equal since both rely on the same process of eddy diffusion.[13] It is also frequently assumed that $\partial T / \partial z / (\partial e / \partial z)$ can be replaced

---

13 See for example Verma, S. B. *et al.*, *J. appl. Meteorol.*, 1978, **17**, 330.

by the ratio of the differences between the mean air temperature and vapour pressure at the absorbing surface, T_o and e_o, and in the air at some selected height, z, above the surface, T_a and e_a. Hence for non-advective conditions, that is, assuming no horizontal transport of energy downwind, the Bowen ratio may be expressed as:

$$\beta = \frac{\gamma \, (T_o - T_a)}{(e_o - e_a)} \qquad [11.8]$$

Substituting $C = \beta \, \lambda \, E$ from 11.7 into 11.4, ignoring M and J and solving for $\lambda \, E$ gives:

$$\lambda E = \frac{R_n - G}{1 + \beta} \qquad [11.9]$$

Typically, over well-watered short grass or a wet soil surface β is about 0 to 0.2. As the surface dries, β becomes larger because the proportion of sensible heat increases and it approaches infinity for very dry surfaces. In arid regions, β may become negative over crops as heat is extracted from the air to evaporate water. Such regional advection results in lower temperatures nearer the crop than in the air above it (i.e. $T_o - T_a$ becomes negative). At around midday on a typical sunny day, R_n may be 600 W m^{-2} of which 60 W m^{-2} warms the soil; if β is 0.2 then the rate of evaporation will be 0.19 g m^{-2} s^{-1} or, assuming stable conditions over one hour, 0.67 mm h^{-1}.

Several comparisons of the Bowen ratio method with measurements from lysimeters show good agreement between the two techniques in non-advective conditions.[14] Figure 11.2 compares measured evaporative losses from a lysimeter with those calculated using the Bowen ratio method; in advective conditions the Bowen ratio generally underestimates the measured evaporation. However, despite the use of the technique by micrometeorologists, there is a problem in its more widespread use because the temperature and humidity at two levels are required.

Penman[15] simultaneously solved the heat and mass balance equations to eliminate surface temperature. Rearranging equation [11.4] (there is no M or J) gives:

$$R_n - G = C + \lambda E \qquad [11.10]$$

At this point it is convenient to rewrite equations [11.5] and [11.6] in terms of the resistance nomenclature now widely used.[16] The sensible heat and latent heat equations can be written:

14 Tanner, C. B., *Soil Sci. Soc. Am. Proc.*, 1960, **24**, 1;   Blad, B. L. and Rosenberg, N. J., *J. appl. Meteorol.*, 1974, **13**, 227.
15 See ref. 8.
16 See for examples Monteith, J. L., *Principles of Environmental Physics*, Arnold, 1975, and Campbell, G. S., *An Introduction to Biophysics*, Heidelberg Science Library, 1977.

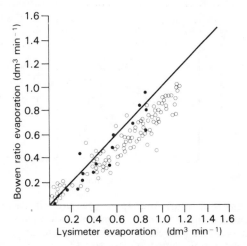

Fig. 11.2 Comparison of evaporation estimated by the Bowen ratio method with that measured using a lysimeter under advective (o) and non-advective (●) conditions. (From Blad, B. L. and Rosenberg, N. J., *J. appl. Meteorol.*, 1974, **13**, 227; measurements for 6–10 July and sensors at 125–150 cm.)

$$C = \frac{\rho c_p\,(T_o - T_a)}{r_h} \qquad [11.11]$$

and

$$\lambda E = \frac{\rho c_p}{\gamma}\frac{(e_{so} - e_a)}{r_v} \qquad [11.12]$$

where ρc_p is the volumetric specific heat capacity of air and T_o and e_{so} are the temperature and saturated vapour pressure at the surface, T_a is the temperature at a reference height in the air, and e_a is the actual vapour pressure at the same height. The terms r_h and r_v are resistances to sensible heat (convection) and vapour transfer respectively between the wet surface and the air at reference height. Penman linearized the relation between the saturated vapour pressure and the temperature of the air (Fig. 11.3) to estimate the vapour pressure difference between the surface, assumed saturated, and the air $(e_{so} - e_a)$ from the temperature difference and the saturation deficit $(e_{sa} - e_a)$. The linearization and substitution shown in Fig. 11.3 allow $e_{so} - e_a$ to be written as:

$$e_{so} - e_a = (e_{sa} - e_a) + \Delta\,(T_o - T_a) \qquad [11.13]$$

where Δ is the slope of the saturation vapour pressure curve at $(T_o + T_a)/2$. In practice, Δ is usually determined at T_a, so a small error results but this

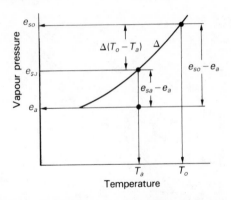

Fig. 11.3   The change of saturation vapour pressure with temperature showing the approximation used by Penman to estimate the difference in vapour pressure between the surface and the air $(e_{so} - e_a)$; $(e_{sa} - e_a)$ is the saturation deficit of the air, $(T_o - T_a)$ is the difference in temperature between the surface and the air, and Δ is the slope of the saturation vapour pressure curve at $(T_o + T_a)/2$.

is not usually serious. By substituting for $(T_o - T_a)$ from equation [11.13] into equation [11.11], then using equation [11.12] to eliminate $(e_{so} - e_a)$ from the resulting expression for C, and finally substituting for C in equation [11.10] gives:

$$\lambda E = \frac{\Delta \ (R_n - G) + (\rho \ c_p \ (e_{sa} - e_a)/r_h)}{\Delta + \gamma \ (r_v \ / \ r_h)} \qquad [11.14]$$

This equation is known as the Penman equation and describes evaporation in terms of radiation, saturation deficit, temperature, and diffusion resistances.

To describe evaporation from a crop (transpiration), G can be ignored but an additional resistance, r_c, must be added because the air at the leaf surface is not saturated[17]; r_c represents the resistance to water transfer from within the substomatal cavities, where air is saturated, to the leaf surface. So, for crops:

$$\lambda E = \frac{\Delta \ R_n + (\rho c_p \ (e_{sa} - e_a)/r_h)}{\Delta + \gamma \ (r_v + r_c)/r_h} \qquad [11.15]$$

The canopy resistance, r_c, depends on the degree of stomatal opening, except when the leaves are wet with dew, rain or irrigation water when it is close to zero. Many crops have a value of r_c between 0.3 and 0.5 s cm^{-1} when

17 Monteith, J. L., *Soc. Exp. Biol. Symp.*, 1965, **19**, 205.
18 See ref. 17.
19 Szeicz, G. *et al.*, *Water Resources Res.*, 1969, **5**, 380.

well supplied with water and when their leaves are dry,[18] rising to 1.0 to 1.5 for forests[19]; but if the stomata begin to close due to water shortage r_c will rise to higher values, up to 3.5 to 7.5 s cm^{-1} for orange orchards in Arizona[20] and up to 15 for a prairie in Nebraska subjected to drought.

Since water on the outside of a leaf evaporates more easily than water inside a leaf, under comparable evaporative conditions, the rate of evaporation from the wet leaves of a crop will be 1.3 to 1.6 times as fast as transpiration from the dry canopy. For a forest, the rate of evaporation from a wet canopy will be about 5 times as high as from the dry. This result emphasizes that spray irrigation, by maintaining a wet canopy, will increase evaporation during and just after irrigation. It also shows that dew is a relatively inefficient mechanism for providing a crop with water.[21] More details of the derivation of these equations and the importance of some of the terms are given in recent micrometeorological textbooks and review papers.[22]

Using evaporation equations

To make use of standard meteorological measurements, the Penman equation was first derived in a different form from that shown above. Using measured rates of evaporation, E, from a small tank of water, Penman found a linear relation between r_h^{-1} and the daily run of wind, u_2, at 2 m above the ground. This empirically derived relation, determined for periods of a week or more, gave the original form of the Penman equation as:

$$\lambda E = \frac{\Delta}{\Delta + \gamma} (R_n) + \frac{\gamma E_a}{\Delta + \gamma} \qquad [11.16]$$

where $E_a = 0.26 (1.0 + 6.2 \times 10^{-3} u_2) (e_{sa} - e_a)$, and $(e_{sa} - e_a)$ is the saturation deficit of the air at the height of 2 m. The term E_a is often referred to as the aerodynamic term in contrast to R_n which is a radiation term. In its initial form it is sometimes referred to as an 'empirical' relation but, as shown, it is physically based; the empiricism enters in the relation between r_h^{-1} and u_2.

The aerodynamic and radiation terms are weighted by the factors Δ and γ both of which are temperature dependent. As temperature rises, the factor Δ/γ increases from 0.9 at 5 °C to 1.6 at 15 °C and 2.8 at 25 °C so that in hot conditions the net radiation term is likely to be more important than the aerodynamic term, and in cool conditions the reverse obtains.

20 van Bavel, C. H. M. *et al., Agric. Meteorol.,* 1967, **4**, 27.
21 Waggoner, P. E. *et al., Agric. Meteorol.,* 1969, **6**, 227;   Black, T. A. *et al., Agron. J.,* 1970, **62**, 66.
22 Monteith, J. L. *Quart. J. R. Meteorol. Soc.,* 1981, **107**, 1;   Rosenberg, N. J. *et al., Microclimate: the biological environment* (2nd edn), Wiley, 1983.

This basic form of the Penman equation is rarely used in practice because net radiation values are rarely recorded at meteorological sites and must be calculated from records of the number of hours of bright sunshine. The exact form of the equation used depends upon the type of measurements available and there are now numerous tables and computer programs available to aid what used to be a tedious calculation.[23] In using such tables, however, caution is necessary because of the uncertainties introduced by the various algorithms used to estimate, for example, net radiation and the wind function when they are not directly measured. For example, Batchelor[24] found a 23 per cent difference in the estimate of evapotranspiration from rice growing in north-east Sri Lanka using the different empirical relations used in the FAO modified Penman equation.

The rate of evaporation from an open water surface is often written E_o. Penman[25] compared the evaporation from short, green grass freely supplied with water, E_t, with E_o and showed that the ratio E_t/E_o depended on the season and increased from 0.6 in winter to 0.8 in summer. An important part of the reason that E_t/E_o is less than unity is that E_o is calculated for a surface with a reflection coefficient of 0.05, whereas the reflection coefficient of vegetation is always greater, up to 0.25 for short grass, though approaching 0.05 for high forest. Initially E_t was regarded as the potential rate of evaporation from short green vegetation completely covering the ground and with no shortage of water. However, results from other crops have shown that the relation is not restricted simply to short crops[26] particularly if the lower reflection coefficient of tall vegetation is allowed for.

Table 11.1 illustrates one use that can be made of the Penman equation. It gives the calculated average evaporation from well-watered short grass at Rothamsted and the average rainfall; and it shows that on average a soil moisture deficit of about 75 mm will be established by the end of July, which will not be made good again until early November. In summers and autumns that are drier than average, much higher deficits will be built up: in 1970 the potential deficit was about 150 mm at the end of July, and it was not until early October that it began to be reduced. In years of below-average winter rains, the deficit may not be made good until the following spring.

Table 11.1 also gives the total amount of incoming radiation, expressed as the depth of water that it would evaporate if it was dissipated entirely as latent heat. Maps have been published for England and Wales (Fig. 11.4) in which the average maximum soil moisture deficit for any area can be deduced and from which an estimate of the likely value of irrigation can be made.[27]

23 See for example Doorenbos, J. and Pruitt, W. O., *Irrigation and Drainage*, Paper No. 24, FAO, Rome, 1977.

24 Batchelor, C. H., *Irrig. Sci.*, 1984, **5**, 223.

25 Penman, H. L., *J. Soil Sci.*, 1949, **1**, 74; *Quart. J. R. Meteorol. Soc.*, 1950, **76**, 372.

26 See for examples Glover, J. and Forsgate, J., *Quart. J. R. Meteorol. Soc.*, 1964, **90**, 320; Pereira, H. C. and Hosegood, P. H. *J. Soil Sci.*, 1962, **13**, 299.

27 Hall, D. G. M. *et al.*, *Soil Survey Tech. Monog.*, No. 9, 1977.

Table 11.1 Mean incoming radiation, rainfall and potential evaporation at Rothamsted. Original data supplied by H. L. Penman. The values of E_t have been rounded to the nearest 5 mm

	Incoming radiation (1956–69) (mm)	Potential evaporation E_t 1967–70 (mm)	Rainfall R 1948–69 (mm)	$R - E_t$ (mm)
January	29	5	59	54
February	45	10	51	41
March	97	30	43	13
April	137	50	47	−3
May	194	70	54	−16
June	213	90	56	−34
July	193	80	60	−20
August	157	65	66	1
September	117	45	60	15
October	68	20	63	43
November	30	5	71	66
December	19	0	68	68
Total	1300	470	698	

Computer-based models are also now available to calculate soil moisture deficits taking account of meteorological, crop and soil variables. For example, simple functions of crop growth, calculations of potential evaporation, and estimates of water stored in various soil types have been used successfully to calculate the soil moisture deficits under cereal crops in Britain.[28] The model accurately reflected changes in soil water deficit as measured with a neutron probe in six out of eight test crops representing a range of soil types and weather patterns.

The second important aspect of the Penman equation is that the net radiation is divided into a latent heat and a sensible heat term, which emphasizes that sensible heat is an additional method of dissipating absorbed energy. Sensible heat transfer depends on the temperature difference between the leaves of the crop and the air, and leaf temperature is controlled by the heating of the leaf due to its absorbing solar energy and its cooling due to transpiring water. Thus, sensible heat transfers are intimately bound up with transpiration rates. It may happen that transpiration rates are sufficiently high for the leaf to be cooler than the air, and so there will be transfer of sensible heat from the air to the latent heat. This effect will be particularly noticeable when warm, dry air is blowing over a rough crop, because transpiration will be encouraged, which will cool the leaves, thereby cooling the air as well as adding water to it.

28 Francis, P. E. and Pidgeon, J. D., *J. agric. Sci.*, 1982, **98**, 651, 663.

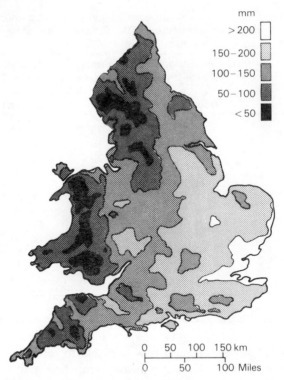

mm

> 200

150 – 200

100 – 150

50 – 100

< 50

0   50   100   150 km

0   50   100 Miles

Fig. 11.4   Average maximum potential soil moisture deficit (mm) for England and Wales. The values are most relevant to perennial crops (e.g. grass or trees) and adjustments are needed for arable crops with shorter growing seasons. (From Hall, D. G. M. *et al., Soil Survey Tech. Monog.*, No. 9, 1977.)

The result of these processes is that for crops growing in arid locations, the measured rate of evaporation may exceed the open pan rate (E_o). Rijks,[29] working in the Gezirah region of the Sudan, found that the latent heat of evaporation from actively growing crops of cotton was often 1.8 times the amount of net radiation available. Similarly, evapotranspiration from an irrigated crop of alfalfa grown in Nebraska ranged from 4.8 to 14.2 mm day^{-1} whereas R_n provided energy for no more than 7 mm day^{-1}; on all days λE exceeded $R_n + G$.[30] Nevertheless, with appropriate modifications the Penman equation can be used in such environments. For example, for northern New South Wales (a sub-humid area with uncertain rainfall), in a season when large-scale advection was significant, evaporation from cotton plants was accurately predicted using the Penman equation containing a modified wind function.[31]

29 Rijks, D. A., *J. appl. Ecol.*, 1971, **8**, 643.
30 Rosenberg, N. J. and Verma, S. B., *J. appl. Meteorol.*, 1978, **17**, 934.
31 Cull, P. O. *et al., Irrig. Sci.*, 1981, **2**, 141.

In addition to the effects of regional advection, there are also important effects resulting from local advection. The effect is most pronounced at the leading edge of the crop, that is, where the wind first strikes it, and it rapidly falls to a fairly constant level within 60 to 80 m from the leading edge,[32] and the effect decreases slowly over distances measured in tens of kilometres.[33] This means that the crop at the leading edge should be irrigated more frequently than the crop further in, though there seem to be no examples where this is done in practice. The effect can even be seen to a small extent in England, so that wind blowing over a dry barley stubble, reaching a field of sugar beet, caused increased transpiration at the leading edge of the crop and neither the air temperature nor the saturation deficit had reached their stable values 70 m behind the leading edge.[34]

Many experimenters have sought to define the relation between the actual rate of evaporation for an annual crop and either the potential rate of evaporation calculated from the Penman equation or the rate of evaporation from an open water surface measured with a class A evaporation pan. This has been made possible with the introduction of weighing lysimeters[35] holding many tonnes of undisturbed soil and of neutron moisture meters.[36] Figure 11.5 shows a typical result for the way in which evaporation changes for two annual crops. For winter wheat, the soil surface is covered by vegetation by mid-April but for spring barley, the soil is bare, so that the measured evaporative flux is very dependent on the moisture content of the soil surface; it is small when the surface is dry but close to E_t when wet. As the barley crop grows, both its leaf area and the aerodynamic roughness of the surface increase and the measured rate of evaporation becomes increasingly independent of the soil surface and more dependent on the store of water in the soil. Evaporation in both crops reaches a plateau when leaf growth is complete but declines rapidly as a result of depletion of the stored water and because of leaf senescence. Similar patterns have been determined by numerous other workers.[37]

It is clear from such studies and from Table 11.1 that during the summer months in the United Kingdom, and for longer periods of the year in other climates, most crops and forests will be dependent on moisture stored in the soil. The size of the store and its accessibility to roots will therefore be important in determining the length of time for which the actual and potential rates of evaporation remain close to unity.

[32] For examples from the Gezirah in the Sudan see Hudson, J. P., *Expl Agric.*, 1965, **1**, 23; Head, D. W., *Emp. J. expl. Agric.*, 1964, **32**, 263.
[33] For an example from Texas see Lemon, E. R. *et al.*, *Soil Sci. Soc. Am. Proc.*, 1957, **21**, 464.
[34] Davenport, D. C. and Hudson, J. P., *Agric. Meteorol.*, 1967, **4**, 339.
[35] For example Sammis, T. W., *J. Hydrol.*, 1981, **49**, 385.
[36] Bell, J. P., *Soil Sci.*, 1969, **108**, 160 and *Inst. Hydrol. Rept.* No. 19, 1976.
[37] For an example with maize see Denmead, O. T. and Shaw, R. H., *Agron. J.*, 1962, **54**, 385.

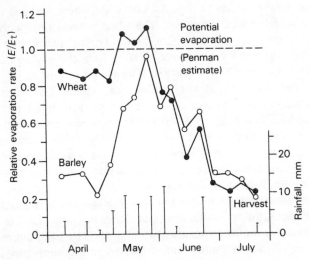

Fig. 11.5   Evaporation from neighbouring crops of winter wheat and spring barley grown on Wick series soil in 1975, measured using a neutron probe. (From McGowan, M. and Williams, J. B., *J. Soil Sci.*, 1980, **31**, 217.)

The available water capacity of a soil

The available water capacity is a soil characteristic related to plant response, and in its widest sense it is the maximum quantity of water that can be extracted from a soil profile by plants. The concept was put forward by Veihmeyer[38] who defined it as the amount of water a soil held between its condition at field capacity and its permanent wilting point. He found that the yields of many irrigated fruit orchards in central California were little affected by drought if the top 1.8 m of soil was allowed to dry to its permanent wilting point before the next irrigation was given.[39] This concept of available water capacity has only proved useful in practice under very limited conditions,[40] and then only for soils which hold most of their available water, as so defined, at suctions close to field capacity. These are typically coarse textured soils, not finer than fine sandy or silt loams; soils with this texture commonly occur on alluvial river terraces, which constitute a large proportion of all

38 Veihmeyer, F. J., *Hilgardia*, 1927, **2**, 125.
39 Veihmeyer, F. J. and Hendrickson, A. H., *Calif. Agric. Exp. Stn., Bull.* 479, 1929; *op. cit.*, 573, 1934 for peach and prune orchards; *op. cit.*, 667, 1942, for apples and pears; *op. cit.*, 668, 1942, for cotton. See *Proc. Hort. Soc. Am.*, 1931, **28**, 151, for grapes; *op. cit.*, 1937, **35**, 289, for walnuts; and for growth rates of fruit on peaches, pears and prunes, *op. cit.*, 1942, **40**, 13.
40 See Hagan, R. M. *et al.*, *Adv. Agron.*, 1959, **11**, 77.

irrigated lands of the world. This concept has never been of any value on irrigated clays.[41]

Problems occur in calculating the magnitude of the available water capacity in a particular soil. The main difficulty is in identifying the field capacity which indicates 'the amount of water held in the soil after the excess . . . water has drained away and after the rate of downward movement of water has materially decreased'.[42] The difficulty posed by this definition is the interpretation of the time when the downward movement has 'materially decreased'. Initially many workers chose a time two to three days after an irrigation or heavy rain but many soils, even with coarse texture, continue to drain for many weeks.[43]

As a practical means of overcoming this difficulty, the moisture content at field capacity has been defined in terms of a matric suction. Colman[44] found 33 kPa (0.33 bar) an acceptable value for his soils, while Haise[45] found that 10 kPa was more appropriate. In England, the suction in the surface horizons of well-drained soils is typically 3 to 7 kPa during winter and spring,[46] so that 5 kPa is a frequently used value. Some soils, however, are imperfectly drained and pose difficulties in defining an appropriate suction corresponding to field capacity. For example, soils developed on clay-shale sediments showed restricted rewetting of subsoils even when the topsoils were waterlogged. In these soils, matric suctions of 12.5 kPa at 1.3 m were common throughout the winter.[47]

The lower limit of available water, the permanent wilting point, was first defined by Hendrickson and Veihmeyer[48] as the moisture content of the soil when the leaves of plants growing in the soil first reached a stage of wilting from which they did not recover when placed in a saturated atmosphere without the addition of water to the soil. This assumes that all plants will act the same way in any one soil; experimental results show this to be invalid.[49] Again, though, a directly measured value of soil matric suction has been taken as the permanent wilting point. Usually 1.5 MPa (15 bar) is the standard, but because of the shape of the moisture characteristic curve the exact value matters less than that for field capacity.

Both soil structure and soil texture influence the amount of available water. The effect of soil structure is most pronounced in determining the water

41 Lewis, M. R., *US Dept. Agric. Tech. Bull.* No. 432, 1934; *Pl. Physiol.*, 1935, **10**, 309; Furr, J. R. and Taylor, C. A., *US Dept. Agric. Tech. Bull.* No. 640, 1939.
42 Veihmeyer, F. J. and Hendrickson, A. H., *Soil Sci.*, 1931, **32**, 181.
43 For examples see van Bavel, C. H. M. *et al.*, *Soil Sci. Soc. Am. Proc.*, 1968, **32**, 310; McGowan, M. in *Isotope and Radiation Techniques in Soil Physics and Irrigation Studies*, p. 435, IAEA, Vienna, 1973.
44 Colman, E. A., *Soil Sci.*, 1947, **63**, 277.
45 Haise, H. R. *et al.*, *Soil Sci. Soc. Am. Proc.*, 1955, **19**, 20.
46 Webster, R. and Beckett, P. H. T., *J. agric. Sci.*, 1972, **78**, 379.
47 McGowan, M., *J. Soil Sci.*, 1984, **35**, 317.
48 See ref. 39.
49 Slatyer, R. O., *Bot. Rev.*, 1957, **23**, 585.

content at field capacity because the presence of cracks, fissures and channels will determine, in part, the amount of water retained against gravitational forces (refer to Ch. 10 for the relation between suction and pore size). This contributes to the greater variations in moisture content at field capacity found within a textural class than those at permanent wilting point.[50]

Soil composition influences the amount of water retained at both field capacity and permanent wilting point. Analysis of over 500 topsoils and subsoils from England and Wales[51] shows that volumetric water content at field capacity is little affected by clay contents > 20 per cent but that the content at permanent wilting point increases linearly at a rate of 0.66 per cent water for each additional 1 per cent clay in the range 0–45 per cent clay. Further analysis of these results (Fig. 11.6) shows that in topsoils, the content of organic carbon also has a marked influence on the water retention at field capacity. Clay content alone accounted for most of the variance at permanent wilting point in both topsoils and subsoils. Similarly, the amounts of silt and clay were highly correlated with laboratory and field determinations of field capacity, permanent wilting point and available water capacity of 11 ferralitic soils in Uganda.[52] Moreover the inclusion of the percentage of organic matter in the regression equations significantly improved the relation in both topsoils and subsoils.

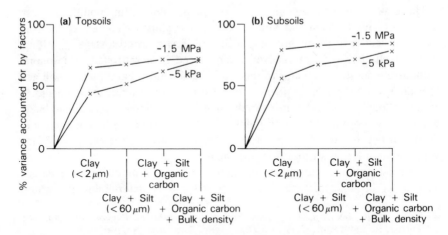

Fig. 11.6   Multiple regression analyses for water retention at different potentials; results from about 520 samples of both topsoils and subsoils. (From Reeve, M. J. and Thomasson, A. J., private communication.)

50 Salter, P. J. and Williams, J. B., *J. Soil Sci.*, 1965, **16**, 310.
51 See ref. 27.
52 Pidgeon, J. D., *J. Soil Sci.*, 1972, **23**, 431.

To avoid the problem of defining the upper and lower limits of available water in relation to a specific value of matric suction, Ritchie[53] has proposed the use of 'extractable water' defined as 'the difference between the highest volumetric water content in the field (after drainage) and the lowest measured water content when plants are very dry and leaves are either dead or dormant'. Extractable water has the advantage of being a field-based measurement and of taking into account the distribution of roots in the soil profile but it is site and season specific. Figure 11.7 typifies many field observations where in the upper soil layers, water is lost beyond the −1.5 MPa limit of available water but, in the subsoil, it is rare for all the available water to be extracted. If the extractable water is measured for several years it can be used for predictive purposes; it is, though, of little use for more general surveys.

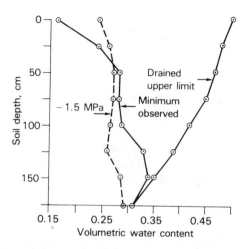

Fig. 11.7 Comparison of extractable soil water and available soil water for a crop of sorghum grown on Houston black clay. (From Ritchie, J. T. in *Soil Water and Nitrogen in Mediterranean-Type Environments* (eds J. L. Monteith and C. Webb), p. 81, Nijhoff/Junk, 1981.)

The soils with the greatest volume of available water are loams containing appreciable quantities of silt. Table 11.2 gives some values for topsoils and indicates the considerable variation within a textural class. However, it is not simply the amount of water in a soil that determines its availability. In the short term, the availability of water to a plant is determined by the rate at which it can be conducted through the soil to the root

53 Ritchie, J. T. in *Soil Water and Nitrogen in Mediterranean-Type Environments* (eds J. L. Monteith and C. Webb), p. 81, Nijhoff/Junk, The Hague, 1981.

Table 11.2   Available water capacity of topsoils in England and Wales. (The author is grateful to M. J. Reeve of the Soil Survey of England and Wales for these figures)

Particle-size class	Mean available water capacity (mm/100 mm)	Standard Deviation	Number of samples
Clay	16.3	3.9	21
Silty clay	18.3	2.5	12
Sandy clay loam	17.1	2.0	23
Clay loam	17.7	4.0	111
Silty clay loam	19.9	5.0	53
Silt loam	22.5	3.3	17
Medium sandy silt loam	20.5	4.3	32
Fine sandy silt loam	22.8	2.0	9
Medium sandy loam	17.8	4.0	71
Fine sandy loam	20.2	3.6	21
Loamy medium sand	17.1	3.8	10
Loamy fine sand	18.7	3.9	11
Medium sand	13.6	2.2	3

surface. Hydraulic conductivity decreases as the matric suction increases, with the result that water held at low suctions is more readily available than water held at high suctions. Figure 11.8 shows typical curves for three contrasting textures of the proportion of the total available water held at different suctions and shows that a much greater proportion of this water is held at low suctions for loamy sands than for silt loams or clays; and this greater

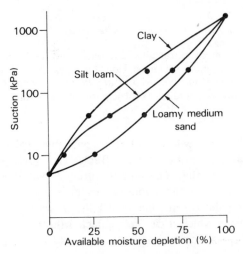

Fig. 11.8   Relation between relative moisture depletion and suction for soils of different textures. (From Reeve, M. J. and Thomasson, A. J., private communication.)

proportional availability at low suctions in the sandy loams counterbalances, to some extent, the advantage of the higher total available water content of the finer-textured loams.

The available water capacity of a soil can be increased marginally by increasing its organic matter content or increasing the proportion of finer-textured particles in a coarse-textured soil. Many laboratory experiments have suggested the benefits of organic amendments for increasing the available water content of soils[54] although the significance of these amendments for farming practice is doubtful.

On the clay loam soil of Broadbalk field at Rothamsted, 93 applications of 35 t ha^{-1} of farmyard manure given most years between 1843 and 1943 increased the available water capacity in the plough layer (0–22.5 cm) by 18 mm from 50 mm to 68 mm and had little or no effect below this depth.[55] This represents an additional 4 to 5 days' water supply at typical potential rates of evaporation during an English summer. Similarly, 50 t ha^{-1} of farmyard manure, given twice a year for five years to the sandy loam soil at Wellesbourne, in Warwickshire, had only increased the available water capacity in the plough layer (0–15 cm) by about 5 mm; and 15 t ha^{-1} annually for nine years on a fine sandy loam at Efford, in Hampshire, had no measurable effect although both field capacity and permanent wilting point were increased a little.[56] Other reviews of the literature have also shown that while additions of organic materials to soils may promote aggregation resulting in increased pore space, the increase in available water is small.[57] However, while the effects on increasing water availability are small, they may be important in dry years in those regions where frequent rain showers can normally be expected.[58]

Clearly, the depth of rooting of a crop will influence how much of the water in a soil profile is available (see Ch. 4 for details of rooting depths). For example, when pearl millet was sown on a sand in Niger, the planting density had a marked influence not only on the depth of rooting and on the total amount of water used but also on the rate at which water became available. These differences in availability and accessibility were reflected in crop dry matter which was 40 per cent less in the densest sown crop as compared with the sparsest.[59]

54 For an example with sewage sludge see Gupta, S. C. *et al.*, *Soil Sci. Soc. Am. J.*, 1977, **41**, 601. For an example using farmyard manure see Kumar, S. *et al.*, *Aust. J. Soil Res.*, 1984, **22**, 253.
55 Russell, E. W. and Balcerek, W., *J. agric. Sci.*, 1944, **34**, 123; Salter, P. J. and Williams, J. B., *J. agric. Sci.*, 1969, **73**, 155, found lower figures than this for samples taken in 1967.
56 Salter, P. J. and Williams, J. B., *J. hort. Sci.*, 1968, **43**, 263; and with Harrison, P. J., *Expl. Hort.*, 1965, **13**, 69.
57 Jamison, V. C., *Soil Sci.*, 1956, **81**, 459.
58 Holliday, R. *et al.*, *J. agric. Sci.*, 1965, **64**, 161.
59 Azam-Ali, S. N. *et al.*, *Expl. Agric.*, 1984, **20**, 203 and 215.

It should also be noted that a recently irrigated crop, with a well-established root system, growing in places where the potential evaporation rate is fast, will use a certain amount of water from its root zone that would have drained out of that zone if the transpiration demand had been low. As a result, the amount of water available in the soil profile to an irrigated crop will depend, to some extent, on both the rate of water use and of downward drainage. It is not unusual for irrigated crops to use up to 50 mm of water that is potentially capable of draining out of the profile.[60] Conversely, water may move back into rooting zones from depths below it as the season progresses.[61] On a deep soil under maize this amounted to 4 mm day^{-1} for several days, although the total upward flux only amounted to 25 mm. This upward water movement into the zone of rooting can also be appreciable (up to 80 mm) on shallow soils overlying chalk in southern England.[62]

In conclusion, the concept of available water capacity and its associated terms remains an inexact part of soil science. It is an equilibrium concept and therefore inadequate to describe the rate at which water becomes available over short periods, but for a crude estimate of water balance for irrigation purposes over long periods, it has some advantages. For example, an irrigated area of uniform soils with an available water capacity of 100 mm m^{-1} soil, which grows crops with an average rooting depth of 0.8 m, will if transpiring at a rate of 4 mm day^{-1} deplete all of the water in 20 days.

Dynamic aspects of water availability

The classical concept of water availability led to a variety of hypotheses as to the effects of soil water depletion on plant growth (see Fig. 11.9). A considerable number of experiments were conducted over many years to define such relations but inevitably they possessed elements of site, season and crop specificity. As the concept of potential evaporation determined by meteorological factors developed, it became increasingly clear that uptake of water by plants was a dynamic system involving fluxes of water through the soil to the root, across the root, from the roots to the leaves, and thence into the atmosphere. This soil–plant–atmosphere system[63] forms the basis of most recent expositions of soil water availability.

The importance of considering water uptake dynamically, as opposed to the static concepts of availability, can be illustrated using the results of

60 For example see Nielsen, D. R. *et al.*, *Soil Sci. Soc. Am. Proc.*, 1959, **23**, 408;   Wilcox, J. C., *Canad. J. Soil Sci.*, 1962, **42**, 122.
61 van Bavel, C. H. M. *et al.*, *Soil Sci. Soc. Am. Proc.*, 1968, **32**, 317.
62 Wellings, S. R., *J. Hydrol.*, 1984, **69**, 259.
63 For an introduction see Slatyer, R. O. and Taylor, S. A., *Nature, Lond.*, 1960, **187**, 922;   Cowan, I. R., *J. appl. Ecol.*, 1965, **2**, 221.

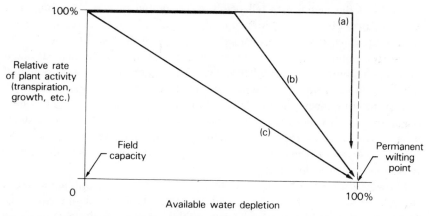

Fig. 11.9   Three classical hypotheses regarding the availability of soil water to plants: (a) equal availability from field capacity to wilting point, (b) equal availability from field capacity to a 'critical moisture' content beyond which availability decreases, and (c) availability decreases gradually as soil moisture content decreases. (From Hillel, D., *Applications of Soil Physics*, p. 218, Academic Press, 1980.)

Denmead and Shaw.[64] They measured the rates of transpiration of maize plants subjected to different conditions of irrigation and atmospheric demand for water by growing plants in columns and then transferring them to the field. Figure 11.10 shows that the value of soil water suction at which actual and potential transpiration rates deviated was dependent upon the atmospheric demand for water. When the potential rate of evaporation was below

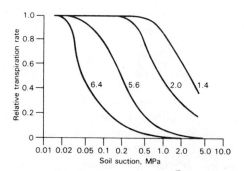

Fig. 11.10   Relative transpiration rate of maize as a function of the soil suction for different potential transpiration rates (mm day^{-1}, shown by the numbers) on a silty clay loam in Iowa.[64]

64   Denmead, O. T. and Shaw, R. H., *Agron. J.*, 1962, **54**, 385.

2 mm day^{-1} the actual rate fell below the potential rate at an average soil water suction of about 0.2 MPa but as the potential rate increased, so the suction at which the soil was capable of supplying this rate decreased. The volumetric water contents at which the measured transpiration rates deviated from the potential rate varied between 23 per cent at the lowest potential rate, and 34 per cent at the highest potential rate. Although this experiment is open to criticism because it did not fully allow the plants to adapt to field conditions,and because the plants were grown in pots so that the root systems were restricted (a crop growing in soil at 0.05 MPa suction would normally be able to extract 6.4 mm day^{-1}), it offers a partial explanation of the various responses shown in Fig. 11.9.

Movement of water through the soil–plant–atmosphere continuum can be thought of as occurring along a gradient of decreasing water potential, ψ. Gradients of potential between parts of the system constitute the driving force for flow within the system. However, at different parts in the system, different components of the total potential will be important to varying degrees. For example, the horizontal movement of liquid water in soil is affected by a gradient of matric potential and little affected by the osmotic potential whereas within the plant, the osmotic and matric potentials both strongly influence the flow. The various processes involved in the movement of water through the system can be visualized as being analogous to the flow of an electric current which obeys Ohm's law, that is,

$$\text{Water flux} = \frac{\text{difference in water potential } (\Delta\psi)}{\text{resistance } (r)}$$

i.e. the flux of water through each part of the system is determined by the operating potential gradient and the resistance in that same segment. In 1948, van den Honert[65] described the steady-state flow of water in the form:

$$\text{Flow} = \frac{\psi_{soil} - \psi_{root\ surface}}{r_{soil}} = \frac{\psi_{root\ surface} - \psi_{xylem}}{r_{root}}$$

$$= \frac{\psi_{xylem} - \psi_{leaf}}{r_{xylem} + r_{leaf}} = \frac{\psi_{leaf} - \psi_{air}}{r_{leaf} + r_{air}}$$

An electrical analogue of the system is shown in Fig. 11.11 and indicates qualitatively the various paths of water movement. Although a simple analogue such as this contains assumptions not strictly applicable to plants, the essential features of the pathway are evident. There are clearly two parallel pathways of vertical water movement to the atmosphere, either through the soil and directly to the atmosphere or through the xylem vessels of the roots, stems, and leaves to the stomatal pores and thence to the atmosphere. Other, similar analogues have been presented[66] highlighting

65 *Disc. Faraday Soc.*, 1948, **3**, 146.

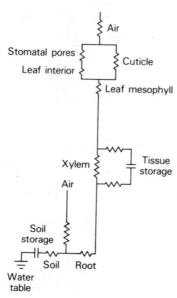

Fig. 11.11 Simplified diagram treating water movement through the soil–plant–atmosphere continuum as analogous to movement of an electrical current through a conducting system containing resistances and capacitances. (Derived from Cowan, 1965, adapted from Kramer and Kozlowski, 1979, *Physiology of Woody Plants*, by permission of Academic Press.)

other possible resistances of the system, e.g. the importance of cuticular versus stomatal resistance. Like many generalizations, though, this analogue is an oversimplification because it assumes a steady state, something which is rarely achieved in plants, and is therefore not applicable in periods shorter than those necessary to reach equilibrium.

Water stored in the parenchyma may act analogously to an electrical capacitor and, therefore, prevent a strict application of Ohm's law. Thus, in trees, for example, the sapwood of the stem may be an important reservoir of water that is normally depleted during the day and replaced at night.[67] The size of the capacitor varies, as might be expected, with the size of the plant so that arable crops generally have smaller capacitance than trees (Table 11.3).

The analogue also assumes that there are no solutes flowing in the system and that the resistances in the various parts of the pathway are constant

66 See for examples Philip, J. R., *Ann. Rev. Pl. Physiol.*, 1966, **17**, 245;   Molz, F. J. and Ferrier, J. M., *Plant Cell Environ.*, 1982, **5**, 191.
67 For apple trees see Landsberg, J. J. *et al.*, *J. exp. Bot.* 1976, **27**, 579.   For Scots pine see Waring, R. H. *et al.*, *Plant Cell Environ.*, 1979, **2**, 309.

Table 11.3   Water capacitance of various trees and crops

Crop	Capacitance (m^3 MPa^{-1})	Source
Potato	1300×10^{-7}	1
Apple tree	2200×10^{-7}	2
Wheat	$(1.1 \text{ to } 14.4) \times 10^{-7}$	3
Wheat	$(2 \text{ to } 12) \times 10^{-7}$	4

1 Whitehead, D., cited by Wallace, J. S. and Biscoe, P. V. (see below).
2 Powell, D. B. B. and Thorpe, M. R. in *Environmental Effects on Crop Physiology* (eds J. J. Landsberg and C. V. Cutting), p. 259, Acad. Press, 1977.
3 Jones, H. G., *J. appl. Ecol.*, 1978, **15**, 613.
4 Wallace, J. S. and Biscoe, P. V., *J. agric. Sci.*, 1983, **100**, 591.

whereas in many cases, the resistance appears to vary with the flow rate.[68] Finally, the analogue is only applicable to the flow of liquid water where the difference in potential (Δ_ψ) is the driving force for water movement. Water is lost from the plant as water vapour and this vapour flow is proportional to the difference in vapour concentration or vapour pressure. Water potential and vapour pressure are related by:

$$\psi_w = \frac{RT}{V_m} \ln e/e_s \qquad [11.17]$$

where R is the universal gas constant, T is the absolute temperature, V_m is the molar volume of water, and e is the vapour pressure and e_s the saturated vapour pressure at temperature T. Typically the potential difference between the bulk soil and the atmosphere will be several tens of MPa and in arid conditions may exceed 100 MPa. Table 11.4 shows some values of water potentials for a crop of maize grown on a sandy loam in South Carolina, USA, where it is clear that the potential difference between soil and leaf is much smaller than that between leaf and atmosphere.

In many actively growing plants, the water potential of the shoot normally decreases during the day particularly as the soil starts to dry. The changes in soil water potential during a single day are usually small and the potential of the shoot generally increases again during the night, implying an appreciable resistance to flow somewhere in the liquid phase. The site of this resistance has been the subject of controversy because, as most of the resistances to water flow of the system are in series, it is usually impossible to measure directly the individual components and they are therefore usually inferred.

Gardner[69] calculated the difference in soil matric potential between the root surface, ψ_r, and a point b, in the soil ($\psi_{so,b}$). Under steady-state

68 For a discussion of these various assumptions and a critique of the Ohm's Law analogy see Passioura, J. B., *Aust. J. Pl. Physiol.*, 1984, **11**, 333. For some results see Passioura, J. B. and Munns, R., *op. cit.*, 1984, **11**, 341, 351.

Table 11.4 Calculated water potentials in the soil–plant–atmosphere system for an unirrigated crop of *Zea mays* L. (Reicosky, D. C. and Lambert, J. R., *Soil Sci. Soc. Am. J.*, 1978, **42**, 221)

Time	Location	Water potential (MPa)
05.20	Air	−6.69
	Leaf	−0.02
	Root	−0.02
	Soil	−0.02
14.00	Air	−66.33
	Leaf	−1.56
	Root	−1.28
	Soil	−0.02

conditions and considering water moving from point b (measured from the root axis) then

$$\psi_{so,b} - \psi_r = \frac{I_w}{4\pi K} \ln \frac{b^2}{a^2} \qquad [11.18]$$

where I_w is the inflow of water, K is the hydraulic conductivity of the soil and a is the radius of the root. To a reasonable approximation b is equal to half the distance between neighbouring roots, i.e. $b = (1/\pi L_v)^{1/2}$, where L_v is the root length per unit volume of soil. Using a value for I_w of 0.1 ml cm^{-1} day^{-1}, Gardner showed that $(\psi_{so,b} - \psi_r)$ can be appreciable in some soils when $\psi_{so,b}$ is less than −0.3 MPa. This implied an appreciable resistance to flow in the soil surrounding the root as it dried. However, rates of inflow are rarely as fast as 0.1 ml cm^{-1} day^{-1} (see Ch. 4), and if slower and more typical rates of inflow are used, calculations suggest that the soil resistance will generally be smaller than the plant resistance until near or beyond −1.5 MPa.[70]

Figure 11.12 shows that onion seedlings grown with their roots aligned in a plane in thin layers of Upper Greensand soil, at an initial matric potential of −0.3 MPa, produced substantial gradients of potential close to the root surface after six days. Calculations with these results support Newman's suggestion that soil resistance is generally small by indicating that if the root–soil geometry had been radial rather than linear, then the potential at the root surface would have been about −0.35 MPa rather than the −2.0 MPa measured.[71]

The analysis and results available suggest that rates of water uptake by root systems are usually slow enough to give only small gradients of water content

69 Gardner, W. R., *Soil Sci.*, 1960, **89**, 63.
70 Newman, E. I., *J. appl. Ecol.*, 1969, **6**, 1, 261.
71 Dunham, R. and Nye, P. H., *J. appl. Ecol.*, 1973, **10**, 585.

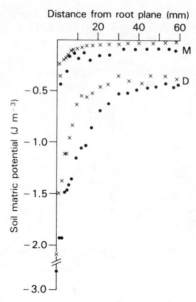

Fig. 11.12 Soil matric potential gradients at $2\frac{1}{2}$ (x) and 6 days (•) in planes of soil close to onion roots. Soil M was initially at a matric potential of about -10 kPa and soil D at about -250 kPa. (Adapted from ref. 71.)

around roots. However, local rates may occasionally be higher, some roots may be in only partial contact with the soil, and salt accumulation may occur around roots; in such circumstances, large gradients might occur.[72] For example, the root tissues of sunflower may contract by as much as 25 per cent of their turgid volume when the water potential of the leaves falls to -1.5 MPa. By either squeezing or vibrating the soil to close vapour gaps around the roots, a temporary reduction in water stress of the plants was achieved.[73]

Within the plant, there may be appreciable resistance to moving water from the root surface to the xylem (radial resistance) or to moving water within the xylem vessels (axial resistance). The Poiseuille–Hagan equation has been used to evaluate the magnitude of axial resistances within the xylem of individual roots. Assuming the xylem consists of a bundle of n capillary tubes of radius r, the relation between the rate of water flow along the root (Q) and the gradient in hydraulic potential along the root axis ($\Delta\psi/L$) is given by:

$$Q = \frac{n\pi r^4 \Delta\psi}{8\eta L} \qquad [11.19]$$

[72] For a review see Tinker, P. B., *Phil. Trans. R. Soc. Lond.*, 1976, **B273**, 445.
[73] Faiz, S. M. A. and Weatherley, P. E., *New Phytol.*, 1982, **92**, 333.

where η is the viscosity of water and $\Delta\psi$ is the difference in potential between the ends of the root of length L. The axial 'resistance' is the gradient in hydraulic potential required to give unit rate of flow along a single root (i.e. $8\eta/n\pi r^4$); axial 'resistances' for cereal roots[74] are generally 0.1 to 1 kPa day cm^{-4}. Direct measurements of axial resistance in roots of loblolly pine and bean were approximately twice the values estimated using equation [11.19]. The axial resistance of bean roots was less than that of pine roots and changes in resistance with temperature were consistent with changes in the viscosity of the solution.[75] There are, unfortunately, few values of this sort available but there are known to be differences between plant species and between different types of root on the same plant.[76] Axial and radial resistances are, though, important terms to quantify since the pattern of water extraction by roots will be substantially influenced by them (see Ch. 4).

Most models of water use by crops now take account of these dynamic processes within the soil and plant and of the diurnal variation in the rate of evaporation, though not of root growth.

Growth and transpiration

The early, pioneering experiments of Briggs and Shantz[77] established that different crops used different amounts of water to produce the same amount of dry matter. Clearly, this is of considerable economic importance in areas where water is in short supply because, other things being equal, the crop that uses water most efficiently will be the one best adapted to such areas. Table 11.5 shows results from a number of crops and indicates the wide range of amounts of water used to produce a unit weight of dry matter. It also shows that the results vary from year to year and from site to site. More recently, the production of 1 g dry weight of barley grown at Rothamsted in experiments where rainfall was excluded required 330 g water in 1976 (which was exceptionally hot) and 230 g water in 1979.

In an extensive review of the literature, de Wit[78] showed that crop growth, N and water use (transpiration, T) could be empirically related by:

$$N = mT/E_o^n \qquad [11.20]$$

where m and n depend mainly on crop and site respectively, and E_o is the mean daily 'free water evaporation'. The analysis was applied to crops over

74 Greacen, E. L. *et al.*, in *Ecological Studies, Analysis and Synthesis*, Vol. 19 (eds O. L. Lange, L. Kappen and E. D. Schulze), p. 86, Springer-Verlag, 1976.
75 Sands, R. *et al.*, *Aust. J. Pl. Physiol.*, 1982, **9**, 559.
76 See for example Taylor, H. M. and Klepper, B., *Adv. Agron.*, 1978, **30**, 99.
77 Briggs, L. J. and Shantz, H. L. *USDA Bur. Pl. Ind. Bull.*, No. 284, 1913.
78 de Wit, C. T., *Versl. Landb. O'nderz.*, No. 64.6, 1958.

Table 11.5   Estimates of the units of water transpired in producing a unit of dry matter for selected crops.

Author	Crop	g water transpired/g dry matter produced	Comments	Source
Lawes, J. B.	Wheat	235	Mean of manured and unmanured crops	1
	Barley	257		
	Beans	214		
	Peas	235		
	Clover	249		
Briggs, L. J. and Shantz, H. L.	Wheat	507	Mean result for all varieties grown at Akron, Colorado, 1911	2
	Barley	539		
	Sorghum	306		
	Peas	800		
	Potato	448		
			Mean of 1910 and 1911	
	Wheat	566	Akron, Colorado	
		763	Dalhart, Texas	
	Sorghum	327	Akron, Colorado	
		336	Dalhart, Texas	
Tanner, C. B.	Potato	182	Hancock, Wisconsin, 1972	3
		278	Hancock, Wisconsin, 1973	
		238	Hancock, Wisconsin, 1976	
Day, W.	Barley	330	Rothamsted, 1976	4
		230	Rothamsted, 1979	

1 *J. hort. Soc. Lond.*, 1850, **5**, 38.
2 *USDA Bur. Pl. Ind. Bull.*, No. 284, 1913.
3 *Agron. J.*, 1981, **73**, 59.
4 *J. agric Sci.*, 1978, **91**, 599 and personal communication.

long time periods (sowing to harvest) and m was shown to be governed predominantly by species and independent of soil nutrition and water deficits unless nutrients or aeration were severely limiting. The constant n was dependent on site and had a value of 1 in dry, high radiation climates and a value of 0 in humid, and more temperate climates. De Wit gave no reason for the difference between climates but speculated that other values of n might be applicable in environments not covered by his review.

A physiologically based analysis of photosynthesis and transpiration allows a clearer understanding of the observations cited earlier.[79] Water vapour is evaporated mainly from surfaces within the substomatal cavity into the air inside the stomata and passes through the stomatal opening and the thin boundary layer of air surrounding the leaf into the ambient air. During photosynthesis, the passage of carbon dioxide is in the reverse direction. The rate of transpiration (T_L) per unit area of leaf is determined by the difference in vapour pressure between the evaporating surface inside the leaf (e_i) and the outside ambient air (e_a) and the resistance to water vapour transfer:

$$T_L = \left(\frac{\rho\epsilon}{p} \frac{(e_i - e_a)}{(r_s + r_a)} \right) \qquad [11.21]$$

where r_s and r_a are the stomatal and boundary layer resistances to diffusion of water vapour, ρ is the density of moist air, ϵ is the ratio of the molecular weight of water vapour to that of air, and p is the atmospheric pressure. Similarly, the rate of photosynthesis, N_L, per unit area of leaf is determined by the difference in carbon dioxide concentration inside the leaf, c_i, and the outside air, c_a, and the resistance to carbon dioxide transfer:

$$N_L = \frac{(c_a - c_i)}{r'_s + r'_a} \qquad [11.22]$$

where r'_s and r'_a are the stomatal and boundary layer resistances to the transfer of carbon dioxide. Therefore the ratio of photosynthesis to transpiration can be written:

$$\frac{N_L}{T_L} = \frac{(c_a - c_i)(r_s + r_a)}{(e_i - e_a)(r'_s + r'_a)} \left(\frac{p}{\rho\epsilon} \right) \qquad [11.23]$$

Equation [11.23] can be simplified in several ways. The stomatal and boundary layer resistances to both CO_2 and water vapour are related to the diffusion coefficients of the two gases which are approximately inversely proportional to the square roots of their molecular weights, i.e. the ratio $(r'_s + r'_a)/(r_s + r_a)$ is approximately constant with a value of about 1.6.

The internal concentration of CO_2 in the intercellular spaces has also been shown to be an almost constant value depending upon plant species. For C_4 plants the value of c_i is about 120 ppm whereas for C_3 plants it is about 210 ppm (see Ch. 2 for more details). This means that $c_a - c_i$ may also be regarded as a constant at normal atmospheric concentrations depending upon the type of plant.

Finally, Bierhuizen and Slatyer[80] approximated e_i by noting that because the daily average temperature of a leaf is frequently close to air temperature,

79 Penman, H. L. and Schofield, R. K., *Symp. Soc. Exp. Biol.*, 1951, **V**, 115. For a more general review, see Fischer, R. A. and Turner, N. C., *Ann. Rev. Pl. Physiol.*, 1978, **29**, 277.
80 Bierhuizen, J. F. and Slatyer, R. O., *Agric. Meteorol.*, 1965, **2**, 259.

then $(e_i - e_a)$ will be almost equal to the saturation deficit of the air $(e_{sa} - e_a)$, where e_{sa} is the saturation vapour pressure and e_a the actual vapour pressure. Using the above approximations, derived from consideration of the behaviour of a single leaf, and further assuming that crops can be described similarly, then:

$$\frac{N}{T} = \frac{k}{e_{sa} - e_a} \qquad [11.24]$$

where k is a crop-specific constant. The general validity of the relation has yet to be fully tested and the constancy of k assessed. However, reviews suggest that k is a very stable term in the cropping system and that the transpiration efficiency, N/T is, therefore, dependent on $e_{sa} - e_a$.[81]

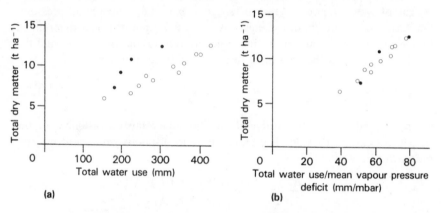

(a)

(b)

Fig. 11.13 (a) Growth and water use by barley given variable amounts of irrigation, 1976 (O) and 1979 (●). (b) Growth and water use when saturation deficit is allowed for. (Original data of Dr W. Day, Rothamsted published by Gregory, P. J., *Outlook on Agriculture*, 1984, **13**, 208.)

The importance of the saturation deficit term can be judged from Fig. 11.13 which indicates how weather can influence the transpiration efficiency of barley crops grown at Rothamsted; similar results have been obtained with potatoes grown in The Netherlands.[82] This analysis provides an explanation for de Wit's findings and for some year to year variations in growth and water use relations. The analysis leads to concepts of practical importance. First, any adaptation of plants to increase growth when water is

81 See reviews by Tanner, C. B. and Sinclair, T. R. in *Limitations to Efficient Water Use in Crop Production*, p. 1, ASA–CSSA–SSA publication, 1983; Feddes, R. A., in *Conférence Internationale de la CIID sur les Besoins en Eau des Cultures*, Paris, 1984.
82 Rijtema, P. E. and Endrodi, G., *Neth. J. agric. Sci.*, 1970, **18**, 26.

in short supply must take account of factors affecting both transpiration and photosynthesis. Second, crops growing in arid regions will generally use more water than the same crops growing in temperate regions to produce the same amount of growth; this is because $e_{sa} - e_a$ is lower in temperate regions than in arid regions. Finally, plants such as maize, sorghum and millet (C_4 plants) can maintain a larger gradient of carbon dioxide concentration between the internal stomatal cavity and the ambient air than plants such as temperate cereals and legumes (C_3 plants); C_4 plants will therefore produce more dry matter per unit of water transpired. For example, when soyabean and sorghum were grown in the same location, sorghum produced almost three times as much dry matter and seed yield as soyabean per unit of water used.[83]

Growth and use of soil water

Crops growing on soil lose water by both transpiration from the leaves and evaporation directly from the soil surface. However, while the production of dry matter per unit of water transpired may be a fairly conservative value for a particular crop species, the production per unit of evapotranspiration will vary considerably more since it depends on numerous crop, climate and soil factors. Obtaining the maximum water use efficiency (defined as units of dry matter per unit of water used) will therefore be dependent on transpiring as high a proportion as possible of a limited water supply. The highest water use efficiency allowing transpiration only can be calculated from the equation in the preceding section. If the leaf and air temperature is 25 °C, the relative humidity of air 50 per cent (saturation deficit 12 mg l^{-1}) and photosynthetic affinity for CO_2 infinitely high, then the water use efficiency would be 30 mg CO_2 g^{-1} H_2O.[84] In practice, the efficiency of use is substantially lower than this except for plants practising CAM (see Ch. 2).

In many places, the loss of water directly from the soil surface represents a major pathway of loss. Cooper[85] has determined that in the Mediterranean climate of northern Syria almost 60 per cent of the total water loss from land cropped with barley is via the soil surface. Adding 27 kg P ha^{-1} and 60 kg N ha^{-1} to these crops almost doubles yields but results in the same total water

83 Teare, I. D. *et al.*, *Agron. J.*, 1973, **65**, 207.
84 Fischer, R. A. and Turner, N. C., *Ann. Rev. Pl. Physiol.*, 1978, **29**, 277.
85 Cooper, P. J. M. in *Proc. 17th Coll. Int. Pot. Inst.* p. 19, Rabat, 1983.
86 See also Gregory, P. J. *et al.*, *J. agric. Sci.* 1984, **103**, 429; Cooper, P. J. M. *et al.*, *Expl Agric.*, 1987, **23**, 113.

use because the fertilizer promotes rapid leaf growth early in the season, thereby covering the soil surface and reducing evaporative losses. In such environments, fertilizer can almost double the overall efficiency of water use.[86] (See also p. 447.)

An important factor influencing the growth of crops and their overall use of water is their pattern of root growth relative to the growth of the shoots. In perennial plants, the root system may be very extensive before drought commences. Dagg[87] showed that the grass *Cenchrus ciliaris* could build up a deficit of 320 mm water in the top 3 m of the profile without its rate of transpiration being appreciably affected, though by the time the deficit was 450 mm, water was being taken from 6 m and the rate of transpiration had fallen well below its potential rate. Nevertheless, it continued to dry the soil slowly, and before the end of the drought had built up a deficit of 540 mm and dried the top 6 m to drier than −1.5 MPa. Similarly, tea at Mlanje, Malawi, dried the top 4 m of soil to −1.5 MPa and took water from 5 m depth during the dry season, but its rate of transpiration decreased when only 40 per cent of the available water (200 mm) had been extracted.[88]

In contrast, the root system of most annual cereal crops ceases to grow substantially after the crop has flowered (see Ch. 4) so that the crop does not have much time to develop a deep root system. This often results in water remaining in the soil profile after the crop has been harvested. For example, in the Canadian prairies, wheat crops often experience drought during grain

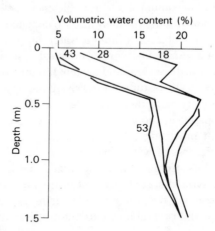

Fig. 11.14 Changes in the volumetric water content beneath a crop of millet growing on stored water. The numbers indicate days after sowing. (From Squire, G. R. *et al.*, *Expl Agric.*, 1984, **20**, 135.)

87 Dagg, M., *J. Hydrol.*, 1969, **9**, 438.
88 Laycock, D. H., *Trop. Agric. Trin.* 1964, **41**, 227;   Willatt, S. T., *Agric. Meteorol.*, 1971, **8**, 341.

filling although there is substantial water in the soil below 1 m.[89] This frequently made observation in other drought susceptible regions has led to suggestions that deeper rooting varieties should be bred to take advantage of this subsoil water. As an example Fig. 11.14 shows the depth from which a rainfed crop of pearl millet took water from an Alfisol at Hyderabad, India. The depth of rooting was restricted to about 1 m although the water potential beneath 1 m was greater than −0.03 MPa throughout the growing period.

In the UK most cereal crops have root systems extending to about 1.5 m and they do not dry the soil to suctions of 1.5 MPa to depths much in excess of 50 cm except on coarse sands. Table 11.6 shows that a winter wheat crop grown on a sandy loam in the East Midlands extracted water to 1.4 m and that during a dry period the soil below 1 m, containing only 3 per cent of the total root weight, was responsible for supplying almost 20 per cent of the water used. Similarly, grass takes most of its water from the surface layers with decreasing amounts from succeeding deeper layers.[90]

Table 11.6 Comparison of the relative water use from different soil depths (cm) with the relative root distribution. (From Gregory, P. J. *et al.*, *J. agric. Sci.* 1978, **91**, 103)

Date	Water uptake (%)				Root dry weight (%)			
	0–30	30–60	60–100	> 100	0–30	30–60	60–100	> 100
9 April–13 May	100	0	0	0	58	22	17	3
13 May–17 June	71	16	13	0	63	22	13	2
17 June–8 July	39	17	25	19	59	21	17	3
8 July–29 July	86	6	2	6	54	22	18	6

The method of cultivation and the degree of drainage both have effects, however, on the amounts of water used and on the patterns of water use through their effects on both root distribution and water distribution within the profile. Work at the AFRC Letcombe Laboratory[91] showed that in a dry year, about 10 per cent more water was stored in clay soils that were direct-drilled compared with soils that were ploughed. This was especially so below 50 cm and allowed winter wheat to extract up to 22 mm more water from the undisturbed soil. On some soil types, direct drilling results in increased soil bulk density and restricted root penetration but in clay soils, the depth of rooting was found to be deeper by early spring, when soils were direct-drilled, for both winter-sown and spring-sown cereals.[92]

89 Hurd, E. A., *Agric. Meteorol.*, 1974, **14**, 39.
90 Garwood, E. A. and Williams, T. E., *J. agric. Sci.*, 1967, **68**, 281.
91 Goss, M. J. *et al.*, *J. Soil Sci.*, 1978, **29**, 475.
92 Ellis, F. B. *et al.*, *J. agric. Sci.*, 1977, **89**, 631; Drew, M. C. and Saker, L. R., *J. agric. Sci.*, 1980, **94**, 411.

In vegetable crops, deep cultivation has been shown to be beneficial by allowing more roots to penetrate deeper into the profile and more water extraction to occur from below the topsoil. Deep cultivation to 45 cm of a sandy clay loam reduced water extraction by potatoes, broad beans, summer cabbage, and red beet from 0–30 cm depth and increased extraction from 30–70 cm depth. Root length of broad beans below about 25 cm was increased by 0.2 cm cm^{-3} to almost 60 cm depth and water use increased by 20 per cent in certain periods; yields were also increased.[93] Similarly, mole drainage has been shown to influence rooting and water use of winter wheat and while in some years the effects are small and transient, generally the maximum depth of water extraction was increased by about 20 cm.[94]

In dry regions, the choice of planting density or the population of plants is often more crucial than in wetter regions because water use must be optimized. If the supply of water is limited and a crop is sown densely, the supply may be used too rapidly and the crop will fail. Conversely, if the crop is sown too sparsely, it may survive to give a utilizable yield but either water will remain in the soil or it will be evaporated directly from the soil surface and lost to subsequent crops. These effects were observed with pearl millet grown on a sand in Niger[95] where a densely sown crop (11.5 plants m^{-2}) initially grew most rapidly, but ran out of water after 40 days and stopped growing; the sparsest sown crop (2.9 plants m^{-2}) grew slowest initially but by harvest at 82 days weighed as much as the crop sown at intermediate density (5.8 plants m^{-2}). The differences in shoot growth were accompanied by differences in root growth and in the amounts and timing of water use from different soil layers. Initially, the sparsest sown crop had a shallower root system and used less water than the other crops. Similarly, sorghum and cowpeas grown on sandy loam soils in Botswana at low populations used slightly less water over the whole growing season but produced higher yields than crops at high populations. The overall effect of the denser population was to decrease the amount of water available per plant.[96] In contrast, on a silt loam soil in Iowa, there was no difference in soyabean yields between crops planted at 25, 50, 75, and 100 cm row spacings except in a wetter than average year when yields were higher in the narrowest row spacing.[97] Agronomic practices for optimizing crop yields in relation to soil water supply will vary according to soil type and, as Anderson[98] has shown for *Triticale*, to the expected water supply during the growing season.

93 Rowse, H. R. and Stone, D. A., *Soil Tillage Res.*, 1980, **1**, 57, 173; Stone, D. A., *J. agric. Sci.*, 1982, **98**, 297 for rooting depths of potatoes.
94 Harris, G. L. *et al.*, *J. agric. Sci.*, 1984, **102**, 561; Ellis, F. B. *et al.*, op. cit., 1984, **102**, 583.
95 See ref. 59.
96 In *ODA, Dryland Farming Research Scheme Phase III Ann. Rept.*, 1982/83; see also Rees, D. J., *Expl Agric.*, 1986, **22**, 153.
97 Taylor, H. M., *Agron. J.*, 1980, **72**, 543.
98 Anderson, W. K., *Field Crops Res.*, 1984, **8**, 281.

Principles of irrigation

The major principle underlying all irrigation systems is to supply crops with sufficient water so that evapotranspiration is close to the rate determined by the atmospheric demand for water; this in turn ensures maximum crop growth. Irrigation may be used for many reasons including low total precipitation and pronounced seasonal or erratic distributions of precipitation. Even when soils are deep, there are few that can sustain crops from sowing until harvest solely from stored water so that even if permanent irrigation systems are not used, supplementary irrigation is often practised for limited periods of time. The continued importance of irrigation as a means of increasing crop production can be gauged from estimates that the irrigated land area worldwide has increased from 16.2×10^6 ha at the end of the nineteenth century to 203.6×10^6 ha in 1975 and will increase to 300×10^6 ha by AD 2000.[99]

Efficient irrigation requires knowledge of crops, soils and weather and of the relations between them, so that both the amount and timing of irrigation can be controlled. For example, it is important to know how the available water content changes as a soil is depleted and the relation between crop growth and available water in order to determine when to irrigate. The growth of some crops is reduced when only a small fraction of the available soil water has been depleted whereas others will suffer no adverse effects until much lower soil water potentials have been achieved. Similarly, it is important to know the depth of rooting so that an estimate can be made of the amount of water available in the soil profile. Some crops also seem to be particularly sensitive to drought at particular growth stages.[100] Generally, the times at which the reproductive organs are developed and fertilized appear to be the most crucial in determining yields since it is at these stages that the potential yields are determined.

Successful irrigation results from supplying sufficient water to prevent plant water stress without the adverse effects caused by excess water. This careful balance can be more readily achieved in arid regions where crop growth is almost wholly dependent on irrigation than in humid regions where sudden rainfall after irrigation can result in waterlogged soils. The timing of irrigation to achieve this desired balance is often achieved by rule-of-thumb methods, but there are three basic approaches.[101] The first method is to measure soil water potential directly but this is often difficult in practice. Tensiometers

99 Gulhati, N. D. in *Irrigation, Drainage and Salinity* (eds V. A. Kovda, C. van den Berg and R. M. Hagan), Hutchinson, 1973.
100 For a review of crop responses to irrigation at different developmental stages see Salter, P. J. and Goode, J. E., *Crop Responses to Water at Different Stages of Growth*, Commonw. Agric. Bur., Hort. Plant. Crops, 1967.
101 For a review of techniques see Hiler, E. A. and Howell, T. A. in *Limitations to Efficient Water Use in Crop Production* (eds H. M. Taylor, W. R. Jordan and T. R. Sinclair), p. 479, Am. Soc. Agron., Madison, 1983; James, D. W., Hanks, R. J. and Jurinak, J. J., *Modern Irrigated Soils*, p. 84, Wiley, 1982.

cover the range of soil matric potential (0 to -0.08 MPa) where growth is best for most crops and although effective in sandy soils, they often become inoperative in clay soils because of shrinkage of the clay during drying. A second method involves the use of both evaporation measurements and a knowledge of the soil water to calculate a water budget.[102] Soil water depletion in the root zone (ΔS_r) may be calculated as

$$\Delta S_r = \sum_{i=1}^{n} (ET - P_e - I_n \pm D_r)$$

where P_e is the effective daily precipitation (i.e. precipitation minus run-off), I_n is the daily net irrigation, D_r is the daily drainage loss from the root zone (positive) or upward movement into the root zone (negative), and n is the number of days. Evapotranspiration (ET) is sometimes obtained from open-water evaporation pans and sometimes calculated from meteorological data (for example using the Penman formula). In the USA the Jensen–Haise formula is widely used to estimate ET.[103] Sometimes, measured open-pan evaporation is multiplied by a 'crop coefficient' based on experimental measurement of the relation between pan evaporation and actual crop evapotranspiration at a particular location.

Finally, the plant itself may be used as the best indicator of the need for irrigation because it integrates the effects of the atmospheric demand and the soil's ability to supply water. Visual indications of water stress include changes in leaf colour, leaf orientation and leaf rolling but these are normally associated with severe stress. Measurements of leaf temperature using an infrared thermometer provide an indirect assessment of water stress because as water supply is reduced, the stomata close, and the lack of evaporative cooling results in higher leaf temperatures.[104] Hiler and Clark[105] developed a quantitative method (the stress day index) for determining the stress imposed on a crop during its growth allowing for differences in stress tolerance between species and at different growth stages. The stress day index (SDI) is the product of the crop susceptibility factor (CS) and the stress day factor (SD). SD is a measure of plant water deficit and is usually the leaf water potential (though it could be the difference between leaf and air temperatures or other plant-based or soil-based measures) and CS is a measure of yield when subjected to water stress. CS is obtained experimentally by subjecting crops to a specified stress at specific developmental stages and determining the yield as a fraction of the yield of unstressed plants. For

102 Jensen, M. E. *et al., Trans. Am. Soc. agric. Eng.,* 1971, **14**, 954; Stegman, E. C. *et al., in Design and Operation of Farm Irrigation Systems* (ed M. E. Jensen), Am. Soc. Agric. Eng., Michigan, 1980.
103 Jensen, M. E. and Haise, H. R., *J. Irrig. Drain. Div. Am. Soc. Civ. Eng.,* 1963, **89**, 15.
104 Idso, S. D. *et al., Agric. Meteorol.,* 1980, **21**, 205.
105 Hiler, E. A. and Clark, R. N., *Trans. Am. Soc. agric. Eng.,* 1971, **14**, 757.

example, CS for maize ranged from 0.25 to 0.50 to 0.21 for stress applied during vegetative growth, silking and tasselling to soft dough and after soft dough respectively.[106] Using this approach, crops would be irrigated when the SDI reaches some predetermined critical value depending on the availability or cost of water or both, and the frequency of irrigation would be greater during periods of high CS. Although this procedure integrates most of the soil, plant and atmospheric factors affecting crop water use, values of CS have to be determined experimentally for different locations and are themselves dependent on the SD used in their determination.

A variety of techniques is used to apply water to crops but the most common are gravity, sprinkler and trickle or drip irrigation. Where the water table is close to the surface then subsurface irrigation may also be practised.[107]

Gravity irrigation has been used for thousands of years. It is a surface system requiring gentle slopes to work effectively and is best on soils with moderate to slow rates of infiltration. Land levelling is often a prerequisite for this form of irrigation which involves spreading water into small basins or in furrows. Furrows are used to irrigate between crop rows and gently sloping land is required to ensure distribution of water along the rows without excessive soil erosion. Water applied at the top of the furrow may exceed crop requirements while insufficient water may reach the bottom of the field if the slope is too shallow. Conversely, if the slope is excessive, erosion may occur. There is, then, an interaction between the amount of water applied and the duration of water application with water infiltration properties of the soil, slope and depth of root zone. With basins, no run-off occurs and the aim is to apply water quickly and evenly.

Sprinkler irrigation is a more recent development and has the advantage over gravity systems in that little land levelling is required. Sprinkler systems generally require less labour than gravity systems and may even involve self-propelled machinery. They generally supply water to the soil at slower rates than gravity systems so that run-off and erosion are minimal and the distribution of water is more uniform. A further advantage is that fertilizers, insecticides and herbicides may be injected into the irrigation water in a controlled manner and spread uniformly. However, if the water has appreciable salt content, leaves may be scorched and the wetting of foliage may encourage the spread of leaf pathogens.

Drip or trickle irrigation supplies water at a slow rate to a limited area of soil close to the plant. The system is particularly advantageous in widely spaced crops (e.g. orchards and vineyards) because losses from seepage and

106 Hiler, E. A. *et al.*, *Trans. Am. Soc. agric. Eng.*, 1974, **17**, 393.
107 For a description of techniques see Israelson, O. W. and Hansen, V. E., *Irrigation Principles and Practices*, Wiley, 1962; James, D. W. *et al.*, *Modern Irrigated Soils*, p. 73, Wiley, 1982.

evaporation are small and as most of the soil surface remains dry, weeds do not grow. Because the rate of water delivery is slow, the water has to be applied frequently and the irrigation system is a permanent installation during the growing period. Also because the water supply to drip systems has to be free of suspended materials – otherwise the drippers block – nearly all such systems have filtration units. Drip irrigation is said to be advantageous where water supplies contain appreciable quantities of dissolved salts because by constant watering, the salt content of the rooting zone is kept low. However, the uneven distribution of salts in the soil may cause problems later if the salt is not leached or if different plant geometry is used.

Irrigation, by whatever means, is not, however, without its problems. Chief among these is the problem of salt accumulation in soils resulting from the level of salts in irrigation water frequently exceeding their uptake by crops (see Plate 11.1). The only solution to this is to supply sufficient water either by precipitation or irrigation to leach away the salts. In humid regions, rainfall normally exceeds evapotranspiration for part of the year and salts are removed by drainage, but in arid environments, there is often insufficient rainfall or irrigation water to achieve this. As a consequence there are vast areas of land where salts have accumulated and crop growth is impossible (see Ch. 27).

A second problem in irrigated cropping, especially in gravity systems, is that of poor aeration (see Ch. 26). Waterlogging of soils in arid regions may

Plate 11.1 Young mango tree with drip irrigation. Note accumulation of salt (white ring).

not only damage plants because of insufficient aeration but may also result in salt damage to the roots. The installation of effective drainage systems (see Ch. 13) is therefore as essential to sustained crop production as the installation of systems for effective water distribution.

12

Soil structure, tilth and mechanical behaviour

The individual particles in soil are not usually randomly arranged but form clods and crumbs, bound together by colloidal material, possessing some characteristic internal organization and external form. The nature and size distribution of aggregates and the converse, that of the pore space, is referred to as soil structure and plays an important part in determining soil physical properties and hence soil fertility. If soil crumbs or the larger pores collapse, this usually has serious consequences for the plants growing there and also on ease of management and on drainage.

There is ambiguity about the exact meaning of the term structure. Every material has some internal arrangement, often on many different levels of scale, and a randomly arranged or close packed mass of sand grains may on one occasion be described as having a single grain structure and on another as 'without structure' or structureless. In this latter sense structure is regarded as an attribute only when it is favourable to plant growth, and it can be lost under the action of various destructive stresses, leading to a structureless or unfavourable physical state of the soil. There is here perhaps an unfortunate confusion between the terms structure and aggregation.

However, whatever the difficulties and ambiguities of definition, soil structure is concerned with the mutual arrangement of the individual soil particles and the stability of the aggregated state, and with the wider range of pore sizes that results.

The geometry of the soil pores to a large extent controls the retention, availability and rate of movement of water, and to a lesser extent the aeration conditions, and the permeability of soil to roots.

If soil structure is of interest mainly because of its effects on soil fertility, it is most usefully analysed in terms of pore characteristics. But since these are themselves controlled by the distribution and stability of the soil solids, then to understand the factors involved in structure development and preservation it is the aggregation of the solids that is important.

Pores and aggregates

Pores

Pores and cracks can be put into several different size classes. Water only moves freely through pores wider than about 0.3 mm, which need suctions of less than 1 kPa to empty them of water, and the young roots of many plants also need pores of about this size for easy entry. Water only drains out of soil under gravity if it can move through pores larger than 60 to 30 μm, that is, pores which need suctions of about 5 to 10 kPa to empty them of water. Root hairs, and the larger members of the soil micro-organisms such as protozoa and fungi with coarse mycelia, need pores larger than about 10 μm to grow into or move in, and the smaller micro-organisms need pores larger than 1 μm for movement, pores which require a suction of 300 kPa to empty them of water.

Water has significant mobility in pores of this size and plants can extract water from them. However, rigid pores smaller than 0.2 μm retain water at suctions exceeding that corresponding to the permanent wilting point, and so water held in these pores appears not to be available to the plant. In many clay soils, as we have seen (p. 316), even the very fine pores may collapse at relatively low suctions releasing water to the root, though probably more slowly than from larger pores.

Pores have been classified into size classes in various ways; for example into capillary and non-capillary with the dividing line at 50–60 μm[1] and more recently[2] into transmission, storage and residual with divisions at 50, 5 and 0.5 μm. It is sometimes necessary to remember that soil pores are not simple tubes, but irregular non-rigid spaces interconnecting through narrow necks. It is the complex geometry of interconnections that largely controls water release and rates of movement, and which may lead to 'islands' of water

1 Schumacher, W., *Die Physik des Bodens*, Berlin, 1864.
2 Greenland, D. J., *Phil. Trans. R. Soc.*, 1977, **B281**, 193.

being cut off from effective hydraulic contact with the rest of the soil and therefore not necessarily in equilibrium with it. Many of these complexities and their effects have been considered by Dullien[3] and some investigated experimentally by Ball.[4]

A soil with a high proportion of pore volume in pores finer than 1 μm necessarily has a large area of pore surface, and this is normally accompanied by plasticity when wet and cohesiveness on drying, with considerable swelling and shrinkage behaviour and often with stickiness. The finer clays have most of their pores less than 0.1 μm in diameter and are difficult to manage, tending to remain wet for long periods, when they are very weak and easily puddled, yet on drying they become very strong and hard.

Aggregates

The soil aggregates can be described and classified in terms of size, shape and the nature of their faces and edges. For example the USDA system[5] uses three main groups of properties: the type of structure or the shape of the peds; the class of the structure or its size; and the grade of the structure or its distinctiveness and durability.

There are four primary types of structure, some of which are illustrated in Plate 12.1 (pp. 381–3): *laminar*, in which the natural cracking is mainly horizontal, such as is often found in the A_2 horizon of leached sols or in soils rich in kaolinite; *prism-like*, in which vertical cracking is better displayed than horizontal and the faces of the peds are fairly smooth and flat, and which is often found in the B horizon of clay loams and clay soils; *blocky*, in which vertical and horizontal cracking are about equally strongly developed, so the peds have roughly equal axes, and the peds may have either flat or rounded surfaces, but when wet they swell and the surfaces of contiguous peds fit exactly one with the other leaving no gaps between them; and *spheroidal* which differs from blocky by the surfaces of the peds being rounded and not fitting the surfaces of their neighbours so, when wet, there are appreciable pores between them.

These types can be further subdivided, as, for example, the prism-like is divided into columnar, when the upper end of the prism is rounded, and prismatic when it is not; the blocky into angular blocky when the surfaces intersect at relatively sharp angles, and subangular when the edges are rounded; and spheroidal into granular, when the ped is relatively non-porous, and crumb when porous.

There are three grades of structure, if one ignores the structureless class, namely, weak, when the peds are barely discernible in the undisturbed soil; strong, when the soil consists of well-defined peds which only adhere weakly

3 Dullien, F. A. L., *Fluid Transport and Pore Structure*, Acad. Press, 1979.
4 Ball, B. C., *J. Soil Sci.*, 1981, **32**, 465, 483.
5 USDA Soil Survey Staff, *Soil Survey Manual, Agric. Handbk.*, 436, 1975.

to one another; and medium for structures intermediate in development between these two.

Soil peds may have a well-developed structure of their own, particularly if the soil is a clay or loam under natural vegetation which is little disturbed by man or his grazing animals. Under these conditions, any large aggregate or clod is seen to be built out of smaller clods or crumbs, and these out of still smaller units. This can often be seen by scratching the surface of a clod with a sharp needle, or putting a dry crumb in water, for both will loosen smaller crumbs from the ped or crumb surface. Crumbs that are built out of definite micro-crumbs are said to possess a well-developed microstructure or marked granularity; and in fine-textured soils, granularity and mellowness of tilth are closely associated. A wet deflocculated or puddled clay or loam soil usually dries out into hard clods devoid of microstructure, but adding a soluble salt, such as gypsum or an adequate amount of sodium chloride to flocculate the soil, will increase the granularity of the clods.

The actual distribution of the micro-crumbs, and of their distribution relative to the sand particles, can be seen by examining thin sections of soil crumbs under a petrological microscope. This technique was introduced by

(a)

Plate 12.1   (a) prismatic aggregates with flat and rounded tops, (b) and (c) fine blocky structure, (d) fine prismatic aggregates; (a), (b) and (c) half natural size, (d) natural size. (From Nikiforoff, C. C., *Soil Sci.*, 1941, **52**, 193).

(b)

(c)

(d)

Kubiena, who named some of the types of microstructure which can be seen in different soils. The classification of these structures has been greatly expanded by Brewer[6] and is now extensively used in soil description.[7] In typical English arable soils, the micro-crumbs lie between sand grains whose surface appears to be clean, and they prevent the sand grains falling into close packing because they are larger than the pores present in such a system. But, in some soils, the sand grains can be seen to have a film of oriented clay particles on their surface, and these films adhere to each other sufficiently

6 Brewer, R., *Fabric Analysis of Soils*, Wiley, 1976.
7 Bullock, P. in *Soil Survey Laboratory Methods*, Soil Survey of England and Wales, Rothamsted, 1974.

strongly to hold the sand particles apart in open packing against moderate mechanical stress.

The individual clay particles present in the micro-crumbs cannot be seen under a petrological microscope, and in thin sections they usually appear as a plasma whose opacity depends on the thickness of the section and on the amount of iron and manganese in or on the clay. The plasma may be composed of films of clay particles, all of which have the same orientation, in which case a thin section of the film will be birefringent; and it sometimes contains birefringent spots, up to about 1 μm in size, but usually it appears to be structureless. The last condition is probably typical of an intimate mixture of clay and humic colloids bonded together. These soils, in which the plasma shows birefringent spots, probably contain small granules or domains of clay particles, in which all the particles have about the same orientation and are held together by di- or polyvalent cations between the layers of adjacent particles.

Clay particles can be resolved using the electron microscope but preparation of soil sections thin enough for examination by the transmitted electron beam is difficult. The scanning electron microscope has been much more widely used[8] but essentially examines the surface of aggregates.

The concept of the domain structure of clay particles in soils was first put forward by Emerson.[9] It was developed by Quirk[10] who studied the fine microstructure of clays and clay soils using their adsorption isotherms for nitrogen at 78 K and tetrachloroethane at room temperature, these liquids causing no swelling of the clay. A plot of the slope of the adsorption isotherm against the relative vapour pressure of the liquid gives the proportion of the total pore volume that is present in pores of different sizes. They found that for the clays and clay soils they used, the pore volumes present were usually concentrated in three zones: one of about 2–4 nm due to spaces between overlapping lattices at the broken edge of the clay particle, one at around 20 nm, probably due to spaces between contiguous clay particles held together in at least partially random orientation, and one at 100 nm or above, probably due to spaces between small packets of clay particles and, again, held together in more or less random arrangement. The word domain is used for the group of clay particles containing pores of about 20 nm but separated from their neighbours by pores around 100 nm. Plate 12.2 gives an electron micrograph of the broken edge of a micaceous clay, showing the parallel channels between groups of overlapping layers.

The principal structural units present in soils can be put into various size groups, each with a definite name. These are:

Domains: groups of clay particles, probably up to 5 μm in size, with a

8 Smart, P., *Soil Sci.*, 1975, **119**, 385.
9 Emerson, W. W., *J. Soil Sci.*, 1959, **10**, 235.
10 Aylmore, L. A. G. and Quirk, J. P., *J. Soil Sci.*, 1967, **18**, 1.

Plate 12.2 Electron micrograph of a replica of a fracture surface of a clay core of Willalooka illite. The arrows show the direction of shadowing.

concentration of pores around 20 nm and interdomain pores of about 100 nm.

Granules: groups of domains and silt and fine sand particles aggregated together in compound particles up to 0.5 mm size. The aggregates in this group are sometimes called microaggregates or microcrumbs, and the upper size is about the largest granule that does not break up into finer granules when wetted with water. They are not much larger than the clay granules Edwards and Bremner[11] found in soils whose clay did not disperse easily, which was up to 0.25 mm.

Crumbs: a collection of granules having a size of several millimetres. There is no particular property which defines an upper size limit to a crumb, but in general aggregates larger than about 5 to 10 mm are too large for a good seedbed.

Clods: soil aggregates larger than about 1 cm in size. They have normally to be broken down into crumbs when preparing a seedbed.

In a well-structured soil, each type of aggregate is built up out of smaller units in a semi-random packing so that there are appreciable volumes of structural pores between each group of units.

11 Edwards, A. P. and Bremner, J. M., *J. Soil Sci.*, 1967, **18**, 47.

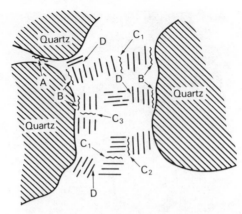

Fig. 12.1 Possible arrangements of domains, organic matter and quartz in a soil
crumb.

Type of bond: A   Quartz–organic matter–quartz
                B   Quartz–organic matter–domain

                                                            C_1 face–face
                C   Domain–organic matter–domain C_2 edge–face
                                                            C_3 edge–edge

                D   Domain–domain, edge–face

Figure 12.1 due to Emerson,[12] summarizes in pictorial form the way
clay particles help to build up structure. The clay particles are shown as
domains consisting of several clay particles held together face to face, and
the domains may be held edge to face by coulombic forces, or edge to face,
edge to edge or face to face through organic polymers. These polymers can
also bond clay particles to the surface of siliceous silt or sand particles.

The stability of soil structure

Soil aggregates can be broken or dispersed and pores compressed or
destroyed by the action of traffic and cultivation operations especially when
the soil is wet. Similar damage can be caused by environmental forces like
raindrop impact, freezing, and by rapid wetting of dry materials. Many of
these processes involve the action of different types of stress which may
interact and which are often difficult to analyse. Consequently there is no
single method for measuring stability of soil structure that is appropriate for
all circumstances. A large number of methods have been suggested,[13] some

12 See ref. 9.
13 de Boodt, M. (ed), *West European Methods for Soil Structure Determination*, Ghent,
   1967.

of them designed to simulate a particular breakdown situation, like raindrop impact, or one particular stress like shear, but many are simply intended to detect differences in soil behaviour without regard to any actual process or field conditions. In the absence of detailed knowledge of field stresses and of the soil moisture conditions on which the effects of stress so much depend, no method can be entirely satisfactory. Some apparently rather arbitrary methods can be useful for detecting changes or differences in structural stability. The results of *all* methods must be interpreted with caution.

Breakdown on wetting

Depending on their 'wet' strength and on the internal forces produced on wetting, different soils show different responses to rapid wetting. Some are completely stable, others show various degrees of swelling and slaking, or even complete dispersion. These differences form the basis of the simple and rapid test devised by Emerson[14] for distinguishing between soils whose reactions to immersion in water differ. Further subdivision is based on soil composition and the enhanced sensitivity to breakdown following manipulation while wet.

Many factors other than the nature of the soil itself affect the extent of breakdown on rapid wetting. These include the initial moisture content, itself dependent on the humidity of the air in which the soil has been air-dried, storage time and temperature, the rate of wetting and the ionic composition of the wetting solution.[15] Wetting in a vacuum, or slow wetting to allow the escape of entrapped air[16] so avoiding the compression of air within the pores, usually results in much less breakdown. Wetting by heavy rain causes additional disintegration because of the impact of the faster large raindrops.

This instability of structure due to slaking, for a given soil, is usually more noticeable the lower its humus content, possibly due to clods with the higher humus content wetting more slowly than those with a lower content, for the humus may make the soil surfaces more hydrophobic, in which case the water films make a positive angle of contact with the soil surface. However, Robinson and Page[17] found that the humic colloids present in a normal mineral soil did not appear to affect the angle of wetting appreciably. But Quirk and Panabokke[18] showed that a more likely reason is that some of the humus strengthens the coarser pores which, in turn, reduces the tendency for cracks to form as the clod wets. Crack formation reduces the rate of wetting

14 Emerson, W. W., *Aust. J. Soil Res.*, 1967, **5**, 47.
15 Kemper, W. D. and Rosenau, R. C. in *Methods of Soil Analysis* (ed A. Klute), p. 425, Am. Soc. Agron., Madison, Wisc., 1986.
16 Kemper, W. D. and Koch, E. J., *USDA Tech. Bull.* No. 1355, 1966.
17 Robinson, D. O. and Page, J. B., *Soil Sci. Soc. Am. Proc.*, 1950, **15**, 25.
18 Quirk, J. P. and Panabokke, C. R., *J. Soil Sci.*, 1962, **13**, 60.

of a clod, for cracks allow water to flow through them more rapidly than through the finer channels constituting the original pore space.

An important group of methods for measuring the stability of soil structure is based on measuring the size distribution of the crumbs that small dry clods break down into on wetting. The methods differ in the initial moisture of the clods that are used, their rate of wetting and the mechanical forces the crumbs are subjected to during the measurement of their sizes. The size distribution can be determined by such methods as separation of the crumbs on a bank of sieves under water – a group of methods that have been very widely used – or the size distribution of the pores between the crumbs can be measured. This is done by determining the moisture content–suction curve for a shallow column of the soil crumbs, after each cycle of drying the column and rewetting it. The slope of this curve, plotted against suction, gives the pore size distribution between the crumbs. Figure 12.2, taken from some work of Childs[19] and designed to determine whether a clay soil is suitable for mole draining or not, illustrates the results obtained on an old pasture soil,

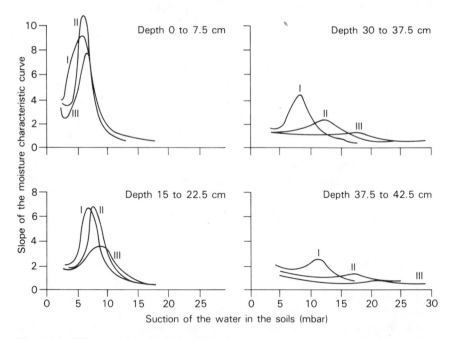

Fig. 12.2   The variation in the stability of the pore space distribution down the profile of an Upper Lias clay. Curve I is for the first, Curve II for the second and Curve III for the third wetting. (Soil initially in crumbs 1–2 mm size.) (1 mbar = 100 Pa).

19 Childs, E. C., *Soil Sci.*, 1940, **50**, 239.

and shows clearly that the surface crumbs are very stable to successive wetting and drying treatments, but the subsoil crumbs are very unstable. This is, in fact, an example of a clay quite unsuited to mole draining because of the instability by slaking of the clay when wetted.

These tests give little or no indication of the strength of soil crumbs that are water stable. Emerson[20] devised a method in which internal swelling pressures are controlled by first leaching a bed of soil crumbs with 1 M sodium chloride to saturate the soil with exchangeable sodium, and then the concentration of salt is reduced until the increasing swelling pressure leads to crumb collapse and a sudden fall in permeability. The concentration of the solution at which this fall occurs gives an indication of the wet strength of the soil crumbs. Soils with very stable structure maintain permeability even in water, while soils with a weak structure become impermeable at concentrations between 10^{-1} and 10^{-2} M.

This test is done under conditions when the crumbs are put under no mechanical force, but in the field this condition often does not apply, for soils must often be cultivated when they are wet. Emerson and Dettman[21] found that crumbs stabilized with organic matter would lose most of their stability if the soil was puddled when wet, showing that the organic matter bonds holding the soil particles could be broken fairly easily; and on the basis of this observation, Emerson[22] suggested a qualitative test based on a more gentle working of the soil at a moisture content corresponding to field capacity and determining the stability of small wet lumps of this reworked material. Soils with strong crumbs, particularly some with calcium carbonate in the clay fraction, gave lumps that remained stable; but if these lumps were then shaken up with water, those from some soils dispersed and from others remained flocculated. Many calcium-saturated soils can be dispersed by this process, for the wet working will pull some of the clay particles apart through distances sufficient to overcome the energy barrier to deflocculation.

Given sufficiently energetic mechanical dispersive treatment, for example by ultrasonics, very few soils resist dispersion even when normally very stable. Attempts have been made to find the energy used in ultrasonic breakdown of structure by measuring the energy input and temperature rise[23] but the results give surprisingly high values, especially so since the soil is saturated with water in ultrasonic treatment and is therefore in its weakest condition.

20 Emerson, W. W., *J. Soil Sci.*, 1954, **5**, 233.
21 Emerson, W. W. and Dettman, M. G., *J. Soil Sci.*, 1959, **10**, 227.
22 See ref. 14.
23 North, P. F., *J. Soil Sci.*, 1976, **27**, 451; Koenigs, F. F. R., *J. Soil Sci.*, 1978, **29**, 117.

Breakdown by mechanical stress

Structural damage is often caused by some form of mechanical impact, by raindrops, implements, tyres or the feet of animals. The stresses imposed are usually not simple and breakdown may be by shear or tensile failure or a combination of the two. Slowly applied and steady loads rarely exceed a few bars (1 bar equals 100 kPa), but impacts can give transient stresses ten-fold larger.[24]

Methods of measuring the resistance of aggregates and pores to mechanical stresses often apply very low but unmeasured stresses, as by sieving, or involve unknown impact stresses as in the drop shatter test[25] and raindrop impact at relatively low velocities.[26] Sieving and raindrop impact usually operate on wet soil whose strength may have been greatly reduced both by the wetting process itself and by the resulting high water content. Drop shatter can be used at any moisture content below the plastic limit but an air. dry state is usually specified. This moisture content is not generally sufficiently precise, especially for clay soils.

Raindrop impact and drop shatter are usually characterized in terms of energy expended although there is no direct evidence that energy is the limiting factor. Over a limited range of heights of fall in the drop shatter test Marshall and Quirk[27] found a straight line relationship between breakdown, expressed as mean size of aggregates, and the logarithm of cumulative energy of impact, that is the product of mass, gravity, height and number of falls. This is claimed to support the importance of energy. The energies involved are between 20 and 200 J kg^{-1}, or about 1 kJ per square metre of new surface produced. These results may be compared with the energies for dispersion by ultrasonics found by North[28] for saturated soil of 10–100 kJ kg^{-1}, which would probably represent a surface energy similar to that found from drop shatter, but the significance of these apparent surface energy results is open to doubt since they are larger by two orders of magnitude than the largest surface energies known.

Raindrop impact can also be expressed in terms of the kinetic energy lost at impact, and this is commonly done as a measure of the erosivity of rain and of simulated rain.[29] Again the real significance of this is not clear.

Static crushing strengths of soil aggregates have been studied by a number of workers.[30] Failure of dry brittle soil is then by tensile fracture and results

24 Ghadiri, H. and Payne, D., *J. Soil Sci.*, 1981, **32**, 41.
25 Marshall, T. J. and Quirk, J. P., *Aust. J. agric. Res.*, 1950, **1**, 266.
26 McCalla, T. M., *Soil Sci.*, 1944, **58**, 117;   Hudson, N., *Soil Conservation*, Batsford, 1971.
27 See ref. 25.
28 See ref. 23.
29 See ref. 26.
30 Rogowski, A. S. and Kirkham, D., *Meded. Fac. Landbouw.*, Ghent, 1976, **41**, 85;
   Braunack, M. V. and Dexter, A. R. in *Modification of Soil Structure* (eds W. W. Emerson *et al.*), p. 119, Wiley, 1976.

can be expressed as approximate tensile strengths. These vary according to the soil composition, structure, and moisture content, typically being a few hundred kPa for air-dry soil, falling to zero at the liquid limit. Failure will only be tensile at moisture contents below the lower plastic limit.

Factors involved in aggregation

Shrinkage of soil on drying, the burrowing action of some of the soil fauna and of plant roots, and the movement and expansion of water on freezing are the principal factors involved in aggregation and the formation of large pores. The relative importance of each of these depends on circumstances. The stability of the resulting structure depends on the presence and distribution of cementing agents such as sesquioxides and polysaccharide gums.

Creation of structural pores in undisturbed soils

The most important group of agents that create structural pores in undisturbed soils are plant roots. The roots of many plants grow by forcing a root tip, only a few tenths of a millimetre in diameter, into soil pores of about their size, and the young root then swells making the pore larger. This swelling involves soil particles, domains or granules being moved relative to their neighbours. This movement causes the structural units to be twisted relative to each other, and since they are not spheres, this will enlarge some of the pores between them, if they were in fairly close packing. Further, one of the functions of a plant root is to extract water from the soil, and if the soil shrinks as it loses water, this drying of the soil will cause it to crack. Thus the more intimately the roots permeate through such a soil, the more intimate will be the system of cracks and pores that will be developed. Cracks produced by shrinkage can be long-lived, for once a loam or clay clod has been dried it may take months before the soil has swollen again to its full volume. In cracking soils, roots help to create their own spatial environment for growth; as the root dries the soil, cracks develop around the root into which a new young root can grow.

The other group of agents that can create structural pores are some of the larger members of the soil fauna. Earthworms are the most important group in many temperate soils, for they make their burrows through a considerable depth of soil, and they make worm casts, which are fine ribbons of soil excreted by the worm and left in a small heap .The larvae of a number of insects, particularly beetles, make channels in the soil as they move about in search of food; and ants make numerous channels in the upper part of the soil. Green and Askew[31] found on some Romney Marsh pastures that the ants *Lasius flavus* and *L. niger* made numerous channels about 1 mm in diameter which penetrated to depths of 1.5 m, and that a proportion of the soil in the anthills came from depths of 70 to 100 cm. In the tropics, ants and termites are the principal animals that create pores, channels and burrows in the soil.

Stabilization of structural pores

The structural pores in a soil can only be stable against wetting if there is some agent holding the soil particles apart against the surface tension forces set up as it is wetted, for these forces pull the soil particles together. The most important soil component resisting the collapse of structure is clay. A pure fine sand is a most unfavourable environment for root growth, for any channels or cracks made in such a soil are filled up by collapse of the sand grains on wetting. Clay particles, by being in domains and granules, help keep sand particles apart on wetting, provided they retain their individuality, that is, provided they are flocculated.

All clays are flocculated if the concentration of salts in the soil solution is high enough. (This must be very high for many clays containing a large proportion of exchangeable sodium ions, and in normal circumstances such clays usually deflocculate if they are wetted by rain.) Clays with a high proportion of exchangeable calcium ions form metastable domains and granules, for they are stable in water if not subjected to a force sufficiently large to pull the individual clay particles apart; but once they have been separated from each other, many types of clay form a deflocculated paste which needs a calcium ion concentration of over 10^{-3} M for flocculation.

Soils containing calcium carbonate, particularly if finely divided, are usually flocculated because the calcium ion concentration is generally higher than 10^{-3} M. Calcium carbonate also appears to have a direct cementing effect and this also contributes to structural stability.[32] Nevertheless some calcareous clays do give rise to soils which are difficult to manage and whose structure can be easily damaged by untimely cultivation. They might well be worse in the absence of calcium carbonate.

Clays differ in the ease with which they can be deflocculated, depending on the number of positive charges they carry. Soils and clays containing appreciable amounts of aluminium hydroxide give very stable granules, for they are difficult to disperse even if they contain an appreciable proportion of exchangeable sodium ions. Deshpande[33] found by selective removal of

31 Green, R. D. and Askew, G. P., *J. Soil Sci.*, 1965, **16**, 342.
32 Rimmer, D. and Greenland, D. J., *J. Soil Sci.*, 1976, **27**, 129.
33 Deshpande, T. L., Greenland, D. J. and Quirk, J. P., *J. Soil Sci.*, 1968, **19**, 108.

aluminium oxide from a number of Australian soils that this was more important than iron oxide, and further that iron oxides in the clay fraction occurred in discrete microcrystals. Hydrated ferric oxides are probably much less important than aluminium hydroxide, because of their much smaller surface area in most soils, though ferric hydroxides may be as efficient. Most workers who have studied the effects of the ferric oxides in soils,[34] however, have not distinguished between their effect and that of the associated aluminium hydroxides on structural stability. There is also evidence indicating that aluminium oxide may be involved in the bonding of organic matter to clay particles.[35] Less is known about the stability due to broken edge positive charges and exchangeable aluminium ions, but field experience is that clay domains and granules that contain such charges and ions have similar stability to calcium-saturated clays.

There is much evidence that soil organic matter stabilizes aggregates and structural pores. Although only a small fraction of the total organic matter is involved (and there is evidence that some organic anions favour the dispersion of clays) many studies have shown good correlations between total organic matter and structural stability. In Britain a level of 5 per cent of organic matter appears to be necessary for soils to be structurally stable.[36] There is some evidence concerning the nature of the more active fractions of soil organic matter. A variety of organic and inorganic materials, natural and synthetic, have been introduced or extracted from soils and various tests for stability then made.[37]

Organic binding agents have been classified by Tisdall and Oades[38] into transient, temporary and persistent depending on their longevity. *Transient* includes some microbial and root exudate polysaccharides and polysaccharide gums whose effect lasts only a few weeks; *temporary* is associated mainly with roots and fungal hyphae which last a few months; and *persistent* includes aromatic organic compounds linked by polyvalent metal cations to clay surfaces and also other strongly adsorbed polymers. The effectiveness of different polysaccharide gums varies widely, but is usually greater the higher the molecular weight. Some can be effective at a concentration of 0.02 per cent in the soil. The actual disposition of material also appears to be as important as its exact nature.

From a study of their own and earlier results on the mechanism of binding within and between aggregates of a red–brown earth, Tisdall and Oades

34 Lutz, J. F., *Soil Sci. Soc. Am. Proc.*, 1936, **1**, 43.
35 Hamblin, A. P. and Greenland, D. J., *J. Soil Sci.*, 1977, **28**, 410.
36 Greenland, D. J., *et al.*, *J. Soil Sci.*, 1975, **26**, 294; Hamblin, A. P. and Davies, D. B., *J. Soil Sci.*, 1977, **28**, 11.
37 Greenland, D. J., *Soils Fertil.*, 1965, **28**, 415; 521; Baver, L. D., *et al.*, *Soil Physics*, Wiley, 1972; Theng, B. K. G., *The Chemistry of Clay Organic Reactions*, Hilger, London, 1974.
38 Tisdall, J. M. and Oades, J. M., *J. Soil Sci.*, 1982, **33**, 141.

concluded that different mechanisms operate for different sizes of soil aggregate, larger aggregates usually being made of the smaller and more stable
ones.

They found that aggregates with high organic matter content and larger
than 2 mm are predominantly held together by fine roots and hyphae, and
are hence much affected by agricultural practices. They are not affected by
treatment with periodate which oxidizes polysaccharide. Larger aggregates
low in organic matter (<1 per cent organic C) are bound by transient agents
only, probably low molecular weight polysaccharides. Aggregates 20–250 μm
diameter are usually more stable although containing less than half the
organic matter concentration of the larger crumbs. The binding here appears
to be by persistent organic materials, crystalline oxides and disordered
aluminosilicates. Aggregates 2–20 μm often have a high concentration of
organic carbon and may resist dispersion even by ultrasonics. Material
<2 μm consists of orientated clay particles held together by inorganic and
organic cements and by electrostatic bonding.

While treatment with periodate destroys the material that stabilizes structure in many arable soils, it does not destroy the structure in many grassland
or forest mull soils, but sequential treatment with periodate followed by
sodium pyrophosphate removes the stabilizing agent.[39] However, treatment
with the pyrophosphate alone does not destroy the stability of these soils.
These observations are interpreted as showing that in pasture and forest mull
soils there are two groups of stabilizing agents – polysaccharides and true
humic colloids. The interpretation is uncertain because pyrophosphate
removes not only an active humus fraction low in polysaccharides, but iron,
aluminium and calcium ions, all of which are active in structure stabilization.

The result that polysaccharide gums are the principal constituent of humus
that stabilizes the structure in agricultural soils has also been found by Chesters[40] who showed that crumb stability, in the soils he worked with, was much
better correlated with the polysaccharide content of the soil than with the
total humus content. This conclusion implies that the polysaccharides are
active primarily through the large number of weak van der Waals' bonds
between the uncharged polymer and the surface of the mineral soil particles.

The reason why only the polysaccharide fraction of the humus appears to
be responsible for stabilizing the structure of arable soils may be that in these
soils there is no adequate mechanism for the other fractions of the humus
to become sufficiently intimately mixed with the soil particles. This intimate
mixing may be a very slow process brought about in undisturbed pasture and
forest mull soils through the action of earthworms. Humic colloids do,

39 Greenland, D. J. *et al.*, *Soil Sci. Soc. Am. Proc.*, 1962, **26**, 366; Clapp, C. E. and
   Emerson, W. W., *Soil Sci. Soc. Am. Proc.*, 1965, **29**, 130; Stefanson, R. C., *Aust. J.
   Soil Res.*, 1971, **9**, 33–41.
40 Chesters, G. *et al.*, *Soil Sci. Soc. Am. Proc.*, 1957, **21**, 272.

however, have a small effect in stabilizing the structure of arable soils, as determined from the effect of extracting the soil with pyrophosphate.[41]

The polysaccharide polymers certainly, and probably the other humic colloids, are present as ropes or nets of polymers, not as individual molecules.[42] Thus a rope or net can have a large number of clay particles attached to it, or it can cover the surfaces of a number of contiguous clay domains and hold them apart against surface tension forces. These ropes do not enter the clay domains but coat the outside of domains and granules sealing a number of pores by covering them over. Thus they only reduce the interdomain or intergranule swelling, they do not affect the intradomain swelling.[43] But they cannot cause the domains or clay particles to increase their separation, that is, they can stabilize but not create structure. The structure produced by these polymers is stronger, the greater the area of sesquioxide surfaces, for these bind carboxyl groups strongly. As a result, soils high in sesquioxide surfaces can have a very stable structure due to the very stable and strong bonds between the humic colloids and these surfaces.

The nature and origin of some of the polysaccharides involved has been investigated by Cheshire[44] and their part in aggregation is discussed by Oades.[45]

Stabilization by the decomposition of organic matter

If decomposable organic matter is added to a soil, there is a rapid improvement in the stability of the soil structure. This is brought about partly by the production of polysaccharide gums by the soil bacteria, and partly by the growth of fungal hyphae over soil particles, for some of these hyphae are adsorbed on to or stick on to the surfaces over which they grow. Many of the soil bacteria produce gums, either as extracellular or capsular gums or as cell-wall gums, and it is probable that the bulk of the polysaccharides are of bacterial origin. Jackson,[46] using an electron microscope, showed that a proportion of soil crumbs in the soil he examined appeared to be composed of clay particles clustered around a bacterium, as is illustrated in Plate 12.3[47] or a group of bacterial cells. Little is known about the factors that affect the amount of high molecular weight gums produced by bacteria when fresh

41 See ref. 39.
42 Rawlins, S. L. *et al.*, *Soil Sci. Soc. Am. Proc.*, 1963, **27**, 354.
43 Theng, B. K. G. *et al.*, *Aust. J. Soil Res.*, 1967, **5**, 69.
44 Cheshire, M. V., *Nature and Origin of Carbohydrates in Soil*, Acad. Press, 1979.
45 Oades, J. M., *Pl. Soil*, 1984, **76**, 319.
46 Jackson, M. L. *et al.*, *Soil Sci. Soc. Am. Proc.*, 1947, **11**, 57.
47 Rope, M. M. and Marshall, K. C., *Microbiol Ecol.*, 1974, **1**, 1.

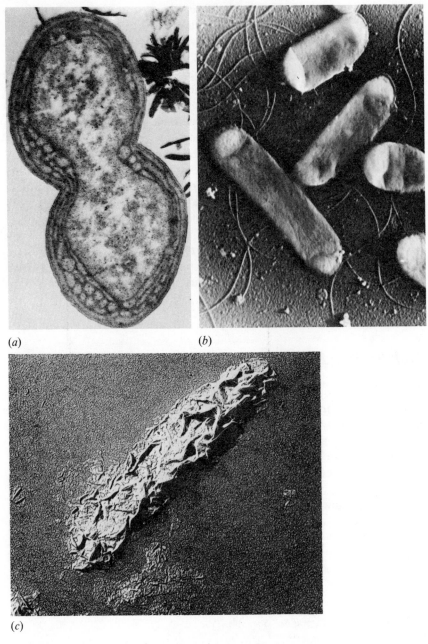

(a)   (b)

(c)

Plate 12.3   Electron micrographs ot illite clay particles absorbed on the surface of the bacterium *Escherichia coli* (× 34 500).

organic matter is added to a soil, but a low level of available nitrogen in the immediate neighbourhood of the bacteria may be favourable.[48]

Filamentous fungi, which multiply when decomposable organic matter is added to a soil, contain species whose mycelia bind clay or soil particles to their surface and will stabilize soil pore surfaces over which they grow. Fungi differ markedly in the laboratory in their power to do this. Thus Swaby[49] found that fungi with fast-growing hyphae were better aggregators of loam and clay soil than those with slow-growing or smooth hyphae. Using ultrasonic vibrations to disperse soil crumbs, Aspiras[50] found great differences between the strength of the fungal bonds of different fungi but this depended to some extent on the soil. He also found that the fungal bond did not appear to involve polysaccharides, and in fact fungi are poor producers of high molecular weight gum.[51] Those fungi whose mycelia bound soil particles strongly to their surface produced more stable crumbs than those stabilized by bacterial gums.

The importance of fungi in the stabilization of soil structure in the field has been difficult to establish, a problem not helped by the difficulty of estimating the number and length of individual bits of hyphae in a soil. In so far as they hold soil particles on their surface and run between a number of soil particles, they will help stabilize structure in the same way as polysaccharide gums, but will be more effective stabilizers, as they actively grow through the wider pore spaces touching a number of soil particles and clay along their length. Bond[52] in Australia has shown that some fungi stabilize sandy soils by forming a web of hyphae covering sand grains, which holds them together, and by waterproofing the weak sand structure makes it difficult to wet.

There is some evidence that the decomposition of fungal hyphae yields products that are more effective in stabilizing structure than the hyphae themselves. This possibility was first suggested by Geltser[53] but it has been little investigated[54] and the conditions under which it occurs and the reasons for it are not known. Plate 12.4 illustrates some experimental results of Swaby[55] for structure build-up by different groups of micro-organisms and the effect of different micro-organisms on structure stabilized by fungal hyphae. In these experiments fungi definitely appear to be the most effective structure stabilizers of the Rothamsted clay loam, and they give no support to Geltser's

48 Acton, C. J. *et al.*, *Canad. J. Soil Sci.*, 1963, **43**, 141, 201;   Schwarz, S. M. *et al.*, *Soil Sci. Soc. Am. Proc.*, 1958, **22**, 409.
49 Swaby, R. J., *J. Soil Sci.*, 1950, **1**, 182.
50 Aspiras, R. B. *et al.*, *Soil Sci. Soc. Am. Proc.*, 1971, **35**, 283.
51 Stacey, M., *J. Chem. Soc.*, 1947, 853.
52 Bond, R. D., *Nature, Lond.*, 1959, **184**, 744.
53 Geltser, F. Y., *The Significance of Micro-organisms in the Formation of Humus and the Stability of Soil Structure*, Moscow, 1940.
54 Griffiths, E. and Jones, D., *Pl. Soil*, 1965, **23**, 17.
55 Swaby, R. J., *J. gen. Microbiol.*, 1949, **3**, 236.

(a) The water-stable soil structure built up by mixed populations of different types of micro-organisms (Rothamsted soil and glucose).

(b) The break-up of the water-stable structure built up by the mycelia of the fungus *Aspergillus nidulans* (initial), after it had been heat-killed and the soil inoculated with a number of species of other organisms.

Plate 12.4 The effect of microbial action on soil structure.

hypothesis that stabilization is due to substances produced by the bacteria decomposing fungal tissue.

The effect of plant roots on aggregation and structural stability is generally found to be favourable, especially with grass, although the traffic and cultivations and exposure of bare soil that usually accompany arable farming may lead to an overall deterioration. However, there is evidence[56] for a specific decrease in soil structural stability around maize and wheat roots, as shown by increasing ease of clay dispersion, so that the addition by roots of organic materials and the stimulation of microbial activity normally associated with plant roots may not always be beneficial. Evidence indicated some removal of iron and aluminium by chelating agents from organic matter–sesquioxide –mineral particle linkages. These effects were not found with ryegrass roots, which always increased structural stability.

Many plant roots exude a gum or mucigel some distance behind the young tip. Webley[57] found the rhizosphere of grasses that he examined was rich in gum-producing bacteria and he also noted that fine dead roots appeared to hold fine crumbs together around them. This could be due to the roots growing through existing crumbs or to fungal hyphae and bacterial gums binding fine material into crumbs. Old root channels are often fairly stable after the decay of the root and if undisturbed may provide channels for rapid drainage and space for new roots to occupy. Under direct drilling these channels may persist whereas their continuity is broken by cultivation.

Roots differ considerably in their ability to stabilize their channels on their death. Cooke and Williams[58] found that when a three-year lucerne ley, on a poorly structured boulder clay soil at Saxmundham in Suffolk, was ploughed out in autumn, the soil had an excellent structure compared with old arable as is shown in Plate 12.5 and it could be worked down to a seedbed very easily, even in wet weather. Yet in the following year, the whole benefit had disappeared. They also noted that the lucerne roots, at the time of ploughing out, contained 25 per cent of water-soluble carbohydrates, so the excellent structure on ploughing out may have been produced from these by the soil population earlier on; but all these stabilizing substances were decomposed after 12 months under winter wheat. This soil contains between 35 and 40 per cent of coarse sand and only about 20 per cent of clay, and forms very compact clods which do not crack on drying, and into which the roots of most crops will not penetrate. Normal additions of farmyard manure do not affect their microstructure appreciably, probably because microbial activity is at a very low level within the clods due to the lack of any mechanism for transferring decomposable organic matter into their interior.

Undisturbed soils are also a favourable habitat for earthworms in

56 Reid, J. B. *et al.*, *J. Soil Sci.*, 1982, **33**, 397.
57 Webley, D. M. *et al.*, *J. Soil Sci.*, 1965, **16**, 149.
58 Cooke, G. W. and Williams, R. J. B., *Rothamsted Ann. Rept.* 1971, Part II, p. 122.

temperate regions, if the soil is not acid or too sandy. These animals are important agents in structure stabilization, partly because they exude a mucus from their skin as they move through their burrows, which may stabilize the burrow walls,[59] though its stabilizing power has not been studied in detail, and partly because they produce wormcasts (see p. 513). These consist of soil which has passed through the earthworm's gut and has had the undigested portion of any plant remains eaten by the worm thoroughly macerated and mixed with it. As the cast ages the stability of its structure increases, presumably due to microbial activity, and the stabilized large granules or small crumbs can last a very long time.

(a)

Plate 12.5   Structure of a chalky boulder clay soil, Saxmundham, East Anglia, shown in a furrow slice (a) after a three-year lucerne ley, and (b) on old arable land (G. W. Cooke).

59 Jeanson, C., *Geoderma*, 1967, **1**, 325.

Swaby[60] found that the increase in the stability of soil structure in a worm-cast depended on the soil it was ingesting. Thus, the crumb stability of casts of *Allolobophora nocturna*, the most common casting species at Rothamsted, was appreciably greater than the soil from which it had been derived in pasture soils, but little different on arable soils with a much less stable structure. He also showed that there were very high numbers of gum-producing bacteria in the casts produced from pasture soils. Evans[61] also showed that these casts had a higher proportion of structural pores finer than 50 μm which increased the volume of readily available water the soil could hold.

(b)

60 Swaby, R. J., *J. Soil Sci.*, 1950, **1**, 195.
61 Evans, A. C., *Ann. appl. Biol.*, 1948, **35**, 1.

Soil conditioners

Much work has been carried out with synthetic organic polymers, both to elucidate their interaction with soil constituents and to investigate their ability to stabilize soil aggregates and so act as soil conditioners. The first products to be introduced commercially were developed by the Monsanto Chemical Company and were sold under the trade name Krilium. This was a vinyl acetate–maleic anhydride copolymer (VAMA) and later a hydrolysed poly-acrylonitrile (HPAN) was also used. Both contained carboxylic groups spaced along the polymer chain and were considered to resemble the naturally occurring stabilizing polysaccharides and polyuronides. In the pH range found in soil they are long chain poly-anions and stabilize clay domains through electrostatic attraction with positive charges on clay particles. Consequently they are more efficient in acid soils because of the strength of the polymer–aluminium oxide–clay bond.[62] They can probably be bonded on to uncharged silica surfaces by hydrogen bonding through the silanol group.

Following Krilium a large number of other materials have been tested. Although many were very effective, various drawbacks eventually outweighed any advantages for agricultural use, though some were successful for high cost soil engineering projects. Expense was one problem, but other difficulties included incorporation, the formation of impermeable layers and the mistaken belief that these materials could ameliorate poor structural conditions rather than simply stabilizing existing aggregates or sometimes reinforcing compacted layers.

Recently a greater variety of cheaper materials have been found to be effective in low concentrations (down to 0.02 per cent). Considerable interest has been shown in the uncharged polymers derived from vinyl alcohol, the polyvinyl alcohols (PVA), which can be obtained with a range of molecular weights and hence of chain length. The higher the molecular weight the more effective per unit mass, but the less mobile the polymer. These PVA polymers are water soluble and are very strongly adsorbed on to clay surfaces by hydrogen bonding, being ineffective in the absence of clay. For binding sandy soils the insoluble polyvinyl acetates (PVAc) in aqueous emulsion are effective, the polymer forming films and bridges at points of contact as the soil dries. Consequently the exact water content at application is important and also the subsequent drying conditions if the polymer is to have its maximum effect.[63]

Incorporation of conditioners is still a major problem, the main successful application at present being surface treatment of seedbeds against capping and erosion. The attempts at deep incorporation to stabilize mole channels

62 Emerson, W. W., *J. agric. Sci.*, 1956, **47**, 117;   Emerson, W. W. and Dettman, M. G., *J. Soil Sci.*, 1960, **11**, 149.
63 Rigole, W. and De Bisschop, F., *Meded. Fac. Landbouw.*, Ghent, 1972, **37**, 938.

and other deep cultivation channels has not yet been successfully achieved. There is also the danger of stabilizing unfavourable structure. Synthetic conditioners are resistant to microbial breakdown, so that effects may be expected to be durable provided the conditioned layers are not broken up or ploughed under, or lost in other ways.

Effect of wetting and drying cycles

As mentioned earlier a number of processes may promote or damage a favourable soil structure and pore characteristics. Alternate wetting and drying, especially with clay soils, causes swelling and shrinkage stresses which result in the weakening of the larger clods and the development of shrinkage cracks in clods and in fresh sediments. Many clay subsoils when brought to the surface and exposed to the weather will more or less rapidly break down into a mass of aggregates or granules, presumably by alternate wetting and drying. This weathering process is usually a one-way action, and if it continues, as it does with some soils, eventually leads to a homogeneous mass with very unfavourable results. Clay soils whose clods do not break up on weathering remain very cloddy and are usually difficult to manage, especially when preparing a seedbed. Generally, soils high in clay tend to weather easily, and it is the clay loams and silty clays that are most difficult. In the laboratory repeated wetting and drying usually lead to increasing break-down, but drying is more severe than in the field, and wetting more rapid.[64]

Frost action

Frost can also have an important effect on soil structure. On sandy soils it tends always to be excessively disruptive, but on clays it can be very beneficial and is traditionally held to be so. One aim of autumn cultivations in suitable climates is to put the soil into a condition where it will gain maximum benefit from frost by exposing the maximum clod surface and at a suitable moisture content. Too fine a tilth in autumn tends to lie wet in winter and then frost is liable to cause excessive breakdown instead of producing the desired 'frost tilth'.

The effect of freezing on soil depends on the rate of freezing, on the moisture content and on the pore characteristics. There is always some unfrozen water present which becomes less and less mobile as the temperature falls, and as ice forms in ever finer pores with the films of liquid water on pore walls decreasing in thickness.[65]

64 Richardson, S. J., *J. Soil Sci.*, 1976, **27**, 287.
65 Homshaw, L. G., *J. Soil Sci.*, 1980, **31**, 399.

Rapid freezing, when little water movement can occur, necessarily implies low temperatures and then water freezes within pores before it can migrate out, leading to considerable and usually excessive breakdown which is later revealed on thawing.[66] Under natural field conditions freezing is not rapid and there is considerable movement of water upwards towards the lower temperatures. and also towards any ice formed as the potential of the water in contact with it decreases. There is probably always supercooling before ice is formed, certainly by one or two degrees according to field temperature records, with ice forming initially in lenses or successive horizontal bands separated by unfrozen soil. The ice bands grow at the expense of the water in the unfrozen soil, which itself contracts in three dimensions, forming a pattern of vertical cracks often fairly close together and dividing the soil into small roughly cubic blocks. These are compressed and dehydrated by the growing ice bands, with ice also forming in the vertical cracks. As water moves up into the frozen layers from below in response to the potential gradients the surface may be pushed upwards, giving one contribution to frost heave. Very large pressures may occur, with damage to structures built on and in the soil, especially if the unsaturated conductivity of the subsoil is relatively high.[67] On susceptible soils a layer of coarse material of low unsaturated conductivity is often used in foundations to reduce water conduction and hence reduce heave. Surface heave in arable soils may amount to several centimetres, damaging plant roots and leaving a hollow layer beneath the surface on thawing, which slowly subsides but which may benefit from rolling in the spring. Some chalk soils are particularly prone to 'heave'.

As the temperature falls, ice forms between the peds of a clay soil with shrinkage of the unfrozen clay until at a certain temperature ice begins to form within the pores and shrinkage stops. The temperature at which ice begins to form within pores depends on pore size, and appears to be consistent with the entry of an ice–water interface governed by the same principles that hold for the entry of an air–water interface,[68] with the interfacial tension of ice–water being about 30 mN m^{-1} and the pressure difference given by the capillary equation.[69] Reviews of frost action in soils are given by Miller[70] and by Yong and Warkentin.[71]

Mechanical stress and aggregation

Another process leading to aggregation or to breakdown is compaction and

66 Miller, R. D. in Hillel, D., *Applications of Soil Physics*, Acad. Press, 1980.
67 Taber, S., *J. Geol.*, 1930, **38**, 303; Beskow, G., *Swedish geol. Soc.*, Series C, 375. Trans. by Tech. Inst. North Western Univ., Evanston, Illinois, 1947.
68 Everett, D. H., *Trans. Faraday Soc.*, 1961, **57**, 1541.
69 Schofield, R. K., *Trans. 3rd Int. Cong. Soil Sci.*, 1935, **2**, 37.
70 See ref. 66.
71 Yong, R. N. and Warkentin, B. P., *Soil Properties and Behaviour*, Elsevier, 1975.

shear. If a dry soil is stirred its crumbs tend to become ground down by attrition and fracture. With a very wet soil, again stirring tends to reduce aggregates to a uniform paste which may dry to a hard cloddy mass. But at intermediate moisture contents some aggregation can occur on stirring or working. If a dry soil is stirred and water added slowly, there usually comes a point at which aggregates form, and increasingly so up to a critical moisture level; beyond this, slaking and puddling occur.

At the optimum moisture content for aggregation not only is the aggregation greatest but the aggregate stability is also highest.[72] This optimum moisture content is near to that at which the soil just becomes sticky (see below), that is, when the pores are nearly water filled, and decreases as the pressure of working the soil increases, when also the range of the optimum increases.[73] Cultivation implements have been designed to exploit this effect[74] but power consumption is high, and if applied outside the appropriate moisture range then serious damage can be done to the soil structure. The aim of cultivation as far as soil structure is concerned should be to exploit the effects of weathering and to put the soil into a condition such that it will gain maximum benefit from the weather.

The mechanical behaviour of soils

The resistance of soil to deformation and breakdown is important in cultivation and in the growth of plant roots, and also in relation to compaction, surface poaching, erosion and bearing capacity for traffic and animals. When dry, soil crumbs behave as brittle solids, fracturing with negligible deformation, but as water content increases breaking strength usually decreases and plastic deformation increases. At still higher water contents as plastic yield strength decreases the soil eventually becomes liquid, flow occurring under gravitational forces. The water content at which the soil changes from brittle to plastic is called the *lower plastic limit*, and at which it just becomes liquid, the liquid limit or *upper plastic limit*. The difference in water contents between the upper and lower plastic limits is the *plasticity index* or *plastic number*. These limits were first proposed as important soil characteristics by Atterberg[75] and are often referred to as the Atterberg limits.

These limits, and the plasticity index, increase with increasing clay content and are commonly used by civil engineers as a better guide than clay content itself to the mechanical behaviour of soil in the field. For aggregated topsoils

72 Vilenskii, D. G., *Trans. 1st Comm. Int. Soc. Soil Sci.*, 1934, p. 17.
73 Henin, S., *Ann. Agron.*, 1936, **6**, 455.
74 Vilenskii, D. G., *Aggregation of the Soil*, Acad. Sci., Moscow, 1945 (trans. from Russian by Howard, A., CSIRO, Melbourne, 1949).
75 Atterberg, A., *Int. Mitt. Bodenk.*, 1911, **1**, 10.

the relevance to field behaviour of these characteristics may be reduced because of the preliminary destruction of any structural organization before their determination. Engineering problems have arisen with very strongly aggregated soils whose field behaviour does not correspond with that predicted from such measurements.

Effect of stress

The effect of a force applied to soil depends on the area over which it acts, the important factor causing deformation or breakdown being force divided by area, or stress. There are three types of stress: compressive, which is an overall pressure such as hydrostatic pressure; tensile, which is a pulling apart with no lateral constraint; and shear, which is the sliding of one layer over another[76] (see Fig. 12.3).

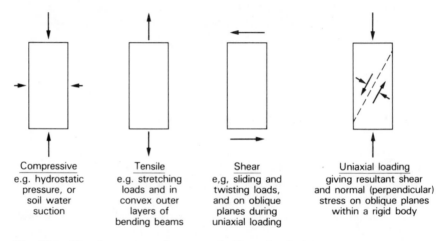

Compressive	Tensile	Shear	Uniaxial loading
e.g. hydrostatic pressure, or soil water suction	e.g. stretching loads and in convex outer layers of bending beams	e,g, sliding and twisting loads, and on oblique planes during uniaxial loading	giving resultant shear and normal (perpendicular) stress on oblique planes within a rigid body

Fig. 12.3   The three types of stress and effect of uniaxial loading.

The nature and magnitude of the reaction of soil to mechanical stress may depend not only on its composition, structure and moisture content but also on the scale of the stress or size of sample involved, and on the duration or rate of application of stress. The effect of size or scale depends on the scale of heterogeneity of strength, that is, the size of aggregates and the distribution and scale of planes of weakness. In general, the smaller the scale or sample size the greater the probability of not including a weak boundary in the sample and hence finding higher and more variable strength than is appropriate for the bulk soil.

When soil yields under stress it most commonly fails in shear, tangentially

76 Marshall, T. J. and Holmes, J. W., *Soil Physics*, Ch. 9, Cambridge Univ. Press, 1979.

acting forces causing it to deform and possibly to fracture. Under some circumstances tensile failure may occur. Cultivation operations and traffic loads generate mainly compressive and shear stresses, and the effect of shear depends partly on the load perpendicular to the shear stress, the 'normal' load. Shear strength increases nearly linearly with increasing normal load (Fig. 12.4), that is:

$$S = C + N \tan \emptyset \qquad \text{(Coulomb equation)}$$

where S is maximum shearing stress, C is shear stress at zero load (called cohesion or apparent cohesion), N is the normal load (stress perpendicular to the shearing stress) and $\emptyset$ is called the angle of internal friction. This normal load includes a component due to soil water suction, which is pulling the soil particles together, as well as components due to external stresses on the soil. Sand grains, unlike clays, interlock as the load increases, resulting in a considerable increase in shear strength and a large angle of internal friction.

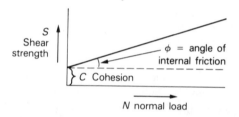

Fig. 12.4   Relation between shear strength and normal load.

Cohesion, the shear strength at zero load, is small for sands, where there is no significant attraction between the grains, and high for clays, where the individual particles are held together by surface tension when moist and strong electrostatic and other bonding when dry. In aggregated soils the crumbs may be relatively strong, resist breakdown and interlock, so that the bulk soil may have a high angle of internal friction and low cohesion irrespective of the properties of the individual aggregates. When the aggregate strength is greater than that of the bulk soil, failure during cultivation occurs between and not within the crumbs and clods, and so there is little breakdown, only movement of intact aggregates within the weaker bulk soil.[77]

Water content can be important because as the soil dries the clod strength increases continuously, whereas the strength of the assemblage of clods, the bulk soil, generally rises at first as soil water suction increases and then falls again as water content falls further, just as happens with sand (see Fig. 12.5).

77 Spoor, G., *Soil Physical Conditions and Crop Production*, MAFF Tech. Bull., No. 29, 128, London, 1975.

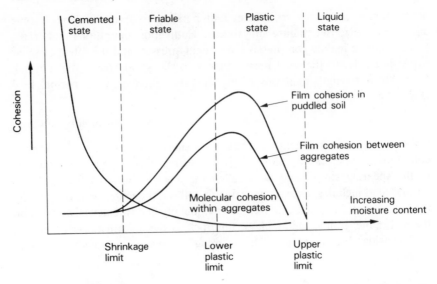

Fig. 12.5   Variation in cohesion with moisture content. (Adapted from Spoor, G., *Soil Physical Conditions and Crop Production, MAFF Tech. Bull.,* No. 29, 128, HMSO, 1975.)

This means that there is a fairly small range of water content when the crumbs are weaker than the bulk soil and yet not plastic, when the soil crumbs are most readily broken, and the soil is said to be friable. At higher water contents the soil smears, at lower ones the stronger aggregates are displaced but remain intact on cultivation.

The pulling together of soil particles as the soil dries causes shrinkage, and with soils high in clay there is usually a wide range of water content over which the volumetric shrinkage is equal to the volume of water lost.[78] This is called normal shrinkage and it continues until the solid components of the system come into contact and resistance to further shrinkage increases, so that air begins to enter as more water is removed. With many clay soils this only occurs when the soil is considerably drier than the permanent wilting point, though cracks may develop between the shrinking blocks of soil.

The effect of rate of application of stress is not usually large, but may have to be taken into account when comparing the soil reaction to stresses applied at very different rates. Root growth rates are about 10^{-7} m s^{-1}, implement blade and raindrop impact rates are eight orders of magnitude faster, and durations correspondingly shorter. As with most materials, the strength of soil increases as rates increase. With soil the increase is greater the lower the hydraulic conductivity, because that part of the load borne by the water is less rapidly transmitted by the less mobile water. Other factors complicate

78 Haines, W. B., *J. agric. Sci.,* 1923, **13**, 296.

the picture. For example, stresses of long duration may cause 'creep'. Impacts can generate shock waves whose resultants may reinforce one another and lead to breakdown some distance from the point of impact, which are not simply related to the initial impact force.

Tensile strength is involved in the failure of loaded beams and hence in the emergence of seedlings through soil crusts. If the crust has too high a strength the plant shoot fails to emerge. Experimentally the crust strength, or that of a remoulded soil block, is found by mounting it on two supports, a distance *l* apart, and measuring the load *F* at the centre which causes it to break, when the modulus of rupture is given by

$$3Fl/2bd^2$$

where *F* and *l* are as stated, *b* is the breadth of the beam or piece of crust used and *d* the thickness. The strength is not called 'tensile strength' here because the necessary theoretical conditions are probably not met.[79] Critical values for modulus of rupture vary with plant and other factors, emergence being prevented in some cases at 20 kPa (0.2 bar) in others at over 100 kPa (1 bar).

Another property of soil of relevance to cultivation is its adhesion to implements. It is the tangential adhesion, analogous to friction, which is important here, and this is not necessarily closely related to 'normal' or 'perpendicular' adhesion, the stress necessary to pull the soil off a surface at right angles perpendicularly. Tangential adhesion increases with normal load at the interface in a way similar to the effect on shear strength:

$$S' = C_\alpha + N\tan \delta$$

where S' is tangential adhesion, C_α adhesion at zero load, and δ the angle of soil–metal friction.[80]

The normal load at the interface includes the product of the soil water suction and the fractional areas of solid–liquid contact, and any other components of stress produced by external forces. The angle of soil–metal friction tends to be low for sands and high for clays, and is not very sensitive to the nature of the non-soil surface.

The relative magnitudes of adhesion and shear strength control whether or not a soil sticks to implements, that is whether the soil fails internally or at the soil–metal interface. If the angle of soil–metal friction is lower than the angle of internal friction, then with increasing normal load adhesion rises less rapidly than shear strength, and a high normal load favours scouring, failure occurring at the soil–metal interface rather than in the body of the soil.

79 Richards, L. A., *Soil Sci. Soc. Am. Proc.*, 1953, **17**, 321.
80 Fountaine, E. R., *J. Soil Sci.*, 1954, **5**, 251.

Soil compaction

Soil compaction and consolidation, accompanied by the loss of the wider pores, results from deformation and breakdown of soil aggregates and pores under load, and leads to a loss in permeability to water and roots. At high water contents when the soil is usually easily deformed, little compaction can occur unless there is time and opportunity for water to escape. So with wet soil under the transient loads produced by traffic and some implements there may be puddling, deformation, loss of aggregates and perhaps some dispersion, but little loss of soil volume. As the soil dries it becomes intrinsically stronger, but susceptibility to compaction may increase as the large pores which empty first are relatively weak. With further drying the increase in soil strength becomes dominant. Measurements of the extent and the distribution of compaction show that its effect is significantly increased by simultaneous shear, as caused by the slipping of driving wheels,[81] and that maximum effect is produced some distance below the surface of contact between wheel and soil.[82] Under field conditions effects are further complicated by the variation with depth of soil water content and hence of strength.

Soil strength and root growth

Soil strength is involved in resistance to root growth. For roots to extend freely there must be sufficient pores of diameter equal to or greater than the root tip (about 200 μm) otherwise the soil must deform easily and aeration must not be limiting.[83] The mechanics of root penetration have been studied over many years,[84] (see Scott-Russell)[85] but the exact mechanism remains obscure, and presumably differs according to conditions. Pfeffer[86] measured axial root pressures of about 1 MPa (10 bar), less than the resistance offered by many compacted soils. Many similar measurements have been made since, though most attention seems to have been paid to soil, rather than to the root, and to its resistance to various penetrometers.[87] Root growth correlates closely with penetrometer resistance whatever the combination of compaction and moisture content involved unless aeration becomes limiting. But measurements of the axial stress experienced by roots give results 4 to 10 times

81 Davies, D. B. *et al.*, *J. Soil Sci.*, 1973, **24**, 399.
82 Soane, B. D., *Soil Physical Conditions and Crop Production, MAFF Tech. Bull.*, No. 29, 160, 1975.
83 Wiersum, L. K., *Pl. Soil*, 1957, **9**, 75.
84 Eavis, B. W., *Pl. Soil*, 1972, **36**, 613.
85 Russell, R. Scott, *Plant Root Systems: their functions and interaction with the soil*, McGraw-Hill, 1977.
86 Pfeffer, W., *Abh. Sach. Ges. (Akad.) Wiss.*, 1893, **33**, 235.
87 Davies, C., *J. S-E agric. Coll.*, Wye, 1931, **28**, 237; ibid., 1933, **32**, 84; Culpin, C., *J. agric. Sci.*, 1934, **26**, 22; Eavis, B. W. and Payne, D. in *Root Growth* (ed. W. J. Whittington), p. 315, Butterworth, 1969.

less than those on metal probes, however closely they resemble root tips in size and geometry.

The effect on root growth of soil resistance appears to have been first investigated by Pfeffer. He found that the root growth of a bean was reduced to half at a penetrometer resistance of about 500 kPa (5 bar). Later studies confirmed this result for a number of species, with limiting resistances of between about 0.5 and 4 MPa. The different approach of measuring the stress needed to compress a pack of glass spheres so that root growth into them was reduced gave much lower stresses.[88] In this case growth was halved at 20 kPa (0.2 bar) and stopped at about 100 kPa (1 bar).

A considerable difficulty in all work on the effects of soil strength arises from the rapid change in strength with change in soil water content, so that only under very strictly controlled conditions can the strength be kept constant in space and time, and hence known. This difficulty is of course compounded by the sensitivity of root growth to other factors such as aeration and other nutrient and inhibitory substances, all of whose presence or availability may be altered by the same factors that alter soil strength characteristics.

88 Russell, R. Scott and Goss, M. J., *Neth. J. agric. Sci.*, 1974, **22**, 305; Abdalla, A. M. *et al.*, *J. agric. eng. Res.*, 1969, **14**, 236.

13

Management of soil physical properties

This chapter deals with the management of the physical properties of agricultural land within the limitations of the prevailing climate, the soil and the terrain. As a general principle, a farmer should aim to leave the land in at least as productive a condition as when it was acquired, the principle being broadly the same if the land is used instead for recreation or nature reserves.

The inherent physical defects which are usually too expensive to rectify are restricted soil depth, and excessive stone, clay or sand content. Excessive soil wetness can be rectified by drainage and its principles are discussed here. Also discussed are soil erosion, which can be controlled or prevented by suitable planning and conservation methods; cultivations to produce seedbeds and improve crop rooting depth; and the effects of grass leys, mulches and organic and inorganic additives. The control of physical conditions interacts with other aspects of soil fertility, for example, drainage affects the nitrogen cycle, cultivations increase the mineralization of soil organic matter, and grass leys may reduce the incidence of some soil-borne pests and increase the amount of nitrogen available to subsequent crops. Thus the control of soil physical conditions is a part of successful soil management, the chemical and biological aspects of which are dealt with in other chapters.

Principles of soil cultivation

Soil cultivations have a number of functions. They may be used to obtain a seedbed, to kill weeds, to undo the damage done by previous traffic over the land or previous cultivations, to increase the permeability of the surface soil or subsoil to water, which will allow better drainage and aeration in the soil so improving root penetration, and to incorporate crop residues.

The basic requirements for a good seedbed are that the seed should be placed at a uniform depth, in good contact with the soil so it can take up water easily, yet the soil must be well aerated. The soil above the seed must remain sufficiently loose for the seedling to grow up through the soil, and the pore space around the seed must contain sufficient wide pores to maintain good aeration and to allow the easy growth of the young roots. There must also be an adequate supply of essential nutrients close to the seed without the osmotic pressure of the soil solution being raised appreciably. Further there must be an absence of weeds in the seedbed, for the young crop is usually very easily affected, and the final yield reduced, by even a small amount of competition at this period.[1]

Seedbed preparation

The traditional method of obtaining a seedbed in Western European agriculture is to plough the land with a mouldboard plough, which turns a furrow slice, then to work this furrow slice down to a suitable tilth for the seedbed with cultivation implements such as cultivators, harrows and rolls, taking advantage of the amount of break-up caused by the weather. The plough, if properly used, is a very efficient implement for levelling land rutted by harvesting equipment, for killing many perennial weeds, and for burying surface trash so that subsequent cultivation implements and drills can work properly. It also allows the soil within the depth of ploughing to be mixed with an efficiency dependent on the type of plough used and the way it is set, and so allows fertilizers and lime to be distributed throughout this depth.

The mouldboard plough has, however, many limitations. If used when the soil is wet, the mouldboard is likely to compress some or all of the soil in the furrow slice leaving it in large clods which require extra work to break them down into a seedbed. An example is illustrated in Plate 12.5b (p. 401). With the plough, as with most implements, the soil will be moulded into clods if it is wetter than its lower plastic limit during cultivation. This can seriously

1 Kasasien, L. and Secyave, J., *PANS*, 1969, **15**, 208;   Neto, J. H. *et al.*, *op. cit.*, 1968, **14**, 159.   For comparison between different systems of cultivation, see Moffatt, J. R., *J. Inst. agric. Engnrs.*, 1971, **25**, 161;   Russell, E. W. and Keen, B. A., *J. agric. Sci.*, 1941, **31**, 326.   For vegetable crops in England see Wolfe, J. S., *Expl. Hortic.*, 1965, **12**, 72.

restrict the period when the soil is suitable for ploughing because the soil is often wetter at field capacity than at the lower plastic limit. Thus with many soils after they have been thoroughly wetted and even though they are well drained, a period of dry weather is needed before they are suitable for ploughing. Under wet conditions the ploughshare also compresses the soil below it, causing a plough pan which can seriously restrict drainage and root penetration. Even more important under wet conditions, the driving wheel of the tractor running on the furrow bottom will compact the soil and, if it is slipping, the wheel will smear its surface and block the coarser pores. By restricting drainage, anaerobic conditions can be created which do further damage to soil structure.

Ploughing can also be harmful in dry weather if it brings moist soil to the surface where it rapidly dries out. This can be particularly serious in semi-arid regions where water needs to be conserved, and even in wetter regions where crops have to be planted in a dry soil profile which is wetted at the start of a rainy season by spasmodic showers.

There has been considerable interest in developing methods of cultivation that do not have the drawback of the traditional plough. Rotary cultivators were used on clay soils early in the twentieth century, originally powered by steam engines, but they had the disadvantage that the rotating tines tended to compact and smear the soil just below the depth of cultivation and to leave the cultivation layer too loose, and liable to slump in wet weather. The soil moisture level for cultivation was even more critical than with traditional ploughing, commonly resulting in very severe puddling and loss of aggregation.

More recently various types of rigid or spring tine and chisel cultivators which do not invert the soil have proved successful, especially on heavy soils, reducing the cost and time needed to produce a seedbed. There is often difficulty with residues of previous crops interfering with seed sowing, though in erosive climates these residues protect the soil surface and help to maintain surface permeability to rainfall. There is generally less subsoil compaction compared with conventional ploughs and discs. If weed control is incomplete herbicides can be used. Shallow tine cultivators with wide sweeps to cut and loosen the soil at 5 or 10 cm depth are a variation of these methods and are used in areas where water conservation is important; this is called stubble-mulch cultivation. They leave the old stubble as protection against wind and water erosion, and help maintain surface infiltration properties. There may be less rapid warming and drying of the soil in spring, and the anaerobic decomposition of residues may give products which are toxic to seedlings.

Reduced tillage

Direct drilling of seeds into undisturbed soil after killing existing vegetation with herbicides has proved successful on many soils in temperate and tropical areas.[2] This is sometimes called zero tillage, and leads to soil conditions which differ markedly from those under more traditional arable systems, but with little effect on final crop yield.

Compared with ploughing, higher bulk densities and penetration resistance[3] occur in the soil surface layers, with a reduction in the wider pores but generally with no change, or even increases, in the water infiltration rates. The increase in soil strength can lead to restricted root growth but this does not necessarily affect yield. In many soils conditions appear to improve after a few years under zero tillage with an accumulation of organic matter in the surface horizons and an increase in the earthworm population[4] together with an increase in the number of continuous vertical pores and cracks no longer disturbed by cultivation.

Some soils do not respond well, especially when compaction and impedance to subsoil drainage occur in poorly aggregated sandy soils.[5] An occasional deep loosening cultivation may then become necessary but such soils are not generally regarded as suitable for zero tillage systems. In the absence of cultivation some nutrients, for example phosphorus and nitrogen, tend to accumulate in the surface soil, and mineralization of nitrogen may be slower, with less nitrate in the drainage water but a requirement for extra nitrate to maintain yields.

Direct drilling can have several advantages over more traditional methods. There is greater flexibility introduced by the reduction in workload in the autumn and much more rapid and flexible seeding which can be carried out more easily at the most suitable time. Although germination and early growth may be slower this may be compensated for by earlier sowing. Under conditions where soil and/or water conservation is important, direct drilling without disturbance of the soil surface, and effectively with a surface mulch, is often an advantage compared with traditional cultivation.

An important benefit of direct drilling is energy conservation shown by the normal range of energy inputs in England[6]:

2 Examples of results of experiments on some clay soils in England are given by Davies, D. B. and Cannell, R. Q., *Outlook on Agriculture*, 1975, **8**, 216; Cannell, R. Q., *Proc. 9th Conf. ISTRO*, 1985, 85. For results from the tropics see Mensah-Bonsu and Obeng, H. B. in *Soil Physical Properties and Crop Production in the Tropics* (eds R. Lal and D. J. Greenland), p. 509, Wiley, 1979; *Soil Tillage and Crop Production* (ed. R. Lal), Int. Inst. Trop. Agric. Proc. Ser. 2, Ibadan, Nigeria, 1979.
3 Soane, B. D. *et al.*, *Outlook on Agriculture*, 1975, **8**, 221; Ehlers, W. *et al.*, *Soil Tillage Res.*, 1983, **3**, 261.
4 Edwards, C. A., *Outlook on Agriculture*, 1975, **8**, 243; see also Chapter 16.
5 Cannell, R. Q. *et al.*, *Outlook on Agriculture*, 1978, **9**, 306.
6 Patterson, D. E., *Outlook on Agriculture*, 1975, **8**, 236.

	Energy input (MJ ha^{-1})	Man hours ha^{-1}
Direct drilling	35–80	0.5
Reduced cultivation	130–230	0.8–1.2
Traditional cultivation based on mould- board ploughing	200–360	2.4–3.7

However, the economy in cost and fuel consumption with direct drilling is at least partially offset by the cost and indirect energy used in the manufacture of herbicides, plus the small amount of extra nitrogen needed to maintain yields.

A major problem with direct drilling is the plant debris from the previous crop and from weeds. Ploughing incorporates these in the soil where they break down slowly and harmlessly. With direct drilling the residues remain on or near the surface, sometimes interfering with drilling and later with germination and early growth. Anaerobic fermentation of straw under wet conditions releases substances which are toxic to seedlings,[7] reducing plant populations and rates of growth sometimes rather severely. In some countries stubble has been normally disposed of by burning *in situ*, a very effective way of disposal, but as this becomes socially unacceptable much effort is being put into alternative methods of meeting the problem.

Subsoil cultivations

Deep cultivations (subsoiling) are used to loosen the subsoil, and break any plough pans or compacted layers within the root range of the young crop. Deep cultivations can be done by deep ploughs, deep cultivators or subsoilers. It is still difficult to predict when subsoiling will be helpful in British conditions, probably because the effectiveness of subsoiling in breaking up pans and loosening compacted layers is very dependent on the moisture content and its distribution in the profile. Deep tine cultivations can cause compaction. For subsoiling to be successful, the subsoil shoe must lift the soil above it in such a way that the force of lifting produces the maximum number of cracks spreading out as wide as possible, as is illustrated in Plate 13.1. Further, if the benefit of subsoiling is to last, the cracks must be stable. This means that subsoiling is best carried out when the soil profile is just moist or not too dry, which means that it is best done in years when the autumn is dry.

Table 13.1 gives the results of a large series of experiments made in Great Britain between 1945 and 1951 on the benefit obtained by subsoiling land to a depth of 15 to 25 cm below the normal depth of ploughing (20 cm), or by

7 Harper, S. H. T. and Lynch, J. M., *J. Soil Sci.*, 1981, **32**, 627.

Plate 13.1 The system of cracks produced by a subsoiler working in compacted soil at a suitable moisture content.

Table 13.1   Per cent distribution of crop responses to deep tillage compared with 20 cm ploughing by soil textural classes

	Benefit		No	Harm	
	High	Medium	effect	Medium	High
Clay soils (about 30 results):					
Subsoiling	19	34	41	6	0
Deep ploughing	29	24	26	12	9
Both together	27	23	27	15	8
Loams (40 results):					
Subsoiling	17	28	35	17	3
Deep ploughing	22	15	40	18	5
Both together	25	20	35	15	5
Light loams (about 30 results):					
Subsoiling	9	42	37	9	3
Deep ploughing	12	30	37	12	9
Both together	33	17	47	0	3
Sands (about 20 results):					
Subsoiling	16	16	42	5	21
Deep ploughing	5	26	27	21	21
Both together	20	13	40	27	0

High: 3.7 t ha^{-1} or over for potatoes and sugar beet, 250 kg ha^{-1} or over for cereals.
Medium: between 3.5 t ha^{-1} and 1.2 t ha^{-1} for roots, 240 kg ha^{-1} and 120 kg ha^{-1} for cereals.
No effect: between 1.0 t ha^{-1} and -1.0 t ha^{-1} for roots, 110 kg ha^{-1} and -110 kg ha^{-1} for cereals.

ploughing the land to a depth of about 35 cm, or by ploughing it to this depth and subsoiling to an additional 10 or 15 cm.[8] More recent experiments on subsoiling have been reviewed.[9] Although some large increases in crop yield have been reported, responses tended to be inconsistent and on average small. It has not been possible to identify the reasons why crops responded to deep tillage on some soils and not on others. Deep tillage has, however, been shown to improve drainage conditions and depth of rooting, and in consequence the amount of water in the profile available to the crop, particularly on heavy clay soils; but these improvements do not necessarily have any large effect on yield.[10] As discussed in Chapter 25, deep incorporation of lime can be of great benefit where subsoils are very acid.

8 Russell, E. W., *J. agric. Sci.*, 1956, **48**, 129. See also Hull, R. and Webb, D. J., *op. cit.*, 1967, **69**, 183.
9 Swain, R. W., *MAFF Tech. Bull.*, No. 29, 189, 1975. HMSO, London. Farrar, K. and Marks, M. J., *Exp. Husb.*, 1978, **34**, 26.
10 Hobbs, J. A. *et al.*, *Agron. J.*, 1961, **53**, 313 (Kansas);   Burnett, E. and Tackett, J. L., *Trans. 9th Int. Congr. Soil Sci.*, 1968, **3**, 329; Unger, P. W., *Soil Sci. Soc. Am. Proc.*, 1970, **34**, 492 (Texas);   Thomasson, A. J. and Robson, J. D., *J. Soil Sci.*, 1967, **18**, 329 (Nottinghamshire).

Removal of excess water by drainage

Excess water in soil excludes air, which leads to anaerobic conditions that restrict root growth and interfere with some of the beneficial activities of the soil population.[11] Water also reduces soil strength, so high water contents limit opportunities for cultivation operations and can make their results less favourable. There are several factors which contribute to excess water and which make it necessary to facilitate the removal of water from a soil.

In humid climates, topography and underlying impermeable layers can cause water tables to be too near the surface especially in low lying or basin-shaped areas. Many soils intrinsically have such low permeability and high affinity for water that drainage is too slow to prevent prolonged and excessive surface wetness, and in other cases soils lie wet below spring lines or from surface run-off upslope. In arid climates irrigation waters may introduce salts or raise the level of saline groundwaters into the surface soil, which would lead to the accumulation of salts in the surface soil unless they can be leached out as required (see Ch. 27). In all these situations it may be necessary to improve drainage in order to accelerate water removal, to divert it, or to encourage leaching.

Usually in temperate regions cultivation is difficult if the water table is less than 60 cm from the surface, although the critical depth depends on soil texture and structure. Rather higher water tables are acceptable for grassland unless there is a poaching problem with grazing animals. The optimum depth of the water table, however, is sometimes influenced by other considerations. On peat and Fen soils which waste by oxidation it is often considered desirable to have as high a water table as possible, both to slow the rate of oxidation of the peat and, if acid sulphate conditions are present, to minimize acidification.[12] However, English Fen farmers prefer the water table deep in order to maintain trafficability for harvesting late-season crops and to reduce the danger of waterlogging in the plough layer where pumping capacity is limited.

Drainage improves the fertility of a medium to heavy soil for many reasons. It improves aeration if the soil contains coarse pores that are emptied as a consequence of draining, so removing stagnant water from the crevices and cracks in the soil, and allows oxygenated rain water to penetrate rapidly into the subsoil. In consequence of this improved aeration of the subsoil and removal of stagnant unoxygenated water, crops develop a deeper root system, and this is particularly important during the winter when transpiration is low. A young plant of winter wheat can, for example, carry a well-

---

11 For a general account of drainage requirements, see *Drainage of Agricultural Lands* (ed J. N. Luthin), *Am. Soc. Agron. Monog.* No. 7, Madison, 1957. For a review of drainage in Britain see Trafford, B. D., *J. R. agric. Soc. Engl.*, 1970, **131**, 129.
12 See Segeren, W. A. and Smits, H. in *Drainage Principles and Applications*, Vol. 4, Publ. 16, Int. Inst. Land Reclamation and Improvement, Amsterdam, 1974.

developed root system in these cracks on a heavy soil, so that as soon as the warmer weather of spring comes, it càn start growing quickly and can tap a large volume of soil for nutrients. The stronger plant on the drained soil will also be able to withstand periods of rapid transpiration during a drought much better than the weaker plant on the undrained, because of its much more extensive root system at the onset of the dry period. This effect of deeper rooting and greater vigour of the crop will also allow it to dry out the soil more during the drought, hence increase the amount of cracking that occurs in the subsoil, and hence again allow water to drain through the soil more easily in the following wet weather.

Draining can have one other important effect for autumn-sown crops or for leys and grassland. The surface of a waterlogged soil tends to heave, or be lifted upwards, when it becomes frozen, with the consequence that if any plants are growing in the soil they get lifted up with the surface and may have all their roots broken in the process. Heaving of soil in frosty weather is almost confined to soils containing an appreciable volume of water held in the coarser pores, hence draining a soil always reduces the liability of the surface to heave, and its magnitude if it does.

Probably the most important single benefit of draining land farmed by intensive mechanized systems is the resultant increase in bearing capacity. A high proportion of the damage sustained by undrained soils is caused by rutting and smearing under driven wheels. This damage leads to yet slower drainage, poorer crops, less transpiration and therefore less cracking and restructuring.

Having established the need for drainage, the next step is to consider how any water removed can be safely and effectively disposed of. This means a study of the natural and other drainage systems of the local catchment and possibly beyond to ensure their capacity is sufficient and with no danger of flooding or causing excessive salinity further down the system. It is also important to identify the origin and extent of the excess water or saline drainage as on this depends the possible solutions. The foreign water problem may often be dealt with by an interceptor system at the top of the slope. The high water table problem and saline irrigation may require a more thorough system of ditches and/or piped drains, and the intrinsically impermeable soil a combination of deep cultivation and a drainage system.

The exact details of a drainage system, the drain type, locations, sizes, depth, fall and spacing can only be decided after survey and measurement of relevant factors, particulars of which are discussed in many standard texts.[13] In some cases reasonably satisfactory mathematical analyses can be made of the effect of different drain geometries on the water table,[14] but in most situations with variable soil properties and weather patterns, and

13 See, for example, Smedema, L. K. and Rycroft, D. W., *Land Drainage*, Batsford, 1983.
14 Schilfgaarde, J. van in *Drainage of Agricultural Lands* (see ref. 11, p. 79).

unhelpful topography, empirical relationships supported or modified by experience and expediency are resorted to.

With many clay and clay loam soils, shrinkage behaviour and the thickness of the capillary fringe (the depth of soil above the water table that remains saturated but under tension), ensure that the soil aggregates remain saturated throughout the year with air entering only in the cracks, fissures and old root and worm channels. These spaces also provide the only pathways for rapid movement of water. To prevent any prolonged surface waterlogging it is often necessary that these larger pores are drained. Since very large suctions would be necessary to remove any water from within the body of the clay peds, no attempt is made to do so, and under these circumstances drains tend to be laid as shallow as possible consistent with protection from cultivation implements and with ensuring a satisfactory fall. The exact spacing of drains proves to be relatively unimportant compared with the need to ensure good conduction of water to them from the drainable spaces. Hence the great importance here of the so-called secondary treatments that is, above the drain permeable fill (PF in drainage jargon) and cross-moling or subsoiling under conditions that will lead to some cracking of the subsoil and so join up as many routes as possible for the rapid removal of water.[15] This subsoiling or moling is then repeated at intervals of a few years, the frequency depending on the stability of the soil structure.

Blockages and other forms of failure arise from silting up, encouraged by inadequate or uneven falls; build-up of gelatinous iron hydroxide deposits ('ochre') especially at outfalls[16] (p. 926); movement of pipes sometimes caused by back-pressure; and mechanical damage. The weakest and most critical part of any system is usually the main outfalls which should therefore be strongly built and accessible.

Soil erosion and conservation

Soil particles can be lost by three processes: they may be blown away, they may be washed away, or the whole soil may slide or slump down a hillside.[17] Soil erosion can cause trouble over large areas: dust storms, once started, may travel great distances; and conditions producing water erosion can then lead to extensive flooding of valleys, to silting-up of rivers, valleys and

15 Trafford, B. D. in *Soil Physical Conditions and Crop Production, MAFF Tech. Bull.*, No. 29, 1975.
16 Bloomfield, C., *J. Soil Sci.*, 1973, **24**, 453; Vaughan, D. *et al.*, *J. Soil Sci.*, 1984, **35**, 149 describe the use of conifer bark to prevent deposition of ochre.
17 For a general account of erosion and its control, see Hudson, N. W., *Soil Conservation*, Batsford, 1971; Troeh, F. R. *et al.*, *Soil and Water Conservation*, Prentice-Hall, 1980. See Rose, C. W. *Adv. Soil Sci.*, 1985, **2**, 2 for an account of erosion models.

reservoirs, and to the great impoverishment of the land above the valley floor.

Erosion can occur without the intervention of man, as shown by the great depths of sedimentary rocks laid down throughout geological times, and the widespread deposition of loess in the northern hemisphere during the Quaternary period. However, destruction by man of the natural vegetation cover and unsuitable methods of agriculture cause accelerated erosion which is a worldwide problem with serious economic and social effects. Only the principles underlying erosion and its control will be discussed here, but it should be recognized that control measures are often costly and may interfere with the farmers' livelihood. Control measures may require catchments to be planned as a whole and cultivations to be limited to gently sloping land; and they may require reduction in the number of grazing animals especially in semi-arid regions. Such measures cause difficult social problems, interfering with traditional rights of land tenure and grazing, and so create issues which are more complex than the control measures themselves.

Wind erosion and soil drifting

High winds can blow much material out of some bare soils, so that the wind itself becomes a dust storm, and soil material drifts across the land, forming dunes, filling up hollows and drifting against farm buildings and hedges. The physics of this action is now fairly well understood, as the principles have been clearly stated by Bagnold[18] for sands, and in considerable detail for soils by Chepil and Woodruff.[19]

Winds move soil and sand particles by three distinct processes: the finer particles (0.05 to 0.1 mm or smaller) are carried in suspension and may be transported as fine dust over very great distances; the coarse particles (over 0.5 mm) are rolled along the surface of the soil; and the medium-sized particles (0.1 to 0.5 mm) move by saltation. Bagnold has shown that saltation is, in fact, the primary process responsible for soil movement. The process of saltation is as follows. A strong eddy of wind at the soil surface picks up a sand grain and carries it up a few centimetres in the air where the wind has a much stronger horizontal component than at the soil surface itself. This wind then gives the sand grain a horizontal acceleration, and as the eddy which picked it up becomes dissipated, the sand grain falls back to the ground after having acquired considerable momentum. On impact it may cause some other sand grains to be shot a little way up in the air, and these, in turn, acquire momentum from the wind, and on hitting the ground may throw up

---

18 For a summary see Bagnold, R. A., *The Physics of Blown Sand and Desert Dunes*, London, 1941.
19 Chepil, W. S. and Woodruff, N. P., *Adv. Agron.*, 1963, **15**, 211.

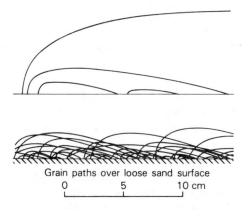

Grain paths over loose sand surface

0　　　　5　　　　10 cm

Fig. 13.1　*Upper:* Typical paths of sand grains moving by saltation. *Lower:* Paths of sand grains moving over a loose soil surface. (From Bagnold, R. A., *The Physics of Blown Sand and Desert Dunes*, Chapman and Hall, 1941.)

other grains. Typical paths for these grains are shown in Fig. 13.1. Thus, the dust storm is partly due to this stream of sand particles which throw up others as they hit the ground. Soil movement by saltation thus requires a source of sand grains of a suitable size, a wind sufficiently strong to give eddies capable of picking up sand grains, and a clear length of run for the wind to build up a sufficient density of sand grains moving in this way.

Soils containing over 60 per cent of unaggregated sand grains and individual granules in the size range 0.1 to 0.5 mm, are very liable to blow, but those with less than 40 per cent do not usually blow easily. Such blowing soils are either fine sands and loamy fine sands or calcareous clays, for these clays tend to have granules of the appropriate size, or peat and fen soils with over 35 per cent of well-rotted organic matter.

Movement of soil particles by saltation causes a fractionation of these particles by size. The finest particles are removed from the area altogether in the form of a dust storm, the larger ones between 0.07 and 0.02 mm settling as loess and the finer often being blown over very large distances and being brought down in rain.[20] The particles moving in saltation, principally in the size range of 0.5 to 0.1 mm, form dunes or drifts up against any obstructions. The particles that are just too large to move by saltation will creep along the soil surface due to the impact of the saltating grains, the larger will remain behind, eventually covering the surface. Thus wind blowing typically increases the content of coarse sand or gravel in the soil surface, unless the soil itself is a loess or a clay or peat in fine tilth, when the whole of the soil surface will be removed.

20 Prospero, J. M. *et al.*, *Nature, Lond.*, 1981, **289**, 570.

The essential condition for wind blowing to occur is that there should be a supply of particles that can move by saltation. This will not happen if the soil is moist, for the water films between the particles will hold them together too firmly for the wind to dislodge any, or if the soil surface has a compact dry crust, such as may be produced by heavy rain on the bare surface or by a heavy roller, or if the soil has a cloddy structure. On the other hand, a supply of these particles can be created by cultivating the soil surface, particularly when dry or nearly dry, or by wetting and drying, or freezing and thawing a cloddy tilth.

Wind blowing can be reduced by the use of surface mulches. Dead vegetation lying on the soil surface, such as the remains of the previous crop, will reduce the build-up of saltation by ensuring that a proportion of the falling sand grains hit vegetation rather than bare soil. Bitumen emulsions sprayed on the soil surface will also reduce saltation very considerably, and can be of very great help in stabilizing blowing sand dunes so a vegetation cover can become established. Saltation can also be reduced by laying the land up in ridges and furrows running across the direction of the wind, for if the ridge height and spacing are correct, few of the particles moving in saltation will jump the ridge.[21] But such methods can only be of temporary help as the furrows will slowly fill up with sand.

A distance of between 50 and 500 metres, depending on conditions, is needed for saltation to build up to its maximum; so if the soil can be trapped within this distance, wind blow can be kept down to an acceptable figure. A thick stubble, such as a tall wheat stubble from the previous year, or a strip of tall grass, is an excellent trap for sand, while a thin maize or sorghum stubble or short grass is relatively ineffective, the latter because its surface on the windward side will soon be covered with sand. In areas where wheat is grown in alternation with fallow, the land can be laid out in strips 100 to 200 m wide with wheat and fallow on alternate strips. An example of this from Alberta is illustrated in Plate 13.2. This method is of only limited value if the wind direction is variable.

Strips of a standing crop also act as windbreaks and at the same time stabilize the soil surface. Farmers on sandy soils in Holland and on peats in the Fens of East Anglia often use strips of spring barley or autumn-sown rye to protect young sugar beet and vegetable crops during critical periods in the spring. In a similar way American farmers sometimes use strips of sorghum to protect autumn-sown wheat until it is established. Another practice in the East Anglian Fens is to 'plant' straw in strips before sowing spring crops particularly onions. Provided these strips are close enough and established across the direction of the prevailing wind the technique provides satisfactory control without the competition caused by living nurse crops. The effect of

21 Armbrust, D. V. *et al.*, *Soil Sci. Soc. Am. Proc.*, 1964, **28**, 557; Chepil, W. S. and Milne, R. A., *Soil Sci.*, 1941, **52**, 417.

(a)

(b)

Plate 13.2   (a) Wheat stubble after being cultivated with a Noble cultivator.
(b) Alternate strips of wheat and fallow near Monarch, Alberta.

a windbreak depends on many factors, including wind velocity and direction, and the shape, width, height and permeability of the barrier.

The fraction by which the wind velocity is reduced remains fairly constant for rigid barriers but for flexible porous ones it improves slightly with increasing wind velocity and falls off less rapidly with distance. Large gaps may cause severe local funnelling and so increase the danger of a 'blow-out', an area from which erosion starts and intensifies.

In general windbreaks are effective for a distance equal to about five times their height upwind and about twenty to thirty times downwind, but the distance depends on various factors including wind velocity and erodibility of soil. In some cases they give protection for much smaller distances. Windbreaks cause shading, and may compete with the crop for water, so reducing crop yield for some distance on both sides. The method is used in Honan Province, China to stabilize sandy soil.[22] Single belts of willow, which grow to about 3.5 m high, are planted in strips 15 to 18 m apart across the direction of the prevailing wind and in strips 30 to 150 m apart at right angles to this direction, leaving only small rectangular strips of field for cultivation.

Farmers can also reduce the liability of land to blow by using appropriate methods of cultivation. One group of these is known as stubble mulch cultivation. It involves leaving the previous year's stubble anchored in the surface, but sticking out above it, as is shown in Plate 13.2. Thus, weeds can be kept in control without the soil surface being left bare. Typically, in the wheat-growing areas where these methods were first developed, a long stubble is left and the straw coming from the combine is also left on the ground. Thus, the wind velocity at ground level is kept very low and the straw and stubble together reduce the chance of a moving sand grain throwing up others by saltation. Chepil[23] found, in fact, that as little as 0.6 t ha^{-1} of straw had an appreciable effect in reducing soil drifting.

The use of suitable herbicides reduces the need for cultivations, making it possible to drill the seed into the undisturbed soil. As a short-term emergency measure, windblow can often be reduced by ploughing the land, for this will bring up clods from the subsoil which will not blow until they have been broken down into finer crumbs by abrasion.

Under English conditions, it may be economic to apply several hundred t ha^{-1} of material containing particles finer than 0.05 mm, such as a clay marl or a power station fly ash. The fly ash is very high in silt-sized particles which are effective in reducing the movement of fine sand particles by saltation.

Increasing the humus content of a blowing fine sand may increase its liability to blow, for the humus will tend to weaken the soil clods when they are dry, so causing them to break down by abrasion more easily. However, decomposing plant debris gives compounds of the polysaccharide type (see

22 Chepil, W. S., *Agron. J.*, 1949, **41**, 127.
23 Chepil, W. S., *Sci. Agric.*, 1944, **24**, 307.

Table 13.2   Influence of wheat straw at different stages of decomposition on the erodibility of a soil by wind. Chopped straw added 29 September 1951. Relative erodibility: control soil = 100

	Months after straw incorporated in soil			
Soil type	7	17	29	36
Loamy sand	0.33	7.4	238	157
Sandy loam	0.08	10.1	200	827
Silt loam	0.17	20.2	255	712
Clay	15.4	9.6	156	651

p. 571) that strengthen soil structure temporarily, but which later on decompose; so that only if organic matter is added at frequent intervals may it reduce the liability to wind blow. This is illustrated in Table 13.2[24]. Under conditions of prolonged drought there are probably no effective methods of control, for then all surface cover will tend to disappear.

Water erosion

Soil erosion by water involves two distinct processes – the break-up of clods into fine granules or particles, and the transport of these fine particles away from the clods.[25] The most important cause of the break-up of clods is the impact of the fast-falling raindrops in a severe storm, for they possess very considerable kinetic energy and momentum. The greater the intensity of the storm, the larger the drops are likely to be and the faster they will fall; and their velocity may even exceed that for free fall because of the air turbulence in the storm. A drop of 1 mm diameter, typical of fairly light rain, will have a terminal velocity of almost 3.8 m s^{-1} and a drop of 4.5 mm in diameter, typical of a heavy storm, will have one of 9 m s^{-1}, but its kinetic energy will be 500 times, and its momentum 200 times greater than that of the smaller drop. Hence, the more violent the storm, the greater is the shattering effect of the raindrops, and the amount of kinetic energy involved is very considerable. Thus, a storm at the rate of 5 cm h^{-1} is dissipating energy at the rate[26] of 5.6 kW ha^{-1}, but one of 7.5 cm h^{-1} is dissipating energy at about 520 kW ha^{-1}, nearly 100 times as fast.[27] Raindrops falling in storms of low

24 Chepil, W. S., *Soil Sci.*, 1955, **80**, 413.
25 For a general account of water erosion and methods for its control in different parts of the world, see *Soil Erosion by Water*, FAO Agric. Devel., Paper 81, 1965 and FAO Soils Bull. No. 50, 1983.   For American work see Smith, D. D. and Wischmeier, W. H., *Adv. Agron.*, 1962, **14**, 109.   For work in the tropics see Lal, R., *Adv. Agron.*, 1984, **37**, 183.
26 Nichols, M. L. and Gray, R. B., *Agric. Eng.*, 1941, **22**, 341.
27 Ellison, W. D., *Emp. J. expl. Agric.*, 1952, **20**, 81.

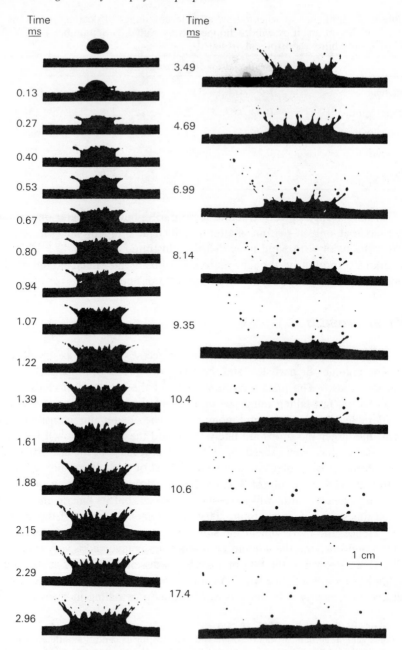

Fig. 13.2 Impact on saturated paste of soil Emerson class 7 by 126 mg drop at 5.9 m s^{-1} velocity (2 m height of fall)

intensity, less than 2 or 3 cm h^{-1}, do not usually have enough velocity to shatter clods, though they may cause the dry clod to slake on wetting. Most of the rain in the United Kingdom falls in storms of low intensity, but in areas subject to heavy rainstorms over half of the annual rainfall may fall at high rates. Hence, liability to water erosion is conditioned by the frequency of serious storms, which is greatest in the tropics.

Raindrop impact shatters clods and causes splash, and some of the droplets splashed up carry fine soil granules and sand grains, most of which are smaller than 0.2 mm.[28] Figure 13.2 shows silhouettes of splash from a raindrop on wet soil, and demonstrates the distribution of droplets, many of which carry fine soil particles a distance of up to a metre. The proportion and range of splash are greater downslope than upslope. As a result of splash, some of the finer particles, which contain much of the soil humus, disperse in the water and run off the land, leaving the coarser sand particles behind.[29] Thus erosion due to splash, and run-off, tends to coarsen the texture of the soil that remains. The finer dispersed particles will also clog coarser pores in the soil surface and this, coupled with the levelling of the soil surface due to break-up of isolated clods and its compaction by the impact of the raindrops, causes a surface cap to be formed, which may become much more resistant to break-up by impact and splash.

Water, if running over the surface of a soil, can also bring soil into suspension. The steeper the slope and the greater the length of run, the thicker will be the sheet of water, the faster the flow rate and the greater the turbulence, and the greater will be its power to disperse and transport soil. However, water running off a slope tends to concentrate into rills, which tend to coalesce, resulting in torrents capable of cutting deep gullies into the land, and carrying or floating large lumps of soil and stones, and even boulders with it. This type of erosion is known as gully erosion, and leaves the land surface uneven, as more soil and subsoil are removed from the rills and gullies than from the areas between them. It can cause spectacular erosion and can only be controlled by preventing the water concentrating into channels. Although water erosion is more severe in the tropics because of the greater frequency of intense storms, it occurs in all countries. Plate 13.3 is an example from southern England. It shows the pale soil on the upper parts of the slope where, possibly because of erosion, ploughing has exposed the subsoil with a high content of calcium carbonate. Water erosion of an Oxisol in Brazil is shown in Colour Plate 2.

28 Laws, J. O., *Agric. Eng.*, 1940, **21**, 431; Ellison, W. D., *op. cit.*, 1944, **25**, 131, 181; 1947, **28**, 7 papers. Ekern, P. C., *Soil Sci. Soc. Am. Proc.*, 1950, **15**, 7.
29 Stoltenberg, N. L. and White, J. L., *Soil Sci. Soc. Am. Proc.*, 1953, **17**, 406; Forrest, L. A. and Lutz, J. F., *op. cit.*, 1944, **9**, 17; Morgan, R. P. C., *Prog. phys. Geog.*, 1980, **4**, 24.

Plate 13.3 Exposure of subsoil, possibly due to erosion, on the upper slopes of the Berkshire Downs, southern England (D. L. Rowell).

The amount of soil lost by splash and rill erosion has been described by the empirical equation, sometimes known as the universal soil loss equation (USLE)[30]:

$$A = RKLSCP$$

A is the average soil loss, R is the erosivity of the rainfall, that is, the product of the kinetic energy of the rain storms and their maximum 30-minute intensity, K is the erodibility of the soil and equals the soil loss per unit value of R from a standard plot 22.1 m long on a 9 per cent slope in clean-tilled continuous fallow, L and S allow for the length and steepness of the slope, C takes account of the cropping system and soil management, and P allows for practices which control erosion such as contouring and strip cropping. The equation was developed for use in the United States and the variables in the equation were averaged from many hundreds of measurements on field plots. For use elsewhere the values may need to be changed. The greatest contribution of the equation is in providing guidance on the conservation practices required to keep soil loss below a prescribed tolerance level. It is, however, strictly empirical and its authors have emphasized that care is required in its use and interpretation.

30 Wischmeier, W. H. and Smith, D. D., *Trans. 7th Int. Congr. Soil Sci.*, 1960, **6**, 418; Wischmeier, W. H., *J. Soil Water Conserv.*, 1976, **31**, 5.

Conservation measures to protect the soil from water erosion usually require planning to be done on a catchment basis.[31] This is because erosion on farmland towards the bottom of a slope can be caused by run-off water upslope, possibly from roads or cattle tracks, over which the farmer may have no control. Aside from such factors, the three principles for reducing erosion are: (i) to minimize the direct impact of raindrops on the soil surface; (ii) to ensure that water moves only slowly over the soil surface; and (iii) to increase the resistance of the soil to erosion.

The value of protecting the soil from raindrop impact is well established. For example, protection with fine netting[32] and straw mulches[33] has been shown to be effective. The leaves of a crop also give protection, although if the leaf or stem causes the rain to collect in large drops, this may initiate run-off. In general a crop cover should be maintained during the period when the rains are most erosive.

Methods which maintain permeability or slow the rate of run-off help to reduce the erosive power of the water. Thus, if a wide-spaced crop is grown in strips with grass in between and along the contour, water running off the relatively unprotected land will be slowed down by the grass, and will drop its soil in the grass sward at the top end of the strip. The water will then have an opportunity of percolating into the soil before it reaches the lower end. However, the width of the erosion-sensitive strips must not be too great otherwise the water will not be flowing uniformly off the strip, and it will also be carrying too large a load of silt for the grass strip to accept. The flow of water can be slowed by other cultivation techniques, such as drawing contour ridges across the slope at frequent intervals, or by using the technique of tied ridges or basin listing (see Plate 13.4).[34] This method, however, is dangerous if the soil has too low a permeability, for if heavy rainstorms follow each other too closely, the basins may be holding water from a previous storm when the next one arrives, or the soil may still be wet; for this method will allow fine silt and clay to accumulate on the bottom of the basins which reduces their permeability until they have been dried and cracked. Also water standing on the soil surface, whether in furrows or basins, will lead to reducing conditions developing in the soil which will harm or kill the crop.

When surplus water must be removed from the land, this can sometimes be done by growing the crop on ridges with the furrows drawn just off the

31 Stocking, M. A. in *Assessment of Erosion* (eds M. de Boodt and D. Gabriels), p. 505, Wiley, 1980.
32 Hudson, N. W., *Rhodesian agric. J.*, 1957, **54**, 297.
33 Mannering, J. V. and Meyer, L. D., quoted by Smith, D. D. and Wischmeier, W. H., *Adv. Agron.*, 1962, **14**, 109.
34 See Cashmore, W. H. and Hawkins, J. C., *J. Inst. Brit. agric. Eng.*, 1957, **13**, 3. For equipment see Pereira, H. C. *et al.*, *Expl. Agric.*, 1967, **3**, 89. For examples from East Africa see Dagg, M. and Macartney, J. C., *op. cit.*, 1968, **4**, 279.

Plate 13.4   Maize growing on ridges with the furrows tied to conserve water, on the Cotton Research Station, Namulonge, Uganda.

contour, so the surplus water will run off along them. If the erosion hazard is only small, the furrows can be ten or more metres apart, leaving 'broadlands' between them, which are wide enough to allow mechanical cultivation.[35] But in general the slopes in a field are too uneven to allow this, so terraces are thrown up just off the contour to collect the water running down the slope and make it to run off slowly along the back of the terrace. The terrace length, the spacing between the terraces, and the steepness of fall of the channel behind the terrace must all be designed so that the water running off does not scour the base of the terrace, nor pond up behind the terrace so it overtops or breaches it. The water that runs off the terraces or furrows can be discharged into grassed waterways which lead it into natural drainage channels such as stream beds.

If the soil is left in large clods after cultivation with some of the stubble left on the surface, the rate of flow of water running off the surface is reduced. Also, the soil aggregation built up under grass, and the retention in the surface soil of roots and grass stems when the grassland is cultivated, increase permeability and may increase aggregate strength. There is,

35 Hill, P. R., *Emp. J. expl. Agric.*, 1961, **29**, 337 (example from Ghana).

however, evidence that a high organic matter content can weaken soil clods, making them more liable to break up during a rain storm.[36] Soil conditioners (see Ch. 12) can stabilize soil aggregates and hence increase their resistance to erosion. One of the most effective agents is sesquioxide cement which give aggregates a considerable wet strength. It is effective in soils derived from basalts and other basic igneous rocks and stabilizes the soil aggregates under raindrop impact. In soils derived from granite and other low-iron rocks, and which also tend to have high contents of fine sand, the aggregates have a low wet strength, are easily broken by raindrops and the soils are very susceptible to erosion.

The effect of cropping sequence on reducing soil erosion losses is now well understood in practice, though the full explanation of the results is not always clearly established. Table 13.3 illustrates the effect of cropping sequences on run-off and soil loss for a series of run-off plots at Zanesville, Ohio.[37] The soil is a silt loam on a 12° slope, and the average annual rainfall is 960 mm. The table shows the magnitude of the reduction of soil and water loss which a two-year grass ley can bring about when the land is subsequently cropped to maize; but the experiment cannot reveal what particular effect the ley had on the soil to reduce its liability to erosion. The table also shows a well-known general result that it is often easier to reduce soil loss than water loss.

The translation of these general principles to detailed planning of farm systems for erosion control depends very much on local circumstances. Plate 13.5 gives an example of an area of rather steep slopes that has been carefully planned to allow cultivation but to minimize soil erosion. This area is in central Kenya and can only be cultivated safely because the soils are derived from basic lavas, so have a very strong stable structure. But two points should

Table 13.3 The effect of crop rotation on the erodibility of a soil. Zanesville, Ohio: 9 year mean

Crop rotation	Average run-off of water		Soil loss, $t\ ha^{-1}\ a^{-1}$
	cm	As % rainfall	
Continuous maize	38	40.3	246
Crops in rotation			
Maize	23	23.7	105
Wheat	23	24.8	28.2
1st year ley	18	17.7	1.5
2nd year ley	13	12.8	0.5
Permanent pasture	4	4.3	0.05

36 McIntyre, D. S., *Soil Sci.*, 1958, **85**, 261.
37 *US Dept. Agric., Tech. Bull.* No. 888, 1945.

Plate 13.5   Example of a well-planned soil conservation scheme in an area of small farms on hill slopes in the Central Province of Kenya. The soils are deep tropical red loams derived from basalt.

be made here. In the first place it has been assumed that fields do not receive water from higher up the slope, particularly in the form of flash floods. The second point is that erosion can occur, and be of agricultural significance on very gentle slopes if the soil is impermeable. Thus self-mulching black vertisol clays on valley floors in the tropics can lose their loose surface by erosion, since water falling in heavy rainstorms must run off the surface. Again, some fine sandy soils will lose permeability easily and can give serious run-off and erosion even on slopes as gentle as 1 per cent.

Although erosion by water is not a severe problem in British agriculture, local problem areas have been recognized for many years. These have tended to on sandy and fine sandy soils growing root crops, field vegetables, soft fruit and hops. Extensive aerial surveys have concluded that erosion losses are more widespread than previously realized and are aggravated by newer farming practices. In particular larger fields without strategically placed hedgerows encourage run-off as does the use of semi-permanent wheelways or tramlines through crops. Control in autumn-sown crops can be fairly easily achieved by ensuring seedbeds are not too fine, by earlier sowing to provide rapid crop cover and by delay in establishing tramlines. However, erosion control on sands growing root crops tends to be unsatisfactory without resorting to techniques such as contour terracing which are incompatible with current farming practice.

Stabilizing the structure of soils

The difficulties of defining and measuring soil structure are outlined in Chapter 12. To improve soil structure, it has often been claimed that an efficient method is to put land into grass for three to four years. Figure 13.3 illustrates the increase of water-stable crumbs on a clay–loam soil in Northamptonshire under a grass–clover ley[38]; and shows that there is a rapid improvement in stability during the first three or four years, and then any further improvement takes place slowly; but even after 12 years the stability of the structure is much less than for the soil under permanent pasture. This benefit from a ley is not always found, and Low has given examples where the benefit was very much less. It is not known what factors are important, but Low concluded that the better the soil structure initially, the more rapidly it improves; that grazing tends to give a slower improvement than taking hay crops due to the treading of the animals; and that improvement takes place

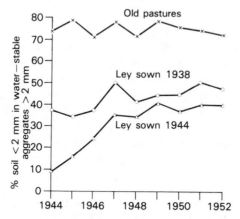

Fig. 13.3 Annual changes in water stability of a clay-loam under grass pastures of different ages. Each point represents the mean of a spring and autumn sampling.

38 Low, A. J., *J. Soil Sci.*, 1955, **6**, 179. For American data showing the rapid build-up of structure in arable soils due to grass or grass–clover leys, see van Bavel, C. H. M. and Schaller, F. W., *Soil Sci. Soc. Am. Proc.*, 1950, **15**, 399, and Wisniewski, A. J. *et al.*, *op. cit.*, 1958, **22**, 320.

more slowly in soils high in fine sand and silt, which are soils in which it is most difficult to maintain good structure.

The type of grass, and perhaps its management, may influence soil structure. One reason for differences between species might be that the greater the tensile strength of the root[39] and the longer the dead root lasts until it is decomposed, the more noticeable will be the build-up of structural units. The number of roots per unit volume of soil might also be important.

The depth of soil in which the structure is improved by the grass has not been studied in detail. Four-year leys at Rothamsted improved the stability of the structure at 10 to 20 cm depth, a result also found on two Welsh fine sandy loam soils,[40] while three-year leys in southern England gave a marked improvement in the top 2 cm, but little below.[41] Thus, when the ley was ploughed out using a 20 cm ploughing depth, the structure of the ploughed surface was no better than that of an older arable soil. This result is also found when the subsurface soil is so dense that grass roots cannot penetrate it, and it is then usually desirable to plough out the ley, loosen the subsoil and reseed; this will often allow deeper rooting of the grass and a much better build-up of the earthworm population.

The structure built up by three to four years of a grass crop is lost fairly rapidly after the ley is ploughed out. On a sandy loam soil in southern England, a three-year ley gave a marked improvement in structure, much of which was still present after a year's cropping to kale, but almost the whole effect had disappeared after the following wheat crop.[42] After old grassland the effects are longer lasting.

The cause of the improvement in structure is only partially understood. There is no evidence that the type of humus produced from grass differs from the bulk of the soil humus, and it has approximately the same ratio of polysaccharide to non-polysaccharide material, and the same distribution of fractions in the different molecular weight ranges. An important aspect of structure in some soils is the presence of channels down which rain or irrigation water can move from the surface to the deeper subsoil, and in pasture soils dead roots produce such channels when they have been decomposed.[43] Ploughing out a grass ley will cut all the channels at the depth of ploughing and will often seal their tops by smearing soil over them. If the grass is killed

39 Stevenson, T. M. and White, W. J., *Sci. Agric.*, 1941, **22**, 108. For examples of effects of different grass species on structure see Pavlychenko, T., *Natl. Res. Counc. Canada, Publ.*, No. 1088, 1942; Pereira, H. C. *et al.*, *Emp. J. expl. Agric.*, 1954, **22**, 148; Emerson, W. W., *Trans. 5th Int. Congr. Soil Sci.*, 1954, **2**, 64; Clement, C. R. and Williams, T. E., *J. Soil Sci.*, 1958, **9**, 252; Pringle, J. and Coutts, J. R. H., *J. Brit. Grassland Soc.*, 1956, **11**, 185.
40 Troughton, A., *Emp. J. expl. Agric.*, 1961, **29**, 165.
41 Clement, C. R. and Williams, T. E., *Int. Symp. Soil Structure*, Ghent, 1959, 166.
42 Low, A. J. *et al.*, *J. agric. Sci.*, 1963, **60**, 229. For similar results at Woburn see Williams, R. J. B. and Cooke, G. W., *Soil Sci.*, 1961, **92**, 30.
43 Emerson, W. W., *J. agric. Sci.*, 1954, **45**, 241; Barley, K. P. and Sedgley, R. H., *Soils Fertil.*, 1959, **22**, 155.

by a herbicide, and a shallow cultivation only is used to produce a seedbed for the following crop these channels can be maintained, which will often give a marked improvement in the ease with which the surface soil will drain.

There has been much discussion on how far the improved yield of arable crops after a grass ley is due to the improved structure it has brought about, and how far it is due to the increased level of organic nitrogen, and therefore of nitrogen supply. This is a very difficult question to answer because the consequences of poor structure can to some extent be ameliorated by additional fertilizers.[44] The fact that, given a high enough level of fertilizer, it is often possible to get high yields of crops with very restricted root systems on soils of poor structure complicates the interpretation of experimental results. A further complication is that leys may only improve the structure of the top few centimetres of soil during a two-or three-year period. When ploughed out, soil with unimproved structure, liable to slump or cap badly, is likely to be turned up on to the surface.[45] The effect on the supply of nitrogen to arable crops after a grass ley is discussed on p. 682.

Value of farmyard manure

Crumb stability in arable soils can usually be increased if regular applications of farmyard manure are given, though the amounts required may be very large. Annual applications of 35 t ha^{-1} for a century have made a measurable increase in the crumb stability of the Rothamsted soil, increasing the proportion of water-stable crumbs over 0.5 mm in size from 28 to 55 per cent on Broadbalk, which is in continuous wheat, and from 54 to 70 per cent on Barnfield, in continuous root crops.[46] Yet 70 annual dressings of 15 t ha^{-1} on the Saxmundham boulder clay had no measurable effect on structure stability,[47] and 18 annual dressings of 75 t ha^{-1} on the Woburn sandy loam cropped to an intensive vegetable rotation,[48] and 28 annual dressings of 40 t ha^{-1} on a loam soil in Ohio, only gave small improvements.[49] On a very unstable fine sandy loam soil, 75 t ha^{-1} annually did not affect the stability of the structure appreciably, but it gave a very large earthworm population which maintained aeration and drainage by the number of their burrows (Low, A. J., unpublished).

44 For examples see Clement, C. R., *J. Brit. Grassland Soc.*, 1961, **16**, 194; Boyd, D. A., *Rothamsted Ann. Rept. for 1965*, p. 216.
45 For an example see Clement, C. R., *Pl. Soil*, 1964, **20**, 265.
46 Russell, E. W., unpublished. Dettman, M. G. and Emerson, W. W., *J. Soil Sci.*, 1959, **10**, 215 established the same result using a different technique. Russell, M. B. *et al.*, *Soil Sci. Soc. Am. Proc.*, 1952, **16**, 156, found 25 annual applications of 25 t ha^{-1} to a Sassafras silt loam gave a small and 100 t ha^{-1} a considerable improvement of structure.
47 Cooke, G. W. and Williams, R. J. B., *Rothamsted Ann. Rept. for 1971*, Part 2, p. 122.
48 Williams, R. J. B. and Cooke, G. W., *Soil Sci.*, 1961, **92**, 30.
49 Havis, L., *Ohio Agric. Expt. Stn.*, *Bull.*, No. 640, 1943.

Mulches

The practice of applying a layer of dead vegetable waste material, such as straw, hay or old grass, composts or farmyard manure, to the surface of the soil around trees and bushes has been prevalent for a very long time in many parts of the world. These surface mulches can have very important effects on the conditions in the surface layers of the soil and in consequence on crops, especially those with shallow root systems. Thus, mulching has been widely used for many fruit trees and bushes and for tropical plantation crops with superficial root systems such as coffee and tea.

A surface mulch has two types of effect on the soil: a specific effect, from its being on the surface of the soil; and a general effect, which it would equally well have if it were ploughed into the soil, due to the plant nutrients set free as it decomposes. The primary specific effects of the mulch are confined to the superficial soil layers, which it keeps at a more even temperature, and moister and more permeable to water than the unmulched soil (p. 289).

A surface mulch affects both the diurnal and seasonal fluctuations in soil temperature. Its principal effect on diurnal temperature is to reduce the midday maximum temperature of a bare soil under hot and dry conditions, without having any large effect on its minimum temperature. Since a bare dry soil under bright sunshine in a continental area can reach temperatures in excess of 40 °C, which is too high for the normal functioning of plant roots, this cooling effect can be very valuable.[50]

In areas with a cold winter a mulch increases the soil temperature during winter, but reduces the rate at which the temperature rises during spring and early summer. This can be particularly serious in regions where maize is an important crop, for maize needs a fairly high soil temperature if it is to give a high germination and fast early growth. American experience is that the germination of their varieties of maize is harmed by mulches in springtime if soil temperatures are below about 23 °C, but is not affected if above.[51]

The mulch slows down the rate of evaporation from a bare, wet soil very considerably. Whereas a wet, bare soil can lose 12 mm of water in three to five days, it takes a mulched soil several weeks to lose this amount, so that in hot climates a mulched soil can retain against evaporation much more of the water falling as rain than an unmulched soil.

This effect can be important in saline soils, for mulches will reduce the rate at which salts accumulate in the surface by evaporation from the soil surface,

50 For results from Nebraska see McCalla, T. M. and Duley, F. L., *J. Am. Soc. Agron.*, 1946, **38**, 75.   See also Harrison-Murray, R. S. and Lal, R. in *Soil Physical Conditions and Crop Production in the Tropics* (eds R. Lal and D. J. Greenland), p. 285, Wiley, 1979. For general effects of a mulch, see Smika, D. E. and Unger, P. W., *Adv, Soil, Sci.*, 1986, **5**, 111.

51 Burrows, W. C. and Larson, W. E., *Agron. J.*, 1962, **54**, 19; *Trans. 7th Int. Congr. Soil Sci.*, 1960, **1**, 629; *Soil Sci. Soc. Am. Proc.*, 1959, **23**, 428.

and this will allow the osmotic pressure of the soil solution around the roots of a young plant to remain low enough for them to function long enough to let the crop become established. If the soil is sodic, a mulch will either delay the onset of the hard surface crust, which is a typical feature of these soils, or prevent it from hardening.

An organic mulch will increase the permeability and aggregation of a poorly structured surface soil,[52] partly by protecting it from raindrop impact, and partly because it serves as a food supply to the larger soil animals such as earthworms, termites and ants. These incorporate some of the organic matter into the soil, creating large channels for air and water movement and, in the case of earthworms, improving aggregation. Thus, not only is evaporation from the surface of the mulched soil reduced, but the amount of water infiltrating into it is increased. Hence, the water-supplying power of a soil can be considerably increased by mulching as is well illustrated by some results on the mulching of coffee with banana leaves in Tanzania,[53] given in Table 13.4. The effect was large enough for the trees to be wilting on the clean weeded plots, but growing well on the mulched plots, at the time the first samples were taken.

Table 13.4 The effect of mulch on conserving and maintaining the water supply in the soil. Percentage moisture in the soil on the oven-dry basis

Depth of soil in cm	After prolonged drought			After a rainy period		
	0–15	15–30	30–45	0–15	15–30	30–45
Mulched plots	27.5	35.0	37.5	41.0	39.0	38.0
Clean weeded plots	15.5	22.0	26.0	30.0	33.0	32.5

A heavy grass or trash mulch can have an important effect on weeds, for weed growth is slow through the mulch. If mulching is done at rather long intervals, the surface soil has usually a sufficiently good structure, when the mulch has nearly all decomposed, to withstand some weed-killing cultivations. This is particularly useful when the use of herbicides is not justified or practicable.

The nutrients in the mulch are returned to the soil as it decomposes, but the effect on the level of ammonium and nitrate cannot always be predicted. In so far as it keeps the soil moist for a long period of time, it encourages the slow decomposition of the soil organic matter which releases ammonium ions; but in so far as the mulch has a low nitrogen content, due to it consisting of mature plant residues, its decomposition will lock up ammonium and

52 For an example from Kenya see Pereira, H. C. *et al.*, *Emp. J. expl. Agric.*, 1964, **32**, 31.
53 Gilbert, S. M., *E. Afr. agric For. J.*, 1945, **11**, 75. For anaerobic conditions under mulches see Tukey, R. B. and Schoff, E. L., *Am. Soc. hort. Sci. Proc.*, 1963, **82**, 68; Carpenter, W. J. and Watson, D. P., *Soil Sci.*, 1954, **78**, 225.

nitrate, so lowering the level of available nitrogen.[54] The effect of the mulch on the potassium supply is usually considerable because the materials used for mulch, such as mature dried grass, grain straw or stover, and dried banana trash, are high in potassium. Mulching also increases the phosphate uptake by crops, partly because it contains phosphates but principally because it encourages the surface rooting of the crop, and keeps the surface soil moist for a longer time. For these reasons, it increases the length of time the roots can take up phosphate from the surface soil,[55] which is where the decomposable organic and fertilizer phosphates are concentrated.

Two types of synthetic mulching material are used: clear and black polythene and bitumen-in-water emulsions.[56] The polythene sheet is laid on the soil surface between the rows of vegetable crops, or of strawberries, melons or pineapples, and the bitumen emulsion is sprayed over the rows after seeding. Both these materials raise the soil temperature, but the effect of polythene is greater than for the bitumen, and both reduce water losses by evaporation from the soil surface. The mulches may also increase the nitrate content of the soil due to the higher soil temperatures.

Effect of lime and gypsum

Spreading chalk on some loam and clay soils at rates in excess of 50 t ha^{-1} was traditional in many parts of England, and this was considered, among other things, to make these soils more easy to cultivate, but there was and still is no real knowledge about the conditions under which this happens. Liming an acid soil could affect the stability of its crumbs in two ways. It will convert exchangeable aluminium ions into polymerized hydroxy-aluminium ions or aluminium hydroxide films on the surface of clay particles and exchangeable calcium ions will take their place. There is no experimental evidence that this affects the stability of the structure or the ease of working the soil.[57]

Liming an acid soil will also increase the concentration of calcium ions in the soil solution, particularly if sufficient lime is applied for some free calcium carbonate to be left in the soil, and this will reduce the tendency for clay particles to deflocculate. A Rothamsted soil which had been in arable culti-

---

54 For an example for the Great Plains in the US see McCalla, T. M. and Duley, F. L., *Agron. J.*, 1943, **35**, 306.

55 For an example with cocoa in Trinidad see Wasowicz, T., *Trop. Agric. Trin.*, 1952, **29**, 163. For arabica coffee in Kenya, see Robinson, J. B. D. and Hosegood, P. H., *Expl. Agric.*, 1965, **1**, 67.

56 For a description and use of some of these materials see Lang, E. W., *J. Inst. agric. Engnrs.*, 1968, **23**, 183.

57 See Emerson, W. W. and Dettman, M. G., *J. Soil Sci.*, 1960, **11**, 149.

vation for a long time and which was low in organic matter began to defloc-culate if the calcium ion concentration fell below 10^{-3} M.[58] It was only necessary for the soil solution to be in equilibrium with air containing 0.1 per cent CO_2 for the calcium ion concentration in a soil containing free calcium carbonate to be above this. It has also been noted[59] that drainage water from a limed clay soil had less suspended silt and clay particles than that from an unlimed soil. Regular annual additions of calcium carbonate to some plots of a brick earth soil at Versailles, which are very low in organic matter, have also given a remarkably good surface structure compared with neighbouring plots not receiving these additions.[60]

Heavy applications of calcium carbonate, of the order of 50 t ha^{-1}, have sometimes given a marked visible improvement in the structure of soils low in organic matter,[61] even if they were neutral, and there is evidence that excess calcium carbonate may reduce the cohesion between clay particles in moist clods[62] and reduce the size of the water-stable crumbs in a soil.[63] On the other hand, the traditional view could be based on the slow stabilization of structure brought about by these heavy applications if the calcium carbonate dissolves and is reprecipitated as films around the clay particles. The Rothamsted soil is naturally acid but the old fields may still contain over 2 per cent $CaCO_3$ in the plough layer, although no chalk has been given for over a century, due to these old heavy applications.

Some interest has been taken for many years in eastern England in the possibility of improving the structure, or easing the problem of cultivation, of some of the more difficult clay soils in this region, by adding gypsum to them. Field evidence is that one year after its application, 25 t ha^{-1} of gypsum to the Saxmundham soil reduced its bulk density and the force needed to push a penetrometer probe into the soil.[64] Presumably its effect will be short-lived as it is slowly soluble in the soil solution. The use of gypsum in sodic soils is discussed in Chapter 27.

Soil management: three examples

Although the principles that underlie soil management are the same throughout the world, their successful application must take account of the

58 Emerson, W. W., *J. Soil Sci.*, 1954, **5**, 233.
59 Childs, E. C., *J. agric. Sci.*, 1943, **33**, 136.
60 Hénin, S. *et al.*, *Le Profil cultural*, p. 293, Paris, 1960.
61 Leenheer, L. de, *Trans. 8th Int. Congr. Soil Sci.*, 1964, **2**, 561.
62 Russell, E. W. and Basinski, J. J., *Trans. 5th Int. Congr. Soil Sci.*, 1954, **2**, 166.
63 Williamson, W. T. H. *et al.*, *Int. Symp. Soil Structure*, p. 176, Ghent, 1958.
64 Cooke, G. W. and Williams, R. J. B., *Rothamsted Ann. Rept. for 1971*, Part 2, p. 122.

local soil properties, climate and cropping systems. Numerous failures, and even disasters, have occurred when soil management methods have been transferred to a new region without taking into account these local conditions, and this requires prior investigations.

The three examples that follow are from widely differing physical environments and illustrate the application of principles described earlier in this chapter. The emphasis is on the management of soil physical properties although, as will be seen, these interact with their nutrient supplying properties. The management of paddy soils is dealt with in Chapter 26.

Arable cropping in eastern England

East Anglia has a mean annual rainfall of 550–650 mm which, on average, is fairly uniformly distributed through the year, but the summer deficit averages almost 200 mm. Winter air temperatures reach freezing point or below, and in the summer reach 20–25 °C. Much of the land is low lying and some has been reclaimed from the sea during the last 300 years. The soils are variable and include peats, sands and clays some of which are of marine origin and some are derived from calcareous and non-calcareous glacial drifts.

From an agricultural industry originally based on mixed cropping, frequently involving livestock, specialist farming systems have evolved during the latter third of the twentieth century concentrating on fewer crops grown on a much larger scale. This specialization has been possible because of three main factors: increased power available for cultivation; abundant supplies of relatively cheap nitrogen fertilizers; and the development of chemicals for the control of weeds, pests and diseases. The change has been accompanied by dramatic increases in average crop yields (Table 1.6 gives average crop yields in the United Kingdom since 1885). However, the change has not been without some penalties, including increased erosion, pollution and chemical residue problems. There has also been concern that decreasing levels of soil organic matter, resulting from a change to continuous arable farming, were leading to deterioration of soil structure.[65] Silts are the worst affected, but with judicious cultivations, the extra root growth provided by the use of high rates of fertilizer nitrogen, and the increasing practice of straw incorporation, the structure of most soils remains satisfactory. The discussion that follows is limited to the management of clay and sandy soils.[66]

The main problem with the clay soils is that at field capacity they are too wet to cultivate and need to lose water by evaporation before they become

65 *Modern Farming and the Soil*, Report of the Advisory Council on Soil Structure and Soil Fertility, MAFF, 1970.
66 For further information see Davies, D. B. *et al., Soil Management*, Farming Press, 1982.

tractable. Some have swell/shrink cycles which lead to surface tilth formation but most have subsoils with low permeability. The problem of cultivation is minimized by sowing crops in autumn rather than in spring, and avoiding root crops which have to be harvested late and from within wet soil. Drainage is required on nearly all the soils. Cultivation has been by mouldboard plough but tined cultivations or discs have been used instead since the 1960s. High crop yields are obtained, given adequate fertilizer use, when the problems of drainage and cultivation are overcome.

The sands have a comparatively low available water capacity and have physical problems different from the clays. They are tractable for much of the year, but because of their weak structure they need regular loosening, especially after compression by tractor wheels. As with clays, autumn-sown crops are preferred but the reason is different: on sands it encourages deeper rooting and so reduces the effect of drought. Another problem is the creation of plough pans, and either regular close-spaced subsoiling is required to loosen them, or short tines are attached to the plough to prevent the compression caused by the tractor wheel in the open furrows. The plough can, however, lead to inadequate consolidation. There is now widespread use of a furrow press attached to the plough, which allows the crop to be drilled directly into the pressed land if the ploughing has left uniform furrows. The soils are suitable for a wide range of crops including potatoes, sugar beet and vegetables especially if irrigation is available. They are less suitable for intensive cereal growing because of the build-up of soil-borne pathogens.

The humid tropics: southern Nigeria

The traditional method in the humid tropics is for the farmer to cultivate a small patch of land for about three years and then allow it to return to forest for about eight years before again bringing it into cultivation. This system of alternating fallow with a cropping period is still widely practised, but the greater food requirements from increasing populations have put the system under pressure. More intensive systems are therefore being investigated.

The characteristics of the physical environment for which a more intensive agricultural system must be suitable are: (1) annual rainfall from 1200 to 2500 mm with one or two peak periods, followed by a dry season of three to four months' duration; (2) heavy and hence erosive storms; and (3) air temperatures of 25–30 °C throughout the year. The soils are strongly weathered and contain predominantly kaolinite in the clay-fraction. The texture and depth of the soils vary with their position on the landscape. Problems in their management arise because on the one hand the rains are erosive (about 14 per cent of the storms have intensities greater than 100 mm h^{-1}) and the soil aggregation is quickly lost under cultivation, and on the other, the high rainfall leads to strong leaching. There is also the restriction on animal husbandry caused by tsetse fly which limits the range of agricultural systems that can be developed.

Following earlier work in the humid tropics, it has been shown that the physical deterioration of the soil can be controlled by minimum or zero tillage and with the use of mulches of straw or weeds. With this system the infiltration rate can be maintained as high as under forest cover, and loss of soil by erosion from slopes up to 15 per cent is minimal, as shown in Table 13.5. The mulch has other effects. It reduces the day-time temperature of the soil surface, returns nutrients to the soil, and maintains a high level of soil organic matter and soil biomass activity, all of which benefit the crop. Together with a no-till system it makes long-term cultivation a possibility.[67]

Table 13.5 Soil erosion (t ha^{-1} a^{-1}) under no-tillage residue mulch and conventional ploughing. Data are means of results for 2 years (4 crops) obtained at Ibadan, Nigeria. (From Lal, R. in *Characterization of Soils in Relation to their Classification and Management for Crop Production* (ed. D. J. Greenland), p. 188, Clarendon Press, 1981)

	Soil erosion under different slopes (%)							
	1		*5*		*10*		*15*	
Crop	NT	P	NT	P	NT	P	NT	P
Maize	0.0	1.4	0.2	7.0	0.1	7.8	0.1	27.7
Soya bean	0.0	0.4	0.0	21.9	0.0	66.1	0.0	18.0

NT No-tillage-residue mulch.
P   Ploughing, residues incorporated.

The particular problems for the control of nutrient supplies arise because organic matter is rapidly mineralized and there is strong leaching. In one investigation, land under a forest fallow was brought into cultivation and the half-life of the soil organic matter, assuming a single rate factor for decomposition, was only three to four years[68]; in 22 months the percentage of nitrogen in the top 10 cm of the soil fell from 0.17 to 0.12. Soil organic matter has been shown to be conserved through the use of crop residues as a mulch,[69] and crop yields have indicated the importance of maintaining the level of soil organic matter as high as possible. There are probably several reasons for the beneficial effect, including increased water permeability and lower erosion losses, better nutrient supplies resulting from mineralization of organic matter, and its ability to hold cations. The latter is particularly

67 Lal, R., *IITA Monog.*, No. 1, 1975, Ibadan, Nigeria.
68 Mueller-Harvey, I *et al.*, *J. Soil Sci.*, 1985, **36**, 585. For similar values see Lathwell, D. J. and Bouldin, D. R., *Trop. Agric. Trin.*, 1981, **58**, 341.
69 Juo, A. S. R. and Lal, R., *Pl. Soil*, 1977, **47**, 567.

important because the components of clay fractions commonly have very low cation exchange capacities.[70]

Once the organic matter built up under the fallow has been mineralized, the natural supply of nutrients needs to be supplemented by the use of fertilizers, deficiencies of macro- and micronutrients being widely reported.[71] Leaching of nitrate is more serious than in temperate countries because crops are grown when there is excess rainfall and the use of split applications of nitrogen fertilizer seems the only way at present to ameliorate the problem. Some soils are also acid, with high contents of exchangeable aluminium, low pH values and little exchangeable calcium.

To summarize: great caution is needed when agricultural development is planned in the humid tropics. Cropping systems suitable for more intensive agriculture are being developed, with more emphasis given to agroforestry, and to systems based on root crops in the uplands and rice-based agriculture in the wetlands.[72]

The cool subtropics: northwest Syria

The cool subtropics, otherwise known as Mediterranean-type environments, are characterized by cool or cold wet winters and hot arid summers. Such a climate predominates in countries bordering the Mediterranean Sea (north Africa, southern Europe and west Asia) and is typified by that found in northwest Syria. South Australia, parts of South Africa and the Pacific North West of the USA also have a similar environment.

In such regions rainfall is the dominant factor influencing crop production and seasonal rainfall totals are low with highly erratic distribution both between and within years. Typically, the long-term seasonal average rainfall near coastal regions is relatively high (600 mm) but decreases rapidly inland to levels at which crop production is marginal (200 mm) or where natural steppe land or desert predominates (<150 mm). Analysis of rainfall data has shown that, as might be expected, there is increasing risk of drought and crop failure as mean rainfall totals decrease, and this is reflected in the choice of crops and farming systems currently practised in these contrasting locations.

Temperature also plays an important role in determining the choice of crops. In the winter months (December, January, February) temperatures are low and frost commonly occurs. Between October and March potential evapotranspiration is low ($1-2$ mm day^{-1}), rainfall exceeds evapotranspi-

70 Juo, A. S. R. in *Characterization of Soils* (ed. D. J. Greenland), p. 51, Oxford Univ. Press, 1981.

71 Kang, B. T. and Fox, R. L. in *Characterization of Soils* (ed D. J. Greenland), p. 202, Oxford Univ. Press, 1981; Kang, B. T. and Osiname, O. A., *Fertil. Res.*, 1985, **7**, 131.

72 Greenland, D. J., *Science*, 1975, **190**, 841; Juo, A. S. R., *Entwicklung Landlicher Raum*, 1985, **2**, 14.

ration, and excess rainfall is stored in the soil profile. In spring (March and April) temperatures rise rapidly, evapotranspiration exceeds rainfall, and depletion of soil moisture reserves occurs which continues until crop maturity (May/June) where average temperatures reach 25–30 °C and potential evapotranspiration can rise as high as 15–20 mm day^{-1}.

The soils are variable, but calcareous heavy-textured soils are widespread and dominant.[73] The clay fraction is largely montmorillonite and severe swelling and shrinking occur in wetting/drying cycles.

In wetter areas deep cracking is characteristic of both Vertisols and Vertic Luvisols. During soil drying, self-mulching is observed with the formation of a robust subangular blocky structure. Rainfall infiltration is good and run-off is seldom observed. Due to their high moisture-holding capacity, these soils are seldom recharged with moisture below 150 cm, and nutrient leaching out of the rooting zone is not a problem. Soil depth varies greatly with topography, with deep soils (>2 m) forming in valley bottoms, but shallower soils (0.3–0.5 m) on slopes and hilltops. The amount of moisture which can be stored during the winter months (which is directly related to soil depth) influences the effective length of growing season and thus the choice of crops.

At drier locations Calcic Xerosols dominate, with a smaller clay content and greater fine sand and silt fraction. Soil capping, which is common, provides a barrier to rapid infiltration, and run-off is not uncommon during

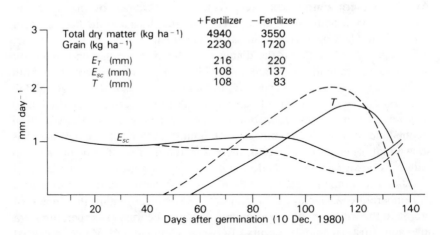

	+ Fertilizer	− Fertilizer
Total dry matter (kg ha^{-1})	4940	3550
Grain (kg ha^{-1})	2230	1720
E_T (mm)	216	220
E_{sc} (mm)	108	137
T (mm)	108	83

Fig. 13.4   Season changes in rates of evaporation of water from soil (E_{sc}), and barley crop (T) with (---) and without (——) fertilizer from experiments in northwest Syria; inset shows components of crop yield and water use (E_T is total evapotranspiration).

73 Harmsen, K. *et al.*, *Proc. 17th Coll. Int. Potash Inst.*, Berne, 1983, 223.

early winter and late spring when high intensity rainstorms (60 to 90 mm h^{-1}) are more frequent. Such soil capping can also result in poor seedling emergence. In these dry locations, soil moisture recharge seldom occurs below 60–75 cm depth.

In all soils, secondary deposition of calcium carbonate is evident within the soil profile at the average depth of profile recharge. In some profiles these deposits are substantial and gypsiferous, causing an effective barrier to root penetration. Organic matter levels are low and both nitrogen and phosphorus deficiencies are widespread. In general terms, nitrogen responses are greatest in the wetter areas and phosphate responses are greatest in the drier areas, but in many instances the greatest crop responses are observed when both nutrients are applied together.[74] Substantial and economic increases in crop production can be achieved through application of fertilizer containing nitrogen and phosphate without a corresponding increase in crop transpiration. Increased leaf area production results in greater and more rapid soil shading which reduces soil evaporation and correspondingly increases crop transpiration (see Fig. 13.4).[75]

Cropping systems

In Mediterranean environments, cropping systems are strongly influenced by the length of the growing season. In both the wetter and drier areas, the start of the rainy season occurs at approximately the same time. During the latter part of the growing season, crops largely rely upon moisture which has been stored in the soil profile during the winter months, and thus the length of the growing season is determined by the amount of seasonal rainfall.[76]

In the wettest areas, winter-sown bread wheat is grown in rotation with spring planted chickpea (a food legume). Summer crops (water melon, cotton, sesame) which are planted in late spring grow on stored moisture following a short winter fallow. Livestock (sheep and cattle) are an integral part of the farming system, and forages (barley and vetch) are often grown on the shallower soils as a source of feed. All crop residues are grazed by animals, and thus there is little direct return of organic matter to the soil.

In the slightly drier areas, durum wheat predominates due to its shorter growing season, and winter planted lentil replaces chickpea. Summer crops are still grown, but in some years farmers will judge that insufficient water has been stored in the profile during the winter and will leave the land fallow.

In areas receiving less than 300 mm of rainfall, barley is grown as a forage (grain and straw) for sheep feed, and these systems are dominated by

74 ICARDA, *Annual Rept. for 1983*, Aleppo, Syria, 1984.
75 Cooper, P. J. M. *et al.*, *Field Crops Res.*, 1983, **7**, 299.
76 Kassam, A. H., *Pl. Soil*, 1981, **58**, 1.

barley/sheep production. Commonly, a two-year rotation of barley/fallow is practised, but following increased mechanization in land preparation and population growth (human and sheep) in the last 30 years, farmers are adopting the practice of barley mono-cropping with resultant decline in soil fertility and crop yields. Farmers estimate that their yields have halved in the last 20 years. In such dry areas, it has been shown that little moisture is stored in the fallow year, but available nitrogen accumulates in the soil profile. Use of fertilizer containing phosphate and nitrogen, and the introduction of forage legumes to replace the fallow (or to break the continuous barley cropping) have both been shown to give substantial increases in crop production.[77]

To summarize, crop production in Mediterranean-type environments occurs against a background of moisture limitation. Any changes in production practices which allow more efficient water use, such as timely sowing, adapted cultivars, correct fertilizer use and good weed control will result in substantial yield increases.[78] Recent survey work[79] has shown that in the wetter, wheat-based farming systems, farmers in Syria are indeed adopting such practices. For example, 90 per cent of farmers use nitrogen fertilizer, 75 per cent use phosphate fertilizer and 50 per cent use herbicide. But in the drier barley/livestock farming systems, the greater risk associated with crop production currently discourages farmers from investing in costly inputs.[80]

77 ICARDA, *Annual Rept. for 1984*, Aleppo, Syria, 1985.
78 Cooper, P. J. M., *Proc. 17th Coll. Int. Potash Inst.*, Berne, 1983, **63**.
79 ICARDA, *Annual Rept. for 1985*, Aleppo, Syria, 1986.
80 For further information on soil management in this region, see *Pl. Soil*, 1981, **58**, also published as *Soil Water and Nitrogen* (eds J. L. Monteith and C. Webb), Nijhoff, 1981, and Cooper, P. J. M. *et al.*, *Expl Agric.*, 1987, **23**, 113.

14

The microbial population of the soil

To a casual observer the most obvious properties of a soil are those which result from its physical constituents, the amounts of sand and clay and organic matter which are present. If a biological element is considered at all, it frequently does not extend beyond consideration of the level of earthworm activity or the frequency of root-devouring beetle larvae. The microscopic nature of most of the soil inhabitants is the primary reason for the ease with which the diversity, complexity and continual activity of the living organisms present in all soils can be overlooked. The small size of soil organisms is, however, compensated by their presence in very large numbers (10^7 to 10^9 bacteria per gram soil are common numbers, and see Ch. 15 for other micro-organisms) and the resulting levels of biological activity are essential to the functioning of the soil. Without the intermeshing vital processes of the bacteria, actinomycetes, fungi and algae (the microflora), the protozoa and nematode worms (the microfauna), and the collembola, mites and other associated meso-fauna, the soil would become a repository of dead plant remains with no facility for the recycling of vital nutrients such as carbon, nitrogen and phosphorus for plant growth. Plants can be grown in totally sterile soils but only when steps are taken to replenish nutrients as they become exhausted by root uptake. A naturally fertile soil is one in which the soil organisms are releasing inorganic nutrients from the organic reserves at a rate sufficient to sustain rapid plant growth. All organisms in soil demonstrate the fundamental biological attributes of feeding, growth, reproduction and death and it is the aggregate result of all these processes, the biological activity of soils, which provides higher plants with a suitable environment for their own development.

Soil micro-organisms demonstrate what is probably a unique diversity of nutrition and physiology and many of the agronomically important properties of soils can be related to this diversity.

Nutrition of the microflora: autotrophic and heterotrophic organisms

Organisms need food or nutrients for two distinct purposes: to supply energy for their physiological processes (catabolism) and to supply the necessary elements to build up their cell tissues (anabolism). Some of the microflora can use entirely different substrates (foods) for these two purposes, whereas for others and for most animals the same material serves both purposes.

Micro-organisms can be divided into four general classes in terms of their energy and carbon sources:

1 *Chemoheterotrophic* (or *chemo-organotrophic*) organisms, which require preformed organic nutrients such as glucose to serve as sources of energy and carbon,
2 *Photoheterotrophic* (or *photo-organotrophic*) organisms which use organic substrates but can obtain some of their energy from light,
3 *Photoautotrophic* (also known as *photosynthetic* or *photolithotrophic*) organisms which obtain all their energy from sunlight,
4 *Chemoautotrophic* (or *chemolithotrophic*) organisms which obtain energy from the oxidation of inorganic compounds.

Both types of autotrophic organism obtain the carbon they require for cell synthesis by the assimilation of atmospheric carbon dioxide. Fungi, most bacteria, protozoa and all animals are heterotrophs whereas algae, the higher plants and a few groups of bacteria are autotrophic. Chemoautotrophy is restricted to a few types of bacteria although the agronomic significance of their activities, such as the oxidation of ammonium to nitrate, is of considerable importance. Some autotrophs such as the ammonium oxidizers may be regarded as strict or obligate autotrophs and are quite incapable of any other form of nutrition, others are best regarded as facultative autotrophs and can grow by using organic substrates for part or all of their needs under certain circumstances. Many algae can be very successfully cultivated in the dark growing heterotrophically by using organic growth media, and certain chemoautotrophic bacteria will respond to additions of preformed vitamins to their growth environment.[1] The latter group are sometimes referred to as mixotrophic.

1 Rittenberg, S. C., *Adv. Microbiol. Physiol.*, 1969, **3**, 159.

Photoautotrophic organisms occur in soils in two distinct groups. In one group are the algae and the cyanobacteria (previously called the blue–green algae) which photosynthesize using chlorophylls related to those of higher plants and release oxygen from the photolysis of water under conditions of adequate oxygen supply. The other group contains certain photosynthetic bacteria which do not produce oxygen and depend on substrates such as hydrogen sulphide, thus restricting their activities to waterlogged conditions such as paddy soils. Many of the photoautotrophic bacteria can fix atmospheric nitrogen whereas the other photoautotrophs require ammonium or nitrate salts to satisfy their nitrogen requirements.

Heterotrophic organisms range in their nutritional requirements from the simple to the complex. The unspecialized organisms such as the Pseudomonad group of bacteria or fungi of the *Mucor* group can obtain all their respiratory energy and cell carbon from simple sugars such as glucose and their nitrogen from inorganic nitrogen salts – ammonium being often preferred to nitrate. These organisms can successfully produce all necessary enzymes and cell constituents from simple nutrients and therefore possess formidable powers of synthesis. It is then possible to ascend a scale of increasing complexity as more and more preformed nutrients such as amino acids and vitamins are needed to support growth. Since the underlying biochemistry of most organisms is very similar, the fact that a particular bacterium or fungus can grow without a preformed supply of, say, the vitamin biotin, should not be interpreted as meaning that biotin does not feature in its metabolism, but that the organism can synthesize it for itself.

The loss of power of synthesizing critical metabolic substances is scattered over many groups of the microflora in such an erratic manner that it has no systematic significance. On the whole, the fastidiousness of the organism is a reflection of the usual food supply: the more specialized the supply, as, for example, in a well-adapted parasite or symbiont, the more likely the organism is to have lost the power of synthesizing some of the enzymes regularly produced by related organisms utilizing a less specialized food supply. Even among free living soil organisms there is a surprisingly high incidence of organisms which require preformed vitamins for growth. Rouatt[2] reported that 54 per cent of the bacteria investigated needed one or more vitamins for growth and showed that there were some 37 per cent of the bacteria which excreted vitamins under the conditions tested suggesting an intricate web of nutritional interdependencies within the soil.

All micro-organisms need minerals as well as sources of carbon, hydrogen, oxygen and nitrogen, though the relative amounts needed vary widely with different types. As with higher plants, potassium, phosphate and sulphate are needed in considerable quantities, typical $C : N : S$ or $C : N : P$ ratios of

2 Rouatt, J. W. in *The Ecology of Soil Bacteria* (eds T. R. G. Gray and D. Parkinson), Liverpool Univ. Press, 1967.

microbial tissue being about 100 : 10 : 1. Sodium, magnesium, calcium, potassium and iron are required in smaller amounts by most organisms and traces of manganese, copper, zinc, boron, molybdenum and cobalt are usually needed, although in laboratory culture they are often present in sufficient quantities as impurities in other media constituents. The function of most of the metals is to act as an enzyme activator or as part of the prosthetic group of an enzyme, iron and copper, for example, being involved with respiratory enzymes among others. Nitrogen fixing organisms such as *Azotobacter* and *Rhizobium* provide examples of organisms needing a trace metal only when certain enzymes are required, both being unable to fix atmospheric nitrogen in the absence of adequate molybdenum, although vanadium can substitute to some extent.

Requirements for energy: aerobic and anaerobic organisms

All organisms require energy for synthetic and other important cell processes. For heterotrophic organisms the energy is derived from the controlled enzymic oxidation of organic molecules – a process referred to as respiration in the presence of oxygen and fermentation in the absence of oxygen. Although energy generation is an essential life process, soil organisms show a range of adaptations which reflect the different environmental conditions which can occur in soils, particularly with respect to oxygen availability. Some organisms, including most soil fungi and actionomycetes and many soil bacteria, are strict or obligate aerobes and must have access to oxygen in order to respire. At the other extreme, obligate anaerobes, which in soils are mainly bacteria, will not grow unless oxygen is absent and may even be killed by oxygen. However, a number of soil bacteria appear to have the best of both worlds by being able to thrive under conditions of both oxygen availability and in its absence.

Three basic mechanisms are used for the liberation of energy from organic substrates and some organisms may have access to more than one depending on conditions. First, aerobic respiration, which is the process used by obligate aerobes and facultative anaerobes in the presence of oxygen, is a straightforward series of oxidation steps with electrons being transported by a chain of electron carriers such as cytochromes until oxygen is utilized as the terminal electron acceptor. The passage of electrons, usually in pairs, through the electron transport chain results in the production of energy-rich phosphate bonds as adenosine diphosphate (ADP) is converted to adenosine triphosphate (ATP) and the ATP can then be used in a wide variety of metabolic processes.

Second, although oxygen must be regarded as the most effective terminal electron acceptor, under conditions where it is not available some organisms are able to use alternative electron acceptors such as nitrate, sulphate or even carbon dioxide. The inorganic acceptors are reduced: nitrate is reduced to nitrogen or nitrous oxide, sulphate to sulphide, and carbon dioxide to methane. The initial steps in the aerobic and anaerobic respiration pathways are identical – the difference lies in the terminal electron acceptor used. Another difference is that the amount of ATP generated is dependent on the difference in redox potential between the organic substrate being respired and the electron acceptor. For this reason, anaerobic respiration using inorganic electron acceptors is less efficient in terms of energy yield to the organism than aerobic respiration using oxygen.

The third type of process, found in the facultative and obligate anaerobic bacteria and in yeasts, is known as fermentation and is distinguished from the previous mechanisms described in that no inorganic terminal electron acceptor is required, an organic substrate acting as both electron donor and acceptor. The usual situation is that the products of fermentation result in some materials being more oxidized than the original substrate and others are more reduced, for example, the carbon dioxide and ethanol which result from the fermentation of glucose by yeasts. Since part of the substrate needs to serve as an electron acceptor it is clear that it cannot be oxidized itself and therefore its potential energy yield to the organism is lost. Because of this diversion of possible energy source to other purposes, anaerobic growth based on fermentation is much less efficient in terms of energy yield than aerobic respiration. The microbial oxidation of glucose to carbon dioxide and water in the presence of oxygen yields about 2800 kJ mol^{-1} of glucose whereas the fermentation of glucose to ethanol and carbon dioxide yields a mere 75 kJ mol^{-1}.

The organisms composing the soil population

Soils contain a wide range of biological forms varying in size from the submicroscopic virus and bacteriophage particles, through bacteria and fungi, the small springtails and mites of the meso-fauna to large earthworms and, of course, the living roots of higher plants. The group known as the microorganisms are rather arbitrarily regarded as those organisms which are invisible as individuals to the naked eye and need the aid of an optical microscope for their visual examination. Within the group are found the bacteria, actinomycetes, fungi, algae and protozoa, and it is those organisms which will be discussed further in this chapter. The larger soil animals are dealt with in Chapter 16.

The types of soil bacteria

Bacteria are the most numerous organisms in soil, with viable populations estimated at up to 200 million individual cells per gram of soil. They are also the most diverse in terms of physiology and nutrition. The system used in Bergey's *Manual of Determinative Bacteriology* is generally accepted as the basis for the taxonomy and identification of the bacteria and actinomycetes. Unfortunately, the identification and classification of a bacterial isolate requires information based on its morphology, nutrition, physiology and in some cases its immunological reactions and even the bases in its constituent nucleic acid. As a result, the identification of all the isolates possible from even a single small sample of soil is a Herculean task. Despite the period during which soil microbiology has been practised it is almost certainly true to say that no soil has yet been completely characterized in terms of the genera and species of all the bacteria present. Although this may appear to be a glaring omission, there are a number of problems apart from the simple question of time and expertise that would be required to conduct such a characterization. First there is the question of whether all soil bacteria are amenable to isolation in the laboratory, for without isolation identification is impossible. Second there is the fact that soil populations are very variable both in space and in time. There is little point in expending a great deal of time, effort and resources on the detailed description of a population if a change in location of a few centimetres or a slightly different sampling date would yield a different population. Improved rapid methods for the identification of bacteria are constantly being developed, ranging from devices which permit up to twenty or so nutritional tests to be completed from a single inoculation[3], to very sophisticated methods of bacterial identification based on pyrolysis–mass spectroscopy,[4] but these methods are, as yet, confined to organisms of medical or industrial importance.

Given that total characterization is impracticable, most of the knowledge that has accumulated concerning the soil bacteria has come from surveys of a wide range of organisms using simplified means of categorization such as simple nutritional classes or has resulted from examination of many soils for a limited range of organisms, for example, anaerobes or cellulose decomposers. The nutritional classification approach is exemplified by that of Lochhead and Chase[5] who devised a series of seven media of increasing complexity into which a selection of bacteria, usually of the order of 100 or so per soil, were inoculated. Responses to the media were recorded and the organisms classified on the basis of the simplest medium in which good growth occurred. Hence it was possible to classify organisms as having simple

3 An example is the API system, manufactured by API Laboratory Products Ltd, Basingstoke, Hampshire, UK.
4 Weiten, G., *et al.*, *J. gen. Microbiol.*, 1981, **122**, 109.
5 Lochhead, A. G. and Chase, F. E., *Soil Sci.*, 1943, **55**, 185.

or complex nutritional requirements and the soil could be scored according to the preponderance of types. Although this approach had a following during the 1950s it did not reveal many significant differences between soils. It emerged as a general pattern that highly organic soils tended to have more organisms with complex requirements and that the soil associated with plant roots (rhizosphere soil) had an enhanced level of organisms with simple requirements, but further application seemed limited.

Other broad surveys have been based on morphological classifications, either of isolates or of the organisms as seen in soil preparations.[6] Typical groupings are shown below:

Morphological classes used in surveys

Small and large cocci	Rods: Gram negative and positive
Spore forming rods	Short rods / long rods
Pleomorphic bacteria	Branched rods
Actinomycetes	Thin flexible rods
Curved rods (vibrios)	

A problem arises in comparing results of the two methods in that it is known that the size and even the shape of some bacteria can differ in their natural habitat compared with artificial culture conditions. Furthermore, many soil bacteria are pleomorphic, that is, they can appear in a number of forms depending on conditions or age. Members of the *Arthrobacter* group, which are very common in soils, may change in morphology from Gram negative rods when young to Gram positive cocci when old. Morphological classification has little systematic significance and there is no correlation between morphology and nutritional classification, for members in any morphological group may occur in virtually any nutritional group. Attempts have been made to apply the approaches of numerical taxonomy to soil bacteria and actinomycetes.[7] The basis of the method is to specify qualitatively a selected, usually large, number of different characteristics of the organisms, and group together those organisms showing the maximum number of similarities. The results from this approach cannot be said to have been very striking with perhaps one exception, and that was to show how little similarity may exist between fresh soil isolates and 'type cultures' which have been resident in

6 Burges, A., *Micro-organisms in the Soil*, Hutchinson, Lond., 1958; Lochhead, A. G., *Canad. J. Res.*, 1940, **18**, 42.

7 For example see a group of papers in *The Soil Ecosystem*, The Systematics Assoc. Publ. 8, 1969, by Gray T. R. G., p. 73; Goodfellow, M., p. 83; Williams S. T. *et al.*, p. 107.

laboratory culture collections for many years and have become domesticated rather than wild.

Direct observation of the bacterial flora in soils shows that small cocci and short rods predominate in most soils (Plates 15.1 and 15.3). Clumps of cocci or coccoid rods can sometimes be seen embedded in slime and some of the organisms are surrounded by gummy capsules as in Plate 15.1a and c. In films removed from soil aggregate surfaces,[8] large homogeneous colonies of cells can often be observed although these natural colonies will be fragmented by most methods of soil preparation.

Where the morphology of isolated bacteria is examined, preferably when isolated on a nutrient-poor non-selective medium, there is a tendency for rod shaped bacteria to be most common. Table 14.1 shows typical results based on a variety of media and sampling dates for two soils in New Zealand.[9]

Table 14.1   *Morphology of bacteria from two New Zealand soils*

Morphology	Percentage of total isolates	
	Hastings soil	*Napier soil*
Non-sporing rods		
Gram negative	19–24	17–29
Gram positive	2–6	3–7
Spore forming rods	12–18	10–15
Pleomorphic bacteria	36–46	31–41
Actinomycetes	11–25	12–32

The largest proportion can be seen to be the pleomorphic bacteria which covers the range from the genus *Arthrobacter*, through the typical *Corynebacterium* and *Mycobacterium* to the actinomycete genus *Nocardia*. The Gram negative rods would include the *Pseudomonas*, *Flavobacterium* and *Achromobacter* genera together with the *Agrobacterium-Rhizobium* group. The Gram positive non-sporing rods will contain some of the members of the *Corynebacterium* and *Mycobacterium* group, while the spore formers are likely to be predominantly *Bacillus* species together with the anaerobic *Clostridium* group.

All the bacteria so far mentioned, and there are many more genera which occur in smaller numbers, are capable of a saprophytic existence and play their part in the general decomposition processes in the soil. A useful conceptual distinction was made by Winogradsky[10] in the early days of soil

8 Harris, P. J., *Soil Biol. Biochem.*, 1972, **4**, 105.
9 Loutit, M. and Loutit, J. S., *New Zealand J. agric. Res.*, 1966, **9**, 84.
10 Winogradsky, S., *Comp. Rend.*, 1924, **178**, 1236.

microbiology. Organisms which relied on humified organic matter and which had a rather low but steady level of activity he named the 'autochthonous population', and the portion of the population with an ability for rapid growth and exploitation of fresh substrates he named the 'zymogenous population'. He envisaged the autochthonous population as providing a level of background biological activity in the soil, with the organisms having complex nutritional requirements; the *Arthrobacter* would be typical members. The opportunist zymogenous population would be subject to more drastic fluctuations in numbers as exploitation was followed by quiescence and the *Bacillus* and *Pseudomonas* group would typify this category. The grouping proposed by Winogradsky is by no means sharply defined and it is possible that an organism could be regarded as a member of both categories depending on the soil and the circumstances.

There are other important groups of soil bacteria which do not appear on non-selective agar and our knowledge of these has developed from specialized culturing methods. Many of the autotrophic bacteria such as the ammonium and nitrite oxidizers, *Nitrosomonas* and *Nitrobacter* and the sulphur oxidizing bacteria *Thiobacillus* can only be revealed by the use of specialized media.

There are also bacteria in the soil that feed on some common species of soil bacteria and possibly on other microbial cells. They do this by the secretion of extracellular enzymes that dissolve, or lyse, the bacteria and then absorb the liquid cell contents. Some belong to a group of very small and motile organisms known as *Bdellovibrio* and others to the myxobacteria or slime bacteria typified by the genera *Myxococcus* and *Polyangium*.[11] The ecological significance of the bdellovibrios is as yet uncertain and they are usually only found in small numbers in soil. The myxobacteria have been isolated from arable and grassland soils quite regularly and Singh[12] reported counts varying from 2000 to 80 000 per gram in the single soil in which he tried to count them.

These predacious bacteria are not the only members of the myxobacteria present in soil, as many cellulose and chitin decomposers, such as the Cytophaga, belong to this group,[13] but it has not yet been conclusively shown that they can form the characteristic raised fruiting bodies of the predacious species.

An important group of nitrogen fixing photoautotrophic bacteria, the Cyanobacteria, occur in soils and these have long been regarded as algae (blue–green algae). Although they should now be considered alongside other bacteria they will for consistency be described with the true algae in a later section.

11 McCurdy, H. D., *Canad. J. Microbiol.*, 1969, **15**, 1453.
12 Singh, B. N., *J. gen. Microbiol.*, 1947, **1**, 1.
13 Stanier, R. Y., *Bact. Rev.*, 1942, **6**, 143; *idem, J. Bact.*, 1947, **53**, 297.

Actinomycetes in the soil

Despite the filamentous nature of most actinomycetes they have few other similarities to the fungi and they are now classified alongside the bacteria, their familiar name now being something of a misnomer. In fact many of the actinomycetes of the genus *Nocardia* are very difficult to distinguish from bacteria as their filaments fragment very readily into cells resembling rod shaped bacteria. The more typical filamentous actinomycetes belong largely to the *Streptomyces*, *Micromonospora* and *Thermoactinomyces* groups. They form very fine, often much branched, filaments with a diameter of approximately 1 μm. The filaments form into conidia by the tip producing one or more spores or else by a length of filament, often twisted into a coil, dividing up into a line of spores. In the soil these characteristic coils are only produced if the organism has free access to air, so they are characteristically borne in the surface pores of a moist soil which is shaded from direct sunlight.[14]

The soil actinomycetes are nutritionally a very adaptable group; they are without exception heterotrophic, and can use a wide range of carbon and nitrogen compounds such as polysaccharides, lipids, saturated hydrocarbons, phenols, protein and chitin. Chitin hydrolysis is a particularly frequent characteristic and good selective media for *Streptomyces* species can be based on chitin. They are not as a group notably dependent on growth promoting substances, though many may excrete a range of vitamins, growth substances and antibiotics into the surrounding medium. The overwhelming majority of soil actinomycetes are saprophytic, with a few species causing plant diseases such as potato scab (*Streptomyces scabies*) and the *Frankia* group, which can develop nitrogen fixing associations with non-leguminous plants, being the notable exceptions.

Soil actinomycetes are typically aerobic organisms and are uncommon in soils that are frequently waterlogged. They are more common in warm than cool soils and are largely intolerant of acidity. Few *Streptomyces* species are active below pH 5 and the actinomycete contribution to the soil population will be low in acid soils. The most significant environmental adaptation of the *Streptomyces* group appears to be their ability to withstand desiccation. Although the organisms require moist conditions for growth, the conidia can withstand prolonged droughts and can outlast other micro-organisms to the point where they can dominate the soil population. Meiklejohn[15] recorded a dramatic rise in the proportion of actinomycetes isolated from a Kenyan soil following a severe drought and similar effects were seen in southern England following the very dry years of 1975 and 1976.

A curious property of certain species of *Streptomyces* is the production of a volatile metabolite with the odour of freshly turned moist soil. The

14 Erikson, D., *J. gen. Microbiol.*, 1947, **1**, 45.
15 Meiklejohn, J., *J. Soil Sci.*, 1957, **8**, 240.

compound has been characterized chemically and has been named 'geosmin' by Gerber.[16] It can be detected with ease when the appropriate organism is growing in a culture plate.

Soil fungi

The fungi form the second of the two major groups of soil micro-organisms, and whether they or the bacteria-actinomycete group predominate depends on local conditions, particularly pH and ambient moisture content. The mycelium of soil fungi cannot be seen with either the naked eye or with a hand lens in normal arable soils, although in woodland soils white mycelial strands called rhizomorphs, consisting of masses of individual filaments (hyphae) matted together, can often be seen beneath the uppermost layer of leaf litter. Slide preparations of soil suspended in a gel[17] or resin[18] will reveal hyphae and spores.

These methods of observing the soil fungi show that the hyphae are commonly 5 to 20 μm in diameter, with a few being as fine as 2 μm and some up to 20 or 30 μm. Most spores have diameters similar to the hyphae although certain chlamydospores, sclerotia and mycorrhizal spores may be over 100 μm in diameter.

The hyphae are often of irregular shape, possibly due to growing through constrictions in the soil. The hyphae may be seen to be subdivided by cross-walls (septa) and they may be colourless or pigmented. Where a soil preparation is stained with cotton blue or trypan blue, the hyphae may be deeply stained indicating the presence of protoplasm or may appear essentially empty. Fungal spores are commonly seen, but the sporing structures are frequently damaged during slide preparation. In many species spores only form at irregular intervals when all the environmental conditions are favourable. An extreme example is shown by many Basidiomycete or mushroom fungi which can be active in soils yet only form their large fruiting bodies at intervals of several years.

A contribution to our knowledge of the condition of fungal hyphae in soil was made by Warcup[19] who isolated fungi by the careful removal of hyphal fragments and transfer to nutrient media. He found that the success rate in terms of subsequent growth ranged from 75 per cent for isolates from an autumn wheatfield after rains to only 3 to 15 per cent when the same soil was sampled in drier conditions. A neighbouring grassland soil gave a lower success rate of 1 to 25 per cent of attempted isolations resulting in growth. Unfortunately it is not possible to distinguish between hyphae that were

16 Gerber, N. N., *Tetrahedron Letters*, 1968, 2971.
17 Jones, P. C. T. and Mollison, J. E., *J. gen. Microbiol.*, 1948, **2**, 54.
18 Nicholas, D. P., *et al.*, *J. Soil Sci.*, 1965, **16**, 258.
19 Warcup, J. H., *Trans. Brit. Mycol. Soc.*, 1957, **40**, 237.

already dead on transfer and those which were unable to respond to the cultural conditions. Since it is at present impossible to culture the vesicular–arbuscular mycorrhizal fungi, the presence of hyphae from these fungi in quantities in the soil[20] would lead to many unsuccessful transfers. The evidence presented by Warcup suggested that the delicate, transparent (hyaline) hyphae were short-lived in moist soils and soon lost their cell contents if not growing actively, but that pigmented hyphae were longer lived.

The evidence available suggests that it is still difficult to obtain a quantitative estimation of the amounts of viable mycelium and fungal spores in soil, though techniques have been developed which separate spores from soil by a flotation technique based on their hydrophobic properties.[21]

Most soil fungi belong to groups that form hyphal filaments although some species of acellular slime moulds (myxomycetes) and unicellular yeasts are also found. The filamentous fungi belong to the Phycomycetes, which have branched unseptate mycelia; to the Deuteromycetes (Fungi Imperfecti) or, where the perfect or sexual spore stage is known, to the Ascomycetes, which have branched septate mycelia; and to the Basidiomycetes. The common soil Phycomycetes include members of the Saprolegniales including the genus *Pythium*, and of the Mucoraceae including the genera *Mucor*, *Rhizopus* and *Zygorrhynchus*. The mycorrhizal fungus *Glomus* is also regarded as a member of the Phycomycetes. The common soil Deuteromycetes include the genera *Trichoderma*, *Penicillium*, *Cephalosporium* and *Fusarium*. These are probably present in the form of spores in soils to which fresh organic material has not recently been added. The soil Basidiomycetes present problems of identification but species of *Rhizoctonia* are common and pasture and forest soils contain members of the Hymenomycetes or mushroom-like fungi together with the Gasteromycetes or puff-ball fungi. Descriptions of the most commonly occurring fungi in agricultural soils are given by Domsch and Gams.[22] Many of the fungi forming mycorrhizas on forest trees as well as the 'fairy ring' fungi in old pastures belong to the Basidiomycetes. In addition there are fungi which are known as zoophagous fungi which prey on small soil animals. Those which catch and feed on nematode worms (eelworms) belong mainly to the Hyphomycetes of which the genera *Arthrobotrys*, *Dactylaria* and *Dactyella* are the most common, whereas the forms which prey on protozoa belong mainly to the Phycomycetes.

Direct observation of fungi in soils shows that there are a number of patterns of mycelial growth which imply differences in feeding strategy:[23]

1   The *Penicillium* pattern, in which the hyphae colonize densely a piece of

20 Jackson, R. M., *New Zealand J. agric. Res.*, 1965, **8**, 865.
21 Williams, S. T. *et al.*, *Pl. Soil*, 1965, **22**, 167.
22 Domsch, K. H. and Gams, W., *Fungi in Agricultural Soils*, Longman, 1972.
23 Hepple, S. and Burges, A., *Nature, Lond.*, 1965, **177**, 1186.

substrate but do not grow out into the surrounding medium, spore formation occurring profusely on the substrate surface.

2 The *Mucor ramannianus* pattern, in which the hyphae colonize a substrate but also grow out into the surrounding medium forming chlamydospores well away from the original food source.

3 The Basidiomycete pattern in which the long-lived mycelium will grow and colonize a substrate and then grow to another substrate well separated from the first. Fruiting bodies are produced from rhizomorphs.

4 The *Zygorrhinchus* pattern in which the fungus seems to grow at random through the soil as dispersed hyphae and not be associated with any particular substrate. This growth habit may be made possible through the use of dissolved organic material in the soil solution.

5 The 'fairy ring' pattern in which the fungus migrates in a well-defined mycelial zone, but the hyphae do not appear to be associated with any particular substrate. The mycelia cause profound microbiological changes in the zone in which they are active, for at least two-thirds of the normal fungal flora appears to be killed and most of the fungi that are active in this zone are not commonly active away from it.[24] The 'fairy ring' fungi are Basidiomycetes of the family Agaricales and belong to the genera *Marasmius*, *Psalliota* and *Tricholoma*.

A number of different types of hyphae can be seen in soil preparations. Thus, Nicholas and his colleagues[25] recorded various distinct morphological types which included thin hyaline septate hyphae, broad aseptate hyaline and broad septate brown stained hyphae, and some sparsely septate purple black hyphae.

The mechanisms whereby fragments of hyphae and especially fungal spores are stimulated into active growth are complex and to a degree still obscure. For soil-borne plant pathogens with poor competitive abilities the germination of spores in response to plant root exudates would appear to have useful survival implications. However, for many fungal spores germination appears to be retarded when in contact with soil although the same spores germinate readily in laboratory media. The term 'soil fungistasis' was used by Dobbs and Hinson[26] to describe this effect and they regarded it as a widespread phenomenon. Explanations of fungistasis fall into two major categories: those which attribute poor spore germination to an absence of nutrients in the soil leading to nutrient loss by the spores,[27] and those which postulate the presence of fungal inhibitors. The inhibitor hypothesis was for some time disregarded for want of demonstrable inhibitors in fungistatic soils but the detailed examin-

24 Warcup, J. H., *Ann. Bot.*, 1951, **15**, 305.
25 Nicholas, D. P., *et al.*, *J. Soil Sci.*, 1965, **16**, 258.
26 Dobbs, C. G. and Hinson, W. H., *Nature, Lond.*, 1953, **172**, 197.
27 Lockwood, J. L., *Ann. Rev. Phytopath.*, 1964, **2**, 341.

ation of soil gases made possible by gas chromatography has revealed volatile inhibitors of microbial origin including ethylene.[28]

Despite a general agreement that bacteria and fungi are important members of the soil microflora, information regarding the relative contribution made by the two groups to soil biomass and soil activity is surprisingly difficult to obtain. The older work on fungal numbers was based on counting fungal colonies on plates by a similar method to that used for viable counts of bacteria. By making conditions in the plates unsuitable for bacteria by additions of streptomycin and by acidification, and by preventing any of the fungi from growing too quickly it is possible to obtain growth from hyphal fragments and spores although it is very difficult to know which form predominates. Numbers counted vary from a few thousands to over a million per gram of soil. Estimates of biomass have suggested that fungi may contribute as much as 70 per cent[29] of the microbial tissue in soil although 50 per cent is a more common figure. A measure of microbial tissue or biomass is not, however, the same as a measure of activity since microbial cells may be alive but dormant. Attempts to distinguish bacterial activity from fungal activity have been attempted by the use of differential inhibitors such as streptomycin for bacteria and actidione for fungi and although initial results suggest a slight balance in favour of the fungi[30] the general picture is by no means clear.

The soil fungi are all heterotrophic and, as a eukaryotic group, are now regarded as being devoid of nitrogen fixing powers, needing a supply of nitrogen as either mineral nitrogen salts or organic nitrogen compounds. The species in soil have a wide variety of carbon substrate requirements varying from those fungi which can utilize simple carbohydrates, alcohols and organic acids, through those capable of the decomposition of polymers such as cellulose and lignin, to those which are obligate parasites of plants, or which have formed an obligate symbiotic relationship with plants as is the case with the mycorrhiza.

The ordinary soil saprophytic fungi can be very efficient converters of substrate into microbial tissue; some can assimilate 30 to 50 per cent of the carbon in the organic matter which they are decomposing into their cell substance, a much higher conversion rate than the 5 to 20 per cent commonly achieved by bacteria. This high efficiency of conversion means that rapidly growing fungi can make considerable demands on the available nitrogen in the soil, although this is mitigated to some degree by the fact that fungal tissue tends to have a higher carbon : nitrogen ratio than is found in bacterial tissue.

A nitrogen containing polymer which is characteristic of fungi (and also

28 Smith, A. M., *Nature, Lond.*, 1973, **246**, 311.
29 Anderson, J. P. E. and Domsch, K. H., *Soil Biol. Biochem.*, 1978, **10**, 207.
30 Anderson, J. P. E. and Domsch, K. H., *Canad. J. Microbiol.*, 1975, **21**, 314.

insect exoskeletons) is chitin which is composed of acetyl glucosamine units. The chitin is an important constituent of fungal cell walls and may be involved in a mechanism whereby microbial growth can result in the immobilization of plant nutrients for a considerable period of time. If a soil is analysed for acetyl glucosamine[31] it is found that the total amount present is far in excess of the amount that can be related to existing living microbial tissue and suggests that fungal cell walls can remain intact in soil long after the fungus has died.[32] Together with the actinomycetes, some fungi can synthesize polyphenolic compounds which resemble forms found in the soil humic fraction and may therefore be contributing to this residual and resistant fraction of the soil organic matter.

The filamentous fungi generally need aerobic conditions to flourish although they may not need aerobic conditions all along the length of their hyphae; they are capable of sending hyphae into poorly aerated pockets of soil, but only if much of the mycelium is growing in well-aerated conditions. Spore formation is particularly dependent on good aeration and sporangia and conidiophores are produced in the larger air-filled soil pores. There is some variation in the tolerance of fungi to restricted aeration and this may be as much a response to enhanced carbon dioxide levels as to lowered oxygen availability. Thus Burges and Fenton[33] found that *Penicillium nigricans*, a fungus usually restricted to the upper 50 mm of the soil, is less tolerant of a high carbon dioxide concentration than *Zygorrhinchus vuellemini*, a species which is usually more abundant below 100 mm. Typically, soil fungi are more common near the surface of the soil than lower down, and they are more common in light well-aerated soils than in heavier soils, as shown in Table 14.2.[34]

Fungi can tolerate a wide pH range, and although they flourish under acid soil conditions such as occur in heaths and forests, this is not so much because of a preference for such conditions but because reduced competition from bacteria and actinomycetes allows them to prevail. It should not, however, be assumed that all fungi are equally common in all soils, for example it has been shown for a group of grassland soils in Suffolk, England that most of the Phycomycete and Ascomycete species inhabiting the acid sandy soils were distinct from those occurring in nearby neutral and calcareous soils.[35] The dominant species in the acid soils belonged to the free sporing penicillia and it is possible that the high numbers recorded reflected the sporing characteristics of the organisms rather than their metabolic activity.

31 Parsons, J. W. and Tinsley, J. in *Soil Components*, Vol. 1 (ed J. E. Gieseking), p. 263, Springer-Verlag, 1975.
32 Jenkinson, D. S. and Ladd, J. N. in *Soil Biochemistry*, Vol. 5 (eds E. A. Paul and J. N. Ladd), Dekker, 1981.
33 Burges, A. and Fenton, E., *Trans. Brit. Mycol. Soc.*, 1953, **36**, 104.
34 Jensen, H. L., *Soil Sci.*, 1931, **31**, 123.
35 Warcup, J. H., *Trans. Brit. Mycol. Soc.*, 1951, **34**, 376.

Table 14.2   Effect of CaCO$_3$ on numbers of fungi, bacteria and actinomycetes as found by plating methods. Number per gram dry soil.

	pH	Fungi × 10³	Bacteria × 10⁶	Actinomycetes ×10⁶
Heath soil				
Untreated	3.7	610	0.84	0
+ CaCO$_3$	7.5	393	314	84
Sandy soil				
Untreated	4.7	341	2	3
+ CaCO$_3$	7.6	365	15	8
Light loam				
Untreated	5.8	127	5.1	2.9
+ CaCO$_3$	7.6	120	13.6	3.4

An ecological classification of soil fungi has been developed by Garrett[36] based on the principal food supply used in the highly competitive arena of the soil. The true saprophytes range from the 'sugar fungi' which are restricted to relatively simple and easily decomposable plant constituents, through the cellulose decomposers to the lignin decomposers. These decomposers of dead organic matter are distinct from the predacious zoophagous fungi and the more obligate root inhabiting fungi which are further dealt with in Chapter 17.

The sugar fungi are typically Phycomycetes, and since there are a large number of different organisms in soil which can use the same simple compounds for food, they must be adapted to exploit these nutrients ahead of their competitors. Their primary source of nutrients are injured, moribund or recently dead plant tissues. They are widely distributed through the soil, often as spores, though sometimes as short lengths of hyphae: the spores germinate immediately a suitable food source is nearby, and the hyphae are characterized by rapid growth and so dominate the population and achieve success by rapid exploitation. It may be significant that a typical Phycomycete, *Mucor hiemalis*, has been found capable of the synthesis of ethylene, a gas implicated in fungistasis.[37]

The cellulose decomposing fungi, most of which belong to the Ascomycetes, Deuteromycetes or Basidiomycetes, come intermediate between the sugar and lignin fungi in their rate of growth. Many are widely distributed throughout the soil as spores and some of them are copious producers of

36 Garrett, S. D., *New Phytol.*, 1951, **50**, 149.
37 Lynch, J. M. and Harper, S. H. T., *Soil Biol. Biochem.*, 1980, **12**, 363.

antibiotics although only the *Penicillium* and *Cephalosporium* species have produced antibiotics of pharmaceutical significance. When woody tissues are decomposed the lignin will remain intact showing the original structure of the tissue and the colour of this lignin gives rise to the term 'brown rot fungi'.

The lignin fungi are generally species of the higher Basidiomycetes and are characterized by very slow growth rates. Slow growth is not an ecological disadvantage as there is little competition for their food supplies and few other organisms are known to possess significant lignoclastic powers. The lignin fungi decompose their substrate more easily if there is a reasonably high cellulose content still associated with the lignin which is not readily accessible to other organisms. Once established, the fungus seeks new sources of food by sending out rhizomorphs into the soil, apparently so that the point on the hyphal strand at which an attack on a new piece of lignified tissue is initiated can be well supplied with energy. When woody plant tissue is attacked by lignoclastic organisms both cellulose and lignin are destroyed and little of the original structure remains apart from a delicate membranous replica. The removal of pigmented material results in a colourless residue which gives the group the name of 'white rot fungi'.

This strategy of using one nutrient base to support an attack on another is also shown by the higher Basidiomycetes which form parasitic associations with tree roots; their rhizomorphs can usually gain entry to a root if they arise from a root which is already heavily infected.

Soil algae

The soil algae are photosynthetic micro-organisms which contain chlorophyll and in some cases other pigments. The term, algae, should now be reserved for organisms which have discrete rather than diffuse nuclei (eukaryotic) and in which the photosynthetic pigments are in specific organelles or chloroplasts. The most common eukaryotic algae in soils are shown below with alternative terminologies:[38]

Common name	Alternative names	
Green algae	Chlorophyceae	Chlorophycophyta
Diatoms	Bacillariaceae	Chrysophycophyta
Yellow–green algae	Xanthophyceae	Chrysophycophyta

38 Metting, B., *Bot. Rev.*, 1981, **47**, 195.

The position of the blue–green algae, known variously as the Cyanophyceae or Cyanochloronta, is now regarded as being systematically closer to the bacteria on account of the prokaryotic nature of the cells, with no discrete nuclear apparatus and with the photosynthetic pigments associated with the cell membrane. The fact that nitrogen fixation is found only in the blue–green group has made the need for their separation from the eukaryotic algae more desirable and they are now known as the cyanobacteria.

The soil algae typically comprise smaller and simpler species than the aquatic forms and tend to have simple morphologies. They occur either as simple unicellular organisms or simple filaments or colonies. Many of the soil algae have their walls covered with a thick layer of gummy material while the cell wall of most diatoms contains a rigid skeleton of silica which persists after the death of the organism.

The fundamental requirements for the algae as a whole are a supply of mineral nutrients together with light and adequate moisture. They are found in the surface of soils throughout the world although little can be said about regional distribution beyond a suggestion that the cyanobacteria are more common in tropical regions. The green algae are more tolerant of acid conditions whereas the diatoms and blue-greens are favoured by neutral to alkaline conditions and the latter group appear to tolerate some degree of salinity. Despite the need for adequate illumination for photosynthesis, algae are regularly recovered from depths of several centimetres in the soil and questions have been raised concerning their capability for a heterotrophic mode of life. Members of all the groups mentioned have been grown in the dark when supplied with organic substrates and should thus be regarded as facultative photoautotrophs. Suitable organic substrates have been found to be starch, sucrose, glucose, glycerol and citric acid.[39] The question of whether these organisms can actually compete for organic substrates and grow heterotrophically in soil is still unresolved. The occurrence of algae below the soil surface can be explained by water movement, soil cultivation and faunal activity and, at present, a significant heterotrophy seems unlikely. In the rhizosphere, where organic nutrients are more abundant than in the soil as a whole, the algae are one of the few groups to decrease in numbers.[40] As many of the soil algae are either motile themselves, for example the diatoms, or have motile zoospores, they can presumably migrate from the subsurface locations back into the surface layers where illumination will be greater.[41]

Numbers of algae can be determined by dilution of soils and the incubation of ten-fold dilution series in mineral salts media in the presence of light. Alternative procedures include the direct counting of algae in soil preparations using ultraviolet light which causes the algal chlorophyll to autoflu-

39 van Baalen, C. and Pulich, W. M., *CRC Crit. Revs Microbiol.*, 1973, **2**, 229.
40 Rouatt, J. W., *et al.*, *Soil Sci. Soc. Am.Proc.*, 1960, **24**, 271.
41 Burges, A. and Raw, F., *Soil Biology*, Acad. Press, 1967.

oresce. A rough estimate of algal abundance can be obtained by extracting soils for chlorophyll with organic solvents and estimating algal abundance from the chlorophyll obtained. Because the algae are concentrated at the soil surface, the significance of numbers of algae per gram of soil is questionable as the result can be influenced by the depth to which the soil is sampled. A better method is to refer to algal counts on an area basis. In general, surface soil samples give counts between 100 and 100 000 per gram but most counts are below 20 000 unless an obvious bloom of algae has occurred, in which case populations exceeding a million cells per gram of surface soil may be encountered.

Algae develop most readily in damp soils exposed to the sun, hence they usually have their maximum development in spring and autumn when the soil is damp, the sunlight is not too fierce and the vegetation is sparse. Excessive sunlight in the tropics with accompanying high temperatures can be deleterious. The location of algae at the soil surface makes them very susceptible to moisture fluctuations and the less motile forms need to be able to withstand desiccation, some species having been recovered from dry soils after storage periods of 10 years.[42] This ability to survive drying, coupled with a rapid response to added moisture means that algae can make a significant contribution to microbial activity, even in the extremes of the desert environment.

The algal population in the soil cannot be regarded as so universally necessary to the functioning of the soil as the bacteria or fungi. In many soils the algae make no significant impact and would scarcely be missed. The importance of the group is to be found in certain specific situations and in these they may have a unique role. In pioneer situations, where raw mineral materials are exposed, where drastic erosion has occurred or where severe burning has taken place the photosynthetic ability of algae permits their growth as primary colonizers, retaining inorganic nutrients that might otherwise be leached away, and bringing to the matrix the initial input of organic material necessary for the transformation into a soil to begin. All algae can contribute organic carbon and the cyanobacteria can also acquire nitrogen from the atmosphere. In the most severe conditions where unprotected algae might be unable to survive, the association with fungi into the lichen symbiosis allows initial colonization, the photosynthetic partner providing carbon substrates while the fungal partner provides adhesion, protection and enhanced inorganic nutrient release from the substratum.[43]

The surface development of algal films can have an influence on soil stability and the reduction of soil loss by erosion. The filamentous nature of many algae, together with the production of polysaccharide gums, aids in the entrapment and cementing of soil particles. The development of surface

42 Trainor, F. R., *Phycologia*, 1970, **9**, 111.
43 Williams, M. E. and Rudolph, E. D., *Mycologia*, 1974, **66**, 648.

blooms of algae during periods of rainfall can later contribute to the stability of the surface layer during periods when the soil may otherwise be very susceptible to wind erosion. The most crucial contribution to soil fertility and crop production must, however, remain the nitrogen fixation by the cyano-bacteria. This may occur in temperate conditions but is at its most important in flooded rice cultivation when the biological fixation can be a central part of the nitrogen economy of the agricultural system. The cyanobacteria may function either as independent organisms or they may contribute to the nitrogen needs of the crop through forming symbiotic associations with the tiny water fern *Azolla*. The nitrogen fixing contribution of this group of prokaryotic algae is dealt with in Chapter 19.

Soil protozoa

The soil protozoa are mostly amoebae (Sarcodina) and flagellates, with the ciliate forms, so common in waters, forming a much smaller fraction of the population. The amoebae include the naked amoebae, of which the genera *Naegleria* and *Hartmanella* are typical representatives, and the testaceous amoebae, or thecamoebae, which are similar but have a hard shell protecting parts of their fragile bodies, representative genera of these forms being *Difflugia* and *Euglypha*. The soil flagellates have one or more flagella to allow motility and are usually quite small organisms, the common soil genera including *Bodo*, *Cercomonas* and *Heteromita*. The soil ciliates, which have short bristles or cilia covering the body (sometimes these may be restricted to one side of the cell) are represented by genera such as *Colpoda*, *Leptopharynx* and *Oxytricha*. Average body volumes for the amoebae, ciliates and flagellates are reported to be 3000 μm³, 400 μm³ and 50 μm³ respectively and all types can feed on bacteria.[44] Provided that adequate supplies of bacteria are available all groups, except perhaps the testaceous amoebae, can double their populations in a few hours.

The soil protozoa are smaller than those indigenous to aquatic situations although in areas subjected to periodic flooding there may be a higher proportion of truly aquatic forms.[45] The aquatic forms of protozoa have a large volume of space in which to move and feed whereas the soil dwellers are restricted to moving and feeding in soil pores and then only in those which contain adequate water.[46] The need for water for activity is accompanied by an ability to survive periods of dryness and many protozoa are capable of forming resistant cysts, in which form desiccation can be withstood and followed by rapid excystment, or hatching, when tolerable condi-

44 Stout, J. D. and Heal, O. W. in *Soil Biology* (eds A. Burges and F. Raw), Acad. Press, 1967.
45 Stout, J. D., *Soil Biol. Biochem.*, 1984, **16**, 121.
46 Darbyshire, J. F., in *Soil Microbiology* (ed N. Walker), Butterworths, 1975.

tions return. Encystment can be regarded as a natural part of the life cycle of many soil amoebae and the number of active amoebae in a soil at any time is dependent on the rates of encystment and excystment.

Protozoa, or protozoa-like organisms, can feed in three ways: some possess chlorophyll and are photoautotrophic, these are confined to a few genera of phytoflagellates such as *Euglena*; some, mainly zooflagellates, can feed saprophytically, absorbing nutrients from solution in pure culture although what proportion do this in the competitive soil environment is unknown; most protozoa, however, feed by capturing, ingesting and then digesting solid particles such as bacteria, a form of nutrition known as phagotrophic.

Bacteria form the staple diet of the protozoan population, though small algae, fungal spores, yeasts, flagellates and small amoeba are sometimes ingested. Not all bacteria are taken with equal readiness; encapsulated bacteria, mycobacteria and bacterial endospores are often avoided as are most streptomycetes. It is likely that different protozoa have different feeding preferences and generalizations concerning what bacteria are edible by all protozoa are probably premature. It is a reasonable assumption that protozoan feeding will affect not only the size of the soil bacterial population but also the balance of its components. Since 1977 there has been increasing interest in mycophagous protozoa,[47] particularly since it was found that certain giant amoebae of the family Vampyrellidae were capable of attacking the spores and hyphae of the take-all fungus *Gaeumannomyces graminis* var. *tritici* as well as such species as *Penicillium cinnamonii* and *Cochliobolus sativus*.[48]

The difficulties of measurement of protozoan populations in soils has always been an obstacle to the investigation of this group, as the predatory mode of nutrition requires the provision of suitable organisms for their growth. The direct counting of protozoa in soils is not a feasible method except for the testaceous amoeba.[49] The classical counting technique devised by Singh[50] in which serial soil dilutions are presented with edible bacteria is still the basis of current approaches although the use of modern containers and micro-dilution apparatus[51] has meant that samples can be handled more rapidly. The population estimates are based on the most probable number technique and considerable replication is required for even a modestly precise value.

The effect of protozoa in the soil was originally thought to be entirely harmful to bacterial life, it being assumed that protozoan feeding reduced the numbers of beneficial bacteria. It is possible to show that, in simple systems, protozoa will indeed reduce bacterial numbers simply by feeding on them,

47 Old, K. M., *Trans. Brit. Mycol. Soc.*, 1977, **68**, 277.
48 Chakraborty, S. and Old, K. M., *Soil Biol. Biochem.*, 1982, **14**, 247.
49 Lousier, J. D. and Parkinson, D., *Soil Biol. Biochem.*, 1981, **13**, 209.
50 Singh, B. N., *Ann. appl. Biol.*, 1946, **33**, 112.
51 Darbyshire, J. F. *et al.*, *Rev. Ecol. Biol. Sol.*, 1974, **11**, 465.

although when the complexity of the system is increased the simple relationship is soon lost. Some early work with soils showed inverse relationships
between bacterial and protozoan numbers[52] whereas later work was less
conclusive.[53]

In the last ten years interest in microbial biomass in the soil and particularly in the cycling of nutrients through the soil biomass has brought a revival
in the attention paid to the protozoa.[54] As the most numerous and widely
distributed predators in the soil, the potential significance of the protozoa in
stimulating microbial turnover is clear. Recent studies in the field and in pot
cultures have shown that the relation between protozoan numbers and
bacterial populations can be related to the biomass and feeding efficiency of
the naked amoeba in particular.[55] Of greater agricultural significance are the
reports of a positive protozoan influence on rates of nutrient mineralization[56]
and nitrogen uptake by plants[57] in soil microcosm systems, in some instances
by as much as 75 per cent.[58]

Although the biomass represented by the protozoa is small compared with
the bacteria and fungi, and is estimated as only one-hundredth that of the
commoner saprophytes, the unique role of these predators may mean that
they have a catalytic influence on the activities of the other groups by maintaining a physiologically younger population and by the rapid cycling of the
nutrients contained within their prey.

Amoeboid and flagellate stages of other organisms

A number of organisms which look like amoebae and flagellates may not be
true protozoa at all but stages in the life cycle of organisms of other groups,
especially the fungi. The Acrasiomycetes or cellular slime moulds are a group
which are variously regarded as protozoa or fungi and are rarely found other
than in the soil. The vegetative form is an amoeba which moves and feeds
like a true protozoan. At intervals the individual amoebae are attracted
together by the influence of a chemotactic agent, acrasin, form a multinu-
cleate mass or pseudoplasmodium (also known as a 'grex' or 'slug') and
finally produce sporing bodies elevated above the pseudoplasmodium. These
organisms are less numerous in arable and grassland soils than in forest soils[59]
and are typical inhabitants of the litter layer. Typical genera are *Acrasis* and

52 Cutler, D. W. *et al.*, *Phil. Trans. R. Soc. Lond. B*, 1922, **211**, 317.
53 Singh, B. N. and Crump, L. M., *J. gen. Microbiol.*, 1953, **8**, 421.
54 Stout, J. D., *Am. Zool.*, 1973, **13**, 193.
55 Clarholm, M., *Soil Biol. Biochem.*, 1985, **17**, 181.
56 Elliot, E. T., *et al.*, *Int. J. Environ. Stud.*, 1979, **13**, 169.
57 Elliot, E. T. *et al.*, in *The Root Soil Interface* (eds J. D. Harley and R. Scott Russell),
   Acad. Press, 1979.
58 Clarholm, M., *Swedish Univ. Agric. Sciences Dept. Microbiol. Rep.*, 17, Uppsala, 1983.
59 Cavender, J. C., *Canad. J. Bot.*, 1972, **50**, 1497.

Dictyostelium. Other fungi may, as part of their life cycle, produce motile spores (zoospores) which are equipped with flagella. In most respects these resemble flagellate protozoa, and since some notable plant pathogenic fungi such as species of *Phytophthora* and *Pythium* produce zoospores capable of migration through the soil towards plant roots, they have been investigated to determine the specificity of their motile response, although it appears to be non-selective.[60]

60 Hickman, C. J. and Ho, H. H., *Ann. Rev. Phytopath.*, 1966, **4**, 195.

15

Ecology of the soil population

The microbial population

In general terms ecology is the study of organisms in relation to their environment. For the soil, although advances have been made in understanding the
physical and chemical nature of the environment there are still formidable
obstacles to be overcome in relating biological information to that available
from other disciplines. Ideally, a detailed study of soil microbiology would
provide information on what organisms are present; in what numbers; where
the organisms are located; what they are doing (if anything) and how they
are interacting with other biological forms, both large and small, in the soil.
    Unfortunately, the inherent nature of soil, particularly its heterogeneity
and opacity, coupled with the limitations of the methods available, means that
it is usually possible to study only one or sometimes two of the above
biological parameters of a soil sample at a time. The destructive nature of
many of the methods ensures that the same sample can only be subjected to
a limited examination. Generally, the study of the soil in biological terms has

been restricted to certain approaches or aspects related to the particular problem in hand and using a limited methodology. It is not an exaggeration to suggest that the complete biological characterization of a soil, as described in the first paragraph, has yet to be accomplished for a single soil sample. A brief survey of the most common methods available is given in the following sections. The excellent review by Burns and Slater[1] should be consulted for detailed information.

Identification of soil micro-organisms

The identification of a soil micro-organism *in situ* is limited to the observation of certain algal forms such as diatoms and the occasional and fortuitous identification of characteristic fungal sporing structures. For most purposes, the identification of an organism requires its removal from the soil and hence its separation from its natural habitat, a process known as isolation. This separation may be purely mechanical, as in the removal of vesicular–arbuscular mycorrhizal fungi in the form of spores from sieved soil fractions, or it may be a combination of a physical disruption of the soil followed by the cultivation of isolated propagules. The use of micromanipulation to extract fragments of fungal hyphae is mentioned on p. 459 but most isolation follows a more drastic soil dispersion. If an organism is suspected of being present in considerable numbers and is not fastidious in its growth requirements it may be sufficient to spread a small sample of soil on the surface of a nonselective nutrient medium to ensure growth. If the organism is present in small numbers or has specific growth requirements, selective culture methods may be adopted whereby a nutrient regime is designed to favour the desired organism and inhibit, or at least not encourage, the growth of other microbes. Once isolation in pure culture has been achieved, identification is based on morphological characteristics for fungi, algae and protozoa, and on physiological and biochemical characteristics for yeasts, bacteria and actinomycetes.

An exception to the need for isolation before identification is through the use of antibodies labelled with a fluorescent dye such as rhodamine B or fluorescein isothiocyanate (FITC). The antibodies are raised by animal inoculation, removed from the blood serum, conjugated with the fluorochrome and then used to stain specifically the desired organisms. This can be carried out directly on soil particle surfaces or in soil smears. It has been used with some success for *Bacillus*, *Rhizobium* and for actinomycetes,[2] and also for certain soil-borne plant pathogens although non-specific reactions can cause problems and must be guarded against.

1 Burns, R. G. and Slater, J. L., *Experimental Microbial Ecology*, Blackwell, 1982.
2 Bohlool, B. B. and Schmidt, E. L., *Adv. Microbial Ecol.*, 1980, **4**, 203.

Numbers of organisms in the soil

The methods available for enumerating soil organisms fall into two broad categories, those which rely on a count carried out by the use of a microscope, usually known as direct counting; and those which are based on growing the organisms, known as viable counts or indirect counting procedures.

Most direct counting procedures are based on the Jones and Mollison[3] method where a known quantity of soil suspended in agar is transferred to a microscope slide, stained and the number of organisms seen in a series of calibrated eyepiece fields is recorded. The original stain employed was phenol aniline blue but various others have been suggested since, including fluorescent stains such as acridine orange, FITC[4] and a combination of a fluorescent brightener and a europium chelate.[5] The autofluorescence of chlorophyll can be exploited in the counting of algae by this method. The use of the transmission electron microscope (TEM) for quantitative estimates of soil organisms was suggested by Nikitin[6] but does not appear to have been developed further, perhaps due to fears of excessive sampling errors because of the very small samples used.

Apart from providing an estimate of numbers, the direct count can also give information on cell morphology and cell sizes, and alternative measurements can be made such as hyphal length of fungi.[7] These direct measurements are valuable in assessing total cell volumes or cell mass (biomass) in soils.

Viable counting techniques for bacteria and fungi are usually based on the dilution plate technique in which a known weight of soil is dispersed in a sterile diluent, and the suspension, after dilution, is either mixed with a suitable nutrient medium in a petri dish or is spread on a previously poured and solidified agar surface. After incubation the number of microbial colonies developed on the plate is counted and the number of organisms present in the original sample calculated. For fungi and actinomycetes the method is severely biased towards those organisms which spore freely since most colonies arise from spores rather than filament fragments. In the case of fungi the method is further discredited due to the widely varying size of the propagules that can give rise to colony formation. For bacteria the population estimate relies on the following assumptions:

1   That each colony on the plate developed from one bacterial cell.
2   That all the bacteria in the soil are brought into suspension.

3 Jones, P. T. C. and Mollison, J. E., *J. gen. Microbiol.*, 1948, **2**, 54.
4 Babuik, L. A. and Paul, E. A., *Canad. J. Microbiol.*, 1970, **16**, 57.
5 Johnen, B. G. and Drew, E. A., *Soil Biol. Biochem.*, 1978, **10**, 487.
6 Nikitin, D. I., *Bull. Ecol. Res. Comm. (Stockholm)*, 1973, **17**, 85.
7 Thomas, A. *et al.*, *Nature, Lond.*, 1965, **205**, 105.

3 That all the bacteria in the suspension can grow on the chosen medium under the conditions of incubation selected.

None of these assumptions is, in fact, justified. Many bacteria regularly exist as clusters of cells held together by gummy substances, in some instances firmly enough to withstand prolonged ultrasonic treatment, and so are counted as one organism instead of many. Most bacteria live on the surfaces of clay or organic matter particles,[8] and it may be very difficult to dislodge them without causing their death.[9] Further, when the soil is shaken in the dispersing solution, bacteria may be adsorbed by small soil crumbs settling through the suspension. Once on the plates, there are problems arising from antagonism between adjacent organisms, the development of spreading colonies which overwhelm other growths, general changes in the medium resulting from microbial metabolism which can affect the pH of the substrate, and finally, there can be no single nutrient medium on which all soil bacteria can develop. It is generally found that the highest microbial counts are obtained on media that are low in readily available nutrients, examples being soil extract media or 0.1 per cent tryptic soy extract agar. Even with such media many of the other problems remain and the method has been criticized for providing 'a mirage of quantification'.[10]

The dilution plate method may have a useful place where strictly comparative measures are being made, or where dilution plating is being used for specific physiological groups, or for specific indicator organisms with selective media. In absolute terms, however, the results from dilution plates are often 100 times and sometimes 1000 times smaller than those obtained from direct counts. It is probable that most of the difficulties with the dilution plate method arise from inadequate dispersion and separation of cells, as it was found [11] that the two methods gave results for a dune sand under forest which only differed by a factor of two or three. Where the isolation of bacteria prior to culturing is the main concern, the dilution plate is still of great value.

The other main counting technique used for specific groups of soil organisms, such as nitrifiers, algae or protozoa, is the dilution to extinction or 'most probable number' method. In this technique, tubes of media, chosen for their selectivity for the organism in question, are inoculated with increasingly dilute soil suspension. After incubation, which may be prolonged in some cases, the number of tubes showing evidence of the presence of the organism sought are recorded, and the number of organisms which would cause the appropriate pattern of results is read off from statistical tables. The problems of dispersion and separation still exist for this method and considerable replication is necessary to obtain a worthwhile estimate of numbers.

8 Alexander, F. E. S. and Jackson, R. M., *Nature, Lond.*, 1954, **174**, 750.
9 Stevenson, I. L., *Pl. Soil*, 1958, **10**, 1.
10 Schmidt, E. L., *Bull. Ecol. Res. Comm. (Stockholm)*, 1973, **17**, 453.
11 Gray, T. R. G. *et al.* in *The Ecology of Soil Bacteria* (eds T. R. G. Gray and D. Parkinson), Liverpool Univ. Press, 1967.

Whatever method is adopted for microbial enumeration there are two aspects of the organism which are not covered by estimates of numbers. One of these is the size of the organism and hence the quantity of metabolizing tissue it might represent. Although the organisms under consideration are all 'micro-organisms' this can easily obscure the fact that an amoeba contains at least 1000 times the protoplasm of an average bacterium. Comparisons between bacteria and fungi are difficult to make because of differences in size and morphology. A second aspect of numerical assessments is that they convey nothing about activity. Many soil organisms produce specific resting structures, and even those which do not are often capable of prolonged dormancy when metabolic activity is at a minimum necessary to sustain viability. These two facts have brought a realization that, even if enumeration techniques could be dramatically improved, they would be limited to showing what was in the soil but would tell little about what was happening. As a result there has been an increasing interest in measurements of total microbial cell tissue (biomass) and microbial activity.

Biomass determinations

The soil biomass can be defined[12] as the living part of the soil organic matter, excluding plant roots and soil animals larger than the largest amoeba (about $5 \times 10^3 \ \mu m^3$). It can be estimated directly by a combined assessment of numbers and sizes of organisms in soil and a calculation based on density and assumed chemical composition. Measurements of this kind are extremely tedious and not without contention but are the nearest to absolute measures at present available.[13] Interest has therefore tended to concentrate on indirect measures of biomass which are capable of good correlation with direct microscopic assessments.

One of the most prominent indirect methods of biomass estimation is the chloroform fumigation technique.[14] This relies on the premise that when a soil is sterilized with chloroform and then re-inoculated with a small amount of fresh soil, the killed microbial cells will act as a carbon substrate and during decomposition will release, as carbon dioxide, a reasonably constant proportion of their constituent carbon. By following the carbon release of a variety of added microbial cells, and by standardizing conditions, a factor can be found which allows the calculation of biomass carbon from measurements of respired carbon dioxide. The method is applicable to a range of

12 Jenkinson, D. S. and Ladd, J. N. in *Soil Biochemistry*, Vol. 5 (eds E. A. Paul and J. N. Ladd), p. 415, Dekker, 1981.
13 Jenkinson, D. S. *et al.*, *Soil Biol. Biochem.*, 1976, **8**, 189;   Paul, E. A. and Johnson, R. L., *Appl. environ. Microbiol.*, 1977, **34**, 263.
14 Jenkinson, D. S. and Powlson, D. S., *Soil Biol. Biochem.*, 1976, **8**, 209.

soils, with the exception of extremely acid solids, and does not require expensive equipment.

Biomass carbon (BC) is expressed as:

(fumigated soil CO_2 per 10 days – control soil CO_2 per 10 days)/0.41, where 0.41 is the proportion of biomass carbon assumed to be mineralized under the conditions of the experiment. An alternative equation which is used in the absence of control measurements is given as[15]:

$$BC = (0.673 \text{ (fumigated soil } CO_2 \text{ per 10 days)} - 3.53)/0.41$$

An alternative method is based on the assumption that micro-organisms contain a relatively constant proportion of adenosine-5-triphosphate (ATP) in their biomass under most conditions. Methods for the estimation of ATP based on the light emitting reaction of the luciferin–luciferase system of the firefly are extremely sensitive, the lower limit being about 10^{-14} g ATP. Providing the highly labile ATP can be liberated from the cells and successfully extracted from the soil the method has considerable attraction. Problems arise mainly from premature destruction of ATP and the highly absorptive characteristics of certain soils.[16]

A third method relies on the initial respiratory response of the existing biomass to a readily available carbon substrate (e.g. glucose).[17] It requires the addition of glucose to the soil at a level which permits the maximum respiration rate over the initial hour of incubation. The respiration rate should be correlated with the biomass value obtained by chloroform fumigation. It is essential to determine the appropriate glucose supplementation level for each soil being investigated as some soils show an anomalous glucose response.

Other attempts at measuring constituents of microbial cells and deducing biomass from the soil content have proved disappointing. Hexosamines (fungal cell wall polymers), muramic acid (a cell wall constituent of bacteria) and nucleic acids have all been tried, but at present the results are equivocal or, at best, of restricted application. The most promising recent development has been to extend the chloroform fumigation technique to investigate other biomass fractions such as phosphorus.[18]

An attraction of the chloroform fumigation methods is that they offer a possible approach for following carbon and other elemental fluxes through the biomass in the course of organic matter decomposition. This may lead to insights into the functioning of the soil as the 'engine' of plant productivity.

15 Lynch, J. M. and Panting, L. M., *Microbial Ecol.*, 1982, **7**, 229.
16 Jenkinson, D. S. and Ladd, J. N. in *Soil Biochemistry*, Vol. 5 (eds E. A. Paul and J. N. Ladd), p. 415, Dekker, 1981.
17 Anderson, J. P. E. and Domsch, K. H., *Soil Biol. Biochem.*, 1978, **10**. 215.
18 Brookes, P. C. *et al.*, *Soil Biol. Biochem.*, 1982, **14**, 319.

Spatial distribution of soil micro-organisms

Although ecology aims to relate the organism to its environment there is, as yet, little clear idea of the nature or extent of the habitat as it relates to soil micro-organisms. Methods are lacking for the detailed characterization of nutrient distribution, aeration and pH at a microscopic level, and information on the distribution of bacteria, actinomycetes, fungi and protozoa is equally limited. Soil can be regarded as a matrix of mixed organic and inorganic composition which is permeated with macro- and micropores. The soil micro-organisms must inhabit these pores, in most cases by developing on the pore surfaces, since we can assume that relatively few exist free in the soil solution. In spite of the large numbers of organisms in soil a considerable proportion of the pore surfaces remain unoccupied. Part of the reason for this is that many of the pores are so narrow that they are inaccessible to the organisms themselves. Thus, there are few bacteria smaller than 0.5 μm and most are about 1 μm in diameter, although investigations of the pore size distribution of soils shows that much of the total pore volume is made up of pores smaller than 1 μm in diameter. It is also possible that micro-organisms can only be really active if they are in pores somewhat larger than 1 μm so that, except for sandy soils or soils in very good tilth, most of the pore surface is unsuited for high microbial activity. Even when the larger pores are considered, there is scant evidence to describe microbial distribution within them, although, since large pores may be equated with soil crumb or aggregate surfaces, it is likely that these are important sites of both colonization and activity. Colonization of soil crumb surfaces can be demonstrated for nitrifying organisms,[19] and since plant roots tend to penetrate between rather than through soil crumbs it can be assumed that rhizosphere activity (see pp. 527–30) will also coincide with crumb surfaces.

Methods for the examination of the spatial distribution of micro-organisms are limited. The Rossi–Cholodney technique, whereby a clean microscope slide is buried in close contact with soil and subsequently removed and stained, allows for only limited interpretation. The glass surface is continuous, impermeable and likely to encourage water to condense. It can reveal the morphology of organisms which are growing on natural soil nutrients, and since it must be in contact with aggregate surfaces, may mimic to some extent the surface conditions. It is not a quantitative method and can have only little spatial relevance.

Other methods adopted to examine the surface colonization of soil particles or aggregates include the use of scanning electron microscopy (SEM)[20] of which examples are reproduced in Plate 15.1 and the surface film removal technique,[21] examples of which are shown in Plate 15.2. Using the SEM it

19 Quastel, J. H., *Proc. R. Soc. Lond.*, 1955, **B143**, 159.
20 Gray, T. R. G., *Science*, 1967, **155**, 1668.
21 Harris, P. J., *Soil Biol. Biochem.*, 1972, **4**, 105.

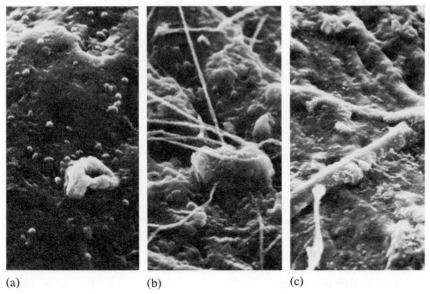

(a)   (b)   (c)

Plate 15.1 Scanning electron microscope photographs of: (a) bacteria,
(b) actinomycetes and (c) fungal hyphae, growing in a chitin
enriched sand dune soil under *Pinus nigra laricio*. (a) and
(b) × 2000, (c) × 1000. Bacterial cells are also visible in both
(b) and (c).

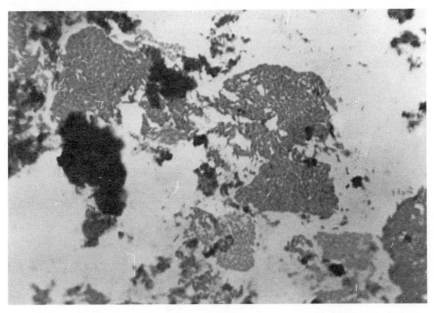

Plate 15.2 Bacterial colonies from removed surface film (× 1000).

is possible to see bacteria as well as fungal and actinomycete filaments, the bacteria often appearing in small groups or singly. Where surface films are removed from aggregates, although single cells are seen, the most striking feature is the occurrence of large colonies of bacteria, usually of single cell thickness which show a great degree of morphological uniformity within a single colony. When such large colonies are examined by SEM they can be difficult to resolve due to enveloping layers of microbial gum which obscures the surface outline. The existence of such colonies has been shown by transmission electron microscopy.[22]

The size of a microbial colony, especially a bacterial colony, may have ecological significance since the combined enzymic activity of many cells in close proximity might be expected to have far more biochemical and degradative significance than a single cell in isolation. The existence of large colonies may also be regarded as a reflection of the availability of nutrients, since the development of a colony of many hundreds of cells can only occur when the nutrient resources are available to permit that amount of growth. The concept of bacterial colony size in soil tends to have been based on results from techniques such as the Jones and Mollison soil slide, where more than half of the bacteria seen are in small colonies of three cells or less (see Plate 15.3). This may be an artefact due to the disruption of larger colonies during the soil preparation and may not reflect the situation in undisturbed soil. When colony size has been assessed in SEM preparations the colonies are usually less than 100 cells with the majority containing fewer than 10 cells.

Provided that microbial cells can be stained and that soil aggregates can be prepared without distortion of either organisms or pores, the use of thin sections of soil could yield valuable information on spatial distribution. The method relies on the impregnation of a dried soil with synthetic resins, followed by slicing and polishing to produce sections with thicknesses of 20 to 30 μm. Such techniques have been used for bacteria[23] and fungi.[24] Although observation is limited by the minimum thickness possible with the sectioning technique, it is possible to see fungal filaments and bacterial colonies if not individuals. It is to be regretted that few investigations of soils have been attempted by these methods although more recent work[25] has used TEM rather than optical microscopy and has revealed the positions of individual cells within soil pores and at the root–soil boundary.

Promising developments are being made in other ecological fields in the study of microbial colonization of interfaces[26] and it is to be hoped that these will cast light on the activities of soil micro-organisms which are very much in an environment consisting of interfaces.

22 Waid, J. S., *Bull. Ecol. Res. Comm. (Stockholm)*, 1973, **17**, 103.
23 Jones, D. and Griffiths, E., *Pl. Soil*, 1964, **20**, 232.
24 Williams, S. T. and Parkinson, D., *J. Soil Sci.*, 1964, **15**, 331.
25 Kilbertus, G., *Rev. d'Ecol. Biol. Sol.*, 1980, **17**, 543.
26 Marshall, K. C. in *Contemporary Microbial Ecology* (eds D. C. Ellwood *et al.*), Acad. Press, 1980.

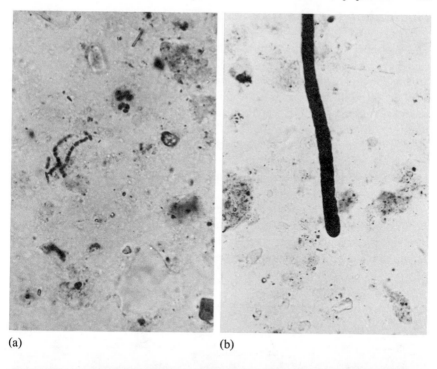

(a)   (b)

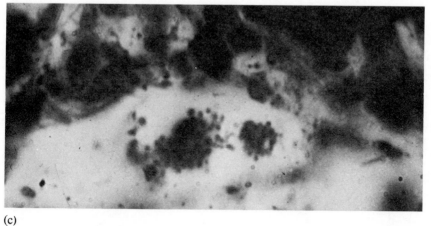

(c)

Plate 15.3   Micro-organisms in the soil:
   (a) Bacteria, a bacterial colony and actinomycetes (× 1000).
   (b) Fungal hypha, displayed by Jones and Mollison's technique (× 1000).
   (c) Bacteria in the pore of a red loam, thin soil section (× 2400).

Fluctuations in the numbers of soil bacteria

The numbers of bacteria that can be detected in soils are reported as changing dramatically when examined over intervals of 24 hours or even as little as 2 hours. This fact was known in the 1920s[27] and 1930s.[28] An example of short-term fluctuations is shown in Fig. 15.1. Attempts were made to relate changes to various environmental factors such as temperature and moisture, although such variations can occur when temperature and moisture are kept constant. Various hypotheses were advanced including the effects of grazing protozoa, but only in a few instances were relationships established. Sample heterogeneity and technical limitations were often invoked to explain the observed results.

More recently, in connection with biomass studies, similar fluctuations have been found in soils using improved methods and much greater replication.[29] The short-term fluctuations were demonstrated over 24 h intervals, and even larger changes of a seasonal nature were shown to occur when sampling was

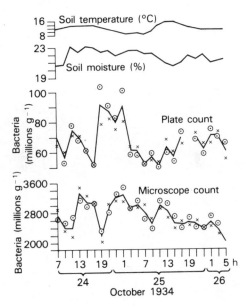

Fig. 15.1   Total cell and plate counts of bacteria from two-hourly samples of fallow garden plot soil. In the curves shown here, the dots and crosses represent numbers in duplicate samples; all curves show mean values.

27 Cutler, D. W. *et al.*, *Phil. Trans. R. Soc. Lond.*, 1922, **211B**, 317.
28 Taylor, C. B., *Proc. R. Soc. Lond.*, 1936, **B119**, 269.
29 Clarholm, M. and Rosswall, T., *Soil Biol. Biochem.*, 1980, **12**, 49.

made at monthly intervals. In the recent work, marked changes of cell size and morphology were shown to accompany the numerical changes and more emphasis was placed on the energy implications of such population fluctuations. In controlled pot and field experiments evidence has indicated that the naked amoebae may be the only predators capable of causing such bacterial fluctuations.[30]

Energy requirements of the soil population

The overall requirements of soil micro-organisms for energy, nutrients, water, suitable temperatures and the absence of harmful conditions are the same as those of plants. The great difference between them is the source of energy. Green plants derive their energy directly from sunlight, whereas the soil organisms, apart from the autotrophs, obtain theirs either directly or indirectly from the photosynthetic activity of the higher plants. The size of the soil population is thus partly controlled by the rate at which energy-containing material synthesized by plants is added to the soil.

The material synthesized by plant photosynthesis forms the initial step in a food chain of considerable complexity. It may be used while the plant is actively growing, as a result of parasitism or symbiosis, as material exuded or sloughed off from roots, or even in small quantities as canopy drip or leaf washings. As plant senescence approaches, greater amounts are introduced as dead or dying roots and upper organs of the plant.

The initial colonizers themselves provide sources of food to subsequent predators or to saprophytes when they die. The number, or activity, of the micro-organisms is partially controlled by the amount of energy that can be released by the decomposition of the organic matter and no matter how many stages or what organisms are involved in its degradation, only a finite amount of energy can be extracted. This amount cannot exceed the energy set free when the organic matter is completely oxidized and this is dictated by its chemical composition. In this sense the energy content of organic matter differs from the content of nutrients such as nitrogen or sulphur or phosphorus.

The nutrients can be used by a succession of organisms: an atom of nitrogen might in the course of a single day form part of a fungus, a bacterium which decomposed it, an amoeba which ate the bacterium, and another bacterium which decomposed the dead amoeba. By contrast, although energy is as indestructible as matter, once transformed into heat it cannot be used by micro-organisms or any other living thing; whatever energy is dissipated by one organism is out of reach of all others. It follows, therefore, that no factor affecting life in the soil which does not add to the stock

30 Clarholm, M., *Microbial Ecol.*, 1981, **7**, 343.

of energy substrates can increase the numbers of all heterotrophic microbial groups; if one group increases, others necessarily decrease.

Interest has centred on the flow of energy through the biomass, both with regard to the source of the energy and the division of its use within the population. Organisms are known to differ in their efficiency of energy use, and this use can be further divided into energy needed for synthesis and growth and that required for maintenance. The latter item in the soil energy budget, the maintenance energy, is the subject of controversy and a variety of values has been suggested.[31] It is significant since only after the maintenance requirements of the population are satisfied is there any energy left for growth. As predation is a constant fact of life for most soil organisms it is clear that growth must occur if organisms are to survive at all. At present the situation is far from satisfactory as there is little information for micro-organisms in soil on their synthetic efficiencies, maintenance energy requirements and the degree of variability between different groups.

It is possible that certain groups of soil micro-organisms, those with specialized resting stages such as bacterial endospores or protozoan cysts, have negligible maintenance needs. Studies using the soil bacterium *Arthrobacter* have demonstrated low maintenance requirements and it has been suggested that the organism is adapted to a carbon limited environment.[32] Where

Table 15.1   Approximate energy changes in Broadbalk soil in relation to micro-organisms.[33]

	Farmyard manure	No farmyard manure
	Energy equivalents, $(10^9 \, J \, ha^{-1} a^{-1})$	
Added in manure	15	0
Added in stubble	2	0.3
Total		
	17	0.3
Net soil change	0.5–1	0.5–1
Energy dissipated	16	1
	Micro-organism numbers, (g^{-1})	
Mean bacteria over 6 months ($\times 10^9$)	2.9	1.6
Fungal length over 6 months (m)	47	38
Mean protozoa over 6 months ($\times 10^3$)	72	17

31 Paul, E. A. and Voroney, R. P. in *Contemporary Microbial Ecology* (eds D. C. Elwood *et al.*), Acad. Press, 1980.
32 Chapman, S. J and Gray, T. R. G., *Soil Biol. Biochem.*, 1981, **13**, 11.
33 Data for protozoa from Singh, B. N., *J. gen. Microbiol.*, 1949, **3**, 204. Other observations from unpublished data by permission of P.C. T. Jones *et al.*

different amounts of organic matter are added to soil, as in Table 15.1, it appears that despite the considerable differences in energy flow through the manured and unmanured soils, the populations of bacteria and fungi are not so markedly different – the differences are much more notable in the protozoa which, being predators, represent a higher trophic level.

The question of trophic level is only now re-emerging as an item of interest in soil ecology but is logically of fundamental interest.To judge a soil by the size of its bacterial population alone is rather like assessing the fertility of a pasture by measuring the height of the grass while ignoring the number of cows in the field.

Results based on short-term studies when planted soils are compared with plant-free soils show the importance of the growing plant to population level.[34] The response of the higher (protozoan) trophic level is greater than the bacterial level especially when biomass values are substituted for numbers. Heterotrophic soil populations appear to be limited more often by a shortage of available energy yielding carbon substrates than by any other nutrient. Apart from in the rhizosphere, or following organic matter additions from dead plants, they are living up to their income in terms of energy supply and respond rapidly to energy supplements.

Activity of the soil population

The numbers of micro-organisms in a soil cannot be regarded as an index of the activity of the microbial population as many organisms may be dormant at the time of sampling. The activity of the soil population is not a concept that can easily be given a quantitative definition, but for many purposes it can be measured by the amounts of either CO_2 or heat evolved by the population, that is, by the rate at which either oxidizable carbon substrate, or the energy it contains, is being dissipated. A variety of respirometers of the Warburg, Gilson and oxygen replacement type have been employed in soil studies. The effects of particular substrates can be examined by measuring their effects on the overall respiration rate of the soil.

A technique for following the fate of specific materials added to soils in small quantities is offered by the use of radiorespirometry. In this method, the substrate, which may be a natural compound or a synthetic pesticide, is labelled with radioactive carbon,[14]C. If the labelled carbon is metabolized, a proportion will be released as labelled carbon dioxide and this can be trapped and measured by scintillation counting with great sensitivity. By careful soil analysis the fate of the labelled substrate in the soil can also be monitored. Heat output, as measured by microcalorimetry,[35] correlates quite

34 Clarholm, M., *Microbial Ecol.*, 1981, **7**, 343.
35 Sparling, G. P., *Soil Biol. Biochem.*, 1981, **13**, 93.

well with respiration measurements and can be regarded as an additional, though more complex, method of assessing microbial activity.

There are reports of a correlation between bacterial numbers and respiration rates for field soils[36] but in general the two parameters show little relation. The major exceptions to this lack of correlation are where readily available organic substrates are added to soils, or where soils are remoistened after drying, and in these cases the various indices of microbial activity tend to increase in unison for a time.

Although the release of CO_2 is usually linked to oxidative degradation of soil organic matter, an interesting complication is provided by the hyphae of mycorrhizal fungi. If these hyphae are a significant proportion of the soil biomass, as has been suggested,[37] they are respiring the products of photosynthesis from their hosts and not using soil organic matter at all.

It is possible to isolate many soil micro-organisms and study their metabolic processes in pure culture. However, the complex nature of the soil environment, including the physical, chemical and biological components, cannot be reproduced in such experiments and extrapolation to the soil must be cautious. Some biological transformations may be influenced by the adsorptive surfaces in soil, for example nitrification, or the degradation of the herbicide paraquat. Other processes, including ammonium and sulphide oxidation and in some circumstances manganese oxidation, may require the activities of more than one organism to proceed to completion. In these circumstances, the whole soil can be studied with the substrate added and products monitored but with less concern for the identity of the individual organisms involved in the various steps.

The perfusion technique, in which a column of soil is percolated slowly by solution drawn from a relatively large reservoir to which the percolate is returned, has proved useful in whole soil studies.[38] Two major advantages are provided by the perfusion technique. First, the substrate can be presented to the soil at low concentrations, but with sufficient supplies in the reservoir to remove the need for constant attention. Second, the bulk solution can be sampled at intervals without disturbing the soil sample. The major disadvantages are the limitations of some soils with poor structure for column percolation due to slumping and column blocking, and particularly with such soils, the difficulty of ensuring adequate aeration which may be essential for the process under examination.[39]

An additional argument for conducting experiments on whole soil samples rather than isolates is that many reactions in soil are reversible but rely on different organisms to drive the reaction in each direction. Soils normally

36 Gray, P. H. H. and Wallace, R. H., *Canad. J. Microbiol.*, 1957, **3**, 191.
37 Jackson, R. M., *New Zealand J. agric. Res.*, 1965, **8**, 865.
38 Quastel, J. H. and Scholefield, P. G., *Soil Sci.*, 1953, **75**, 279.
39 Greenwood, D. J. and Lees, H., *Pl. Soil*, 1960, **12**, 175.

contain populations capable of converting all the usual inorganic constituents to that state of oxidation or reduction which would be predicted if they were in equilibrium with the oxidation–reduction potential of the soil. Typical reversible oxidation–reduction processes are the nitrate–nitrite, sulphate–sulphide, manganic–manganous and ferric–ferrous ion systems.The organisms responsible for these transformations are usually bacteria, the oxidations being carried out by specialist autotrophs whereas the reductions are brought about by relatively unspecialized heterotrophs. The heterotrophs are often of the facultative anaerobe type and in some cases the reductions are indirect and caused by the production of organic metabolites which actually cause the chemical reduction. Oxidation–reduction reactions are discussed more fully in Chapter 26.

Effect of soil moisture content

Moisture content affects the activity of the soil population in a number of ways: through the thickness of the water films in the soil and consequently its aeration; through the reduction in the free energy of the water as the soil becomes drier and the films become thinner, so affecting water availability and the cell osmotic relationships; and finally through the effects that can be demonstrated when soil moisture fluctuates from the dry state to that of adequate moisture. Bacteria, protozoa and nematodes can move only in water films and for bacteria, although many are smaller than 1μm in diameter, they appear to be readily motile only in films appreciably thicker than this although they may well be capable of growth at lower moisture contents. Since the protozoa and nematodes are predatory, a moisture level which prohibits movement is more significant to them and potentials of -0.5 bar are likely to be sufficient to stop activity.[40] For all motile micro-organisms, the rate of movement is dependent on the moisture content of the soil and is thus dependent on soil structure. Movement becomes restricted as the soil becomes drier, partly because the water film pathway between two points in the soil becomes more tortuous[41] and partly due to organisms becoming trapped in moisture filled pores occluded by narrow necks. The filamentous fungi, and to a lesser extent the actinomycetes, differ from bacteria in that their hyphae need not grow in a continuous water film but can cross an open air space. Thus the reports that they can function in drier conditions, even as low as -100 bar $(-10$ kPa),[42] than bacteria[43] can be explained on morphological grounds.

Soil fungi and bacteria must take up their nutrients from the water films

40 Darbyshire, J. F. in *Soil Microbiology* (ed N. Walker), Butterworth, 1975.
41 Hamdi, Y. A., *Soil Biol. Biochem.*, 1971, **3**, 121.
42 Griffin, D. M., *The Ecology of Soil Fungi*, Chapman & Hall, 1972.
43 Dommergues, Y., *Ann. Agron.*, 1962, **13**, 265.

in which they are growing, but these films can be thinner than the dimensions of the organism, even if the organism must maintain a water film around its cell wall. The effect of the thickness of the film on the metabolic rate of the organism depends on the source of its food supply. If the nutrients are being derived from the decomposition of a solid, insoluble material, such as a fragment of dead root, the humidity of the soil air, which remains high until the soil is quite dry, will be sufficient to ensure enough water films for activity, at least until the water tension exceeds -15 bar. On the other hand, if the necessary nutrient is a soluble substance, such as ammonium or nitrate, the rate of diffusion of these ions to the microbial surface will become slower as the water films become thinner. Thus, one would expect to find that the production of ammonium from insoluble organic residues – the process of ammonification – should continue in a drying soil well beyond the point at which the oxidation of ammonium to nitrite ceases – a result that has been observed.[44]

A point that appears to have received scant attention is that microbial activity always involves the excretion of by-products that may be toxic, or at least inhibitory, to the growth of the organism, and as the soil becomes drier these will diffuse away from the organism with increasing difficulty. This may provide a further explanation of the decrease in metabolic activity which is noted when moisture content decreases, even before extreme conditions are reached.

Drying a soil and then rewetting it results in a flush of CO_2 production, a flush of ammonium and then of nitrate, providing that a proportion of the original population survives. The CO_2 flush may last for five to ten days. The more intensively the soil is dried, the longer it is kept dry and the higher the organic matter content of the soil, the larger is the magnitude of the flush.[45] Further, a soil can be dried and rewetted a number of times, and on each rewetting another flush occurs of only slightly less magnitude than the previous one, except that the first flush is usually appreciably greater than the subsequent ones. Freezing and thawing can also give a flush of decomposition. The causes of this rewetting phenomenon are thought to include the desorption of organic matter from clay surfaces, the rupturing of microaggregates to reveal previously protected organic substrates and the rapid decomposition of microbial cells killed by drying.

There are two practical considerations which follow from the rewetting phenomenon: one is that it may play a part in the timing of crop planting in tropical areas where rains follow prolonged dry seasons (see p. 659); the other relates to laboratory experiments where soil drying is a common pretreatment prior to experimentation. On rewetting such experimental soils, the microbial activity will not reflect the situation in the field, and the results

44 Robinson, J. B. D., *J. agric. Sci.*, 1957, **49**, 100.
45 Birch, H. F., *Pl. Soil*, 1960, **12**, 81.

of short-term experiments where microbial activity may be of significance must be interpreted with caution. As an example, it is known that denitrification rates measured in recently rewetted soils are far in excess of those that can be obtained from soils that have been allowed to equilibrate for three weeks.[46]

The effect of high soil moisture contents on microbial activity is through the effect of thick water films and filled pores on the oxygen supply to micro-organisms and the carbon dioxide concentration around them (see also p. 307). Oxygen diffuses some 10 000 times more slowly through water than through air and hence the rate of supply can more easily fall short of demand. Calculation shows[47] that anaerobic zones are likely within the centres of 10 mm crumbs in summer and 20 mm crumbs in winter. Where organic substrates are plentiful, for example near roots, the critical size may be reduced.

The yardstick by which aerobic conditions are judged is that put forward by Greenwood[48] who suggested that, due to the efficient terminal oxidases possessed by most micro-organisms, the soil could be considered aerobic as long as the soil solution could sustain an oxygen concentration of at least 5 μM, corresponding to one hundredth of the oxygen concentration in air-saturated water. Although many soil fungi have efficient cytochrome oxidase systems it is likely that their poor tolerance of low oxygen conditions is due to the adverse effects of high CO_2 levels,[49] and this may also be the reason why many fungi are restricted to the superficial layers of the soil or to the surface cracks or channels in the soil.

Effect of temperature

For a given organism, metabolic activity will begin when a threshold minimum temperature is exceeded, the activity will increase as the temperature rises up to a certain optimum value and then decline at a rapid rate as the temperature increases further. Different organisms have their maximum activity at different temperatures, and the temperature at which any organism has its maximum metabolic rate may be different from the temperature at which it is most abundant, because rapid growth and survival often require slightly different conditions. Within restricted groups of organisms such as the nitrifiers there is evidence of geographically distinct strains with different temperature optima.[50]

46 Galsworthy, A. M. and Burford, J. R., *J. Soil Sci.*, 1978, **29**, 537.
47 Currie, J. A., *Soil Sci.*, 1961, **92**, 40.
48 Greenwood, D. J. in *The Ecology of Soil Bacteria* (eds T. R. G. Gray and D. Parkinson). Liverpool Univ. Press, 1967.
49 Burges, A. and Fenton, E., *Trans. Brit. Mycol. Soc.*, 1953, **36**, 104.
50 Mahendrappa, M. K. *et al.*, *Soil Sci. Soc. Am. Proc.*, 1966, **30**, 60.

Activity in temperate soils during the winter with temperatures in the region of 0 to 5 °C is lower than during the summer. For example, in soils under kale and soils free from plants, the rate of oxygen consumption was ten times lower than when the soil temperature was around 17 °C during the summer months.[51] This ten-fold drop in consumption over a 14 °C decrease is much more than would be expected for micro-organisms where a two-fold change in activity for a 10 °C change is more usual. It would suggest that a large proportion of the soil organisms were below their minimum temperature for activity. However, seasonal sampling of soils involves many other factors than simply temperature; moisture content and availability of decomposable substrates being the most obvious. Further work is necessary on temperature responses of soil organisms for a number of reasons: processes such as nitrification can take place at low temperatures with economic and environmental consequences; the modelling of organic matter turnover in soils demands better field results to determine the accuracy of the simulations; and the interactions between soil-borne pathogens and potentially antagonistic saprophytes are likely to be influenced by temperature. A final area that has been little examined compared with the analogous soil moisture effect is the impact of fluctuating soil temperature. All soils show diurnal temperature changes and even in soils of temperate regions these, at the surface, may be in excess of 20 °C in a single day.

Partial sterilization of soils

Although sterility is an absolute condition meaning the absence of all life, the term partial sterilization has been in use for many years to describe soil treatments which drastically reduce, albeit temporarily, the microbial population of the soil with the total eradication of certain groups. The aim usually is to eliminate soil-borne plant pathogens, especially root infecting fungi and nematodes. The cost is high and partial sterilization is restricted by expense to soils in protected structures such as glasshouses and to a limited extent in forestry nursery beds and citrus orchards.

The commonest method at one time was to inject low pressure steam into the soil until the temperature was raised to between 60 and 70 °C, and leave the soil for a period of time to recover from any phytotoxic effect of heating before replanting. With increasing labour and fuel costs, partial sterilization is now more commonly carried out by incorporating volatile fumigants such as methyl bromide, formalin or 1,3-dichloropropane, covering the soil surface with polythene sheeting to retain the fumigant, then allowing the fumigant to volatilize before planting the next crop.

Steam and chemical sterilization treatments share certain influences on the

51 Currie, J. A., *Soc. Chem. Ind., Monog.*, No. 37, **152**, 1970.

soil such as the initial decline of populations followed by a rapid increase in bacterial numbers when the sterilizing influence is removed. Protozoa recover more slowly and in the case of steam treatments fungal re-establishment may be very slow.[52] Chemical fumigation may encourage the rapid multiplication of the fungus *Trichoderma viride* and related species[53] which are known to be antagonistic to many plant pathogens. Some of the effect of chemical treatment may be due to this alteration of the microbial balance in favour of antagonists, a form of chemically stimulated biological control.

All sterilization treatments lead to an initial increase in ammonium concentrations, possibly due to the autolysis of killed cells, and since the nitrifying organisms responsible for conversion of ammonium to nitrate are eliminated by most treatments, whereas ammonifying organisms survive, the ammonium concentrations in partially sterilized soils often continue to increase for some time, often weeks, until recolonization by nitrifiers takes place. Steam sterilization can lead to phytotoxic effects, though not as severe as can result from dry heating which is never recommended. The reasons suggested for post-heating toxicity have included organic toxins,[54] nitrite and excessive available manganese.[55] Some alleviation of phytotoxicity can be achieved by phosphate fertilization prior to treatment.

Where field experiments have included sterilization treatments with agents such as formaldehyde, a common observation is a stimulation of the crop in the year of treatment, partly due to pathogen control and partly due to enhanced nitrogen mineralization. The year following treatment is often marked by a yield depression. Typical results are shown in Table 15.2.[56]

One explanation for yield reduction in the year following partial sterilization is the reduction of natural antagonists to plant pathogens which may themselves multiply on roots and stubble. In the experiments illustrated in Table 15.2 the principal pathogens were the cereal cyst eelworm, *Heterodera avenae*, and the take-all fungus, *Gaeumannomyces graminis*, and both were more severe on the second crop after treatment than on the untreated crops. The effect is similar to the inclusion of a crop resistant to take-all in a run of successive susceptible cereal crops. The first cereal crop after the resistant crop is usually much less severely attacked, and the following one is more severely attacked, than if the resistant crop had not been included in the rotation.[57]

Even when pathogens are not involved, there can be a yield decrease in the year following partial sterilization which appears to be related to biomass and nitrogen mineralization rates. This suggests that the sterilized soil had

52 Brind, J. E., *Forestry Comm. Bull.*, 1965, **37**, 206.
53 Saksena, S. B., *Trans. Brit. Mycol. Soc.*, 1960, **43**, 111.
54 Rovira, A. D. and Bowen, G. D., *Pl. Soil*, 1966, **25**, 129.
55 Messing, J. H. L., *Pl. Soil*, 1965, **23**, 1.
56 Widdowson, F. V. and Penny, A., *Rothamsted Ann. Rep. for 1969*, Part 2, p. 113.
57 Cook, R. J., *Phytopathology*, 1981, **71**, 189.

Table 15.2   The effect of partial sterilization with formalin on the yield of spring wheat.[56] Yields in t ha^{-1}

Cropping year	Fertilizer N applied (kg ha^{-1})	Formalin treatment			
		None	Crop year	Previous year	Both
Woburn (crop irrigated)					
1964	75	1.45	3.76	—	—
1964	150	2.29	4.68	—	—
1965	75	0.90	3.64	0.75	2.64
1965	150	1.38	4.68	1.92	3.74
Rothamsted					
1965	0	2.13	3.00	—	—
1965	62	3.55	4.38	—	—
1965	124	3.34	4.44	—	—
1966	0	2.50	3.66	1.63	3.46
1966	62	3.88	4.54	2.75	4.81
1966	124	5.04	5.33	3.54	4.77

achieved two years' mineralization in one year, giving a crop increase in the first year but a decrease in the following year. Jenkinson and Powlson[58] found that, although the microbial population in soils treated with formaldehyde and untreated respired at the same rate five and a half years after treatment, the treated soils were measurably lower in biomass carbon and nitrogen, than the untreated soils despite the time lapse.

Soil enzymes

The metabolic processes of all living material in the soil, whether micro-organisms, animals or plant roots, depend on the catalytic activity of enzymes. It is therefore possible to regard soil enzymes as being the sum total of all the enzymes possessed by the soil biomass. For many enzyme classes, such as those concerned with energy generation (the respiratory enzymes) or those concerned with cell synthesis, the enzymes are usually incapable of functioning or even surviving outside intact living cells. Many enzymes of this type act in sequences, chains or cycles of activity and it is the end result of concerted activity that is measured, for example respiration as carbon dioxide release. However, some enzymes have an extracellular function because micro-organisms, with the exception of protozoa, rely on soluble nutrients passing through the protoplasmic membrane into the cell, whereas many of

58 Jenkinson, D. S. and Powlson, D. S., *Soil Biol. Biochem.*, 1970, **2**, 99.

the substrates on which the organisms feed are insoluble. Hence, for micro-organisms to obtain the nutrients they need in a form they can use, they must produce collectively a battery of mainly hydrolytic enzymes which will degrade and solubilize the organic substrates in the soil such as proteins, cellulose, chitin, lignin and many others.

Some enzymes must be released from the protoplasm and be able either to diffuse into the immediate vicinity of the cells or remain bound to the exterior of the cell, and are referred to as extracellular or exoenzymes. From necessity most extracellular enzymes must be capable of functioning inde-pendently and may continue to function after the death of the organism which synthesized them. For this reason, the enzymic activity of a soil may not be directly related to the microbial population found at the time of measure-ment, although this is more often the case for exoenzymes than endoen-zymes. A notable example appears to be the endoenzyme urease, which catalyses the hydrolytic breakdown of urea to ammonium carbonate. Although normally functioning within the cell, most soils contain high levels of urease activity which cannot be related to ureolytic organisms, and a urease active organic fraction containing no clay has been separated from soil, [59] which has the property of resistance to attack by the protein degrading enzyme pronase. Since all enzymes are protein, this resistance is of signifi-cance in understanding the persistence of enzymes in soil. The explanation offered is based on the urease enzyme being located in porous organic particles whose pores are large enough to permit the passage of the small urea molecules, but are too fine to allow the potentially destructive pronase to reach the urease.

Enzymes can, however, be adsorbed by clay and soil organic matter. The effective pH environment of the enzyme attached to a clay or organic matter surface can be quite different from the bulk soil pH as usually measured, and the pH dependence and optima can differ appreciably for an enzyme in soil compared with one in solution.[60]

When compared with the activity of pure enzyme preparations, the enzyme activity of soil can be relatively low, and prolonged exposure of soil to a substrate may be necessary to obtain a measurable change in either substrate or product concentration. There is always a danger that population growth may occur, thereby altering the enzyme level during the course of an exper-iment. A variety of growth inhibitors such as toluene, chloroform, sodium azide and ionizing radiations (gamma rays) have been used to allow enzyme determination without microbial proliferation but none can be regarded as wholly satisfactory. Some may only inhibit part of the population, some may actually reduce enzyme activity while others may increase membrane perme-ability and hence increase access of substrate to enzyme. The best solution

59 Burns, R. G. *et al.*, *Soil Biol. Biochem.*, 1972, **4**, 107.
60 Burns, R. G., *Soil Enzymes*, Acad. Press, 1978.

is to devise enzyme determinations which can be carried out in less than one or two hours and thus avoid the likelihood of microbial growth. The use of radiorespirometry may be of value but most developments are likely to be in the area of substrate analogues, especially fluorogenic analogues, which can be detected at very low concentrations.[61]

By-products of microbial metabolism

Micro-organisms, like larger organisms, take in food and release by-products, which include products of respiration or components in the food supply that either cannot be assimilated or are in excess of requirements. Non-photosynthetic autotrophs must always excrete the oxidized product of the substrate from which they have derived energy, while heterotrophs must always excrete carbon dioxide.

Many micro-organisms can multiply in a well-aerated carbohydrate solution and excrete little except carbon dioxide if the solution contains adequate quantities of ammonium or nitrate. If the conditions are altered, for example if the nitrogen supply or the degree of aeration is greatly reduced, these organisms may excrete a variety of energy-rich compounds in conjunction with the carbon dioxide. Thus, if growth under aerobic conditions is limited by a reduced nitrogen supply, for example when organisms are decomposing plant debris with a high carbon : nitrogen ratio exemplified by some tree litters, the material may be metabolized but less efficient use is made of the carbon fraction. Under these circumstances, by what is sometimes called a 'shunt metabolism', the organisms will excrete partially oxidized compounds such as citric and oxalic acids with a relatively high energy content.[62]

The products of metabolism from heterotrophic organisms growing under limiting conditions consist of two distinct types of compounds: (i) complex carbohydrate gums typically produced by bacteria and yeasts under aerobic conditions[63]; and (ii) a variety of simple soluble compounds such as aliphatic acids from methanoic to butyric or the derived alcohols, aldehydes or ketones; the dibasic acids succinic and oxalic, and hydroxyacids such as citric, tartaric and lactic, the latter three being particularly common products of certain groups of fungi such as the Aspergilli.

Micro-organisms usually excrete excess nitrogen from organic sources as ammonium ions under aerobic conditions although the soil fauna may excrete urea or uric acid. If, however, aeration is reduced, or anaerobic conditions set in, a number of foul-smelling amines may be produced such as the aromatic amines indole and skatole and the aliphatic amines cadaverine and

61 Darrah, P. R., Ph.D. Thesis, Univ. Reading, 1983.
62 Foster, J. W., *The Chemical Activities of Fungi*, Acad. Press, 1949.
63 Hepper, C. M., in *Soil Microbiology* (ed. N. Walker), Butterworth, 1975.

putrescine, the latter compounds being largely produced by the anaerobic spore-forming Clostridia.

Micro-organisms also excrete small quantities of compounds that are fairly specific to a certain microbial group. Some of these substances are antibiotics, such as penicillin, which have the potential to influence the microbial population in the immediate vicinity of the originating organism. Other compounds include vitamins, growth factors and amino acids which may be needed by other members of the soil population which are unable to synthesize them for themselves.

Of considerable interest in recent years has been the possible significance of microbial excretions with the potential to influence plant growth. These include compounds such as ethanoic acid,[64] ethylene[65] and various plant growth regulating compounds with gibberelin-like[66] or indole-acetic acid-like properties.[67] The production by bacteria of powerful iron-chelating compounds called 'siderophores' have been described which result in the decline of root pathogens, the organisms responsible being called plant growth promoting rhizobacteria or PGPRs.[68]

Dissolution of soil minerals by soil micro-organisms

The normal organic matter decomposition processes of soil organisms result in the eventual liberation of soluble nutrients for plant uptake. There has always, however, been interest in the possible solubilization by micro-organisms of insoluble minerals in the soil, especially the sparingly soluble phosphates. If rock phosphate and sulphur are mixed and kept moist, the sulphuric acid resulting from autotrophic oxidation of sulphur will increase phosphate availability and similar mechanisms have been postulated for soils involving organic acids or chelating agents such as 2-ketogluconic acid. Soil inoculation with strains of *Bacillus* sp. known as 'Phosphobacterin' has been proposed at various times. Unfortunately, although it is relatively simple to demonstrate phosphate solubilization in the confines of a petri dish in the laboratory with isolated organisms and ample carbon substrates, there is little acceptable evidence of such activity in the field.[69] As a result, interest in this microbial activity as a means of improving plant nutrition has declined.

Better evidence exists for the improved uptake of iron and manganese through microbial intervention and here the mechanism is thought to involve

64 Lynch, J. M. *et al.*, *Pl. Soil*, 1980, **56**, 93.
65 Lynch, J. M. and Harper, S. H. T., *Soil Biol. Biochem.*, 1980, **12**, 363.
66 Brown, M. E., *Ann. Rev. Phytopath.*, 1974, **12**, 181.
67 Wainwright, M. and Pugh, G. J. F., *Soil Biol. Biochem.*, 1975, **7**, 287.
68 Emery, T., *Nature, Lond.*, 1980, **287**, 776.
69 Tinker, P. B., *Pl. Soil*, 1984, **76**, 77.

chelating agents.[70] Other microbial activities may include the enhanced weathering of aluminosilicates through the removal of divalent cations and the solubilization of silica. Both processes have been observed on newly exposed rock faces but no evidence is available regarding the rate at which such processes might occur in soil, or even whether they take place at all.

Breakdown of toxic chemicals in soil

Provided that extremes of pH, moisture and temperature are excluded, soils possess remarkable powers of decomposition of organic matter. Most natural organic products are decomposed within a few weeks or months and only the more intransigent polymers such as lignin persist for significantly more than a year. Soils can also acquire the ability to decompose compounds not normally present in them (exobiotic compounds), but which may be toxic to the microbial population if used in high enough concentrations. Thus, soils were known to be able to oxidize phenol, cresols and hydrocarbons some 70 years ago.[71]

Developments in pesticide chemistry have produced a wide range of compounds capable of killing weeds or insects or fungi and concern has grown at the extreme persistence shown by some of the chemicals. Alexander,[72] in a short table lists eight compounds with persistences in excess of 15 years. Our understanding of pesticide decomposition has improved over the years, helped by the insistence of various environmental agencies that persistence information be part of the case put forward by manufacturers for pesticide registration.

Environmental factors are important for the degradation of organic exobiotics and conditions which would restrict the decomposition of normal plant debris will usually restrict the degradation of synthetics. For microbial degradation, through the activities of enzymes, it is also necessary for the pesticide to be available or accessible. Adsorption on to or into colloidal material can seriously retard decomposition. Finally, the chemical structure of the synthetic compound will dictate whether enzymic degradation is likely or unlikely.

Most synthetic pesticides are heavily substituted with either chlorine, bromine, fluorine, nitro or sulphonate groups in structures which are not found in biological tissues. It may be the unfamiliarity of shape which renders the compounds difficult to decompose and become known as recalcitrant or refractory. The higher the level of substitution the more recalcitrant the compound. For example, 2,4-D (2,4-dichlorophenoxyacetic acid) is much

70 Barber, D. A., in *Interactions between Non-pathogenic Soil Micro-organisms and Plants* (eds Y. Dommergues and S. Krupa), Elsevier, 1978.
71 Wagner, R., *Ztschr. Gärungsphysiol.*, 1914, **4**, 289.
72 Alexander, M., *Biotechnol. Bioeng.*, 1973, **15**, 611.

more readily decomposed than 2,4,5-T (2,4,5-trichlorophenoxyacetic acid)[73] and certain substitutions, for example at the *meta* position on an aromatic ring, convey more recalcitrance than *ortho* or *para* substitutions.

Pesticide degradation is now recognized to take place by two mechanisms. In the first, the compound is degraded by the organisms with specific enzymes, which may be adapted to deal with the compound, and the organism normally derives some nutrient or energy benefit from the degradation. In the second form of degradation, the synthetic compound is degraded, or sometimes only partly degraded, by metabolic pathways which exist for other purposes, that is, the synthetic compound is metabolized alongside the normal functioning of the cell virtually by accident, the process being known as co-metabolism.

Obviously soil micro-organisms can have no detoxifying effect on radioactive materials, nor can they detoxify poisonous elements such as cadmium or mercury. With mercury there is even a danger that microbial activity may render it more dangerous through the formation of the more soluble monomethyl mercury[74] (see p. 808).

Symbiotic and antibiotic relations between the microflora

The various groups of soil organisms do not live independently of each other, and many have developed a series of symbiotic relations with their neighbours. It seems to be a characteristic of most micro-organisms that they excrete complex organic compounds[75] when growing actively, usually only in very small amounts, and most of the symbiotic and antibiotic interactions take place through the action of these substances on other organisms in the immediate vicinity. These inter-relations are, however, rarely very specific.

There are several non-specific types of beneficial interaction which may occur in soils. In the first place, a soil contains a wide variety of microhabitats, having a range of oxidation–reduction potentials even if well drained. A number of aerobic bacteria are known to produce peroxidases and catalases which prevent the build-up of metabolically produced peroxides, which are strong inhibitors of anaerobic bacteria. It is therefore possible that some anaerobic or microaerophilic bacteria can grow in the neighbourhood of these aerobes protected by excess peroxidases. In the second place, two organisms may be living in association because one needs an excretion product such as an amino acid or growth factor. Again, two organisms in combination may

73 Audus, L. J., *Pl. Soil*, 1951, **3**, 170.
74 Bekert, W. F. *et al.*, *Nature, Lond.*, 1974, **249**, 674.
75 Lynch, J. M., *CRC Crit. Revs Microbiol.*, 1976, **5**, 67.

be able to effect a decomposition which neither can achieve when growing alone. Although proof of such associations in soil is difficult, there are numerous laboratory examples of two soil organisms carrying out a decomposition more quickly when growing together than when separated.[76]

Some soil organisms develop antibiotic relations with other organisms based on materials excreted into the soil solution[77] that prevent the affected organisms from growing naturally, or may even cause them to lyse by dissolving their outer membrane. The fungus *Penicillium urticae* which excretes patulin in stubble mulched soil is such an organism,[78] but in most soils the fungus *Trichoderma viride* may be a more important example, for it excretes two antibiotics, gliotoxin and viridin, which inhibit many soil saprophytes and plant pathogens in culture media.

The ability to produce antibiotics is not confined to fungi, for it is also strongly developed by many actinomycetes and bacteria. At least half the soil fungi and actinomycetes produce such substances in the laboratory under suitable conditions, the proportion of bacteria being somewhat smaller. The substances so far recognized act mainly on the three producer groups, although there are some bacteria such as the Pseudomonads which synthesize metabolites toxic to ciliate, flagellate and amoeboid protozoa.[79]

Antibiotics display a wide variety of chemical structures and mode of action and should never be regarded as an homogeneous group of substances. Furthermore, just because an organism can produce an antibiotic when grown on suitable media in the laboratory, it does not follow either that it behaves similarly in soil, or if it does, that the concentration or persistence of the antibiotic in the soil solution is high enough to have a significant effect on nearby organisms. The zone of activity of an antibiotic is probably confined to a region very close to the producer organism. Since antibiotics are apparently products that accompany active growth, they are most likely to be concentrated in habitats where nutrients, especially energy substrates, permit such growth. In practice it has proved very difficult to prove the existence of antibiotics in soil by the use of extraction techniques unless energy substrate supplementation is practised.[80] Indirect evidence of antibiotic presence in soils is available from the widespread possession of active enzymic defence mechanisms, such as the β-lactamase enzyme of *Bacillus cereus*[81] which destroys penicillin.

Bacteria can also be parasitic on soil fungi and indeed on each other.[82] Rossi–Cholodney slides buried in the soil frequently show portions of fungal

76 Enebo, L., *Nature, Lond.*, 1949, **163**, 805.
77 Brian, P. W. in *Microbial Ecology* (eds R. E. D. Williams and C. C. Spicer), Cambridge Univ. Press, 1957.
78 Norstadt, F. A. and McCalla, T. M., *Soil Sci.*, 1969, **107**, 188.
79 Singh, B. N., *Nature, Lond.*, 1942, **149**, 168.
80 Wright, J. M., *Ann. appl. Biol.*, 1956, **44**, 461.
81 Hill, P., *J. gen. Microbiol.*, 1972, **70**, 243.
82 Stolp, H., *Ann. Rev. Phytopath.*, 1973, **11**, 53.

mycelium being attacked by mycolytic bacteria; one end of the hypha appears free from bacteria, then comes a large concentration of small rods and cocci on what appears to be an almost empty hypha, and behind this the original growth line of the fungus is outlined by the remaining bacteria. The mechanism of fungal lysis is not always simply explained by enzymic attack[83] although some bacteria possess suitable chitinases. Nutrient depletion and parasitism may also precede fungal lysis.[84]

Interactions between the soil microflora and microfauna

Apart from the straightforward predation by protozoa on bacteria and fungi, the soil fauna can affect the other soil micro-organisms in several ways. The larger saprophytic soil animals, by being mobile, distribute both the decaying organic matter and some of the microbial population throughout the soil layer in which they are active, for they will be ingesting food at one place, mixing it with bacteria in their gut, and excreting it at another location. This process of comminuting and distributing dead plant litter throughout the surface layers of the soil is the major contribution of the larger soil invertebrates. Micro-organisms are brought into intimate contact with their substrate and a large surface area is presented for decay. Smaller soil animals, such as the saprophytic mites and Collembola, are known to feed largely on fungal hyphae and will cause an increased turnover in this portion of the biomass. The effects of the microfauna on the cycling of plant nutrients within the soil is becoming a subject of increasing attention. Significant contributions are being made through the use of controlled microcosms in which respiration and nutrient release from organic debris can be followed and the influence of the microfaunal component investigated.[85]

A further interaction of importance, at a symbiotic level, exists between the soil invertebrates and the protozoa. Some invertebrates carry a protozoan population in the gut which help to digest some of the more resistant plant polymers such as cellulose that may form a significant part of their diet. The digestive tracts or organs of these invertebrates therefore constitute an important environment of selected soil protozoa. The final indication of the complexity of microbial interactions in the soil is the suspicion that the protozoan component just described may, in its turn, be dependent on cellulytic bacteria within their own cells for their enzymic activity.

83 Ko, W-H. and Lockwood J. L., *Phytopathology*, 1970, **60**, 148.
84 Scott, P. R., *Ann. appl. Biol.*, 1969, **63**, 27.
85 Anderson, J. M. and Ineson, P., *Soil Biol. Biochem.*, 1982, **14**, 415.

16

Soil fauna other than protozoa

In comparison with microbial studies, the fauna of the soil has, until recently, received little attention. Excellent general accounts of our knowledge on this subject are available in books by Burges and Raw[1] and Wallwork,[2] and a recent account of the present state of knowledge on the quantitative ecology of animals in soil and litter has been edited by Petersen and Luxton.[3]

The relative neglect has been due partly to the difficulty of isolating the animals quantitatively from the soil mass, and partly to the very great problems of their systematic classification when they have been separated out. Again, until recently very little was known about the actual food of many members of this population, and even now our knowledge is very incomplete. Techniques have now been developed, however, for fixing and preparing casts and sections of the soil without disturbing its structure,[4] and for identifying food sources by examination of the gut contents of at least the larger

1 *Soil Biology* (eds A. Burges and F. Raw), Acad. Press, 1967.
2 Wallwork, J. A., *The Distribution of Soil Fauna*, Acad. Press, 1976.
3 Petersen, H. and Luxton, M., *Oikos*, 1982, **39**, 288–388.
4 Rogaar, H., *Neth. J. agric. Sci.*, 1974, **22**, 143.

species, using microscopical and immunological methods. These advances are giving a more complete picture of the location and activity of this population than we have had before.[5]

The size of the soil faunal population depends both on the food supply and on the physical condition of the soil. The soil fauna need a fairly well-aerated soil for active growth: they cannot thrive in waterlogged soils nor in wet soils which have been puddled by the trampling of cattle. The soil invertebrates reach their greatest diversity and highest populations in the soil surface layers of forests or old pastures. In such situations, if the soil becomes temporarily waterlogged by heavy rain or flooding from above, enough air is likely to be entrapped in the pores of the fairly open soil for the smaller animals to survive for considerable periods of time. This entrapping of air does not take place to anything like the same extent if the water table rises to the soil surface, and this has a far more serious consequence for the more aerobic members of the population. The population is, on the whole, more tolerant of dry conditions, and a large proportion of the animals tends to congregate in the top 20 to 50 mm. In many soils only a few groups, such as the earthworms, are common in the deeper layers. The variety and taxonomic complexity of the soil fauna are very great. Table 16.1 gives population densities of invertebrate groups for a faunistically rich site on old grassland. Table 16.2 gives biomass estimates for important groups in a number of ecologically differing locations. Typical members of the fauna of English soils are shown in Plates 16.1 and 16.2.

The complex decomposer ecosystem of the soil subsists primarily on the input of dead plant tissue to the surface of the soil, through such processes as leaf fall and the incorporation of crop residues and manure. This makes it difficult to define what is meant by a soil animal, for many which are normally included are inhabitants of the surface litter layer as well as of the soil. Under forest the litter is fairly evenly distributed, but elsewhere, for example in a tussocky pasture, or in arid semi-desert, the input of organic matter is very unevenly distributed over the soil surface. In such situations there may be a number of quite different environments in and above the soil, each with its characteristic population, so that it would be misleading to lump them all together as if there were a uniform population. In general, animals of a given species are not distributed uniformly or at random throughout the soil, but always show a tendency to congregate in some places and be rare in others.[6] One example is that in the Antarctic tundra of South Georgia the numbers and species of microarthropods recovered from adjacent localities among the sparse vegetation were highly variable.[7] Another example is that grass root samples from Irish turf contained higher

5 Miedema, R. *et al.*, *Neth. J. agric. Sci.*, 1974, **22**, 37.
6 Salt, G. and Hollick, F. S. J., *J. exp. Biol.*, 1946, **23**, 1.
7 West, C. C., *Oikos*, 1984, **42**, 66.

Table 16.1 Total animal populations recorded from old grassland soil at Lyons Hill, Ireland, based on totals from 96 samples, each 7.5 cm deep and of surface area 25 cm². (From Curry, J. P., *Soil Biol. Biochem.*, 1969, **1**, 221)

Group	Population m^{-2}	Percentage of total Arthropoda	Percentage of total Acari
Collembola	105 360	47.4	
Acari			
Mesostigmata	28 240		26.6
Prostigmata	32 800		30.9
Cryptostigmata	42 120		39.7
Astigmata	2 960		2.8
Total Acari	106 120	47.8	
Other arthropoda			
Protura	6 680		
Symphyla	800		
Coleoptera	608		
Araneae	12		
Diptera	1 280		
Chilopoda	120		
Pauropoda	1 024		
Tardigrada	21		
Thysanoptera	21		
Hemiptera	40		
Hymenoptera	4		
Psocoptera	16		
Lepidoptera	8		
Others	4		
Total other arthropoda	10 600	4.8	
Total arthropoda	222 080		
Other invertebrates			
Nematoda	34 320		
Enchytraeidae	5 760		
Lumbricidae	364		
Mollusca	71		

Table 16.2 Typical biomass estimates of important soil fauna groups in main biomes. Figures without brackets represent median values of five or more independent mean biomass estimates. Figures in parentheses are more or less tentative values generally based on less than five biomass estimates. Values are in mg dry wt m^{-2}. (From Petersen, H., *Oikos*, 1982, **39**, 330)

	'Tundra'	Temperate 'grassland'	Tropical 'grassland'	Temperate coniferous forest	Temperate deciduous forest (mor)	Temperate deciduous forest (mull)	Tropical forest
Protozoa		(200)					
Nematoda	160	440	(50)*	120		330	(50)
Enchytraeidae	1800	330	(20)	480		430	(20)*
Collembola	150	90	(10)	80	(130)	110	(20)
Cryptostigmata	60	110	(20)	450	(700)	180	–
Mesostigmata	20	(60)	(10)	(80)		40	–
Prostigmata	10	(40)	(50)	(30)		(10)	–
Acari (*in toto*)	90	(120)	(80)	500	(900)	(300)	(100)
Large Oligochaeta (empty gut)	330	3100†	170	450	200	5300	340
Diplopoda	(~0)*	1000	(10)	50		420	20
Diptera larvae	470	60	(10)	260		330	(~0)
Isoptera	0	(~0)	(1000)	0		0	(1000)
Chilopoda	(20)	140	(5)	70		130	5
Carabidae + Staphylinidae	(50)	(80)	(10)*	120		90	(10)*
Araneae	10	(30)	(30)	50		40	20
Gastropoda	(~0)*	(100)*	(10)*	(20)		270	(10)
Formicoidea	(~0)*	100	(300)	(10)		(10)	(30)
Total	3300	5800	1900	2400	3500	8000	1800

* Estimate with practically no data support.   † Excl. North American prairie sites.

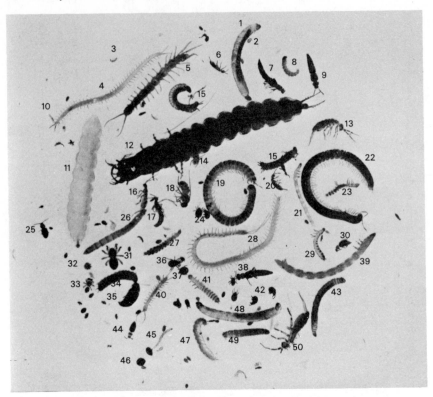

Plate 16.1 Typical members of the soil fauna (natural size).
Chilopoda: Centipedes 4, 5, 28, 40. *Diplopoda:* Millipedes 19, 22, 41.
Arachnida: (a) Araneae: Spiders 31, 33; (b) Acarina, Mites: Gamas-
idae 2, 14, 36; Oribatidae 24, 32, 37, 45; Tyroglyphidae 42.
Insecta: (a) Collembola: Springtails 3, 6, 10, 13, 16, 20, 50. (b)
Lepidoptera (larva) 8. (c) Coleoptera: Staphylinoidea (adults) 7, 9,
17, 30, 35, 38, 44; Staphylinoidea (larvae) 15, 23, 27, 29, 34, 41;
Carabidae (adult, small species) 25; Carabidae (larva, large species)
12; Elateridae (larvae): Wireworms 21, 26, 39, 45. (d) Diptera:
Bibionidae (larvae) 1, 43, 48, 49; Cyclorrapha (larva) 11; Antho-
myidae (adult) 18.

numbers of soil arthropods than did weed root samples.[8] This aggregated
distribution must be taken into account in the design and analysis of
experiments.

The food supply of the invertebrate population is not restricted to dead or
living plant material. The saprophagous forms which feed on dead material
and the associated microflora are perhaps the most important. In addition

8 Curry, J. P. and Ganley, J. in *Ecol. Bull.* Vol. 25 (eds U. Lohm and T. Person) p. 330,
Stockholm, 1977.

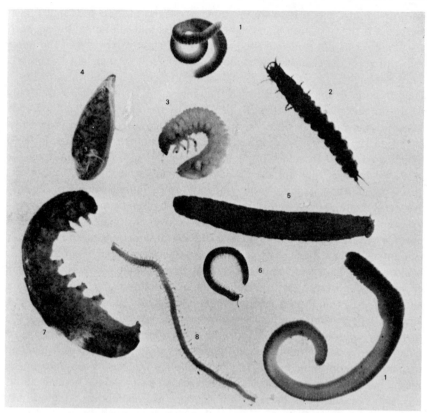

Plate 16.2 1. Earthworm, *Lumbricus* sp. (Oligochaeta); 2. Beetle larva, Carabidae (Coleoptera); 3. Chafer grub, *Phyllopertha* sp. (Coleoptera); 4. Slug, *Agriolimax* sp. (Mollusca); 5. Leatherjacket, Tipulidae (Diptera); 6. Millipede (Diplopoda); 7. Cutworm, Agrotidae (Lepidoptera); 8. Centipede (Chilopoda).

there is a coprophagous population feeding upon the excreta of the primary group, and a diverse predator population subsisting upon the whole animal community. Knowledge of the types of material in the plant debris which are digestible by the different species of the saprophagous fauna is limited. Soil detritus feeders have been shown to possess a wide range of carbohydrase enzymes,[9] and molluscs, tipulid larvae, termites, some millipedes and some lumbricid earthworms are all considered to digest plant structural polysaccharides.[10] In some cases, as in the termites, this ability is related to the presence of symbiotic cellulose-decomposing bacteria in the gut. Most of the soil fauna, however, appear to have no power of digesting cellulose, hemicellu-

9 Nielsen, C. O., *Oikos*, 1962, **13**, 200.
10 Luxton, M., *Oikos*, 1982, **39**, 340.

loses or lignins, and it is likely in many forest and pasture soils that fungal mycelium is the transition material between plant and soil faunal tissue. Such mycelium forms the principal food supply of a large proportion of the mites and springtails, as is shown by study of the contents of their gut.[11] This has the consequence that the soil fauna living on plant remains only oxidize a small fraction of the material which they ingest, and it is probable that only rarely do they use more than 10 per cent of the energy in the food they consume. Further, they differ from most soil micro-organisms in using most of this energy for their vital processes, and only convert a small fraction into their own tissues. In so far as most of these animals will probably be the primary source of food for the predatory population in the soil, little if any of the carbon they assimilate will contribute to the humus supply of the soil.

Measurements of respiratory metabolism have often been used in attempts to assess the importance of the soil fauna in overall organic matter decomposition. In these terms the influence of the soil fauna appears to be slight, amounting to less than 5 per cent of the total decomposer respiration.[11] It is now clear, however, that this measurement totally obscures the true ecological importance of the soil fauna in decomposition processes. Soil animals play an important role in breaking up plant remains such as leaves, and the larger forms, particularly earthworms, are effective in mixing the surface organic matter into the soil. Material which has passed through the gut of soil animals is more readily attacked by the microflora and the rate of mineralization is increased. Experiments in which leaves are enclosed in plastic bags made from mesh of various sizes, so as to exclude animals above a certain size, show that even small forms such as Collembola and mites have an important effect on the rapidity with which decomposition proceeds.

Microarthropods

In numbers, the microarthropod population of the soil is dominated by the springtails (Collembola) and the mites (Acari), as indicated by the figures in Table 16.1 relating to samples from an old grassland soil. The microarthropods may be considered in two ecological groups. Those occurring principally on the surface or in the litter layer are hemi-edaphic forms, characterized by well-developed eyes, long appendages, well chitinized and with marked pigmentation, and are clearly adapted to an active life in a relatively open habitat. In contrast, eu-edaphic forms are more specialized deep soil animals, often lacking pigment and eyes, soft bodied and with short appendages. The springtails are entirely detritus or fungal feeders, as are many of the mites, but the mites include many and varied predatory forms which subsist upon other members of the soil animal community.

11 Forsslund, K. H., *Medd. Skogsförsöksanst*, 1945, **34**, 1.

The investigation of microarthropod populations is dependent on methods for separating them quantitatively from the soil and associated organic debris. Two types of technique have been developed: dynamic methods, in which the living animals are persuaded to leave the soil sample, and mechanical methods, based upon mechanical separation of the animals from the soil. The dynamic methods have their origin in the Berlese funnel, a copper funnel fitted with a jacket filled with hot water, into which the soil sample is placed, with the aim of driving the animals out of the lower end of the funnel. This method was improved in the Tullgren funnel, in which the heat is applied from above by an electric lamp bulb. Dynamic techniques are represented today by various controlled gradient extractors,[12] in which the soil samples, as far as possible undisturbed cores, are exposed to a controlled gradient, from top to bottom, of high to low temperature and light, and low to high humidity, so that the animals are driven gradually downwards and out into suitable collecting jars. The mechanical methods depend upon wet sieving to break up the soil gently and free animals from entanglement, followed by flotation in solutions of appropriate density to separate organic matter, both plant and animal, from the mineral part of the soil. In more recent techniques, this is followed by procedures to separate the microfauna from the plant debris. Flotation was used by Berlese many years ago, but the procedures have been much developed and improved.[13] The separation of arthropod and plant material has been effected by exploiting their differing wettability in aqueous solutions and in organic solvents.[14] The choice of technique is dependent on the physical nature of the soil samples and on the purpose for which the animals are required. The dynamic methods are excellent for loose litter samples, where the animals can readily migrate downwards, and these methods produce good specimens for taxonomic purposes. They are less suitable for clay soils, where smearing tends to occur on sampling and the sample can dry hard before the animals have been able to escape. Furthermore, the methods are dependent on the animals being in a healthy state and able to migrate, and so may be unsatisfactory if samples have to be transported a long distance before extraction.

Microarthropod populations attain their greatest abundance and species diversity in the soil and litter of deciduous forests and old grasslands. Studies on the fauna of an old grassland site in Ireland (Table 16.1) showed a mean population density of 220 000 arthropods m^{-2} in the top 75 mm. This community was dominated by four genera or species which accounted for one-third of the population. Different grassland sites, however, differ considerably in the composition of the microarthropod populations, and those

12 Macfadyan, A., *J. anim. Ecol.*, 1953, **22**, 65.
13 Salt, G. and Hollick, F. S. J., *Ann. appl. Biol.*, 1944, **31**, 53; Raw, F. in *Soil Zoology* (ed. D. K. McE. Kevan), p. 341, Butterworths, 1955; Raw, F. in *Progress in Soil Zoology* (ed. P. W. Murphy), p. 199, 1962.
14 Newman, J. F. in *Crop Loss Assessment Methods*, FAO, 1970.

in woodland litter are again very different. There is evidence that microarthropod numbers are associated with particular species of plants in the
ground cover.[15]

The effect of soil microarthropods on the decomposition of organic matter
and the cycling of nutrients has been the subject of much speculation and
study. Effects arising from their comminution of plant remains have been
suggested, as has an effect produced by feeding on fungal hyphae. In experiments in which ^{14}C-labelled fungal mycelium was placed in soil and exposed
to mites, six out of thirteen species or groups were consistently labelled, but
only a small proportion of the food supplied was taken up by the mites,
indicating a minor role in the grazing of the microflora.[16]

Earthworms: Lumbricidae

The probable importance of earthworms was noted in 1777 by Gilbert
White.[17] He observed that worms seem to be great promoters of vegetation,
by boring, perforating and loosening the soil, by drawing leaves and twigs
into it, and by throwing up worm casts. More than a century later, Charles
Darwin[18] wrote *The Formation of Vegetable Mould through the Action of*
Worms, one of the classics of soil science, and at about the same time the
importance of worms was stressed by Henson.[19] Henson called Darwin's
attention to the work of Müller on humus[20] but Darwin had no opportunity
to consult Müller's work, which was not published in German until later.[21]
In spite of this early start, and of the obvious importance of worms in soil,
for a long time little work was done on their ecology. Morris, working with
Rothamsted arable soils,[22] and Bornebusch, working with Danish forest
soils,[23] found that where earthworms flourished they constituted between 50
and 75 per cent of the total weight of animals present. There are recent
accounts of earthworm biology and ecology.[24]

Earthworms appear to attain their highest populations in the pastures and
deciduous woodlands of Europe and the temperate regions of Asia. In tropical areas, and in arid areas of the subtropics, earthworms are scarce, their

15 Curry, J. P., *Soil Biol. Biochem.*, 1972, **5**, 645.
16 Coleman, D. C. and McGinnes, J. T., *Oikos*, 1970, **21**, 134.
17 and later published: White, Gilbert, *The Natural History of Selborne*, 1789.
18 Darwin, C., *The Formation of Vegetable Mould through the Action of Worms*, Murray,
   1881 and reissued as *Darwin on Humus and the Earthworm*, Faber & Faber, 1945.
19 Henson, V., *Ztschr. wiss. Zool.*, 1877, **28**, 354.
20 Müller, P. E., *Tidsskrift for Skroubrug*, 1878, Band **3**, Heft 1 & 2, Copenhagen.
21 Müller, P. E., *Studien über die natürlichen Humusformer*, Berlin, 1887.
22 Morris, H. M., *Ann. appl. Biol.*, 1922, **9**, 282, 1927, **14**, 442.
23 Bornebusch, C. H., *Forstl. Forsoegsv, Danmark*, 1930, **11**, 1.
24 Satchell, J. E. (ed), *Earthworm Ecology*, Chapman & Hall, 1983. Wallwork, J. A.,
   Earthworm Biology, Institute of Biology Studies in Biology, No. 161, Arnold, 1983.

roles in the soil being taken over by termites and to some extent by ants. The earthworms are composed of various families of the order Oligochaeta. The earthworms of the Palaearctic region – the temperate areas of Europe and Asia – are almost entirely of the family Lumbricidae, and the natural distribution of this family is essentially confined to this region. The earthworm fauna of the British Isles[25] is somewhat impoverished in numbers of species in comparison with that of continental Europe,[26] a consequence of their elimination during ice ages and subsequent isolation from Europe. The forms which have re-established are the so-called peregrine species, which readily colonize new areas and are relatively tolerant of agricultural disturbance. With the spread of agriculture, these species of lumbricids have established themselves in many other parts of the world, both in North America and in the temperate areas of the southern hemisphere,[27] where they have often displaced the native species in agricultural soils.

The weight of earthworms present in favourable conditions can be impressive. In two productive forest soils in Denmark, 2.5 to 3.5 million worms per hectare were recorded, weighing 1700–2000 kg.[23] This is equivalent to the weight of livestock carried on a comparable area of pasture. The earthworms, however, would only be respiring about one-tenth the amount of carbon dioxide that the livestock would. About 6 million earthworms per hectare, weighing nearly 1700 kg, were found on productive four-year-old ryegrass/white clover leys at Wye in Kent in the UK.[28] The most productive pastures at the Grasslands Station, Palmerston North, New Zealand carried numbers up to 7 million per hectare, weighing 2400 kg[29] and a close correlation was noted between the productivity of the pasture and the weight of earthworms, there being about 170 kg of worms for each 1000 kg of annual dry matter production, averaged over the year. Observations on the numbers and weights of worms on a number of orchard and arable sites in Britain[30] showed that the greatest populations were in grassed orchards, the largest being over 8 million worms per hectare, weighing about 2300 kg. Populations in arable sites were much lower, ranging up to a quarter of these values.

In an attempt to classify earthworms on an ecological basis, Bouche[26] has divided them into three types, based upon lifestyle and burrowing habit. Epigeal forms, including such species as *Lumbricus rubellus* and *Eisenia foetida*, hardly burrow in soil at all, but inhabit decaying organic matter on the surface, or in an agricultural context, manure or compost heaps.

25 Gerard, B. M., *Synopses of the British Fauna*, No. 6, Linnean Society, London, 1964.
26 Bouche, M. B., *Lumbriciens de France*, Institut Nationale de la Recherche Agronomique, 1972.
27 Lee, K. E., *The Earthworm Fauna of New Zealand*, New Zealand Dept. Sci. Ind. Res. Bull., No. 130, 1959.
28 Davis, A. G. and Cooper, M. M., *J. Brit. Grassland Soc.*, 1953, **8**, 115.
29 Sears, P. D. and Evans, L. T., *New Zealand J. Sci. Tech.*, 1953, **31A**, Suppl. 42; Waters, R. A. S., *New Zealand J. Sci. Tech.*, 1953, **36A**, 516.
30 Edwards, C. A. in *Earthworm Ecology* (ed J. E. Satchell), Chapman & Hall, 1983.

Endogenous species, such as *Allolobophora chlorotica* and *A. caliginosa*, produce shallow branching burrows in the organo-mineral layers of the soil. Anectic forms are the deep burrowing species, producing channels to a depth of 1 metre or more. These include the giant megascolicid worms of Australia and other southern latitudes, and are represented in Britain by *Lumbricus terrestris*, *Allolobophora longa*, *A. nocturna* and *Octolasion cyaneum*.

The quantitative sampling of earthworms is a matter of some difficulty. The more shallow dwelling species are perhaps best sampled by digging and hand sorting, which may be assisted by washing and flotation procedures. Digging is less satisfactory for the anectic species, as the approaching disturbance causes them to retreat into the deep burrow system. More quantitative samples may be obtained by the application of irritant chemical solutions, such as a 0.2 per cent solution of formaldehyde in water, to measured areas. Electrical methods, in which a current is passed between electrodes pushed into the soil, have been tried and can expel many earthworms. There are, however, paths of widely varying conductivity in the soil, so that the method is unsatisfactory as a quantitative technique.

Food and nutrition

The principal food of earthworms is dead or decaying plant remains, including both leaf litter and dead roots, but there are differences in the acceptability of litter from different plant species. Forest leaf litter with a nitrogen content exceeding 1.4 per cent, as from *Alnus glutinosa*, *Fraxinus excelsior* or *Sambucus nigra* was taken more readily by *L. terrestris* than litter from *Fagus sylvatica* or species of *Betula* and *Quercus*, where the nitrogen content was under 1 per cent, though since there was a close correlation between nitrogen content and soluble carbohydrate in the leaves, it was not possible to decide which of these two constituents was the more important.[31] Roughly similar results were obtained with other forest trees,[32] but in addition it was shown that items of the typical ground flora of mull forest soils – *Mercurialis perennis*, *Urtica dioica* and *Oxalis acetosella* – were even more strongly preferred. Litter high in polyphenols, such as the needles of coniferous trees and fresh oak leaves, is eaten relatively little.

The level of available nitrogen is probably an important factor in controlling the size of the earthworm population. Animal dung is an attractive food for some species of earthworm, and large populations may be found under cow pats in pastures. Additions of farmyard manure to soils will favour earthworm populations. At Rothamsted, the long-term experiments include

31 Wittich, W., *Schr. Reihe forstl. Fak. Univ. Gottingen*, 1953, **9**, 5.
32 Bornebusch, C. H., *Dansk Skovforen. Tidsskr.*, 1953, **38**, 557.

Broadbalk which has grown continuous wheat since 1843, Barnfield, continuous root crops since 1843 and Park Grass, continuous grass since 1836. Annual treatments include farmyard manure (48 and 96 kg N ha^{-1}), various forms of inorganic nitrogen (48, 96, 144 and 192 kg ha^{-1}), and liquid and solid sewage sludge in various doses. In the arable experiments, all species of earthworms are more numerous in plots treated with organic manures than in untreated plots and there is a strong positive correlation between amounts of inorganic nitrogen applied and populations of earthworms. The effects of nitrogen on earthworm populations in grassland are less than those in arable experiments.[33]

Satchell has attempted to draw up a nitrogen balance sheet for *L. terrestris* in an ash–oak wood at Merlewood in the English Lake District.[34] He estimated the annual return of nitrogen in the leaf litter at about 50 kg ha^{-1}, the amount of nitrogen returned as dead earthworm tissue at about 60 to 70 kg ha^{-1}, and the amount of metabolic nitrogen excreted at about 30 kg ha^{-1}, making a total turnover for this one species of worm of about 100 kg ha^{-1} or twice the amount of nitrogen in the leaf fall. These estimates may not be entirely accurate but they do emphasize what a severe competition for nitrogen there must be both among the worms themselves and between the worms and other members of the soil population. It may be that earthworms feed selectively upon organic matter in which the C : N ratio has already undergone some reduction. In experiments upon the cycling of nitrogen by surface casting earthworms, predominantly *L. rubellus*, in a pasture in New Zealand, it was found that the C : N ratio of casts (10.7) was consistently lower than that of underlying soil material (15.0 and 14.2 for the 0–5 and 18–22 cm depths respectively) and it was suggested that this was due partly to the loss of carbon as carbon dioxide from plant-derived organic matter during passage through the earthworm and partly due to the utilization of low C : N ratio litter.[35]

While both Gilbert White and Charles Darwin noted the importance of earthworms in incorporating organic matter into the soil, it has been difficult to obtain reasonably accurate quantitative measurement of this effect. Stout has summarized recent work on organic matter incorporation at various undisturbed sites in the USA, in England and in the North and South Islands of New Zealand.[36] At various depths in the soil, measurements were made of the isotopic composition of the organic matter carbon. Information on the rate of incorporation of organic matter into the soil is provided by the concentration of the long-life radioactive carbon isotope ^{14}C. ^{14}C occurs naturally in low concentration, but the release of this isotope from atomic bomb explosions produced a sudden increase in the concentration in the

33 Edwards, C. A. and Lofty, J. R., *Soil Biol. Biochem.*, 1982, **14**, 515.
34 Satchell, J. E. in *Soil Biology* (eds A. Burges and F. Raw), Acad. Press, 1967.
35 Syers, J. K. *et al.*, *Soil Biol. Biochem.*, 1978, **11**, 181.
36 Stout, J. E. in *Earthworm Ecology* (ed. J. E. Satchell), Chapman & Hall, 1983.

atmosphere. The concentration of ^{14}C at various depths in the soil thus provides a guide to the relative age of different organic materials and to rates of turnover. The use of this technique to compare the rates of soil mixing on sites differing widely in earthworm populations showed that mixing rates were higher where earthworms were numerous. This work again indicates that the role of earthworms is an important one.

Apart from its importance as a mixer of soil, the role of the earthworm as an actual decomposer of organic matter is more difficult to evaluate. Earthworms may contribute only around 4 to 5 per cent of the total soil metabolism[37] but by providing a well-comminuted medium for microbial activity they may well have a much greater influence on total soil metabolism.

Numerous enzymes have been reported to be present in the gut of earthworms. These include proteases, lipase, amylase, cellulase and chitinase.[38] It is not clear, however, which of these enzymes are directly secreted by the gut and which are produced by the diverse symbiotic flora and fauna of micro-organisms which are commonly present in the earthworm.

Burrowing activity

The habits of the different species of earthworms differ appreciably. Some feed on surface litter, coming to the surface to collect it and then dragging it down into their burrows. Those that are truly inhabitants of the soil typically consume soil along with the plant debris. Their excreta, in the form of worm casts, therefore contain a high proportion of soil. Most species void this in the body of the soil, in their burrows, but some make casts on the surface. At Rothamsted only *Allolobophora longa* and *A. nocturna* have been proved to make casts regularly on the surface, but other species such as *Lumbricus terrestris* sometimes do.

The burrowing activities of earthworms clearly have some importance in relation to soil drainage and aeration. The introduction of earthworms to pastures where they are deficient has been seen to produce a substantial increase in water infiltration while conversely, it was reduced three-fold when earthworms were eliminated.[39] Channelling by earthworms and the opening of burrows to the surface appear to be the main explanation. Numerous field observations of this effect are supported by laboratory experiments in which it was found that water passed through containers of light sandy soil in two days when the soil was worked by worms, and in eight days in the absence of worms.[40] In fields of irrigated lucerne up to 15 million channels per hectare

---

37 Phillipson, J., Abel, R. *et al.*, *Oecologia, Berlin*, 1978, **33**, 291.
38 Laverack, M. S., *The Physiology of Earthworms*, Pergamon Press, 1963.
39 Syers, J. K. and Springett, J. A. in *Earthworm Ecology* (ed J. E. Satchell), Chapman & Hall, 1983.
40 Guild, W. J. McL. in *Soil Zoology* (ed D. K. McE. Kevan), Butterworths, 1955.

were found coming to the surface, allowing the irrigation water to penetrate rapidly.[41] In Holland where gang mowing was practised in a grassed orchard on a clay loam soil there were 5 million large worm channels and many more smaller ones per hectare.[42] In Germany up to 10 million were found in the subsoil of well-manured arable fields and 1 million were common on normal arable soils.[43] In an apple orchard in Cambridgeshire, an average of 18.5 *L. terrestris* burrows opening at the surface per quadrat of 0.37 m^2 were counted. If the average diameter of such burrows is reckoned as 6.4 mm, this would be equivalent in cross-section to a drainage pipe 44 mm in diameter in each metre square.[44]

The effect of earthworms on soil aeration is less clear. Data obtained at Rothamsted suggest that the volume of soil ejected annually on to the surface by earthworms amounts to only about 0.5 to 6 per cent of the soil in the top 10 cm, the total pore space amounting to 40 to 59 per cent of its volume. Since the rate of casting on the pastures concerned seems to be about average for Western Europe, it seems that the effect of earthworms on soil aeration is of only minor importance.[34] The overall effects of earthworms in the soil become apparent when comparisons are made between soils in which they are absent and otherwise similar soils in which they are present. In young polder soils recently reclaimed from Lake Ijssel in Holland earthworms are absent, although the ecological conditions are favourable. Pastures produced on these areas, in the absence of worms, are subject to various problems, including detachment of the sward by grazing cattle or sheep, and surface compaction by treading. Experimental plots were inoculated with earthworms, batches of about 3000 worms being distributed over plots at a density of 10 m^{-2}. The problems were alleviated in the inoculated areas, although the effort involved was considerable in relation to the slow rate of dispersal of the worms, the full benefits not being realized for a number of years. It is of interest to note that although the plots were not visible on standard colour aerial photographs of the area, they became visible on infrared thermal image scans, being warmer by night and cooler by day than the surrounding worm-free areas. This is thought to be related to the absence of a mat of dead organic debris on the surface in the worm-bearing plots.[45]

Effect of cultivations

Arable soils usually contain a much smaller weight of earthworms than pastures, often of the order of 100 kg ha^{-1} less, and typically composed of

41 Dimo, N. A., *Pedology*, 1938, **4**, 494.
42 Edelman, C. H., *Trans. 5th Int. Congr. Soil Sci.* (Leopoldville), 1954, **1**, 119.
43 Finck, A., *Z. Pfl, Ernähr.*, 1952, **58**, 120.
44 Raw, F., *Nature, Lond.*, 1959, **184**, 1661.
45 Hoogercamp, M. *et al.*, in *Earthworm Ecology* (ed J. E. Satchell), Chapman & Hall, 1983.

smaller species, unless the land is given large regular treatments with farmyard manure. It seems, therefore, that cultivation is in some way damaging to earthworm populations. It is important to note, however, that the type of shallow cultivation practised in some orchards does not adversely affect earthworms. At least as high numbers of *L. terrestris* were found in apple orchards which received regular shallow cultivations as under a gang-mown grass sward, but most of these earthworms would have been below the layer of soil that was loosened by the cultivation.[46] The two cultivated orchards examined contained nearly 2 t ha^{-1} of *L. terrestris* which was as high as the highest grassed orchard. These worms buried over 90 per cent of the leaf fall, which contained just over 1 t ha^{-1} dry matter, during late autumn and winter. It seems probable that any adverse effect of shallow cultivation in orchard soils is more than compensated for by this regular high input of organic matter to the soil surface, so that the soil can support a larger earthworm population than when under annual crops.

The practice of direct drilling, in which the pasture or other existing vegetation is destroyed by the use of a non-persistent herbicide such as paraquat, and the new crop is slit-seeded into the soil without further cultivation, appears to be beneficial to earthworm populations. With direct drilling, populations under arable crops can be maintained at a similar level to those under grassland. This effect of direct drilling has been studied by a number of authors, and the situation is well summarized by Edwards.[47] Figures 16.1 and 16.2 show earthworm populations in several experimental situations where comparisons have been made between direct drilling and various degrees of cultivation.[47] The nature of the adverse effect on earthworm populations produced by traditional cultivations on arable land has been investigated in some detail.[48] It is commonly noted that large numbers of birds, particularly gulls, follow the plough to feed on material turned up from the soil. The situation was studied in arable fields in Switzerland, particularly in relation to earthworm predation by the Black-headed Gull (*Larus ridibundus* L.). Earthworm populations were assessed in the fields by the standard procedures of formaldehyde expulsion combined with digging, flotation and hand sorting, and earthworm biomass was found to vary between 1000 and 2000 kg ha^{-1}. Samples of exposed earthworms were collected from behind the plough and mortality due to mechanical damage was estimated to be about 25 per cent. Predation by gulls was studied by direct observation through binoculars and by analysis of stomach contents. While it was clear that earthworms formed an important part of the diet of the gulls, it was concluded that they eat only a small fraction, a quarter to a third, of the worms made available by cultivation, and a quarter of these

46 Raw, F., *Ann. appl. Biol.*, 1962, **50**, 389.
47 Edwards, C. A. in *Earthworm Ecology* (ed J. E. Satchell), Chapman & Hall, 1983.
48 Cuendet, G. in *Earthworm Ecology* (ed J. E. Satchell), Chapman & Hall, 1983.

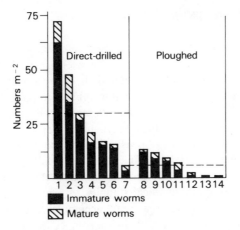

Fig. 16.1 Populations of *L. terrestris* in seven direct-drilled and seven ploughed fields in Sussex.

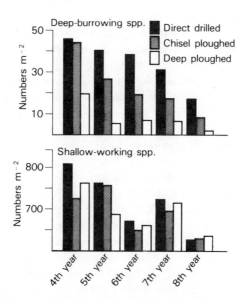

Fig. 16.2 Populations of earthworms in plots that had been direct-drilled, chisel-ploughed or deep ploughed for eight years.

would die anyway from mechanical injury. Furthermore, the exposed worms represent only about 10 per cent of the total biomass, and hence it is most unlikely that bird predation is an important factor in the population decrease in arable land.

Earthworms and pollution

Earthworms feed on plant remains, in some cases taking the material from the soil surface, while the earthworms themselves are taken as food by numerous species of birds and small mammals. They are thus well placed to ingest pollutant chemicals which reach the soil and to pass on such chemical residues to higher forms of life. The degree to which accumulation and concentration of chemical residues takes place depends on the chemical stability of the material and also on its solubility properties. Organochlorine pesticides and related industrial chemicals such as polychlorinated biphenyls (PCBs) have a very low solubility in water but are readily soluble in oils and fats, and also possess great chemical stability. A low level of contamination by such substances in the soil may thus accumulate in the fat deposits of earthworms and be passed on, with further concentration, into earthworm predators. Where large applications of organochlorine insecticides were used, as in orchards or forests, considerable concentration into earthworms has been recorded. In Michigan, DDT was applied to elm trees to combat Dutch Elm Disease at an application rate of 11 to 13 kg ha^{-1} annually for three years. The soil was found to contain 6.3 ppm of DDT, while samples of *Lumbricus terrestris* and *Allolobophora caliginosa* carried residues of 11.3 and 28.0 ppm respectively.[49] Accumulations in normally treated arable land are more modest, the residues being highest in the small species of *Allolobophora*, such as *A. caliginosa* and *A. chlorotica* and lower in the deep burrowing *Lumbricus* species. The organophosphorus insecticides which have now largely replaced the organochlorine compounds in agriculture tend not to accumulate in earthworms. In general, they have little effect on earthworms, although some of the more toxic compounds may produce some initial direct kill of worms.

The accumulation into earthworms of tetrachloro dibenzodioxin (TCDD) in the soil contaminated by a chemical plant explosion at Seveso in Italy in 1976 has been investigated.[50] TCDD is not a pesticide, but has comparable solubility and accumulative properties to organochlorine insecticides. Earthworms accumulated TCDD in their tissues and concentrated it, relative to the mean level in the soil, by about 14-fold. Their burrowing and casting activity tended to bring the TCDD back to the surface. It was noted that moles, earlier present, were absent from the most highly contaminated area. Earthworms being an important part of the diet of moles, it was presumed that the predators were affected by TCDD.

The uptake of heavy metal compounds by earthworms has been often reviewed.[51] Heavy metals may reach the soil in a variety of ways. There are

49 Brown, A. W. A., *Ecology of Pesticides*, Wiley, 1977.
50 Martinucci, G. B., Crespi, P. *et al.* in *Earthworm Ecology* (ed J. E. Satchell), Chapman & Hall, 1983.
51 O'Connor, F. B. in *Soil Biology* (eds A. Burges and F. Raw), Acad. Press, 1967.

natural levels of these materials, which may vary in relation to the geology in different localities. Levels of man-made pollution may be high in old industrial areas where mining or smelting has been carried on for many years, in times when there was less concern than today in relation to the disposal of wastes. In an agricultural context, massive soil contamination may occur in orchards and vineyards with a long history of spraying with copper fungicides, where levels of copper in the leaf litter may reach 2500 mg kg^{-1} of the dry weight. In such situations few earthworms can survive and there is marked inhibition of litter decomposition.

Soil contamination by heavy metals at a generally lower level may occur where sewage sludge is used as a fertilizer. If the sludge is derived from sewage from industrial areas, it is likely to contain residues of a number of toxic metals, including cadmium, arsenic, lead and zinc. Earthworms do not in general accumulate heavy metal residues, although there is evidence that cadmium is accumulated to some extent. Problems are more likely to arise from toxic effects upon plant growth and the presence of residues in crops (see Ch. 23).

Enchytraeid worms

The oligochaete group of Enchytraeidae have received less attention than the Lumbricidae, probably because they are of greatest relative importance in soils of low agricultural value. A valuable introduction to the study of the Enchytraeidae, including guidance on taxonomy and identification, together with information on their ecology, is provided by O'Connor.[51] They have been studied in some Danish pastures,[52] in a coniferous forest in Wales[52] and in Pennine moorlands.[53] These soils contain between 10 000 and 300 000 m^{-2}, with a biomass of between 30 and 100 kg ha^{-1}. Information on the structure and size of their populations has been reviewed more recently.[54] A number of tundra sites appear to have the highest enchytraeid biomass values. This is a result of high population densities and also of high average individual weights in such areas. The Enchytraeidae appear to be intolerant of drought. Comparison of population densities at various sites in northwestern Europe show that humid habitats favour high population densities due to relatively low summer mortality.

Enchytraeidae may be extracted from the soil by exposing cores to

52 Nielsen, C. O., *Oikos*, 1955, **4**, 187; 1955, **6**, 153.
53 Peachy, J. E. in *Progress in Soil Biology* (ed P. W. Murphy), Butterworths, 1962; *Pedobiologia*, 1963, **2**, 81.
54 Petersen, H., *Oikos*, 1982, **39**, 311.

temperature and humidity gradients, and two different techniques have been described.[51] Information on the feeding habits of the Enchytraeidae is inadequate. They appear to feed on rotting material with some fungal and bacterial content.

Nematodes

The soil nematodes, commonly known as eelworms, are small non-segmented worms with thin spindle-shaped bodies. Many, when adult, are between 0.5 and 1.5 mm in length and 10 to 30 μm in width. Most plant-feeding nematodes, of which there are many species, live in association with roots or other underground structures as ecto- or endoparasites. These plant-feeding species are characterized by the possession of mouth spears or stylets, used to penetrate the cells of the host structures. Many of these plant-feeding species are the cause of particularly intractable problems as pests of economic crops. Species of *Heterodera*, the cyst-forming eelworms, produce egg-filled cysts upon the attacked roots. The cysts can persist in the soil for seven years or more, until the eggs are triggered to hatch in response to root exudates from susceptible plants. The economically important potato root eelworm is included in this group. Species of *Meloidogyne*, the root knot eelworm, attack and distort the roots of a wide range of herbaceous plants, while species of *Ditylenchus* infest cereal crops.

The soil also supports many groups of free-living eelworms. Bacterial feeders are often abundant in decaying organic matter and in wounds or lesions caused by other pests or diseases. There are also numerous predatory forms, subsisting on other soil animals, from protozoa to other nematodes. These predators often carry an impressive toothed armature around the jaws.

The general biology of soil nematodes is discussed by Nielsen[55] while the plant parasitic forms are well presented by Jones and Jones.[56]

While various specialist techniques have been developed to sample infestations of plant parasitic nematodes, free-living forms may be extracted from soil samples by the use of variants of the Baerman funnel. The soil sample is placed in a mesh bag or on a wire grille across the neck of a water-filled funnel, the funnel being closed by a short length of plastic tube and a clip at the lower end. Nematodes migrate from the soil into the water and fall down to the bottom of the funnel, where they may be collected by releasing the clip. Several variations of this method have been devised.[57]

55 Nielsen, C. O. in *Soil Biology* (eds A. Burges and F. Raw), Acad. Press, 1967.
56 Jones, F. G. W. and Jones, M., *Pests of Field Crops*, Arnold, 1964. For effects of eelworms on specific crops see Webster, J. M., *Economic Nematology*, Acad. Press, 1972; Nickle, W. R. (ed) *Plant and Insect Nematodes*, Dekker, 1984.
57 Southwood, T. R. E., *Ecological Methods*, Methuen, 1966.

Nematodes live in the soil water films, and they have very specific space requirements if they are to be numerous. Most are strictly aerobic, so although they may be able to stay alive in flooded soils, they are only active when it is drained.[58] Space and moisture in soils are critical factors for nematode movement. Except in coarse textured soils, nematodes cannot penetrate the spaces between particles but are confined to those between aggregates. In an untilled woodland soil containing clay, nematodes were seen in macropores between impenetrable aggregates, and the amount of usable space available was estimated from the moisture characteristics of samples taken down the soil profile. At 2, 6, 18 and 30 cm, 13, 8, 6 and 5 per cent of the soil volume was available and the decrease with depth corresponded with increasing amounts of clay.[59] Pore space requirements are not as important for the gall-forming or root-inhabiting nematodes, because they remain in the root channels and can move through them to attack new roots of a susceptible crop. They can therefore be serious pests in all soils, but since most only attack a relatively narrow range of crops, they may be susceptible to control by suitable crop rotations. Some nematodes, however, produce eggs in highly resistant cysts which can persist and remain infective in the soil for a number of years, making these species difficult to control by crop rotation.

The feeding habits of free-living nematodes are not known in detail. Studies have been made on the feeding relationships of common soil nematodes from subalpine tussock grassland soil of the Southern Alps of New Zealand.[60] Twenty-three species of stylet-bearing and eighteen species of non-stylet-bearing species were observed separately in laboratory feeding trials. Twenty-one of the stylet-bearing species fed on plant material or on other nematodes, but none of the twenty-three fed on bacteria. Sixteen species of the non-stylet-bearing nematodes fed and reproduced on mixed bacterial plates and eleven were established on cultures of single species of bacteria.

Nematodes are preyed upon by other nematodes and also by some species of fungi which have developed trapping mechanisms. About fifty species of such fungi are known, and although some are quite widely distributed in field soils, they do not appear to effect much natural control of nematodes. Active nematodes are trapped by sticky hyphae or contractile rings which hold them while trophic hyphae penetrate their integument and ramify through their bodies. Unfortunately, there are not yet any methods available for stimulating these nematode predators to help control the numbers of plant parasitic nematodes in soil.

58 van Gundy, S. H., *Phytopathology*, 1962, **52**, 628.
59 Jones, F. G. W. *et al., Soil Biol. Biochem.*, 1969, **1**, 153.
60 Wood, F. H., *Soil Biol. Biochem.*, 1973, **5**, 593.

It is possible that in some pasture soils nematodes parasitic on plant roots may constitute about half the nematode population, while in forest soils nearly all will feed on soil organisms. The dry weight of nematodes may be as high as 440 mg kg^{-1} on suitable temperate grassland.[61] Nematodes have a high metabolic rate when active and will then have an oxygen demand per unit weight appreciably higher than other members of the soil fauna.

The larger arthropods

Soils normally support a diverse assemblage of larger arthropods, as distinct from the microarthropods, which are normally considered to include the Collembola and Acari. The larger arthropods include terrestrial Isopoda (woodlice), Chilopods and Diplopods (centipedes and millipedes), larvae and sometimes adults of a number of Orders of insects, of which the Coleoptera (beetles), the Diptera (flies), the Isoptera (termites) and the Hymenoptera (ants) are the most abundant. Some larger members of the Class Arachnida (spiders and related forms) may also be included. The extent to which these larger arthropods can be considered to be soil animals is often somewhat doubtful. The larvae of many flies and some beetles possess the character-istics of true soil animals, such as poorly developed appendages and eyes, and lack ability for rapid movement. They can be important burrowers and chan-nellers in the soil. The active predacious larvae of carabid and staphylinid beetles, however, use the soil, and particularly the litter layer, as a refuge and do most of their foraging for prey on or above the soil surface. Similar considerations apply to the spiders. Many beetle and fly larvae are sapro-phagous, but some will attack the roots of agricultural crops or the young plants just above the soil surface. Among the principal agricultural pests on living crops are wireworms, which are the larvae of elaterid beetles, predomi-nantly of the genus *Agriotes*, and leatherjackets, which are the larvae of craneflies (Tipulidae).

Both millipedes and centipedes are present in soils, often in considerable numbers. In general the food of millipedes is decaying plant litter, although some will damage living plant roots and are recognized pests, particularly in greenhouses. Most of the centipedes are active predators, subsisting upon the other soil fauna. While some millipedes have been shown to be able to digest cellulose,[62] many cannot and so must consume a very large volume of litter to extract sufficient carbohydrates for their requirements. They can thus be important as mechanical comminuters and mixers of litter with soil. It has been estimated, for example, that the millipede *Glomeris marginata* feeding

61 Petersen, H., *Oikos*, 1982, **39**, 330.
62 Hartenstein, R., *Soil Biol. Biochem.*, 1982, **14**, 387.

on ash (*Fraxinus excelsior*) litter only utilized 6 to 10 per cent of the dry matter which it ingested and only converted about 0.3 to 0.5 per cent into its body tissues.[63] In both millipedes and centipedes there are species adapted for burrowing in the soil and species that mainly feed and live above the soil surface. The humidity of the air is an important environmental factor for both these groups. Many have little ability to control water loss from their body in a dry atmosphere and equally cannot control water intake if immersed in water. The surface layers of forest soils, with much leaf litter, thus form a favourable environment for many species. Forest soils which are not too acid, but are for some reason not suitable for large earthworm populations, such as acid sandy soils, often have high millipede populations.

The woodlice – isopods – are predominantly saprophagous or phytophagous, but their relative importance in the decay of organic matter is not known. Investigations on soil respiration in the presence and absence of the woodlouse *Oniscus asellus* indicated that litter comminution by the animals was the main factor contributing to the enhancement of organic matter mineralization.[64] Woodlice may have some importance in sites periodically too dry for earthworm activity. They may be very active tunnellers and burrowers during the summer in semi-desert soils when earthworms are aestivating. They were found burrowing down to 60 or 90 cm and bringing up about 5 t ha^{-1} of soil during the season.[65]

The extraction of these larger forms from the soil, for population sampling purposes, is accomplished through a variety of techniques.[57] Wireworms have been sampled by floating them from crumbled soil samples in salt solutions of appropriate density. Tipulid larvae may be expelled from the undisturbed soil by watering with an emulsion of o-dichlorbenzene. The larger beetles and their active larvae are best sampled by the use of pitfall traps containing a preservative liquid. The choice of the preservative is important, in that it must not be repellent to the forms it is required to sample nor too attractive to other animals which are not required. Dilute formaldehyde solution seems the best material for beetles. It must be remembered that no trapping method alone is quantitative, in that all traps are selective. There may be specific attractive or repellent effects, and in any event, traps select for high activity, as active animals are more likely to encounter the trap. Quantitative population estimates may only be made through the use of a mark and recapture technique.

63 Bocock, K. L. in *Soil Organisms* (eds J. Doeksen and J. van de Drift), North Holland, 1963.
64 Hanlon, R. G. D. and Anderson, J. M., *Soil Biol. Biochem.*, 1980, **12**, 255.
65 Dimo, N. A., *Pedology*, 1945, **2**, 115.

Ants and termites

The ants (Order Hymenoptera) and the termites (Order Isoptera) belong to distinct insect Orders which are not closely related. Their ecological similarity, as insects living in nests and having a complex social structure, are the result of parallel evolution. Numerous genera of both ants and termites make nests in the soil. Some make passages and chambers for their brood without making any mound, while others make mounds out of surface soil or subsoil without, however, mixing in any organic matter in the way earthworms do. This channelling and burrowing in the soil can have an important effect on the aeration and drainage of soils, and they can be as effective as earthworms in improving soil conditions in this way.[66] They are very active insects and forage over considerable distances for their food. They can thus survive in land carrying sparse vegetation, such as the fringes of deserts. There are genera in each group which harvest green vegetation and carry it back to special chambers in their nests where it is used as a food base on which they cultivate fungi upon which they feed. Some ants collect the seeds of grasses. In semi-arid country, particularly if the grass is already damaged due to overgrazing, these insects can remove almost all the remaining grass. This greatly complicates the problems of soil conservation because it leaves the soil exposed to rainfall impact, run-off and soil erosion.

Ants differ from termites in that they are not confined to equatorial and subtropical regions, but are common inhabitants of temperate soils. They can also differ in their feeding habits, for many genera of ants are predacious, feeding on the mesofauna and on the smaller or more immature members of the soil macrofauna, including termites, of which they are one of the principal predators. They are not restricted to the soil fauna but can also prey on insects feeding on plants and crops. An example of the control of an insect pest by a species of ant is that of the coreid bug *Pseudotherapsis wayi* causing damage to coconuts in East Africa which can be largely controlled by the ant *Oecophylla longinoda*. Unfortunately this ant is itself preyed upon by three other species which do not feed on the bug, and we have not yet learned how to control the fortunes of battle between warring species of ants.

Ants also differ from termites in that some genera protect aphids or coccids within their nests and feed on their exudates. These may either be cultivated on the roots of plants or may be tended on the leaves of herbaceous plants or trees, where the ants protect them from predators. The size of the ant colony may be controlled by the size of the colonies it can tend, and the damage done to crops by ants may be greater than any benefit which might accrue from predation on insects.

A number of genera of termites have developed the use of wood as a substrate for their fungal gardens. Termites are also able to utilize wood

66 Hopp, H. and Slater, C. S., *Soil Sci.*, 1948, **66**, 421.

directly, the cellulose being degraded through the activities of symbiotic micro-organisms in the termite gut. Termites are thus the dominant insects in many of the woodland and forest soils of the tropical and subtropical regions, the ants of these areas being predacious upon them.

Termites are the dominant animals in many tropical soils in that they probably consume more organic matter per hectare per year than any other group, although this is a difficult measurement to make with any accuracy. Certainly in the drier regions of the tropics the termites become the main component of the soil fauna biomass. Investigations on the role of termites in an equatorial rainforest ecosystem in western Malaysia concluded that they consumed 22–32 per cent of the total forest floor leaf litter, although the fraction of this assimilated was much lower.[67] The role of termites as soil animals has been reviewed by Lee and Wood[68] and much information on the ecology of ants and termites is available in reports, edited by Brian,[69] on work done under the International Biological Programme established in 1964.

The termites can be classified into a number of groups on the basis of their nesting and feeding habits. The dry-wood termites live in trees or in the timber of buildings, they need no contact with the soil and therefore do not affect it. Wood-feeding termites in general are probably of more importance in the damage they do to buildings than in their effects on the soil. Little is known of the effects of the so-called humus-eating termites on soil organic matter. They pass humus-rich topsoil through their gut, though whether they break down humus or partially decayed plant remains or fungi or micro-organisms is not clear. They do not enrich the soil they excrete as earthworms do. Some species of termites have very considerable ability to penetrate hard pans and possibly laterite (plinthite) crusts. Termite activity may have been a factor in causing the particular structure of the vesicular or vermicular laterite crusts which are common on the Miocene peneplains of central Africa.

The termites with the most spectacular effect on the soil are the large mound-builders of Africa and northern Australia. In Africa they seem to be confined to areas having an adequate rainy season and a long dry one. They cover hundreds of thousands of square kilometres of equatorial and southern Africa. The mounds can be 12 to 18 m in diameter and up to 7.5 m high, and their remains may persist long after a change of climate has altered the vegetation. The mounds are not inhabited all the time by the termites which built them, but serve to shelter other species of ants and termites, and some of these may inhabit parts of the mound even when the building species is in occupation. These mounds may contain hundreds of tons of earth above soil level, and commonly occur at a spacing of about two per hectare, which

67 Matusomoto, T. and Abe, T., *Oecologia, Berlin*, 1979, **38**, 261.
68 Lee, K. E. and Wood, T. G., *Termites and Soils*, Acad. Press, 1971.
69 Brian, M. V. (ed), *Production Ecology of Ants and Termites*, Cambridge Univ. Press, 1978.

gives some measure of the foraging range of these termites. The important factor for most of these mound-building termites is an adequate wood supply. The exact shape and architecture of the mound depends somewhat on the species present. The mound consists of two parts, a compact outer casing and an inner nest. The soil forming the mound is subsoil which may have come up from as deep as 3 m, and the channels from which the subsoil is derived can extend far beyond the mound itself. In an area of rather small mounds, it was estimated[70] that the average transfer of subsoil to the surface was just over $1 \, t \, ha^{-1} \, a^{-1}$. This figure should be compared with the estimate of $50 \, t \, ha^{-1}$ of worm casts produced during the six months of rain by the worm *Hippopera nigeriae* on a neighbouring area under bush vegetation which shaded the ground and had an abundant leaf fall.

There has been much discussion on the fertility of soils from a termite mound compared with the soil around it. If the soil around the mound is acid and low in calcium, and the crop being grown requires a large amount of calcium, such as sisal, one would expect a better crop on the mound, and on mounds which have been more or less levelled: this effect can be seen quite strikingly on some Tanzanian sisal estates. On the other hand, if the subsoil is poor in nutrients, most of the soil in the mound will be poor, particularly in available nitrogen, and levelling the mound can distribute a nutrient-poor subsoil over a wider area.

Areas where active mound-building *Macrotermes* are common can be difficult to cultivate both because the mounds themselves are difficult to cultivate, and also because, if knocked down by a bulldozer, they may be built up again surprisingly quickly. Since these termites are primarily wood eaters, they would be expected to die out quickly when forest is cleared for cultivation. This is seen in some areas of East Africa, but some will attack annual crops and then cause serious trouble. It is also difficult to maintain mulches on the surface of soils where soil-inhabiting termites of any kind are active.

Ants affect soil in ways similar to termites, particularly those species which are mound builders, although anthills are much smaller than most termite mounds. In and close to the hill, the soil is full of their burrows, the bulk density of the soil is reduced, and the potassium and phosphate status is frequently higher. Much of the soil forming the hill itself may have come from the subsoil. Some species of ants use a hill for a limited number of years, and these may then slowly become levelled but they are not normally used by other species of ants.

70 Nye, P. H., *J. Soil Sci.*, 1955, **6**, 73.

Gasteropod molluscs

Slugs and snails are the two groups of Gasteropods which may to some extent be regarded as soil animals. A general account of these terrestrial molluscs is provided by Newell.[71] In the case of some of the smaller slugs, a considerable proportion of the population may be sheltering beneath the soil surface at any given time. Some species of snails have abundant cellulase enzymes in their gut which allow them to digest cellulose. It seems likely that these enzymes are secreted by the digestive gland of the snail, but that bacterial cellulases also make a contribution.

In agricultural soils it has been shown that the food of slugs is preferentially dead or damaged vegetation, such as freshly fallen leaves and old grass. They are typically scavengers, rendering plant tissues more readily usable by other organisms. The populations of slugs and snails can be very large. The slug *Agriolimax reticulatus* can cause extensive damage to seedling crops in damp conditions, and populations have been estimated as high as 1.4×10^6 ha^{-1}. Slug damage has been a factor of some importance in relation to direct drilling, particularly in heavy clay soils. In such soils the seeding slots may remain open in damp conditions, providing shelter for slugs in a position where they can readily attack the germinating seeds.

Soil-inhabiting mammals

Various small mammals – mice, voles, shrews, rabbits, moles, gophers, prairie dogs – are present in appreciable numbers in some undisturbed soils and some may cause damage, usually minor, in pastures and arable land. Though their total weight per hectare is usually small, probably under 5 kg ha^{-1} in forest soils and up to 10 kg ha^{-1} in open ranges, they can cause loosening of the surface layers of the soil by their often extensive excavations.

71 Newell, P. F. in *Soil Biology* (eds A. Burges and F. Raw), Acad. Press, 1967.

17

Interactions between plant roots and micro-organisms

The rhizosphere

The influence of plant roots on the microbial population in the surrounding soil is considerable, for plant roots and root hairs may carry large
concentrations of micro-organisms on their surfaces. Not all the roots or root
hairs carry these concentrations, and those that do so are not necessarily
uniformly covered. The interfacial volume between the plant root and the
bulk of the soil, in which this microbial population lives, is called the rhizosphere.[1] The boundaries of the rhizosphere are not exact; for example, some
fungal hyphae live both in the soil and penetrate the outer cortical cells of
the root (the endorhizosphere). There is also no sharp boundary between the

---

1 Hiltner, L., *Arb. deut landw. Ges.*, 1904, **98**, 59, first used this word.

rhizosphere external to the root surface (the ectorhizosphere) and the bulk of the soil. The rhizosphere may be up to 1 to 2 mm thick[2] and, if so, nearly all the surface soil may be considered as rhizosphere soil for crops such as some pasture grasses which have a very extensive root system in the surface soil. Some workers have also introduced the concept of the rhizoplane[3] population, which is defined as the microbial population living on the outer surface of the root. There is, in fact, a continuous gradation in the characteristics of the microbial population living in the various surface regions of the rhizosphere and in the soil well away from the root.[4]

Direct microscopical examination of plant roots taken from a soil shows that the growing root tip is typically free of micro-organisms, but that the zone of root elongation behind the tip carries a bacterial population, and sparse clumps of bacteria near the tip become almost a sheath further back. The older part of the root may carry a considerable fungal and bacterial population, the fungal population becoming of increasing importance as root cells become moribund.[5] There can also be protozoa and nematodes in the rhizosphere which are predatory on the rhizosphere population,[6] but some of the nematodes are parasitic on the root itself. Plate 17.1 shows the bacterial colonization of cereal root rhizoplanes.

The size of the bacterial population has usually been investigated through the use of the plating technique, which is a technique known to underestimate bacterial numbers. The results obtained by a number of workers for a range of crops in a range of soils indicate that the soil adhering fairly tightly to the root surface may contain up to 200 times as many bacteria as in the body of the soil, with ratios between 10 and 50 being common.[3] Louw and Webley[7] found that for the oat rhizospheres they examined, the ratio of bacterial numbers as determined by the plating technique, using a relatively nonselective medium, and the direct cell count was between 1:1 and 1:5 in the rhizosphere and between 1:8 and 1:11 in the soil. Thus, a part of the apparent concentration of bacteria reported by most workers in the rhizosphere compared with the body of the soil, could be due to their counting technique. From measurements of the rate of microbial respiration,[8] the activity of the population in the rhizosphere soil appears to be between two and four times

2 Rovira, A. D., *Aust. J. agric. Res.*, 1961, **12**, 77.
3 Clark, F. E., *Adv. Agron.*, 1949, **1**, 241.
4 For reviews of this subject see Newman, E. I., *Biol. Rev.*, 1978, **53**, 511; Bowen, G. D. in *Contemporary Microbial Ecology* (eds D. C. Ellwood, *et al.*), 1980, Acad. Press; Lynch, J. M., *Soil Biotechnology*, 1983, Blackwell.; Whipps, J. M. and Lynch, J. M., *Adv. Microbial Ecol.*, 1986, **9**, 187.
5 Parkinson, D., *et al.*, *Pl. Soil*, 1963, **19**, 332.
6 See Henderson, V. E. and Katznelson, H., *Canad. J. Microbiol.*, 1961, **7**, 162; Katznelson, H., *Soil Sci.*, 1946, **62**, 343; Clarholm, M., *Microbial Ecol.*, 1981, **7**, 343.
7 Louw, H. A. and Webley, D. M., *J. appl. Bact.*, 1959, **22**, 216.
8 See Katznelson, H. and Rouatt, J. W., *Canad. J. Microbiol.*, 1957, **3**, 673; Reuszer, H. W., *Soil Sci. Soc. Am. Proc.*, 1949, **14**, 175; Vancura, V. and Kunc, F., *Z. Bakt. Abt. II*, 1977, **32**, 472.

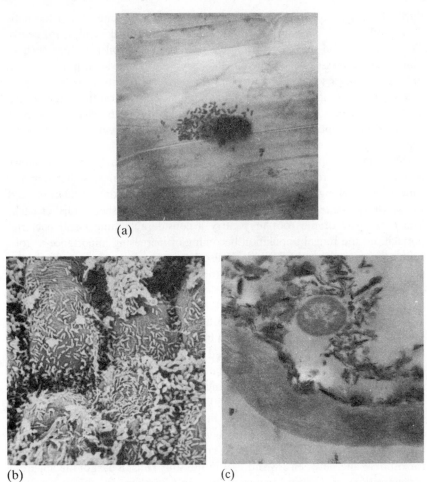

(a)

(b)   (c)

Plate 17.1   Bacterial colonization of cereal root rhizoplanes. (a) Light micrograph of bacterial colony on wheat root in solution culture. Note
colony development and the difficulty of counting the number of
cells. (b) Scanning electron micrograph of bacteria on maize root in
solution culture. Note accumulation of bacteria in the intercellular
spaces of the convex epidermis root cells. (c) Transmission electron
micrograph of bacterium on wheat root surface surrounded by clay
particles.

as high as in the bulk soil, but the reliability of the techniques used and the
interpretation of the results is uncertain. Other measurements on soils in the
field have shown an increase in the soil microbial biomass measured by the
fumigation respiration technique when roots are growing.[9]

9 Lynch, J. M. and Panting, L. M., *Soil Biol. Biochem.*, 1980, **12**, 29.

The bacterial population in the rhizosphere differs from that in the bulk soil in that it contains larger proportions of Gram-negative non-spore-formers such as *Agrobacterium radiobacter* (a group of bacteria in the same family (Rhizobiaceae) as root nodule bacteria), *Pseudomonas* spp., and myco-bacteria and corynebacteria. While most *Bacillus* spp. and Gram-positive cocci are reduced, *B. polymyxa*, *B. circulans* and *B. brevis* may be stimulated.[10] The population also contains a higher proportion of chromogenic and motile forms and more ammonifiers, nitrifiers, denitrifiers and aerobic cellulose decomposers, but a lower proportion capable of using aromatic acids such as benzoic.[11] The essential nutrient requirements of the rhizosphere bacterial population also differ from those of the general soil population; the rhizos-phere population containing a lower proportion of bacteria requiring complex growth factors such as are present in soil extract and yeast extract, but a higher proportion needing simple amino acids.[12] This is what would be expected if the principal sources of nutrients for this population are root exudates, cell lysates, sloughed off cells from the growing root cap, moribund root hairs and moribund cortical cells in the older root region. It is still tech-nically difficult, or not yet possible, to make any quantitative estimate of the relative importance of these sources, but root excretions are probably the principal source in the younger regions of the root, and moribund tissues become of increasing importance as the root ages. Root excretions contain a wide variety of substances, including sugars, amino acids, organic acids, nucleotides and enzymes. The zone of elongation behind the root cap is a zone where excretion and particularly amino acid excretion[13] appears to be active.[14] This could account for the bacterial population in this region being characterized by simple nutritional requirements and high metabolic activity. Of unknown importance is the mucilaginous sheath or mucigel which may surround the roots of some plants under some conditions and in which bacteria can sometimes be seen embedded.

The only reliable method for investigating the amount of carbon released by roots is to grow the plant on a source of $^{14}CO_2$ and measure the ^{14}C released by the roots into a sealed container of soil.[15] The total ^{14}C loss meas-ured includes CO_2 due to microbial respiration of the carbon derived from

10 Clark, F. E., *Trans. Kansas Acad. Sci.*, 1940, **43**, 75.
11 See for example Rouatt, J. W. and Katznelson, H., *J. appl. Bact.*, 1961, **24**, 164; Rouatt, J. W. *et al.*, *Soil Sci. Soc. Am. Proc.*, 1960, **24**, 271; Sperber, J. I. and Rovira, A. D., *J. appl. Bact.*, 1959, **22**, 85. For older work see Starkey, R. L., *Soil Sci.*, 1929, **27**, 355; Lochhead, A. G., *Canad. J. Res.*, 1940, **18C**, 42.
12 Lochhead, A. G. and Rouatt, J. W., *Soil Sci. Soc. Am. Proc.*, 1955, **19**, 48.
13 Pearson, R. and Parkinson, D., *Pl. Soil*, 1961, **13**, 391; Schroth, M. N. and Snyder, W. C., *Phytopathology*, 1961, **51**, 389; Frenzel, B., *Planta*, 1960, **55**, 169.
14 For reviews see Rovira, A. D., *Soils Fertil.*, 1962, **25**, 167; with McDougall, B. M. in *Soil Biochemistry* (eds A. D. McLaren and G. H. Peterson), Arnold, 1967.
15 Barber, D. A. and Martin, J. K., *New Phytol.*, 1976, **76**, 69.

Table 17.1   Percentage distribution of photosynthetically fixed ^{14}C in wheat and barley plants grown in soil for 21 days under different temperature and daylength regimes in an atmosphere containing 0.03–0.10% constant specific activity $^{14}CO_2$ (70–90 $\mu Ci\ g^{-1}C$)

| | 15 °C constant | | 18 °C day–14 °C night | | | |
| | 16 h daylength | | 16 h daylength | | 12 h daylength | |
	Barley	Wheat	Barley	Wheat	Barley	Wheat
Shoot	45.5	47.6	48.1	41.2	73.1	48.6
Root	17.1	19.0	17.1	18.7	9.2	26.8
Respired CO_2 from roots + micro-organisms	25.9	21.6	18.9	23.1	9.4	12.9
Total ^{14}C loss (including respired CO_2)	37.4	33.5	34.7	40.0	17.7	24.6

roots. Recent results[16] indicate that up to 40 per cent of the plant's dry matter production can be released by roots (Table 17.1).

Effects of micro-organisms on plants

Plant nutrition

Although the rhizosphere population is dependent on the plant for its principal source of energy and nutrients, it is not certain how far the growth of the plant is affected under normal conditions. There is no evidence that the rhizosphere population weakens a healthy plant, but under conditions of very low nutrient supply in the soil the rhizosphere population may compete for any nutrient in short supply so reducing the supply available to the crop. Plate 17.2 reproduces a radioautograph of a section of a sterile and non-sterile root growing in a solution labelled with ^{32}P. It shows clearly the concentrations of ^{32}P-labelled material on the outside of the non-sterile root, a feature that is absent from the sterile root.[17] Barber and colleagues[18] have shown that the rhizosphere population around barley roots growing in solution culture, can either reduce or increase the amount of phosphate taken up by the plant compared with a sterile root system (Table 17.2). Under some conditions the rhizosphere may have microsites which are anaerobic

16 Whipps, J. M. and Lynch, J. M., *New Phytol.*, 1983, **95**, 605;   Whipps, J. M., *J. exp. Bot.*, 1984, **35**, 767; but see also Lambers, H. in *Root Development and Function* (eds P. J. Gregory *et al.*), p. 125, Cambridge Univ. Press, 1987.
17 Barber, D. A. *et al.*, *Nature, Lond.*, 1968, **217**, 644.
18 Barber, D. A. *et al.*, *Aust. J. Pl. Physiol.*, 1976, **3**, 801.

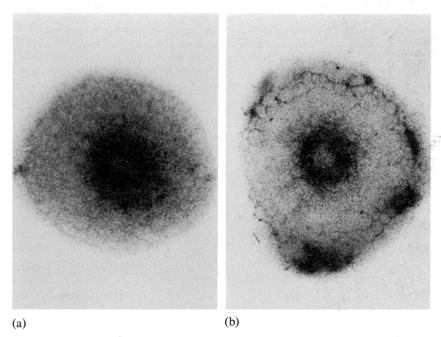

(a)   (b)

Plate 17.2   Distribution of ^{32}P in (a) sterile root; (b) non-sterile root. Three-week-old barley plant growing in a culture solution containing 3 × 10^{-6}M phosphate.

and where there is active denitrification, even if the soil as a whole is aerobic.[19]

The population may also reduce the availability of a nutrient to a crop by converting it to an insoluble form just outside the root surface. Thus, Timonin[20] found that the microbial population in the rhizosphere could induce manganese deficiency in some varieties of oats by oxidizing manganous ions to manganese dioxide, and he also found that this harmful effect of the population could be aggravated by applying a straw mulch. However, in solution culture the rhizosphere bacteria can produce iono-phores which promote manganese uptake by roots.[21]

One particular aspect of the rhizosphere–plant interaction which has received a great deal of attention is seed inoculation. Russian agronomists for several decades inoculated the seed of wheat with a culture of *Azotobacter* before drilling, and found that sometimes this practice increased yields up to about 10 per cent. They also found some evidence that inoculating the seed

19 Woldendorp, J. W., *Meded. Landb. Hoogesch, Wageningen*, 1963, **63**, 1.
20 Timonin, M. I., *Soil Sci. Soc. Am. Proc.*, 1947, **11**, 284; see also Gerretsen, F. C., *Ann. Bot.*, 1937, **1**, 207.
21 Barber, D. A. and Lee, R. B., *New Phytol.*, 1974, **73**, 97.

Table 17.2 Effect of micro-organisms on the uptake by barley seedlings of phosphate from ^{32}P-labelled KH_2PO_4 (5 μM); the seedlings had been grown under sterile or non-sterile conditions in complete nutrient solution with or without phosphate. Note that micro-organisms depressed phosphate uptake except for short periods of uptake with young seedlings. (Adapted from ref. 18)

Pretreatment conditions	Age of seedlings (days)	P uptake by plants	
		(period)	(nmol)
Complete nutrient			
Sterile	8	30 min	3.1
Non-sterile	8	30 min	3.2
Phosphate-free			
Sterile	8	30 min	3.8
Non-sterile	8	30 min	4.1
Complete nutrient			
Sterile	8	24 h	22.2
Non-sterile	8	24 h	14.8
Phosphate-free			
Sterile	8	24 h	67.7
Non-sterile	8	24 h	67.8
Phosphate-free			
Sterile	21	30 min	86.7
Non-sterile	21	30 min	77.9
Phosphate-free			
Sterile	21	24 h	757.0
Non-sterile	21	24 h	580.2

with *Bacillus megatherium* increased the phosphate supply to the crop though the evidence for this is less convincing.[22] *Azotobacter* is not an organism that is normally concentrated in the rhizosphere but if the wheat seed has a sufficiently large population of *Azotobacter*, these organisms grow along, or are carried along, the root surface as it grows into the soil, and give a number of large colonies on the roots.[23] Their beneficial effect is due not to an increase of the nitrogen supply to the crop but to the excretion of growth substances, such as some gibberellins and β-indolyl acetic acid[24] which, when taken up by the young plant, modify its growth and subsequent yield.

22 For a review of the Russian work see Cooper, R., *Soils Fertil.*, 1959, **22**, 327.
23 Jackson, R. M. and Brown, M. E., *Ann. Inst. Pasteur*, 1966, **111**, Suppl. 112. For a photograph of this development on barley roots see Rovira, A. D. and McDougall, B. M. in *Soil Biochemistry* (eds A. D. McLaren and G. H. Peterson), Arnold, 1967.
24 Brown, M. E. and Burlingham, S. K., *J. gen. Microbiol.*, 1968, **53**, 135; *J. exp. Bot.*, 1968, **19**, 544; with Walker, N., *Pl. Soil*, 1970, **32**, 250.

Associative nitrogen fixation

Studies originated by Döbereiner[25] in Brazil have demonstrated that tropical grasses have the nitrogen-fixing bacteria *Azotobacter paspali* and *Azospirillum* spp. associated with their roots, which might provide a useful source of nitrogen for the plant by converting atmospheric dinitrogen gas into amino acids. However, theoretical and practical studies[26] make it seem unlikely that nitrogen fixation is very great because nitrogenase needs more energy from carbohydrate to produce a significant amount of nitrogen for the plant than the root system will secrete. Again it seems possible that beneficial effects of inoculants may be due to the production of plant growth regulators by the bacteria and perhaps these could stimulate uptake of soil nitrogen by roots.

Symbiotic nitrogen fixation in non-leguminous plants

Many non-leguminous families and genera of dicotyledonous angiosperms which are phylogenetically unrelated, possess root nodules with nitrogen fixing capacity. They include shrubs and trees and are generally of little agricultural value, being more important as pioneer colonizers. They generally do not feature in climax vegetation.

An actinomycete of the genus *Frankia* is the microbial partner in the symbioses described by Becking[27] (Table 17.3). There are two types which are distinct morphologically and anatomically. A coralloid root nodule is formed by dichotomous branching in the *Alnus* type, and the nodule lobes are in the axil of lateral roots whose apical meristem grows slowly or not at all. In the *Myrica/Casuarina* type, the apex of each nodule lobe produces a normal but negative geotropic root, the nodule being covered by upward growing roots. Figure 20.1 shows the effectiveness of alder (*Alnus*) in increasing the amount of soil nitrogen on glacial moraines.

Of great agricultural significance on a global scale is the *Azolla–Anabaena* symbiosis.[27] *Azolla* is a genus of heterosporous floating aquatic ferns and has long been used as an alternative nitrogen source for rice in Vietnam and the People's Republic of China. *Anabaena*, a blue–green alga (see pp. 465–468), is contained endophytically in ellipsoidal cavities of aerial dorsal lobes of the fern. Algal spores germinate when the female gametophytes form within the megasporocarp, producing undifferentiated filaments. The filaments become

25 Döbereiner, J. in *Biological Nitrogen Fixation for Tropical Agriculture* (eds P. H. Graham and S. C. Harris), Centro International de Agricultura Tropical Colombia, 1982.

26 Brown, M. E., *J. appl. Bact.*, 1976, **40**, 341. Barber, D. A. and Lynch, J. M., *Soil Biol. Biochem.*, 1977, **9**, 305; Lethbridge, G. and Davidson, M. S., *Soil Biol. Biochem.*, 1983, **15**, 365.

27 Becking, J. H. in *Advances in Agricultural Microbiology* (ed N. S. Subba Rao), p. 89, Butterworths, 1982.

Table 17.3 Classification of non-leguminous dinitrogen-fixing angiosperms with *Frankia* symbiosis. (Adapted from ref. 27)

Family	Genera	No. of nodulated spp. (in parentheses total no. of spp.)
Casuarinaceae	*Casuarina*	24 (45)
Myricaceae	*Myrica, Comptonia*	27 (36)
Betulaceae	*Alnus*	33 (35)
Elaeagnaceae	*Elaeagnus, Hippophae, Shepherdia*	20 (51)
Rhamnaceae	*Ceanothus, Discaria, Colletia, Trevoa*	41 (88)
Coriariaceae	*Coriaria*	13 (15)
Rosaceae	*Rubus, Dryas, Purshia, Cercocarpus*	10 (276)
Datiscaceae	*Datisca*	2 (2)

associated with the shoot apex of the developing sporophyte to perpetuate the symbiosis.

The rate of non-leguminous dinitrogen fixation is compared with that of legumes in Table 17.4.[28] The range of values is wide, but the high values are much greater than normally achieved.

Table 17.4 Examples of rates of symbiotic N_2 fixation in terrestrial habitats

Vegetation	$(kg\ ha^{-1}\ a^{-1})$	Vegetation	$(kg\ ha^{-1}a^{-1})$
Tropical/sub-tropical		Temperate	
Pulse legumes			
Soyabean	1–168	Broad bean	45–552
Groundnut	77–124	Garden pea	52–77
Pigeon pea	168–280	Lupin	145–208
Forage legumes			
Stylosanthes	34–320	Lucerne	184–463
Centrosema	126–395	Perennial clover	45–673
Tick clover	897		
Non-legumes			
Casuarina	52	*Alnus rubra*	139
Azolla	83–125	*Hippophae rhamnoides*	3–58

28 Postgate, J. R. and Hill, S. in *Microbial Ecology: a conceptual approach* (eds J. M. Lynch and N. J. Poole), Blackwell, 1979.

Symbiotic nitrogen fixation in leguminous plants

Nitrogen fixation in legumes depends upon a highly coordinated sequence of interactions between plants of the family Leguminosae and soil micro-organisms belonging to the genera *Rhizobium* and *Bradyrhizobium* (see p. 645) which results in the formation of root nodules. Only 48 per cent of the genera of the Leguminosae have been examined for nodulation to date, and of these, 86 per cent were found to be nodulated.[29] Nodules on young lucerne roots are illustrated in Plate 17.3. The association is symbiotic, with the plant leaves supplying carbohydrate for the bacteria which fix atmospheric dini-trogen gas and supply amino acids for both organisms. The root nodule bacteria may become antagonistic to the plant if for any reason the carbo-hydrate supply is restricted, as, for example, when the plant is in the dark or at the time of flowering or setting of seed.[30]

Legumes provide the most important nitrogen-fixing association in agri-culture, and world production is dominated by soyabeans (annual production of almost 10^8 tonnes). As a consequence, much of the research work on

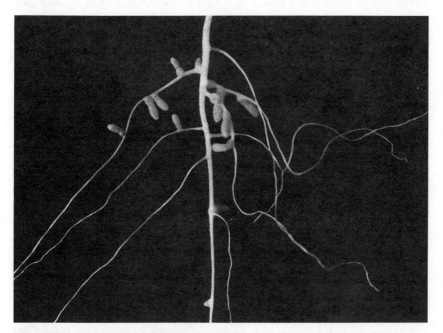

Plate 17.3 Nodules on young roots of lucerne, *Medicago sativa* ($\times$ 5).

29 Allen, O. N. and Allen, E. K., *The Leguminosae: a source book of characteristics, uses and nodulation*, Macmillan, 1981.
30 Pate, J. S., *Aust. J. biol. Sci.*, 1958, **11**, 366, 496.

legumes has been carried out on soyabean and other temperate legumes, with breeding programmes often selecting for increased yield without considering nitrogen-fixing ability.

Norris[31] has emphasized that any general discussion on symbiotic nitrogen fixation must be based on the fact that the great majority of leguminous genera and species are tropical, and that the ancestral ecological niche was probably on fairly strongly leached soils in tropical rainforests. The temperate and subtropical legumes developed from these. Some of them, particularly certain species belonging to the Trifolieae, which includes the genera *Trifolium*, *Medicago*, *Melilotus* and *Trigonella*, and the Viciae, which includes the genera *Vicia*, *Pisum*, *Lens*, *Cicer* and *Lathyrus*, have become adapted to neutral or calcareous soils of higher nutrient status. A high proportion of the leguminous crops cultivated in temperate agriculture belong to these groups, and it is only recently that the role of legumes in tropical and subtropical soils has been more extensively studied.[32]

Legume crops can be broadly divided into those grown for their protein-rich seeds (grain legumes such as soyabean and chickpea) and those grown for forage (pasture legumes such as white clover and *Stylosanthes*). A major advantage of including legumes in farming systems is that they need not depend on the soil supply of mineral nitrogen. When effectively nodulated, legumes may also increase the nitrogen status of the soil when the plant residues are returned to the soil. This is particularly true for pasture legumes, where the role of the grazing animal is important in cycling nitrogen, but it is less true for grain legumes where the seeds and much of the vines or straw are removed from the land. Many legume crops may fail to nodulate when introduced into an area with no previous history of cropping those particular legumes. This is well illustrated by the failure of US soyabean varieties to nodulate when introduced into African countries.[33] The problem may be overcome either by selecting legume varieties which will nodulate with indigenous rhizobia, or by inoculating the introduced variety with one or more suitable strains of *Rhizobium* (see p. 650).

Mineral nutrition

Nodules can only fix nitrogen actively if the plant is adequately supplied with all the mineral elements essential for active growth. In addition, all nodulated plants require small quantities of cobalt,[34] which is required for the formation

31 Norris, D. O., *Emp. J. expl. Agric.*, 1956, **24**, 247.
32 Andrew, C. S. and Kamprath, E. J., *Mineral Nutrition of Legumes in Tropical and Subtropical Soils*, CSIRO, Melbourne, 1978; Döbereiner, J. *et al.*, *Limitations and Potentials for Biological Nitrogen Fixation in the Tropics*, Plenum Press, New York, 1978.
33 Kueneman, W. R. *et al.*, *Pl. Soil*, 1984, **82**, 387.
34 See for example Delwiche, C. C. *et al.*, *Pl. Physiol.*, 1961, **36**, 73; Hallsworth, E. G. *et al.*, *Nature, Lond.*, 1960, **187**, 79.

of leghaemoglobin, and have a high requirement for molybdenum, a constituent of nitrogenase. Some species have a high calcium or pH requirement, and a few a high boron[35] or sulphur requirement; for example, lucerne and groundnuts may have a higher sulphur demand than many other legumes. Magnesium, which facilitates ATP supply to nitrogenase, is essential for nitrogen fixation. Although many legumes have an appreciable calcium demand,[36] there is no evidence that calcium plays any role in the fixation process.

Acid tolerance of legumes depends partly on their tolerance to aluminium and manganese ions and partly on their calcium requirement. Lucerne and red clover, for example, only grow well on soils of fairly high pH or calcium status, and in general they need soils of pH 6 or above for optimum growth. On the other hand, some white clovers and subterranean clovers can fix nitrogen in soils of pH 4.2 to 4.5[37] although they usually respond to liming if the soil is as acid as this. Many tropical legumes can fix nitrogen actively at these low soil pH values, provided they can extract calcium from the soil.[38] A legume relying on nitrogen fixation is more sensitive to acidity than the same legume using soil mineral nitrogen.[39] Tolerance of acidity depends on the strain of *Rhizobium*. For example, inoculation with strains of *Rhizobium japonicum* selected for symbiotic ability in acid soils resulted in yields of soya beans up to 2 t ha^{-1}, and liming had no effect on the response.[40]

The reason for this difference in pH or calcium requirements of the various legumes is not completely known. It is partly due to the toxic effect of aluminium ions,[41] for legumes[42] and rhizobia[43] vary in their sensitivity to acidity and aluminium. Solution culture techniques have enabled the important acidity factors limiting the legume symbiosis to be identified and tolerant strains of rhizobia to be selected.[44] Nodulated legumes vary greatly in their response to calcium, although this may only be a reflection of the differing abilities of legumes to take up calcium from soils poorly supplied with this element.[45]

35 Brenchley, W. E. and Thornton, H. G., *Proc. R. Soc.*, 1925, **B98**, 373.
36 Norris, D. O., *Aust. J. agric. Res.*, 1959, **10**, 651;　Vincent, J. M. and Colburn, J. R., *Aust. J. Sci.*, 1961, **23**, 269.
37 See, for example, Jensen, H. L., *Proc. Linn. Soc. NSW.*, 1947, **72**, 265.
38 Albrecht, W. A., *J. Am. Soc. Agron.*, 1933, **25**, 512;　Albrecht, W. A. and Harston, C. B., *Soil Sci. Soc. Am. Proc.*, 1942, **7**, 247.
39 de Carvalho, M. M. *et al.*, *Agron. J.*, 1981, **73**, 261.
40 Bromfield, E. S. P. and Ayanaba, A., *Pl. Soil*, 1980, **54**, 95.
41 Bryan, O. C., *Soil Sci.*, 1923, **15**, 37. Norris, D. O., *Emp. J. expl. Agric.*, 1956, **24**, 247; *idem, Nature, Lond.*, 1958, **182**, 734.
42 Andrew, C. S. *et al.*, *Aust. J. agric. Res.*, 1973, **24**, 325.
43 Keyser, H. H. and Munns, D. N., *Soil Sci. Soc. Am. J.*, 1979, **43**, 519; Lowendorf, H. S. and Alexander, M., *Appl. environ. Microbiol.*, 1983, **45**, 737.
44 Keyser, H. H. *et al.*, *Soil Sci. Soc. Am. J.*, 1979, **43**, 719;　Wood, M. *et al.*, *Pl. Soil*, 1984, **78**, 367;　Wood, M. and Cooper, J. E., *Soil Biol. Biochem.*, 1985, **17**, 493.
45 Norris, D. O. in *Nutrition of Legumes* (ed E. G. Hallsworth), Butterworths, 1958.

The molybdenum requirements of leguminous plants are almost entirely for nitrogen fixation, for the plants will generally grow well if they have a supply of mineral nitrogen, and their roots will be well nodulated, in soils very low in molybdenum, but the nodules will fix no nitrogen[46]. In Australia and New Zealand there are some soils on which subterranean clover will not grow unless it is given molybdenum (Colour Plate 4). Applications of 20 to 50 g ha^{-1} of Mo as sodium molybdate[47] are common, but if the concentration of molybdenum in the soil is too high, its accumulation in the leaves of leguminous plants can cause toxicity to grazing animals[48] (see Ch. 23).

The growth of an effectively nodulated legume may be independent of the concentrations of mineral nitrogen in the soil, but high concentrations of mineral nitrogen in the soil solution inhibit nodulation and nitrogen fixation. In contrast, low rates of fertilizer nitrogen application at sowing often stimulate plant growth and may enhance nodulation and nitrogen fixation.[49] High concentrations of nitrate inhibit nearly all stages in the development of the symbiosis.[50] The nodulation process may be inhibited at the stage of attachment of rhizobia to the host root,[51] root hair curling and infection thread formation.[52] Variation exists in the ability of different symbioses to fix nitrogen in the presence of mineral nitrogen,[53] offering the possibility of selecting improved combinations of legume variety and *Rhizobium* strain for use in agriculture.

Carbon economy

If they are to fix nitrogen the bacteria in the nodules must be supplied with energy, and hence with oxygen for the oxidation of the carbohydrate. Bond[54] has estimated that, in some of his experiments with soyabeans, 16 per cent of the total carbohydrates synthesized by the plants was respired by the nodules. Of the total respiration from the plant, 57 per cent was from the tops, 18 per cent from the roots and 25 per cent from the nodules. If soil aeration is reduced, the growth of soyabeans given nitrate or ammonia is not

46 Anderson, A. J., *Aust. Counc. Sci. Ind. Res. Bull.* No. 198, 1946. Jensen, H. L., *Proc. Linn. Soc. NSW*, 1945, **70**, 203.
47 Trumble, H. C. and Ferres, H. M., *J. Aust. Inst. agric. Sci.*, 1946, **12**, 32.
48 For the margin between optimum molybdenum contents for nitrogen fixation and the minimum for toxicity to animals, see Jensen, H. L., *Proc. Linn. Soc. NSW*, 1945, **70**, 203.
49 Haystead, A. and Marriott, C., *Grass Forage Sci.*, 1979, **34**, 241.
50 Munns, D. N. in *A Treatise on Dinitrogen Fixation*, Section IV (eds R. W. F. Hardy and A. H. Gibson), Wiley, 1977.
51 Dazzo, F. B. and Brill, W. J., *Pl. Physiol.*, 1978, **62**, 18;   Dazzo, F. B. and Hrabak, E. M., *Pl. Soil*, 1982, **69**, 259.
52 Munns, D. N., *Pl. Soil*, 1968, **29**, 33.
53 Manhart, J. R. and Wong, P. P., *Pl. Physiol.*, 1980, **65**, 502;   Nelson, L. M., *Canad. J. Microbiol.*, 1983, **29**, 1626.
54 Bond, G., *Ann. Bot.*, 1941, **5**, 313.

much affected, but it is significantly reduced if they are nodulated plants receiving no nitrogen fertilizer.[55] Hence a well-nodulated leguminous crop in the field needs a better oxygen supply to its roots, which respire more carbon dioxide than does a cereal or root crop.[56] Wilson[57] calculated that for a crop fixing 1 kg dinitrogen per hectare per day in summer, nodular respiration would amount to 3.3 g m^{-2} per day of carbon dioxide.

Nitrogen fixation, however, is a hydrogenation reaction, and must therefore take place in an environment low in oxygen. Nodules that are fixing nitrogen actively contain leghaemoglobin (related to the haemoglobins present in animal blood), the function of which appears to be the active transport of oxygen to the bacteria while maintaining a low oxygen tension in the bathing solution. The pink colour due to this leghaemoglobin is characteristic of nodules fixing nitrogen; if it is absent it is usually a sign that nitrogen fixation is not taking place within the nodules. Legumes may also control the concentration of oxygen inside the root nodule by varying the resistance to oxygen diffusion through the nodule tissue.[58]

Rhizobium nodules are more efficient fixers of nitrogen than the non-symbiotic bacteria, for *Rhizobium* fixes 1 g of nitrogen for every 15 to 20 g of carbohydrate oxidized, whereas the non-symbiotic bacteria use about 2.5 times as much carbohydrate. This is probably because the non-symbionts have to divert a substantial proportion of the carbon uptake to growth whereas the symbiont is non-growing when the nodule has formed. A healthy lucerne nodule can fix up to 100 mg of nitrogen daily per gram of nodule dry matter, which corresponds to 1.36 times its own nitrogen content. Nodules of subterranean clover can fix up to 50 mg g^{-1} nodule daily, though for each crop the average fixation per unit of dry matter in the nodules is usually about half of these rates.[59]

Legumes vary markedly in their rates of nitrogenase activity expressed in terms of nodule mass, which suggests that the proportion of nodule respiration supporting maintenance and growth of uninfected cells in the nodule also varies.[60] Although it is impossible to measure the individual costs of nodule function, theoretical estimates can be made (Table 17.5). Nitrogenase activity is the major respiratory cost in the nodule,[60] and part of this cost is due to the reduction of protons to hydrogen gas by the enzyme while reducing nitrogen gas.[61] The enzyme continues to allocate at least 25 per cent

55 Bond, G., *Ann. Bot.*, 1950, **15**, 95. For the corresponding work with red clover, see Ferguson, T. P. and Bond, G., *Ann. Bot.*, 1954, **18**, 385.
56 See for example Hasse, P. and Kirchmeyer, F., *Ztschr. Pflanz. Düng.*, 1927, A**10**, 257.
57 Wilson, P. W., *The Biochemistry of Symbiotic Nitrogen Fixation*, Univ. Wisconsin Press, Madison, 1940.
58 Witty, J. F. *et al.*, *Ann. Bot.*, 1984, **53**, 13.
59 Jensen, H. L., *Proc. Linn. Soc. NSW*, 1947, **72**, 265.
60 Atkins, C. A., *Pl. Soil*, 1984, **82**, 273.
61 Schubert, K. R. and Evans, H. J., *Proc. Natl. Acad. Sci. USA*, 1976, **73**, 1207.

Table 17.5   Comparison of theoretical carbon costs for the processes of nitrogen fixation by legume nodules with nitrogen assimilation from soil nitrate. (Adapted from ref. 60)

Item of cost	g C/g N
Biological fixation	
Nitrogenase/hydrogenase	1.7–3.5
Ammonia assimilation and ancillary C metabolism	0.4–0.5
Transport of N	0.3
Growth and maintenance of nodule	0.5–1.8
Total	2.9–6.1
Nitrate assimilation	
Nitrate uptake	0.1
Nitrate + nitrite reductases	0–1.5*
Ammonia assimilation and ancillary C metabolism	0.4–0.5
Transport of N	0.3
Total	0.8–2.4

* 0 = Complete reduction in shoots with direct utilization of photosynthetic reductant.
1.5 = Complete reduction in roots at the expense of respired assimilate.

of its reductant to hydrogen production even under a nitrogen pressure of 50 atmospheres.[62]

Some strains of *Rhizobium* possess a hydrogen uptake (Hup) system in the bacteroid form which oxidizes hydrogen gas to water. Dixon[63] suggested three possible advantages in possessing an uptake hydrogenase system: (i) utilization of oxygen, thereby protecting nitrogenase; (ii) removal of hydrogen gas which might otherwise be inhibitory to nitrogenase; and (iii) recovery of at least some of the energy used by nitrogenase in producing hydrogen. Theoretical estimates suggest that complete recycling of evolved hydrogen could result in a saving of 11–26 per cent of the energy required for nitrogenase function[60] and could therefore produce an increase in the efficiency of use of photosynthate in nodules possessing a Hup system. The practical significance of the presence of an uptake hydrogenase system remains uncertain, however, with some reports indicating an increase in nitrogen fixation and yield[64] and others showing no benefit.[65]

62 Simpson, F. B. and Burris, R. H., *Science*, 1984, **224**, 1095.
63 Dixon, R. O. D., *Arch. Microbiol.*, 1972, **85**, 193.
64 Albrecht, S. L. *et al.*, *Science*, 1979, **203**, 1255.
65 Rainbird, R. M. *et al.*, *Pl. Physiol.*, 1983, **71**, 122;   Cunningham, S. D. *et al.*, *Appl. environ. Microbiol.*, 1985, **50**, 791.

Although nitrogen fixation represents an economy in terms of the costs of agricultural production, a legume fixing dinitrogen pays a price in terms of the respiratory costs involved. A legume obtaining most of its nitrogen from fixation may require more energy for nitrogen assimilation than the same legume utilizing mainly soil nitrogen.[60] The comparison in Table 17.5 shows that although the most efficient symbiosis would be energetically similar to the most expensive example of nitrate assimilation, many symbioses would probably require a greater allocation of carbon. This is supported by experimental measurements for lupins[60] and cowpeas.[66]

The infection process

The nodule bacteria are now classified as two genera, *Rhizobium* and *Bradyrhizobium*.[67] The bacteria are typically rod shaped when grown on suitable media and when actively growing in healthy nodules, but they may have a shape like an X, Y, T or club if growing in unfavourable conditions in media or in the nodule, where they assume the bacteroid form (p. 544). They have characteristic banded and branched shapes in the older cells of the nodule. The classification of nodule bacteria is discussed on pp. 643–647.

Despite a great deal of research into the infection of legumes by rhizobia, the mechanisms involved remain unclear.[68] Particular attention has been focused on the specificity shown by *Rhizobium* strains for the species of legume that they infect. Actively growing roots of leguminous plants secrete substances into the soil which may stimulate the multiplication of nodule bacteria,[69] but there is no direct evidence for specific stimulation of those rhizobia which can enter into symbiosis. The bacteria demonstrate a chemotactic response to amino acids produced by the roots, but as *Escherichia coli* can move as rapidly as *Rhizobium* to amino acids[70] there appears to be little ecological advantage to the nodule-forming bacteria.

Curling and sometimes branching of root hairs is the first visible response by the host legume to the presence of rhizobia.[71] The production of β-indoleacetic acid (IAA) by rhizobia, possible from tryptophan excreted by legume roots, has been suggested as the cause of root hair deformation.[72] However,

66 Atkins, C. A. *et al.*, *Pl. Physiol.*, 1980, **66**, 978.
67 Jordan, D. C. in *Bergey's Manual of Systematic Bacteriology* (eds J. G. Holt and N. R. Kreig), Williams and Wilkins, 1984.
68 Schmidt, E. L., *Ann. Rev. Microbiol.*, 1979, **33**, 355. Bauer, W. D., *Ann. Rev. Pl. Physiol.*, 1981, **32**, 407.
69 Van Egeraat, A. W. S. M., *Mededelingen Landbouwhogeschool Wageningen*, 1972, 72–77, Netherlands.
70 Gaworzewska, E. T. and Carlile, M. J., *J. gen. Microbiol.*, 1982, **128**, 1179.
71 Ward, H. M., *Phil. Trans. R. Soc. Lond.*, 1887, **B178**, 539.
72 Thiman, K. V., *Proc. Natl. Acad. Sci. USA*, 1936, **22**, 511; Chen, H. K., *Nature, Lond.*, 1938, **142**, 753.

IAA does not cause the curling characteristic of infected root hairs.[73] A marked degree of curling by root hairs (360° or greater) appears to be the form of curling which is characteristic of an interaction between rhizobia and their homologous host plant.[74]

The work of Dazzo and co-workers[75] suggests that adhesion of the bacterial symbiont to the root surface constitutes one of the critical 'recognition' interactions prior to successful infection and development. In the early recognition a unique, immunologically cross-reactive antigen on the cell walls of clover and *Rhizobium trifolii* is responsible for the attachment (Fig. 17.1).[76] The antigen contains receptors which bind to a lectin called trifoliin. In the model system trifoliin recognizes similar saccharide residues on *R. trifolii* and clover. There is a second stage of adherence (phase II) which firmly anchors bacteria to the root hair surface, possibly by the production of extracellular cellulose microfibrils.[76] However, although there is substantial experimental evidence to support this hypothesis, other workers have been unable to demonstrate specific attachment.[77]

Rhizobia attached to the surface of the root hairs are often enclosed as the root hair curls, and this may lead to a localized increase in the concentration of chemical signals and effectors from the rhizobia and subsequent responses

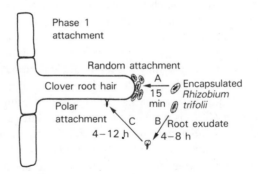

Fig. 17.1   Proposed sequential steps of phase 1 attachment of *Rhizobium trifolii* to the surface of clover root hairs. The inoculum consists of fully encapsulated cells which bind trifoliin A uniformly. The sequence is phase A, B, and C.

73 Kefford, N. P. *et al., Aust. J. biol. Sci.*, 1960, **13**, 456;   Sahlman, K. and Fahraeus, G., K. *Lantbrukshögsk. Ann.*, 1962, **28**, 261.
74 Yao, P. Y. and Vincent, J. M., *Pl. Soil*, 1976, **45**, 1.
75 Dazzo, F. B. in *Microbial Adhesion to Surfaces* (eds R. C. W. Berkeley, J. M. Lynch *et al.*) Ellis Harwood, 1980;   Dazzo, F. B. and Gardiol, A. E. in *Genes Involved in Microbe–Plant Interactions* (eds D. P. S. Verma and Th. Hohn), Springer–Verlag, 1984.
76 Dazzo, F. B. *et al., Appl. environ. Microbiol.*, 1984, **48**, 1140.
77 Badenoch-Jones, J. *et al., Appl. environ. Microbiol.*, 1985, **49**, 1511.

from the plant.[78] Infection of the root hair by rhizobia appears to be associated with curled or branched root hairs.[79] Only a small and variable proportion of root hairs become infected, and many infections (68–99 per cent for *Trifolium* species) abort before they reach the base of the root hair.[80]

Infection threads are tubular cellulose structures that carry *Rhizobium* cells, often in single file, from the point of entry into the root hair to the inner cells of the root cortex.[81] Nutman[82] proposed that when infected by rhizobia the growth of the host cell wall is reorientated so that the root hair becomes invaginated. There is an alternative mechanism[83] in which the rhizobial exopolysaccharide stimulates an increase in plant pectic enzyme (polygalacturonase) activity, which causes a softening of the root hair cell wall and allows penetration by rhizobia. Recent evidence from light and electron microscopy[84] lends support to the polygalacturonase hypothesis.

When the rhizobia penetrate the inner cortical cells they cause a proportion of the cells they penetrate to start proliferating. They themselves proliferate in these cells until they have almost filled them. With red clover (*T. pratense*) only a tetraploid cell can begin proliferation and the proliferating cells are predominantly tetraploid, but it is not known how general this feature is. These proliferating cortical cells remain within the root endodermis and form the nodules.

Nodule formation

Nodules themselves differ in shape and size, partly as a response to soil conditions and partly as a characteristic of the particular bacterial strain–plant variety interaction. They may be roughly spherical, cylindrical, flattened and often bidentate or with coralloid branching, or they may have an entirely irregular shape. Their size may vary from that of a pin head to over 1 cm. The larger nodules are never spherical but have shapes giving a high ratio of surface area to volume, possibly to ensure an adequate supply of nitrogen gas to the active nodule cells and an adequate means of disposal of the carbon dioxide produced in the nodule. In general, a nodule will only contain a single strain of *Rhizobium*, although Johnston and Beringer[85] found

78 Bauer, W. D., *Ann. Rev. Pl. Physiol.*, 1981, **32**, 407; Ridge, R. W. and Rolfe, B. G., *Appl. environ. Microbiol.*, 1985, **50**, 717.
79 Li, D. and Hubbell, D. H., *Canad. J. Microbiol.*, 1969, **15**, 1133; Yao, P. Y. and Vincent, J. M., *Aust. J. biol. Sci.*, 1969, **22**, 413.
80 Nutman, P. S., *J. exp. Bot.*, 1959, **10**, 250.
81 For electron microscope photographs of an infection thread in soyabeans see Goodchild, D. J. and Bergersen, F. J., *J. Bact.*, 1966, **92**, 204.
82 Nutman, P. S. in *Ecology of Soil-Borne Plant Pathogens* (eds K. F. Baker and W. C. Snyder), Univ. California, 1965.
83 Ljunggren, H. and Fahraeus, G., *J. gen. Microbiol.*, 1961, **26**, 521.
84 Callaham, D. A. and Torrey, J. G., *Canad. J. Bot.*, 1981, **59**, 1647.
85 Johnston, A. W. B. and Beringer, J. E., *Nature, Lond.*, 1976, **263**, 502.

that when pea plants were inoculated with mixtures of *R. leguminosarum* and either *R. trifolii* or *R. phaseoli*, at least one nodule on more than half of the plants contained the heterologous strain together with the homologous strain.

Once the bacteria have filled a proliferating cell, they change their form into a bacteroid. Plate 17.4 shows the bacteria, which have become enlarged, in the nodules of subterranean clover during this stage. Only the bacteroid possesses the nitrogen fixing enzymes, but it has lost the ability to multiply and it also appears to have lost most of the ribosomes the cell originally contained. It may therefore have lost its ability to synthesize proteins. It is usually larger than the actively multiplying forms, and it usually contains granules of poly-β-hydroxybutyric acid, which are probably food reserves for the cell.

The bacteroids in the cortical cells are themselves surrounded by membranes which are part of the host cell. The number of bacteroids enclosed within a membrane depends on the plant species. There is only one bacteroid within each membrane in the clovers, pea, medics and groundnuts; while in the cowpea,, soyabean and acacia there are several, usually much smaller bacteroids, within each membrane.[86] Thus in soyabean, there are four to six within each membrane,[87] and there are about 10^5 such groups within a single cell. If the nodule is a few millimetres in size, it may contain 35 000 cells, giving something over 10^{10} bacteroids per nodule. The bacteroids within their envelopes are bathed in a solution containing leghaemoglobin, which they synthesize; this allows them to function with oxygen concentration of only 10^{-8} M in the bathing solution.[88] As the nodule ages the cells lose their leghaemoglobin. They become brown in colour and a large vacuole appears in each, crowding the bacteroids and the cell contents into a dense mass against the periphery of the cell wall. Finally the bacteroids break up into bacteria which digest the remaining cell contents and attack the cell wall. The nodule now becomes necrotic, and the bacteria are released into the soil. The period of maximum development and leghaemoglobin content in the nodules of annual legumes is probably just before flowering.[89]

This description of infection and nodule formation only applies if the bacterial strain can form an effective or fully healthy nodule. But this process can fail to be completed due to incompatibility at any of the following stages of development[90]:

1   At the primary infection of the root hair;
2   In the growth of the infection thread;

86 Dart, P. J. with Mercer, F. V., *J. Bact.*, 1966, **91**, 1314; with Chandler, M., *Rothamsted Ann Rept. for 1971*, Part 1, p. 99.
87 Bergersen, F. J. and Briggs, M. J., *J. gen. Microbiol.*, 1958, **19**, 482.
88 Bergersen, F. J., *Trans. 9th Int. Congr. Soil Sci.*, 1968, **2**, 49.
89 Jordan, D. C. and Garrard, E. H., *Canad. J. Bot.*, 1951, **29**, 360.
90 Bergersen, F. J., *Aust. J. biol. Sci.*, 1957, **10**, 233;   Nutman, P. S., *13th Symp. Soc. gen. Microbiol.*, 1963, 51.

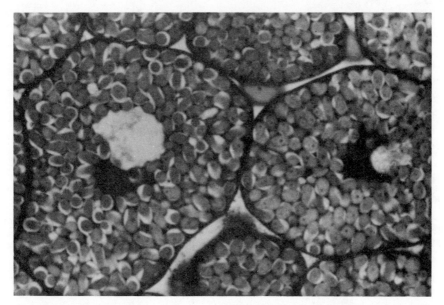

Plate 17.4 The bacteroid zone in subterranean clover (× 1600).

3 In the release of the bacteria from the infection thread;
4 In their multiplication within the cytoplasm of the host cell;
5 In bacteroid formation;
6 In bacteroid persistence;
7 In the functioning of the bacteroids.

In some of these, the cause has been traced to a bacterial defect, in others to a hereditary defect in the host plant, and in still others it involves a specific interaction between the bacteria and the host plant.

The lack of bacteroid persistence appears to be the most common cause of the nodules being ineffective. The ineffective nodules are not only often short-lived, but are also much smaller than effective ones. Although they are typically far more numerous than effective ones, the total volume of bacteroid, tissue per plant is very much smaller and the colour is less pink. However, the rate of nitrogen fixation per unit volume of pink tissue appears to be about the same for this type of ineffective nodule as for an effective one.[91]

Other types of ineffective nodules which can occur in the field have not been fully studied. In equatorial Africa, lucerne at Muguga (Kenya) and groundnuts at Yangambi (Zaïre) sometimes have swellings on their roots which look like nodules, but their colour is white or pale green, and they

91 Chen, H. K. and Thornton, H. G., *Proc. R. Soc.*, 1940, **B129**, 208.

appear to be a proliferation of cortical parenchyma without the cellular differentiation of the normal nodule.[92]

The root can normally carry only a limited number of nodules per unit length; hence, if root growth ceases fairly early in the season, as it may do with peas, for example, the root system can become saturated with nodules. Once this has happened no further bacteria of any other strain can produce additional nodules on the root. Root saturation will not be shown by plants such as clovers whose root system continues to develop through much of the growing season.

The number of nodules produced on a unit length of root depends on the bacterial strain, the genetic constitution of the plant and the density of plant roots in the soil. Nutman[90] showed that clovers possess several genes which determine the number of nodules a given bacterial strain can produce and the period in the growth of the plant when they are produced. He also found that clovers selected for abundant nodulation produce a larger number of lateral roots than those selected for sparse nodulation, whether or not the plants are nodulated.[93] The addition of charcoal or bentonite to the rooting medium may enhance nodulation,[94] possibly by adsorbing toxic compounds produced by other soil micro-organisms.[95]

Nodules already present on the roots of clover can affect the number of new nodules produced. Roots infected by ineffective strains of bacteria carry far more, but much smaller, nodules than those infected by effective strains. Initial nodulation is as rapid with either strain, but the effective nodules inhibit further nodulation as soon as nodule growth is properly started.[96] Recent work has shown that only certain parts of legume roots are susceptible to infection at a particular time and that root cells in these infection zones remain susceptible for only 12–24 h.[97]

The longevity of nodules

The amount of nitrogen fixed by a leguminous crop depends very largely on the longevity of the nodules on its roots. Four factors affect longevity: the physiological condition of the plant; the moisture content of the soil; parasites in the nodule; and the strain of bacteria forming the nodule.

The effect of physiological condition applies particularly to annual plants whose nodules tend to die at flowering and seed set. This is presumably

92 Bonnier, C., *INEAC Sci. Ser. Publ.*, 1957, 72.
93 Nutman, P. S., *Ann. Bot.*, 1948, **12**, 81; *ibid.*, 1949, **13**, 261; *idem, Heredity*, 1949, **3**, 263; *idem, Proc. R. Soc.*, 1952, **B170**, 176.
94 Anderson, A. J. and Spencer, D., *Aust. Inst. agric. Sci.*, 1948, **14**, 39. Nutman, P. S., *Nature, Lond.*, 1951, **167**, 288; *idem, Ann. Bot.*, 1953, **17**, 95.
95 Hely, F. W. *et al., Aust. J. agric. Res.*, 1957, **8**, 24.
96 Nutman, P. S., *Ann. Bot.*, 1949, **13**, 261; *ibid.*, 1952, **16**, 79.
97 Bhuvaneswari, T. V. *et al., Pl. Physiol.*, 1981, **68**, 1144.

because at this time the flowers and developing seeds are drawing on the carbohydrate reserves of the plant very heavily, and the young seeds may also be drawing on the nitrogen compounds in the nodules and leaves.[98] A recent study by Sheehy[99] suggests that senescence and withdrawal of nitrogen from the vegetative portion of the crop may not be essential for seed production. By selecting for maintenance of 'green' leaf colour at pod maturity, soyabean lines have been selected for delayed leaf senescence and these lines continue to fix nitrogen throughout seed filling.[100] The cutting or hard grazing of clovers may also cause the death of the nodules, presumably because the carbohydrate supply to the nodules is interrupted. Perennial legumes differ appreciably in the longevity of their nodules. Perennial clovers, for example, have nodules that are normally only short-lived and they are shed in winter. Legumes that are shrubs or trees may, however, carry nodules for several years, and this group may include some herbaceous legumes such as lupins.[101]

Nodules only seem to remain on the roots of many leguminous crops if the soil is kept moist. The first effect of the onset of drought is for the crop to shed its nodules, though unfortunately no systematic work has been done on the moisture deficit in the plant, or the tension of water in the soil, at which shedding is severe. In the laboratory, Sprent[102] found that soyabean nodules cease to fix nitrogen when their fresh weight falls below 80 per cent of their fully turgid value. This effect is very noticeable on introduced legumes in parts of Africa, for even quite moderate breaks in the rains can cause their nodules to be shed.

Nodules can also be short-lived through being parasitized by the larvae of insects. It has been found that the nodules of pea and field bean crops can be heavily attacked by the larvae of the pea weevil *Sitonia lineata*.[103] The severity of this attack can be reduced by irrigation in dry seasons,[104] which thereby prolongs the life of the nodule.

Measurement of nitrogen fixation by legumes

To assess the value of legumes in agriculture, information is needed on the amounts of nitrogen fixed by these crops. However, despite much effort, there is not a single legume crop for which we have valid estimates of the

98 See Pate, J. S., *Aust. J. biol. Sci.*, 1958, **11**, 366, 496, for a description of this on the roots of field peas and field vetch.
99 Sheehy, J. E., *Ann. Bot.*, 1983, **51**, 679.
100 Abu-Shakra, S. S. *et al.*, *Science*, 1978, **199**, 973.
101 Harris, J. O., *Am. J. Bot.*, 1949, **36**, 650.
102 Sprent, J. I., *New Phytol.*, 1971, **70**, 9.
103 Mulder, E. G., *Pl. Soil*, 1948, **1**, 179; Masefield, G. B., *Emp. J. expl. Agric.*, 1952, **20**, 175.
104 Masefield, G. B., *Emp. J. expl. Agric.*, 1961, **29**, 51.

amount of nitrogen fixed under field conditions.[105] Rates of fixation have been reported for many legumes, see for example Table 17.4, but these estimates vary considerably, for example 45 to 673 kg ha^{-1} a^{-1} for white clover,[106] and depend greatly upon the method used to measure fixation.[107]

The only direct measurement of nitrogen fixation is given by $^{15}N_2$ reduction in which a legume is incubated in an atmosphere enriched with $^{15}N_2$ and ^{15}N fixed in the harvested plant is measured by emission or mass spectroscopy. This method is expensive and is not suitable for use in the field, but all other methods should be referred to it for calibration if they are to be quantitative.[108]

Nitrogenase catalyses the reduction of acetylene to ethylene (p. 626) and this reaction has been used to estimate nitrogen fixation in the field.[109] Plants are incubated in an enclosed atmosphere containing acetylene and the rate of ethylene production is measured by gas chromatography. Problems associated with this method include endogenous ethylene production by soil micro-organisms,[110] inhibition of hydrogen evolution from nodules by acetylene, and reduction of acetylene varying with time.[111] The decline in nitrogenase activity is caused by the acetylene and is only detected by using a continuous gas flow system rather than the sealed systems commonly used. Measurements by acetylene reduction provide only short-term estimates of nitrogenase activity; large errors are associated with field measurements and a large number of samples are required if site variation, diurnal trends and day-to-day changes are to be accounted for in a seasonal estimate of fixation.

The amount of nitrogen fertilizer which must be added to a pure stand of grass to produce the same yield as a mixed grass and legume sward may be used as an estimate of the nitrogen fixed by the legume. This method assumes that nitrogen fixed by a legume is transferred to a companion crop within the same season. Further, the estimate is not of nitrogen fixation but of the fertilizer equivalence of fixation.

Nitrogen fixation may also be estimated from the balance of nitrogen entering the soil after all gains and losses of nitrogen to and from the system have been measured. Accurate seasonal measurements of nitrogen fluxes such as leaching and denitrification are required, and these are difficult to achieve. A nitrogen balance also requires the use of lysimeters (p. 664) and is not therefore suitable for routine measurements in the field.

105 LaRue, T. A. and Patterson, T. G., *Adv. Agron.*, 1981, **34**, 15.
106 Nutman, P. S. in *Symbiotic Nitrogen Fixation in Plants* (ed P. S. Nutman), Cambridge Univ. Press, 1976.
107 Knowles, R. in *Current Perspectives in Nitrogen Fixation* (eds A. H. Gibson and W. E. Newton), Aust. Acad. Sci., Canberra, 1983.
108 Bergersen, F. J. in *Methods for Evaluating Biological Nitrogen Fixation* (ed F. J. Bergersen), Wiley, 1980.
109 Mahon J. D. and Salminen, S. O., *Pl. Soil*, 1980, **56**, 335; Sinclair, A. G. *et al.*, *New Zealand J. agric. Res.*, 1976, **19**, 451.
110 Witty, J. F., *Soil Biol. Biochem.*, 1979, **11**, 209.
111 Minchin, F. R. *et al.*, *J. exp. Bot.*, 1983, **34**, 641.

Attention has recently been focused on the use of ^{15}N-labelled nitrogen fertilizer to measure nitrogen fixation. The method is indirect, but provides an integrated measurement over a long period of time such as a season. [15]N is applied as small amounts of ammonium or nitrate fertilizer which becomes diluted by the soil mineral nitrogen; the enrichment ([15]N : [14]N ratio) is estimated by growing a non-fixing plant near to the fixing plant. The legume assimilates some soil nitrogen, but the ^{15}N is also diluted by ^{14}N obtained from nitrogen fixation. Nitrogen fixation is then calculated from the relative enrichment in ^{15}N of the legume and non-fixing plant.[112] Because ^{15}N occurs naturally at low concentrations, analysis of the natural abundance of ^{15}N in fixing and non-fixing plants may also be used to estimate nitrogen fixation.[113] Problems associated with the use of ^{15}N isotope dilution include the choice of a suitable non-fixing reference plant which will assimilate soil mineral nitrogen similarly to the fixing plant, and declining levels of ^{15}N in soil during the measurement period.[114]

Availability of nitrogen from legumes

A major advantage of using legumes, particularly pasture legumes, in agriculture is that they can increase the nitrogen content of the soil and thereby increase the yield of either companion or subsequent crops. Crops which produce an extensive rooting system which can be nodulated throughout the season, for example clovers and lucerne, leave more nitrogen than those with a restricted root system particularly when most of the nodules are formed during a relatively short part of the growing season, for example peas and beans. Furthermore, up to 90 per cent of the nitrogen in nodules of peas and vetches is transferred to the tops before the plant dies,[115] and may therefore be removed from the land.

Nitrogen may be released from legumes in a number of ways. Soluble organic compounds such as amino acids may be excreted from legume roots and this nitrogen could be available to a companion crop during the same season. Evidence has been given to support direct transfer of legume nitrogen by using ^{15}N isotope dilution studies,[116] but is not conclusive because of different nitrogen uptake patterns by the non-fixing plants when grown in mixtures and pure stands. Subterranean clover and lucerne labelled with ^{15}N by foliar absorption showed no direct transfer of nitrogen to ryegrass in the field.[117]

112 McAuliffe, C. *et al., Agron. J.,* 1958, **50**, 334.
113 Turner, G. L. and Bergersen, F. J., *Soil Biol. Biochem.,* 1983, **15**, 525.
114 Witty, J. F., *Soil Biol. Biochem.,* 1983, **15**, 631; Chalk, P. M., *Soil Biol. Biochem.,* 1985, **17**, 389.
115 Pate, J. S., *Aust. J. biol. Sci.,* 1958, **11**, 366, 496.
116 Eaglesham, A. R. J. *et al., Soil Biol. Biochem.,* 1981, **13**, 169; Broadbent, F. E. *et al., Agron. J.,* 1982, **74**, 625.
117 Ledgard, S. F. *et al., Soil Biol. Biochem.,* 1985, **17**, 575.

Studies in South Australia have shown that only 10.9 to 17.3 per cent of the nitrogen in medic (*Medicago littoralis*) is taken up by a following wheat crop after 15 months, indicating that the main value of legumes is in their capacity to maintain or improve the concentration of soil organic nitrogen which is decomposed at relatively slow rates in the following years.[118] The rate of mineralization of nitrogen in root residues depends largely upon the crop species and the environment; nitrogen is available more rapidly from legumes in the tropics.[119] The breakdown of nodules is also accelerated by chilling at the onset of winter, shading by a vigorous companion crop, or defoliation by cutting or grazing.

The largest flux of fixed nitrogen in a legume pasture is through the grazing animal which returns nitrogen to the soil in urine and faeces. One estimate is that about 80 per cent of the fixed nitrogen which is returned to the soil passes through the grazing animal and of this about 70 per cent is urine nitrogen[120] (see also Ch. 20).

The data in Table 17.6[121] illustrate the value of legume leys in a crop rotation. The yields of three test crops, potatoes followed by wheat followed by barley, are compared when grown in rotation with a three-year ley of lucerne, grass–white clover or grass plus manure, or with a one-year grass–white clover ley followed by two years of arable. The two three-year legume leys apparently increased the nitrogen content of the soil for at least three years. The use of legumes in low input systems incorporating rotations offers an alternative to systems based upon crop monoculture and high rates of nitrogen fertilizer.

Table 17.6 Residual effects of three-year grass and legume leys. Rothamsted: mean of experiments in 2 fields for 2 years (1969–70). Yield in t ha^{-1}

Crop after ley	First: potatoes				Second: wheat			Third: barley	
Nitrogen added kg ha^{-1} N	0	75	150	225	0	50	100	0	50*
Lucerne	40.0	45.5	45.9	47.9	6.4	7.5	7.8	4.7	5.7
Grass–clover	40.0	45.5	47.4	46.4	6.4	7.4	7.7	4.7	5.5
Grass	37.5	43.5	45.5	46.2	5.9	6.7	7.7	4.4	5.3
Arable	30.3	40.0	42.0	45.5	5.3	6.7	7.5	3.9	5.3

* 40 kg N given in 1969.

118 Ladd, J. N. *et al.*, *Soil Biol. Biochem.*, 1981, **13**, 251.
119 Vallis, I., *Aust. J. agric. Res.*, 1983, **34**, 367.
120 Newbould, P. and Haystead, A., *Hill Farming Research Organisation 7th Rep.*, p. 49, 1977.
121 *Rothamsted Ann. Repts for 1969 and 1970.* See also Ch. 20, p. 682.

Plant root pathogens

A plant root, during its life in the soil, will exist in four conditions: it begins as an actively growing organism, it then becomes mature and ceases growth, later becoming senescent, and finally it becomes diseased and dies. In each of these stages it usually has a characteristic fungal flora. Further, the rhizosphere fungi may be growing on the root surface, or they may penetrate the outer or inner, cortical cells, or the stele, and each of these zones again has its characteristic fungal flora.

The relative proportions of the species present on or in the root depends on the plant and the soil conditions. Thus, *Fusarium* tends to be more common in acid soils and *Cylindrocarpon* in neutral soils. *Rhizoctonia*, *Gliocladium*, *Trichoderma*, *Penicillium* and *Mortierella* are usually confined to the root surface or, if they penetrate the root, to the outer cortical cells; while *Fusarium* and *Cylindrocarpon* have a greater ability to penetrate into the inner cortical cells or even the outer cells of the stele.[122] As the root ages, the fungal hyphae tend to penetrate deeper into the root tissues,[123] and in mature or senescent roots which have lost their cortical cells the stele is much more exposed to infection. The activity of the fungi, as measured by their respiration, or by the length of mycelia in the soil, increases as the young plant grows; it reaches a maximum when the vegetative growth rate of the crop is at its maximum.[124]

Root-rot fungi

Many of the fungi which develop on the young root can be weak unspecialized parasites, for they can enter and kill juvenile but not mature roots. The attack is usually only serious if seedling growth receives a check so that the root remains juvenile for an extended period. This can occur if the soil temperature is high enough to allow the seed to germinate, but not high enough for rapid root growth. Thus, fungi are the cause of the characteristic damping off of seedlings growing under unfavourable conditions.[125] The specialized *Fusarium* and *Verticillium* vascular wilt fungi fall into the same group, and they are long-lived in the soil. They can only attack juvenile roots, but once in the root they invade the vascular system. Initially they are confined there by the active resistance of the living cells to infection, though once a

122 Taylor, G. S. and Parkinson, D., *Pl. Soil*, 1965, **22**, 1; Gadgil, P. D., op. cit., 1965, **22**, 239; Waid, J. S., *Trans. Brit. Mycol. Soc.*, 1957, **40**, 391.
123 Waid, J. S., *Trans. Brit. Mycol. Soc.*, 1962, **45**, 479.
124 Parkinson, D. and Thomas, A., *Canad. J. Microbiol.*, 1969, **15**, 875.
125 For a fuller discussion of this see Leach, L. D., *J. agric. Res.*, 1947, **75**, 161. For an example with maize see Harper, J. L., *Ann. appl. Biol.*, 1955, **43**, 696; *idem, New Phytol.*, 1955, **54**, 107, 119; 1956, **55**, 35.

cell dies the fungus enters. Plant resistance to these vascular wilts thus occurs entirely in the root system.

The ability of a plant root to withstand attack by these specialized root-rot fungi depends partly on the health and vigour of the root at the point of attack and partly on the vigour of the fungus growing in the soil or on the root surface before it penetrates into the root cells, and this may depend on the conditions in the rhizosphere. As the rhizosphere may contain fungi such as *Trichoderma* spp., which are antagonists to the root infecting fungi, altering the root environment to encourage these antagonists reduces the susceptibility of the crop roots to attack. *Trichoderma viride*, for example, is more resistant to chemical sterilants than most pathogenic fungi, and treating a soil with one of these chemicals encourages its development. It grows more strongly in acid than calcareous soils, as has been shown[126] in a study of the susceptibility of pine roots to attack by *Heterobasidon (Fomes) annosus* on some sandy East Anglian soils; for the *Fomes* attack was much more serious, and the growth of *Trichoderma* was much less on the calcareous than on the acid soils. It has also been found[127] that root-rots of strawberries and tobacco, caused by a complex of weak unspecialized root-parasitic fungi, could be reduced by ploughing in soyabean tops, which was attributed to a protective effect of the decomposition products of carbohydrates but might have been due to a change in pH value.

Other organisms which are good producers of powerful antibiotics may protect plant roots against the root-rot fungi in the same way as *T. viride*. Thus, varieties of cotton resistant to *Phymatotrichum* root-rot had roots carrying a much higher number of blue–green fluorescent bacteria of the *Pseudomonas–Phytomonas* group in their rhizosphere than had the roots of susceptible varieties.[128]

The antagonistic action of the bacteria may, however, be indirect. For example, it has been found[129] that certain members of the rhizosphere population produced antibiotics which were taken up by the roots of the bean *Phaseolus vulgaris* and were distributed throughout the plant, thus protecting it from attack by *Xanthomonas phaseoli*.

The rhizosphere environment can be altered by adding different types of decomposable organic matter to the soil. Thus, adding 560 kg ha^{-1} of chitin to a soil two weeks before planting beans (*Phaseolus vulgaris*) markedly reduced the severity of attack by the root-rot *Fusarium solani* f. *phaseoli*.[130] By contrast, adding the soluble products from decomposing rye or grass to a soil increased the severity of attack on tobacco roots by *Thielaviopsis*

126 Rishbeth, J., *Ann. Bot.*, 1950, **14**, 365; 1951, **15**, 1, 221.
127 Hildebrand, A. A. and West, P. M., *Canad. J. Res.*, 1941, **14C**, 199.
128 Eaton, F. M. and Rigler, N. E., *J. agric. Res.*, 1946, **72**, 137.
129 Krezel, Z., *Ann. Inst. Pasteur*, 1964, **107**, 168.
130 Mitchell, R. and Alexander, M., *Soil Sci. Soc. Am. Proc.*, 1962, **26**, 556.

basicola.[131] The inoculation potential of a root-rot fungus may also be markedly affected by changes in the oxygen or carbon dioxide content of the soil air.[132] Thus, the pathogenic activity of *Rhizoctonia solani* on radish and sugar beet seedlings was reduced by raising the CO_2 concentration in the soil air, and a similar result was found for the *Fusarium* wilt in flax (*F. oxysporum* f. *lini*)[133]; for its virulence was much reduced if the soil aeration was poor. It has also been shown that *Streptomyces scabies*, the cause of potato scab, can only cause appreciable infection of potato tubers if it can gain entry to the tubers during the three weeks after tuber initiation, but it can only do this if the soil is dry.[134] By maintaining the soil moisture during this critical period by irrigation, a good quality marketable crop can be obtained on moderately infected land.

It is sometimes possible to weaken the root-rot fungus during its resting phase in the soil. For instance, some parasitic strains of *Fusarium oxysporum* can be killed or severely weakened if subject to anaerobic conditions. Infected land can therefore be much improved by being flooded in warm weather, particularly if it contains much decomposable organic matter. Moreover, land heavily infected with banana wilt, *Fusarium oxysporum* f. *cubense*, can have its infection very considerably reduced by waterlogging the soil for several months. This method has the limitation that the surface layer of the soil will remain aerobic, due to algal growth, so the fungus can maintain itself in this layer.[135] The soil minerals can also have a major effect on the spread of this and other pathogens.[136]

The specialized root-rot fungi have considerable power to decompose lignocellulose and to attack mature root cells, with which they show a degree of symbiosis, for they do not initially kill or seriously disorganize the cells they are attacking. They differ from the mycorrhizal fungi, which will be considered in the next section, for these enter only juvenile roots or the growing points of laterals and they form a stable symbiosis with the cells they enter. The root-rot fungi typically grow on the root surface before they penetrate into the root cells, and for fungi which attack mature tree roots such as *Armillaria mellea*, and *Fomes annosus*, their mycelia may run along a root for several metres before they send branches into root cells themselves. Further, in undisturbed equatorial forests, *A. mellea* grows over the surface of roots of some species of trees as a weft without penetrating

131 Patrick, Z. A. and Koch, L. W., *Canad. J. Bot.*, 1963, **41**, 747.
132 Papavizas, G. C. and Davey, C. B., *Phytopathology*, 1962, **52**, 759.
133 Lachance, R. O. and Perrault, C., *Canad. J. Bot.*, 1953, **31**, 515; Stover, R. H., *op. cit.*, 693.
134 Lapwood, D. A. *et al., Ann. appl. Biol.*, 1966, **58**, 447; 1970, **66**, 397; *idem, Pl. Path.*, 1967, **16**, 131; Lewis, B. G., *Ann. appl. Biol.*, 1970, **66**, 83.
135 Stover, R. H. *et al., Soil Sci.*, 1953, **76**, 223; 1954, **77**, 401; 1955, **80**, 397.
136 Stotzky, G. in *Microbial Adhesion to Surfaces* (eds R. C. W. Berkeley *et al.*), Ellis Harwood, 1980.

into the roots, provided the tree is healthy. Only if the forest tree is weakened, by a severe drought, for example, or if the tree is cut down, can the fungus enter the root and start growing actively inside it. Even a plantation crop such as tea, which is usually considered to be susceptible to *A. mellea*, may carry rhizomorphs on the surface of its roots without suffering any serious attack if the crop is growing vigorously.

Plant roots have a certain resistance to attack by these fungi, which the fungus must overcome before it can enter. A fungus with a high inoculum potential will have the energy or ability to overcome a high resistance to attack by the host, and one with a low inoculum potential will only be able to enter a root with a very low resistance. For a given fungus, the better the food supply at the point of attack, the greater is its inoculum potential. Inoculum potential is difficult to quantify and in an attempt to remedy this the term colonization potential (the biomass of colonist per unit length or weight of root) has been used.[137]

The concept of inoculum potential can be illustrated by three different fungi which attack roots in different ways. *Gauemannomyces graminis* only sends out very fine hyphae into the soil and can only infect the fine roots of susceptible Gramineae. The fungus of cotton root-rot, *Phymatotrichum omnivorum*, sends out mycelial strands from the root in which it is living and it can infect the thicker roots of herbaceous plants. Fungi causing the root-rot of trees, such as *A. mellea*, send out rhizomorphs starting from a large piece of tree root. If the root of the tree is relatively resistant to attack, its resistance can only be overcome if the root grows fairly close to the source of infection. After the fungus has built up a weft of mycelia on the root it sends a large number of hyphae into the root more or less simultaneously, which increases the inoculum potential of the fungus very considerably.

A point of great importance in the control of the specialized root-disease fungi on herbaceous and tree crops in the field is that when the crop is harvested the roots do not die immediately. For a considerable time the root will maintain its immunity to saprophytic and weakly parasitic fungi, while becoming a better food source for the specialized parasite. Roots of some tree species that were resistant to infection by these specialized parasites when the tree was alive can lose their resistance and become attacked once it has been felled. Such roots will carry the parasite in the soil for a considerable time, which may be measured in years in the case of trees, and are a source of infection for any new susceptible roots growing in their neighbourhood for this period. This source of infection is of great importance when one is converting natural forest, whose roots may carry a heavy population of *A. mellea* to plantations of susceptible crops such as tea or rubber.

137 Bennett, R. A. and Lynch, J. M., *Curr. Microbiol.*, 1981, **6**, 137.

Biological control of plant root pathogens

The natural control of a plant pathogen by a microbial antagonist is termed biological control. There is scope for the inoculation of such antagonists to elevate their natural populations and increase the likelihood of bringing about biological control.[138] There is also the half-way stage of integrated control in which inoculation with an antagonist and a low rate of fungicide are used. As yet there have been few successful commercial developments but with the increasing interest being shown in this subject by agricultural biotechnology companies there is likely to be an increased number. One bacterial disease already being controlled is crown gall of fruit trees[139] which is caused by *Agrobacterium radiobacter* var. *tumefaciens* which has a tumour-inducing (Ti) plasmid. Control is brought about by a non-pathogenic strain of the bacterium *A. radiobacter* var. *tumefaciens* strain K84 which produces an anti-biotic, agrocin 84, against the pathogen.

One of the major fungal diseases of wheat, take-all (*Gauemannomyces graminis*) is the target of much current attention. One approach has been to isolate pseudomonad bacteria antagonistic to the pathogen from soils which suppress the disease.[140] Other work has demonstrated the capacity of *Bacillus cereus* var. *mycoides* to control the disease.[141]

Another example of the commercial exploitation of biocontrol agents is the use of *Peniophora gigantea* to control *F. annosus*. Asexual oidia of *P. gigantea*, distributed as dehydrated tablets or in fluid suspensions, are painted on the surface of tree stumps immediately after felling. The treatment has been used in England since 1963.[142] The antagonism is by pre-emption (establishment before *F. annosus*) and by hyphal interference.

Mycorrhizal symbiosis

Symobiotic associations between plant roots and specialized fungi are normal for the great majority of plant species both in natural environments and in cultivation. These associations are termed mycorrhizas[143] and may be simply classified as ecto- or endomycorrhizas depending on the physical relationship of the fungus with the root. In the former a distinct fungal sheath or mantle

---

138 For interesting examples see Baker, K. F. and Cook, R. J., *Biological Control of Plant Pathogens*, W. H. Freeman, 1974; *idem, The Nature and Practice of Biological Control of Plant Pathogens*, Am. Phytopath. Soc., St. Paul, 1983.

139 Kerr, A., *Pl. Disease*, 1980, **64**, 24.

140 Shipton, P. J. *et al.*, *Phytopathology*, 1973, **63**, 511.

141 Faul, J. L. and Campbell, R., *Canad. J. Bot.*, 1979, **57**, 1800.

142 Rishbeth, J., *Eur. J. For. Path.*, 1979, **9**, 331.

143 For a general account see Harley, J. L. and Smith, S. E., *Mycorrizal Symbiosis*, Acad. Press, 1983.

develops on the root surface and penetration of the cortex by the fungus is intercellular. Root hair formation is usually suppressed and fine mycorrhizal roots may be stimulated to branch, sometimes repeatedly. With endomycorrhizas the fungal hyphae penetrate the cells of the root cortex and the sparse mycelium external to the root cannot be observed by the unaided eye. Fungi from different families form the two types of mycorrhizas, but their beneficial effects on plants are similar.

Many, but not all, of the fungi forming ectomycorrhizas can be grown in pure culture but require association with the appropriate host plant in the natural environment. Endomycorrhizal fungi have not yet been grown axenically and have little if any capacity for saprophytic growth in soil. Both groups are of ecological and economic importance.[144]

Ectomycorrhizas

Ectomycorrhizas are typically formed on the roots of trees and shrubs of temperate and tropical regions. The fungi which form these mycorrhizas are principally Basidiomycetes and Ascomycetes, and they infect a very wide range of plant families and genera. The symbiosis is of considerable ecological significance through its influence on the mineral nutrition of the tree or shrub.

The fungal hyphae form a sheath round the roots (Plate 17.5) which is 20 to 100 μm thick and can be readily observed by a magnifying glass. Some of the hyphae grow between the cells in the epidermis and root cortex, forming what is called the 'Hartig net' (Plate 17.6). This provides a large surface area for the interchange of material between the host and the fungus. The fungus benefits by receiving from the plant carbon compounds and specific growth substances, for example, thiamine. The supply of carbon compounds used for the synthesis of mycelium and respiration, implies a cost to the host in terms of energy supply. It has not been established, however, what proportion of the carbon compounds are a drain on the metabolic resources of the plant, and what proportion comes from the mucigel, the root cap and associated cells, which would be lost from the root whether or not it was mycorrhizal.

The fungal sheath which forms round the root provides a greater absorbing surface for nutrients than does the root alone, which is of benefit to plants growing in soils low in nutrients. Thus, *Pinus radiata* was found to grow better with low phosphate in the soil when it was mycorrhizal than when it was non-mycorrhizal.[145] The absorbed phosphate moved more slowly to the shoot in the mycorrhizal plants, suggesting that it accumulated in the sheath before being transferred to the host. This accumulation of phosphate in the sheath may be of ecological significance in providing a steady supply to the

144 Gerdemann, J. W., *Ann. Rev. Phytopath.*, 1968, **6**, 397.
145 Morrison, T. M., *Nature, Lond.*, 1957, **179**, 907; *idem, New Phytol.*, 1962, **61**, 10.

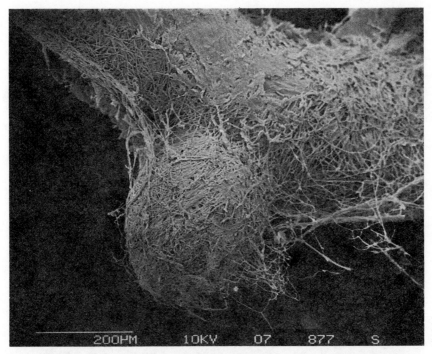

Plate 17.5 Young root lateral of *Larix kaemferi* (Japanese larch) emerging from mother root. Both are covered with a sheath of ectomycorrhizal hyphae. (From Duddridge, J. A., Oxford Univ.)

host plant when transport to the root is limited, as by a period of drought. The fungus also appears to confer a degree of drought resistance on the host, and although the explanation is not agreed, it might be that the fungus in some way improves the conduction of water between soil and root. It has been suggested that a further effect of the fungus is to hydrolyse organic phosphorus compounds present in soil organic matter. Infected roots have been shown to have greater phosphatase activity than uninfected roots,[146] but it has not yet been demonstrated that this results in increased hydrolysis of soil organic phosphorus compounds.

Most species of ectomycorrhizal fungi are not host specific, but attempts to establish tree plantations in new areas very commonly require the introduction of an effective fungus. With many tree species the ectomycorrhiza is probably essential for normal plant development and there are numerous examples of crop failure in the absence of the fungus.[147] The most common method of introducing the required fungus has been to use soil or humus

146 Woolhouse, H. W. in *Ecological Aspects of Mineral Nutrition of Plants* (ed I. H. Rorison), *Symp. Brit. ecol. Soc.*, 1969, **9**, 357; Bartlett, E. M. and Lewis, D. H., *Soil Biol. Biochem.*, 1973, **5**, 249; Alexander, I. J. and Hardy, K., *op. cit.*, 1981, **13**, 301.
147 Marx, D. H. in *Tropical Mycorrhiza Research* (ed. P. Mikola), Clarendon Press, 1980.

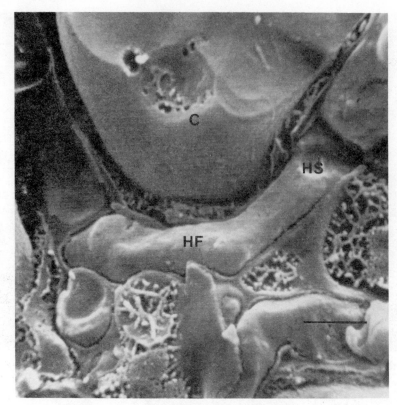

Plate 17.6 Scanning electron micrograph of *Picea sitchensis* (spruce) showing Hartig net hyphae in surface view (HF) and in cross-section (HS) adjacent to root cortical cell (C); bar = 4 μm. (From Jones, D., McHardy, M. J. and Alexander, C., Macaulay Institute for Soil Research.)

from an established forest or nursery to inoculate the soil in which the seedlings are to be grown. This provides a mixed inoculum, and the most desirable fungus is not necessarily dominant. Trials have therefore been made with selected pure cultures of certain fungi, which have sometimes been more successful than a soil inoculum. There seems to be considerable potential for inoculation with selected fungi in establishing plantations of new tree species, especially in the tropics.[147]

Endomycorrhizas

Endomycorrhizas can be subdivided into ericaceous, orchidaceous and vesicular–arbuscular mycorrhizas (VAMs); growth characteristically occurs

within the plant root, no sheath is formed and the external mycelium is relatively sparse. Only the morphologically distinct VAM is important in most
agricultural situations and further discussion is limited to this group.[148]

VA mycorrhizas have been found in the roots of a wide range of plants,
including those that form ectomycorrhizal associations. The intensity of infection varies, and in the family Cruciferae it is either very weak or non-existent.
VA mycorrhizas result from the colonization of young roots by fungi of the
family Endogonaceae. The main genera are *Gigaspora*, *Glomus*, *Acaulospora*
and *Sclerocystis*. The characteristic structures of a generalized VA infection
are shown in Plate 17.7, and Colour Plate 3 shows the vesicles and arbuscules
in a stained onion root.

Infection begins with hyphae from a germinating fungal spore forming an
appressorium on the root surface from which hyphae penetrate the epidermal
cells. The hyphae spread inter- and intracellularly through the outer cortex
and proliferate in the inner cortical layer to form arbuscules (Plate 17.7). An
arbuscule develops within a host cell by repeated dichotomous branching of
the invading hypha. The arbuscules are short-lived and disintegrate within the
soil when the root dies. It is believed that they have special significance in
transfer processes between host and fungus.[149] Following arbuscular development, the structures known as vesicles develop inter- or intracellularly as
swellings, along or at the tips of hyphae. They contain many lipid droplets
and probably function as temporary storage organs.[144] Despite the density of
internal fungal development, there is little effect of VAM infection on root
morphology. The internal structure of VAMs can, however, be easily
observed in cleared and stained root samples under the microscope.

Externally the VAM fungi form a loose network in the soil around the
roots. The hyphae are dimorphic and essentially non-septate. There is a
dichotomously branched, non-septate, coarse hyphal network which is thick-
walled and 'knobbly'; and tufts of much finer, repeatedly branched, thin-
walled and ephemeral hyphae which become septate as they die. The large
resting spores are produced on the coarse external hyphae. They range from
about 100 to 600 μm in diameter and are borne singly in the soil or aggregated into sporocarps.

The intensity of a VA infection may be affected by various factors such
as plant nutrition and fertilizers, pesticides, light intensity, soil moisture, pH
and plant susceptibility.[150]

Although ectomycorrhizas and VA mycorrhizas are different in
morphology, they have similar physiological roles. The most striking feature
of VA mycorrhizal infection is that infected plants tend to grow better than

148 For a good general reference see *Endomycorrhizas* (eds F. E. Sanders *et al*)., Acad.
Press, 1975.
149 Cox, G. and Sanders, F. E., *New Phytol.*, 1974, **73**, 901.
150 Hayman, D. S. in *Interactions between Non-Pathogenic Soil Microorganisms and Plants*
(eds Y. R. Dommergues and S. V. Krupa), Elsevier, 1978.

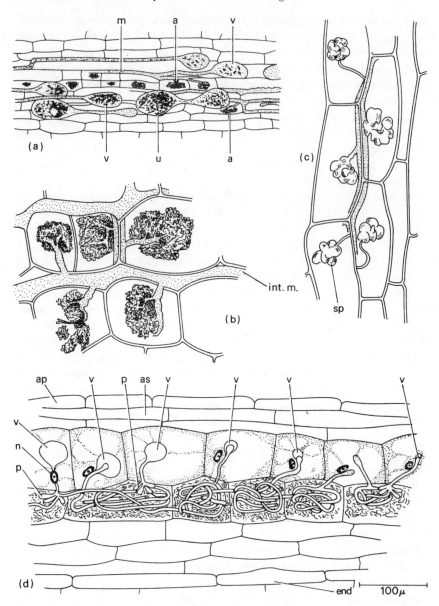

Plate 17.7   (a) Vesicles in a strawberry rootlet (× 60); (b) and (c) Arbuscules
in strawberry rootlets (× 600); (b) well developed; (c) partially
decomposed; (d) Vesicles and pelotons in autumn crocus roots
(× 210). m = mycelium; int. m = intercellular mycelium; v = vesicles;
a = arbuscules; sp = sporangioles; end = endodermis; ap = as =
epidermis.

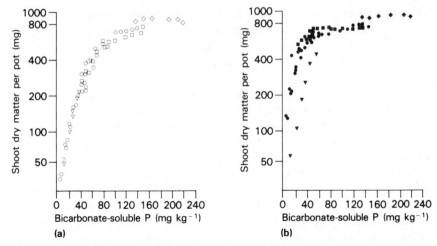

Fig. 17.2 Effect of mycorrhizal infection on the relationship between shoot dry matter and bicarbonate-soluble phosphate in soil: (a) soils sterilized by γ irradiation; (b) as for (a) but plants infected with *Glomus mosseae*. Symbols refer to soils at Rothamsted, all of which were close to pH 7 except (▽) and (▼) which were acid (pH 5.4).

non-mycorrhizal plants in phosphate deficient soils.[150] With a high level of soluble phosphate in soil, roots carry little or no VA mycorrhizas. The effect is shown in Fig. 17.2[151] which compares the growth of leeks in soil after γ-radiation to kill the native fungi; in (a) the plants were not infected and in (b) they were infected with *Glomus mosseae*. In the infected plants the dry matter yield reached a maximum in most soils at a bicarbonate-soluble P level of about 60 mg kg^{-1} soil, but it required about twice this value in the uninfected plants. The exceptions were plants growing in a soil of pH (in water) 5.4. An interesting observation was that in the same soils where the plants were naturally infected, the relation between yield and bicarbonate-soluble P was more variable. Under these conditions, where the plant depends on mycorrhiza for its phosphate supply, a close relationship between growth and soluble phosphate should therefore not be expected, because growth will depend on the amount and type of fungi in the soil and the speed and extent of their spread.

The effect of mycorrhizal infection is probably greater in phosphate deficient soils than under conditions where other nutrients are limiting because phosphate is relatively immobile in soil (see Ch. 21). Using ^{32}P it has been shown that mycorrhizal plants use the same pool of soil phosphate as non-mycorrhizal plants, indicating that the fungus extends the volume of soil from

151 Stribley, D. B. *et al.*, *J. Soil Sci.*, 1980, **31**, 655.

which phosphate is drawn, rather than making it more soluble.[152] Thus, the process is of benefit to the plant because it increases the rate of supply of phosphate to the roots. The mycelium might have other effects, possibly increasing the phosphatase activity, but as yet there is no convincing evidence of its importance.

Much of the work on growth responses to VA mycorrhizal infection has, however, been done on phosphate deficient soils which have been partially sterilized, and is not necessarily applicable under field conditions, where responses have been few. In Western agriculture where any phosphate deficiency is obviated by fertilizer application, mycorrhizas may not have an important effect. Mycorrhizas can also increase uptake of other nutrients such as potassium, sulphate, copper and zinc although data on increases in nitrogen uptake are more equivocal.[153] Interestingly, VA mycorrhizas also enable plants to establish and grow in harsh conditions such as heavily polluted or eroding sites, and areas with high variability of temperature or adverse pH.[154] There are also several reports that mycorrhizal plants are less sensitive to periods of drought than non-mycorrhizal plants and that mycorrhizas play a role in the water relations of plants.[155] Mycorrhizas may also increase disease resistance in plants by several possible mechanisms, one of which (hormone production) may also stimulate plant growth directly.[156] In legumes, infection with VAM increased nitrogen fixation.[157] A synergistic effect on yield and thus protein nitrogen could have great significance for agriculture, emphasizing the need for further research on mycorrhizal–legume relationships.

In return for all these potentially beneficial effects, the plant supplies most if not all the carbon and vitamins needed by the mycorrhizal fungus. Enhanced photosynthesis or increased 'leafiness' may offset the effect of the carbon drain on VA mycorrhizal plants.[158] If the mycorrhizal association is to be exploited a better understanding is, however, needed of the source–sink relationships, and especially of the carbon cost.

The present account of mycorrhizas outlines their role in the soil ecosystem. The beneficial activities of mycorrhizal fungi are areas of potential exploitation in the field of agricultural biotechnology, but before exploiting them, it must be confirmed that the symbiont will not induce a net drain on

152 Tinker, P. B. in *Endomycorrhizas* (eds F. E. Sanders *et al*)., Acad. Press, 1975. Owusu-Bennoah, E. and Wild, A., *Pl. Soil*, 1980, **54**, 233.
153 Tinker, P. B., *Pl. Soil*, 1984, **76**, 77.
154 Menge, J. A., *Canad. J. Bot.*, 1983, **61**, 1015.
155 Safir, G. R. *et al.*, *Pl. Physiol.*, 1972, **49**, 700.
156 Slankis, V. in *Ectomycorrhizae* (eds G. C. Marks and T. T. Kozlowski), Acad. Press, 1973.
157 Shenck, N. C. and Hinson, K., *Agron. J.*, 1973, **65**, 849.
158 Kucey, R. M. N. and Paul, E. A., *Soil Biol. Biochem.*, 1982, **14**, 407; Snellgrove, R. C. *et al.*, *New Phytol.*, 1982, **92**, 75.

the plant's photosynthate. Thereafter, using suitable fermentation technology, development of ectomycorrhizal inocula[159] should present no major problems but until the VAM fungi can be produced in culture, their economic use will be greatly limited.[160]

159 Molina, R. and Trappe, J. M. in *Advances in Agricultural Microbiology* (ed N. S. Subba Rao), Butterworths, 1982.
160 Hayman, D. S. in *Microbiological Methods for Environmental Biotechnology* (eds J. M. Grainger and J. M. Lynch), Acad. Press, 1984.

18

Soil organic matter and its dynamics

Soil organic matter

Arable topsoils commonly contain 1 to 3 per cent of organic carbon, with
grassland and forest soils often containing somewhat more, particularly if
poorly drained. This organic matter consists of a whole series of products
which range from undecayed plant and animal tissues through ephemeral
products of decomposition to fairly stable brown to black material bearing
no trace of the anatomical structure from which it was derived; it is the latter
material that is normally defined as 'humus'.[1] Soil organic matter, defined as
the totality of the organic matter in soil, also includes the organisms that live
in soil, the soil biomass, although they usually account for less than 5 per cent
of the soil organic carbon.[2]

1 For reviews of early work, see Waksman, S. A., *Humus*, Baillière, Tindall and Cox,
  London, 1936; Kononova, M. M., *Soil Organic Matter*, Pergamon, Oxford, 1966.
  More recent books on soil organic matter include Allison, F. E., *Soil Organic Matter
  and its Role in Crop Production*, Elsevier, Amsterdam, 1973; Schnitzer, M. and Khan,
  S. U., *Soil Organic Matter*, Elsevier, Amsterdam, 1978; Stevenson, F. J., *Humus
  Chemistry, Genesis, Composition, Reactions*, Wiley, New York, 1982;   Aiken, G. R.,
  McKnight, D. M. *et al.*, *Humic Substances in Soil, Sediment and Water*, Wiley, New
  York, 1985; Vaughan, D. and Malcolm, R. E., *Soil Organic Matter and Biological
  Activity*, Nijhoff, Dordrecht, 1985; Orlov, D. S., *Humus Acids of Soils*, Balkema,
  Rotterdam, 1985.
2 Jenkinson, D. S., and Ladd, J. N. in *Soil Biochemistry*, Vol. 5 (eds E. A. Paul and
  J. N. Ladd), p. 415, Dekker, New York, 1981.

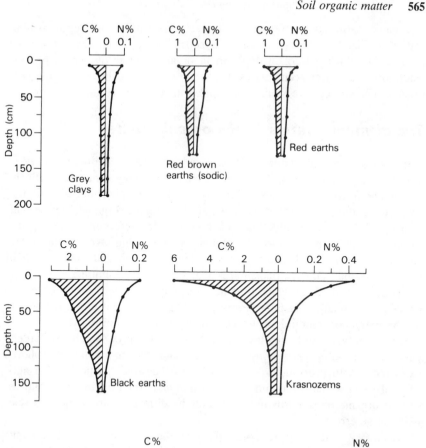

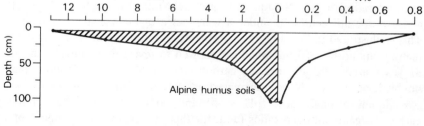

Fig. 18.1 Distribution of organic carbon and nitrogen with depth in selected Australian soils.[3] For a similar diagram for European and Asiatic soils see Parsons and Tinsley.[4]

3 Spain, A. V. *et al.*, in *Soils – an Australian viewpoint*, p. 551, CSIRO, Melbourne/Acad. Press, London, 1983.
4 Parsons, J. W. and Tinsley, J. in *Soil Components*, Vol. 1 (ed J. E. Gieseking), p. 263, Springer, Berlin, 1975.

Organic matter concentration usually decreases rapidly with depth, as illustrated in Fig. 18.1 for six Australian soils. The principal exceptions are podzols, in which a horizon depleted in organic matter overlies a horizon where organic matter accumulates (the B horizon) and palaeosols, where old organic-rich surface layers are often buried under more recent deposits.

The elemental ratios in soil organic matter

Although soil organic matter is composed of carbon, hydrogen, oxygen, nitrogen, sulphur and phosphorus, most studies are only concerned with the carbon and nitrogen (and, less commonly, sulphur and phosphorus) contents, because of the analytical difficulties of determining organic oxygen and hydrogen in soil. Acceptable methods now exist for determining soil organic carbon, nitrogen and sulphur, the organic content being taken as the difference between the total quantity of the element in soil and that present in inorganic forms.[5] Organic phosphorus is also determined indirectly, either by the ignition method, in which organic phosphorus is taken as the difference between the quantity of phosphorus extracted before and after ignition, or by the extraction method, in which the organic phosphorus extracted by a sequence of acid and alkaline treatments is measured as the difference between the total phosphorus extracted and the inorganic phosphorus extracted.[6] Both procedures leave much to be desired, particularly in agricultural soils, which often contain more inorganic than organic phosphorus, so that organic phosphorus measurements by difference are subject to relatively large errors.

Truchot[7] was the first to point out that the organic carbon to organic nitrogen ratio, the C : N ratio expressed as a ratio by weight, is relatively constant for different soils under a wide range of management conditions. This is illustrated in Fig. 18.2 for a Rothamsted soil that had been arable for many years up to 1883 but was then allowed to revert to woodland.[8] In the 81 years after 1883 the organic carbon content of the topsoil more than doubled, but the C : N ratio increased only slightly, from 9.0 to 10.5.

Excluding strongly acid (pH < 5) and poorly drained soils, organic carbon : organic nitrogen ratios (and, for that matter, organic carbon : total nitrogen ratios) of topsoils fall within surprisingly narrow limits, most lying

5 Nelson, D. W. and Sommers, L. E. in *Methods of Soil Analysis*, Vol. 2 (eds A. L. Page, R. H. Miller and D. R. Keeney), p. 539 Am. Soc. Agron., Madison, 1982; Bremner J. M. and Mulvaney, C. S. *op. cit.*, p. 595; M. A. Tabatabai, *op. cit.*, p. 501.
6 Olsen, S. R. and Sommers, L. E. in *Methods of Soil Analysis*, Vol. 2 (eds A. L. Page, R. H. Miller and D. R. Keeney), p. 403, Am. Soc. Agron., Madison, 1982; see also Anderson, G. in *Soil Components*, Vol. 1. (ed J. E. Gieseking), p. 305, Springer, Berlin, 1975.
7 Truchot, M., *Ann. Agron.*, 1875, **1**, 535.
8 Jenkinson, D. S., *Rothamsted Ann. Rept. for 1970*, Part 2, p. 113, 1971.

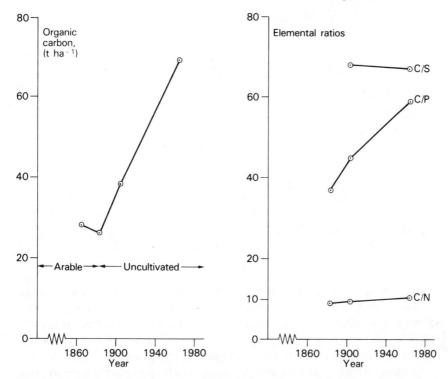

Fig. 18.2 Changes in carbon content (t ha $^{-1}$) and in elemental ratios of organic carbon, nitrogen, phosphorus and sulphur in a soil (0–23 cm) that had been arable for many years up to 1883 and thereafter allowed to revert to woodland (Broadbalk Wilderness).[8]

between 10 and 14. This is much less than in mature plant material and indeed a C : N ratio greater than 14 or so is a strong indication that a soil contains much partially decomposed plant material. Decomposition, often referred to as 'humification', is slowed under acid or under anaerobic conditions so that acid soils and poorly drained soils have the widest organic carbon : total nitrogen ratios.[3,4] No one yet knows why the humified organic matter in well-drained topsoils has a carbon : nitrogen ratio in the 10–14 region, rather than say 5 to 7: presumably the explanation will be found when more is known about the chemical structure of the humic substances and of how their carbon and nitrogen are bonded. Soils can develop under the most diverse conditions – temperate grassland, deciduous forest, tropical rainforest, and yet show similar organic carbon : nitrogen ratios.

Organic carbon : total nitrogen ratios often decrease down the soil profile (Fig. 18.3)[9] and values of 5 or thereabouts are not uncommon at depths of

9 Dyer, B., *US. Dept. Agric. Off. Exp. Stn Bull.* No. 106, 1902.

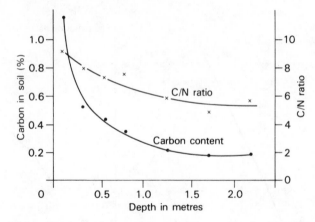

Fig. 18.3 The decline in carbon content and C : N ratio with depth in Broadbalk soil.[9]

a metre or so. Part and sometimes all of this decrease is due to the presence of fixed NH_4^+. The organic carbon content of most soils decreases with depth but the content of fixed NH_4^+ remains constant or increases, so that the organic carbon : total nitrogen ratios narrow sharply.[10] Very narrow carbon : nitrogen ratios are also found in marine sediments.[11]

Organic nitrogen : organic sulphur ratios are also remarkably constant over a wide range of climates and soils,[12] values in the range 7 to 8 being common.[13] Because most of the sulphur in topsoils is organic (the principal exceptions being pyritic soils[14] and gypsiferous soils), as is most of the nitrogen, the same constancy holds for the total nitrogen : total sulphur ratio. This constancy can be seen in Fig. 18.2 where the organic carbon : organic sulphur ratio of a soil remained the same, despite a very large increase in carbon content. As with nitrogen, this constancy might be a reflection of similarities in the underlying chemical structure of the humic substances formed in different soils.

There is much evidence that soil organic carbon is less intimately linked with soil organic phosphorus than with nitrogen or sulphur.[13] Again this can be seen in Fig. 18.2, where both carbon : nitrogen and carbon : sulphur ratios changed little as organic matter accumulated, whereas the organic carbon : organic phosphorus ratio widened, with organic phosphorus lagging behind.

10 Stevenson. F. J., *Soil Sci.*, 1959, **88**, 201.
11 Müller, P. J., *Geochim. Cosmochim. Acta*, 1977, **41**, 765.
12 Williams, C. H., *Trans. Int. Soc. Soil Sci.*, Comm. II & IV, Aberdeen, p. 93, 1967.
13 Williams, C. H. *et al.*, *J. Soil Sci.*, 1960, **11**, 334.
14 Bloomfield, C. and Coulter, J. K., *Adv. Agron.*, 1973, **25**, 265.

The extraction of organic matter from soil

Alkali is the traditional reagent for extracting organic matter from soil, first used by Achard in 1786 and later by many others.[1] Figure 18.4 shows a typical fractionation scheme based on extraction with alkali. The major materials thus separated, 'humic acid' (extracted by alkali and precipitated by acid), 'fulvic acid' (extracted by alkali and not precipitated by acid) and 'humin' (not extracted by dilute acid or alkali) cannot be defined in precise chemical terms, despite the efforts of generations of chemists. They are complex mixtures, no two humic molecules being exactly alike, and cannot be crystallized or otherwise separated into classes of homogeneous molecules. There are no sharp distinctions between humic acid, fulvic acid and humin; for example the proportion of the humic matter that is precipitated by acid from an alkaline extract depends on the type of acid used and its concentration.

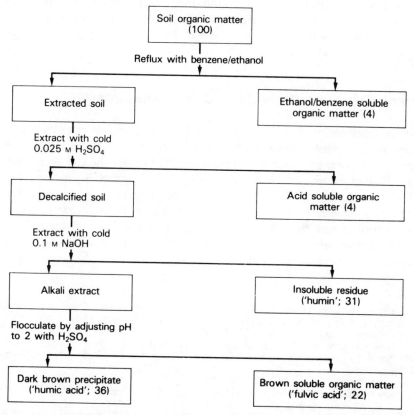

Fig. 18.4 Fractionation scheme for soil organic matter, after Kononova.[1] The figures in parentheses show the distribution of carbon between fractions from the surface layer of an uncultivated chernozem.

Many other reagents have been used to extract organic matter from soil,[15] the aims usually being to obtain more complete extraction than is possible with alkali and to avoid the possibility of alkali-induced chemical degradations and rearrangements. Bremner showed that humic colloids in alkaline solution take up atmospheric oxygen[16] and this autoxidation results in an increase in the acidity of the polymers.

Among the more effective inorganic reagents is sodium pyrophosphate.[17] Hot anhydrous formic acid has been used to extract organic matter from a wide range of soils,[18] and organic reagents that complex with iron and aluminium can effectively extract the organic matter from podzolic B horizons, a particular form of soil organic matter that is relatively easily and completely extracted.[15]

Organic fractions can also be isolated from soil on the basis of their density; the light, largely non-humified, fraction being of particular interest.[19] Soils also contain a small fraction of organic matter that can be extracted by organic solvents (see Fig. 18.4); again this largely comes from the non-humified part of the soil organic matter, for example, from fresh plant remains and living organisms, and contains a wide range of lipids, many of them chemically identifiable.[20]

Identifiable constituents of soil organic matter

All the constituents of plants, animals and micro-organisms enter the soil at some stage, however transient their existence once there. Some constituents are more resistant to microbial degradation than others and can accumulate to a limited extent, although the presence of the carbon-limited and biochemically-flexible soil population ensures that no great accumulation of biodegradable material ever occurs. Schreiner and Shorey (see footnote[1]) were the first to isolate and identify organic substances from soil in a systematic way and since then hundreds of organic compounds have been isolated and characterized. Soils contain traces of free sugars,[21] amino acids,[22] organic acids[23]

15 For a review see Hayes, M. H. B. in *Humic Substances in Soil, Sediment and Water* (eds G. Aiken, D. McKnight *et al.*), p. 329, Wiley, New York, 1985.
16 Bremner, J. M., *J. Soil Sci.*, 1950, **1**, 198.
17 Bremner, J. M. and Lees, H., *J. agric. Sci.*, 1949, **39**, 274.
18 Tinsley, J. and Walker, C. H., *Trans 8th Int. Cong. Soil Sci.*, Vol. 3, p. 149, Bucharest, 1964.
19 Ford, G. W. *et al.*, *J. Soil Sci.*, 1969, **20**, 291.
20 See for example Morrison, R. I. and Bick, W., *J. Sci. Fd Agric.*, 1967, **18**, 351; Butler, J. H. A. *et al.*, *Aust. J. Chem.*, 1964, **17**, 817.
21 Cheshire, M. V., *Nature and Origin of Carbohydrates in Soil*, Elsevier, London, 1979. For other reviews see Finch, P. *et al.*, in *Soil Biochemistry*, Vol. 2 (eds A. D. McLaren and J. J. Skujins), p. 257, Dekker, New York, 1971; Greenland, D. J. and Oades, J. M. in *Soil Components*, Vol. 1 (ed J. E. Gieseking), p. 213, Springer, Berlin, 1975.

and rather greater quantities (1 to 5 per cent of the soil carbon) of lipids.[23]

However, polymeric constituents comprise a much more important part of soil organic matter than transient monomers. Carbohydrates usually account for 10 to 20 per cent of the carbon in soil.[21] Some of these carbohydrates are detrital, in that they originally entered the soil in plant debris, but most are thought to be products of microbial synthesis *in situ*. When soils are subjected to acid hydrolysis, the polysaccharides are degraded to their constituent monomers. The dominant monomer in soil hydrolysates is glucose (Table 18.1) but representatives of all the major classes of carbohydrate are present – pentoses, hexoses, deoxyhexoses, hexosamines and uronic acids. The commonest polysaccharide linkage in soil is thought to be between carbon atoms 1 and 4 of the monomers, as in cellulose, but other linkages are known and indeed a feature of the polysaccharides isolated from soil is their highly branched nature.[21]

Although free amino acids are only present in minute concentrations, amino acid polymers that can be hydrolysed by acid are ubiquitous in soil and commonly make up 30 to 45 per cent of the nitrogen in surface soils, the proportion decreasing with depth.[22] Amino sugars account for a further 5 to 10 per cent of the soil nitrogen. Most of the amino acids in soil hydrolysates are those that are normally found in proteins but a few, for example diaminopimelic acid, which is a common constituent of bacterial cell walls, are not. Acidic amino acids (aspartic acid, glutamic acid) account for 15 to 25 per cent of the total amino acid N, basic amino acids (arginine, hystidine, lysine and ornithine) 20 to 25 per cent, neutral amino acids 50 to 55 per cent and sulphur-containing amino acids some 2 to 3 per cent.[24] Infrared studies show that peptide links are present in soil organic matter extracts[25] but the preponderance of evidence is that free proteins, as such, have a very transient existence in soil. Most of the amino acids in soil are thought to exist as peptide chains, stabilized against enzymic hydrolysis by linkage to other fractions of the soil organic matter, by reaction with clay or by reaction between amino acid containing organic matter and polyvalent metal cations such as iron or aluminium.[22]

As well as amino acids and amino sugars, soil hydrolysates contain unidentified acid-soluble organic nitrogen, usually accounting for 10 to 20 per cent

22 For reviews see Stevenson, F. J. in *Nitrogen in Agricultural Soils* (ed. F. J. Stevenson), *Agron. Monog.*, No. 22, 67, 1982; Kowalenko, C. G. in *Soil Organic Matter* (eds M. Schnitzer and S. U. Khan), p. 95, Elsevier, New York, 1978; Parsons, J. W. and Tinsley, J. in *Soil Components*, Vol. 1 (ed J. E. Gieseking), p. 263, Springer, Berlin, 1975; Bremner, J. M. in *Soil Biochemistry*, Vol. 1 (eds A. D. McLaren and G. H. Peterson), p. 19, Dekker, New York, 1967.

23 For reviews see Stevenson, F. J. in *Soil Biochemistry*, Vol. 1 (eds A. D. McLaren and G. H. Peterson), p. 119, Dekker, New York, 1967; Braids, O. C. and Miller, R. H. in *Soil Components*, Vol. 1 (ed. J. E. Gieseking), p. 343, Springer, Berlin, 1975.

24 Sowden, F. J. *et al.*, *Geochim. Cosmochim. Acta*, 1977, **41**, 1524.

25 Stevenson, F. J. and Goh, K. M., *Geochim. Cosmochim. Acta*, 1971, **35**, 471.

Table 18.1 Structural units of the major soil polysaccharides

Saccharide class	Structure of monomer	Sugars	Relative proportion in soil*
Pentoses	CHOH —— O / \\ HOCH CHOH \\ / CHOH —— CHOH	Arabinose Ribose Xylose	12 2 10
Hexoses	C^6H_2OH | C^5H —— O / \\ HO-C^4H C^1HOH \\ / C^3HOH —— C^2HOH	Galactose Glucose Mannose	15 35 17
Deoxyhexoses	CH_3 | CH —— O / \\ HO-CH CHOH \\ / CHOH —— CHOH	Fucose Rhamnose	3 6
Uronic acids	COOH | CH —— O / \\ HO-CH CHOH \\ / CHOH —— CHOH	Galacturonic acid Glucuronic acid	
Hexosamines	NH_2 | CH —— O / \\ HO-CH CHOH \\ / CHOH —— CHOH	Galactosamine Glucosamine	

* Average proportions, as calculated by Cheshire[21] from analyses done on numerous soils.

of the soil nitrogen. Some 20 to 35 per cent of the nitrogen in soil hydrolysates is ammonium, much of which is thought to be formed during hydrolysis by the breakdown of certain amino acids, amides and amino sugars.[22] A further 20 to 35 per cent of the soil nitrogen is not hydrolysed by acid and remains in the soil residue: little is known about its chemical structure.[22]

Although a cultivated soil normally contains much less organic nitrogen than the corresponding uncultivated soil from which it came, there is relatively little change in the distribution of nitrogen between the different hydrolysis products, so that cultivated and uncultivated soils contain much the same proportions of amino acids, amino sugars, ammonium released on hydrolysis, etc. In contrast, there is a marked difference in distribution of hydrolysis products from long-stabilised organic nitrogen in soil ('old' nitrogen) and from organic nitrogen that has only recently been stabilised ('new' nitrogen), with the new nitrogen containing a higher proportion of amino acids and amino sugars.[26] Presumably this is because the 'new' nitrogen contains a greater proportion of living microbial biomass.

Organic phosphorus compounds

Humus contains appreciable amounts of organic phosphates, mostly present as complex organic esters, in which phosphorus is bonded to carbon via oxygen. Recent NMR studies by Hawkes and co-workers[27] have shown that alkaline extracts of soil organic matter contain orthophosphate diesters, in which two of the oxygen atoms in a phosphate group are esterified, in addition to the orthophosphate monoesters which predominate. In addition, a small part of the organic phosphorus in certain highly organic acid soils is present as phosphonate, in which phosphorus is directly bonded to carbon. Polyphosphates are also present. Unlike nitrogen and sulphur, which occur predominantly in the organic form, organic phosphorus makes up a very variable part of the total soil phosphorus, most of the phosphorus in old grassland being organic, whereas in arable soils as little as 10 to 20 per cent can be organic.

Organic phosphorus differs from organic carbon and nitrogen in that it can be extracted almost quantitatively by relatively mild treatments.[28] Thus Halstead[29] was able to disperse almost all the organic phosphate from some soils using ultrasonic dispersion in acetyl acetone. Omotoso and Wild[30] dispersed about 90 per cent in the soil they used by a pretreatment with acid, followed by dispersion using a Na^+-saturated exchange resin and then extraction with acetyl acetone: most (but not all) of the organic phosphorus thus extracted was in the fulvic acid fraction.

26 Allen, R. W. *et al.*, *J. environ. Qual.*, 1973, **2**, 120. See also Kai, H. *et al.*, *Soil Sci. Pl. Nutr. (Tokyo)*, 1973, **19**, 275.
27 Hawkes, G. E. *et al.*, *J. Soil Sci.*, 1984, **35**, 35.
28 Barrow, N. J., *Soils Fertil.*, 1961, **24**, 169.
29 Halstead, R. L. *et al.*, *Nature, Lond.*, 1966, **211**, 1430.
30 Omotoso, T. I. and Wild, A., *J. Soil Sci.*, 1970, **21**, 216. See also Williams, J. D. H. *et al.*, *Soil Sci.*, 1970, **110**, 13.

The first phosphate compounds to be recognized in soil organic matter were the inositol phosphates and in particular phytic acid or myoinositol hexaphosphate. The inositol phosphates do not occur free in soil but are probably associated with proteins or polysaccharides. A range of inositol phosphates is present, from the mono to the hexa, but the penta- and hexaphosphates are the predominant forms. Commonly, 10 to 50 per cent of the organic phosphate in soil occurs as inositol penta- and hexaphosphates.[31] Several stereoisomers of the D-inositol phosphates are found, mostly based on myoinositol but other isomers are also present, some being found only in soils.[31] Inositol hexaphosphate forms highly insoluble salts with ferric and aluminium ions and it may well be that many of the organic phosphates owe their persistence to the formation of such salts, particularly in podzolic soils.

Other organic phosphates have been found in soil extracts but usually only in very small amounts. Thus Anderson[32] found about 1 or 2 per cent of the organic phosphorus as nucleotides, but most of this could well be present in the RNA and DNA of the living micro-organisms. Brookes and co-workers[33] showed that up to a quarter of the organic phosphorus present in old grassland and woodland soils was in the microbial biomass, although most soils contained a very much smaller proportion. DNA has been isolated from soil, but in very small quantity.[34] Phospholipids have also been isolated, but as presently determined, these comprise less than 1 per cent of the total organic phosphorus in soils.[35]

Organic sulphur compounds

About as much organic sulphur as organic phosphorus is present in humus but it is much more closely associated with the organic nitrogen than is the phosphorus, and, as already pointed out, the organic nitrogen : organic sulphur ratio in humus from a given group of soils is usually less variable than the organic nitrogen : organic phosphorus ratio.

The amino acids cystine and methionine and their derivatives such as methionine sulphoxide and cysteic acid have been isolated from soil hydrol-

31 For reviews on soil organic phosphorus see Kowalenko, C. G. in *Soil Organic Matter* (eds M. Schnitzer and S. U. Khan), p. 95, Elsevier, Amsterdam, 1978;   Cosgrove, D. J. in *Soil Biochemistry*, Vol. 1 (eds A. D. McLaren and G. H. Peterson), p. 216, Dekker, New York, 1967;   Halstead, R. L. and McKercher, R. B. in *Soil Biochemistry*, Vol. 4 (eds E. A. Paul and A. D. McLaren), p. 31, Dekker, New York, 1975; Anderson, G. in *Soil Biochemistry*, Vol. 1 (eds A. D. McLaren and G. H. Peterson), p. 67, Dekker, New York, 1967; Dalal, R. C., *Adv. Agron.*, 1977, **29**, 83; Cosgrove, D. J., *Adv. Microbial Ecol.*, 1977, **1**, 95.
32 Anderson, G., *Soil Sci.* 1961, **91**, 156; ibid., 1970, **110**, 96.
33 Brookes, P. C. *et al.*, *Soil Biol. Biochem.*, 1984, **16**, 169.
34 Torsvik, V. L. and Goksoyr, J., *Soil Biol. Biochem.*, 1978, **10**, 7.
35 See for example Kowalenko, C. G. and McKercher, R. B., *Soil Biol. Biochem.*, 1971, **3**, 243;   Dormar, J. F., *Soil Sci.*, 1970, **110**, 136.

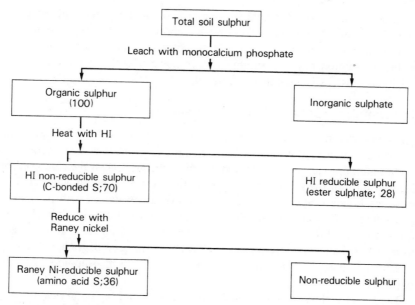

Fig. 18.5 Fractionation scheme for soil organic sulphur. The figures in parentheses show the distribution of sulphur between fractions from the surface layer of a yellow podzolic soil under subterranean clover.[37]

ysates.[36] However, most of the organic sulphur in soil has not been identified, although some progress has been made in subdividing it into broad categories. Figure 18.5 shows a fractionation scheme[37] used to resolve soil organic sulphur into ester sulphate and carbon bonded sulphur, of which the part reducible with Raney nickel is considered to consist mainly of amino acid sulphur. In this fractionation, ester sulphate is defined as the organic sulphur that can be reduced to H_2S by a reducing agent containing hydriodic, formic and hypophosphorus acids; it will include both O-bonded sulphate and N-bonded sulphate. Kowalenko[36] assembled data on the fractionation of organic sulphur in a wide range of surface soils and showed that 40 to 60 per cent of the organic sulphur was in the ester form.

Alkali extracts much, but by no means all, of the organic sulphur in soil,[38]

36 For reviews of the chemistry of organic sulphur in soil see Freney, J. R. in *Soil Biochemistry*, Vol. 1 (eds A. D. McLaren and G. H. Peterson), p. 229, Dekker, New York, 1967; Anderson, G. in *Soil Components*, Vol. 1 (ed J. E. Gieseking), p. 333, Springer, Berlin, 1975; Kowalenko, C. G. in *Soil Organic Matter* (eds M. Schnitzer and S. U. Khan), p. 95, Elsevier, Amsterdam, 1978; Probert, M. E. in *Soils: an Australian viewpoint*, p. 583, CSIRO, Melbourne/Acad. Press, London, 1983.
37 Freney, J. R. *et al.*, *Soil Biol. Biochem.*, 1971, **3**, 133; *ibid.*, 1975, **7**, 217.
38 Williams, C. H. and Steinbergs, A., *Aust. J. agric. Res.*, 1959, **10**, 340 used 0.1M NaOH. For recommended methods see Tabatabai, M. A. in *Methods of Soil Analysis*, Part 2 (ed A. L. Page), p. 501, Am. Soc. Agron., Madison, 1982.

so that in this respect sulphur resembles organic carbon and nitrogen. However, organic sulphur, like organic phosphorus, is more completely extracted by dispersion in acetyl acetone solution than is organic nitrogen,[29] so the association between sulphur and nitrogen in soil organic matter may not be as close as is sometimes assumed.

The chemical nature of humic substances

The composition of humic materials found in mineral soils, peats and sediments has exercised the minds of soil chemists from the very beginning.[1] The traditional approach of the organic chemist, chemical cleavage, followed by identification of the fragments, has been much used[39-41] and a wide range of products isolated, almost invariably in low yield. Oxidation with permanganate releases fatty acids, especially the n-C_{16} and n-C_{18} acids, dicarboxylic acids and tricarboxylic acids. Several benzene carboxylic acids are released, as are phenolic acids.[39] Humic substances have also been degraded by reductive techniques, by heating with phenol or sodium sulphide, and by pyrolysis.[40]

The ferocity of the degradation techniques used and the poor yield of identifiable compounds, make it difficult to come to firm conclusions about the original structure of the humic substances, other than that they contain linked aromatic rings, mostly heavily substituted. In general, physical methods of analysis have been more helpful than degradation studies in elucidating the structure of humus, although the chemical structure of the humic substances is still poorly understood. Indeed the phrase 'chemical structure of the humic substances' is probably misleading, in that it implies that there is one chemical structure to be discovered, however complex and difficult to establish. It is much more likely that a wide range of structures is present, formed by random reactions between molecules in a way that is quite different from the formation of polymers in plants and animals, which are constructed in a systematic manner by enzymes operating under genetic control.

Infrared spectroscopy and electron spin resonance spectrometry (ESR) have been repeatedly used to examine humic substances.[39,40] ESR spectra show that these substances contain numerous stable free radicals (i.e. groups containing unpaired electrons), presumably trapped in polymers containing semi-quinone systems. Infrared spectroscopy shows the presence of hydrogen-bonded OH groups, aliphatic C–H bonds, carboxyl groups, amide links and, arguably, of aromatic C=C bonds.[40]

More recently, nuclear magnetic resonance spectrometry (NMR) has been

39 Schnitzer, M. in *Soil Organic Matter* (eds M. Schnitzer and S. U. Khan), p. 1, Elsevier, Amsterdam, 1978.
40 Hayes, M. H. B. and Swift, R. S. in *The Chemistry of Soil Constituents*, Vol. 1 (eds D. J. Greenland and M. H. B. Hayes), p. 179, Wiley, Chichester, 1978.
41 Flaig, W. *et al.*, in *Soil Components*, Vol. 1 (ed. J. E. Gieseking), p. 1, Springer, Berlin, 1975.

used to examine humic substances, as yet mainly in soil extracts.[42] Proton NMR of humic substances has revealed the presence of aromatic protons; for example a humic acid isolated from a New Zealand silt loam contained 16 per cent of aromatic protons, with a further 27 per cent attached to oxygen, presumably mainly as OH groups in polysaccharides. Signals were also obtained from protons in methylene ($-CH_2-$) groups and methyl ($-CH_3$) groups. Although ^{12}C cannot be examined by NMR spectrometry because it does not possess the necessary nuclear spin, the isotope ^{13}C, which does, is present to the extent of 1.11 per cent in naturally occurring carbon. This means that ^{13}C NMR spectra can be obtained from the various humic substances found in nature. As pointed out by Wilson and co-workers;[43] '^{13}C NMR has enormous potential in soil science. No other analytical technique reveals such structural detail on the nature and forms of organic carbon.' They illustrated this for a New Zealand soil by identifying polymethylene chains, aliphatic carbon-to-oxygen bonds (as in the $-C-OH$ group in sugars), carboxylic carbon (about 10 per cent of the total) and aromatic carbon (about 15 per cent of the total). They concluded that some of the carboxylic groups in humus are bonded to aliphatic carbon, and some to aromatic carbon. It should be noted that this information was obtained on whole soils – the organic matter was not extracted, and thus possibly altered, prior to spectroscopic examination. Other workers have shown that aromaticity (i.e. the fraction of the carbon which is aromatic) varies greatly in humic preparations, being low in humic acids from marine environments and high in terrestrial soils.[44]

Functional groups in humus

Although mineral soils commonly contain only a few per cent of humus carbon, the humic substances have a very large influence on the cation exchange capacity of the soils, often contributing half or more. Most of the measurements of the exchange capacity of humus have been made in one of two ways: by measuring the cation exchange capacity before and after destruction of the organic matter by treatment with hydrogen peroxide[38] or from multiple regressions between the exchange capacities of a series of soils and their clay and organic carbon contents.[45]

42 For reviews on the NMR spectroscopy of humic substances see Wilson, M. A., *J. Soil Sci.*, 1981, **32**, 167; Wershaw, R. L. in *Humic Substances in Soil, Sediment and Water* (eds G. R. Aiken, D. M. McKnight *et al.*), p. 561, Wiley, New York, 1985.
43 Wilson, M. A. *et al.*, *Nature, Lond.*, 1981, **294**, 648.
44 Hatcher, P. G. *et al.*, in *Aquatic and Terrestrial Humic Materials* (eds R. F. Christman and E. T. Gjessing), p. 37, Ann Arbor Science, Ann Arbor, 1983.
45 Hallsworth, E. G. and Wilkinson, G. H., *J. agric. Sci.*, 1958, **51**, 1; Yuan, T. L. *et al.*, *Soil Sci.* 1967, **104**, 123; Martel, Y. A. *et al.*, *Soil Sci. Soc. Am. J.*, 1978, **42**, 764.

The exchange capacities of humus determined by these methods for different soils varies widely, although some of this variation may have arisen because different authors used different methods for determining cation exchange capacity. But the general result is that the exchange capacity of humus in well-drained near-neutral soils lies mostly between 300 and 550 meq per 100 g organic carbon. The values are lower for more acid or for poorly drained soils and figures between 50 and 120 have been found for acid soils. The probable reason for such low values is that much of the organic carbon in these acid soils is not fully humified, since the exchange capacity of the fulvic and humic acid fractions isolated from them is about the same as for neutral soils. Humus becomes increasingly important as the seat of cation exchange as the pH of a soil rises;[46] for each unit increase in pH, the change in cation exchange capacity of organic matter is several times greater than that of clay (see Ch. 7).

Of the functional groups in humus, undoubtedly the most important are the carboxylic acid groups, because of their contribution to the cation exchange capacity. The R–COOH groups are mostly dissociated to the carboxylate anion at the pH values found in all but the most acid soils. Phenolic hydroxyls are also present, although these are normally not substantially dissociated when the pH is less than 9. However, deactivating ring substituents can lower the pK_a values of phenolic hydroxyls, thus allowing them to contribute to the soil cation exchange capacity at lower pH values.

The determination of functional groups in complex polymeric materials like the humic substances is not easy and the results obtained are often influenced by the methods used.[47] Some results obtained by current methods are shown in Table 18.2, constructed from data assembled by Schnitzer.[48] This table gives the elemental composition and functional group content of humic and fulvic acids isolated from a wide range of soils from different climatic zones. Although the analytical values spread over a wide range, as would be expected, it can be seen that the humic acids contained more carbon and less oxygen than the fulvic acids and that both total acidity and carboxylic acid content tend to be greater in fulvic acids than in humic acids. A fulvic acid sample studied by Hatcher and co-workers[49] contained either a carboxyl group or an oxygen containing substituent (carbonyl, alcohol or ether) for

46 See for an example for Wisconsin soils, Helling, C. S. *et al.*, *Soil Sci. Soc. Am. Proc.*, 1964, **28**, 517.

47 For useful discussions of the measurement of functional groups in humus see Hayes, M. H. B. and Swift, R. S. in *The Chemistry of Soil Constituents* (eds D. J. Greenland and M. H. B. Hayes), p. 179, Wiley, Chichester, 1978; Stevenson, F. J., *Humus Chemistry, Genesis, Composition, Reactions*, p. 221 Wiley, New York, 1982. For details of the experimental methods see Schnitzer, M. in *Methods of Soil Analysis*, Vol. 2 (eds A. L. Page, R. H. Miller and D. R. Keeney), p. 581, Am. Soc. Agron., Madison, 1982.

48 Schnitzer, M. in *Proc. Int. Symp. Soil Organic Matter*, Vol. 2, p. 117, IAEA, Vienna, 1977.

49 Hatcher, P. G. *et al.*, *Soil Sci. Soc. Am. J.*, 1981, **45**, 1089.

Table 18.2 Elemental composition and functional group composition of humic and fulvic acids isolated from soils taken from different climatic zones[48]

Elemental composition (%)	Humic acids	Fulvic acids
C	53.6–58.7	40.7–50.6
H	3.2–6.2	3.8–7.0
N	0.8–5.5	0.9–3.3
S	0.1–1.5	0.1–3.6
O	32.7–38.3	39.7–49.8
Functional group (meq g^{-1})		
Total acidity	5.6–7.7	6.4–14.2
—COOH	1.5–5.7	5.2–11.2
Phenolic—OH	2.1–5.7	0.3–5.7
Alcoholic—OH	0.2–4.9	2.6–9.5
Quinoid and ketonic C=O	0.1–5.6	1.2–4.2
Methoxyl (—O—CH$_3$)	0.3–0.8	0.3–1.2

every 1.8 non-carboxyl carbons. Most of the carboxyl groups are therefore sterically close to each other or to other polar groups, so that the dissociation of one group will influence the dissociation of its neighbours. Titration curves of humic substances are best interpreted as arising from a continuous distribution of acidic functional groups, ranging from strongly acidic carboxyl groups to weakly acidic groups, such as phenolic hydroxyls – although the number of phenolic hydroxyls may not be as large as once thought, being limited by the amount of aromatic carbon present.[44,48,50]

Interaction of humus with metallic ions

Metallic ions, particularly iron and aluminium, are tenaciously held by humic preparations and can only be removed with difficulty. The reaction with metallic ions can lead to increased metal solubility; for example, ferric ions can be held in solution by low molecular weight fulvic acids at pH values far above those at which precipitation would have taken place in the absence of the fulvic acids.[50] However, it is much more common for the metal concentration in solution to be diminished by reaction with humus, leading to the formation of insoluble metal-organic compounds.

The reaction between a metal ion (M) and an organic ligand (A) can be described by the equation[51]

50 For a discussion of the acidity of humic substances see Perdue, E. M. in *Humic Substances in Soil, Sediment and Water* (eds G. R. Aiken, D. M. McKnight *et al.*), p. 493, Wiley, New York, 1985.

51 For a discussion of the stability constants in metal–humus reactions and how they are measured, see Stevenson, F. J., *Humus Chemistry, Genesis, Composition, Reactions*, p. 355, Wiley, New York, 1982.

$$jM + iA \rightleftharpoons M_j A_i$$

with the stability constant (K) being given by

$$K = \frac{(M_j A_i)}{(M)^j (A)^i}$$

In general, the stability constants for reaction between humus and metal increase according to the Irving-Williams series for the stability of metal complexes, with trivalent iron and aluminium being most strongly held, followed by the divalent metals in the sequence

$$Cu > Ni > Co > Zn > Fe(II) > Mn(II) > Mg > Ca > Ba$$

However, this divalent metal sequence is by no means always observed and, as pointed out by Stevenson,[51] serious problems are encountered in the determination of stability constants for reactions between metal ions and humic or fulvic acids.

In general the capacity of humic acids to bind a metal ion is, at most, equal to the number of titratable H^+ ions, divided by the valency of the particular metal ion. The exact nature of the chemical bonds between humic substances and metal ions is still uncertain[51] but at least some of the linkages are of the chelate type, with the metal linked to two adjacent functional groups, for example.[52]

However, it is unlikely that a large proportion of cations are held by a specific arrangement of functional groups like this. In all probability there is a range of stability constants for each metal, the magnitude of the constant for a particular site depending on the exact environment of the reacting functional group. Functional groups containing oxygen are not the only ones capable of reacting with metal ions,[51] copper in particular being able to react with the $=N-H$ groups in the humic peptides.

52 van Dijk, H., *Geoderma*, 1971, **5**, 53.

Interaction of humus and clay

Humic material has for a long time been known to absorb, or be absorbed by, clay particles. Schloesing[53] wrote in 1874 that *'L'argile possède une certaine tendance à s'unir aux humates du terreau pour former probablement une de ces combinaisions entre colloïdes signalées par Graham.'* He attempted to separate the humus from the clay by suspending the mixed soil in ammonia and then adding ammonium chloride which flocculated the clay, but left much of the humus in suspension.

Several lines of evidence have been advanced to show that much or most of the humus in a soil can be in close association with the clay particles. Direct microscopic examination of thin sections of most mull and arable soils, using the techniques pioneered by Kubiena, is consistent with a close association of humus and clay particles, for they reveal a plasma in which these two components cannot be distinguished; nor can most of the humus in these soils be seen in electron microscope photographs.[54] Again, if a soil is dispersed ultrasonically in a heavy liquid, with a specific gravity as high as 2, most of the humic fraction settles with the mineral fraction, for the organic matter that floats is primarily unhumified or partially humified and is only lightly contaminated with dark-coloured material. The proportion of the total organic matter in this dense fraction varies from soil to soil, as illustrated in Table 18.3,[19] where, of the soils tested, it was least in the podzol, with little of the organic matter complexed to clay, and greatest in a grey clay soil in which almost all the organic matter was thus complexed.

Table 18.3 Proportion of soil organic carbon in the clay–organic complex[19]

	Per cent organic carbon in soil	Per cent soil carbon in clay–organic complex*
Podzol	1.5	18
Solonized brown soil	1.5	77
Grey clay soil	1.1	91
Terra rossa	2.8	82
Ground water rendzina	5.4	69
Black earth	1.8	82
Krasnozem	4.9	90
Red brown earth	2.5	66

* Defined as the material sinking when the soil was ultrasonically dispersed in an organic liquid of density 2 g cm^{-3}.

53 Schloesing, T., *Compt. Rend.*, 1874, **78**, 1276.
54 See for example some of the photographs in Kubiena, W., *Soils of Europe*, Murby, London/Madrid, 1953; Babel, U. in *Soil Components* (ed J. E. Gieseking), p. 369, Springer, Berlin, 1975.

The linkage between inorganic and organic colloids in soils

Since the polyanionic humic colloids are normally repelled by the negatively charged clays, absorption of humic substances on clay surfaces only occurs when polyvalent cations are present. There is no evidence to suggest that a sodium saturated clay will absorb any appreciable amount of a sodium saturated 'humate'. Theng[55] suggests that there are two main binding mechanisms[56] between humic substances and inorganic colloids, viz.:

1 *Ion dipole interactions.* These form the strongest links, and can be of two sorts: links formed by polyvalent cations between clay surfaces of the mica type and humic polyanions; and links formed between positively charged hydrous oxide surfaces and the humic polyanions.[56] Edwards and Bremner[57] proposed that organic matter is linked to clay particles through polyvalent cations so that a clay–organic matter aggregate would be held together by numerous clay–(cation)–organic matter links. This linkage is illustrated in Fig. 18.6 which also shows an alternative, and probably more realistic, formulation in which the cation is associated with both organic anion and clay surface via water bridges. Scharpenseel[56] showed that the extent to which a humate or fulvate salt was adsorbed by a montmorillonite clay depended on the valency of the bridging cation and that

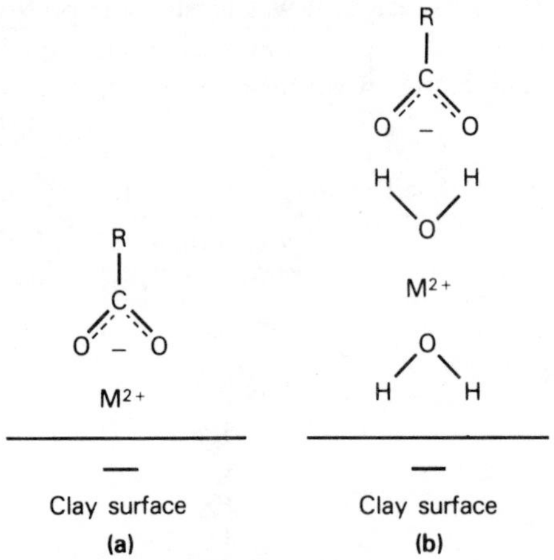

Fig. 18.6 Diagram illustrating cation bridging: (a) in its simplest form; and (b) a more realistic model in which the organic anion is coordinated to the cation through water molecules from the hydration shell of the cation. R is a polyanionic humic colloid.

55 Theng, B. K. G. *Formation and Properties of Clay-Polymer Complexes*, Elsevier, Amsterdam, 1979.
56 Scharpenseel, H. W., *Trans. Int. Soc. Soil Sci.*, Comm. II & IV, p. 41, Aberdeen, 1967.
57 Edwards, A. P. and Bremner, J. M., *J. Soil Sci.*, 1967, **18**, 47.

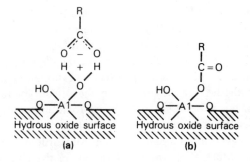

Fig. 18.7 Diagram illustrating anionic bridging: (a) anion exchange; and (b) ligand exchange. R is a polyanionic humic colloid.

adsorption generally rose with an increase in cation valency. Figure 18.7 shows how organic polyanions can be linked to positively charged hydrous oxide surfaces by anion exchange. The clay minerals found in the surface layers of soil are frequently coated with a skin of poorly crystalline hydrous oxides of iron and aluminium, so that there are numerous opportunities for the formation of this sort of linkage. Anion exchange gives a relatively weak bond and the organic anions can be displaced by other anions in solution, such as Cl^- or NO_3^-. Fig. 18.7 also shows bonding by ligand exchange, which gives very strong bonding, the organic anions only being desorbed by even more strongly adsorbed anions such as hydroxyl or pyrophosphate. According to Greenland,[58] ligand exchange occurs when the organic anion penetrates the coordination shell of Fe or Al and is incorporated into the surface OH layer.

2   *Hydrogen bonding and van der Waals' forces.* The bonds formed by both those mechanisms are short range, weak but additive, in that more are formed if the organic polyanion collapses on to the clay surface when the system is dehydrated. Several of the functional groups in humus, for example carboxyl, hydroxyl and amino, can be involved in hydrogen bonding with inorganic colloidal surfaces. Adsorption of humic particles on clay surfaces by the even weaker van der Waals' forces is through the attraction that develops between atoms when brought close (but not too close) together. Thus if a humus polymer comes close to a clay surface, forces of attraction may become important, particularly if the polymer is flexible and can bend to touch a large area of the clay. These forces become stronger and the adsorption firmer as the system is dried, for this removes water from between adjacent surfaces, allowing closer contact to be made.

The effects of sodium hydroxide in releasing humic materials can be explained by the effects of the alkali on the polyvalent bridges. The Na^+ ion

58 Greenland, D. J., *Soil Sci.*, 1971, **111**, 521;   Greenland, D. J., *Soils Fertil.*, 1965, **28**, 415; *ibid.*, 1965, **28**, 521.

replaces bridges formed by polyvalent cations, leading to the disruption of the clay–(polyvalent cation)–humus linkage and solubilization of the adsorbed humate. At the same time, the added OH⁻ ions increase the negative charge on the humic molecule, thus increasing the repulsion between clay and organic matter. In addition to the effect on cation bridging, the OH⁻ ions also compete with the anionic groups on the organic matter for positively charged sites on the mineral surface, again promoting desorption.[55]

Organic molecules can occupy the interlayers of expanding 2.1 clay minerals but it is generally thought that relatively little organic matter is thus retained in soils. Most of the humic materials in the clay–organic complex are thought to be held on external clay surfaces, being too large to penetrate interlamellar spaces.[59]

Humus adsorbed on the clay usually changes the colour of the clay, but this change need not be large. In general the adsorbed organic matter gives clay a dark brown or black colour. Some red and orange-coloured clays, particularly in tropical soils, may have quite an appreciable content of humic material, whereas calcium-saturated humus–montmorillonite complexes are black in colour, although their humus content may be quite low.

Colloidal properties of humic substances

Particles too small to settle under the influence of gravity, say less than 1 μm in diameter, and too large to be in true solution, say 0.001 μm in diameter, are conventionally described as 'colloidal'. Although relatively little is known about the size of humic particles as they exist in soil, much attention has been given to the molecular size of humic substances in aqueous extracts, usually as obtained from soil by the traditional alkaline extraction procedure. None of the methods that have been used to measure molecular size (gel permeation chromatography, ultrafiltration, light scattering, small angle X-ray scattering) or molecular weight (ultracentrifugation, viscometry or colligative property measurement) give unequivocal results with humic substances, and different methods usually give different results. Nevertheless, it is now accepted that most of the humic substances in aqueous soil extracts are polydisperse colloids, with a wide range of molecular sizes.[60] Molecular weights start from 500 to 1000 in the fulvic acids and rise, in a more or less

59 For a discussion see Tan, K. H. and McCreery, R. A. in *Proc. Int. Clay Conf.*, Mexico City, 1975, p. 629; Theng, B. K. G., *Geoderma*, 1976, **15**, 243; Stevenson, F. J., *Humic Chemistry, Genesis, Composition, Reactions*, p. 374, Wiley, New York, 1982.

60 For reviews see van Dijk, H. in *Soil Biochemistry*, Vol. 2 (eds A. D. McLaren and G. H. Peterson), p.16, Dekker, New York, 1972; Orlov, D. S. *et al.*, *Geoderma*, 1975, **13**, 211; Stevenson, F. J., *Humus Chemistry Genesis, Composition, Reactions*, p. 285, Wiley, New York, 1982; Wershaw, R. L. and Aiken, G. R. in *Humic Substances in Soil, Sediment and Water* (eds G. R. Aiken, D. M. McKnight *et al.*), p. 477, Wiley, New York, 1985.

continuous fashion, through the fulvic acids into the humic acids, with values of 10^6 being measured for some of the largest humic acid 'molecules'. Molecular size and weight measurements are frequently complicated by aggregation between individual particles, leading to an apparent increase in molecular size.

There is strong evidence to show that soil organic colloids in solution (and possibly, in soils) are not the compact globular particles seen in dried preparations under the electron microscope[61] but are flexible, three-dimensional chains, with a relatively open structure. A random coil model for humic acid molecules has been suggested in which charged humic acid strands coil and wind randomly.[62] Some of the strands are branched and these give rise to more compact structures than would be obtained with a linear molecule of equivalent molecular weight. These colloidal humic particles are perfused with solvent molecules which are able to exchange with bulk solvent molecules, provided the overall particle size is not too great. In large particles, the solvent can flow freely through the periphery but is effectively trapped within the central region. The negative charge on the particles arises largely from the ionization of –COOH groups distributed along the humic acid strands. It has also been suggested that both humic and fulvic acids behave as rigid 'spherocolloids' at high organic concentrations, high electrolyte concentrations and at low pH.[63] At low organic matter concentrations, low electrolyte concentrations and at pHs of 7 or above, the ionization of the –COOH groups leads to mutual repulsion and the humic molecule expands to give flexible linear tangles.[62]

Reactions between soil organic matter and pesticides

Field experiments have repeatedly shown that the amount of pesticide that must be added to a soil to control a particular pest is strongly influenced by the organic matter content of the soil; in general the more organic matter, the more pesticide that must be used.[64] Pesticides applied to soil include herbicides, fungicides, insecticides, nematocides and can also be taken to include chemicals used to manipulate the nitrogen cycle, such as nitrification inhibitors and urease inhibitors.

61 See for example Flaig, W. *et al.*, in *Soil Components*, Vol. 1 (ed J. E. Gieseking), p. 1, Springer, Berlin, 1975.
62 Cameron, R. S. *et al.*, *J. Soil Sci.*, 1972, **23**, 394.
63 Ghosh, K. and Schnitzer, M., *Soil Sci.*, 1980, **124**, 266.
64 For reviews of the role of organic matter in modifying the behaviour of pesticides in soil see Adams, R. S., *Residue Reviews*, 1973, **47**, 1;   Hayes, M. H. B., *Residue Reviews*, 1970, **32**, 131;   Stevenson, F. J., *ACS Symposium Series*, 1976, **29**, 180;   Weed, S. B. and Weber, J. B. in *Pesticide–Organic Matter Interactions* (ed. W. D. Guenzi), p. 39, ASA, Wisc., 1974;   Stevenson, F. J., *Humus Chemistry, Genesis, Composition, Reactions*, p. 403, Wiley, New York, 1982.

Both organic matter and clay contribute to the colloidal properties of a soil and only rarely can the sorption of a particular pesticide be attributed exclusively to organic matter, although in most cases organic matter dominates the process. Walker and Crawford[65] examined the adsorption of several of the *s*-triazine herbicides in a range of soils and showed that both organic matter and clay participated, up to an organic matter content of about 8 per cent; above this, adsorption occurred mainly on organic surfaces. Stevenson[66] showed that the adsorption of a range of different herbicides was almost always more closely correlated with the organic matter content of a soil than with the clay content.

The organic compounds used as pesticides are conveniently classified as ionic, including cationic, basic and acidic compounds and non-ionic, a class that includes most of the pesticides currently in use. The surface of soil organic matter contains both hydrophobic and hydrophilic areas, so that both ionic and non-ionic molecules can be adsorbed. Non-ionic compounds are considered to be mainly adsorbed on the hydrophobic areas. The fats, waxes and resins in soil organic matter provide hydrophobic surfaces, as do the extensive aliphatic chains that are nowadays thought to be an important part of humus.

Indeed, the partition of a non-polar compound between water and octanol gives a very good guide to its sorption by soil, compounds that strongly favour the octanol phase being the most strongly absorbed by soil organic matter.[67]

Of the ionic pesticides, the cationic herbicides paraquat and diquat are strongly bound to organic matter in soil, presumably by ion exchange between the herbicide cation and the humus carboxylate polyanion (Fig. 18.8), just as these herbicides are strongly held on negatively charged clay surfaces by cation exchange.[68] Because of the large size of the paraquat and diquat cations, their hydration shells are likely to be small, so that adsorption by short range van der Waals' forces will also contribute to the uniquely strong bonding observed with these compounds.

The sorption of basic and acidic organic pesticides is strongly dependent on soil pH, as well as on the structure of the organic molecule being sorbed.[68] Soil pH governs not only the ionization of acidic functional groups on the humic colloids but also the proportion of the herbicide that occurs in the ionized form. This can be seen with the weakly basic *s*-triazine herbicides, which are most strongly adsorbed by soil organic matter at pH values close

65 Walker, A. and Crawford, D. V. in *Isotopes and Radiation in Soil Organic Matter Studies*, p. 91, IAEA, Vienna, 1968.

66 Stevenson, F. J., *J. environ. Qual.*, 1972, **1**, 333.

67 Khan, A. *et al.*, *Soil Sci.*, 1979, **128**, 297;   Briggs, G. C., *Proc. 7th Brit. Insecticide and Fungicide Conf.*, *Brighton*, 1973, **2**, 83.

68 For detailed discussions of these processes, see Burchill, S. *et al.*, in *The Chemistry of Soil Processes* (eds D. J. Greenland and M. H. B. Hayes), p. 221, Wiley, Chichester, 1981;   Khan, S. U. in *Soil Organic Matter* (eds M. Schnitzer and S. U. Khan), p. 137, Elsevier, Amsterdam, 1978.

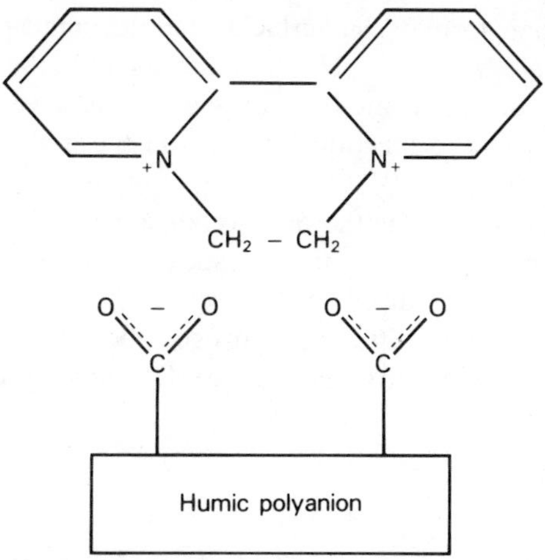

Fig. 18.8 Diagram illustrating the adsorption of the herbicide diquat on a humic polymer.

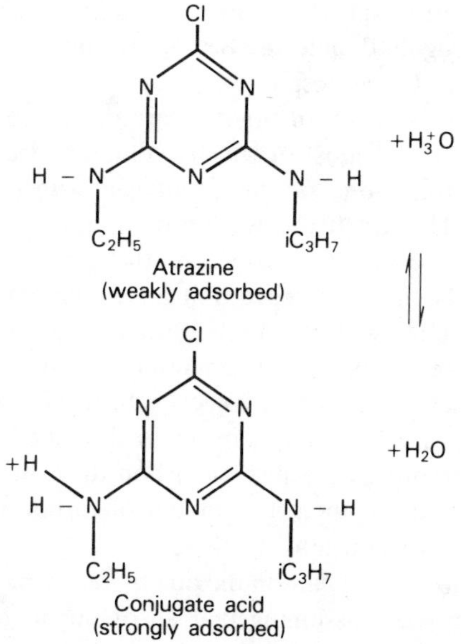

Fig. 18.9 Formation of the conjugate acid from the herbicide atrazine.

to the pK_a of the conjugate acid of each particular herbicide.[69] As the pH rises above the pK_a of the herbicide, the proportion of the herbicide present as the conjugate acid decreases (Fig. 18.9) so that binding by ion exchange

[69] Weber, J. B. *et al.*, *Weed Sci.*, 1969, **17**, 417.

between the cationic form of the herbicide and the humic polyanions becomes weaker. Conversely, when the pH falls below the pK_a, the carboxylate groups on the humic polymers become less dissociated, again weakening adsorption, even though most of the functional groups on the herbicide are now in the conjugate acid form.

Anionic pesticides, such as the phenoxyalkanoic acids, can presumably be held by salt bridges between ionized pesticide and humic polyanions at pH values above their pK_a; at pH value below the pK_a, such binding as is observed is presumably through hydrogen bonds between the ionized carboxyl groups in the pesticide and carboxyl or amide groups on the organic matter.

The turnover of organic matter in soil

Over the world as a whole, there is substantially more carbon in the top metre of soil than in the vegetation it carries or in the atmosphere. It is estimated[70] that terrestrial soils now contain some 1500×10^9 tonnes of organic carbon, roughly double the 800×10^9 tonnes in green plants or the 700×10^9 tonnes in the atmosphere.

The net annual production of organic carbon by photosynthesis on land is about 55×10^9 tonnes,[71] most of which eventually decomposes on or in the soil. In the long run, none of this input can withstand decomposition to carbon dioxide and water. If this were not so, any completely resistant fraction would by now cover the surface of the earth.[72] When a geologically stable surface has been under vegetation for a long time, the soil approaches steady-state conditions, with the annual loss of organic matter balancing the annual input. Thereafter the organic matter content remains constant, until the next vegetational change or geological catastrophe. This process, in which losses and gains proceed simultaneously, is described as turnover and may be defined (for carbon) as the flux of carbon through the organic carbon in the soil. Turnover time is then the amount of carbon in the soil, divided by the annual input of carbon into the soil.

Taking the net fixation of carbon dioxide by terrestrial plants to be 55×10^9 tonnes per annum[71] and assuming that all of this carbon is decomposed in the soil, then the global mean turnover time of soil organic matter is 27 years. In calculating the global turnover time in this way, it is assumed that all the

70 Buringh, P. in *The Role of Terrestrial Vegetation in the Global Carbon Cycle*, Scope 23 (ed G. M. Woodwell), p. 91, Wiley, New York, 1984. See also Schlesinger, W. H., *op. cit.*, p. 111.

71 Bolin, B., *Carbon Cycle Modelling*, Scope 16, Wiley, New York, 1981.

72 For a critical examination of this argument, see Alexander, M., *Adv. appl. Microbiol.*, 1965, **7**, 35.

primary production is decomposed in the top metre of soil. The more that is decomposed above or below this layer, the longer will be the turnover time in the top metre.

During the last 100 years, the carbon dioxide concentration of the atmosphere has risen from about 290 to about 330 ppm. Part of this increase has come from the combustion of fossil fuel but part from the decline in soil organic matter level brought about by the clearing of forests and the cultivation of virgin prairies. Buringh[70] has suggested that soils now contain about three-quarters of the soil organic matter that was present before the spread of civilization.

Effects of management on the amount of organic matter

Changes in management affect the quantity of organic matter in two ways, by altering the annual input of organic matter from dead plants and animals and by altering the rate at which this organic matter decays. It is usually impossible to separate these two processes in analysing the results of a particular change in management.

Figures 18.2, 18.10, 18.11[73] and 18.12, all taken from experiments at Rothamsted, give examples of how management alters the amount of organic matter in a soil – examples for other climates and other farming systems can be found in a comprehensive review by Campbell.[74] Figure 18.2 shows how

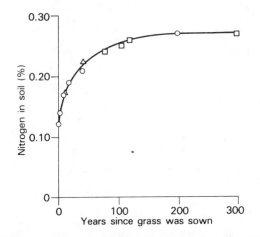

Fig. 18.10 The accumulation of nitrogen in soil (0–23 cm) at Rothamsted when old arable land is laid down to grass. J. B. Lawes and J. H. Gilbert's results, ○; results provided by A. E. Johnston, △ and □.

73 Jenkinson, D. S. and Johnston, A. E., *Rothamsted Ann. Rept. for 1976*, Part 2, 1977, p. 87.
74 Campbell, C. A. in *Soil Organic Matter* (eds M. Schnitzer and S. U. Khan), p. 173, Elsevier, Amsterdam, 1978.

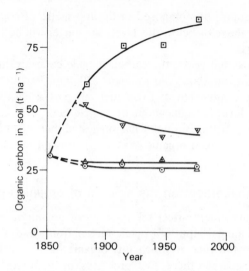

Fig. 18.11 The effects of long-continued manuring on organic carbon in soil (0–23 cm, adjusted for changes in soil bulk density) from four plots on the Hoosfield Continuous Barley Experiment at Rothamsted.[73] The plot receiving farmyard manure every year since 1852, □; plot receiving farmyard manure 1852–71 and nothing thereafter, ▽; plot receiving NPK every year, △; unmanured plot, ○.

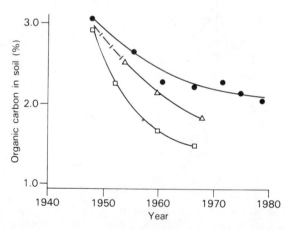

Fig. 18.12 The effects of different farming systems on the carbon content of Rothamsted soils ploughed out from old grass. Results provided by A. E. Johnston. Six-course rotation (two roots, three cereals, one year ley), ●; three-course rotation (two roots, one cereal), △; bare fallow, □.

organic carbon and nitrogen accumulate when a soil that had been arable since Roman times was allowed to revert to woodland in 1883; in less than 100 years the carbon and nitrogen contents of the soil more than doubled. Although cultivation in itself can cause an increase in the mineralization of soil organic carbon, this effect is transient; almost certainly the major reason for the accumulation of soil organic matter in this experiment is that much more plant debris enters the soil each year under woodland than under arable cultivation, rather than that cultivation ceased in 1883. Figure 18.10 shows the corresponding situation when an old arable soil is put down to grass – soil organic nitrogen reaches equilibrium after some 200 years, about 25 years being needed to move halfway to the final equilibrium nitrogen content.

Manuring can also affect the amount of organic matter in a soil. Figure 18.11 shows the changes brought about in organic matter levels by different systems of manuring in a Rothamsted soil that has been under continuous barley since 1852. When no manures were given, there was an initial decline in organic matter content but little further fall over the next 100 years. The amount of organic matter entering the soil each year must now be very similar to the amount lost by mineralization to carbon dioxide. Plots given nitrogen, phosphorus and potassium produced more grain and straw and presumably returned more organic matter to the soil each year in stubble and roots. This extra input of plant material is almost certainly the reason why the plot receiving inorganic nitrogen, phosphorus and potassium every year now contains about 15 per cent more soil organic matter than the unfertilized plot. However, when farmyard manure, containing 3 t C ha^{-1}, was applied annually, the carbon content of the soil more than trebled in 125 years, although equilibrium had still not been reached. A particularly interesting plot in this experiment received farmyard manure between 1852 and 1871 and nothing thereafter. One hundred and four years after the last addition, this plot still contained more organic carbon than plots that did not receive organic manures, illustrating the great stability of humic material in soil.

The effects of cropping system on the amount of organic matter in soil can be seen in Fig. 18.12, rotations that return more organic matter to the soil over a run of years tending towards a higher equilibrium value than rotations that return less. Not surprisingly, organic matter declines most rapidly under bare fallow, when returns of plant material are substantially zero.

A decline in soil organic matter almost inevitably accompanies the introduction of arable agriculture. The experiment illustrated in Fig. 18.11 was done on a soil that had originally been cleared from primeval forest some 2000 years ago, and which contained perhaps only one-third of its original organic matter when the experiment started in 1852. Yet, despite this decline in organic matter, good yields are still obtained – provided the necessary fertilizers are applied each year. When tropical soils are cleared and cultivated, soil organic matter declines dramatically.[75] When forest in Ghana was cleared and cropped to a maize/cassava rotation for eight years, the carbon content of the top 15 cm of soil fell from 2.19 to 1.50 per cent.

By and large, and excluding soils of extreme textures, modern temperate agriculture works satisfactorily at a soil organic matter content that can be maintained by the crop residues. This is not the case in many parts of the tropics, where production cannot be maintained under continuous cropping and recourse must be made to shifting cultivation, with its attendant build-up of organic matter and soil nutrients during the fallow stage, if yields are to be maintained at acceptable levels.

The introduction of improved methods of farming can of course increase soil organic matter. The accumulation of organic matter under improved pasture is an example. Thus Russell[76] showed that the nitrogen content of the top 15 cm of an unimproved virgin pasture in Australia changed little in 38 years, increasing from 0.9 t ha^{-1} to 1.1 t. However, in an adjacent plot receiving superphosphate and lime, the soil nitrogen increased from 0.9 to 2.6 t ha^{-1}. Plant production in this soil was limited by phosphate deficiency and when this was remedied, plant production, sheep carrying capacity, and soil organic matter all increased.

Decomposition of plant material

Given time, all naturally occurring compounds, with the possible exception of carbonized materials like charcoal, can be decomposed in moist aerobic topsoils. Characteristically this attack is in two stages, illustrated in Fig. 18.13[77] for cellulose, the major constituent of plants, and in Fig. 18.14[78] for whole plant material. During the rapid phase the new substrate is broken down: secondary products, the soil microbial biomass and its metabolites, are synthesized concomitantly. This new biomass and its metabolic products are themselves the substrates for the much slower second phase. Sorensen's results[77] illustrate the two phases: when labelled cellulose decomposed in soil, more than half the cellulose carbon was evolved in the first 30 days. At the same time, labelled carbon derived from the cellulose appeared as amino acid carbon, presumably through the activities of the soil microbial population, although the quantities formed were too great for all the newly synthesized amino acids to be held in the biomass alone. After three months, the rate of evolution of cellulose-derived carbon dioxide decreased markedly to a

75 For extended discussions of this see Sanchez, P. A., *Properties and Management of Soils in the Tropics*, Wiley, New York, 1976; Nye, P. H. and Greenland, D. J., *The Soil Under Shifting Cultivation*, Tech. Comm., No. 51, CAB, Harpenden, 1960.

76 Russell, J. S., *Aust. J. agric. Sci.*, 1960, **11**, 902.

77 Sorensen, L. H., *Soil Biol. Biochem.*, 1975, **7**, 171.

78 Jenkinson, D. S. and Ayanaba, A., *Soil Sci. Soc. Am. J.*, 1977, **41**, 912.

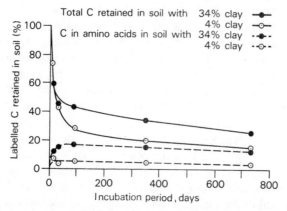

Fig. 18.13 Decomposition of labelled cellulose and concomitant formation of labelled amino acid carbon in two soils of contrasting texture, one with 34 per cent clay and the other with 4 per cent. (Redrawn from Sorensen's data.[77])

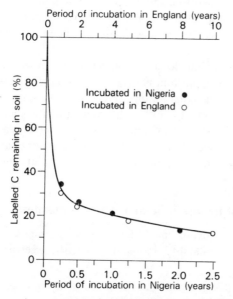

Fig. 18.14 Decomposition of uniformly labelled ryegrass in soils of similar clay contents in England (Rothamsted) and in southern Nigeria.[78]

level that remained substantially constant over the remainder of the two-year incubation period.

In general, plant materials containing large quantities of lignin and other polyphenols are more resistant to decomposition than materials low in these compounds. Thus oak and beech litter, both with wide C : N ratios and rich in polyphenols, decompose relatively slowly compared to leaf litter from ash, elm, alder or hazel, all of which are rich in nitrogen and soluble carbohydrates but poor in polyphenols.[79] Lignin is a heterogeneous cross-linked polymer made up of substituted phenyl propane units, mainly derived from coniferyl, sinapyl and coumaryl alcohol monomers.[80] There are relatively few micro-organisms that can degrade lignin and they are exclusively aerobic: indeed the lignases they produce are thought to need hydrogen peroxide for their operation.[81] Not only is lignin difficult to decompose in itself but it encrusts the cellulose and hemicelluloses of plant walls, decreasing the rate at which these materials are decomposed. Nevertheless lignin is steadily if slowly degraded under aerobic conditions in soil: Martin and Haider found that 30 per cent of the carbon in labelled lignin was evolved as carbon dioxide when the lignin was mixed with soil and allowed to decompose for a year.[82]

Although the aromatic plant constituents, lignin and the other polyphenols, undoubtedly contribute to the formation of resistant forms of organic matter in soil, they are not, as has sometimes been thought, the only source. Thus although glucose added to a soil disappeared in a matter of days, some 6 per cent of the glucose-derived carbon was still present in the soil organic matter six years later: the corresponding figure for cellulose-derived carbon was 12 per cent.[83]

The proportion of the total carbon remaining in the soil once the rapid initial phase of decomposition is over is surprisingly similar for a wide range of crop residues: thus ryegrass tops, ryegrass roots, green maize and mature wheat straw all left about one-third of their carbon in the soil after one year in the field.[84, 85] This is not of course true of the initial phase itself: long ago Waksman and Tenney[86] showed that mature plant material decomposed much more slowly over a 28-day period than immature material with a large water-soluble fraction.

Environmental factors influencing decomposition

The factors that govern microbial activity – temperature, pH, moisture content, oxygen availability, inorganic nutrients, accessibility, etc., all influ-

79 Jensen, V. in *Biology of Plant Litter Decomposition*, Vol. 1. (eds C. H. Dickinson and G. J. F. Pugh), p. 69, Acad. Press, New York, 1974.
80 For an account of lignin and its biodegradation see Crawford, R. L., *Lignin Biodegradation and Transformation*, Wiley, New York, 1981; Zeikus, J. G., *Adv. Microbiol. Ecol.*, 1977, **5**, 211.
81 Crawford, R. L. and Crawford, D. L., *Enzyme Microbiol. Technicol.*, 1984, **6**, 434.
82 Martin, J. P. and Haider, K., *Appl. environ. Microbiol.*, 1979, **38**, 283.
83 Sorensen, L. H., *Soil Biol. Biochem.*, 1972, **4**, 245.
84 Paul, E. A. and van Veen, J. A., *Trans. 11th Int. Soil Sci. Congr. Symp. Papers*, Vol. 3, p. 61, Edmonton, 1978.
85 Jenkinson, D. S., *Soil Sci.*, 1971, **111**, 67.
86 Waksman, S. A. and Tenney, F. G., *Soil Sci.*, 1927, **24**, 317.

ence the decomposition of both freshly added organic matter and humified organic matter.[87]

In general, the van't Hoff rule (that Q_{10}, the ratio of the velocity constants at temperatures T and $T + 10$, where T is the temperature in degrees Celsius, has a value of between 2 and 3) is approximately valid for the decomposition of organic matter in soil over the temperature range 10 to 40 °C.[87] This effect of temperature can be seen in an experiment in which the same plant material was allowed to decompose under temperate and tropical conditions (Fig. 18.14). The decomposition *pattern* was similar under temperate conditions (mean annual temperature 8.9 °C) and under tropical conditions (mean annual temperature 26.1 °C) but the whole decomposition process was four times faster under tropical conditions. Other things being equal, soils from warmer climates contain less organic matter than those from cooler areas. Thus Jenny[88] showed that the relationship

$$N = c\,e^{-kT}$$

held for well-drained cultivated soils from the semi-humid region of the US Mid-West, where N was the total nitrogen content of the surface soil, c and k constants, and T the mean annual temperature.

Microbial activity (as measured by carbon dioxide evolution) is strongly influenced by water potential. Soils dried to a water potential of -10 MPa evolved carbon dioxide at about half the rate observed if the soils were incubated at the optimal water content, usually in the range -20 to -50 kPa.[89] As water potentials reach large negative values, microbial activity ceases.

Decomposition is less complete and slower under anaerobic conditions than under aerobic conditions.[87] When soils become so wet that the larger pores are water filled, decomposition of organic matter is limited by the rate at which oxygen can diffuse to the sites of microbial activity. The diffusion coefficient for oxygen in water is 10 000 times smaller than that in air, so that even quite a modest oxygen demand cannot be met if the larger soil pores are water filled. The Michaelis constant of cytochrome oxidase, the most important terminal oxidase of aerobic soil organisms, is about 8×10^{-10} g cm^{-3} (2.5×10^{-8} M), so that the oxygen concentration at the site of the enzyme has to fall to this value to halve the rate of oxygen uptake. This concentration is much below the oxygen content of water in equilibrium with the atmosphere, which is about 10^{-5} g cm^{-3} at 20 °C.[90] In theory, the partial

87 Jenkinson, D. S. in *The Chemistry of Soil Processes* (eds D. J. Greenland and M. H. B. Hayes), p. 505, Wiley, Chichester, 1981.

88 Jenny, H. *The Soil Resource*, p. 318, Springer, New York, 1980. See also Jenny, H., *Factors of Soil Formation*, McGraw-Hill, New York, 1941.

89 Sommer, L. E. *et al.*, in *Water Potential Relations in Soil Microbiology*, Vol. 9, Soil Sci. Soc. Am. Spec. Publ., 1981, p. 97. See also Boddy, L., *Soil Biol. Biochem.*, 1984, **15**, 501.

90 Griffin, D. M. *Ecology of Soil Fungi*, Chapman & Hall, London, 1972.

pressure of oxygen in the soil atmosphere can therefore fall to about a ten-thousandth of that in air before the respiration rate is halved, assuming, firstly, that the soil solution in contact with the organism is in equilibrium with the soil atmosphere and secondly, that oxygen consumption at the enzyme site is not limited by the rate at which oxygen can move through the cell wall. In practice, oxygen concentration in the soil air becomes limiting at levels far above this, although the limiting oxygen partial pressures are usually well below that of the atmosphere.[87]

Although it can be confidently assumed that no organic plant or animal constituent is completely resistant to decomposition under moist aerobic conditions, this assumption does not hold under anaerobic conditions. Lignin, in particular, is not attacked under anaerobic conditions and the partly humified organic debris described as peat can accumulate indefinitely, as long as conditions remain anaerobic.[91]

A mat of humified organic matter frequently develops on the surface of strongly acid soils: thus the plot on the Park Grass Experiment at Rothamsted that has received ammonium sulphate every year since 1856 now has a pH of 4.0 and is covered by a mat of humified organic matter lying on top of the mineral soil. This mat arises from the acidifying effects of the ammonium sulphate: when lime is added at regular intervals no mat develops. However, although the *distribution* is different, the overall quantities of organic matter in the plot receiving ammonium sulphate and the control unfertilized plot are similar (Table 18.4).[92] The difference in distribution is almost certainly related to the activities of earthworms in the unfertilized plot, which distribute the organic debris through the soil profile. In their absence, it accumulates on top. However, there is ample evidence that fresh organic matter *initially* decomposes more slowly in strongly acid soils than in soils with pH values of 5 and above. When ryegrass was allowed to decompose under field conditions in a soil of pH 3.7, 42 per cent of the ryegrass-derived carbon still remained after one year: the corresponding figure for a soil of pH 4.8 was 31 per cent and for a soil of pH 6.9 also 31 per cent.[93] However, by the end of five years this difference had largely disappeared, suggesting that only the early stages of decomposition were retarded by acidity.

When organic matter is added to soil, the micro-organisms decomposing it can obtain the necessary inorganic nutrients (nitrogen, phosphorus, sulphur, calcium, etc.) for growth from two sources: those already present

91 For a review of the geochemistry of peat see Mathur, S. P. and Farnham, R. S. in *Humic Substances in Soil, Sediment and Water* (eds G. R. Aiken, D. M. McKnight *et al.*), p. 53, Wiley, New York, 1985.   For a review on its biochemistry and microbiology see Given, P. H. and Dickinson, C. H. in *Soil Biochemistry*, Vol. 3 (eds E. A. Paul and A. D. McLaren), p. 123, Dekker, New York, 1975.

92 Warren, R. G. and Johnston, A. E., *Rothamsted Ann. Rept. for 1963*, p. 240, 1964.

93 Jenkinson, D. S., *J. Soil Sci.*, 1977, **28**, 424.

Table 18.4 The effects of acidity on the quantity of organic carbon in a grassland soil[92]

	Treatment, 1856–1959*		
	Unmanured	$(NH_4)_2SO_4,$ 48 kg N ha^{-1}a^{-1}	
	Unlimed	Unlimed	Limed
pH f mat	—	3.7	—
pH of 0–23 cm layer	5.2	4.0	7.2
pH of 23–46 cm layer	5.3	5.2	6.6
C in mat, 4–0 cm layer (t ha^{-1})	0	21	0
C in soil, 0–23 cm layer (t ha^{-1})	81	74	86
C in soil, 23–46 cm layer (t ha^{-1})	47	32	47
Total C in profile(t ha^{-1})	128	127	133
Offtake of hay (t ha^{-1} a^{-1})	1.5	1.7	2.4

* Sampled in 1959 from plots 3 unlimed, 1 unlimed and 1 limed on the Park Grass Continuous Hay Experiment at Rothamsted.

in the soil in available form and those in the added plant material itself. The nutrient required in greatest abundance is nitrogen and it is not surprising that this is the inorganic element that most often limits microbial activity in soil. Lack of phosphorus, sulphur, etc., can restrict microbial activity *in vitro* but it is unusual in agricultural soils for nutrient elements other than nitrogen to limit the decomposition of plant or animal material.

Some of Allison's experiments illustrate how nitrogen need can slow the decomposition of organic matter in soil.[94] He incubated a mixture of sawdust and soil with and without inorganic nitrogen, the quantity of inorganic nitrogen being sufficient to reduce the initial C : N ratio of the sawdust from 346 to 64. Without nitrogen, 19 per cent of the sawdust carbon was mineralized to carbon dioxide in 40 days; with nitrogen 45 per cent. However, by 160 days the difference had narrowed: 52 per cent for the soil–sawdust mixture by itself and 57 per cent for the mixture with nitrogen. This pattern is characteristic of the effect of nitrogen deficiency on the decay of organic matter in soil; once the initial rapid phase of decomposition is over, the amount of humified organic matter remaining from a given addition of plant material is similar, whether nitrogen is limiting or not. Thus Sauerbeck, incubating labelled straw and soil with and without added inorganic nitrogen,

94 Allison, F. E., *US Dept Agric. Tech. Bull.* No. 1332, 1, 1965.

showed that the nitrogen usually increased the retention of carbon in prolonged incubations but the effect was small, and sometimes not observed.[95]

There is also strong indirect evidence from field experiments that, in the long run, the addition of inorganic nitrogen has little effect on the retention of carbon in soil. Table 18.5 shows the effects of long-continued nitrogen applications on the amounts of organic matter in two soils, one under grass and the other under arable cropping. The arable soil that had received fertilizers for 130 years contained slightly more organic matter than that not receiving fertilizers. The C : N ratio of the arable fertilized soil was, if anything, slightly greater than that of the unfertilized soil. Thus there was neither accumulation of carbon-rich organic matter in the unfertilized soil, nor did fertilizer cause a decline in the amount of organic matter in the soil. The additional organic matter in the fertilized arable soil is mainly, and quite possibly entirely, the result of larger yields and larger crop residues.

The old grassland that had received fertilizer nitrogen every year (Table 18.5) also contained a little more soil organic matter than the corresponding unmanured grassland. Thus, the large quantities of inorganic nitrogen received over the years by both the arable soil and the soil under grass have

Table 18.5   The effects of long-continued applications of inorganic nitrogen on the carbon and nitrogen contents of soils from two of the Rothamsted Classical Experiments[87]

Experiment	Fertilizer treatment	Inorganic N applied during the whole experiment (kg ha⁻¹)	Organic C (%)	Organic N (%)	C : N ratio
Continuous wheat since 1843*	Unmanured	0	0.90	0.098	9.2
Continuous wheat since 1843*	NPK Mg annually	18 000	1.08	0.112	9.6
Continuous grass since 1856†	Unmanured	0	3.0	0.26	11.6
Continuous grass since 1856†	NPK Mg annually	12 000	3.3	0.29	11.3

* Sampled (0–23 cm) in 1972 from Broadbalk plots 03 and 08.
† Sampled (0–23 cm) in 1976 from Park Grass plots 3d and 9a, with pH values of 5.3 and 6.4 respectively.

95 Sauerbeck, D., *Landw. Forsch.*, 1968, **21**, 91.

had remarkably little effect on both the amount of organic matter in the soil and its C : N ratio.

The effects of clay on decomposition

The amount of biomass and biomass metabolites formed during the first phase of the decomposition process is greatly influenced by soil texture, clay soils retaining much more organic matter than sandy soils. This is illustrated in Fig. 18.13; a soil containing 34 per cent clay retained about twice as much cellulose-derived carbon than a soil with 4 per cent clay. More than twice as much of the cellulose carbon was converted to amino-acid carbon in the clay soil than in the sandy soil. This stabilizing effect of clay is well known: Jenny showed that, other things being equal, clay soils contain more organic matter than sandy soils.[88] Similar observations have been made for tropical soils.[3,96] Clay can protect organic matter in a number of ways, by adsorbing otherwise readily available substrates, making them less available to the soil population, by stabilizing the newly formed metabolites, and by increasing the longevity of soil organisms. In general, montmorillonitic clays give more protection than illites or kaolinites.[97] Protein present as a monolayer in the interlamellar spaces of montmorillonite was almost completely resistant to microbial attack over a period of four weeks: when more protein was present than could form a monolayer, the protective action of the clay was very much less.[98] There is also good evidence that poorly crystalline aluminium silicate ('allophane') is particularly effective in stabilizing organic matter.[99]

Insoluble substrates such as cellulose and whole plant material cannot react with clay surfaces. With such substrates it is probable that the organic material itself is not protected but that the secondary products of decomposition are.[77] It is also possible that clay or other minerals present in soil can catalyse the formation of humic polymers. For example, it has been shown that the condensation of phenols such as hydroquinone to dark polymers is catalysed by manganese dioxide.[100]

Clay is not the only colloid capable of slowing the decomposition of organic matter in soil. It has been shown that the loss of carbon from algal protein could be reduced merely by mixing it with a humic acid preparation.[101] In contrast, free amino acids were not appreciably protected, suggesting that the

96 Jones, M. J., *J. Soil Sci.*, 1973, **24**, 42; for similar results see Kadeba, O., *Soil Sci.*, 1978, **125**, 122.
97 Lynch, D. L. and Cotnoir, L. J., *Soil Sci. Soc. Am. Proc.*, 1956, **20**, 367.
98 Pinck, L. A. *et al.*, *Soil Sci.*, 1954, **78**, 109.
99 Broadbent, F. E. *et al.*, *Soil Sci.*, 1964, **98**, 118; Zunino, H. *et al.*, *Soil Biol. Biochem.*, 1982, **14**, 37. For reviews see Fox, R. L. in *Soils with Variable Charge* (ed. B. K. G. Theng), p. 195, New Zealand Soc. Soil Sci., Palmerston North, 1980; Tate, K. R. and Theng, B. K. G., *op. cit.*, p. 225; Goh, K. M., *op. cit.*, p. 373.
100 Shindo, H. and Huang, P. M., *Soil Sci. Soc. Am. J.*, 1984, **48**, 927.
101 Verma, L. *et al.*, *Soil Sci. Soc. Am. Proc.*, 1975, **39**, 279.

protein was protected by being sorbed on the humic acid in a way that made it less accessible to proteolytic enzymes. Thus both organic and inorganic colloids can stabilize organic matter in soil.

Accessibility of organic matter in soil   The effects of clay on the decomposition of organic matter, just considered, are one aspect of the accessibility of substrates in soil. Another is the effect of comminution: finely divided organic matter usually decomposes more quickly than coarse. For example, when finely ground and coarsely ground rye straw were incubated in soil the finely ground material lost 61 per cent of its carbon in 448 days, but the coarse only 52 per cent.[102] Similar results have been obtained with sawdust[103] and with whole and finely ground oak leaves.[104] Coarse organic material is more resistant, presumably because grinding creates extra surface and thus exposes more substrate to attack. Soil animals play a major role in comminuting organic matter[105] and those large enough to transport organic matter (certain earthworms, termites, etc.) can also accelerate the decomposition of plant debris by moving it to an environment more propitious for biological attack (see Ch. 16). Litter in the soil surface is more subject to unfavourable conditions for decomposition than is material buried in the soil: straw has repeatedly been shown to decompose more slowly on the surface than when incorporated.[106]

The stability of humified organic matter

Radiocarbon dating has shown that some at least of the organic matter in soil is extremely stable.[107] This is illustrated in Table 18.6 which shows the radio-carbon ages of soil from the Broadbalk experiment at Rothamsted. The topsoil dates at 1450 years, even though conditions for decomposition are favourable, the freely draining soil having developed under a moist temperate climate. The radiocarbon age increases and the carbon content decreases with soil depth, an effect repeatedly observed in different soils.[107] Surface soils

102 Cheshire, M. V. *et al.*, *J. Soil Sci.*, 1974, **25**, 90.

103 Allison, F. E. and Cover, R. G., *Soil Sci.*, 1960, **89**, 194.

104 van der Drift, J. and Witkamp, M., *Archs Neerl. Zool.*, 1960, **13**, 486.

105 For reviews of the role of the soil population in comminuting and transporting organic matter see *Biology of Plant Litter Decomposition*, Vol. 2 (eds C. H. Dickinson and G. J. F. Pugh), Acad. Press, New York, 1974.

106 Shields, J. A. and Paul, E. A., *Canad. J. Soil Sci.*, 1973, **53**, 297; Harper, S. H. T. and Lynch, J. M., *J. Soil Sci*, 1981, **32**, 627; Sain, P. and Broadbent, F. E., *J. environ. Qual.*, 1977, **6**, 96; Christensen, B. T., *Soil Biol. Biochem.*, 1986, **18**, 523.

107 For reviews of work on the radiocarbon dating of soils see Scharpenseel, H. W. and Schiffmann, H., *Z. Pflanzenernaehr. Bodenkd.*, 1977, **140**, 159; Goh, K. M. in *Soils with Variable Charge* (ed B. K. G. Theng), p. 373, New Zealand Soc. Soil Sci., Palmerston North, 1980; Stout, J. D. *et al.*, in *Soil Biochemistry*, Vol. 5 (eds E. A. Paul and J. N. Ladd), p. 1, Dekker, New York, 1981; O'Brien, B. J. and Stout, J. D., *Soil Biol. Biochem.*, 1978, **10**, 309.

collected before contamination with radiocarbon from the thermonuclear bomb tests of the early 1960s frequently date at 500 to 2000 years, with the older dates tending to be associated with chernozemic soils high in organic matter. Radiocarbon ages as large as these can be explained if the soil contains a small, biologically inert fraction of great age or, alternatively, if the major part of the organic matter has an age of the order of that measured for the soil as a whole. Different fractions of the soil organic matter do indeed have different radiocarbon ages. Thus the humic acid isolated from a Rothamsted topsoil had a radiocarbon age of 750 years, the fulvic acid 420 years and the humin 2400 years (Table 18.6),[108] suggesting that at least some of the soil organic matter must be present in a form that is almost totally biologically inert.

Fresh decomposable organic matter sometimes stimulates the decomposition of the organic matter already in the soil, an effect usually described as a 'priming action'.[109] However experiments with labelled glucose,[83] cellulose,[110], lignin[82] and whole plant material[111] have shown that priming actions, even when they occur, are transient and negligibly small compared to the total amount of organic matter in soil. The great stability of soil organic matter in soil is due to the physical and chemical structure of the humic molecules, not from the lack of decomposable substrates to 'operate as a forced draught on the smouldering bacterial fires'.[112]

Table 18.6   Radiocarbon age of organic matter in soil collected during the last century from the unmanured plot of the Broadbalk Continuous Wheat Experiment at Rothamsted[108]

Sampling depth (cm)	Organic C (%)	Radiocarbon age (years)	
0–23	0.94	1 450	750 (humic acid fraction) 420 (fulvic acid fraction) 2 400 (humin fraction)
23–46	0.61	2 000	
46–69	0.47	3 700	
114–137	0.24	9 900	
206–229	0.20	12 100	

108 Jenkinson. D. S., *Welsh Soils Discussion Group Rep.*, 1976, **16**, 91.   For comparable results on Canadian soils see Campbell, C. A. *et al.*, *Soil Sci.*, 1967, **104**, 81; *ibid.*, 217.
109 For reviews on the priming action see Jenkinson, D. S., *Soil Sci.*, 1971, **111**, 67; Sauerbeck, D. in *The Use of Isotopes in Soil Organic Matter Studies*, p. 209, Rept. FAO/IAEA, Braunschweig, Pergamon, Oxford, 1966.
110 Jansson, S. L., *Trans. 7th Int. Congr. Soil Sci.*, Vol. 2, p. 635, Madison, 1960.
111 Nyhan, J. W., *Soil Sci. Soc. Am. Proc.*, 1975, **39**, 643.
112 Broadbent, F. E. and Norman, A. G., *Soil Sci. Soc. Am. Proc.*, 1946, **11**, 264.

The reasons why the particular chemical and physical structures found in humified organic matter are resistant to microbial attack are still not fully understood. Kleinhempel[113] pointed out that it is unlikely that humified substances owe their stability to a particular linkage that is inherently resistant to microbial attack. It is much more likely that their resistance stems from a multiplicity of different linkages, each differing from the others to an extent that a single enzyme (or group of enzymes) cannot break down more than a very few of them. An organism must expend energy to produce and excrete an exocellular enzyme. Only if the energy gained by the organism through the activity of the enzyme is greater than the energy needed for its production, can the organism use the substrate degraded by that enzyme for growth. Thus it is much more profitable for organisms to use biopolymers with molecular regularity, such as starch, cellulose or chitin, than to use the irregular humic polymers. Cellulose, for example, is a regular polymer of glucose, formed under genetic control in living organisms and a single molecule of cellulase can hydrolyse many β-glucosidic links in cellulose. If the humic substances are products of *random* condensation reactions, only a relatively few links will have the necessary configuration for rupture by a particular enzyme.[113] Another factor leading to stability in the cross-linked humic polymers is steric hindrance, which will also have the effect of reducing the number of bonds that can be hydrolysed by a particular enzyme.

Neither the time, place nor mode of formation of the humic substances is known with any certainty. It is generally agreed that the lignoprotein theory,[114] which postulates reaction between lignin and protein to form 'lignoprotein' complexes that are resistant to microbial attack, is incorrect: infrared and NMR spectroscopy show that soil organic matter does not contain enough unaltered lignin.[44] Kononova[115] pointed out that dark humic polymers were formed at a relatively early stage in the decomposition of plant material, before much lignin had been decomposed. Likewise, labelled recalcitrant organic matter was formed within a very few days of the addition of labelled acetate to soil.[116] The lignoprotein theory has been supplanted by the polyphenol theory,[115] in which plant and animal products are first degraded, more or less completely, to soluble low molecular weight products, often monomers. The phenolic degradation products formed from lignins are then oxidized to quinones, which react with themselves, or with other degradation products, to give the dark-coloured humic polymers. Phenolic

113 Kleinhempel, D., *Pedobiologia*, 1971, **11**, 425.
114 Waksman, S. A., *Humus*, Baillière, Tindall and Cox, London, 1936.
115 For reviews on theories about humification see Kononova, M. M., *Soil Organic Matter*, Pergamon, Oxford, 1966; Flaig, W. in *Soil Components*, Vol. 1 (ed J. E. Gieseking), p. 1, Springer, Berlin, 1975; Stevenson, F. J., *Humus Chemistry, Genesis, Composition, Reactions*, Ch. 8, Wiley, New York, 1982; Haider, K. *et al.*, in *Soil Biochemistry*, Vol. 4 (eds E. A. Paul and A. D. McLaren), p. 195, Dekker, New York, 1975; Anderson, D. W., *J. Soil Sci.*, 1979, **30**, 77.
116 Sorensen, L. H. and Paul, E. A., *Soil Biol. Biochem.*, 1971, **3**, 173.

substances can also be synthesized from non-phenolic substrates by the microbial population and then undergo polymerization to humic substances in the same way as lignin-derived phenolics. A scheme for the oxidative reaction of phenolic substances with themselves or with peptides and amino acids to give humic polymers is shown in Fig. 18.15. Swaby and Ladd[117] suggested that the most likely site for the formation of humic polymers was in moribund or dead microbial cells, when degradative enzymes were still active but before lysis occurred. Under these conditions, high concentrations of amino acids and quinones could build up, favouring rapid (and largely chemical) polymerization. The newly formed humic polymers would be released into the soil when the cell wall was finally lysed.

With the realization, from NMR studies, that humic substances contain a substantial proportion of aliphatic carbon, heavily substituted with carboxyl groups, the polyphenol theory as set down in Fig. 18.15 is in need of change and revision. The fulvic-acid-like properties of polymaleic acid[118] may perhaps provide an indication of the direction of change.

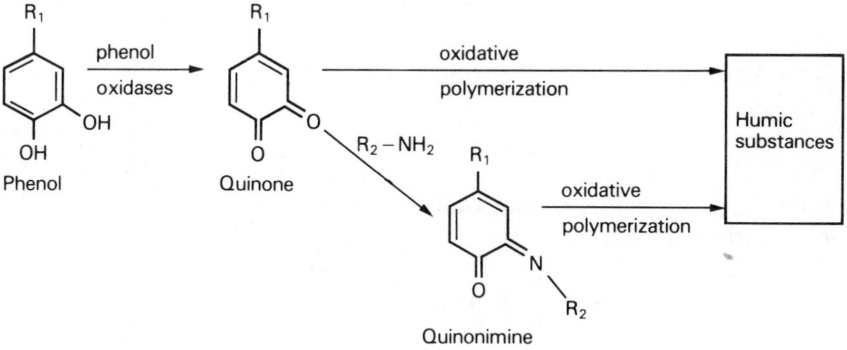

Fig. 18.15 The polyphenol theory for the formation of humic substances.

Modelling the turnover of organic matter

Soil organic matter contains fractions of widely different turnover times: fresh undecomposed plant and animal debris, the soil microbial biomass and humified materials of varying ages. Multicompartmental or multisequential models are therefore necessary if the turnover processes are to be represented in a realistic way. However, before considering these complex models it is worth examining a simple two-compartment model that has been

117 Swaby, R. J. and Ladd, J. N. in *The Use of Isotopes in Soil Organic Matter Studies*, p. 153, Rept. FAO/IAEA, Braunschweig, Pergamon, Oxford, 1966.
118 Anderson, H. A. and Russell, J. D., *Nature, Lond.*, 1976, **260**, 597.

widely used in studying the dynamics of organic matter in agricultural soils. In this model (Fig. 18.16) compartment P, containing undecomposed plant and animal remains, feeds compartment Q. containing the remainder of the soil organic matter – the various humic fractions, plus the soil microbial biomass. Let H be the quantity of organic carbon in compartment P, and C be the quantity in compartment Q, both expressed per unit area of soil to a fixed depth. Let A be the annual input of fresh plant and animal carbon, of which a fraction f (the 'isohumic coefficient')[119] enters compartment Q each year. Assume that $C \gg H$, so that undecomposed plant and animal tissues make up a negligibly small part of the soil organic matter at any particular time. The 'isohumic coefficient' is basically a device for splitting the decomposition process into a fast and a slow part. The justification for this can be seen in Fig. 18.14, which shows that there is a marked decrease in the rate of decomposition once about two-thirds of the plant carbon has been mineralized. Assuming that all parts of C are equally decomposable, then if r is the fraction of C decomposed per year and t is time, in years:

$$\frac{dC}{dt} = fA - rC \qquad [18.1]$$

The solution of this differential equation is

$$C = \frac{fA}{r} + \left(C_0 - \frac{fA}{r}\right)e^{-rt} \qquad [18.2]$$

where C_0 is the initial carbon content of compartment Q, effectively the initial carbon content of the soil. At equilibrium

$$\frac{dC}{dt} = 0 \text{ and } \frac{fA}{r} = C_E$$

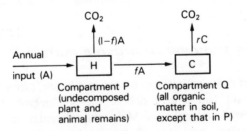

Fig. 18.16   Two-compartment model for the turnover of organic carbon in soil. The fraction of A entering compartment Q each year is f. H and C are the quantities of organic carbon in compartments P and Q respectively and r is the fraction of C decomposed each year.

119 Hénin, S. and Dupuis, M., *Ann. Agron.*, 1945, **15**, 17.

where C_E is the equilibrium organic matter content of compartment Q, again effectively that of the soil. It follows that

$$C = C_E + (C_o - C_E) e^{-rt} \qquad [18.3]$$

The turnover time, t_E, for a soil in equilibrium is defined as the amount of carbon in the soil, divided by the annual input (or output, since input is equal to output when equilibrium has been reached). So, for this model

$$t_E = \frac{C_E}{rC_E} = \frac{1}{r}$$

At equilibrium, the average age of the organic carbon in compartment Q can be shown[120] to be the same as the turnover time, t_E. Another term used in discussions of the turnover of organic matter in soil is the 'half-life' $(t_{\frac{1}{2}})$, given for compartment Q by

$$t_{\frac{1}{2}} = 0.693\, t_E$$

Jenny[88] developed an analogous model for the turnover of organic nitrogen in soil:

$$N = N_E + (N_o - N_E) e^{-rt} \qquad [18.4]$$

in which N is the organic nitrogen content of the soil at time t, N_o the starting content and N_E the equilibrium content. This nitrogen model is, however, simpler than the carbon model in that it contains only one compartment and does not involve an isohumic coefficient. The justification for this difference is that most plant materials entering the soil have wide carbon : nitrogen ratios. During the early stages of decomposition carbon is rapidly mineralized by microbial activity: in contrast, nitrogen is retained, so that there is no need to postulate the rapid initial loss that occurs with carbon.

Turnover times and decomposition constants are found by fitting equations [18.3] or [18.4] to data from field experiments in which a change in management has brought about a considerable (and accurately known) change in soil organic carbon or nitrogen level, such as those illustrated in Fig. 18.11. Table 18.7 shows some turnover times calculated in this way for soils from different parts of the world: they lie in the 20 to 40 year range. In general, these simple exponential models give reasonably satisfactory representations of reality over the 10 to 100-year period, and have been successfully used to predict the effects of shifting cultivation on organic matter in tropical soils.[121]

Despite their value over the intermediate time span, these one or two compartment exponential models are not useful over short periods, when

120 Bartholomew, W. V. and Kirkham, D., *Trans. 7th Int. Congr. Soil Sci.*, Vol. 2, p. 471, Madison, 1960.
121 Nye, P. H. and Greenland, D. J., *The Soil Under Shifting Cultivation*, Tech. Comm., No. 51, CAB, Harpenden, 1960.

Table 18.7 The turnover time of organic nitrogen in soil organic matter

Land use	Soil layer (cm)	Period studied (years)	Rate constant (years^{-1})	Turnover time (years)
Pasture*	0–5	39	0.029	34
Arable[†]	0–18	60	0.055	18
Grassland[‡]	0–23	300	0.028	36
Arable[§]	0–23	123	0.031 (0.024)[π]	32 (42)

* P plots, Kybybolite Experiment, Australia.[76]
[†] Continuous maize plots, Sanborn Experiment, USA.[120]
[‡] Fig. 18.10.
[§] Fig. 18.11; continuous barley experiment, farmyard manure plots.
[π] Figures in parentheses are the corresponding results for organic carbon.

plant composition and microbial colonization influence decomposition rates, or over long periods, in that they predict soil radiocarbon ages that are far too young.

Numerous attempts to construct more realistic models have been made over the last decade or so. The incoming plant material has been placed in a series of compartments, each decomposing exponentially but at its own characteristic rate, the aim being to model the early stages of decomposition, when the decay processes are dominated by the chemistry of the incoming plant debris.[122,123] In another family of models (for an example see Fig. 18.17) the incoming plant material enters a net of interconnected compartments, some in series and some in parallel.[124–127] Again, the organic matter in each compartment decays exponentially with a characteristic rate constant. In the more sophisticated of these models the rate constants are adjusted for temperature and soil moisture content: good fits were obtained to data from a wide range of Canadian soils.[128]

A further group of models are particularly concerned with the microbiology of the early stages of the decomposition processes, when microbial populations are still increasing and Michaelis–Menten kinetics are relevant, rather than with the long-term first-order exponential processes considered so far. The early stages of decay are much influenced by the carbon : nitrogen ratio

122 Minderman, G., *J. Ecol.*, 1968, **56**, 355.
123 Morel, R., *Ann. Agron.*, 1978, **29**, 357.
124 Jenkinson, D. S. and Rayner, J. H., *Soil Sci.*, 1977, **123**, 298.
125 Hunt, H. W., *Ecology*, 1977, **58**, 469.
126 Parton, W. J. *et al.* in *Nutrient Cycling in Agricultural Ecosystems* (eds R. R. Lowrance, R. L. Todd *et al.*), p. 533, Univ. Ga. Coll. Agric. Exp. Stn. Spec. Publ. No. 23, 1983.
127 O'Brien, B. J. and Stout, J. D., *Soil Biol. Biochem.*, 1978, **10**, 309.
128 Van Veen, J. A. and Paul, E. A., *Canad. J. Soil Sci.*, 1981, **61**, 185; Voroney, R. P. *et al.*, *Canad. J. Soil Sci.*, 1981, **61**, 211.

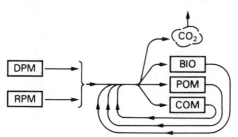

Fig. 18.17 Multicompartmental model for the turnover of C in soil. DPM is decomposable plant material, RPM resistant plant material, BIO the soil microbial biomass, POM the physically stabilized fraction of the soil organic matter and COM the chemically stabilized fraction. For an annual input of 1 tonne plant C ha^{-1}, the equilibrium carbon content of the soil will be 24.2 tonnes C ha^{-1}, of which 0.01 tonnes is in DPM (turnover time 0.2 year), 0.47 tonnes in RPM (turnover time 3.3 years), 0.28 tonnes in BIO (turnover time 2.4 years), 11.3 tonnes in POM (turnover time 71 years) and 12.2 tonnes in COM (turnover time 2900 years).[124]

of the incoming organic debris and the effects of nitrogen deficiency on decomposition are central to this type of model.[129] Another mathematical approach[130] has been to treat the incoming organic matter as of continuously declining 'quality', organic matter moving down a quality scale as it decays, fresh decomposable organic matter having a high quality and completely undecomposable material a quality of zero.

The current interest in modelling the turnover of organic matter should lead to a more integrated understanding of the decomposition process, from start to finish, and also to a better understanding of the long-term effects of agriculture on the amount and behaviour of organic matter in soil.

129 Smith, O. L., *Soil Microbiology: a model of decomposition and nutrient cycling*, CRC Press, Boca Raton, 1982.
130 Bosatta, E. and Agren, G. I., *Soil Biol. Biochem.*, 1985, **17**, 601.

19

Microbial transformations of nitrogen

Micro-organisms are fundamentally involved in important changes to the nitrogen compounds in soil. They are involved in: (i) the release of nitrogen from organic compounds; (ii) the oxidation of the ammonia so released to nitrite and nitrate ions; (iii) the reduction of nitrate to nitrogen and its oxides, principally nitrous oxide; and (iv) the fixation of atmospheric nitrogen. They contribute much to the nitrogen cycle in the soil–plant–atmosphere system and the transformations which are so important to plant nutrition are predominantly microbial.

The basic transformations are those recognized by the following names:

Process	Principal organisms
Ammonification organic nitrogen → ammonium	Saprophytic and predatory heterotrophs, including bacteria, fungi and protozoa
Nitrification ammonium → nitrite → nitrate	Mainly autotrophic bacteria but other groups may contribute
Denitrification nitrate → dinitrogen and nitrous oxide	Mainly heterotrophic bacteria
Nitrogen fixation dinitrogen → ammonium	Bacteria, actinomycetes and blue–green algae (cyanobacteria), both free living and symbiotic

Mineralization of soil organic nitrogen

The term 'nitrogen mineralization' is commonly used to describe the conversion of organically bound nitrogen, mainly as amine groups but not exclusively so, into inorganic forms such as ammonium or nitrate. The conversion to nitrate is a separate step involving specialized organisms and it is best to deal with it separately and concentrate on the first conversion stage and be specific in calling this ammonification.

Heterotrophic micro-organisms, regardless of whether they are bacteria, fungi or protozoa, need organic material as a source of combined carbon from which they can derive energy by respiration and carbon for cell synthesis. To accompany this carbon they also require nitrogen and various other nutrients. As protein is progressively decomposed, ammonium is the endpoint of the degradation and this will be used by the organisms for cell synthesis. If more nitrogen is present in the organic substrate than is required by the feeding organisms the surplus is released as a waste product. If insufficient nitrogen is present in the substrate the organisms will draw on the mineral nitrogen in the soil to make good the deficit and nitrogen immobilization will result. The balance between ammonium release and nitrogen immobilization is a function of the carbon–nitrogen ratio of the cells being synthesized, the energy efficiency of the organisms and the carbon–nitrogen ratio of the material undergoing decomposition. Calculations can be made which show that the latter factor is the most important. This can vary from 5 : 1 for some animal wastes to as much as 100 : 1 for cereal straws. The ratio at which nitrogen is neither released nor taken up is approximately 35 : 1.

Products of animal excretion may contain urea or, in the case of insects, uric acid although nematodes and protozoa excrete nitrogen as ammonia. Urea and uric acid are readily decomposed and have a short life in soil.

The release of ammonia from organic compounds is the result of hydrolytic enzyme action and enzymes may survive the death of the organisms which produced them. Hence, although high temperature sterilization of soil by autoclaving will destroy enzymes and hence prevent further ammonification, treatments such as gamma radiation at 2.5 to 4 Mrad[1] or chemical treatment such as toluene can leave enzymes intact, despite the absence of viable micro-organisms, and will permit limited ammonification. Urease enzymes in particular can survive in soils and levels of urea decomposition often bear little relation to the existing populations of ureolytic organisms.[2]

The ability of a soil to meet the mineral nitrogen requirement of a crop over a growing season is largely dependent on the reserves of organic nitrogen in the soil, the nature of the organic residues and the ammonifying capabilities of the microbial population. Recent work in Sweden[3] has focused on the importance of the protozoa in speeding up the translocation of nitrogen from organic matter to growing plants and has shown that the ability of protozoa to metabolize the proteins of ingested bacteria may be important. Free-living nematodes may perform a similar function. Associated work has also indicated that the plant itself may increase the rate of ammonification by supplying energy sources in the form of root exudates which stimulate the bacterial decomposers into greater activity.

Ammonification is an important microbial process, not only because it supplies the raw material for the nitrification sequence but also because it provides a readily available nitrogen source for cereals and grassland species and could be regarded as the preferred nitrogen supply as it is less readily lost from the soil than nitrate and is used more efficiently within the plant.[4]

Reports on the levels of mineral nitrogen found in forest systems indicate a general trend from very low to low levels in northern forests, rising in temperate climates to the highest levels in tropical systems.[5] Coniferous forests tend to have lower levels of mineral nitrogen than deciduous forests. The nitrogen required is often taken in the form of ammonium from the litter layer and in areas of low nitrogen availability the mycorrhizal associations formed by the trees are regarded as important.

1 Cause, P. A., in *Soil Biochemistry*, Vol. 3 (eds E. A. Paul and A. D. McLaren), Dekker, 1975.
2 Burns, R. G. *et al.*, *Soil Sci. Soc. Am. Proc.*, 1972, **36**, 308.
3 Clarholm, M., *Soil Biol. Biochem.*, 1985, **17**, 181.
4 Reisenaur, A. M. in *Nitrogen in the Environment*, Vol. 2 (eds D. R. Nielsen and J. G. McDonald), Acad. Press, 1978.
5 Greenland, D. J., *J. agric. Sci.*, 1958, **50**, 82.

Both coniferous and deciduous forests have been found to have higher ammonium levels and lower nitrate levels and fewer nitrifiers than an old field site.[6] The coniferous forest showed these trends in the most pronounced form and the addition of lime to the forest soils stimulated nitrification in all cases. Where ammonium is the dominant mineral form of nitrogen, the trees take up this form preferentially having evolved an ammonium based nutrition.[7]

The effect of the low available nitrogen status of many natural forest ecosystems can be self-sustaining through the effect of nitrogen nutrition on plant polyphenol production. A cyclical system has been described whereby low available nitrogen levels led to plant litter with a high polyphenol content.[8] This litter showed low rates of ammonification which reinforced the poor nitrogen nutrient status. The addition of fertilizer nitrogen resulted in the cycle being broken, and increased ammonification rates.

Nitrification in the soil

The organisms involved

Early experiments on sewage purification by soil percolation led Schloesing and Müntz in 1877[9] to the initial discovery that the production of nitrate and nitrite ions from ammonia was a biological process, since chloroform or boiling water brought the transformation to a halt. Attempts to isolate the causative organisms using the organic media pioneered by Koch failed and Winogradsky in 1890 postulated an autotrophic mode of nutrition in which organic substrates played no part.

He developed entirely inorganic growth media based on silica gel saturated with inorganic salts on which organisms with nitrifying ability were isolated. He showed that the process was sequential with *Nitrosomonas* being responsible for the oxidation from ammonium to nitrite and *Nitrobacter* for the stage from nitrite to nitrate. Subsequent work has resulted in the isolation of many other genera of nitrifiers but closer examination has indicated that a proportion of these may have been mixed and the current status is shown below.[10]

6 Montes, R. A. and Christensen, N. L., *Forest Sci.*, 1979, **25**, 287.
7 Bledsoe, C. S., *Pl. Physiol.*, 1976, **57**, 49.
8 Davies, R. I. *et al.*, *J. Soil Sci.*, 1964, **15**, 299.
9 Schloesing, T. and Müntz, A., *Comp. Rend.*, 1877, **84**, 301.
10 *Bergey's Manual of Determinative Bacteriology* (8th edn) (eds R. E. Buchanan and N. E. Gibbons), Williams and Wilkins, 1974.

Autotrophic nitrifying bacteria

Ammonium oxidizers (all obligate autotrophs)

> *Nitrosomonas europaea*
> *Nitrosospira briensis*
> *Nitrosococcus nitro-oxydans*
> *Nitrosolobus multiformis*

Nitrite oxidizer (not all strains are obligate autotrophs)

> *Nitrobacter winogradskii*

Genera such as *Nitrosogloea* and *Nitrosocystis* are now thought to have been morphological variants resulting from cultural conditions and have been relegated to the category *incertae sedis*.

Besides the well-established autotrophic nitrifiers there have been numerous reports of ammonium oxidation, or at least nitrate production, by heterotrophic organisms including bacteria, fungi and actinomycetes. Unlike autotrophic nitrification, which is closely linked to cell growth and is proportional to nitrifier biomass, heterotrophic nitrification is independent of cell yields. As the rates of nitrification reported for heterotrophs are usually less than one-tenth those of the autotrophs it is doubtful whether significant quantities of nitrate are so generated in most natural systems.[11] Addition of an inhibitor which was selective for autotrophic nitrification failed to show the occurrence of significant heterotrophic activity.[12] In extreme alkaline or acid environments where autotrophic nitrification will be reduced, it is possible that the heterotrophs might make a contribution.[13] However, in instances where heterotrophic nitrifiers have been isolated from acid environments the performance of the organism is still much better under neutral conditions.[14]

The biochemistry of nitrification

The transformation of ammonium to nitrate involves the oxidation of nitrogen from its most reduced state with a valency of -3 to its most oxidized state with a valency of $+5$. The process takes place in two stages with different organisms effecting the two transformations. Although there is a considerable change in free energy the change is not equally distributed as can be seen below.

Nitrosomonas stage

$$NH_4^+ + 1\tfrac{1}{2}O_2 \xrightarrow{\text{6e}} NO_2^- + H_2O + 2H^+ \qquad \Delta G = -276 \text{ kJ}$$

11 Focht, D. D. and Verstrate, W., *Adv. Microbial Ecol.*, 1977, **1**, 135.
12 Engel, M. S. and Alexander, M., *J. Bact.*, 1958, **76**, 217.
13 Verstrate, W. and Alexander, M., *Environ Sci. Tech.*, 1973, **7**, 39.
14 Remacle, J., *Ecol. Bull.* (*Stockholm*), 1977, **25**, 560.

Nitrobacter stage

$$NO_2^- + \tfrac{1}{2}O_2 \longrightarrow NO_3^- \qquad \Delta G = -73 \text{ kJ}$$

Despite the limited energy available, autotrophic bacteria expend considerable amounts of it in fixing carbon dioxide for cell synthesis. Furthermore, some of the nitrogen oxidation steps require the use of cytochromes, large quantities of which are characteristic of these organisms. Compared with organic carbon oxidations low cell yields are obtained in terms of the amount of substrate used.

Some detailed aspects of the biochemistry of nitrification still remain obscure. The oxidation catalysed by the nitroso group of organisms almost certainly involves hydroxylamine (NH_2OH) and oxidation of hydroxylamine can be achieved with cell free preparations. The final steps leading to nitrite appearance have yielded traces of nitric oxide and nitrous oxide in the laboratory, and they have also been detected during active nitrification in field soils. These two gases are probably by-products of an unisolated intermediate which is likely to be enzyme bound nitroxyl (HNO). Transient formation of a haem–nitric oxide complex has been reported.[15] Nitroxyl would be the logical two-electron transport product derived from hydroxylamine but its existence has never been proven.

Anaerobically, the nitrosobacteria may produce nitric oxide and nitrous oxide at the expense of nitrite. The exact biochemistry is uncertain and may result from *either* the use of nitrite as an alternative electron acceptor when oxygen is limiting *or* as a means of protecting the cell against toxic effects of accumulated nitrite. It follows that the presence of nitrous oxide in a soil atmosphere is no longer to be explained solely on the basis of the activities of heterotrophic denitrifiers.

The oxidation of nitrite to nitrate by *Nitrobacter* is much better understood and appears to be a cytochrome linked single step hydrolytic oxidation.

Populations of nitrifying organisms in the soil

The numbers of nitrifying organisms normally reported as being found in soils are surprisingly low and there have been suspicions for some time concerning the techniques used. Meiklejohn[16] counted ammonium and nitrite oxidizers on various Broadbalk plots, Rothamsted, on different occasions and usually found a few thousand ammonium oxidizers and only a few hundred nitrite oxidizers per gram of soil. Higher populations were found[17] in a survey of 25 fields over a 27-month period. Ammonium oxidizers were as high as 1 to 10 million per gram and nitrite oxidizers were both much more variable and

15 Hooper, A. B. in *Microbiology-78* (ed D. Schlessinger), ASM Publication, 1978.
16 Meiklejohn, J., *Rothamsted Ann. Rept. for 1968*, Part 2, p. 177.
17 Barkworth, H. and Bateson, M., *Pl. Soil*, 1965, **22**, 220.

only between one-tenth and one-hundredth as numerous.[18] Great care is necessary in assessing populations of nitrifiers, including the need for very lengthy incubations of up to 70 or even 100 days in some cases.[19] Belser[20] describes the outcome of attempts to investigate nitrifier populations using fluorescent antibody techniques and suggests that soils may contain closely related but serologically distinguishable strains of nitrifiers. This population diversity could give rise to a greater flexibility in response to environmental changes but also offers the possibility of competition between similar but not identical organisms. The populations of nitrifiers that are theoretically possible in soils are very much higher than the actual populations measured, as the latter are limited by ammonium substrate availability. This has led to the concept of the 'carrying capacity of the soil for nitrifiers'.[21]

Despite a continuing interest in estimating nitrifier populations, caution is necessary when inferring nitrification rates from measured populations. Belser states:

> The main problem . . . is attempting to infer something about nitrification rates from data that do not necessarily correlate with these rates. Thus, in general, the presence of a high nitrifying population does not mean that nitrification is occurring, only that it occurred sometime in the past. Low counts, on the other hand, do not rule out the possibility of active nitrification taking place since counting may be inefficient. Low nitrate concentrations can occur in an actively nitrifying habitat as a result of leaching, denitrification or rapid plant uptake. It should be stressed that if nitrification rates are the factor of interest they should be measured directly.

Environmental factors affecting nitrification

Transformations in the field

The obvious prerequisite for nitrification to take place in a field soil is the presence of a nitrifying population together with the ammonium substrate. However, because the population is a specialized one, the amount of nitrate formed is subject to the influence of a combination of factors such as temperature, aeration, soil moisture content, pH and the presence of toxic materials.

Nitrifying bacteria have a high optimum temperature for activity and reach a maximum at about 25 to 30 °C. In soils with higher average temperatures the optimum may be higher, indicating an adaptation of the population to

---

18 Techniques for estimating nitrifiers and nitrification are reviewed by Prosser, J. and Cox, D. in *Experimental Microbial Ecology* (eds R. G. Burns and J. H. Slater), p. 178, Blackwell, 1982.

19 Malulevitch, V. A. *et al., Appl. Microbiol.*, 1975, **29**, 265.

20 Belser, L. W., *Ann. Rev. Microbiol.*, 1979, **33**, 309.

21 Ardakani, M. S. *et al., Soil Sci. Soc. Am. Proc.*, 1973, **37**, 53.

climatic conditions. This feature was first reported[22] for soils in the USA and more recently further work[23] has produced evidence of similar adaptations. It is also worth remembering that results under constant laboratory conditions may not relate well to fluctuating temperatures in the field.

Nitrification is slow at temperatures below 4 or 5 °C[24] and ammonium nitrogen applied as fertilizer in late autumn may remain largely intact until spring. Under winter conditions freezing and thawing were considered by Campbell[25] to affect nitrification activity more than ammonification and thus to lead to an increased proportion of ammonium.

Soil moisture has a considerable influence on nitrification both by itself and through its effect on aeration. At low moisture contents microbial activity is depressed and mineralization of organic nitrogen will be slow, hence limiting the amount of ammonium available. Nitrifiers can, however, produce measurable amounts of nitrate even below the wilting point of plants.[26]

The relation between nitrification rate and pF varies for different soils, some authors[27] claiming a linear relation between log nitrification rates and pF whereas others consider that the relationship is only linear over a short range of pF values.[28] Figure 19.1 illustrates the effects of pF on relative ammonium and nitrate accumulation between pF 5.6 and pF 2.7 for a Senegal black clay.[29]

As moisture contents increase, the predominant influence will be through the effects on aeration as the conversion of ammonium to nitrate is an

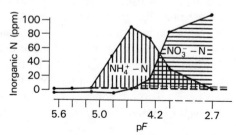

Fig. 19.1 Influence of soil pF on the accumulation of NH_4^+-N and NO_3^--N in a tropical black clay soil of Senegal.

22 Mahendrappa, M. K. *et al.*, *Soil Sci. Soc. Am. Proc.*, 1966, **30**, 60.
23 Myers, R. J. K., *Soil Biol. Biochem.*, 1975, **7**, 83.
24 Anderson, O. E. and Boswell, F. C., *Soil Sci. Soc. Am. Proc.*, 1964, **28**, 525. Addiscott, T. M., *J. Soil Sci.*, 1983, **34**, 343.
25 Campbell, C. A. in *Soil Organic Matter* (eds M. Schnitzer and S. U. Khan), p. 173, Elsevier, 1978.
26 Sabey, B. R., *Soil Sci. Soc. Am. Proc.*, 1969, **33**, 263.
27 Tyllova, A., *Rostliuva Vyroba*, 1981, **27**, 315.
28 Sludnickova, I., *Folia Microbiol.*, 1979, **24**, 478.
29 Dommergues, Y., *Biologie du Sol*, Presses Universitaires, Paris, 1977.

oxidative process and does not take place at redox potentials below +200 mV. Poorly aerated soils cannot nitrify ammonium and when ammonium sulphate amended soil was aerated with various gas mixtures the following results were obtained after 21 days[30]:

% Oxygen in gas mixture	20	11	4.5	2.1	1.0	0.4
% Added N nitrified	46	43	38	28	21	2

The sharp decrease occurring at the lowest level was confirmed using oxygen electrode techniques[31] which showed that the rate of nitrification is not seriously reduced until the dissolved oxygen concentration falls to 1.3×10^{-7} g cm^{-3}, which corresponds to a solution in equilibrium with air containing 0.3 per cent oxygen. Both groups of nitrifiers possess cytochrome-C enzymes with a strong affinity for oxygen. At low oxygen levels[32] the nitrite oxidizers may recycle nitrate to some extent and reports exist of the production of nitrous oxide by pure cultures of *Nitrosomonas* under conditions of restricted aeration.[33]

The effect of alternate wetting and drying of the soil depends to a large extent on the duration and degree of the two phases. In so far as wetting and drying increase mineralization, the nitrification process benefits from extra ammonium substrate. In hot regions having a pronounced dry and wet season, nitrate is produced most rapidly at the commencement of the rains and only slowly during the rainy season itself. Flushes of between 13 and 180 kg ha^{-1} were found in 13 soils from different parts of East Africa shortly after the onset of seasonal rains.[34] Agronomically it is vital to plant as early as is safely possible, taking advantage of early showers so that crops can start growing quickly when the main rainfall arrives and make use of the flush of nitrate before it is leached beyond rooting depth. In tropical fallow soils concentrations as high as 110 ppm nitrate-N in the top 150 mm[35] or 200 ppm in the top 50 mm have been recorded, compared with 30–40 ppm nitrate-N in the top 150 mm of a well-manured Broadbalk field at Rothamsted Experimental Station.

Several other factors can affect the rate of nitrification in wetting and drying cycles. Extreme drying conditions can kill off many nitrifiers, and in the seasonally arid tropics and subtropics there may be a lag following rainfall before nitrification begins. It is possible that, under some circumstances, the nitrite oxidizers are more sensitive than the ammonium oxidizers to desiccation, so that nitrite sometimes accumulates in the soil for a number of days. Assessment of the progress of nitrification following the rewetting of a dried

30 Amer, F. M. and Bartholomew, W. V., *Soil Sci.*, 1951, **71**, 215.
31 Greenwood, D. J., *Pl. Soil*, 1962, **17**, 365.
32 Belser, L.W., *Ann. Rev. Microbiol.*, 1979, **33**, 309.
33 Goreau, T. J. *et al.*, *Appl. environ. Microbiol.*, 1980, **40**, 526.
34 Semb, G. and Robinson, J. B. D., *East Afr. agric. For. J.*, 1969, **34**, 350.
35 Griffith, G.ap., *Emp. J. expl. Agric.*, 1951, **19**, 1.

soil is further complicated by the fact that the soil, if approaching a water-logged condition, may be capable of denitrification at a much enhanced rate. Much higher denitrification rates lasting for one to three days have been found after rewetting a dried soil compared with the denitrification in a soil maintained in a moist condition.[36]

In soils subjected to heavy dews, it is possible that enough water may condense in the surface of dry soil to allow localized nitrification.[37] Some Canadian soils have shown that the heterotrophic soil population can respond to very small water additions in the form of simulated dew.[38]

Very heavy rainfall in the field may lead to nitrate leaching to depth in the profile and sampling depth and sampling time need to be carefully selected if an adequate picture of the processes involved is to emerge. If the profile is dry to depth, and the rainfall limited, nitrate will accumulate lower down the profile and these accumulations can be considerable. As much as 400 ppm nitrate-N was found in a soil at 900 mm depth.[39]

In the extreme case of waterlogging, the activity of nitrifiers is likely to be restricted to the air–water or soil–water interface, although the populations seem quite capable of survival. No significant changes in the nitrifying populations in organic soils during waterlogging have been reported.[40]

All the carbon utilized by the autotrophic nitrifiers is supplied by carbon dioxide in the dissolved state. There is no evidence to suggest that carbon dioxide ever limits the activity of nitrifiers in the field through lack of availability, although the situation in the laboratory can be very different. The paper by Clark[41] on the effect of soil sample size on nitrification rates (in laboratory incubations) due to varying carbon dioxide levels seems to have made insufficient impact on experimental procedures.

Effect of soil pH

Nitrification does not take place readily in very acid soils with the possible exception of limited heterotrophic nitrification.[42] There is, however, no clear relationship between soil pH and the rate of nitrification. There are examples of soils with a pH between 4 and 5 which will nitrify slowly[43] and others with a pH over 5 which will only nitrify after liming.[44] In general over the pH range of 5.5 to 8.0 there is usually little effect of pH and there is evidence

36 Galsworthy, A. M. and Burford, J. R., *J. Soil Sci.*, 1978, **29**, 537.
37 Drouineau, G. and Lefèvre, G., *Ann. Agron.*, 1951, **21**, 1.
38 Beiderbeck, V. O. *et al.*, *Canad. J. Soil Sci.*, 1977, **57**, 93.
39 Mills, W. R., *East Afr. agric. For. J.*, 1953, **19**, 53.
40 Terry, R. E. and Tate, R. L., *Soil Sci.*, 1980, **129**, 88.
41 Clark, F. E. in *The Ecology of Soil Bacteria* (eds T. R. G. Gray and D. Parkinson), Liverpool Univ. Press, 1967.
42 Remacle, J., *Ecol. Bull. (Stockholm)*, 1977, **25**, 560.
43 Weber, D. F. and Gainey, P. L., *Soil Sci.*, 1962, **94**, 138.
44 Millbank, J. W., *Pl. Soil*, 1959, **11**, 293.

that nitrifiers in soils of low pH are adapted to those conditions. A better understanding of how nitrification proceeds in the heterogeneous environment of the soil has been provided by the identification of localized clusters of nitrifiers of perhaps 100 cells associated with soil aggregates.[45] Nitrifying activity by these clusters was very intense but short-lived and declined sharply. The nitrification process may occur in pulses, spatially separated and making up an overall rate when sufficient soil is considered at the same time. This aspect of spatial separation permits the hypothesis of localized areas of acceptable pH conditions, helped by local buffering.[46] Rapid local decline in activity would result from localized self-acidification.[47]

Substrate availability

Since the sole source of energy for nitrifiers comes from the oxidation of ammonium and subsequently nitrite, it is obvious that ammonium concentration and availability are crucial to the nitrifying processes providing that other environmental factors are not limiting. It follows that a soil will show the highest rates of nitrification if well provided with ammonium, either from the ammonification of organic matter or from fertilizers containing urea or ammonium. Very high rates of nitrification, up to 350 mg of nitrate-N per kilogram of soil per day are possible in soils with adequate buffering capacity. Nitrate concentrations in arable soils are typically lowest in winter when temperature limits nitrification and rainfall tends to leach away any accumulations previously formed. Concentrations increase in spring with the commencement of mineralization. In summer the crop will lower nitrate concentrations due to uptake and finally in autumn the concentration will increase as the senescent plant materials decay.

The relationship between fertilizer type and soil pH can influence nitrification rate. In an acid soil, alkaline hydrolysing materials such as urea and diammonium phosphate usually nitrify at a much faster rate than acid hydrolysing materials such as ammonium sulphate or monoammonium phosphate. In alkaline soils the reverse may be true. Hauck regards the critical microsite pH as about pH 7.5. The form in which the ammonium exists in soil is important, whether in solution, on exchange sites or fixed. Fixed ammonium is only slowly available to nitrifiers.[48]

An extreme case of substrate availability is where either urea or anhydrous ammonia or highly nitrogenous organic manures are applied to alkaline soils. Apart from the potential wastefulness of such practices due to ammonia loss by volatilization, ammonium itself is toxic to the two groups of nitrifiers and especially the nitrite oxidizing *Nitrobacter* group. Thus when high levels of

45 Molina, J. A. E. *et al.*, *Soil Sci. Soc. Am. J.*, 1979, **43**, 728.
46 Kunc, F. and Stotzky, G., *Folia Microbiol.*, 1980, **25**, 106.
47 Bazin, M., *CRC Crit. Revs Microbiol.*, 1976, **4**, 463.
48 Sowden, F. J., *Canad. J. Soil Sci.*, 1976, **56**, 319.

ammonia occur in neutral or alkaline soils of low cation exchange capacity, nitrite may accumulate to levels causing phytotoxicity, particularly during cool weather.[49] Free calcium carbonate, by buffering the soil and so reducing the rise in pH, will reduce the likely nitrite concentration and soils with considerable clay or organic matter content will be less likely to give rise to problems because more ammonium ions can be adsorbed.

Purchase[50] showed that slow nitrification response to added ammonium sulphate could be overcome by the addition of phosphate and that the nitrite oxidizers were very sensitive to soil phosphate deficiency. The interaction of phosphate availability and soil pH might be an interaction worth further investigation with respect to nitrification rates.

Natural and artificial inhibition of nitrification

The nitrifying bacteria are among the most delicate of soil micro-organisms and are susceptible to the influence of toxic materials. As nitrate can be lost from field soils by both leaching and denitrification it is not unreasonable to regard nitrification as a process to be controlled if not eradicated and there have been a number of attempts to find nitrification inhibitors. Specific nitrification inhibitors such as Nitrapyrin (2-chloro-6-trichloromethyl pyridine), marketed as N-Serve, quite effectively inhibit the oxidation of nitrifiable nitrogen in the field. However, even for Nitrapyrin, which is the best inhibitor known at present, the inhibition and duration of effect depend to a considerable degree on factors such as soil type and climatic conditions.

Although positive crop responses have been reported following the use of nitrification inhibitors there have also been instances of crop yield depression.[51] Comprehensive reviews of the literature on nitrification inhibition are available.[52]

The effects of toxic agrochemicals other than specific nitrification inhibitors have received attention and a comprehensive review of pesticide influence on nitrification is available.[53] At field levels of application the effects of insecticides and herbicides tend to be short-lived if indeed detectable; fungicides and more notably fumigants such as methyl bromide have more pronounced effects. It has been suggested that there can be important interactions with soil pH which can influence the toxicity of pesticides to the nitrifying population.[54]

49 Chapman, H. D. and Liebig, G. F., *Soil Sci. Soc. Am. Proc.*, 1952, **16**, 276.
50 Purchase, B. S., *Pl. Soil*, 1974, **41**, 541.
51 Warren, H. L. *et al.*, *Agron. J.*, 1975, **67**, 655.
52 Meisinger, J. J. *et al.*, *Nitrification Inhibitors – Potentials and Limitations*, Am. Soc. Agron. Special Publ., 1980, Madison;   Sahrawat, K. L., *Pl. Soil*, 1980, **57**, 335.
53 Goring, C. A. I. and Laskowski, D. A. in *Nitrogen in Agricultural Soils* (ed F. J. Stevenson), Am. Soc. Agron. Publ. No. 22, 1982.
54 Wainwright, M., *J. Soil Sci.*, 1978, **29**, 287.

The adverse effects of salinity on nitrification have been recognized for some time, part of the effect being due to osmotic stress and part to specific toxicities. When sodium sulphate is the dominant salt little effect can be seen until the osmotic pressure reaches about 5 bar (0.5 MPa), followed by severe inhibition at 20 bar (2 MPa) and complete cessation at 30 bar (3 MPa).[55] With sodium chloride, there is evidence that the chloride ion has a particular toxicity and an osmotic pressure of 10 bar (1 MPa) caused by sodium chloride is sufficient to inhibit nitrification, and the addition of 10 mg of sodium chloride per gram of soil completely inhibited nitrification during a 14-week study.[56] Most studies of salinization effects on nitrification processes are conducted by adding salts to non-saline soils and there is a possibility that although nitrifiers are adversely affected by chloride ions, they may be able to adapt to some degree though not to the extreme levels of salinity found in saline alkaline soils.[57]

There is a possibility that certain cropping systems may have an inhibitory influence on nitrification. The original finding by Richardson[58] that the ratio of ammonium to nitrate was higher in grassland soils compared with similar arable situations, has been repeated often enough[59] for it to suggest that the nitrogen economy of grassland is distinctive and that nitrification is in some way depressed beneath a continuous sward. Apart from some natural grassland soils, particularly savanna soils, where the soil may be too acid for nitrifying populations to exist, the lack of actively nitrifying organisms appears to be an effect of the grass crop itself. Ploughing out grassland usually results in ample nitrate production indicating that there is no inherent soil problem and the only cause other than extreme pH is where the C : N ratio of the organic material incorporated by ploughing results in excessive immobilization and an inadequate supply of ammonium for the nitrifiers.

Various explanations have been suggested to explain the effects of the growth of the grass crop, with its dense rooting pattern, on nitrification or nitrate concentrations, such as:

1 Rapid root uptake of mineralized ammonium by grasses reducing availability of ammonium for nitrifiers.
2 Rapid root uptake of nitrate resulting in an overall low nitrate content.
3 Water removal by the crop creating conditions which are less conducive to nitrification.
4 Release of carbon-rich root exudates which encourage microbial immobilization of ammonium.

55 Johnson, D. D. and Guenzi, W. D., *Soil Sci. Soc. Am. Proc.*, 1963, **27**, 663.
56 McCormick, R. W. and Wolf, D. C., *Soil Biol. Biochem.*, 1980, **12**, 153.
57 Painter, H. A., *Water Res.*, 1970, **4**, 393.
58 Richardson, H. L., *J. agric. Sci.*, 1938, **28**, 73.
59 Clark, F. E. and Paul, E. A., *Adv. Agron.*, 1970, **22**, 375.

5   Release of exudates which encourage denitrification in the rhizosphere.[60]
6   Release of nitrification inhibitors by grass roots.

The last of these hypotheses has received considerable attention starting with the work of Theron[61] who obtained material from plants which inhibited pure cultures of nitrifiers. Unfortunately, the nitrifying autotrophs are readily inhibited in pure culture and drastic plant extraction methods may well create inhibitory substances. The use of root washings has shown that exudates inhibitory to nitrification in soils as opposed to pure cultures can be obtained from grasses and other species,[62] but some grasses do not appear to be active.[63] So far nothing is known about whether there are a number of inhibitory substances excreted or whether plants differ only in the amounts of a single substance. A further possibility is that when grasses are using ammonium as a nitrogen source they may liberate hydrogen ions in sufficient quantities to influence the localized pH and hence depress nitrification.

Loss of inorganic nitrogen

The growing plant is dependent on supplies of inorganic nitrogen in the form of ammonium or nitrate from the soil unless it is capable of nitrogen fixation to meet its growth needs. Micro-organisms influence the availability of inorganic nitrogen by two major processes: the assimilation of nitrogen into their cell material, a process known as immobilization; and the conversion of nitrogen compounds into volatile products which can be lost into the air. Gaseous loss as ammonia from ammonium ions present in soil is not directly effected by microbial activity and is discussed in Chapter 20; here, gaseous loss will be discussed only in relation to denitrification.

Immobilization

Immobilization can influence the nitrogen nutrition of plants severely if there is an imbalance of carbon and nitrogen in the organic materials being decomposed by micro-organisms. High carbon : nitrogen (C : N) ratio plant residues such as cereal straw can cause limitation of available nitrogen as the micro-organisms draw on soil reserves to satisfy their own nutritional needs. Sometimes the situation can be sufficiently severe to require fertilizer supplementation to overcome the deficiency. A combination of high C : N ratio

60 Woldendorp, J. W., *Meded. Landbouwhog. Wageningen*, 1963, **63**, 13.
61 Theron, J. J., *J. agric. Sci.*, 1951, **41**, 289.
62 Moore, D. R. E. and Waid, J. S., *Soil Biol. Biochem.*, 1971, **3**, 69.
63 Neal, J. L., *Canad. J. Microbiol.*, 1969, **15**, 633.

plant residue and low soil nitrogen may restrict microbial growth and reduce the rate of breakdown of the organic matter. As a general guide, organic residues with C : N ratios less than 30 : 1 are unlikely to make demands on soil nitrogen during the course of decomposition. Immobilization does not constitute a permanent loss of nitrogen and subsequent death of the micro-organisms will result in the mineralization and return to the soil pool of the immobilized nitrogen. Attention is currently being given to the dynamics of nitrogen turnover through the microbial population[64], and the role of the predatory fraction of the population, especially the protozoa and nematodes, is seen as having considerable importance. A recently discovered phenomenon involving the rapid, but short-term, immobilization of added fertilizer nitrogen[65] may also be connected with the microbial population.

Microbial denitrification

One of the most important causes of soil and fertilizer nitrogen loss as gaseous products is microbial denitrification, that is, the reduction of nitrate to nitrous oxide (N_2O) and dinitrogen gas (N_2). Field losses due to denitrification have usually been estimated by the difference between nitrogen additions and recoveries where leaching loss is measured or controlled. A large survey of soil nitrogen balance sheets[66] suggests that up to 30 per cent of fertilizer nitrogen can be lost by denitrification with an average in the range of 9 to 15 per cent. The authors place emphasis on the very variable nature of denitrification losses. Experiments using ^{15}N to monitor fertilizer nitrogen loss have indicated losses of the same magnitude.[67] Losses from arable soils are higher than from grassland soils since the latter tend to maintain lower nitrate levels.

In the simplest terms, nitrate will be lost from soil most rapidly when it is warm, wet and well supplied with easily decomposable organic matter. It can be a very active process[68] and in the laboratory a soil containing 300 ppm nitrate-N can lose almost the whole amount in 28 to 96 hours. The reduction to gaseous forms of nitrogen can be brought about by numerous groups of bacteria which are predominantly facultative in that they use the oxygen of nitrate, nitrite or oxides of nitrogen as a hydrogen acceptor only in the absence of free oxygen.[69] A comprehensive list of organisms[70] shows that the most common are the heterotrophs such as *Pseudomonas* and *Alcaligenes*

64 Coleman, D. C. *et al.*, *Ecol. Bull.* (*Stockholm*), 1977, **25**, 299.
65 Seneviratne, R., Ph.D. Thesis., Univ. Reading, 1983.
66 Legg, J. O. and Meisinger, J. J. in *Nitrogen in Agricultural Soils* (ed F. J. Stevenson) Am. Soc. Agron. Publ., No. 22, Wisconsin, 1982.
67 Colbourn, P. and Dowdell, R. J., *Pl. Soil*, 1984, **76**, 213.
68 Cooper, G. S. and Smith, R. L., *Soil Sci. Soc. Am. Proc.*, 1963, **27**, 659.
69 Valera, C. L. and Alexander, M., *Pl. Soil*, 1961, **15**, 268.
70 Knowles, R., *Ecol. Bull.* (*Stockholm*), 1981, **33**, 315.

although certain autotrophs such as *Thiobacillus denitrificans* can reduce nitrate in the course of oxidizing sulphur compounds. Denitrification ability of bacteria is invariably established under laboratory conditions on artificial media and there are difficulties in extrapolating laboratory rates to the soil.

The sequence of reductions can be represented as a series of steps:

Nitrate → Nitrite → Nitric oxide → Nitrous oxide → Dinitrogen

NO_3^- NO_2^- (NO) N_2O N_2

in which the nitrate is gradually reduced while acting as a hydrogen acceptor in dissimilatory nitrate reduction. All nitrate which is taken up for the purpose of cell synthesis by micro-organisms or plants, must be reduced to the ammonium form if it is to be converted into amino acids and subsequently protein. This latter process is known as assimilatory nitrate reduction. The possibility exists for ammonium to be formed as a dissimilatory product but it is considered that such a pathway would only contribute about 2 to 4 per cent of the total dissimilatory nitrate reduction.[71]

It is certain that nitrate reduction goes through the nitrite stage, but nitrite does not normally accumulate in soil and the most commonly detected products are nitrous oxide and nitrogen gas. The ratio between the latter two gases can be very variable and seems to depend on the relative abundance of available nitrate and organic substrate. High nitrate concentrations relative to the amount of organic substrate lead to a greater proportion of nitrous oxide. Nitric oxide is rarely detected and is thought to remain enzyme bound during the reduction.

Nitrate reduction only takes place under conditions of low oxygen supply. The enzymes responsible for the reduction are inhibited by oxygen and even their synthesis is repressed. The later steps of the reduction are most sensitive and nitrous oxide is a more common product as oxygen concentrations rise.

Greenwood[72] found that the concentration of dissolved oxygen in the solution surrounding the bacteria must fall to about 4×10^{-6} M before reduction begins. However, soils which appear well aerated may yet reduce nitrate to nitrous oxide and nitrogen, particularly if the organic matter status is high, because of the existence of microsites where the oxygen demand by the microbial population has exceeded the supply from the soil atmosphere. Hence the interior of soil aggregates or crumbs may become anaerobic and sites for denitrification. Greenwood[73] has suggested that aggregates which exceed 9 mm in radius are likely to have such sites within them even in soils in which aeration around the aggregates appears satisfactory. The situation is complicated by the fact that the requirements for active denitrification, that is, available organic substrate, nitrate, suitable organisms and low

71 Tiedje, J. M. *et al.*, *Ecol. Bull.* (*Stockholm*), 1981, **33**, 331.
72 Greenwood, D. J., *Pl. Soil*, 1962, **17**, 365.
73 Greenwood, D. J., *Min. Agric. Fish Fd. Tech. Bull.*, No. 29, p. 261, HMSO, London, 1975.

oxygen concentrations, may be distributed throughout the soil in an irregular fashion and must all coincide for denitrification to occur. The potential complexity of soil oxygen distribution is demonstrated by models of oxygen concentration which allow for inter- and intra-aggregate porosities.[74]

In earlier work, Smith[75] had found large variations in oxygen and nitrous oxide concentrations in the soil atmosphere over small distances at the same depth in soils and had remarked on the transient nature of the nitrous oxide concentrations measured. In California, after measuring denitrification in a variety of well-characterized soils, workers were forced to conclude that soil microstructure may be critical in determining nitrate loss rates.[76] It emerges clearly from much of the work on denitrification that there is a demand for a reliable means of establishing the aeration status of soils, not as an overall level but at a resolution that might have some relevance to microsite behaviour.

Denitrification is typically a rapid process under warm conditions in soils containing adequate organic matter, and hence nitrate in the deep subsoil suffers little denitrification because of a lack of bacterial activity.

Ploughing in easily decomposable organic matter may encourage denitrification during a period of poor aeration and it has been suggested[77] that the roots and root exudates of plants may contribute sufficient organic matter to stimulate denitrification.

In general, organic additions, by providing a substrate for bacterial growth, are liable to enhance denitrification providing that nitrate concentrations are high. Where nitrate concentrations are low, and especially if the organic amendments have a wide C : N ratio, the nitrate may be immobilized and assimilated by the micro-organisms before significant denitrification loss has taken place. The importance of the easily available nature of the organic matter is indicated by the relations between denitrification potential and water-extractable organic carbon in a range of soils.[78] Air drying, which is known to increase organic matter availability and microbial activity on rewetting, increases denitrification.[79] It is possible that many of the rates of denitrification reported in the literature may be overestimates due to soil treatment prior to experimentation.

The effect of temperature on denitrification is very marked and nitrate loss can double with a temperature increase of 10 °C over the range from 10 °C to 35 °C.[80] In the range from 0 °C to 5 °C denitrification is much reduced but

74 Smith, K. A., *J. Soil Sci.*, 1980, **31**, 263.
75 Smith, K. A. and Dowdell, R. J., *J. Soil Sci.*, 1974, **25**, 231.
76 Gilliam, J. W. *et al.*, *Soil Sci. Soc. Am. J.*, 1978, **42**, 61.
77 Woldendorp, J. W., *Meded. Landbouwhog. Wageningen*, 1963, **63**, 13.
78 Burford, J. R. and Bremner, J. M., *Soil Biol. Biochem.*, 1975, **7**, 389.
79 Galsworthy, A. M. and Burford, J. R., *J. Soil Sci.*, 1978, **29**, 537.
80 Stanford, G. *et al.*, *Soil Sci. Soc. Am. Proc.*, 1975, **39**, 867.

still measurable and the proportion of nitrous oxide to dinitrogen gas increases.

There is a positive relation between denitrification and pH, the process taking place more readily in neutral and calcareous soils than in acid situations with a peak in the region pH 7 to pH 8. The proportion of nitrous oxide to dinitrogen increases with reduced pH as the reduction of nitrous oxide is the most acid sensitive step and there is some tendency for nitrite to accumulate.

That pH is a major limiting factor is shown by the considerable rates of denitrification obtained when an acid peat soil of pH 3.5 is amended with lime and nitrate.[81] At low pH values any nitrite which occurs in the soil solution may undergo chemical reactions, particularly in the presence of organic matter. Nitrite can, under acid conditions, react with amino acids, polyphenols, lignin and humic acids and release nitrogen-containing gases.[82]

Methods for measuring denitrification losses have improved in recent years. The methods themselves and the results they have yielded are discussed in Chapter 20.

The potential scale of economic loss is indicated by a survey of 34 sets of experiments,[83] where the use of ^{15}N techniques had revealed that up to 30 per cent of added nitrogen was unaccounted for and had presumably been lost as nitrous oxide or dinitrogen gas. In terms of current fertilizer application this could amount to a very considerable emission of gas, at least some of which is likely to be nitrous oxide. A further consideration is that the higher the level of nitrogen fertilizer application the higher is the proportion of nitrous oxide in the products of denitrification.

Gains of nitrogen

A number of attempts have been made to estimate the natural gains of nitrogen by the soil on a global scale. Burns and Hardy[84] suggest an average input of about 13 kg N ha^{-1} a^{-1} from atmospheric deposition and biological fixation but affirm that this will be subject to very considerable regional variation. Desert areas remote from industrialization may be much lower than the average whereas temperate areas with active leguminous crops will far exceed the average level.

81 Klemedtsson, L. *et al.*, *Swedish J. agric. Res.*, 1977, **7**, 179.
82 Nelson, D. W. and Bremner, J. M., *Soil Biol. Biochem.*, 1969, **1**, 229. Knowles, R. lists possible reactions and products in *Soil Biochemistry*, Vol. 5 (eds E. A. Paul and J. Ladd), Dekker, 1979.
83 Hauck, R. D., *Ecol. Bull. (Stockholm)*, 1981, **33**, 551.
84 Burns, R. C. and Hardy, R. W. F. in *Nitrogen Fixation in Bacteria and Higher Plants*, Springer-Verlag, 1975.

On a broad scale the most important natural process for increasing the nitrogen content of soils is through the agency of nitrogen fixing micro-organisms, a group collectively known as the diazotrophs. Among the most important are those micro-organisms which form symbiotic associations with plants such as the legume–*Rhizobium* symbiosis and the nodule-forming symbioses caused by the actinomycete group *Frankia*. Free-living organisms may not readily fix nitrogen at the rates common among symbionts but may still make a significant contribution as in the case of the blue–green algae (cyanobacteria) in rice cultivation. The last 12 years has seen a new area of interest in microbial nitrogen fixation which falls between obligate symbiosis and the independence of free-living fixation. Firm evidence now exists for enhanced nitrogen fixation in the rhizosphere of a wide range of plants, variously called associative nitrogen fixation or rhizocoenoses. The remainder of this chapter will concentrate on various aspects of biological nitrogen fixation.

The diversity of biological nitrogen fixation

The expansion of our understanding of biological nitrogen fixation which has occurred over the last two decades has been based to a considerable degree on new and improved techniques. The development of methods permitting the greater use of the radioactive and stable isotopes of nitrogen, ^{13}N and ^{15}N respectively, has meant that nitrogen fixation can be identified and quantified with more certainty than ever before. The acetylene–ethylene reduction assay has meant that nitrogen fixing systems can be investigated under more natural conditions with less expensive equipment. Finally, the progress made in understanding the biochemistry of nitrogen fixation and the genetics underlying its occurrence has revealed that it is a process that is basically similar wherever it occurs, but one in which the organisms demonstrate great ingenuity in the means employed to permit the relevant enzymes to function in a variety of ecological situations.

All forms of biological nitrogen fixation require a micro-organism, either free living or as a partner in an association or symbiosis. The micro-organisms are all simple prokaryotic types, either bacteria, actinomycetes or blue–green algae. The latter group are now classified with the bacteria under their alternative name, the cyanobacteria. No thermophilic nitrogen fixers have yet been discovered and an upper limit for the biological fixation of nitrogen has been tentatively placed at 40 °C.[85]

Previous claims for nitrogen fixation by filamentous fungi, yeasts and mycorrhizas are now thought to be incorrect. A brief survey of nitrogen fixing micro-organisms, or diazotrophs as they are now known, is given below and they will be examined in greater detail in succeeding sections.

85 Postgate, J. R., *Nitrogen Fixation*, Studies in Biology, No. 92, Arnold, 1978.

A variety of bacteria and cyanobacteria are free-living nitrogen fixers. Both aerobes and anaerobes are known and in both categories there are both heterotrophs and photosynthetic autotrophs. Most are single celled organisms although some of the cyanobacteria are filamentous. At present, organisms from 33 separate genera are recognized as diazotrophs with the largest number of genera coming from the bacteria.

Associative nitrogen fixation by micro-organisms

A number of situations are now recognized where nitrogen fixation is enhanced through the association of nitrogen-fixing organisms with the roots or other organs of higher plants or in some cases with other organisms. The micro-organisms concerned are invariably free-living nitrogen fixers which are capable of a higher degree of activity because of particular ecological features of the association. In most cases, such as the rhizosphere association between *Paspalum notatum* and *Azotobacter paspali*, the improved nitrogen fixation is a result of improved carbon substrate supply to the micro-organisms. These loose associations in the root zones of a variety of plant species are known as rhizocoenoses. In associations between micro-organisms the non-fixing partner may contribute by reducing oxygen tension.

Symbiotic nitrogen fixation by non-leguminous plants

The most common type of symbiosis in this category involves the actino-mycete genus *Frankia*, members of which have only recently been grown in pure culture.[86] Symbiotic fixation takes place in nodules on the roots of a range of plants and some 17 genera are now known to have members capable of symbiotic fixation. Apart from a type of raspberry (*Rubus*) none of the non-legumes is of agricultural or horticultural significance but the contri-bution made by this symbiotic form to natural plant communities may be considerable. Under this heading might also be considered the leaf–nodule associations which have been recorded for a variety of plants. Present evidence suggests that these latter have little importance in nitrogen fixation. Free-living surface dwelling organisms are known to fix nitrogen but the inhabitants of the various leaves are not always nitrogen-fixing types and even where the organism is potentially capable of fixation, this function appears to cease when the organism enters the leaf.

Finally, a very different form of symbiosis, that which forms between the water-fern *Azolla* and members of the cyanobacteria, can be an efficient nitrogen-fixing system and is significant in certain areas of Asia where it is already exploited.

86 Callaham, D. *et al., Science,* 1978, **199**, 899; Tjepkema, J. D. *et al.,* 1980, *Nature, Lond.,* **287**, 633.

Symbiotic nitrogen fixation by leguminous plants

Of all the biological nitrogen-fixing systems known, the symbiotic association of the plants of the family Leguminosae and the bacteria of the family Rhizobiacae is the most significant and the most extensively investigated. Many of the plants are raised as crops producing valuable protein and the fact that the bacteria are relatively easy to culture has meant that both their study and their exploitation have been comparatively easy. The quantities of nitrogen fixed, often exceeding 100 kg ha^{-1} a^{-1}, are of great significance, and the improvement that can result in soil fertility was pronounced enough for it to have been recorded by writers on agricultural matters in Roman times.

The biochemistry of nitrogen fixation

Although the brief survey above indicates a considerable diversity in the types of organisms and in the levels of sophistication of their associations with higher plants, the outstanding advances made in the study of nitrogen fixation biochemistry have pointed to a remarkable similarity in the basic biochemical process. All biological nitrogen fixation is brought about by the same mechanism using very similar enzymes. The differences between the nitrogen-fixing systems are in the way in which the organisms overcome environmental limitations on enzyme function.

The basic problem faced by an organism when it attempts to tap the large atmospheric reserves of elemental nitrogen is that the dinitrogen molecule is extremely stable. Industrial fixation requires high temperatures, high pressures and the use of catalysts whereas the micro-organism must function at low temperatures, atmospheric pressure and in an aqueous medium.

For many years the major obstacle to biochemical understanding of nitrogen fixation was an inability to isolate the nitrogenase enzyme complex, a name given to the system when it was still a concept rather than a reality. The general use of radioisotopes which had proved so useful in studying carbon metabolism was precluded by the unhelpful fact that the ^{13}N isotope has a half-life of 10 minutes. The breakthrough came in 1960 when workers at the American laboratories of Du Pont, in attempting to extract the nitrogen-fixing complex from the anaerobic bacterium *Clostridium pasteurianum* took great precautions to exclude oxygen from all steps of their procedure.[87] The resulting successful extraction of a functioning enzyme, and the subsequent discovery of the extreme sensitivity of the nitrogenase complex to inactivation by atmospheric oxygen, stimulated other workers to adopt similar rigorous oxygen exclusion techniques. The approach worked, and permitted the isolation of nitrogen-fixing enzymes from other diazotrophs, regardless of whether they were ostensibly anaerobic or not.

87 Carnahan, J. E. *et al.*, *Biochim. Biophys. Acta*, 1960, **42**, 530.

The success of the oxygen exclusion procedure can be judged from the fact that functioning nitrogenase complexes have now been isolated from a wide variety of organisms, free living and symbiotic, heterotrophic and autotrophic. Furthermore, the similarity in biochemistry that has emerged is quite striking.

All nitrogenase systems comprise two water-soluble proteins. The larger molecule known as dinitrogenase or P1 is a molybdenum- and iron-containing protein with a molecular weight of about 200 000 to 220 000 Daltons, while the smaller partner is an iron protein with a molecular weight of 55 000 to 65 000 Daltons. This second protein is known as dinitrogenase reductase or P2 and is the most oxygen sensitive of the two with a half-life in air of less than a minute. Both enzymes are required for nitrogen fixation and a ratio of one or two P2 molecules for each P1 molecule is required. The extreme similarity of the nitrogen-fixing systems from different organisms is shown by the fact that the P1 and P2 elements can be interchanged and still yield working complexes. When workers cross-combined the nitrogenase fractions from eight different organisms they found that of the 55 non-homologous crosses tested 45 yielded active nitrogenases, suggesting a high degree of similarity.[88] Where non-active combinations occurred it was found that one of the constituent partners had been derived from a *Clostridium*. Further investigation indicated that the dinitrogenase reductase from *Clostridium* was bound very tightly to non-homologous dinitrogenases and could not be released. More recently the nitrogenase from the actinomycete *Frankia* on *Alder* has been added to the homologous list.[89]

Dinitrogenase reductase requires the supply of both protons and electrons together with considerable amounts of energy. Another strict and general requirement of a functioning nitrogenase complex is that the energy which comes from adenosine-5-triphosphate (ATP) must be supplied by a magnesium salt of ATP.

Apart from the important function of reducing dinitrogen to ammonia, the dinitrogenase complex is a versatile reductase and will reduce other substrates. In the absence of dinitrogen, hydrogen gas is evolved, cyanide can be reduced to ammonia and methane, and, although it is a non-physiological reaction, there is the very conveniently detected reduction of acetylene to ethylene.

The extreme sensitivity of the nitrogenase complex to oxygen might be expected to pose problems for nitrogen-fixing organisms with the exception of strict anaerobes which do not function in the presence of oxygen. When the organisms or symbioses are examined a variety of devices are found to overcome the problem. They include respiratory protection, conformational protection and the use of oxygen scavengers or buffers. These protection

88 Emerich, D. W. and Burris, R. H., *J. Bact.*, 1978, **134**, 936.
89 Benson, D. R. *et al., Science*, 1979, **205**, 688.

mechanisms will be discussed in more detail in connection with specific groups of organisms.

Although the reaction brought about by nitrogenase:

$$3H_2 + N_2 = 2NH_3$$

is exothermic, biological nitrogen fixation requires a high energy input and in this is no different from the industrial Haber–Bosch process. Photosynthetic organisms such as the cyanobacteria and other photoautotrophs, together with the symbiotic organisms *Rhizobium* and *Frankia*, rely directly or indirectly on the products of photosynthesis. Free-living organisms are likely to be energy limited, and this explains the interest shown in the associative forms of nitrogen fixation where the plant, albeit rather more loosely, can still make the products of photosynthesis available in the form of root exudates.

An important feature of biological nitrogen fixation is the non-obligate nature of the process. All organisms which are capable of fixation are also capable of growth on combined nitrogen and will do so if it is supplied in adequate amounts. Ammonia, or more correctly, an amino product of ammonium assimilation, is known to inhibit the synthesis of the enzymes necessary for fixation. Nitrate is a less effective inhibitor but, in the case of the legume–*Rhizobium* symbiosis, can affect nitrogen fixation by inhibiting infection and nodule formation. At present, no naturally occurring nitrogen-fixing system can be expected to function where the ambient levels of soil mineral nitrogen are already high, although organisms have been produced by mutation in the laboratory which are not so sensitive to nitrogen fixation inhibition by high surrounding nitrogen levels.[90]

The genetic manipulation of biological nitrogen fixation has become a major goal in agricultural biotechnology.[91] The primary aims have been to make a formal genetic analysis of the genes involved in dinitrogen fixation (*nif* genes), to determine their function, their interaction with other genes and to characterize the gene products. This information would assist in the genetic manipulation of nitrogen-fixing bacteria to produce improved strains for use as inoculants on crop plants.[92] The organisms studied have included *Klebsiella*, *Rhizobium*, *Azotobacter* and the cyanobacteria. Use has been made of both spontaneous and induced mutation and the genes have been mapped using the techniques of transformation, transduction, conjugation and protoplast fusion.

90 Brill, W. J. in *The Biology of Nitrogen Fixation* (ed A. Quispel), p. 639, North Holland, 1974.

91 Brill, W. J., *Sci. Am.*, 1981, **245**, 199.

92 Beringer, J. E. in *Advances in Agricultural Microbiology* (ed N. S. Subba Rao), Butterworth, 1982.

Free-living nitrogen fixers

Nitrogen-fixing micro-organisms, or diazotrophs, which do not require the presence of a living plant in either loose or intimate association can be defined as free-living. It has been established that all organisms involved in nitrogen fixation, including the symbionts *Rhizobium* and *Frankia*, can fix nitrogen to a limited extent under highly specialized artificial culture conditions, so the term free-living should perhaps take on an ecological rather than physiological connotation.

Using this restricted category, the free-living diazotrophs can be divided into heterotrophs and autotrophs and further subdivided on the basis of the aeration status of the environment in which fixation can take place. The heterotrophic bacteria can be separated into aerobes, facultative anaerobes and anaerobes.

Aerobic heterotrophic nitrogen fixation

The principal group of aerobic diazotrophs are the members of the family Azotobacteriaceae which includes the genera *Azotobacter*, *Azotococcus*, *Azomonas*, *Beijerinckia* and *Derxia*. All are large, ovoid to rod-shaped organisms, superficially similar to yeasts, and are unable to grow in the absence of oxygen. When provided with ample organic matter they will grow readily in the laboratory on nitrogen-free media. The azotobacters are typically residents of neutral to alkaline soils with a few exceptions. *Azotobacter beijerinckia* is known to occur over a wider than normal pH range along with the genus *Derxia*. The *Beijerinckia* group, once part of the *Azotobacter* genus, are common in the tropics but largely absent from temperate soils. None of the azotobacters are thermophiles although the optimum temperature for fixation in the laboratory is around 30 °C. It is possible to outline general soil conditions favourable to the *Azotobacter* group, but the organisms are not always present in soils that would appear to be suitable and often when they do occur they are present in small numbers, maybe only a few hundred to a few thousand cells per gram of soil.

As the *Azotobacter* group are obligately aerobic in their respiration requirements they are faced with the necessity of protecting the oxygen-sensitive nitrogen-fixing enzyme system. Two strategies are adopted which can be regarded as complementary. First, the group has a very high respiration rate, some 10 to 50 times that of normal aerobes on a cell weight basis. This high respiration rate, although undoubtedly wasteful in terms of carbon substrate, serves to keep oxygen tensions low at the site of nitrogen fixation, the oxygen being consumed before it can cause damage. If the oxygen tension in the surrounding medium is reduced, the organisms do not need to respire as quickly to protect the nitrogenase and their efficiency improves providing that enough oxygen remains to support respiration.[93] This process is known as respiratory protection.

A further protective mechanism, conformational protection, has also been developed by the *Azotobacter* group. Although the purified nitrogenase complex is easily damaged by oxygen, crude cell extracts contain the enzyme in a protected or aggregated form in which oxygen-sensitive sites appear to be masked by combination with a protein which is not part of the nitrogenase complex. When oxygen concentrations increase to a point where respiratory protection is inadequate, perhaps due to a lack of carbon substrate, the nitrogenase complex can be protected by association with the protein and becomes temporarily inactive.

Other aerobic bacteria exist, such as *Xanthobacter flavum* and *Azospirillum lipoferum*, which although capable of excellent growth under full aeration in the presence of combined nitrogen, can fix nitrogen only when oxygen concentrations are severely reduced but not absent. This mode of life is known as microaerophily and indicates a limited ability to protect the nitrogenase complex by means of respiratory protection.

A rather special case of free-living microaerophily is found in the rhizobia of the slow-growing 'Cowpea group'. Although these bacteria are symbiotic nitrogen fixers, nitrogen fixation has also been shown to take place under controlled culture conditions but at a very low rate[94]: an indication that there are rarely any absolute rules in soil microbiology.

Fixation by facultative anaerobes

Heterotrophic bacteria of the facultative anaerobe type are capable of good growth in either the presence or absence of air. Many are common constituents of the soil population and the group includes such genera as *Bacillus*, *Klebsiella*, *Enterobacter* and *Citrobacter* although by no means all species or even all strains of a single species fix nitrogen. The important feature of this group is that they are only capable of nitrogen fixation in the absence of oxygen, which suggests that their ability to protect the oxygen-sensitive enzyme system is very limited. In the case of *Klebsiella*, the presence of oxygen inhibits the synthesis of nitrogenase.

Anaerobic nitrogen fixation

Although the facultative group only fix nitrogen under anaerobic conditions the distinction in the case of truly anaerobic nitrogen-fixing bacteria is that they are unable to grow under any other aeration regime. Thus nitrogen fixation is potentially possible under any environmental condition that permits growth. The majority of anaerobic nitrogen fixers belong to the genus

93 Parker, C. A., *Nature, Lond.*, 1954, **173**, 780.
94 Pagan, J. D. *et al.*, *Nature, Lond.*, 1975, **256**, 406.

Clostridium, a group quite closely related to the genus *Bacillus* mentioned in the previous section. Both *Clostridium* and *Bacillus* produce very resistant spores and are thus capable of surviving adverse conditions such as extreme heat, desiccation or in the case of an obligate anaerobe, an abundance of oxygen.

The first micro-organism to be identified as being capable of nitrogen fixation, *Clostridium pasteurianum*, was isolated by Winogradsky in 1893, the same species that first yielded up the nitrogenase enzyme complex some 60 years later. Many of the clostridia are capable of nitrogen fixation although it is not a universally distributed property within the group. The bacteria are common throughout the soil, inhabiting microsites where temporary anaerobiosis may occur and are relatively easy to isolate from virtually any location.

Despite the contribution made over geological time by the free-living heterotrophic nitrogen fixers it seems clear that they make little contribution to the nitrogen economy of agricultural soils at present. They need considerable energy supplies in the form of combined carbon and many will only function when the microenvironmental conditions are propitious. Estimates of fixation for aerobes and anaerobes are of the order of 0.5 to 1 kg N ha^{-1} a^{-1}, rates that are below the inputs from rainfall. Their importance lies in the scientific information that they can yield concerning a process that is of increasing economic importance and, for the future, the possibility that they represent a wide gene pool which can be drawn upon in attempts to establish biological nitrogen fixation on a wider basis than at present.

Autotrophic nitrogen fixation

The microbiology of free-living autotrophic diazotrophs is simplified in a soil context because one of the two groups that might be considered, the photosynthetic sulphur bacteria, are essentially restricted to aquatic situations where sulphide is abundant, and apart from a minor significance in rice paddy soils at the anaerobic mud surface,[95] they can largely be disregarded.

Of far wider distribution, and potential ecological and agronomic importance, are the cyanobacteria. Because the cyanobacteria are photosynthetic they do not suffer from the shortage of energy substrate suffered by heterotrophic organisms which must compete with other soil micro-organisms for the limited amount of decomposable organic matter in the soil.

The most common cyanobacteria are those like *Nostoc* and *Anabaena* which are filamentous forms and, when fixing nitrogen, show a morphological feature that is now known to be the site of nitrogen fixation. These are cells called heterocysts which are less pigmented than normal and appear at regular intervals along the filament. If the cyanobacteria are supplied with adequate ammonium nitrogen the formation of heterocysts is suppressed. The

95 Kobayashi, M. and Haque, M. Z., *Pl. Soil*, Spec. Vol., 1971, 443.

heterocyst represents an adaptation which permits the functioning of the nitrogenase system in an organism which relies on photosynthesis and the consequent release of oxygen. By separating the oxygen-producing photosynthetic activities from the oxygen-sensitive nitrogen-fixing process the organisms can reconcile the conflicting requirements. Some 18 genera of heterocyst-forming cyanobacteria are known, the majority of which fix nitrogen in air at normal illumination levels.[96] There are also some ten genera of non-heterocyst-forming filamentous and unicellular cyanobacteria, some members of which are known to fix nitrogen. The exact nature of the protective mechanism for the nitrogenase complex in these organisms is not fully established, but it may consist of a combination of conformational protection and a temporal mechanism whereby the incompatible demands of photosynthesis and nitrogen fixation are separated in time.

The primary requirements of the cyanobacteria are those of light, moisture and a pH of about 7. Although the organisms are photosynthetic it does not follow that nitrogen fixation is greatest under high light intensities. Intense sunlight has been shown to reduce nitrogen fixation in rice fields[97] and optimum growth and fixation occur at less than 10 per cent of full incident light. The importance of plant cover in reducing light intensity and thus improving performance has also been shown to occur where cyanobacteria are functioning under a grass leaf canopy. Here the cyanobacteria are protected from excessive illumination and from desiccation. Although the most obvious environment for the cyanobacteria is in aquatic or semi-aquatic situations such as rice cultivation, they may contribute more to non-flooded soils than has been appreciated even in temperate climates. Soil drying will inactivate cyanobacteria but the majority of the organisms are well adapted to survive desiccation and respond rapidly to the return of moisture. For this reason they can make a contribution even in desert areas where the rainfall is restricted and sporadic.[98]

Although the cyanobacteria of temperate zones can tolerate a range of pH conditions, they grow best in neutral to slightly alkaline conditions. Nitrogen fixation is at a maximum at around pH 7. Cyanobacteria are present in highest numbers in alkaline soils, sparse in acid forest soils and largely absent from soils below pH 5. Table 19.1[99] shows that for temperate soils the occurrence of cyanobacteria was greatest in clay and calcareous sites.

The cyanobacteria are very common in tropical soils and can develop heavy populations in waterlogged fields. The genera found vary with geographical location but *Anabaena*, *Nostoc* and *Calothrix* are widely distributed throughout India, Japan and Southeast Asia generally. There seems to be a

96 Fogg, D. E. *et al.*, *The Blue–Green Algae*, Acad. Press, 1973.
97 Reynaud, P. A. and Roger, P. A., *Ecol. Bull. (Stockholm)*, 1978, **26**, 148.
98 Stewart, W. D. P., *Pl. Soil*, 1970, **32**, 555.
99 Granhall, U. in *Nitrogen Fixation by Free-living Micro-organisms* (ed W. D. P. Stewart), Cambridge Univ. Press, 1975.

Table 19.1 Relationship between soil type and occurrence of cyanobacteria

Soil characteristics	*No. of soils*	*Soils with cyanobacteria (%)*	*Soils with N_2 fixing cyanobacteria (%)*	*Occurrence of heterocystous filaments (%)*	*Total no. of genera*	*No. of N_2 fixing genera*
Sand	6	50	17	0	4	1
Moraine	7	57	0	0	4	0
Clay	16	100	81	31	13	7
Calcareous	15	100	93	40	10	4
Forest soils (humus)	20	25	10	0	3	1
Total	64	66	47	17	17	7

distinction between the pH tolerances of temperate and tropical cyanobacteria which parallels the distinction between *Azotobacter* and *Beijerinckia*. In the tropics, although the activity of the cyanobacteria is still highest under neutral conditions, the range of tolerated pH levels appears to be extended and significant nitrogenase activity has been reported at a pH value as low as 4.[100] The development of cyanobacteria may be significantly influenced by concentrations of nutrients such as phosphate, and by competition with other organisms including photosynthetic algae, as much as by direct effects of pH.

Despite the importance of the rice crop, with which most work on cyanobacteria is associated, the technical difficulties of field measurements of nitrogen fixation mean that relatively few quantitative evaluations of nitrogen fixation in paddy fields exist. An average of 38 field determinations[101] amounts to only 27 kg N ha^{-1} of crop with highest recorded values of 50 to 80 kg N ha^{-1} of crop. The difficulty with most field measurements is that it is virtually impossible to separate the contribution due to cyanobacteria from the fixation that is at least theoretically possible by anaerobic photosynthetic sulphur bacteria and bacteria growing in association with the rice rhizosphere.

The need for improved rice yields has directed investigations towards the inoculation of paddy fields with cyanobacterial cultures. Current developments in this area have been reviewed,[102] with reports of average grain yield increases of 15 per cent in field experiments. Higher increases of up to 28 per cent have been obtained in greenhouse pot experiments. Most field experiments are carried out on a single crop and it is likely that less than half of the nitrogen acquired by the nitrogen-fixing organisms would be transferred to the crop in such a short time. Not all experiments result in crop improvements and some of the best results are only obtained when lime, phosphate fertilizer and occasionally molybdenum supplementation are also made. In some reports an increase in grain yield was obtained with inoculation even when nitrogen fertilizer up to 150 kg N ha^{-1} was applied.[103] This effect suggests that the cyanobacteria may cause growth-promoting effects similar to those found for *Azotobacter*, which are independent of nitrogen fixation.[104]

Dramatic crop yield increases have been achieved in the rice-growing areas of the world through plant breeding and fertilizer technology, but with increasing fertilizer costs it may need intensive investigation of the role of cyanobacteria if the improvements are to be sustained at a cost that can be afforded.

100 Stewart, W. D. P. *et al.* in *Limitations and Potentials for Biological Nitrogen Fixation in the Tropics* (eds J. Döbereiner *et al.*), Plenum Press, 1978.
101 Roger, P. A. and Koolasooriya, S. A., *Blue Green Algae and Rice*, Int. Rice Res. Inst., Los Banos, Philippines, 1980.
102 Roger, P. A. and Koolasooriya, S. A., *Blue Green Algae and Rice, Int. Rice Res. Inst.*, Los Banos, Philippines, 1980.
103 Venkataraman, G. S. in *Nitrogen Fixation by Free-living Micro-organisms* (ed. W. D. P. Stewart), Cambridge Univ. Press, 1975.
104 Brown, M. E. and Burlingham, S. K., *J. gen. Microbiol.*, 1968, **53**, 135.

Associative nitrogen fixation: rhizocoenoses

The need has arisen for a category of nitrogen fixation which falls between the truly independent free-living form and the genetically based and virtually obligate symbiotic form typified by the *Rhizobium*–legume symbiosis. This is mainly because of discoveries of rhizosphere associations made since 1965. The term 'associative nitrogen fixation' can also be conveniently used to cover leaf surface diazotroph populations and also other associations between micro-organisms whereby nitrogen is fixed at rates in excess of those possible by one or both partners functioning alone. In many cases it has been the acetylene reduction assay which has permitted the investigation of these systems and in some instances has been largely responsible for their discovery.

Intensive examination of root environments for active nitrogen-fixing associations was stimulated by the work of Dobereiner who, in the early 1970s, reported on the activities of *Azotobacter paspalum* in the rhizosphere of *Paspalum notatum* cv. batatais, a cultivar of the tropical grass with poor grazing qualities. Nitrogenase assays with acetylene suggested nitrogen fixation rates, based on washed roots, approaching 90 kg N ha^{-1} a^{-1} which are comparable with *Rhizobium*–legume rates. Measurements made on intact soil cores give distinctly lower rates of fixation.[105]

It is possible that the *Azotobacter–Paspalum* grass is a more restricted association than some of the later discoveries in that the bacterium, despite ease of culture in the laboratory, is ecologically restricted to the *Paspalum* rhizosphere in nature and establishes itself poorly, if at all, in the rhizospheres of other, more palatable, cultivars and species.

In temperate regions the stimulation of nitrogen-fixing organisms in the rhizosphere of crop plants is not pronounced although broadleaved weeds at Rothamsted have shown considerable rhizosphere-linked nitrogenase activity,[106] and attention has tended to concentrate on tropical species. Dobereiner and De Polli[107] list the following categories of nitrogen-fixing root associations or rhizocoenoses:

1 *Paspalum notatum–Azotobacter paspali*
2 Sugar cane–*Beijerinckia*
3 Wheat–*Bacillus*
4 Rice–*Achromobacter*-like organisms
5 Various grasses with *Azospirillum*

More than 30 species of grass are now known to carry *Azospirillum* in their rhizospheres although not all of these can be considered to fix agronomically

105 van Berkum, P. W. B., Ph.D. Thesis, Univ. London, 1978.
106 Day, J. M. *et al.* in *Nitrogen Fixation by Free-living Micro-organisms* (ed W. D. P. Stewart), Cambridge Univ. Press, 1975.
107 Döbereiner, J. and De Polli, H. in *Nitrogen Fixation* (eds W. D. P. Stewart and J. R. Gallon), Acad. Press, 1980.

significant amounts of nitrogen. There are suggestions that the C_4 photosynthesis – pathway plants which are more common in the tropics tend to be colonized by *Azospirillum lipoferum* whereas C_3 photosynthesis plants are more commonly colonized by *Azospirillum brasilense*.

More field studies are required of the natural distribution of *Azospirillum* species and their degree of specificity in forming root associations. Some field experiments with millet suggest that significant increases in grain yield can result from the association when inoculation is carried out,[108] but other results have been more equivocal.[109] It is known that *Azospirillum brasilense* can produce plant growth hormones[110] and it is to be hoped that the investigations into *Azospirillum* inoculation prove more rewarding in the long run than the *Azotobacter* inoculation or 'bacterial fertilizer' activities of the 1950s and early 1960s.

A note of caution has been sounded regarding the amount of nitrogen fixation that might be expected in temperate regions.[111] Based on calculations which assumed that all the carbon released by cereal roots is used exclusively by known nitrogen-fixing bacteria functioning at their maximum rates, it can be estimated that only 15 per cent of the cereal nitrogen requirements could be provided by associative nitrogen fixation.

Compared to the root, the development of nitrogen-fixing populations on plant leaves would seem to offer a much simpler situation. There is little doubt, on the basis of both acetylene reduction assays and ^{15}N studies, that there can be an enrichment of nitrogen-fixing organisms associated with the aerial parts of the plant. The organisms seem most often to be enterobacter or bacilli in temperate regions but large numbers of *Azotobacter* or *Beijerinckia* are found in tropical areas. The humid tropics, where moisture is usually present on the leaf surface and photosynthesis is active, give rise to considerable carbon substrates in the leaf surface moisture layer. Leaf leachates containing up to 100 mg l^{-1} of carbohydrates have been reported,[112] and in the water trapped in the leaf sheaths of tropical grasses the level of carbohydrate can rise to 4 g l^{-1}.[113] In such situations the usual energy limitation to heterotrophic nitrogen fixation is eliminated and localized nitrogen fixation may occur. The major problem in assessing such nitrogen fixation is that it may fluctuate with time and Ruinen[114] emphasizes the difficulties of making reasonable extrapolations of the results obtained from short-term assays. A figure of 1 kg N ha^{-1} a^{-1} may be all that can be assigned to this

108 Smith, R. L. *et al.*, *Science*, 1976, **193**, 1003.
109 van Berkum, P. and Bohlool, B. B., *Microbiol. Rev.*, 1980, **44**, 419.
110 Tien, T. M. *et al.*, *Appl. environ. Microbiol.*, 1979, **37**, 1016.
111 Barber, D. A. and Lynch, J. M., *Soil Biol. Biochem.*, 1977, **9**, 305.
112 Ruinen, J., *Pl. Soil*, 1965, **22**, 375.
113 Ruinen, J., *Pl. Soil*, 1970, **33**, 661.
114 Ruinen, J. in *Nitrogen Fixation by Free-living Micro-organisms* (ed W. D. P. Stewart), Cambridge Univ. Press, 1975.

route for cultivated annuals but a higher figure may apply for natural forests.

The last form of associative nitrogen fixation to be considered is that between two types of micro-organism. The literature abounds with reports of enhanced nitrogen fixation by mixed cultures of organisms in the laboratory.[115] In many cases the stimulation can be related to reduction of partial oxygen pressure by the non-fixing member of the mixture which enhances the performance of the diazotroph. In a minority of cases the stimulation relates to the increased availability of either organic metabolites or inorganic nutrients. Although these laboratory experiments show the possibilities for microbial interaction in the soil, proof of such *in situ* association is difficult.

An example of a microbial association is currently being investigated in connection with the subject of straw disposal. A major problem is the high C : N ratio of most cereal straw, a factor which can seriously retard decomposition. A desirable combination would be an active cellulose decomposer working in harmony with a good nitrogen fixer, the first supplying simple carbon substrates to the second in return for combined nitrogen. One such combination, utilizing *Trichoderma harzianum* as the cellulytic fungus and *Clostridium butyricum* as the diazotroph, is under investigation.[116] It has the attraction of not only producing a nitrogen-enriched compost but also, through the antagonistic properties of the *Trichoderma*, of reducing the plant pathogen burden of the resulting compost.

Symbioses with cyanobacteria

Although pride of place in symbiotic nitrogen fixation must go to the *Rhizobium*–legume combination there are a number of other associations between micro-organisms and plants. Some of these already make significant contributions to agriculture or are important in natural ecosystems while others may yield scientific information which improves our knowledge of nitrogen fixation. The symbioses fall into two major categories, those involving cyanobacteria and plants and those in which the microbial partner is an actinomycete. A third and at present very limited category is the aberrant but intriguing symbiotic association between a genuine *Rhizobium* and a non-legume.

The symbioses involving cyanobacteria are essentially tropical or subtropical in distribution. The first of these is the root nodule producing symbioses between genera such as *Nostoc* and the 'seed-ferns' or cycads. The cycads are the primitive gymnosperm relatives of an ancient tropical flora and many

115 Jensen, V. and Holm, E. in *Nitrogen Fixation by Free-living Micro-organisms* (ed W. D. P. Stewart), Cambridge Univ. Press, 1975.
116 Veal, D. A. and Lynch, J. M., *Nature, Lond.*, 1984, **310**, 695.

similar plants are well known from the fossil record. The roots develop coral-like protuberances which reveal dense zones of cyanobacteria. Measurements suggest fixation rates of about 20 kg N ha^{-1} a^{-1} but the plants rarely occur in sufficient density to approach this figure in the field. It is likely that the significance of the cycads to the nitrogen budget of soils lies in the distant past. Of similar status is the restricted example of cyanobacteria forming an association with an angiosperm. In this case the plant is *Gunnera*, a genus with about 40 species in which, again, *Nostoc* forms somewhat unusual 'glands' at the base of the leaf. Within these glands nitrogen is fixed and made available to the plant. Because of the large amount of photosynthetic activity in the close vicinity of the 'glands', *Nostoc* in this situation has an unusually high density of heterocysts. Estimates of nitrogen-fixing activity suggest rates of about 30 kg N ha^{-1} a^{-1} but, as with the cycads, the sporadic distribution of the *Gunnera* plants would mean much lower overall rates in natural communities.

The most significant cyanobacteria symbiosis, the value of which has been exploited for centuries, is that between the tiny water-fern *Azolla* and a cyanobacterium of the genus *Anabaena*. In terms of nitrogen-fixing potential, this symbiosis can be regarded as the paddy field equivalent of the temperate legume, and under conditions of multiple harvest may well exceed the best that can be achieved with legumes. The water-fern is only 2 to 3 mm in diameter and has overlapping lobed leaves. It reproduces mainly by vegetative propagation although a sexual stage exists. The cyanobacteria, a type of *Anabaena* which is difficult to grow in isolation, inhabit cavities in the upper lobes of the *Azolla* leaf structure. High numbers of heterocysts are found in the *Anabaena* cells, presumably to counteract oxygen inhibition of the nitrogenase although the cyanobacteria are known to photosynthesize. It is not clear whether the water-fern contributes energy or reductants to the nitrogen fixation process.

The cyanobacteria are thought to excrete nitrogen as ammonium which the water-fern converts to amino acids. A particularly valuable feature of the *Azolla* symbiosis is its apparent insensitivity to ambient mineral nitrogen levels.

The potential significance of the *Azolla–Anabaena* association can be realized from a report[117] in which a number of field experiments yielded nitrogen fixation rates of 1 to 2.6 kg N ha^{-1} day^{-1}. Where the water-fern is harvested from the water surface regularly and forced to recolonize the amounts of nitrogen fixed per annum run into several hundred kg N ha^{-1}.

The value of *Azolla* has been known for a long time in China and Southeast Asia. It can be grown as a wetland green manure and incorporated with the soil where its rapid decomposition makes it as effective as mineral fertilizers.

117 Watanabe, I. in *Microbiology of Tropical Soils and Plant Productivity* (eds Y. Dommergues and H. G. Dien), Martinus Nijhoff, 1982.

It is also encouraged to grow at the same time as the rice crop although there are prejudices against this in some areas. It has also been harvested and used on dryland crops. Other possibilities are its direct use as animal feed, although it has a low digestibility for animals other than ruminants and may need amino acid supplementation to balance some deficiencies in an otherwise high protein material. The nitrogen-fixing power of the *Azolla–Anabaena* symbiosis is not in doubt, but future developments will lie in its more efficient management and the control of the fungal and insect damage to which it is prone.[118]

Actinomycete symbioses

Nodular nitrogen-fixing symbioses formed by actinomycetes of the genus *Frankia*, for which the term 'actinorhizas' has been suggested, are known to occur in at least 175 plant species and the number can confidently be expected to continue rising for some time. At present some 8 families and 17 genera are represented.[119] The plants involved comprise mainly perennial herbs together with shrubs and trees, annual herbs being poorly represented. Their distribution is widespread in nitrogen-poor habitats in temperate regions and although they occur in warmer climates, most tropical and subtropical actinorhizas are found at higher altitudes. A feature of the ecology of actinorhizal plants is the more frequent occurrence of acid tolerance than is found in the legumes. Alder trees (*Alnus glutinosa*) are capable of growth at pH 4.5 to 5.5 and Bog Myrtle (*Myrica gale*) is a colonizer of acid peat soils with pH values of 3.5 to 5.

In a number of the plant families, e.g. Casuarinaceae, Coriariaceae, Eleagnaceae and Myricaceae, all the species found are nodulated whereas in other families such as the Betulaceae only species of *Alnus* are known to carry nodules on their roots. The Rosaceae have a scatter of genera in which isolated species are found to be nodulated, including members of the genera *Dryas*, *Cercocarpus*, *Purshia* and *Rubus*. The last genus is of interest because it has a large number of species and yet only one, *Rubus ellipticus*, has so far been found which is nodulated. A further unique feature of this plant is that it is virtually the only plant with an actinorhizal symbiosis to have been used in agriculture or horticulture, and it was grown at one time in the United States as a yellow fruited raspberry. One of the main reasons for the lack of attention paid to the actinorhizal symbiosis in the past is undoubtedly the absence of agriculturally exploitable crops, although the value of trees such as the alder has been recognized in forestry.

118 Liu, C. C. in *Nitrogen and Rice*, Int. Rice Res. Inst., Los Banos, Philippines, 1979.
119 Becking, J. H. in *Advances in Agricultural Microbiology* (ed N. S. Subba Rao), Butterworths, 1982.

A serious obstacle to the study of actinorhizal symbiosis until recently has been the inability to grow the microbial symbiont in pure culture. This has meant that inoculation experiments were limited to using crushed nodule material and the detailed identification of the organism was impossible. Since 1978 a number of successful isolations have been made ·from, among others, *Alnus*, *Eleagnus*, *Myrica* and *Comptonia*. Evidence that *Frankia* can grow independently immediately raises the question of whether it has a non-symbiotic phase in the soil but no information is currently available.

The taxonomic position of the species of *Frankia* is at present unsatisfactory as most species are recognized on the basis of the host plant from which they are obtained. Attempts at establishing host ranges by cross-inoculation have been made which indicate that most *Frankia* species have a modest infection ability beyond their immediate host, but definitive work must await further pure culture isolations.

It is now virtually certain that the biochemistry of nitrogen fixation within the actinorhizas is similar to that in other diazotrophs. Problems have been encountered in purifying the dinitrogenase and dinitrogenase reductase due to the high levels of polyphenols encountered in the nodules. Less is known about the internal organization and physiology of actinorhiza nodules than for the *Rhizobium* nodule, and the exact means by which the nitrogenase in the actinorhizal nodule is protected from oxygen is unknown. The localization of reducing power into highly active vesicles, which can be demonstrated by histochemical means, may indicate a morphological protection. Some form of protection is necessary since it has been shown that aerobic respiration within the nodule must take place.

Given that the host plants are perennial trees and shrubs growing in mixed natural vegetation it is not surprising that firm estimates of the amount of nitrogen fixed are hard to obtain. Comparison of actinorhizal nodules with *Rhizobium* nodules using acetylene reduction suggests that they are of similar nitrogen-fixing ability, weight for weight. Estimates have been based on measurement of acetylene reduction for nodules and then extrapolated on the basis of the amount of nodule tissue per unit of soil surface area. A number of assumptions of activity and seasonal variation need to be made and at present the best estimates for *Alnus* species suggest rates between 50 and 200 kg N ha^{-1} a^{-1}, and from negligible to 50 kg N ha^{-1} a^{-1} for other hosts that have been examined.[120] The nitrogen-fixing potential of the *Alnus–Frankia* partnership make it attractive for improving poor soils and for recovering degraded soils or spoil heaps.

120 Akkermans, A. D. L. and van Dijk, C. in *Nitrogen Fixation*, Vol. 1 (ed W. J. Broughton), Clarendon Press, 1981.

A rhizobium–non-legume symbiosis

The last non-legume symbiosis to be described has attracted attention out of all proportion to its frequency of occurrence or ecological significance. In 1973 a *Rhizobium*, which subsequently was shown to behave as a member of the slow-growing cowpea group, was isolated from a plant identified as *Parasponia rugosa*.[121] The initial investigations were hampered because of the close similarity of *Parasponia rugosa* to a plant of wider distribution called *Trema cannabina* which has not yet been shown to form nodules. *Parasponia rugosa* and two other species, *P. parviflora* and *P. andersonii*, are all natives of Southeast Asia and occur as weeds, in tea plantations. Although the rhizobia isolated from *Parasponia* can infect the legume *Vigna*, the form of the nodules on the non-legume is quite different from those on the legume. Firstly, the bacteria within *Parasponia* nodules are contained within highly branched infection threads[122] and are rarely released to form bacteroids. Secondly, the nodules are devoid of haemoglobin[123] which is present in effective legume nodules. Apart from undermining the long-held view that the *Rhizobium*–legume symbiosis is exclusive, the nature of the *Parasponia* nodule also suggests that bacteroid formation and haemoglobin production are a feature of the particular host and are not essential for nitrogen fixation. Although of no immediate practical value the *Rhizobium*–*Parasponia* symbiosis may add greatly to our understanding of the relationship between *Rhizobium* and its host.

Symbiotic nitrogen fixation by leguminous plants

No other area in soil microbiology has received the attention, over the past 100 years of the subject's history, as the nitrogen-fixing symbiosis that exists between plants of the Leguminosae and the heterotrophic bacteria known generally as the rhizobia. The reasons for the effort expended in investigating this symbiosis are four-fold:

1 The family Leguminosae is highly successful and widespread in both tropical and temperate regions.
2 It contains many useful crop and forage plants that have been cultivated for centuries.
3 The symbiosis can be very effective at fixing nitrogen and produces crops of high nutritive value.

121 Trinick, M. J., *Nature, Lond.*, 1973, **244**, 459.
122 Trinick, M. J., *Canad. J. Microbiol.*, 1979, **25**, 565.
123 Coventry, D. R., Trinick, M. J. and Appleby, C. A., *Biochim. Biophys. Acta*, 1976, **420**, 105.

4 The micro-organism responsible is readily isolated from nitrogen-fixing legumes and easily cultured in the laboratory which facilitates selection of improved strains and inoculum production.

The Leguminosae is a large family of plants with some 12 000 species recorded and is subdivided into subfamilies of which there are three main ones:

1 The Papilionaceae, which is the largest subfamily, contains most of the well-known plants of agriculture including beans, peas, lentils, clover and many tropical grains such as chick-peas. Of the Papilionaceae examined, some 80 to 90 per cent are found to form nodules.
2 The Mimosaceae subfamily, containing such plants as the mimosas and wattles, are less nodulated and only about 25 per cent of the species appear to form a nitrogen-fixing symbiosis.
3 Finally, the subfamily Caesalpinaceae contains few nodule-forming species and is notable for just a few plants of economic value such as senna and tamarind.

Large numbers of species have yet to be collected and studied for nodulation and nitrogen fixation. Although the Leguminosae are a large family, only about 50 species have been cultivated and less than 30 of these are cropped to any significant extent.[124]

The microbial partner in the symbiosis, the rhizobia, present a number of problems of classification. In simple terms they are small Gram-negative heterotrophic bacteria capable of producing nodule structures on leguminous plants which may result in nitrogen fixation. Young cells are frequently motile with varying arrangements of flagella.

On yeast-extract–mannitol agar fast-growing rhizobia (mean generation time typically 2 to 4 h) isolated from clovers, peas and beans produce copious quantities of extracellular gum and give rise to characteristic raised glistening colonies in culture plates. Some fast-growing strains, such as those from lucerne, tend to produce less gum. The slow-growing strains, which include many of the tropical rhizobia (mean generation time from 6 to 8 h), form smaller colonies with much less gum. Rhizobia, in common with all micro-organisms, can change by mutation, and repeated plating and re-isolation from isolated colonies rather than general growth is not recommended. Production of acid or alkali during growth on media has been used for broad identification purposes, the fast-growing organisms showing a tendency to produce acid metabolites while the slow-growing tropical types tending towards alkalization of the medium. This change in medium pH has been invoked to explain ecological preferences of rhizobia but since the by-prod-

124 A detailed account of the *Leguminosae* including geographical distribution and nodu-lation records is given by Lim, G. and Burton, J.C. in *Nitrogen Fixation*, Vol. 2 (ed W. J. Broughton), Clarendon Press, 1982.

ucts of metabolism are heavily influenced by the constitution of the medium this approach is now less acceptable than it was.

For many years the rhizobia group has been divided into species based on the cross-inoculation group principle which relies on the assumption that legumes within a particular infection group will be nodulated by one species of nodule bacterium.[125] With repeated evidence of aberrant infections between groups, the organization of the rhizobia into species has been altered in the 1984 edition of *Bergey's Manual of Systematic Bacteriology*.[126] The nodule bacteria are now divided into two genera; *Rhizobium*, which are now distinguished as the bacteria which cause nodule formation on plants predominantly of the temperate zone and which show fast growth on yeast-extract–mannitol media; and a new genus, *Bradyrhizobium*, which grow slowly on yeast-extract–mannitol media and are more important in causing nodules on tropical zone plants although some temperate groups are also infected. The genus *Rhizobium* has three species, *R. leguminosarum*, *R. meliloti* and *R. loti*. Within the species *R. leguminosarum* three biovars (varieties) are recognized namely *viceae, trifolii* and *phaseoli*. The genus *Bradyrhizobium* is at present restricted to a single species, *B. japonicum*, with no biovars at present recognized. It seems inevitable that there will be further modifications in the classification and taxonomy of the rhizobia as more legumes are examined and already the occurrence of fast-growing isolates from tropical plants such as soyabean raises the possibility of transitional groups. Although the previous system based on cross-inoculation groups was known to have drawbacks and inconsistencies, it is not immediately apparent that the new classification is significantly better or clearer. Table 19.2 shows in a simplified form the relation between the new and old classification.

Our understanding of the *Rhizobium*–legume symbiosis, from the nature of the infection specificity to the biochemistry of the nitrogen-fixing process and the energy economics within the plant, has advanced considerably since 1970. A detailed account of the plant–micro-organism relationship is given in Chapter 17. The expression of the symbiosis in the field in terms of plant growth and nitrogen gain is also complex and many questions remain to be answered. Nutman[127] states that 'Over the last 60 years a significant part of the world's primary production has depended on the large changes brought about in the microbial population of soils by the spread of legume crops and by the practice of inoculation . . . the greatest success has normally been with alien plants and alien bacteria.' The potential nitrogen fixation of the *Rhizobium*–legume partnership is very considerable and is only approached by

125 Fred, E. B., Baldwin, I. L. and McCoy, E., *Root Nodule Bacteria and Leguminous Plants*, Univ. Wisconsin Studies in Science, Vol. 5, Madison, 1932.
126 Jordan, D. C. in *Bergey's Manual of Systematic Bacteriology*, Vol. 1 (eds J. G. Holt and N. R. Krieg), p. 234, Williams and Wilkins, 1984.
127 Nutman, P. S. in *Soil Microbiology* (ed. N. Walker), Butterworths, 1975.

Table 19.2   Comparison of old and new *Rhizobium* classification

Old species	Principal plant nodulated	New species
R. leguminosarum	Pisum (pea)	R. leguminosarum
	Lens (lentil)	
	Vicia (vetch)	biovar viceae
R. phaseoli	Phaseolus (bean)	biovar phaseoli
R. trifolii	Trifolium (clover)	biovar trifolii
R. meliloti	Melilotus (sweet clover)	R. meliloti
	Medicago (lucerne)	
	Trigonella (fenugreek)	
R. lupini	Lupinus (lupin)	Bradyrhizobium sp.
	Lotus (trefoils)	R. loti
R. japonicum	Glycine (soyabean)	Bradyrhizobium japonicum
Tropical or cowpea miscellany	Vigna (cowpea)	Bradyrhizobium sp.

that of the *Azolla–Anabaena* association. Nitrogen fixation rates have been quoted for the soyabean *Glycine max* as up to 369 kg N ha^{-1}, the cowpea *Vigna unguiculata* as up to 240 kg N ha^{-1}, *Centrosema pubescens* as up to 395 kg N ha^{-1} and *Sesbania cannabina* as up to 542 kg N ha^{-1}, all measurements being in terms of fixation per crop.[128] They define the potential symbiotic nitrogen fixation of a given legume as the maximum activity when nodulated with the most effective *Rhizobium* strain and grown under the most favourable conditions. Despite the high figures quoted, in reality nitrogen fixation rates may only be 10 to 15 per cent of the potential as judged at present.

The achievement of high nitrogen fixation rates depends on a very complex mixture of interactions including the plant, the *Rhizobium* strain, climatic factors for both infection and plant growth and soil factors such as ambient soil mineral nitrogen concentrations, soil acidity, calcium and molybdenum deficiency and aluminium and manganese toxicity. The effects of environmental factors on the *Rhizobium*–legume symbiosis are dealt with in Chapter 17.

Legumes have been used as green manures and ground cover crops in plantations of tea, cocoa and rubber for well over 50 years, although the plants employed are often native 'weed' species and relatively little has been done to improve their effectiveness compared with the effort expended on crop species. Development of these cover crops may bring considerable benefits as interest moves towards sustainable systems of agriculture which are less dependent on fertilizer inputs. A few legumes, such as *Crotolaria* and

128 Gibson, A. H. *et al.*, in *Microbiology of Tropical Soils and Plant Productivity* (eds Y. Dommergues and H. G. Dien), Martinus Nijhoff, 1982.

Sesbania are of note since they possess the unusual capability of tolerating waterlogged conditions and have been used in rice-growing areas of southern India for green manure purposes resulting in nitrogen enrichments of up to 50 kg N ha^{-1} a^{-1}.

Rhizobium in the soil

Although small numbers of rhizobia may occur as natural contaminants of the seed surface, the usual source of the rhizobia which infect legumes will be the soil unless steps are taken to introduce inoculants. For this reason a large number of surveys have been conducted to establish the normal levels of rhizobia existing in soils.[129] Surveys of rhizobia present certain technical problems as no suitable culture medium exists for their selective isolation and reliance has to be placed on legume infection. Soil dilution assays are conducted whereby the soil is progressively diluted with a growing medium devoid of rhizobia such as sand-vermiculite, and the test legume is introduced as a sterile seedling.[130] In a soil dilution containing at least one *Rhizobium* cell, the micro-organisms can multiply in the rhizosphere and infect the plant which is then examined for nodules. The procedure amounts to a 'most probable number' assay and needs considerable replication if the population estimates are to have any degree of precision. Where known populations of rhizobia have been added to soils, the recoveries suggest that the soil dilution procedure is generally acceptable for *trifolii* biovars and strains from lupins but less good for *R. leguminosarum* and *R. meliloti*.

For the identification of a particular strain of *Rhizobium*, rather than the population in general, there are more sophisticated techniques available. If antibodies can be raised for a particular strain, it may be identified serologically once isolated. The use of fluorescent-labelled antibodies allows direct observation of soil for the occurrence of the test organism although non-specific adsorption of the labelled antibody by soil can be a problem and the fluorescent label is most useful in identifying rhizobia in excised nodules. The enzyme-linked immunoadsorbent assay (ELISA) has been shown to offer some useful possibilities with rhizobia.[131]

Production of antibiotic-resistant mutant strains is useful because the organism can be selectively isolated on media enriched with the antibiotics to which resistance has been acquired. Resistance to one or two antibiotics can be acquired quite readily by *Rhizobium* as single step processes and where double resistance occurs it is very unlikely that natural soil organisms will have the same resistance pattern. It is important that the resistance-

129 Nutman, P. S. in *Soil Microbiology* (ed N. Walker), Butterworth, 1975.
130 Brockwell, J. in *Methods for Evaluating Biological Nitrogen Fixation* (ed. F. J. Bergerson), p. 417, Wiley, 1980.
131 Nambiar, P. T .C. and Anjaiah, V., *J. appl. Bact.*, 1985, **58**, 187.

labelled rhizobia retain their infectivity and effectiveness and this should always be checked. Of the vast range of antibiotics available, streptomycin[132] and spectinomycin[133] appear to cause the least loss of desirable characteristics.

A broad general rule governing rhizobial populations in soils.is that if the host plant is normally found in the location, the rhizobia infecting that host will be present too. The size of the rhizobial population will depend on how recently the host plant was a member of the plant community and in what proportion. Populations of rhizobia in the region of 10^4 to 10^6 per gram of soil appear to be adequate for nodulation. If the host is removed, the rhizobial population will usually decline, the rate of decline depending on how well the rhizobia can tolerate the soil conditions. Where the host plant has never been a natural inhabitant of the area, for example in the case of soyabeans in the Cerrado soils of Brazil, or where continuous cereal growing or long fallows are practised, as in the classical arable fields at Rothamsted, *Rhizobium* numbers may be very low, amounting to a few hundred cells per gram of soil if present at all.[134]

For the rhizobia of the temperate regions, survival in the absence of the host plant appears to depend on soil pH to a large extent and higher numbers persist in limed soils than in those with lower pH levels. After pH the next most important factor appears to be soil moisture content; rhizobia have relatively poor abilities to withstand desiccation and tend to die out more rapidly in sandy soils low in organic matter, than in soils with a high clay or organic content.[135]

Differences between *Rhizobium* types are known to exist and the slow-growing rhizobia typical of the tropics are inherently more tolerant of both acid conditions and desiccation.[136] A feature of acid soils which is still not fully understood is the relative importance of pH and aluminium toxicity (see also Ch. 17). Many organisms seem able to tolerate low pH and high levels of aluminium *in vitro* but suffer difficulties in nodulation. Since the infection process involves a degree of 'recognition' by the plant, in which lectins are now thought to be involved, it may be that pH or aluminium interfere at this crucial stage. A further complication with acid tropical soils is their ability to bind rhizobia very tightly to their surfaces which can result in poor recoveries of added rhizobia under experimental conditions.[137]

The effects of the surface charge of various soil constituents such as clay minerals, iron oxides and organic matter have been discussed[138] in relation

132 Schwinghamer, E. A., *Ant. van Leewenhoek.*, 1967, **33**, 121.
133 Schwinghamer, E. A. and Dudman, W. F., *J. appl. Bact.*, 1973, **36**, 263.
134 Nutman, P. S. and Ross, G. J. S., *Rothamsted Ann. Rept. for 1967*, Part 2, p. 148.
135 Pena-Cabriales, J. J. and Alexander, M., *Soil Sci. Soc. Am. J.*, 1979, **43**, 962; Osa-Afiana, L. O. and Alexander, M., *Soil Sci. Soc. Am. J.*, 1979, **43**, 925.
136 Bushby, H. V. A. and Marshall, K. C., *Soil Biol. Biochem.*, 1977, **9**, 143.
137 Kingsley, M. T. and Bohlool, B. B., *Appl. environ. Microbiol.*, 1981, **42**. 241.
138 Bushby, H. V. A. in *Nitrogen Fixation*, Vol. 2 (ed W. J. Broughton), Clarendon Press, 1982.

to rhizobial growth and heat resistance but no account has been made of the straightforward immobilization of cells. Aluminium at high concentrations may flocculate *Rhizobium* cells and reduce their availability for infection and nodulation. It is known that the poor nodulation of some tropical crops following inoculation results from the apparent inability of the introduced *Rhizobium* to migrate significantly beyond the point of inoculation and cause nodule formation over the root system as a whole. In addition to selecting rhizobia in the laboratory for tolerance to acidity and high aluminium levels it will also be profitable to seek strains that are less tightly bound to soil surfaces.

The importance of salinity to rhizobia in the field is as yet unclear. It is known, for example, that *Rhizobium meliloti* is more tolerant of salinity than *Bradyrhizobium japonicum* and can grow in media containing 340 mM sodium chloride. However, most experiments on rhizobial tolerance to salinity have been conducted in cultures in the laboratory. The significance of the salinity tolerance of the microbial partner may only be secondary if, as has been suggested,[139] the primary effect of salinity is on the host plant. Finally, the report that strains of rhizobia from *Sesbania cannabina*, growing in a saline–alkaline soil, have more tolerance to saline–alkaline conditions than isolates from normal soils suggests that localized adaptation may be just as important in the ecology of rhizobia as for many other micro-organisms.

Other factors which have been investigated with respect to the survival of rhizobia in soils have been biological interactions with other soil micro-organisms and the possible effects of pesticides. Many biological factors could affect the rhizobial population in the soil including antagonisms due to anti-bacterial metabolites and influences such as predation by protozoa or attack by bacteriophages. The outcome of a number of microbial antagonism studies using soil bacteria and fungi suggests that rhizobia are no more susceptible to biological antagonism than any other group of soil heterotrophs. In the case of protozoan predation and bacteriophage attack there is no evidence to suggest that poor *Rhizobium* performance in the field can be ascribed to either of these influences.[140] The only situations where problems from bacteriophage action have been reported are in pot cultures where very high rhizobial populations have been introduced.[141]

Although many tests have been conducted on rhizobia in laboratory cultures to determine their sensitivity to fungicides, herbicides and insecticides, the situation in the soil is so different due to adsorption phenomena and other factors, that it is difficult to predict field behaviour from culture study results. Furthermore, in the field it is often impossible to separate effects on the micro-organisms from those on either the host itself or the symbiosis.

139 Wilson, J., *Aust. J. agric. Res.*, 1970, **21**, 571.
140 Habte, M. and Alexander, M., *Soil Biol. Biochem.*, 1978, **10**, 1.
141 Kleczkowska, J., *Pl. Soil*, 1972, **37**, 425.

Inoculation of legumes with rhizobia

An important aspect of the successful exploitation of the *Rhizobium*–legume symbiosis has been the opportunity afforded for introduction of effective rhizobial strains into the soil. This is usually carried out at seed planting time with the aim of ensuring adequate nitrogen-fixing nodules in the crop.

Successful inoculation depends on the selection of the correct rhizobial strains and their provision in high enough numbers to colonize the developing root system. Inoculation is usually adopted for two different reasons and the choice of strains will vary accordingly. In the first situation, such as when lucerne was first introduced to Great Britain, or when soyabeans were grown in the Mid-West of the United States, the appropriate rhizobia were absent or present in numbers too small to nodulate effectively. A similar situation existed when subterranean clover was introduced to Australia. As long as the rhizobia can tolerate the soil conditions, its choice can be based very largely on its nitrogen-fixing powers in symbiotic combination.

A different, and more difficult problem exists for many legume crops in tropical areas. Here, the soil often contains abundant rhizobia capable of nodulating the crop but the indigenous rhizobia are often of low nitrogen-fixing effectiveness. The normal criteria for *Rhizobium* strain selection are first that it should be infective, i.e. able to enter the plant and form a nodule, and second, it should be effective, i.e. the nodules should be able to fix as much nitrogen as possible. Where there are large indigenous populations of rhizobia with poor effectiveness, it is necessary to add a further criterion to strain selection, that of competitiveness. The latter characteristic is very difficult to define and may vary from soil to soil. There is no correlation between effectiveness and ability to compete with other strains. Beringer and Johnston[142] state that 'competitiveness reflects that particular strain in that particular soil with that particular plant in that particular year'. If any factor is altered the competitiveness may be lessened. Simply increasing the number of inoculant rhizobia is not a guarantee of success, as a good competitive strain, however ineffective, may be able to overcome a numerical disadvantage of 1000 to 1. At least part of the competition phenomenon is related to the pre-infection stage of rapid rhizobial development in the rhizosphere.

Although very useful pointers can be gained from laboratory and greenhouse trials, the eventual performance of an inoculant must be established from field trials in a variety of soils over a number of seasons. Some insurance against poor rhizobial performance in specific situations can be gained by using inoculants which comprise a mixture of strains. Where mixed strain inocula are used the only precaution needed is that the strains are not mutually antagonistic as some rhizobia are known to produce bacteriocins,

142 Beringer, J. E. and Johnston, A. W. B. in *Isotopes in Biological Dinitrogen Fixation*, Proc. Advisory Group Meeting, IAEA, Vienna, 1979.

which are protein-like compounds which can destroy closely related organisms.[143]

Inoculant rhizobia are usually introduced with the seed either as a coating to the seed, known as pelleting, or as a finely divided peat or compost which is drilled along with the seed. In the first method the rhizobia are incorporated with calcium carbonate or finely divided rock phosphate and a binding agent such as gum arabic is used to form a pellet around the seed. The calcium carbonate or rock phosphate is more than just an inert carrier and can bring about a useful localized amelioration to soil conditions for the developing seedling.

Where a peat-based inoculant is produced, calcium carbonate might be needed to raise the pH to around 7. The peat is then sterilized. The rhizobia are grown in a medium containing yeast extract and sucrose with forced aeration until a viable count of about 10^9 ml^{-1} is achieved. The culture is sprayed on to the peat and allowed to mature for 48 to 72 hours before packaging. If the final product is not too moist and is stored at 4 °C it will maintain a high viability for up to 12 months. High temperatures are a major reason for poor inoculant viability and can cause plasmid loss and thus loss of infectivity even in the organisms which survive.[144] The poor performance of inoculants may result as often from inadequacy of inoculant storage facilities as from inappropriate choice of *Rhizobium* strains. In this context, wet inoculants are much more easily damaged than those with a low moisture content.[145] Further developments are needed to overcome the viability problems of inoculants and new materials such as polyacrylamide gels are under investigation.[146] Ideally, pre-inoculated seed with a long shelf-life under a variety of storage conditions would satisfy most agronomic requirements but this ideal has yet to be produced.[147]

Despite the developments that have taken place in the understanding of the legume symbiosis and its exploitation for human needs since the initial isolation of the *Rhizobium* organism by Beijerinck in 1888, there is little doubt that interest will continue unabated. Increased interest in tropical legumes and the need to conserve resources used for fertilizer production will provide an economic stimulus to further research.

143 Schwinghamer, E. A. *et al.*, *Canad. J. Microbiol.*, 1973, **19**, 359.
144 Zurkowski, W. and Lorkiewicz, Z., *Arch. für Microbiol.*, 1979, **123**, 195.
145 Bushby, H. V. A. and Marshall, K. C., *Soil Biol. Biochem.*, 1977, **9**, 143.
146 Dommergues, Y. *et al.*, *Appl. environ. Microbiol.*, 1979, **37**, 779.
147 An excellent review of *Rhizobium* technology and inoculant production is given by Burton, J. C., *Microbial Technology*, Vol. 1, Acad. Press, 1979.

20

Plant nutrients in soil: nitrogen[1]

The top 15 cm of soil profiles of temperate regions usually contains between 0.1 and 0.3 per cent chemically combined nitrogen.[2] With a soil of bulk density of 1.3 g cm^{-3} this represents between 2 and 6 tonnes of N per hectare. Hence the whole soil profile may contain 3 to 20 t N ha^{-1} although under permanent grassland and woodland the amount can be greater and in peat profiles it can be much greater. In contrast, soils of arid regions generally contain less than 0.1 per cent N in the top 15 cm and may contain less than 0.02 per cent N.

Unless the soil profile contains a horizon of eluviated organic matter, ploughed-in organic matter, or a buried A horizon, the per cent N decreases

---

1 For reviews see *Soil Nitrogen* (eds W. V. Bartholomew and F. E. Clark), *Agron. Monog.*, No. 10, Am. Soc. Agron., Madison, 1965; *Nitrogen in Agricultural Soils* (ed F. J. Stevenson), *Agron. Monog.*, No. **22**, Am. Soc. Agron., Madison, 1982; Haynes, R. J., *Mineral Nitrogen in the Plant-Soil System*, Acad. Press, 1986.

2 For examples of total soil N, see Parsons, J. W. and Tinsley, J. in *Soil Components*, Vol. 1 (ed J. E. Gieseking), p. 263, Springer-Verlag, 1975.

with depth. In topsoils more than 90 to 95 per cent of the total nitrogen usually occurs in organic compounds, the rest being present as nitrate and ammonium ions, some of the latter being held within clay minerals. The ratio of per cent organic C to per cent organic N in the organic matter is in the range of 10 to 14 in most arable and grassland topsoils (the ratio is often wider in acid soils and peats, but is rarely narrower). The amount of total nitrogen therefore correlates with the organic carbon content, and a change in the content of organic matter in the soil, whether by addition or loss, is usually accompanied by a change in the content of N. Most of the nitrogen in soil organic matter is present in high molecular weight polymers as amino acids, amino sugars and in heterocyclic ring structures. The chemistry of these compounds is discussed in Chapter 18.

Sources of nitrogen

The combined nitrogen in primary rocks, has been given as 10–12 μg g^{-1}, corresponding to about 0.04 t ha^{-1} for a thickness of 15 cm at bulk density 2.6 g cm^{-3}, and is present as NH$_4^+$ within the structure of the rock silicates.[3] Under natural conditions, additions to soil are through biological fixation of dinitrogen, and wet and dry deposition of nitrogen compounds from the atmosphere. The rates of addition can be high. For example, the nitrogen content of newly vegetated mudflows in California increased by an average of 63 kg ha^{-1} a^{-1} over 60 years,[4] and at Rothamsted Experimental Station by 49 kg ha^{-1} a^{-1} for 81 years on land which had been allowed to revert to woodland after being under arable cultivation for many years.[5] If the soil is not disturbed by man, the nitrogen content increases towards an equilibrim value where rates of additions and losses are in balance. The equilibrium value depends on climate, parent material and topography. Soils that are at equilibrium occur under old woodland or grassland, for example the prairie soils of North America before they were brought into cultivation.

The most important natural process for increasing the nitrogen content of soils is biological fixation of dinitrogen by micro-organisms which are of two types: those that are free living and those that fix dinitrogen symbiotically in root nodules. A good clover or lucerne crop will increase the organic nitrogen in the soil by over 100 kg ha^{-1} a^{-1} through the activity of *Rhizobium*. Symbiotic nitrogen fixation is also brought about by actinomycetes in root nodules on certain trees such as alder, a process that is important in the

3 Stevenson, F. J. in *Soil Nitrogen* (eds W. V. Bartholomew and F. E. Clark), *Agron. Monog.*, No. 10, 1965, 1 (see ref. 1).
4 Dickson, B. A. and Crocker, R. L., *J. Soil Sci.*, 1953, **4**, 142.
5 Jenkinson, D. S., *Rothamsted Ann. Rept. for 1976*, Part 2, p. 103.

colonization of land after it has been freed from ice-sheets and of land left by the retreat of the sea. Nitrogen fixation also takes place on the leaves of some plants. These processes are discussed in Chapters 17 and 19.

Soils also gain small amounts of nitrogen as ammonium and nitrate from the rain and by dry deposition. The annual amounts are usually less than 5 kg N ha^{-1} in parts of the world remote from industrial and urban centres. At Rothamsted Experimental Station, the annual addition was 4 to 5 kg N ha^{-1} from 1888 to 1913 but had increased to about 20 kg N ha^{-1} by 1983,[6] due to the production of oxides of nitrogen, and to a less extent of ammonia, from the combustion of fossil fuels, and ammonia from fertilizer use and livestock production systems. Soils can also gain nitrogen by absorption of ammonia from the atmosphere, and a value of 4 kg N ha^{-1} a^{-1} has been measured at Rothamsted.[7] The amounts may be greater where atmospheric concentrations are higher, for example near feedlots, grazed pastures and industrial centres. Additions of organic nitrogen compounds in rainfall may also occur, but their origin is uncertain. The additions of nitrogen in rainfall and by dry deposition are of particular importance in stable ecosystems and the increase during the last few decades as a result of burning fossil fuels and increased use of fertilizer nitrogen might have long-term ecological effects. Ammonia and presumably ammonium ions in the atmosphere can be used by plants after deposition on the foliage and will later be returned to the soil as organic nitrogen compounds.

An example of the increase in nitrogen on recessional ice moraines in southeast Alaska is shown in Fig. 20.1. After an initally slow increase in nitrogen content when the moraine was becoming colonized by mosses and species of *Epilobium*, *Equisetum* and *Dryas*, there was a more rapid increase when the site was colonized by *Alnus* (alder). During this phase of the ecological succession there was an average annual addition of 49 kg N ha^{-1} to the soil (to 46 cm depth) and the forest floor. Most of the increase in soil nitrogen occurred near the surface but there was also an increase at depths to 46 cm. An interesting observation is that during the alder phase, when there was rapid fixation of nitrogen, the soil pH fell from 8.0 to 5.0, even though the parent material contained calcium carbonate.

Natural colonization of mining waste by vegetation has also provided information on the rate of accession of nitrogen. China clay waste in southwest England colonized by leguminous shrubs (gorse, *Ulex* spp.; broom, *Sarothamnus scoparius* and tree lupin, *Lupinus arboreus*) gained nitrogen much more rapidly than sites containing no legumes. Where legumes were

6 Goulding, K. W. T. and Poulton, P. R., *Soil Use Management*, 1985, **1**, 6;   Eriksson, E., *Tellus*, 1952, **4**, 215 reviewed early data. See also Stevenson, C. M., *Quart. J. R. Meteorol. Soc.*, 1968, **94**, 56;   Jones, M. J. and Bromfield, A. R., *Nature, Lond.*, 1970, **227**, 86 for Nigerian data.
7 Rodgers, G. A., *J. agric. Sci.*, 1978, **90**, 537.

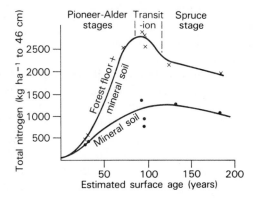

Fig. 20.1 Accumulation of nitrogen in soils after the recession of ice moraines in southeast Alaska. (From Crocker, R. L. and Major, J., *J. Ecol.*, 1955, **43**, 427.)

dominant,[8] the average annual rate of nitrogen accumulation to 27 cm depth was as high as 124 kg N ha^{-1}. From the evidence at this site and elsewhere, Bradshaw has argued that in order to meet the nitrogen requirement of the vegetation through mineralization, a soil nitrogen 'capital' of about 700 kg ha^{-1} is needed to support *Salix* scrub and about 1200 kg ha^{-1} to support woodland.[9]

Gains and losses under cultivation

When soil is brought into cultivation the nitrogen content usually decreases. This is mainly a consequence of increased rates of oxidation of the soil organic matter. Perhaps the best known example is the decrease following the breaking up of the prairie lands of North America. In Saskatchewan, where the rainfall is sufficiently low for there to be very little drainage from the soil, the nitrogen content of the top 22.5 cm of soil fell from 0.37 to 0.25 per cent in the first 22 years of cropping, an average loss of 112 kg ha^{-1} a^{-1} of which 36 kg ha^{-1} a^{-1} was recovered in the crop[10]; the difference, 76 kg N ha^{-1} a^{-1}, must have been lost mainly in gaseous forms. In subsequent years, the annual loss was less and almost all could be accounted for by the nitrogen in the crop. In Kansas the nitrogen content of the soil (0–18 cm) decreased from about 0.23 to 0.13 per cent after 30 years' continuous culti-

8 Dancer, W. S. *et al.*, *Pl. Soil*, 1977, **48**, 153.
9 Bradshaw, A. D., *et al.*, *Phil. Trans. R. Soc. Lond.*, 1982, **B296**, 557.
10 Shutt, F. J., *J. agric. Sci.*, 1910, **3**, 335.

vation, but fell only to 0.12 per cent after cultivation for a further 30 years.[11] Under continuous cereals with no fertilizer nitrogen the uptake by the crops exceeded the loss of soil nitrogen, indicating accession of nitrogen to the soil–crop system from the atmosphere.

There is usually a gain of soil nitrogen if the land receives frequent applications of organic manures or is put down to grass or becomes woodland. Fertilizer alone has little or no effect, as shown by the analyses of soils from Broadbalk field under wheat at Rothamsted given in Table 20.1. The small increases compared with the no manure plot can be attributed to additions of carbon and nitrogen through increased root growth. Where carbon and nitrogen were added in the farmyard manure, the increase in soil nitrogen was very much greater, and from analysis of the rate of increase it was predicted that it had almost reached its equilibrium value for this treatment of about 6.9 t N ha^{-1}.

Table 20.1   Total N in topsoils (0–23 cm), Broadbalk, Rothamsted (t N ha^{-1}). (From *Rothamsted Ann. Rept. for 1976*, Part 2, 103)

Year of sampling	No manure	FYM	PKMg	N_1PKMg	N_2PKMg	N_3PKMg
1881	3.06	5.35	2.97	3.35	3.67	3.81
1966	2.88	6.85	3.11	3.29	3.35	3.43

Treatments started in 1843 and were, annually: FYM 35 t ha^{-1} (about 225 kg N); PKMg, 35 kg P, 90 kg K, 10 kg Mg ha^{-1}; N_1, N_2 and N_3, 48, 96 and 144 kg N ha^{-1}. Sampling depth on the FYM plot was adjusted for decreased bulk density.

The effects of putting old arable land down to grass were shown in Fig. 18.10. Over 200 years the total N content in the top 23 cm of soil increased from 0.11 to 0.25 per cent, and it was estimated that about half the increase occurred in the first 25 years. When a part of Broadbalk (which had long been under arable cultivation) was allowed to develop a mixed herbaceous vegetation, tree seedlings being cut down (stubbed), it showed a mean gain in the soil (0–23 cm) of 49 kg ha^{-1} a^{-1} over 81 years (Fig. 20.2a). On a part that was allowed to become woodland the mean gain was 45 kg ha^{-1} a^{-1}. By contrast, after forest clearance in southern Nigeria the content of N in the top 10 cm of soil fell from 0.17 to 0.12 per cent in 22 months, with an estimated half-life of 3.3 years (Fig. 20.2b). At this site, 622 kg ha^{-1} of N was mineralized over the 22 month period from the top 10 cm of soil.

The gains and losses of nitrogen are not constant with time but change exponentially. They have been described by equations such as $dN/dt = A - rN$ or by its integrated form:

11 Hobbs, J. A. and Brown, P. L., *Kansas Agric. Exp. Stn. Tech. Bull.*, No. 144, 1965. For other examples see F. J. Stevenson in *Nitrogen in Agricultural Soils* (ed. F. J. Stevenson), *Agron. Monog.*, No. 22, 1, Am. Soc. Agron, Madison, 1982.

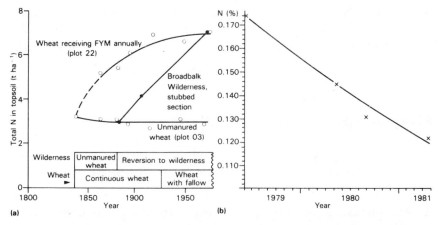

Fig. 20.2 Nitrogen contents of soils: (a) changes in the top 23 cm of soil at Rothamsted on plots which are unmanured, receive farmyard manure or have been allowed to revert from arable to grassland, (b) decrease in top 10 cm of soil under arable cultivation in southern Nigeria after forest clearance. (From (a) Jenkinson, D. S., Rothamsted Rept. for 1976, Part 2, 1977, p. 103, and (b) Mueller–Harvey, I. *et al.*, 1985, **36**, 585.)

$$rN = A - (A - rN_o)e^{-rt}$$

where N is the mass of nitrogen in the soil, A is the annual addition of nitrogen, r is the fraction of nitrogen that is mineralized each year and N_o is the initial nitrogen content of the soil. After a period of years the nitrogen content of the soil will approach the equilibrium value A/r asymptotically. After a period of $0.693/r$ years the nitrogen content will be half-way between its initial and equilibrium values. This period is called the half-life of soil nitrogen.[12] (See p. 605 for discussion of half-life of nitrogen turnover.)

For newly cultivated soil, r is generally in the range 0.01 to 0.03 (1 to 3 per cent) but becomes lower with time; it is higher with recently added plant material. More complex equations which incorporate more than one decomposition constant are referred to by Campbell.[12]

The amounts of nitrogen gained by the soil or lost from it can be very large and they are of economic importance to agriculture throughout the world. Losses as nitrate and as nitrous oxide can also have adverse effects on the environment.[13] Except for the loss of soil by erosion, losses occur only after mineralization of the organic nitrogen compounds to ammonium and nitrate,

12 See Campbell, C. A. in *Soil Organic Matter* (eds M. Schnitzer and S. U. Khan), p. 173, 1978. Greenland, D. J., *Soils Fertil.*, 1971, **34**, 237 gives an application of the equation to agronomic practices in South Australia.
13 Magee, P. N., *Phil. Trans. R. Soc. Lond.*, 1982, **B296**, 543 discusses potential hazards to health from nitrogen oxides and nitrate. See also Bryson, D. D., *Fertil. Soc. Proc.*, No. 228, 1984.

which behave in soil similarly to the same forms when added in fertilizers. It is therefore convenient to describe the processes of loss irrespective of the source of the nitrogen. This will be done after discussing the mineral nitrogen content of soil.

Mineral nitrogen content of soils[14]

Mineralization of organic nitrogen compounds is brought about by micro-organisms to meet their energy and nutritional requirements. As discussed in Chapter 19 there are two steps:

1   Ammonification, whereby organic nitrogen compounds are hydrolysed to NH_4^+, a reaction catalysed by enzymes from many heterotrophic micro-organisms; the NH_4^+ ions are held in the soil solution and also on exchange sites on clays and organic matter where they are oxidized by:

2   Nitrification, (a) $NH_4^+ + 1\frac{1}{2}O_2 \rightarrow NO_2^- + H_2O + 2H^+$
   carried out mainly by a small number of autotrophic bacterial genera of which *Nitrosomonas* occurs most commonly, and

   (b) $NO_2^- + \frac{1}{2}O_2 \rightarrow NO_3^-,,$
   carried out mainly by a single autotrophic bacterial genus (*Nitrobacter*).

The step which limits the rate of mineralization is usually ammonification, even though the enzymes required (proteinases, peptidases, amino acid dehydrogenases, amino acid oxidases, etc.) can be provided by a large proportion of the microbial population of the soil. The limitation appears to be because of the chemical recalcitrance of the organic nitrogen compounds in soil, but limitations of carbon substrates and steric hindrance to enzyme activity may also be factors. One of the effects of earthworms is to mix dead plant material more intimately with the soil, and it has been observed that the mineral nitrogen content of surface casts is higher than in the underlying soil.[15] Because oxidation of NO_2^- is more rapid than that of NH_4^+, NO_2^- seldom accumulates in soil. As a result of the relative rates of these reactions, nitrate is usually the most important source of nitrogen for arable crops even though they can use ammonium and some low molecular weight organic N compounds.

On land which is kept under a bare fallow, where there is no crop uptake

14 Determination of mineral (inorganic) forms of nitrogen is described by Keeney, D. R. and Nelson, D. W. in *Methods of Soil Analysis*, Part 2 (ed A. L. Page) p. 643, Monog. No. 9 (2nd edn), Am. Soc. Agron., Madison 1982.
15 Syers, J. K., *et al.*, *Soil Biol. Biochem.*, 1979, **11**, 181;   Syers, J. K. and Springett, J. A., *Pl. Soil*, 1984, **76**, 93.

of nitrogen, nitrate accumulates in the soil. Four conditions are needed for the accumulation of nitrate under fallow: there must be decomposable organic matter in the soil; the fallow must be kept free from weeds; there must not be too much rain otherwise nitrate is leached out of the soil or lost to the atmosphere through denitrification; and the soil must be moist or subject to alternate wetting and drying.

Fallowing in seasonal rainfall areas

In regions having alternating dry and wet seasons, ammonification can proceed during the dry season when the soil is too dry for nitrification. At the start of the rains the ammonium which has accumulated nitrifies.[16] The alternate wetting and drying which occur during the early rains also cause more rapid oxidation of the soil organic nitrogen compounds than in the rainy season itself. If crops are to benefit from the initial flush of nitrate they must be planted as early as possible, taking advantage of any showers that fall before the main rains. In this way they can start growing quickly when the rains start and use this nitrate before it is leached out of the soil. During a summer fallow, frequent ploughing, a traditional practice in some arid countries, controls weeds which would otherwise take up nitrate. It also dries a greater volume of soil between rain showers and would be expected to increase the nitrate content of the soil by a wetting and drying effect. During a long dry season, upward movement of the soil solution may increase the concentration of nitrate near the soil surface.[17]

Effect of cultivations

Aside from the effects on a bare fallow, cultivations lead generally to an increase of soil mineral nitrogen, most probably because soil disturbance leads to more carbon substrates becoming available to support greater microbial activity. This effect has agronomic importance in that crops which are grown after direct drilling (no primary cultivation) require more fertilizer nitrogen than those grown after conventional cultivation techniques. For example, in a field experiment on a clay soil in southern England growing winter wheat, the nitrate concentrations in the soil solution were as shown in Fig. 20.3. The crop had been sown in October 1972 after either ploughing to a depth of 20 cm or direct drilling. Fertilizer nitrogen was applied at

16 Semb, G and Robinson, J. B. D., *E. Afr. agric. For. J.*, 1969, **34**, 350. See Dommergues, Y. *et al., in Nitrogen Cycling in West African Ecosystems* (ed T. Rosswall) SCOPE/UNEP, 1980 for relationship between ammonification, nitrification and soil water potential. For earlier work on wetting/drying effects see Birch, H. F., *Pl. Soil*, 1958, **10**, 9; 1959, **11**, 262; 1960, **12**, 81; *idem, Trop. Agric. (Trin.)*, 1960, **37**, 3.

17 Simpson, J. R., *J. Soil Sci.*, 1960, **11**, 45; Wetselaar, R., *Pl. Soil*, 1961, **15**, 110, 121; Stephens, D., *J. Soil Sci.*, 1962, **13**, 52; Robinson, J. B. D. and Gacoka, P., *J. Soil Sci.*, 1962, **13**, 133.

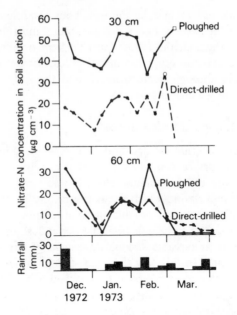

Fig. 20.3 Changes in mean nitrate-N content of samples of soil solution obtained from an Oxford clay soil which had been ploughed or direct-drilled. Open symbols indicate that the mean was calculated from less than four samples. (From Dowdell, R. J. and Cannell, R. Q., *J. Soil Sci.*, 1975, **26**, 53.)

40 kg N ha^{-1} with the seed in both treatments. The nitrate concentration in the ploughed plots was generally between two and four times more than that in the direct drilled plots. Although the seasonal changes are not entirely explicable, leaching loss and denitrification were believed to be small, and the differences were attributed mainly to mineralization, with a small part due to crop uptake. The nitrate concentrations throughout the whole profile

Table 20.2 Yield of winter wheat (t ha^{-1}) after a one-year fallow; Broadbalk 1970–78; plots also received P, K, Mg and Na. (From *Rothamsted Ann. Rept. for 1982*, Part 2, 5)

Nitrogen manuring (kg N ha^{-1}) as ammonium sulphate	Years after fallow		Continuous wheat
	First	Second	
0	3.56	1.53	1.78
48	4.60	3.57	3.52
96	5.10	5.27	4.84
144	4.96	5.66	5.32

(to 80 cm) at the end of the winter showed effects of the cultivation treatments similar to those illustrated for the 30 and 60 cm depths.

The effect of a fallow on subsequent crop yields is illustrated in Table 20.2 which gives yields of winter wheat on Broadbalk, Rothamsted. The data show that with continuous cropping 48 kg N ha^{-1} was required to raise the yield to that obtained in the year after fallow. The effect of fallow was small for the second crop. Results in Table 20.3[18] indicate that some of the nitrate accumulated in the subsoil between 25 and 70 cm, although some was probably leached out of the root zone. In the heavy soils of the Palouse area of Washington State, nitrate produced during the summer fallow also accumulated in the subsoil between 60 and 150 cm during winter and was used by the following wheat crop.[19]

Table 20.3 Subsoil storage of nitrate over winter on Broadbalk, Rothamsted, 1935–36. Amount of (nitrate + ammonium) nitrogen in the soil in μg g^{-1}. Section III fallowed 1934–35, first crop after fallow; section II, fourth crop after fallow; soils sampled in November, April and June

Depth of sampling	November Section		April Section		June Section	
(cm)	III	II	III	II	III	II
Plot 2B (37 t farmyard manure ha^{-1})						
0–25	12	7	7	7	7	6
25–45	16	11	17	8	4	5
45–69	20	10	14	11	3	4
Plot 3 No manure						
0–25	5	5	5	3	4	3
25–45	10	5	6	8	5	4
45–69	9	4	9	6	4	2
Plot 8 145 kg N ha^{-1} as ammonium sulphate 1/4 in October, 3/4 in May						
0–25	10	8	6	6	6	4
25–45	12	10	7	6	5	3
45–69	13	7	11	8	4	2

Effect of crops

Cropping reduces the ammonium and nitrate concentrations in soil through plant uptake, but there is evidence of an additional effect. This is illustrated

18 Gasser, J. K. R., *Pl. Soil*, 1962, **17**, 209.
19 Sievers, F. J. and Holtz, H. F., *Washington Agric. Expt. Stn. Bull.*, No. 166, 1922.

Table 20.4   Nitrate-N in soil, and nitrogen in crop (wheat), in cropped and fallow soil at Rothamsted. (From Russell, E. J., *J. agric. Sci.*, 1914, **6**, 18; and with Appleyard, A., op. cit., 1917, **8**, 385)

Nitrogen (kg ha^{-1})	Hoosfield		Broadbalk		
	June 1911	July 1912	Unmanured July 1915	Dunged 26 July 1915	17 Aug 1915
In top 45 cm of fallow soil	61	51	78	139	168
In top 45 cm of cropped soil	17	15	49	41	77
In crop	26	7	23	78	78
Total	43	22	72	119	155
Deficit in cropped land	18	29	6	20	13
	(mg N kg^{-1} soil)				
Fallow, 0–22.5 cm depth	12	8			
22.5–45 cm depth	9	10			
Cropped, 0–22.5 cm depth	4	2			
22.5–45 cm depth	2	3			

Cropping. Hoosfield: Alternate wheat and fallow, unmanured since 1851.
    Broadbalk: Continuous wheat, but fallow taken over part of this field for the third time since 1843.

by the analyses in Table 20.4 which show that there is more nitrate-N in the fallow soil than the total of nitrate-N in the cropped soil plus nitrogen in the crop. Although the roots and stubble were not included in the analyses, the deficits are too great for nitrogen in these parts to be the whole explanation. It is possible that some of the soil mineral nitrogen was immobilized by micro-organisms which use carbon compounds from the root as a source of energy.

In soils under cut grass, concentrations of nitrate are usually lower than under arable crops and ammonium concentrations may be higher. As Richardson[20] showed, the ammonium-N concentration in some soils at Rothamsted remained fairly constant between 3 and 9 μg g^{-1} throughout the year, yet if an ammonium or nitrate fertilizer was added this mineral nitrogen soon disappeared from the soil, presumably by uptake by the grass. As discussed elsewhere (p. 620), one possible explanation is that inhibition of nitrification occurs under a grass sward.

20 Richardson, H. L., *J. agric. Sci.*, 1938, **28**, 73.

Immobilization

During the decomposition in soil of plant materials with wide C : N ratios, mineral nitrogen concentrations may decrease because of uptake by micro-organisms. Neither the N content of the material nor the C : N ratio is a safe guide to the effect on the mineral nitrogen content of the soil, although there is usually a loss of mineral nitrogen, at least temporarily, if the material contains less than about 1.8 per cent N, corresponding to a C : N ratio of about 30.[21] The effect may last for several weeks, during which period micro-bial respiration results in the loss of carbon as carbon dioxide and a narrowing of the C : N ratio. As the microbial cells die, or act as the food source for predators, some of the nitrogen within them is mineralized. During the first period there is net immobilization and during the second there is net mineralization. These changes (mineralization/immobilization) have been discussed in Chapter 19.

Ammonium fixation

Ammonium ions (ionic radius 0.143 nm) are similar in size to potassium ions (ionic radius 0.133 nm) and can be similarly held as non-exchangeable (fixed) ions within clay minerals. Soils that contain the clay minerals hydrous mica and vermiculite can fix added ammonium ions, and the amount fixed is increased if the soil is subsequently dried. Nitrification rates may be decreased[22] but a lessening of availability of the ammonium ions to plants is generally short-lived.[23] The difference between the effects on nitrification and plant uptake probably arises because plants take up potassium, thereby allowing the interlayer spaces of the clay minerals to expand and the ammonium ions to diffuse more rapidly into the soil solution.

Soils often contain a proportion of their total nitrogen in the form of fixed ammonium. This fraction, 'native' fixed ammonium, is small in most surface soils but may constitute up to half the nitrogen present in the subsoil.[24] It is measured by partly dissolving the clay in hydrofluoric–hydrochloric acid mixture after the organic matter has been extracted with alkali.[25] Most of this fixed ammonium is not accessible to nitrifying bacteria or to plant roots.

21 Bartholomew, W. V. in *Soil Nitrogen* (eds W. V. Bartholomew and F. E. Clark), p. 285, 1965 (see ref. 1).
22 Broadbent, F. E., *Soil Sci. Soc. Am. Proc.*, 1965, **29**, 292.
23 Nommik, H. and Vahtras, K. in *Nitrogen in Agricultural Soils* (ed F. E. Stevenson), p. 123, 1982 (see ref. 1).
24 Rodrigues, G., *J. Soil Sci.*, 1954, **5**, 264.
25 Bremner, J. M., *et al.*, *Soil Sci. Soc. Am. Proc.*, 1967, **31**, 466.

Losses of nitrogen from soils

Erosion can cause the loss of all, or part of, a soil and hence of the soil nitrogen (see Ch. 13 for an account of erosion). Nitrogen can also be lost from the soil by one or more of the following processes:

1 By plant uptake of mineral nitrogen. The uptake generally increases with crop yield, but because the main part of the requirement of legumes for nitrogen can be met by symbiotic fixation their uptake from the soil is less than that of non-legumes. It should also be noted that not all the nitrogen taken up by plants is permanently removed from the soil. It may be returned as excreta directly from the grazing animal or from farmyard manure or slurry[26]; and with all crops, grass in particular, some of the nitrogen is retained in roots in the soil, and some is returned in leaf fall, stubble, and excretion from the roots. Hence some of the nitrogen taken up by plants enters a soil/crop/soil cycle, which might include the grazing animal (and also man if sewage is returned to the land).
2 By leaching, primarily of nitrate.
3 By gaseous loss as ammonia, dinitrogen and nitrous oxide.

There are as yet few results from experiments in which all losses have been measured simultaneously. This is because of the practical difficulties of measuring gaseous losses under field conditions, and of accounting accurately for nitrate which is leached. A compromise with field conditions is to use monolith lysimeters which are wide cores of soil in its natural structured state. As used by the former Letcombe Laboratory[27] these cores of soil, 80 cm diameter and 135 cm deep, are contained in glassfibre cylinders, to the base of which a small suction is applied to simulate field drainage conditions. Drainage water can be collected without loss and then analysed for nitrate and other ions. Loss of nitrogen as nitrous oxide can be estimated by covering the monoliths for short periods and measuring the concentration of nitrous oxide within the cover. A further advantage of lysimeters is that they have a defined soil volume which leads to an accurate estimate of fertilizer nitrogen retained in the soil if this is labelled with ^{15}N.

Table 20.5 shows results from an experiment at Letcombe with lysimeters containing a silt loam overlying Chalk. Each year the lysimeters received fertilizer at a rate equivalent to 80 or 120 kg N ha^{-1}, applied in the first year as calcium nitrate labelled with ^{15}N at an enrichment of 30 atoms per cent, and in the following three years as unlabelled ammonium nitrate. About half the applied ^{15}N-labelled fertilizer was recovered in the four crops of spring barley, a little over one-quarter was retained in the roots and soil, and about 6 per cent was lost in drainage water. The rest was not accounted for, but

26 See Ryden, J. C., *Fertil. Soc. Proc.*, No. 229, 1984 for a comprehensive review.
27 Belford, R. K., *J. Soil Sci.*, 1979, **30**, 363.

Table 20.5 Balance sheet over four years for applied ^{15}N-labelled fertilizer (kg N ha^{-1} and percentage of quantity applied); crop: spring barley on silt loam over Chalk (Andover Series). (From Dowdell, R. J. and Webster, C. P., *J. Soil Sci.*, 1984, **35**, 183)

	Treatment							
	B		C		D		E	
	kg	%	kg	%	kg	%	kg	%
Input								
N applied	80	100	120	100	120	100	120	100
Output								
Recovered in:								
grain and straw	42	52	59	49	69	57	59	50
roots and soil	22	28	30	25	28	23	31	26
drainage water	5	7	8	6	8	7	8	7
unaccounted for	11	13	23	19	16	13	22	18

In first year, N applied to all treatments as calcium nitrate (^{15}N-labelled) and in subsequent years as ammonium nitrate (not labelled); in treatment D the lysimeters were protected from rainfall in April and May, and in treatment E the lysimeters received additional water.

separate laboratory measurements showed that 10–20 per cent of the ^{15}N could have been lost as dinitrogen gas, which was not measured. The high proportion of the applied ^{15}N that was retained in the soil plus roots is of particular interest and is referred to on p. 687. A loss of 6 to 10 per cent of the ^{15}N applied to cut grass was recorded[28] after taking account of the ^{15}N removed in herbage, that left in the soil, and the ^{15}N lost by leaching and denitrification. The lysimeters contained a sandy loam over clay, and received ammonium nitrate supplying 250, 500 and 900 kg N ha^{-1} a^{-1} in split applications, the fertilizer being labelled in the first year only. As an example of the results, the per cent recovery of ^{15}N after three years from the 250 kg N ha^{-1} application labelled on the NH_4^+–N was 43.8, 0.2 and 40.5 in the herbage, leachate and soil, respectively. Denitrification was estimated to account for 6 per cent of the fertilizer nitrogen. The total recovery of 90.5 per cent might have been an underestimate by 2 per cent for analytical reasons. Ammonia was not measured but the loss was probably small because the swards were not grazed and fertilizer nitrogen was applied as ammonium nitrate.

28 Barraclough, D. *et al.*, *J. Soil Sci.*, 1984, **35**, 191; ibid 1985, **36**, 593 and personal communication.

Nitrate leaching[29]

Of the various combined forms of nitrogen only nitrate is washed out of soil in appreciable amounts by percolating water. Ammonium ions are held by cation exchange on clay and humus and, except from sands, are only displaced by application of a salt solution; most soil organic N compounds have a very low solubility. Nitrate ions, on the other hand, are held in the soil solution because they are not adsorbed by most soils,[30] and there are no simple insoluble nitrates. The solution is displaced downwards by rainfall or irrigation and, if sufficient water is added, the dissolved ions such as nitrate that it contains will be displaced below the root zone. The depth of leaching of nitrate is therefore determined, at least in part, by the amount of water passing through the soil, which can be calculated from the components of the water balance equation:

$$Q = W_p - W_r - W_{ev} - W_{tr} - \Delta W$$

where Q is the amount of through-flow, and the other terms are respectively precipitation (plus irrigation), run-off, evaporation from soil, transpiration from crops, and the increase in water content of the soil. Each of the components on the right-hand side of the equation affects the amount of through-flow. The amount of nitrate that is lost also depends on its concentration in the soil solution throughout the soil profile, and on the ability of plant roots to absorb it before it is leached below the limit of the root zone. The solution containing nitrate which passes below the root zone may percolate through porous rocks into an aquifer, or move laterally to rivers through agricultural drains or through a system of naturally occurring channels and fissures.[31]

Effect of soil texture and structure

Soil texture and structure affect the amount of water retained at field capacity (p. 353). If this water content is expressed as a fraction, θ, of the soil volume, and it is assumed that all the incoming water displaces the resident solution, the downward movement is Q/θ. Q is either measured or calculated from the water balance equation, see above.[32]. Thus, nitrate is more readily washed

29 For reviews of nitrate leaching see Wild, A. and Cameron, K. C. in *Soils and Agriculture (ed P. B. Tinker), 1980, 35; Cooke, G. W. in Agriculture and Water Quality, MAFF Tech. Bull.*, No. 32, 5, 1976.

30 See Singh, B. R. and Kanehiro, Y., *Soil Sci. Soc. Am. Proc.*, 1969, **33**, 681; Kinjo, T. and Pratt, P. F., *op. cit.*, 1971, **35**, 722, 725, 728. For examples of adsorption of nitrate by some acid soils in the tropics, see Pleysier, J. L. and Juo, A. S. R., *Soil Sci Soc. Am. J.*, 1981, **45**, 754.

31 Young, C. P. describes water and nitrate transport to aquifers based on borehole analyses in *Water Research Centre Tech. Rept.*, 31, 1976 (with Hall, E. S. and Oakes, D. B.) and *op. cit.*, 69 (with Grey, E. M.).

from a sandy soil which has a value of θ of about 0.1 than from a loam which has a value of 0.35. During the winter months in countries such as Britain, run-off, evaporation and transpiration are low and ΔW tends towards zero when the water deficit developed during the preceding summer has been made good. Under these conditions Q is approximately equal to the rainfall; hence for each centimetre of rainfall nitrate will be displaced downwards by about 10 cm in a sandy soil and 3 cm in a loam.

Effect of macropores

In reality, nitrate movement is more complex because: (i) differences of nitrate concentration within the profile cause diffusive dispersion; and (ii) different rates of water flow within the soil cause hydrodynamic dispersion. Together, they have the effect of spreading out the front between the resident soil solution and the displacing rainfall. In many soils in the field there are cracks and biopores which can be several millimetres across and through which rain-water flows quickly and carries nitrate (and other ions) with it.[33] Tillage, by breaking the continuity of these channels, results in less rapid infiltration and less deep movement of nitrate on land which is ploughed rather than untilled.[34] For the same reason intense autumn rainstorms leach some nitrate rapidly from the soil, as happened after the summer drought in Europe of 1976,[35] but bypass the nitrate held in the soil matrix.

One consequence of the presence of conducting channels is that there is a marked lateral variation of leaching depth. In the absence of such channels, light rainfall may equilibrate with the soil solution, and Burns[36] has used this to give a simple and useful model for nitrate leaching. The model has been modified to include fast flow through channels.[37] Accurate prediction of the rate of leaching of nitrate (or of any ion) through a soil profile is, however, not yet possible. This is because, for a given input of water (through-flow, as referred to above), the leaching rate on many soils depends to a considerable extent on the size and number of transmission pores, and these cannot easily be measured.

32 Rousselle, V., *Ann. Sci. Agron.*, 1913, 4th series, 97.
33 Bouma, J. and Anderson, J. L., *Soil Sci. Soc. Am. J.*, 1977, **41**, 766. Omoti, U. and Wild, A., *Soil Sci.*, 1979, **128**, 98. Smettem, K. R. J., *et al.*, *J. Soil Sci.*, 1983, **34**, 499.
34 McMahon, M. A. and Thomas, G. W., *Agron. J.*, 1976, **68**, 437; Goss, M. J. *et al.*, *J. Soil Sci.*, 1978, **29**, 475.
35 Garwood, E. A. and Tyson, K. C., *J. agric. Sci.*, 1977, **89**, 767.
36 Burns, I. G., *J. Soil Sci.*, 1974, **25**, 165; *J. agric. Sci.*, 1976, **86**, 305. For leaching during crop growth see *J. Soil Sci.*, 1980, **31**, 155, 175, 187.
37 Addiscott, T. M., *J. Soil Sci.*, 1977, **28**, 554; Addiscott, T. M., *et al.*, *op. cit.*, 1978, **29**, 305.

A solution to the problem proposed by Jury[38] is to attribute the dispersion down the profile to a range of water flow velocities. His method is to calculate the mean and variance of the velocities from 15 measurements of the distribution of the solute concentrations at one soil depth, say 30 cm, after a known water input. The mean and variance of the water flow velocities at this depth are then used to calculate the distribution of solute concentrations at a greater depth after a further input of water. A vertically uniform distribution of flow velocities has to be assumed. Under these conditions the method gives good agreement with measured distributions.

Effect of cultivations and fertilizer use

Much nitrate is lost when land is cultivated after being under grass or forest. For example a leaching loss of about 200 kg N ha^{-1} over two years was measured following the cultivation of 3–5-year-old leys for winter wheat.[39]

The effect of rate of application of fertilizer N is shown by experiments at Jealotts Hill Research Station (Table 20.6). Several applications of fertilizer (ammonium nitrate) were applied annually to isolated 0.4 ha grassland plots on a sandy loam overlying impermeable London Clay. Depending on the season, the grass was cut and removed 3, 4 or 5 times, and the volume of drainage water and its nitrate concentration were measured. The average amount of NO_3-N lost by leaching over the three years represented 1.5, 5.4 and 16.7 per cent of the fertilizer applied at 250, 500 and 900 kg N ha^{-1} a^{-1}. In experiments conducted at the same time in monolith lysimeters on an adjoining site the total amounts of nitrate in the leachate were similar. In the lysimeter experiments the fertilizer was labelled with ^{15}N and at application rates of 250, 500 and 900 kg N ha^{-1}, the fertilizer nitrogen represented 9, 39 and 75 per cent of the total nitrogen in the leachate. As the application of fertilizer nitrogen increased so did its leaching loss, and so also did the leaching loss of soil nitrogen. For example, in 1979/80 2.0, 3.9 and 39 kg N ha^{-1} from the soil were leached at the three application rates.

Mineralization/immobilization turnover, discussed on p. 687, is probably the reason for the effect of the fertilizer application on the leaching loss of soil nitrogen. It should, however, be noted that because of the return of nitrogen in excreta, the leaching loss from soil under grazed grass is very much higher than from soil under cut grass: in one experiment it was 5.6 times higher than from comparable cut grass and was higher than has normally been observed from arable land.[40]

38 Jury, A., *Water Resources Res.*, 1982, **18**, 363; with Stolzy, L. H., *op. cit.*, 369. See also White, R. E. *et al.*, *J. Soil Sci.*, 1984, **35**, 159;   White, R. E., *J. Hydrol.*, 1985, **79**, 21, 37.
39 Cameron, K. C. and Wild, A., *J. environ. Qual.*, 1984, **13**, 274.   For loss following deforestation see Bormann, F. H. *et al.*, *Ecol. Monog.*, 1974, **44**, 255.
40 Ryden, J. C. *et al.*, *Nature, Lond.*, 1984, **311**, 50.

Table 20.6 Loss of nitrate into drains from grassland receiving ammonium nitrate fertilizer. (From Barraclough, D. *et al.*, *J. Soil Sci.*, 1983, **34**, 483. See also Barraclough, D. *et al.*, *op. cit.*, 1984, **35**, 191.)

Year	N rate (kg N ha⁻¹)	Drainage flow (mm)	Total NO₃-N leached (kg N ha⁻¹)	Mean NO₃-N concentration (mg l⁻¹)	Range (mg l⁻¹)	Per cent of flow with NO₃-N> 11.3 mg l⁻¹	Per cent of flow with NO₃-N> 22.6 mg l⁻¹
1978–79	250	235.3	4.7	2.0	0.2–12.4	9	0
	500	235.6	54.0	22.9	0.5–57.2	86	35
	900	230.7	152.2	66.0	27.7–127.4	100	100
1979–80	250	163.0	6.3	3.9	0.1–13.3	8	0
	500	160.0	18.6	11.6	0.9–16.2	68	0
	900	158.7	155.8	98.2	1.3–142.9	100	100
1980–81	250	157.9	0.47	0.3	0.1–0.8	0	0
	500	156.6	7.9	5.1	0.9–8.9	0	0
	900	154.5	144.1	93.2	0.5–148.6	98	98

No measurements of comparable reliability have yet been reported under annual grain crops, but reviews of experimental results indicate higher losses than from cut grass, often in the range of 5 to 20 per cent of the fertilizer nitrogen applied.[41] The main reasons for this higher loss are the earlier cessation of nitrogen uptake and transpiration by grain crops compared with grass. In temperate regions, net mineralization often occurs after senescence of the flag leaf of cereals, and is accelerated by post-harvest cultivations. Little nitrogen is taken up by an autumn-sown crop and the nitrate left in the soil is subject to leaching by winter rain.

Gaseous loss of nitrogen

Gaseous losses of nitrogen from soils occur mainly as dinitrogen (N_2), nitrous oxide (N_2O) and ammonia (NH_3). Losses have also been reported as nitric oxide (NO), which are usually attributed to the chemical decomposition of nitrite under acid conditions.[42] The evidence at present is that nitric oxide losses are usually negligible at pH values above 5.0 to 5.5. Gaseous losses have also been shown to occur from plants, the greatest loss occurring between anthesis and maturity, probably as ammonia, nitrogen oxides and possibly as dinitrogen and other volatile nitrogen compounds.[43] These gaseous losses may account partly for the fall in the nitrogen content of crop plants that is commonly observed between anthesis and final harvest. Its contribution to the total loss of nitrogen from crop plants under field conditions is not yet known. Nitrous oxide and ammonia are water-soluble gases and the direction of the flux between the soil–plant system and the atmosphere depends on the partial pressure of the gas in the atmosphere and its concentration in the soil solution. Net gains or losses of nitrogen may therefore occur.

Ammonia volatilization[44]

Ammonia is lost under conditions which can be predicted from the equations:

$$NH_4^+ \overset{K_a}{\rightleftharpoons} H^+ + NH_3(s) \qquad [20.1]$$

$$NH_3(s) \overset{K_H}{\rightleftharpoons} NH_3(g) \qquad [20.2]$$

41 Cooke, G. W., *MAFF Tech. Bull.* No. 32, 5, HMSO, 1976.
42 Fillery, I. R. P. in *Gaseous Loss of Nitrogen from Plant–Soil Systems* (eds J. R. Freney and J. R. Simpson), p. 33, Nijhoff, 1983; Galbally, I. E. and Roy, C. R., *Nature, Lond.*, 1978, **275**, 734.
43 Wetselaar, R. and Farquhar, G. D., *Adv. Agron.*, 1980, **33**, 263.
44 For reviews see Freney, J. R. *et al.*, in *Gaseous Loss of Nitrogen from Plant–Soil Systems* (eds J. R. Freney and J. R. Simpson), p. 1 Nijhoff, 1983; Terman, G. L., *Adv. Agron.*, 1979, **31**, 189.

where $K_a = \dfrac{[NH_3][H^+]}{[NH_4^+]}$ and K_H (the Henry coefficient) $= \dfrac{p[NH_3]g}{[NH_3]s}$

p being the partial pressure of the gas, $[NH_3]_g$, and $[NH_3]_s$ the concentration of NH_3 in solution. It will be seen from equations [20.1] and [20.2] that volatilization of NH_3 from an ammonium source will leave residual acidity.

At pH 7 about 1 per cent of the ammonia/ammonium in solution is present as NH_3, but the percentage increases rapidly as the pH is increased, as shown by equation [20.1]. The partial pressure of NH_3 gas depends on the concentration of NH_3 in solution; it increases with increase of concentration of ammonium ions, pH and temperature. Under field conditions, where the soil surface is open to the atmosphere, volatilization varies with the rate of transport of NH_3 away from the surface, which is determined mainly by wind speed, and with evaporation of water from the surface.

It will be apparent that the amount of ammonia lost from soil depends on properties of the soil and of the microenvironment. Loss occurs from soils, or microsites within soil, which are at pH 7 or above, especially if the ammonium is present in a drying soil surface. It can be severe following applications of urea fertilizer to the soil surface because of the rise of pH when the urea is hydrolysed to NH_4^+ and HCO_3^- (and CO_3^{2-}). This reaction is particularly important in relation to the efficient use of fertilizer nitrogen worldwide, as urea is now the form of nitrogen most commonly applied as fertilizer. When ammonium fertilizers containing sulphate or phosphate are added to calcareous soils ammonia loss is facilitated because precipitation of the sparingly soluble Ca salts shifts the equilibrium towards ammonium carbonate. The cation exchange capacity of the soil also has an effect because the greater the proportion of the ammonium ions held on exchange sites the lower the concentration in solution.

Much ammonia is lost from urine from grazing animals as most of the nitrogen is present as urea. An example is shown in Fig. 20.4. The ryegrass sward received 60 kg N ha^{-1} as ammonium nitrate on seven occasions during the grazing season (April–October) and was grazed on a 28-day rotational basis by steers at an average stocking rate of 11 head per hectare.[45] The ammonia losses recorded in two summer months averaged 0.75 and 1.13 kg N ha^{-1} day^{-1}, and were equivalent to 35 and 53 per cent respectively of the fertilizer N applied. Losses were highest during and immediately after grazing and decreased during periods of rain. No loss was observed from cut swards receiving the same fertilizer N treatments.

In experiments in the Philippines, loss of nitrogen from ammonium sulphate was 30–60 per cent when applied to the flood water of paddy rice, the large loss being attributed to rise in pH to 9–10 caused by algal photo-

---

45 Ryden, J. C. *et al.*, *Grassland Res. Inst. Ann. Rept. for 1982*, p. 28; Ball, P. R. and Ryden, J. C., *Pl. Soil*, 1984, **76**, 23.

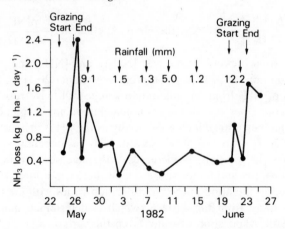

Fig. 20.4   Daily rates of NH₃ loss measured aerometrically during and after two-day grazings by yearling steers within a 28-day rotation: average stocking rate 11 head per ha. The sward received 420 kg N ha⁻¹ a⁻¹ as ammonium nitrate in seven equal applications. (From Ball, P. R. and Ryden, J. C., *Pl. Soil*, 1984, **76**, 23.)

synthesis.[46] With this system of cultivation, as with all others, ammonia loss can be minimized, or avoided altogether, if the fertilizer is placed well below the soil surface, as any ammonia that is released will be absorbed by the soil before it can reach the atmosphere.

The extent of ammonia volatilization has in the past been uncertain because of the lack of suitable methods to measure it, but the introduction of micrometeorological (aerometrical) methods by Denmead in Australia has removed this obstacle.[47] On a global scale the principal sources of ammonia in the atmosphere are animal excreta, decomposition of soil organic matter, fertilizer use, and the burning of fuels.

Ammonia which is volatilized from soil may be absorbed by the plant canopy above the soil or be transported through the atmosphere to other soils, plant canopies, or to bodies of water. The mean concentration of ammonia in the atmosphere over land has been given as 3 ppbv (partial

46 Reported by Watanabe, I. *et al.*, in *Nitrogen Cycling in South-East Asian Wet Monsoonal Ecosystems* (eds R. Wetselaar, J. R. Simpson and T. Rosswall), p. 4, Aust. Acad. Sci., 1981, from unpublished data of Bouldin, D. and Alimagno, B.   See Vlek, P. L. G. and Fillery, I. R. P., *Fertil. Soc. Proc.*, No. 230, 1984 for summary of N loss, including that as NH₃, under wetland rice.

47 Reviewed by Denmead, O. T. in *Gaseous Loss of Nitrogen from Plant–Soil Systems* (eds J. R. Freney and J. R. Simpson), p. 133, Nijhoff, 1983. A method using wind tunnels in the field has been described by Lockyer, D. R., *J. Sci. Fd Agric.*, 1984, **35**, 837.

pressure 3×10^{-9} atmosphere),[48] a value similar to the partial pressure at which there is zero flux of ammonia between plants and the atmosphere.[49] At higher concentrations the plant canopy can be expected to absorb ammonia from the atmosphere and at lower concentrations to lose it. The present evidence is that ammonia loss from plants is greatest during senescence, but the amounts of ammonia lost or gained by plants are not yet known. Absorption of ammonia by the soil depends on its water content, pH, content of organic matter, and content and type of clay minerals. The annual accretion can therefore be expected to depend on soil properties as well as on the partial pressure of ammonia in the atmosphere. Except near ammonia sources the amount is small; for example, at Rothamsted the amount measured in 1976 and 1977 was 4 kg N ha^{-1} a^{-1} (see ref. 7).

The most important reaction of ammonia with soil is to dissolve in the water to form NH_4^+, the ratio of NH_3 to NH_4^+ depending on pH as described above. The NH_4^+ may remain exchangeable or become non-exchangeable, and both NH_3 and NH_4^+ will normally become oxidized by nitrification. In addition, NH_3 reacts with oven-dry (105 °C) soil minerals and organic matter. In the reaction with montmorillonite and vermiculite, NH_3 is adsorbed by coordination to exchangeable cations, which causes loss of water molecules from their hydration shells, and by formation of NH_4^+ with interlayer water.[50] With vermiculite, NH_3 has been shown to react with protons dissociated from the hydroxyl groups in the octahedral layer.[51] The ammonia adsorbed by these reactions is more slowly nitrified than ammonium sulphate.[52] Less is known about the reaction with soil organic matter. Chemical reactions are likely to occur in soil with phenolic constituents and reducing sugars because these reactions have been demonstrated with pure substances. There is probably also a reaction with acid groups, e.g. carboxyl groups by protonation of NH_3, because the extent of the reaction increases at pH values above 7. Although the NH_3 becomes resistant to acid or alkali extraction after reaction with organic matter, its availability to crops appears not to be affected.[53]

48 Galbally, I. E. and Roy, C. R. in *Gaseous Loss of Nitrogen from Plant–Soil Systems* (eds J. R. Freney and J. R. Simpson), p. 265, Nijhoff, 1983.

49 Farquhar, G. D., *et al.*, in *Gaseous Loss of Nitrogen from Plant–Soil Systems* (eds J. R. Freney and J. R. Simpson), p. 159, Nijhoff, 1983.

50 Ahlrichs, J. L., *et al.*, *Clay Miner.*, 1972, **9**, 263; Mortland, M. M. and Raman, K. V., *Clays Clay Miner.*, 1968, **16**, 393; and review by Mortland, M. M. in *Agricultural Anhydrous Ammonia* (eds M. H. McVickar *et al.*), p. 188, Soil Sci. Soc. Am., Madison, 1966.

51 Stone, M. H. and Wild, A., *Clay Miner.*, 1978, **13**, 337.

52 Stone, M. H. and Wild, A., *Pl. Soil*, 1977, **46**, 633, Russell, J. D. and Fraser, A. R., *J. Sci. Fd Agric.*, 1977, **28**, 852.

53 See Nommik, H. and Vahtras, K. in *Nitrogen in Agricultural Soils* (ed. F. J. Stevenson), p. 123, 1982 (see ref. 1).

Denitrification

Denitrification is the reduction of nitrate to nitrous oxide and dinitrogen gas. Under most conditions the process is brought about biologically, but there is also evidence for a purely chemical pathway. In the Soil Science Society of America Glossary of Scientific Terms 'denitrification' is restricted to biological reduction, but the term is used here more generally because, for measurements in the field, although the end products are measured, the mechanism is usually not identified. For example, nitrous oxide may be produced during both nitrification and denitrification.[54]

Reduction of nitrate is brought about by a number of groups of bacteria, some of which may only be able to carry out one stage of the reduction process from nitrate to nitrogen gas. The bacteria are predominantly facultative anaerobes. The reduction is generally thought to proceed via the pathway shown below, although it is still uncertain whether NO is an obligatory intermediate:

$$NO_3^- \rightarrow NO_2^- \rightarrow (NO) \rightarrow N_2O \rightarrow N_2$$

Denitrification is the principal pathway for nitrate reduction and its microbiology and biochemistry are discussed elsewhere (pp. 622–5). Although there is no evidence that nitrate is reduced to ammonium in soils which are only occasionally anaerobic, in long-term anaerobic situations such as paddy soils this reduction has been shown to occur.[55] Its importance is not known. Nitrate can, however, be assimilated by bacteria, although to a less extent than ammonium, and after their death, and decomposition of their protoplasm, the nitrogen may be present as ammonium ions.

There is evidence of a purely chemical pathway involving the conversion of nitrite to nitrogen oxides. Nitrite is unstable in acid solutions, decomposing to nitric oxide, NO, and nitrogen dioxide, NO_2. These gases are soluble in water, but in an open system they can diffuse into the atmosphere. Also, nitrous acid reacts with phenolic constituents of soil organic matter with the liberation of N_2 and N_2O.[56] The contribution of these reactions to total denitrification is likely to be small in most soils, except those which are acid.[57]

There have as yet been relatively few reliable field measurements of denitrification losses. The earliest estimates were from deficits in nitrogen balance sheets from the few experiments in which additions, crop removal, leaching loss and changes in total soil nitrogen were measured over a period

54 See Firestone, M. K. in *Nitrogen in Agricultural Soils* (ed F. J. Stevenson), p. 289, 1982 (see ref. 1).
55 Tiedje, J. M. *et al.*, in *Terrestrial Nitrogen Cycles* (eds F. E. Clark and T. Rosswall), p. 331, Ecol. Bull. (Stockholm) 33, 1981.
56 Nelson, D. W. and Bremner, J. M., *Soil Biol. Biochem.*, 1969, **1**, 229.
57 Nelson, D. W. in *Nitrogen in Agricultural Soils* (ed F. J. Stevenson), p. 327, 1982 (see ref. 1).

of years.[58] The method was refined by measuring the ^{15}N recovery from added labelled sources and, if there was no leaching or leaching was measured, the ^{15}N deficit was attributed to gaseous loss. In soils of pH below 7, the gaseous loss can reasonably be attributed to denitrification. The disadvantage of the method is that accumulated errors can give an unreliable estimate, and these errors are serious unless leaching loss and ammonia volatilization are measured, and changes in total soil nitrogen carefully monitored.[59]

Direct measurements of nitrous oxide flux in the field have been made from the increase in nitrous oxide concentration in covers placed over the soil surface.[60] From a heavy clay soil (50 per cent clay) the highest daily loss was 0.6 to 0.7 kg N ha^{-1} in late autumn and spring, and on this soil amounted to 5.4 and 8.6 kg N ha^{-1} over each of two growing seasons. From a soil with 35 per cent clay and probably less anaerobic in wet weather, the corresponding loss[61] in each of the two growing seasons was 1.5 and 2.1 kg N ha^{-1}. Nitrous oxide is soluble in water and loss of the dissolved gas in drainage water has been measured.[62] Because of its water solubility, the soil may therefore act as a sink as well as a source for atmospheric nitrous oxide.

Direct measurement of the loss of dinitrogen has proved more difficult. One method is to measure the ^{15}N-labelled gases after the addition of ^{15}N-labelled fertilizer. It may be satisfactory for rates of denitrification above about 0.5 kg ha^{-1} day^{-1}, but the assumption has to be made that the fertilizer ^{15}N does not take part in mineralization/immobilization reactions (see p. 687). Another method makes use of low concentrations (0.1–1.0 per cent v/v in soil air) of acetylene (C_2H_2) to inhibit the reduction of nitrous oxide; total denitrification is then equal to the total N_2O-N production on C_2H_2-treated plots.[63] Using this method, Ryden[64] found annual denitrification losses of 1.6, 11.1 and 29.1 kg N ha^{-1} from soil under cut grass receiving 0, 250 and 500 kg N ha^{-1} a^{-1} respectively. The seasonal distribution of these losses (Fig. 20.5) was closely related to fertilizer applications, most of the annual loss occurring within about three weeks after the application if the water content of the soil was over 20 per cent. By contrast with these comparatively low losses under grassland, the same method of acetylene

---

58 Allison, F. E., *Adv. Agron.*, 1966, **18**, 219.   For an account of the development of methods for measuring denitrification and a summary of results see Colbourn, P. and Dowdell, R. J., *Pl. Soil*, 1984, **76**, 213.

59 For a review of more recent methods see Ryden, J. C. and Rolston, D. E. in *Gaseous Loss of Nitrogen from Plant–Soil Systems* (eds J. R. Freney and J. R. Simpson) p. 91, 1983.

60 Burford, J. R. *et al.*, *J. Sci. Fd. Agric.*, 1981, **32**, 219.

61 Dowdell, R. J. *et al.*, *Letcombe Ann. Rept. for 1979*, p. 18.

62 Dowdell, R. J. *et al.*, *Nature, Lond.*, 1979, **278**, 342.

63 For reports on the testing and evaluation of the method see Smith, M. S. *et al.*, *Soil Sci. Soc. Am. J.*, 1978, **42**, 611;   Ryden, J. C. *et al.*, *op. cit.*, 1979, **43**, 104, 110.

64 Ryden, J. C., *J. Soil Sci.*, 1983, **34**, 355.

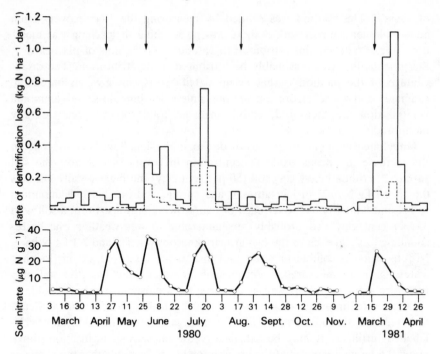

Fig. 20.5   Weekly mean rates of denitrification (solid line), N_2O flux (dashed line) and soil nitrate content (0–20 cm) for a Wickham loam under grass receiving 500 kg N ha^{-1} a^{-1} as ammonium nitrate. The arrows indicate the times when fertilizer was applied at one-quarter of the annual rate.

inhibition showed annual denitrification losses[65] of between 95 and 223 kg N ha^{-1} from irrigated soils in California under vegetables that had received 290 to 665 kg N ha^{-1}. Peak losses occurred at water potentials of −5 to −10 kPa and appeared to be increased by additions of organic manure.

All measurements point to denitrification being an intermittent process, the rate of which is determined mainly by nitrate concentration, temperature, oxygen concentration and readily decomposable carbon compounds. Estimates might therefore be reasonably obtained from algorithms based on field measurements of losses under conditions where these soil properties have been measured.

To summarize: losses of nitrogen from soils and crops are still poorly quantified. New techniques described in previous sections will help to identify the sources of loss and lead to improved management practices, although much

65 Ryden, J. C. and Lund, L. J., *Soil Sci. Soc. Am. J.*, 1980, **44**, 505.   See also Legg, J. O. and Meisinger, J. J. in *Nitrogen in Agricultural Soils* (ed F. J. Stevenson), p. 503, 1982 (see ref. 1).

remains to be done. An example referred to in Table 3.12 illustrates this point. In animal production, only 10 to 16 per cent of the input of nitrogen might leave the farm in the form of milk, meat or wool. Recent measurements of nitrate leaching under urine patches, ammonia volatilization and denitrification have shown that together these losses can account for about 80 per cent of the nitrogen applied (420 kg N ha^{-1}) to grazed pastures.[66]

Nitrogen and crop growth

Uptake of nitrogen and its metabolism

Nitrogen is a constituent of all proteins and nucleic acids and is therefore essential for plant growth. The usual sources of nitrogen are NH_4^+ and NO_3^- ions which are taken up through the roots. Dinitrogen is used by plants such as legumes which have an association with nitrogen-fixing micro-organisms, e.g. *Rhizobium*, and atmospheric ammonia may make a significant contribution if its concentration is high.[67] As nitrate concentrations in the soil are usually much higher than ammonium concentrations, the main source of nitrogen for non-leguminous crop plants is nitrate.

In nutrient solutions most crop plants grow equally well whether supplied with nitrate or ammonium ions, as long as the solution pH is controlled (see below). With a wide range of plant species, calcifuges appear to prefer ammonium and calcicoles to prefer nitrate.[68] However, in an experiment by Cox and Reisenauer[69] wheat plants grew better when supplied with both ammonium and nitrate than when provided with only one source of nitrogen.

Differences in the effectiveness of the two sources which are sometimes reported probably arise indirectly. For example, Cox and Reisenauer found a lower uptake of Ca, Mg and K when the nitrogen was supplied as NH_4^+ than when supplied as NO_3^-. There is also an effect on the pH value of the external solution, an effect also shown when plants are grown in soil. It is generally believed that when the ratio of anions to cations taken up by plant roots is greater than 1, OH^- is excreted and the pH rises, whereas if the ratio is less than 1, H^+ is excreted and the pH drops. It is therefore usual for the pH to increase in nitrate solutions and to decrease in ammonium solutions. This pH change can affect the availability to the plant of other nutrient ions. There may also be an effect of pH on the relative uptake of NH_4^+ and NO_3^- ions because there is some evidence that anions are taken up faster than cations

66 Ball, P. R. and Ryden, J. C., *Pl. Soil*, 1984, **76**, 23, and see also ref. 26.
67 Cowling, D. W. and Lockyer, D. R., *Nature, Lond.*, 1981, **292**, 337.
68 Haynes, R. J. and Goh, K. M., *Biol. Rev.*, 1978, **53**, 465.
69 Cox, W. J. and Reisenauer, H. M., *Pl. Soil*, 1973, **38**, 363.

from acid solutions, and that the reverse holds as the pH is increased.[70] At low root temperatures the uptake of both ions is small, but with ryegrass the uptake of nitrate is depressed more than that of ammonium.[71]

Nitrate ions are reduced to ammonia in the roots and leaves of plants by a two-step process through nitrite. The enzymes (nitrate and nitrite reductase) which catalyse the process contain molybdenum and iron (see pp. 92, 97). In some plants most of the nitrate taken up by the roots is carried through the xylem to the leaves where it is reduced, whereas in others, e.g. *Vicia* and *Pisum*, the reduction may occur mainly in the roots.[72] In cereals and a large number of other species reduction occurs both in leaves and shoots. The end product of biological dinitrogen fixation is also ammonia.

The ammonia produced, whether by nitrate reduction or dinitrogen fixation, has a central role in protein metabolism. It can react with oxoglutarate to give glutamate, and with glutamate to give glutamine. The amino N subsequently reacts with keto acids (oxo acids) to produce the corresponding amino acids and hence proteins. If nitrate reduction occurs in the roots the reactions of NH_3 to produce amides and amino acids also occur there, and the nitrogen is transported to the leaves in these forms.

The conversion of ammonia to amino acids and proteins requires a supply of keto (oxo) acids, and carbon substrates to provide energy for the reactions. Hence the balance between the supply of organic compounds, which are the result of photosynthesis, and the supply of NO_3^- and NH_4^+ is critical for protein synthesis. Protein which is synthesized in the vegetative stage of growth is used to produce new and bigger leaves whereas carbohydrate at the same stage of growth forms mainly cell wall material. The ideal is to have a balance between these two requirements. If there is excess NO_3^- this ion accumulates in plant cells, and although it normally does no harm to the plant it may have a deleterious effect when the plant is used as food (see p. 681). If NH_4^+ accumulates it can be toxic, possibly by inhibiting the Hill reaction in the chloroplasts, or by reducing the proton gradient on which ion transport appears to depend.[73]

Excess or deficient uptake of nitrogen results in changes in plant morphology. The effects of added nitrogen to a crop that would otherwise be deficient, may be summarized as follows:

1   It increases leaf size and therefore the potential for greater photosynthesis, which will increase root growth, total dry matter and yield of useful product.

70 Bowling, D. J. F. *Uptake of Ions by Plant Roots*, Halstead Press, 1976. Results of Blamey, F. P. C. *et al., Proc. 9th Int. Pl. Nutr. Conf.*, 1982, Vol. 1, p. 66 with sunflower do not support the general statement.
71 Clarkson, D. T. and Warner, A. J., *Pl. Physiol.*, 1979, **64**, 557; Clarkson, D. T. *et al.*, Plant, Cell and Environment, 1986, **9**, 535.
72 See review by Pate, J. S., *Ann. Rev. Pl. Physiol.*, 1980, **31**, 313; also Raven, J. A. and Smith, F. A., *New Phytol.*, 1976, **76**, 415.
73 Gregory, R. P. F., *Biochemistry of Photosynthesis*, pp. 66, 77, Wiley, 1976.

2   It increases tiller survival in cereals.
3   It increases the protein content of the storage organs, that is, grain, tubers or roots.
4   It increases the size of plant cells and reduces the thickness of their walls.
5   It increases the proportion of water in the plant fresh weight because of increased plant protoplasm.
6   It darkens the green colour of the leaves.
7   It increases the shoot/root ratio.

If the addition of nitrogen is excessive, the harmful effects that can occur are:

1   A greater susceptibility of the plant to attack by pests and diseases and to damage by drought and frosts, as a result of thinner cell walls.
2   With cereals, the straw is weakened so that the crop is liable to lodge.
3   The quality of crops such as sugar cane, sugar beet and malting barley grown mainly for their carbohydrate is reduced.
4   Weed competition can become severe and reduce crop growth.
5   Where water is in short supply, the greater transpiration loss from the increased leaf area may reduce final yields.
6   A delay in crop maturity.

Effect of supply of nitrogen on crop growth and composition

An example of increasing the supply of nitrogen to cereals through fertilizer additions is given in Table 20.7. The results are from Broadbalk at Rothamsted where wheat has been grown continuously since 1843. In the period 1852–61 when a long-strawed variety of winter wheat (Red Rostock) was grown, application of fertilizer nitrogen increased the yield of grain from 1.20 to 2.51 t ha^{-1}, but because the effect on the yield of straw was greater, the harvest index fell from 0.38 to 0.33. Using a short-strawed variety (Cappelle Desprez) in 1970–78, fertilizer nitrogen again reduced the harvest index but, as has generally been found with short-strawed varieties of cereals, the harvest index was very much higher than that of the old varieties. Table 20.7 also shows that the responses of grain and total dry matter to fertilizer nitrogen were higher in 1970–78 than in 1852–61. The increase of response is attributed partly to a change from autumn to spring application of the fertilizers, and to better disease and weed control. On plots not receiving fertilizer nitrogen, grain yields have increased slightly over the period of the experiment. The table also shows the higher yield potential of the newer varieties.

As illustrated by the fertilizer response curves in Fig. 20.6, similar observations have been made from other experiments. On average in these exper-

Table 20.7 Effects of farmyard manure and fertilizers on yields of winter wheat grown continuously on Broadbalk. Rothamsted Experimental Station. (From *Rothamsted Ann. Rept. for 1982*, Part 2, 20)

Treatment*	Period and variety		
	1852–61 Red Rostock	1902–11 Squarehead's Master	1970–78 Cappelle Desprez
1. Mean yields of grain (t ha^{-1})			
None	1.12	0.80	1.65
FYM	2.41	2.62	5.84
PKMg(Na)	1.29	1.00	1.78
N_1PKMg(Na)	1.91	1.58	3.52
N_2PKMg(Na)	2.42	2.28	4.84
N_3PKMg(Na)	2.52	2.76	5.32
N_4PKMg(Na)	—	—	5.49
2. Mean yields of grain plus straw at 85% dry matter (t ha^{-1})			
None	3.03	2.01	2.86
FYM	6.67	7.76	11.60
PKMg(Na)	3.43	2.56	3.15
N_1PKMg(Na)	5.21	4.39	6.25
N_2PKMg(Na)	6.99	6.74	8.84
N_3PKMg(Na)	7.59	8.49	9.98
N_4PKMg(Na)	—	—	10.34
3. Harvest index†			
None	0.37	0.40	0.58
FYM	0.36	0.34	0.50
PKMg(Na)	0.38	0.39	0.57
N_1PKMg(Na)	0.37	0.36	0.56
N_2PKMg(Na)	0.35	0.34	0.55
N_3PKMg(Na)	0.33	0.33	0.53
N_4PKMg(Na)	—	—	0.53

* Plots 3, 22, 5, 6, 7, 8 and 9 (all periods). In 1970–78 the nitrogen fertilizer was ammonium nitrate/calcium carbonate at 48, 96, 144 and 196 kg N ha^{-1} (N_1, N_2, N_3, N_4). Earlier applications were as ammonium sulphate and chloride supplying N at about the same rates. In 1852–61 all the fertilizer nitrogen was applied in the autumn, in 1902–11 24 kg N ha^{-1} was applied in autumn and the rest in spring and in 1970–78 all N was applied in the spring.

† Harvest index calculated as (mean yield of grain)/(mean yield of grain plus straw).

iments the yield of grain with no fertilizer nitrogen had increased from a little under 3 t ha^{-1} to a little under 5 t ha^{-1} over a 30-year period, perhaps due mainly to a carry-over of nitrogen from the preceding crops.

The increases of crop yield shown in Table 20.7 and Fig. 20.6 are usually accompanied by increased uptake of nitrogen and sometimes by increased concentrations of nitrogen in the plant tissues; Table 3.5 gives an example

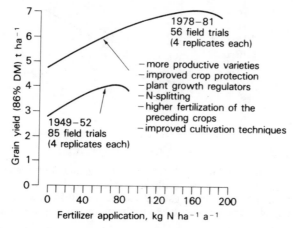

Fig. 20.6 Comparison of average response curves of winter wheat to fertilizer N applications in 1949–52 and 1978–81 in the Federal Republic of Germany. (From Becker, F. A. and Aufhammer, W., *Fertil. Soc. Proc.*, No. 211, 1982.)

of increased N in the grain of wheat with fertilizer N application. However, with crops that would otherwise be severely starved of nitrogen, the extra growth stimulated by the addition of nitrogen can be very low in nitrogen so that the overall concentration of N in the crop is depressed.

Most of the nitrogen taken up by crops is converted to protein, but NO_3^- accumulates if there is excessive uptake. Accumulation in vegetable crops has been the subject of investigation because of the fear that a high nitrate intake in the diet might cause methaemoglobinaemia in infants, or cause the formation of carcinogens such as N-nitrosamines by reduction of nitrate in the body to nitrite which then reacts with amides and amines. Nitrate concentrations tend to be high in leafy vegetables such as spinach, cabbage and lettuce, but measured concentrations vary greatly. They tend to be highest in outer leaves and petioles, but depend also on the conditions under which the vegetables were grown, harvested and marketed. Common values (as NO_3) are 200–2000 mg kg^{-1} fresh weight. Using average values for the nitrate content of an adult diet it is estimated that water will contribute 55 per cent of the nitrate uptake at a concentration of 50 mg NO_3^- l^{-1} (11.3 mg NO_3^- N l^{-1}) and 10 l consumed weekly,[74] and the diet will contribute 45 per cent.

Leafy crops, including grass, are the most responsive to applications of fertilizer nitrogen. At 21 field sites in England and Wales ammonium nitrate fertilizer was applied to perennial ryegrass at annual rates up to

74 Royal Commission on Environmental Pollution, *7th Rept..*, 1979, HMSO; Bryson, D. D., *Fertil. Soc. Proc.*, No. 228, 1984; also Maynard, D. N. *et al.*, *Adv. Agron.*, 1976, **28**, 71 for a review of nitrate concentrations in vegetables.

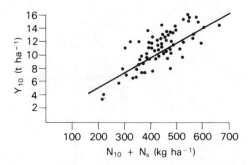

Fig. 20.7   Response of perennial ryegrass in England and Wales to soil and fertilizers N, where fertilizer N (N_{10}) is the optimum application, defined as that which gave a response of 10 kg DM ha⁻¹ at a yield of Y_{10}; N_s is the nitrogen taken up by perennial ryegrass on control plots.

750 kg N ha⁻¹ in six split applications.[75] The optimum application of fertilizer N, defined as the greatest amount to give a response of at least 10 kg DM kg⁻¹ N, was 388 kg ha⁻¹ averaged over four years at the 21 sites. Figure 20.7 shows the relation between the optimum yield (Y_{10}) and the nitrogen supply expressed as available nitrogen from the soil, N_s, plus the optimum fertilizer application, N_{10}. The soil available nitrogen was assumed to be that taken up by ryegrass on plots that received no N. The figure shows the close dependence of the optimum yield on the nitrogen supply. At different sites, however, the response curves to N depended on the summer rainfall and the available water capacity of the soil. Response curves for two contrasting sites are shown in Fig. 20.8. There was a higher response to nitrogen on a clay loam with high summer rainfall than on a sandy loam under low summer rainfall. It is of interest that at nearly all sites there was a linear response to fertilizer at rates up to 300 kg N ha⁻¹, the average response for all sites up to this rate being 23 kg DM kg⁻¹ N.

When land is put under grass or grass–legume mixtures, soil nitrogen increases because the rate of addition from roots and shoots exceeds the rate of loss. At a single site the rate of increase depends on the botanical composition, and especially the legume content of the herbage and its management,[76] and over a range of sites the rate depends also on the soil properties and the climate. At a single site the increase of soil nitrogen, most of which was in the top 2 cm of soil, was found to be much greater under mixed swards that had been grazed than under similar swards that had been cut and the herbage removed, reflecting the return of nitrogen in the excreta, in particular the dung. Fertilizer nitrogen up to 314 kg ha⁻¹ had no

75 Morrison, J. *et al.*, *Grassland Res. Inst. Tech. Rept.*, No. 27, 1980. Hurley, Berkshire. For a discussion of response curves see Cooke, G. W., *Fertilizing for Maximum Yield*, Granada, Lond., 1982.

76 Clement, C. R. and Williams, T. E., *J. agric. Sci.*, 1967, **69**, 133.

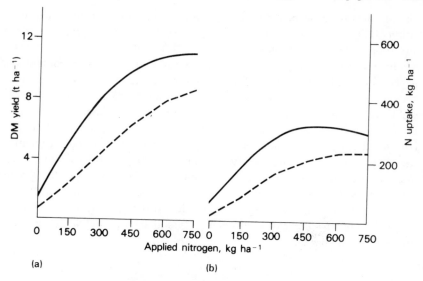

Fig. 20.8 Response of ryegrass to fertilizer N at two sites: ——— DM yield, ------ N uptake. (a) High summer rainfall and (b) low summer rainfall. (From Morrison, J. *et al.*, *Grassland Res. Inst. Tech. Rept.*, No. 27, 1980, Hurley, Berks.)

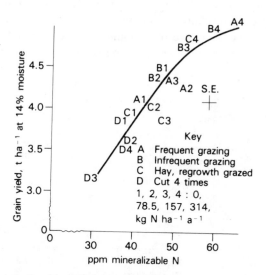

Fig. 20.9 The relationship between mineralizable N in the soil after three years under various swards and the yield of winter wheat. (From Clement, C. R., *Residual Value of Applied Nutrients, MAFF Tech. Bull.*, No. 20, 166,, HMSO, 1971.)

measurable effect on the increase of soil nitrogen for two reasons: (i) accumulation of nitrogen depends on accession of carbon through the death of herbage and roots, which was probably not greatly different at the different rates of fertilizer used; and (ii) fertilizer nitrogen reduced the clover content and hence nitrogen fixation. After three years' growth, the swards were ploughed and winter wheat was grown. Figure 20.9 shows the relation between grain yield and the amount of mineralizable nitrogen (20 days' aerobic incubation at 30 °C) in the soil under four systems of sward management. Grain yield correlated closely with mineralizable nitrogen whatever the management of the sward. Yields were highest after the three grazing treatments and smallest after the herbage had been cut and removed. Grain yields after the grazing treatments generally increased with increasing fertilizer nitrogen, but this did not hold after cutting. The benefit of the ley to the subsequent arable crops was attributed to the extra nitrogen provided by the soil under grass or grass–clover mixtures.

Soil nitrogen available to crops and fertilizer requirement

The principal source of nitrogen for non-leguminous crops growing on land not receiving any nitrogen fertilizer is that which is released as ammonium by the micro-organisms decomposing the soil organic matter and which is later oxidized to nitrate. The organic matter which is decomposed includes recently added crop residues and humus. The amount of nitrogen which is available to a crop is therefore the amount of mineral N (NH_4^+ and NO_3^-) in the soil profile at the start of the growing season, plus the amounts added

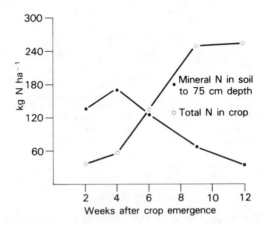

Fig. 20.10 Comparison of mineral N in soil and total N in potato crop. (From Asfary, A. F. *et al., J. agric. Sci.*, 1983, **100**, 87,, and Asfary, unpublished.)

as fertilizer and mineralized in the soil while the crop is growing, and minus that which is lost from the root zone by leaching or in gaseous forms. Figure 20.10 compares crop uptake of nitrogen with the amounts of mineral N (extracted by KCl) within the rooting zone of a potato crop receiving nitrogen fertilizer. The experiment was conducted on a sandy loam after a previous arable crop. During the period of rapid growth (4–9 weeks after emergence) the nitrogen uptake rate of 4–5 kg N ha^{-1} day^{-1} was about three times the rate of mineralization of soil organic nitrogen (1.3–1.7 kg^{-1}ha^{-1} day^{-1}). Consequently, during this period the crop depended largely on the nitrogen applied as fertilizer. The mineral nitrogen that had accumulated in the soil by the date of planting, and that which was mineralized after planting were not sufficient for the crop to reach its growth potential.

The concept of the crop requirement for nitrogen in terms of soil mineral nitrogen and mineralizable nitrogen has been expressed by Stanford[77] as:

$$N_c = N_i + N_m + N_f$$

where N_c = uptake of nitrogen for a maximum or most profitable yield, N_i is the initial quantity of mineral nitrogen in the rooting zone, N_m is the amount of nitrogen which will be mineralized and N_f is the required amount of fertilizer nitrogen. If it is assumed that the fractional use, e, of all three sources of mineral nitrogen is the same, then

$$N_c = e \, (N_i + N_m + N_f)$$

and the fertilizer N requirement is given by

$$N_f = (N_c/e) - (N_i + N_m)$$

In a modified form of the last equation, N_m is divided into the nitrogen which is mineralized from soil organic matter, crop residues and organic manures. The equation requires an estimate to be made early in the season of the expected final crop yield. The uncertainty of this estimate is a problem whatever method is used to advise on the amount of fertilizer N to apply. There are also practical difficulties. First, taking deep samples to the depth of the root zone is laborious. It has, however, been found in Canada that the correlation between nitrogen uptake by barley and the NO_3^- present in soil improved when sampling was to 61 cm rather than to shallower depths, but there was no improvement when sampling was to greater depth. Accordingly, sampling to about 60 cm has become widely used in the Canadian Provinces.[78] A more serious difficulty is estimating subsequent mineralization. The

77 Stanford, G., *J. environ. Qual.*, 1973, **2**, 159. See also Remy, J. C. and Viaux, Ph., *Fertil. Soc. Proc.*, No. 211, 1982, 67, and other papers on European experience in the same number.
78 Soper, R. J. *et al.*, *Canad. J. Soil Sci.*, 1971, **51**, 45. See also Smith, C. M. in *Soil Testing: correlating and interpreting the analytical results* (eds T. R. Peck *et al.*), pp. 85–98, Am. Soc. Agron. Spec. Publ., No. 29, 1977.

large number of incubation and chemical extraction techniques that have been used are described by Stanford.[79] They include aerobic and anaerobic incubation, extraction with mild or stronger oxidizing and hydrolysing agents, and the determination of more specific fractions, e.g. nitrogen in soil biomass and in undecomposed or partially decomposed crop residues. No single method can be expected to be universally successful because of the wide range of conditions under which crops are grown, but can be useful for a limited range of soil and climatic conditions.

The most serious difficulty is, however, the unpredictability of final crop yield at the time when a decision has to be made on the rate of application of fertilizer nitrogen. The prediction of yield has to be based on average yields in earlier years, but drought, low temperatures, pests and other causes of reduced yields will result in low utilization of fertilizer nitrogen. In contrast, in a particularly good growing year an average application will be too low. If the growing conditions are fairly similar between years, and the ratio of grain price to fertilizer cost is high, it is economically sound practice to apply sufficient fertilizer nitrogen to ensure high yields even though the application will be in excess in some years. It is often different in developing countries where the ratio of grain price to fertilizer cost is low and yields vary considerably between years because of variable rainfall or for some other reason. In these countries the optimum application will usually be much less than that required for the occasional high yield.

The difficulty of basing advice on soil nitrogen analyses has led to the use in England and Wales of the Soil Nitrogen Index.[80] This is based on previous land use and is used as a guide for average conditions. If, for example, a cereal crop is to follow a succession of cereal crops the Index is 0, if it follows a grazed forage crop the Index is 1, and if it follows crops which received heavy applications of organic manures the Index is 2. The fertilizer requirement has been established by field trials on several soils and is adjusted for a particular season according to the Index and, where necessary, by rainfall in the previous winter and by soil type.

Measurement of fertilizer nitrogen recovery by crops

The efficiency of use of fertilizer nitrogen can be expressed in various ways. The per cent apparent recovery by the whole crop is one way, although nitrogen in roots is frequently not included. From trials in which increments of fertilizer N are added and the crop at final harvest is analysed:

79 Stanford, G. in *Nitrogen in Agricultural Soils* (ed F. J. Stevenson), p. 651, 1982 (see ref. 1). Methods for submerged rice are reviewed by Sahrawat, K. L., *Adv. Agron.*, 1983, **36**, 415.
80 *MAFF Reference Book*, **209**, London, HMSO 1985. For discussion see Needham, P., *Fertil. Soc. Proc.*, No. 211, 1982.

$$\text{per cent apparent recovery} = \frac{N_x - N_o}{F_x} \times 100$$

where N_o is the total nitrogen in the crop receiving no fertilizer nitrogen, and N_x is the total N in the crop receiving an amount of fertilizer nitrogen, F_x. The term 'apparent recovery' is used because the assumption has to be made that the contribution of soil N to crop uptake is the same for the fertilized and unfertilized treatments. If the crop uptake is linearly related to application rate of nitrogen, the apparent recovery is the same at all rates, but if the response curve is concave to the axis for application rates, as is usually the case, the apparent recovery decreases as application rates increase.

Another method is to use fertilizer enriched (or less commonly, depleted) in ^{15}N, and to calculate the percentage recovery of fertilizer nitrogen in the crop as:

$$\frac{(\text{At. } \% \ ^{15}N_1 - \text{At. } \% \ ^{15}N_0) \cdot \text{total N in crop}}{(\text{At. } \% \ ^{15}N_f - \text{At. } \% \ ^{15}N_0) \cdot F} \times 100$$

The measurements required are the atom (At.) per cent ^{15}N in the fertilized (N_1) and unfertilized (N_0) crops, the total nitrogen in the crop which received an amount of fertilizer nitrogen, F, and the atom per cent ^{15}N in the fertilizer (N_f).[81] The assumption has to be made that the fertilizer nitrogen remains separate from the pool of soil nitrogen (see below) throughout crop growth.

Assumptions must therefore be made in both assessments of recovery, and they often give different results. This is illustrated in Table 20.8 by results from ^{15}N experiments in lysimeters which show higher apparent recoveries than ^{15}N recoveries.

The difference between the two estimates of recovery arises because, as shown in Table 20.8 and commonly reported from other experiments, more soil nitrogen is taken up by the fertilized crop. Reasons that have been given are: (i) nitrogen fertilizer stimulates the growth of roots which can then use mineral nitrogen from deeper in the soil; (ii) fertilizer nitrogen stimulates the mineralization of soil nitrogen, which is sometimes called a priming effect; and (iii) fertilizer ^{15}N enters a pool of soil nitrogen in which there is turnover, due to microbial activity, between mineral N and immobilized N, which has been called mineralization/immobilization turnover or MIT.[82]

Where only MIT occurs, the amounts of ^{15}N immobilized and soil nitrogen released are equal, that is, there is pool substitution. There is now good evidence that much of the ^{15}N not taken up by the crop remains in the soil in organic form, which supports the view that substantial MIT often occurs and might therefore be the main, and in some circumstances the whole, explanation of the uptake of extra soil nitrogen. This extra nitrogen is termed

81 Hauck, R. D. and Bremner, J. M., *Adv. Agron.*, 1976, **28**, 219.
82 Jansson, S. L. and Persson, J. in *Nitrogen in Agricultural Soils* ed F. J. Stevenson, p. 229, 1982. (see ref. 1).

Table 20.8 Lysimeter experiment: uptake of fertilizer N (^{15}N) and soil N (kg ha^{-1}) by crops during year in which fertilizer had been applied, and comparison of ^{15}N recovery and apparent recovery by each crop; fertilizer applied as ammonium nitrate. (Adapted from Dowdell, R. J., *Phil. Trans. R. Soc. Lond.*, 1982, **B296**, 363)

	Ryegrass			Winter wheat
	Sandy loam	*Clay*	*Silt loam*	*Clay*
Total N applied	400	400	400	80
N uptake by fertilized crop				
Labelled N	201	206	247	52
Unlabelled N	143	194	152	87
Total	344	400	399	139
N uptake by unfertilized crop	72	121	138	63
Recovery of labelled N(%)	50	51	62	65
Apparent recovery of fertilizer N(%)	68	70	65	95
Difference in uptake of soil N (fertilized – unfertilized)	71	73	14	24

the 'added nitrogen interaction' or a.n.i. The term implies no mechanism, but provides a distinction between a 'real' a.n.i. where fertilizer nitrogen increases the mineralization of soil organic N, and an 'apparent' a.n.i. where there is only pool substitution.[83]

One consequence of a.n.i. is that the use of ^{15}N to measure the extra nitrogen taken up by a fertilized crop, as described above under methods for measuring recoveries, is often an underestimate. It gives uptake of fertilizer nitrogen *sensu strictu*, but underestimates the benefit to the crop. This conclusion applies equally to the use of ^{15}N to measure gaseous or leaching loss of nitrogen. Only if the measurements of apparent recovery and ^{15}N recovery agree can there be confidence that the recovery of fertilizer nitrogen by the crop is correctly measured.

Nitrogen fertilizers[84]

Traditionally, crops depended for their nitrogen on the use of organic manures and on biologically fixed nitrogen added to the soil through the

83 Jenkinson, D. S. *et al.*, *J. Soil Sci.*, 1985, **36**, 425.
84 For fuller information see Cooke, G. W., *The Control of Soil Fertility*, Crosby Lockwood, Lond., 1967; Cooke, G. W., *Fertilizing for Maximum Yield*, Granada, Lond., 1982; Follett, R. H., Murphy, L. S. and Donahue, R L., *Fertilizers and Soil Amendments*, Prentice-Hall, 1981; Archer, J., *Crop Nutrition and Fertiliser Use*, Farming Press, Ipswich, 1985.

growth of clovers and other legumes but, under most conditions, fertilizer N is required if crops are to reach their yield potential. The requirement has increased with the introduction of improved crop varieties, better cultivations and use of other mineral fertilizers, and better control of weeds, pests and diseases, which together have led to a potential for higher crop yields.

Nitrogen fertilizers were first introduced from a few natural deposits of nitrate salts and as ammonium sulphate as a by-product from gasworks, but the Haber process established industrially in the early 1920s has ensured supplies from atmospheric nitrogen. The products now available are easy to handle and provide much better control of the mineral nitrogen available to crops than do organic manures. Their efficiency of use has improved as a result of many years of field experimentation, of better knowledge of crop requirements, and of the conditions under which losses occur. In the section that follows general comments are made on the best ways of using the fertilizers. (For organic manures see Ch. 3).

Forms of nitrogen fertilizers

Most nitrogen fertilizers are water soluble and derived from ammonia. Worldwide the important ones are ammonium nitrate, ammonium phosphate, urea, ammonium sulphate and ammonia itself either anhydrous or in solution. Because of the microbial oxidation of ammonium to nitrate, most of the fertilizer nitrogen taken up from non-acid soils a few weeks after it has been applied is likely to be in the form of nitrate. In so far as nitrate and ammonium fertilizers differ in effectiveness, nitrates tend to be somewhat quicker acting, but are subject to loss by denitrification and leaching from the time of application. Ammonium fertilizers and urea may lose nitrogen by ammonia volatilization soon after application, but denitrification and leaching loss may occur later when the ammonium has been oxidized to nitrate. One point on which care must be taken in the use of nitrate fertilizers is that, since they are very soluble in water and nitrate is usually not adsorbed by the soil, they may raise the osmotic pressure of the soil solution round seedlings to a damaging level if used during dry weather. Because of loss by denitrification they should not be used in poorly drained soils and particularly not for paddy rice.

Cause of soil acidification

All ammonium salts and urea cause soil acidification and loss of exchangeable cations, with the loss of calcium predominating in most soils, and of calcium from calcium carbonate if it is present.[85] The main cause is the oxidation of NH_4^+ to yield H^+ and NO_3^- (see Ch. 25). This also applies to NH_4^+ formed

---

85 See for example Pierre, W. H., *J. Am. Soc. Agron.*, 1928, **20**, 270;   Leutenegger, F., *E. Afr. agric. For. J.*, 1956, **22**, 81.

from the hydrolysis of urea. Under field conditions some of the nitrate produced by nitrification, in so far as it is not taken up by the crop, is leached out of the soil as calcium nitrate. In addition, calcium is lost because it is leached with the anion in the fertilizer. With ammonium sulphate, for example, calcium is leached with the sulphate which makes it the most acidifying of the commonly used fertilizers. There is a smaller loss of soil calcium from the use of ammonium phosphates because the calcium phosphates that are formed have a low solubility. The loss of calcium following the application of urea and ammonium nitrate is also smaller because, if most of the nitrate which is added or formed by nitrification is taken up by the crop, there is no excess anion to be leached which would require calcium as a counter ion. For ammonium sulphate, each 100 kg fertilizer requires about 100 kg $CaCO_3$ to neutralize the acidity; the requirement is less for the other ammonium fertilizers and urea. The continued use of ammonium fertilizers, and especially of ammonium sulphate has, in the absence of remedial lime applications, led to severe soil acidification in many weakly buffered soils and consequently to reductions in crop yields.[86]

Use of ammonium salts

Ammonium sulphate can also be an inefficient nitrogen fertilizer if broadcast on the surface of alkaline or calcareous soils, as for example those containing more than 10 per cent calcium carbonate.[87] The ammonium sulphate reacts to give calcium sulphate and free ammonia. Ammonium sulphate appears to suffer a greater loss of ammonia in this way than does ammonium nitrate or phosphate.[88] Ammonium sulphate can also be undesirable for use in some paddy rice soils, for the sulphate becomes reduced to sulphide which can be toxic to the rice roots. It is because of this that ammonium chloride is fairly widely used in Japanese rice fields, for it does not suffer this disadvantage. Ammonium nitrate and the ammonium phosphates are widely used nitrogen fertilizers, and in the past the nitrate was usually intimately mixed with calcium carbonate, which reduced its natural hygroscopicity. This carbonate had the incidental effect of neutralizing the acidity caused by the loss of calcium nitrate from the soil by leaching.

86 For an example from northern Nigeria see Bache, B. W. and Heathcote, R. G., *Expl Agric.*, 1969, **5**, 241, with later results by Jones, M. J. and Stockinger, K. R., *op. cit.*, 1976, **12**, 49. Results from the Park Grass Experiment at Rothamsted have been summarized by Johnston, A. E. *et al., Soil Use and Management*, 1986, **2**, 3.

87 Devine, J. R. and Holmes, M. R. J., *J. agric. Sci.*, 1964, **62**, 377; Gasser, J. K. R., *J. Soil Sci.*, 1964, **15**, 258.

88 Larsen, S. and Gunary, D., *J. Sci. Fd Agric.*, 1962, **13**, 566. For conditions leading to ammonia loss see Fenn, L. B. and Kissel, D. E., *Soil Sci. Soc. Am. Proc.*, 1973, **37**, 855; 1974, **38**, 606; 1975, **39**, 631; *Soil Sci. Soc. Am. J.*, 1976, **40**, 394; Fenn, L. B. and Escarzaga, R., *op. cit.*, 1977, **41**, 358; Feagley, S. E. and Hossner, L. R., *op. cit.*, 1978, **42**, 364.

Use of urea

Urea is the most concentrated solid nitrogen fertilizer on the market, containing about 46 per cent N, and in many countries it is the cheapest form of fertilizer nitrogen.[89] It is soluble in water and is deliquescent, but it is produced in a form that stores well by coating urea prills or granules with a suitable resin; also it can be bonded with sulphur, thus becoming a nitrogen and a sulphur fertilizer.[90] Some processes produced urea containing sufficient impurity of biuret, $(NH_2CO)_2.NH$, to be toxic to many crops, but present-day samples of urea contain less than 1 per cent, which is below the danger limit. Urea is only weakly absorbed by soils, and so can be washed out of the soil profile if applied in periods of continuous wet weather. It is quickly hydrolysed by urease to ammonium carbamate and most of the practical problems of using urea as a fertilizer turn on the fate of this ammonium carbamate, for it dissociates and can release free ammonia. If urea is spread on the surface of soil and allowed to become moist through rain or dew formation, much of the nitrogen is liable to be lost to the atmosphere as ammonia. Even if the urea is worked into the soil surface, ammonia may be lost if the soil is sandy and so has only a low ammonia absorbing power.[91] It has also been found to be inefficient when broadcast on pastures because of their high urease activity and loss of ammonia to the atmosphere.[92] Further, the free ammonia formed in the soil is toxic to germinating seeds and young seedlings, so it must not be used in fertilizers which are placed close to the seed. Under some conditions the rise of pH due to hydrolysis can be sufficient to inhibit the oxidation of nitrite by *Nitrobacter* which leads to nitrite concentrations in the soil which are toxic.[93] These harmful consequences are mitigated by using small application rates. The value of urea as a fertilizer would often be increased if its rate of hydrolysis to ammonium carbonate could be decreased. There are a number of substances which inhibit the activity of urease enzymes in soils under laboratory conditions, such as dihydric phenols, quinones, and hydroxamates[94] but none is yet in use commercially.

Since, as already mentioned, urea is both concentrated and cheap, it is most important that the conditions in which it can be used efficiently should be clearly understood. General experience has shown that if used in applications not exceeding 45 to 70 kg N ha^{-1}, it is usually, but not always, as effective as ammonium or nitrate fertilizers, but applications of 100 kg N ha^{-1}

89 For a review of problems using urea see Tomlinson, T. E., *Fertil. Soc. Proc.*, No. 113, 1970.
90 Giordano, P. M. and Mortvedt, J. J., *Agron. J.*, 1970, **62**, 612.
91 Gasser, J. K. R., *J. Soil Sci.*, 1964, **15**, 258; Court, M. N. *et al.*, *J. Soil Sci.*, 1963, **14**, 247; Volk, G. M., *Agron. J.*, 1959, **51**, 746.
92 Simpson, J. R., *Trans. 9th Int. Congr. Soil Sci.*, 1968, **2**, 459.
93 Low, A. J. and Piper, F. J., *J. agric. Sci.*, 1961, **57**, 249; Court, M. N. *et al.*, *J. Soil Sci.*, 1964, **15**, 42, 49.
94 Bremner, J. M. and Douglas, L. A., *Soil Biol. Biochem.*, 1971, **3**, 297.

are usually appreciably less efficient.[95] However, even this generalization can be unsatisfactory, for there is an example of 17 kg ha^{-1} as urea harming the very early growth of wheat seedlings on a sandy soil although 70 kg N ha^{-1} was safe on a clay soil.[96] In general, urea tends to be least suitable as a nitrogen fertilizer for grassland, for it can only be broadcast on the surface of the sward so that ammonia losses to the atmosphere are likely to be large.

An important fertilizer in some developed countries is ammonia itself, either anhydrous or as a saturated or supersaturated solution.[97] This is the cheapest nitrogen fertilizer to manufacture, but it needs special facilities for transport and storage; and to minimize losses to the air it must be injected into the soil, which is often difficult. Further, ammonia is toxic to seedlings. It is also toxic to nitrifying organisms if used in sufficient concentration, so may persist longer than ammonium salts.

Residual value of nitrogen fertilizers

Water-soluble nitrogen fertilizers given to one crop used to be considered to have no residual value for the following crop. Nitrogen fertilizers applied to arable crops have little effect on the level of soil organic matter or organic nitrogen in the soil, and it is not possible to build up the nitrogen reserves in a soil by the use of nitrogen fertilizers (Table 20.1). However, under suitable conditions a heavy application of a nitrogen fertilizer to one crop can have a residual effect on the succeeding crop. This may occur through unused nitrate being stored in some subsoils over the winter, and also through the decomposition of leafy residues with comparatively low C : N ratios left by crops such as potatoes[98] and oil seed rape. Thus, Widdowson and Penny[99] found at Rothamsted that the average residual effect of 190 kg N ha^{-1} as ammonium sulphate given to potatoes, was equivalent to 63 kg N ha^{-1} as a top dressing to the following wheat crop in spring. The residual effect of 125 kg N ha^{-1} given to winter wheat was equivalent to 25 kg N ha^{-1} applied to the following potato crop. This residual effect is influenced by the winter rainfall, being larger in years of dry than of wet winters.[100]

The nitrogen present in a nitrogen fertilizer applied to an established ley or permanent pasture during periods of active growth may be taken up almost completely by the grass, if allowance is made for the nitrogen in the roots

95 See for example Devine, J. R. and Holmes, M. R. J., *J. agric. Sci.*, 1963, **61**, 391; Collier, P. A. *et al. op. cit.*, 1970, **74**, 153.
96 Mees, G. C. and Tomlinson, T. E., *J. agric. Sci.*, 1964, **62**, 199.
97 For reviews on its use in the USA see Hauck, R. D. and Russel, D. A., *Outlook Agric.*, 1969, **6**, 3; and in Europe, van Burg, P. F. J., *op. cit.*, 53.
98 Asfary, A. F., *et al.*, *J. agric. Sci.*, 1983, **100**, 87.
99 Widdowson, F. V. and Penny, A., *J. agric. Sci.*, 1965, **65**, 195. For a review of results in UK see Gasser, J. K. R. in *Residual Value of Applied Fertiliser*, MAFF Tech. Bull. No. 20, 114, HMSO, 1971.
100 van der Paauw, F., *Pl. Soil*, 1963, **19**, 324.

as well as in the shoots; though if allowance is not made for this between two-thirds and three-quarters of the applied nitrogen is usually found in the shoots.[101] Usually up to about half the applied nitrogen is found in the grain and straw of cereal crops at harvest and about 5 per cent is in the roots.

Part of the difference between arable crops and established grass is that the roots of grass permeate the surface soil intimately, whereas those of arable crops are much further apart, so that the nitrate can be washed through the soil without coming into contact with a root. Further, the dense root system under grass will usually take up the nitrate from the soil fairly rapidly, whereas, with arable crops, it may persist longer in the surface soil and in consequence have a greater probability of being denitrified or leached. Another difference is that grass will take up nitrogen throughout most of the year whereas with arable crops there are periods after sowing and close to harvest when there is little or no uptake.

Methods of improving efficiency of use

Considerable interest has been taken in the problem of increasing the efficiency of use of nitrogen fertilizers, because of the very large amounts of applied nitrogen that are not harvested in the crop. First, it is possible to increase the efficiency, particularly when the crop foliage is well developed, by spraying urea on the crop. This will give uptakes of the order of 70 per cent, but it is only possible to apply a relatively small amount (about 10 kg N ha^{-1} per application) if efficiencies as high as this are to be obtained.[102] Each application of urea is limited to 10–20 kg N ha^{-1} to avoid leaf burn, and the method is restricted to fruit crops.

In practice, the correct timing of the application, or of split applications during periods of active growth, and the placement of the fertilizer near the seed, will often increase the proportion of fertilizer taken up. Second, there is also the possibility of reducing the loss of nitrate by leaching through the use of ammonium fertilizers with which a nitrification inhibitor is incorporated.[103] These compounds inhibit the oxidation of ammonium to nitrite, and 2-chloro-6-(trichloromethyl)-pyridine (nitrapyrin) and dicyandiamide are two examples, but field trials have not yet shown an inhibitor to be sufficiently useful to justify its cost.[104] Nitrification inhibitors are most likely to be useful

101 Walker, T. W. *et al.*, *J. Brit. Grassland Soc.*, 1954, **9**, 249; Whitehead, D. C., *Commonw. Bur. Past. Field Crops Bull.*, No. 48, 1970, Farnham Royal.

102 Thorne, G. N., *Rothamsted Ann. Rept. for 1954*, p. 188.

103 Schmidt, E. L. in *Nitrogen in Agricultural Soils* (ed F. J. Stevenson), p. 253, 1982 (see ref. 1).

104 Gasser, J. K. R., *J. agric. Sci.*, 1965, **64**, 299; 1968, **71**, 243; 1970, **74**, 111. Turner, G. O. *et al.*, *Soil Sci.*, 1962, **94**, 270; Goring, C. A. I., *Soil Sci.*, 1962, **93**, 211, 431. For reviews of nitrification inhibitors see Slangen, J. H. G. and Kerkhoff, P., *Fertil. Res.*, 1984, **5**, 1; Gasser, J. K. R., *Soils Fertil.*, 1970, **33**, 547.

where there is a big excess of rainfall over transpiration during the growing season and in making more effective use of the nitrogen in liquid animal manures applied to land.

Slow-acting fertilizers

Sulphur-coated urea (SCU) and urea formaldehydes are two types of slow release nitrogen fertilizers which are manufactured commercially. With SCU the rate of release of urea depends on the thickness of the coating of the granules with elemental sulphur and wax, and on the imperfections in the coating. When released into the soil solution the urea is quickly hydrolysed, as described earlier. The urea formaldehydes used as fertilizers differ from those used in plastics and adhesives in having a higher ratio of urea to formaldehyde. They are usually manufactured to contain 30–38 per cent N, and the nitrogen becomes available to plants only after microbial hydrolysis. The manufacturing cost of both types of fertilizer is higher per unit of N than that of conventional fertilizers. Their use is therefore mainly restricted to high value horticultural crops and to sports turf where they avoid the risk of scorch.[105]

Other synthetic, slow-acting fertilizers include isobutylidene diurea (IBDU), which contains about 32 per cent N and has a low water solubility, and crotonylidene diurea (CDU) with about 30 per cent N.[106] Both are manufactured in Japan and used for the rice crop. Magnesium ammonium phosphate (7.3 per cent N) is another sparingly soluble fertilizer which releases nitrogen slowly at a rate that decreases with increasing granule size.

Slow-acting fertilizers provide a supply of nitrogen over a comparatively long period of time, which is useful for slow-growing plants. Their use is limited, however, by their manufacturing costs, and also because their rate of availability to crops might not match crop requirements. This match is difficult to attain for fertilizers such as the urea formaldehydes which depend on microbial hydrolysis and therefore on soil temperature.[107]

105 Davies, L. H., *Fertil. Soc. Proc.*, No. 153, 1976; Schneider, H. and Veegens, L., *Fertil. Soc. Proc.*, No. 180, 1979.
106 Hamamoto, M., *Fertil. Soc. Proc.*, No. 90, 1966.
107 Gasser, J. K. R., *J. agric. Sci.*, 1970, **74**, 107; Prasad, R. *et al.*, *Adv. Agron.*, 1971, **23**, 337.

Colour Plate 1 Spodosol developed on sandy drift in Norfolk, eastern England; length of rule 0.5 m (S. Nortcliff)

Colour Plate 2 Partial crop failure due to water erosion of an Oxisol in Brazil (J. Newman)

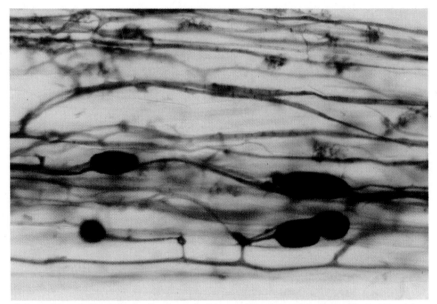

Colour Plate 3 Vesicles and arbuscules in a stained root of onion (P. J. Harris)

Colour Plate 4 Response to molybdenum of *Trifolium subterraneum* grown on an acid soil in Victoria, Australia. LH plot received single superphosphate at 12 kg P ha^{-1}, centre plot received no treatment, RH plot received superphosphate at 12 kg P ha^{-1} plus sodium molybdate at 70 g Mo ha^{-1} (L. H. P. Jones)

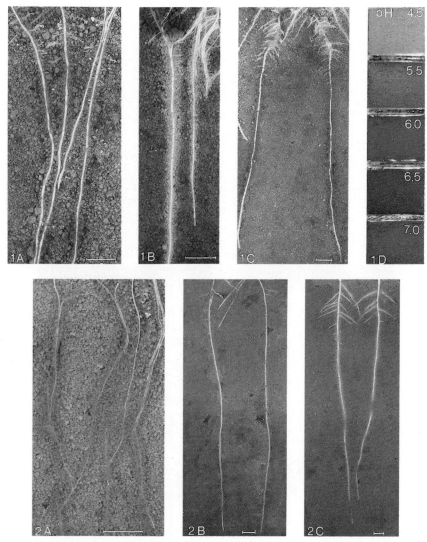

Colour Plate 5 The pH of rhizospheres
1. Effect of source and concentration of nitrogen with maize. A 16.6 mg
   $NO_3^- - N/100$ g soil. B 16.6 mg $NH_4^+ - N/100$ g soil. C 4.2 mg $NO_3^-/100$ g
   soil. Potassium constant at 8.3 mg $K/100$ g soil. D pH calibration standards;
   bar = 1 cm.
2. Comparison of three plant species growing in the same soil (pH 6.0)
   supplied with $Ca(NO_3)_2$ at 8.3 mg $NO_3^-/100$ g soil. A Wheat, B Soyabean,
   C Chickpea; bar = 1 cm. (H. Marschner and V. Romheld)

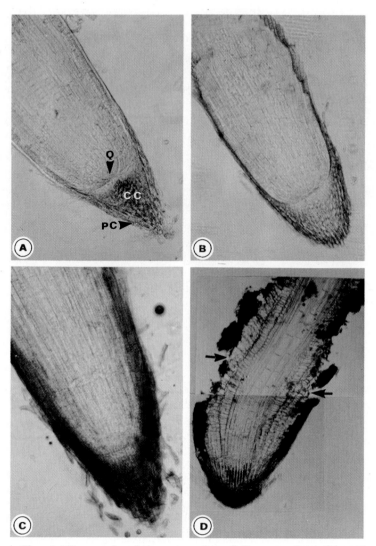

Colour Plate 6 Micrographs of the root apices of *Zea mays* L., cultivar T × 24 after exposure to aluminium solutions of 0.84 mmol dm^{-3} (activity) and stained with haematoxylin. Darkly stained cells indicate the presence of Al.

A (× 100) Control, no Al added. B(×100), C(×100), D(×60) exposure to Al for 30 minutes, 2 hours and 20 hours respectively.

Note staining of peripheral cells of root cap and mucilaginous secretions (B), which extends to the epidermal and outer cortical cells (C), and fracture of the epidermis but with little staining of the cells of the inner cortex, stele and root meristem, as arrowed (D).

PC = cap periphery, CC = cap columella, Q = quiescent centre (R. J. Bennet, C. M. Breen, M. V. Fey)

21

Plant nutrients in soil: phosphate

Before the introduction of phosphate* fertilizers few soils in the world were able to provide sufficient phosphate for a sequence of good crop yields. Where there is no history of use of phosphate fertilizers, deficiency of phosphate is still very common, and under severe conditions the soil may not supply any phosphate at all to a crop. Deficiency is widespread in the world, limiting the production of arable crops and herbage especially in Australia, South America and South Africa. Under extreme conditions, forest trees may be stunted in their growth, and grazing animals may suffer phosphate deficiency.[1] In the British Isles phosphate deficiency has been most severe on the

---

1 For effects on animals see Reid, R. L. in *The Role of Phosphorus in Agriculture* (eds F. E. Khasawneh, E. C. Sample and E. J. Kamprath), p. 847, publ. Am. Soc. Agron., Madison, 1980. An early account of deficiency in Northern Ireland is given by Robertson, G. S., *J. Min. Agric. N. Ireland*, 1927, **1**, 7.

* Throughout this chapter, phosphate refers to orthophosphate, i.e. ions formed by dissociation of phosphoric acid, H_3PO_4, unless stated otherwise.

Jurassic clays of the English Midlands, on the acid Millstone Grits of northern England and in large areas of Northern Ireland.

Soils formed on alluvium and on volcanic lavas are exceptions if the parent material has a high content of phosphate. The most extreme example is probably the soil of the Nile Valley where annual deposits of phosphate-rich alluvium from the basic rocks of the Ethiopian plateau have supported continuous crop cultivation for several millennia. Without such deposits, or additions by man, soils contain only the small amount inherited from their parent material, and usually only a little of it can be readily used by crops. The use of fertilizers containing phosphate has therefore had a big impact on agricultural productivity in nearly all parts of the world. On land with a long history of fertilizer use, however, its frequent use has now reduced the need to that required to balance crop removal. Elsewhere additions are needed to increase crop yields, and the total amounts used are large (Table 1.4).

As a plant nutrient, phosphate differs from nitrogen in several respects. One difference is dependence on the amount provided by the parent material, as mentioned above. Another is that phosphate ions are strongly adsorbed by soil, or are precipitated as products with very low solubility, and so are not lost by leaching except from sands. Another difference is that gaseous loss of phosphorus (as phosphine, PH_3) is rare or non-existent.[2] Finally, because of the low concentration of phosphate in soil solutions its availability to plants depends on rapid replenishment of the solution by desorption of phosphate from the solid phase and by mineralization of soil organic phosphorus compounds, and on diffusion through the solution to the root surface. The relation between the concentration of phosphate in solution and that in the solid phase has therefore been investigated widely. The chemistry of soil phosphate which pertains to this relation will be discussed in the next sections, and this will be followed by a description of the availability of phosphate to plants, and then of the use of phosphate fertilizers.

Amounts of phosphate in soil

The phosphate content of rocks is commonly between 500 and 1400 μg P g^{-1}, with an average of about 1000 to 1200 μg P g^{-1}, depending on the rocks analysed.[3] Of the igneous rocks basalts are usually at the upper end of this range and granites at the lower end, as are many sedimentary rocks. Above this range are deposits of phosphate rock which are exploited commercially, and coprolitic limestones and some basic volcanic lavas, which are of local importance in forming phosphate-rich soil.

2 Burford, J. R. and Bremner, J. M., *Soil Biol. Biochem.*, 1972, **4**, 489.
3 McKelvey, V. E. in *Environmental Phosphorus Handbook* (eds E. J. Griffith *et al.*), p. 13, Wiley-Interscience, NY, 1973.

Table 21.1 Total phosphorus contents (arithmetic means) of soils (usually 0–15 cm depth)

Country or area	No. of samples	Total P ($\mu g \, g^{-1}$)	Source
United Kingdom	—	700	1
Australia	2217	350	2
United States	863	420	3
Ghana (forest)	21	326	4
(savanna)	67	134	4
West African savanna	503	140	5

1 Cooke G. W., *J. Sci. Fd Agric.*, 1958, **9**, 761.
2 Wild, A., *Aust. J. agric. Res.*, 1958, **9**, 193.
3 Shacklette, H. T. *et al.*, *US Geol. Survey Prof. Paper*, No. 574–D, 1971.
4 Nye, P. H. and Bertheux, M. H., *J. agric. Sci.*, 1957, **49**, 141.
5 Jones, M. J. and Wild. A., *Soils of the West African Savanna, Commonw. Bur. Soils Tech. Comm.*, No. 55, 1975.

Very low total phosphate contents have been reported in soils on the old land masses of Australia and Africa (Table 21.1). In unfertilized soils the highest values are associated with parent material formed from basic igneous rocks and soils with high organic matter contents. In general the content in soil is lower than in the lithosphere as a whole. A reduction of the concentration of phosphate occurs as rocks weather probably because apatite, the main phosphate mineral in rocks, dissolves in acidic leaching water before the formation of oxides of iron and aluminium, and clay minerals, which would adsorb the phosphate.[4] Low soil phosphate contents might also be the result of the loss of topsoil by erosion.

In soil profiles most of the phosphate is usually near the soil surface. In cultivated soils this is partly because of the use of phosphate fertilizers; for example, Cooke[5] estimated that nearly half the phosphate content of British topsoils has been supplied by fertilizers. In uncultivated soils phosphate accumulates near the soil surface because of its circulation through vegetation and deposition in leaf fall and litter. Examples are shown in Fig. 21.1.

Part of the phosphate is held in the soil organic matter, but methods for its determination are not entirely satisfactory.[6] One method is to extract a subsample with dilute acid after oxidation by ignition in air or treatment with hydrogen peroxide to destroy the soil organic matter. The increase in the amount of phosphate extracted over a 'control' subsample which has not been oxidized is considered to be the amount of phosphorus held in the organic

4 Wild, A., *Aust. J. agric. Res.*, 1961, **12**, 286.
5 Cooke, G. W., *J. Sci. Fd Agric.*, 1958, **9**, 761.
6 For methods see Hesse, P. R., *A Textbook of Soil Chemical Analysis*, Murray, 1971; Olsen, S. R. and Sommers, L. E. in *Methods of Soil Analysis*, Part 2 (eds A. L. Page, R. H. Miller and D. R. Keeney), p. 403, Am. Soc. Agron., Madison, 1982.

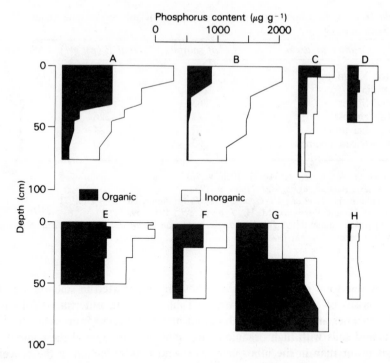

Fig. 21.1 Distribution of phosphorus in various soil profiles (Anderson, G. in *The Role of Phosphorus in Agriculture* (eds F. E. Khasawneh *et al.*), Am. Soc. Agron., Madison, Wisc., 1980.)

    A & B Freely and poorly drained cultivated clay loams of the Insch Association, Scotland.

    C Uncultivated Koputaroa soil developed on windblown sand, New Zealand.

    D Uncultivated Dawes silt loam, Nebraska.

    E Uncultivated Pima calcarous clay loam, Arizona.

    F Cultivated Orthic Deep Black, Melfort, Saskatchewan.

    G Uncultivated *Carex globularis* pine bog, northern Finland.

    H Leached forest soil, Ibadan, Nigeria.

matter. Another method is to extract the organic matter from the soil with alkali. The amount of inorganic phosphate extracted is subtracted from the total amount to give the organic phosphorus content of the soil.

Organically held phosphorus[7]

The amounts of organically held phosphorus vary greatly between soils (Fig. 21.1). They are roughly related to organic matter contents, both between soils and within the soil profile. Hence the circumstances that lead to increased soil organic matter contents (see pp. 588–592) also generally lead to increased organic phosphorus contents. The phosphorus content of soil organic matter is, however, variable. The ratios of per cent carbon and per cent nitrogen to per cent organic phosphorus (C : P and N : P ratios) vary over a wide range, although commonly the ratio of N : org P is in the range 10 : 1 to 10 : 2.[8] The source of the variability arises partly from the analytical methods used, because comparisons have shown disagreement between them. The C : P ratio is, however, more variable than the C : N ratio, possibly because nitrogen atoms are chemically linked within the humic acid structure whereas most of the organically held phosphorus is in structures peripheral to humic acid. Large C : P ratios as, for example those reported from Ghana,[9] might be caused by low phosphorus contents of the vegetation and its litter, suggesting low turnover rates of phosphorus within soil.

The chemical forms of organic phosphorus include a range of inositol phosphate esters, phospholipids, nucleic acids, phosphate linked to sugars and derivatives of a phosphonic acid (containing a C–P bond), probably all occurring in organic complexes. They are discussed in Chapter 18. Their origin is not yet agreed. The inositol phosphates in soil include esters of myo-, chiro-, scyllo- and neo-inositol of which only the first is a common plant constituent; the other isomers might be of microbial origin.[10] The phospholipids may come from plants, animals or microbes, and the nucleic acids are at least partly of microbial origin.[11] It therefore appears that the organic phosphorus compounds found in soil are not simply the result of the accumulation of organic phosphorus compounds from plants that are resistant to decomposition; there must also be biochemical transformations, which are yet unknown. Unfortunately, the organic phosphorus compounds are themselves still incompletely characterized. The use of nuclear magnetic resonance to identify the C–P bonds in phosphonic acids[12] is an indication that new techniques will provide the information that is needed.

It is generally assumed that soil organic phosphorus is derived directly, or after biochemical transformations, from crop residues and leaf litter, in which part of the phosphorus is present in organic compounds. This is supported by the evidence (see below) that phosphate added as fertilizer is converted

7 Anderson, G. in *The Role of Phosphorus in Agriculture* (eds F. E. Khasawneh, E. C. Sample and E. J. Kamprath), p. 411, Am. Soc. Agron., Madison, Wisc., 1980; Dalal, R. C., *Adv. Agron.*, 1977, **29**, 83.
8 Barrow, N. J., *Soils Fertil.*, 1961, **24**, 169.
9 Nye, P. H. and Bertheux, M. H., *J. agric. Sci.*, 1957, **49**, 141.
10 Cosgrove, D. J., *Inositol Phosphates: their chemistry, biochemistry and physiology*, Elsevier, 1980.
11 Anderson, G. and Malcolm, R. E., *J. Soil Sci.*, 1974, **25**, 282.
12 Tate, K. R. and Newman, R. H., *Soil Biol. Biochem.*, 1982, **14**, 191; Hawkes, G. E. *et al.*, *J. Soil Sci.*, 1984, **35**, 35.

to organic forms in soils where organic matter levels are increasing, as under young grassland, as a result of increased root production and leaf decay. There might also be microbial synthesis of organic phosphorus compounds from inorganic phosphate when an energy source is added to soil, because it has been shown that addition of glucose can cause immobilization of phosphate.[13]

The rate of accumulation of organically held phosphorus can be large. Under irrigated pastures of perennial ryegrass and white clover in Australia the phosphate added as superphosphate (174 kg P ha^{-1}) accumulated mainly as calcium phosphates and in organic forms.[14] The highest increase of organic phosphorus was under white clover, the increase being from 168 kg P ha^{-1} one year after establishment of the pasture to 269 kg P ha^{-1} four years later. The addition of superphosphate increased the soil organic phosphorus contents on the Park Grass experiment at Rothamsted[15] in which plots had been kept under grass for more than 100 years, but the effect was small. Only 6 per cent of the accumulated phosphorus was in organic form at pH 4.5 and 1 per cent on limed plots at pH 6.2–6.5. As the total N contents were almost unaffected by fertilizer additions, the ratio N : organic P (by alkali extraction) changed from 15.7 to 9.7 on the acid plots but only from 13.0 to 12.4 on the limed plots. The difference in accumulation rates of organically held phosphorus in these two examples can probably be attributed to the rapid increase of soil organic matter under the new Australian pastures, whereas the content of soil organic matter under the old grassland at Rothamsted was near to equilibrium. The build-up of organic phosphorus can deplete the soil of inorganic phosphate available to plants, as was found on old soils in a chronosequence in New Zealand.[16]

The reverse process, that is, net mineralization of organically held phosphorus, is important in most soils and especially so after a ley (short-term grassland), and bush and forest fallows. After clearing a secondary forest in Nigeria the organic phosphorus content of the top 10 cm of soil fell from 194 to 147 μg g^{-1} over a period of 22 months.[17] The importance for crop nutrition of the mineralization of organic phosphorus compounds is probably universal, but is greatest at high soil temperatures because of high mineralization rates. For example, in Ghana there was a negative regression between organic phosphorus measurements and crop response to phosphate fertilizers.[18] Because adsorption by soil might lessen the availability of organic phosphorus

13 Ghoshal, S. and Jansson, S. L., *Swed. J. agric. Res.*, 1975, **5**, 199. See also Stewart, J. W. B. and McKercher, R. B., *3rd Int. Congr. Phosphorus Compounds*, 1983, p. 551.
14 Rixon, A. J., *Aust. J. agric. Res.*, 1966, **17**, 317.
15 Oniani, O. G. *et al.*, *J. Soil Sci.*, 1973, **24**, 1.
16 Walker, T. W. and Syers, J. K., *Geoderma*, 1976, **15**, 1.
17 Mueller-Harvey, I. *et al.*, *J. Soil Sci.*, 1985, **36**, 585.
18 Smith, R. W. and Acquaye, D. K., *Emp. J. expl. Agric.*, 1963, **31**, 115. For cocoa in Nigeria see Omotoso, T. I., *Soil Sci.*, 1971, **112**, 195.

after mineralization, Friend and Birch in Kenya used the product of organic phosphorus and the phosphate adsorption capacity as a measure of phosphate availability.[19]

As for nitrogen, net mineralization will only occur when the microbial requirements have been met. The critical ratio of C : organic P has been put at 200 : 1 for net mineralization and 300 : 1 for net immobilization, though both ratios are only approximate. The balance between net mineralization and net immobilization in plant material has also been put at about 200 to 300 : 1, which corresponds to about 0.2 per cent P.[20] Enough is already known of the mineralization rates of organic phosphorus to show its importance, but it is often neglected when the phosphate nutrition of crop plants is considered.

Inorganic forms of phosphorus in soil

Primary minerals

Apatite is the most commonly occurring phosphate mineral in rocks.[21] It is present as a primary mineral in the sand fractions of soils especially if these are fairly young and not acidic.[22] Because of its very low solubility and rate of solubilization, primary apatite has little or no significance in supplying phosphate to plants. Plumbogummite minerals, which are hydrated aluminium phosphates containing Pb, Sr, Ca, Ba and Ce, occur in the fine silt and coarse clay fractions of some soils in Australia[23] and elsewhere, and also in. rocks.[24] They are present in soil either as residual primary minerals or are of secondary origin. They also are probably of no direct significance in plant nutrition because of their low solubility. A few secondary minerals have also been identified. They include apatite, identified in calcareous nodules on heavily phosphate-fertilized plots at Rothamsted[25]; crandallite, a calcium aluminium phosphate (a plumbogummite mineral) found in highly phosphatic soil on islands with guano deposits[26]; vivianite, a ferrous phos-

19 Friend, M. T. and Birch, H. F., *J. agric. Sci.*, 1960, **54**, 341.
20 See Dalal, R. C., *Adv. Agron.*, 1977, **29**, 83.
21 Lindsay, W. L. and Vlek, P. L. G. in *Minerals in Soil Environments* (eds J. B. Dixon *et al.*,), p. 639, Soil Sci. Soc. Am., Madison, Wisc., 1977.
22 For examples see Williams, E. G. and Saunders, W. M. H., *J. Soil Sci.*, 1956, **7**, 90, 189; Cescas, M. P. *et al.*, *J. Soil Sci.*, 1970, **21**, 78; Skipp, R. F. and Matelski, R. P., *Soil Sci. Soc. Am. Proc.*, 1960, **24**, 450.
23 Norrish, K., *Trans. 9th Int. Congr. Soil Sci.*, Adelaide, 1968, **2**, 713. For discussion see Norrish, K. and Rosser, H. in *Soils: an Australian viewpoint*, p. 335, CSIRO Div. Soils, Melbourne, 1983.
24 Bain, D. C., *Min. Mag.*, 1970, **37**, 934 for occurrence in some sandstones in Scotland.
25 Nagelschmidt, G. and Nixon, H. L., *Nature, Lond.*, 1944, **154**, 428.
26 Schroo, H., *Neth. J. agric. Sci.*, 1963, **11**, 209.

phate which has a bright blue colour and has been found in some waterlogged soils and peats,[27] and a suite of transitory minerals discussed on p. 731, which are formed at the interface between soil and dissolving granules of phosphate fertilizer.

Chemical forms of soil phosphorus

The primary and secondary minerals, which are crystalline and have been identified by optical and X-ray methods, account for only a small proportion of the inorganic phosphate in soil. The remaining phosphate cannot be identified by these methods. Instead, three groups of chemical methods have been developed which are derived from investigations by Voelcker, Warington and others in the nineteenth century[28] showing that calcium phosphates are soluble in acids, and that iron and aluminium phosphates are hydrolysed in alkaline solution. The methods have been used to characterize the chemical linkage of the phosphate in unfertilized soils and the products of the reactions between soil and added phosphate.

Use of chemical fractionation

Chemical fractionation is used in one group of methods, of which the best known example is that developed by Chang and Jackson.[29] The sample of soil is extracted sequentially with: (i) neutral 0.5 M NH_4F to give aluminium phosphate; (ii) 0.1 M NaOH to give iron phosphate; (iii) 0.25 M H_2SO_4 to give calcium phosphate; (iv) sodium dithionite buffered with citrate to give occluded (reductant-soluble) iron phosphate; and finally (v) the residue is extracted with neutral 0.5 M NH_4F to give occluded aluminium phosphate. There is good evidence that the reagents extract phosphate from chemical linkages other than those specified,[30] and results from the method must therefore be treated with caution.

Use of correlation methods

The second group of methods seeks correlations between the amount of phosphate a group of soils will adsorb and the amount of aluminium or ferric iron soluble in solutions such as Tamm's acid oxalate or neutral pyrophosphate. A strong reducing solution, such as sodium dithionite, is not suitable because it dissolves all the free iron oxide, which usually contains some aluminium, whereas adsorption is a reaction with the mineral surfaces.

   The results of a great deal of work, using both methods, suggest that in

27 Koch, L. E., *Aust. J. Sci.*, 1956, **18**, 200.
28 For this early work see review by Wild, A., *J. Soil Sci.*, 1950, **1**, 221.
29 Chang, S. C. and Jackson, M. L., *Soil Sci.*, 1957, **84**, 133; *J. Soil Sci.*, 1958, **9**, 109.
30 Tandon, H. L. S. *et al.*, *Soil Sci. Soc. Am. Proc.*, 1967, **31**, 168. Bromfield, S. M., *Aust. J. Soil Res.*, 1967, **5**, 225; *idem, Soil Sci.*, 1970, **109**, 388.

well-drained, mildly acid soils of temperate regions, aluminium that is extracted by acid potassium chloride or Tamm's acid oxalate is the principal agent holding phosphate on the soil surface.[31] It has not yet been possible to identify the various sources of the aluminium, whether on the broken edge of a clay lattice, as ions either partially or fully neutralized with hydroxyls,[32] or that which is associated with organic matter.[33] However, in many poorly drained soils of temperate regions, and some soils of tropical regions, the surfaces of ferric oxide or hydrated oxide surfaces are also important.[34] Because of lack of evidence it is not known how far the experimental results merely reflect the relative surface areas of iron and aluminium hydroxides in soils. A considerable proportion of the phosphate adsorbed either by aluminium or iron oxide, as high as two-thirds in some soils, rapidly loses isotopic exchangeability.[35] Some of the phosphate that loses isotopic exchangeability also ceases to be extracted by the solvents mentioned above, and has presumably either become incorporated in stable crystals or has diffused below the active surface film.

Unfortunately, in most work with acid or slightly acid soils, the role of calcium has not been examined. At the same pH value, soils adsorb more phosphate when the dominant exchangeable cation is divalent, e.g. Ca^{2+}, than monovalent, e.g. Na^+. The effect is probably indirect, the exchangeable cation determining the rate and extent of reaction of the phosphate with exchangeable aluminium and mineral surfaces.[36]

Use of solubility products for calcium phosphates

The third method aims to establish the identity of the soil phosphate by assuming compatibility between the phosphate concentration in solution and the solubility product of pure phosphate minerals. This method was first used with the calcium phosphates.[37] Those most likely to occur in soils are, in decreasing order of solubility:

1   $Ca(H_2PO_4)_2.H_2O$, monocalcium phosphate, which is the water-soluble component of superphosphate and reacts in soil to form less soluble products.

31 Bromfield, S. M., *Nature, Lond.*, 1964, **201**, 321.
32 Franklin, W. T. and Reisenauer, H. M., *Soil Sci.*, 1960, **90**, 192.
33 Williams, E. G. *et al., J. Sci. Fd Agric.*, 1958, **9**, 551;   Harter, R. D., *Soil Sci. Soc. Am. Proc.*, 1969, **33**, 630; Le Mare, P. H., *J. Soil Sci.*, 1982, **33**, 691.
34 Chang, S. C. and Chu, W. K., *J. Soil Sci.*, 1961, **12**, 286 (Taiwan soils); Bates, J. A. R. and Baker, T. C. N., *op. cit.*, 1960, **11**, 257 (Nigerian forest soils).
35 Tandon, H. L. S. and Kurtz, L. T., *Soil Sci. Soc. Am. Proc.*, 1968, **32**, 799;   Le Mare, P. H., *J. Soil Sci.*, 1982, **33**, 691.
36 Wild, A., *J. Soil Sci.*, 1953, **4**, 72; Leaver, J. P. and Russell, E. W., *op. cit.*, 1957, **8**, 113.
37 For review of the chemistry of calcium phosphates see Larsen, S., *Adv. Agron.*, 1967, **19**, 151.

2 $CaHPO_4.2H_2O$, dicalcium phosphate (brushite) and the unhydrated form $CaHPO_4$ (monetite).

3 $Ca_8H_2(PO_4)_6.5H_2O$, octacalcium phosphate.

4 $Ca_3(PO_4)_2$, tricalcium phosphate, although it has not been established that it forms by precipitation from aqueous solutions.

5 $Ca_5(PO_4)_3OH$, hydroxyapatite and $Ca_5(PO_4)_3F$, fluorapatite, and substituted apatites, e.g. francolite (see p. 727).

Measurements of phosphate solubility have been represented on a solubility diagram in several ways. For calcareous soils it is convenient to plot the phosphate potential[38] ($0.5\ pCa^{2+} + pH_2PO_4^-$) against the lime potential ($pH^+ - 0.5\ pCa^{2+}$) as in Fig. 21.2. The measurements are made in 0.01 M $CaCl_2$ which has about the same concentration of calcium as is found in calcareous soils. If the concentration remains at 0.01 M, $0.5pCa^{2+}$ is approximately 1.2 when allowance is made for the activity coefficient. The value of $pH_2PO_4^-$ is calculated from the total inorganic concentration of the solution, the dissociation constants of phosphoric acid[39] and the activity coefficients of the phosphate ions. The lime and phosphate potentials of the calcium phosphates in Fig. 21.2 are represented as follows[40]:

diCa phosphate $pH - 0.5pCa = 0.5pCa + pH_2PO_4 + 0.66$

octaCa phosphate $5/3\ (pH - 0.5pCa) = 0.5pCa + pH_2PO_4 + 3.30$ (dissolving) or
$+ 4.13$ (precipitating)

hydroxyapatite $7/3\ (pH - 0.5pCa) = 0.5pCa + pH_2PO_4 + 4.73$ (dissolving) or
$+ 5.06$ (precipitating)

As an example to assist interpretation of Fig. 21.2, if a soil solution has a pH of 7.0 and its calcium concentration is 10^{-2} M, its lime potential is $7.0 - 1.2 = 5.8$. If, at the same pH value, the phosphate concentration is controlled by the solubility of hydroxyapatite its phosphate potential is about 8.5, and pH_2PO_4 is 7.3. Allowing for dissociation constants and activity coefficients this corresponds to a concentration of phosphate of about 10^{-7} M. If, also at pH 7, the concentration is controlled by octacalcium phosphate the phosphate potential is about 6.0, pH_2PO_4 is 4.8, and the concentration is $10^{-4.5}$ M $= 3.2 \times 10^{-5}$ M. As Fig. 21.2 shows, the phosphate concentration in calcareous soils does not correspond to that of any one mineral species. It is consistent with the solubility of octacalcium phosphate in several samples, with a trend

38 Schofield, R. K., *Soils Fertil.*, 1955, **18**, 373.

39 The following values (25 °C) are given by Lindsay, W. L., *Chemical Equilibria in Soils*, 1979, Wiley-Intersci., NY.
$H_3PO_4 \rightleftharpoons H^+ + H_2PO_4^-$ $pK° = 2.15$
$H_2PO_4^- \rightleftharpoons H^+ + HPO_4^{2-}$ $pK° = 7.20$
$HPO_4^{2-} \rightleftharpoons H^+ + PO_4^{3-}$ $pK° = 12.35$
These values, and the solubility products for the calcium phosphates given by Lindsay, are close to the values used by Aslyng and shown in Fig. 21.2.

40 Aslyng, H. C., *R. Vet. Agric. Coll., Copenhagen, Yearbook*, p. 1, 1954.

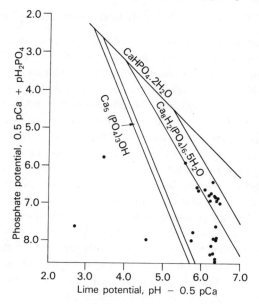

Fig. 21.2 Comparison of phosphate and lime potentials of calcium phosphates with the potentials in soils of Rothamsted Experimental Station. (From Aslyng, H. C., *Roy. Vet. Agric. Coll., Copenhagen, Yearbook,* p. 1, 1954.)

towards hydroxyapatite in soils that had not recently received phosphate fertilizer.

There are several difficulties in using solubility diagrams. First, solubility constants for some phosphates depend on whether they are being precipitated or dissolved. Second, the neutral molecule $CaHPO_4^0$ exists in solution and allowance must be made for it.[41] Third, low soluble phosphates such as hydroxyapatite can adsorb cations, e.g. calcium, or anions, e.g. phosphate, depending on the pH.[42] Perhaps the most important limitation, however, is that soil phosphate is rarely, if ever, at equilibrium and because the ions have low mobility, phosphate concentration (and potentials) vary on a micro scale from site to site.[43] The value of solubility diagrams such as that in Fig. 21.2, is that they show the expected sequence of precipitation in neutral and calcareous soils from soluble monocalcium phosphate to hydroxyapatite.

41 Larsen, S., *Nature, Lond.*, 1966, **212**, 605; improved by Adams, F., *Soil Sci. Soc. Am. Proc.*, 1971, **35**, 420; Bennett, A. C. and Adams, F., *Soil Sci. Soc. Am. J.*, 1976, **40**, 39.
42 Larsen, S., *Adv. Agron.*, 1967, **19**, 151.
43 White, R. E., *Pl. Soil*, 1964, **20**, 1, 184; *idem, Aust. J. Soil Res.*, 1966, **4**, 77; Aslyng, H. C., *Act. Agric. Scand.*, 1964, **14**, 260 discusses the precautions that must be taken. See also Clark, J. S., *Canad. J. Chem.*, 1955, **33**, 1696 for the effect of CO_2 on the solubility constant of hydroxyapatite.

Use of solubility products for iron and aluminium phosphates

Under very acid conditions minerals of the variscite–barrandite–strengite group are precipitated. Variscite has the formula $AlPO_4.2H_2O$, strengite $FePO_4.2H_2O$, and barrandite $(Al, Fe) PO_4.2H_2O$, with an Al : Fe ratio which varies over a wide range. Variscite has been identified as a reaction product of the dissolution of monocalcium phosphate pellets in acid soils, because the pH of the solution diffusing out of the pellet is as low as 1 to 1.5. This, and other reaction products, e.g. taranakites formed under these conditions, are discussed later (p. 731). From the solubility products of variscite and strengite, solubility diagrams have been used to compare concentrations in solution with those expected if these minerals control the concentrations. It was found, for example, that concentrations in some acid soils were compatible with the solubility constant of variscite, but in other soils the concentrations were lower than expected.[44] Solubility constants can, however, only be used if the phosphates dissolve congruently, and the results of Bache[45] in Fig. 21.3 show this to be so for variscite and strengite only at low pH values. Variscite hydrolyses to an increasing extent at pH values above 3.1 and strengite at pH values about 1.4. This evidence strongly indicates that in acid soils, unless they are of exceptionally low pH, the concentration of phosphate is

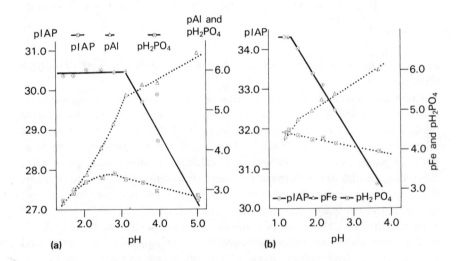

Fig. 21.3   Effect of pH on the solubility of (a) variscite and (b) strengite. pIAP = negative log. of ion activity product. For variscite, pIAP = pAl + pH_2PO_4 + 2POH. For strengite, pIAP = pFe + pH_2PO_4 + 2POH.

44 Clark, J. S. and Peech, M., *Soil Sci. Soc. Am. Proc.*, 1955, **19**, 171. See also Taylor, A. W. and Gurney, E. L., *J. Soil Sci.*, 1962, **13**, 187; *Soil Sci.*, 1962, **93**, 241; 1964, **98**, 9.
45 Bache, B. W., *J. Soil Sci.*, 1963, **14**, 113.

not controlled by the solubility of variscite or strengite. In fact, the only direct identification of variscite in soil is as a transitory reaction product at very low pH values close to phosphate fertilizer pellets, and there is some evidence that strengite can form under the same conditions. As shown later (Fig. 21.6) the solubility of phosphate in acid soils often does not increase as the pH value is increased, as would be expected if variscite and/or strengite controlled the solubility.

Adsorption of phosphate by soil

When soil is shaken with a solution of phosphate, adsorption is initially fast but continues slowly for many days. The rate and extent of the reaction in acid soils probably depend more on the specific surface area of the soil mineral surfaces than on whether the surface is a hydrous oxide of iron or aluminium or clay mineral. Examples of soils which have high phosphate buffer capacities, that is, adsorb large amounts for phosphate for a small increment in solution concentration, are Oxisols[46] which have high contents of disordered (amorphous) oxides and soils containing allophane.[47] Such soils have large areas of surfaces containing iron and aluminium. The chemistry of adsorption is described in Chapter 7.

Adsorption is generally greater at pH 3–5 than at higher pH values unless calcium phosphates are precipitated. It increases with time and temperature, and depends on the soil : solution ratio, on the nature and concentration of salts which are present and on previous treatment of the soil.[48] The plot of amount of phosphate adsorbed against concentration left in solution therefore depends on the conditions of the experiment. The plot is curved but a linear relationship can be found over a limited range of concentration by the use of equations such as those due to Langmuir, Freundlich and Temkin (see Ch. 7). As pointed out by Barrow[48] the two main reasons for using these equations are to understand the processes involved and to summarize many results by a few parameters.

For the purpose of investigating phosphate availability to plants, it is useful to plot the amount of phosphate, Q, desorbed or adsorbed from a phosphate solution over a short period of time, such as one or two hours, against the phosphate potential, I, or concentration in solution. Beckett and White found that for one of the soils they used, derived from the Upper Greensand, the

46 Goedert, W. J., *J. Soil Sci.*, 1983, **34**, 405.
47 Fox, R. L., *Trop. Agric., Trin.*, 1974, **51**, 200.
48 Barrow, N. J., *J. Soil Sci.*, 1978, **29**, 447; Shayan, A. and Davey, B. G., *Soil Sci. Soc. Am. Proc.*, 1978, **42**, 878.   For reviews of the reaction see Mattingly, G. E. G. and Talibudeen, O. in *Topics in Phosphorus Chemistry* (eds M. Grayson and E. J. Griffith), 1967, **4**, 157;   White, R. E. in *Soils and Agriculture* (ed P. B. Tinker), p. 71, Blackwell, 1980.

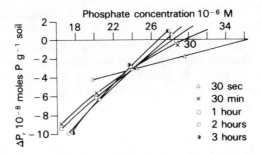

Fig. 21.4 The relation between phosphate desorbed and the phosphate concentration for a field sample of an Upper Greensand soil, after different times of contact between the soil and the solution.

Q/I curve was approximately linear, as is shown in Fig. 21.4.[49] This allowed them to estimate, by extrapolation, a quantity Q_o the amount of phosphate that could be released if the phosphate concentration was reduced to zero, although in this example the range of concentrations is not suitable for extrapolation. The particular samples used for the results illustrated in the diagram were taken straight from the field, and the sample as a whole was clearly not in equilibrium. Q/I curves have proved less reliable for phosphate than for potassium (see p. 746) and have less theoretical justification. They are useful for giving phosphate concentrations in soil solutions (at $\Delta P = 0$), and the slope gives the buffer capacity of the soil for phosphate.

The buffer capacity for phosphate differs widely for different soils and, as indicated earlier, is generally high in soils with high surface areas containing iron and aluminium. In some soils from Britain, the buffer capacity increased with clay content and with the amount of extractable phosphate the soil contained, but decreased with increasing calcium carbonate content.[50] In these experiments the buffer capacities, and phosphate potentials, were determined after short periods of contact (1 to 2 hours) between soils and solutions. Longer term buffering has been observed by depleting the soil of phosphate and then storing the soil moist. In the results shown in Table 21.2 part of the soil phosphate was removed by extraction with an anion exchange resin. After being stored moist for three or four months, the buffer capacity, the concentration of phosphate in solution (expressed as activity), and the value of Q_o, were about the same as for the original soil if up to about half of the original isotopically exchangeable phosphate was removed. Thus, if an appreciable proportion of the isotopically exchangeable phosphate is taken up by a crop, the phosphate buffer capacity and the phosphate potential of

49 Beckett, P. H. T. and White, R. E., *Pl. Soil*, 1964, **21**, 253.
50 See for example Webber, M. D. and Mattingly, G. E. G., *J. Soil Sci.*, 1970, **21**, 111 for soils being depleted of phosphate, and Jensen, H. E., *Pl. Soil*, 1970, **33**, 17 for soils being enriched.

Table 21.2   The effect of removing phosphate on the Q/I curve. Soil: Upper Greensand. Initial isotopically exchangeable P in soil about 80×10^{-8} mol g^{-1}. The Q/I curve was measured by shaking soil samples with 0.01 M $CaCl_2$ for 1 hour

Quantity of P removed (moles/10⁸ g soil)	Period of storage after depletion, days at 25 °C	Activity of H_2PO_4 (μ M)	$\dfrac{\Delta Q}{\Delta I}$	Q_0 (moles/10⁸ g soil)
0	120	42.5	4.3	19.0
8	126	30.0	6.0	18.5
16	117	32.7	5.0	17.0
38	94	38.5	5.5	21.5
95	95	10.0	4.5	5.0

the soil will return close to the original values after a period of moist storage, for example, between cropping seasons.

Experimentally similar to the Q/I curve is measurement of the amount of phosphate a soil must absorb to be approximately in equilibrium with a solution of about 10^{-5} M in phosphate after a fixed period of time such as one day. This measurement identifies soils that adsorb large amounts of phosphate, and so need large applications of phosphate fertilizer. It is discussed on p. 725.

Phosphate available to plants

Phosphate must be in solution before it can be absorbed by plant roots, yet the concentration in the soil solution is usually very small and must be replenished if the supply to plants is to be adequate. The processes by which phosphate moves between solid and liquid phases and is transported to the plant root, are shown schematically in Fig. 21.5, where the numbers in brackets give the order of magnitude of phosphate as μg P g^{-1} in the top 40 cm of a reasonably fertile soil at 30 per cent volumetric water content; other numbers refer to the processes discussed below.

Some of the phosphate ions which are held on the mineral surfaces of the soil are desorbed (process 1) when the concentration in solution falls. Phosphate in solution diffuses to the root surface (process 2), where its concentration is usually low because of adsorption by the root (process 3). Desorption is reversed (process 5) when the concentration of phosphate in solution is raised. Processes 6 and 7 are, respectively, the slow conversion of isotopically exchangeable (that is, exchangeable with phosphate ions labelled with ^{32}P) into non-exchangeable phosphate and its release. Processes

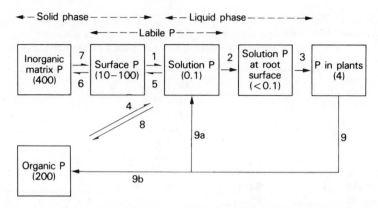

Fig. 21.5   Schematic representation of processes of transfer of phosphate between soil and plants; numbers in parentheses are amounts of phosphate (μg P g^{-1} soil) that might be expected in a reasonably fertile soil at 30 per cent volumetric water content. Processes 1–9 are discussed in the text.

4 and 8 are the mineralization of organic phosphorus and microbial immobilization. Phosphate is returned to the soil from plant roots and tops as inorganic phosphate (process 9a) and as organic phosphorus (process 9b).

The numbers in brackets indicate that the total phosphate in the soil is sufficient for about 150 crops, but even from the early days of crop experimentation it was realized that total phosphate content is a wholly inadequate measure of phosphate availability to plants. As early as 1845 Daubeny[51] distinguished between the 'active' and 'dormant' fractions of soil nutrients. The 'active' fraction was later considered to be easily extracted from soil by various acids, alkalis and salt solutions. Correlations of the amounts extracted with the response of crops to applications of phosphate fertilizer became the basis of methods of advising growers on fertilizer use. These methods are referred to later. The distinction between 'active' (available) and 'dormant' (unavailable) is useful but arbitrary, as indeed is the distinction made in Fig. 21.5 between 'surface' and 'matrix' inorganic phosphate. The distinction made in the diagram is only useful because 'surface phosphate' can be defined by the method of its determination. In principle it is better to regard the relation between solid phase and liquid phase phosphate as a continuous function. This is the basis of the Quantity/Intensity (Q/I) relation in which the quantity of P adsorbed or desorbed is plotted against concentration or phosphate potential as discussed earlier.

51 Daubeny, C., *Phil. Trans. R. Soc.*, 1845, 179.

Phosphate in the soil solution and plant uptake

Although the concentration of phosphate in solution can be below 10^{-8} M in some very poor tropical soils,[52] and is of the order of 10^{-6} M in temperate soils known to be deficient in phosphate, it is about 10^{-5} M in many soils of moderate phosphate status[53] (10^{-5} M corresponds to 0.3 mg P l^{-1} in the soil solution). A soil which has a soil solution 10^{-5}M in phosphate and which, for example, holds 6 cm of water in the top 30 cm will have less than 0.2 kg P ha^{-1} in solution to that depth. If a crop uses 37 cm of water during its growth, there will only be about 1 kg P ha^{-1} dissolved in it, yet it may take up 20 to 40 kg P ha^{-1} during the growing season. Even though it is often very dilute, the solution is able to provide plants with phosphate because roots can absorb it from low concentrations, and phosphate is maintained in solution by desorption from the surface phase. The evidence of Fried and his co-workers[54] is that in soils well supplied with phosphate, the ions can go into solution at a very much higher rate than roots can take them up. This would lead to the expectation that phosphate concentration in solutions of such soils would remain constant throughout the growing season. This has not been thoroughly tested, but at Rothamsted and Woburn phosphate concentrations in calcium chloride extracts of soils were generally highest in spring, decreased to harvest and rose again in the winter.[55] It is not known whether the changes in the summer were due to cropping or to the soil becoming dry.

There is now good evidence that plants can achieve their potential growth rate when grown in solutions of low phosphate concentration. The technique that has been used to measure the minimum concentration of phosphate needed for good plant growth is to grow plants in solutions which are pumped through the root systems (flowing solution culture). Concentrations of 1 to 25×10^{-6} M phosphate were found by Asher and Loneragan[56] to be required for maximum growth by eight different species transferred to the solutions when they were 8 to 13 days old and grown for four weeks. In other similar experiments, the approximate concentrations required by 18 plant species ranged from $>0.25 \times 10^{-6}$ M for stylo (*Stylosanthes guyanensis*), soya bean (*Glycine max*) and maize (*Zea mays*) to $>50 \times 10^{-6}$ M for cassava (*Manihot*

52 *Rubber Res. Inst. Malaya, Ann Rept. for 1967*, p. 70.
53 Analyses are reported by Nelson, W. L. *et al.* in *Soil and Fertilizer Phosphorus* (eds W. H. Pierre and A. G. Norman), p. 153, *Agron. Monog.*, No. 4, Acad. Press, 1953. Wild, A., *Trans. 5th Int. Congr. Soil Sci.*, 1954, **2**, 500, gives references to early analyses. See also Asher, C. J. and Loneragan, J. F., *Soil Sci.*, 1967, **103**, 225; Barber, S. A., *Soil Nutrient Bioavailability*, p. 202, 1984, Wiley-Interscience. For comparison of methods of obtaining soil solutions see Adams, F. *et al.*, *Soil Sci. Soc. Am. J.*, 1980, **44**, 733.
54 Fried, M. *et al.*, *Soil Sci.*, 1957, **84**, 427.
55 Blakemore, M., *J. agric. Sci.*, 1966, **66**, 139; Garbouchev, I. P., *op. cit.*, 399.
56 Asher, C. J. and Loneragan, J. F. (see ref. 53).

esculenta).[57] Lower concentrations were adequate for ryegrass (*Lolium perenne*) grown from seed in the nutrient solution.[58] Young plants needed about 6.4×10^{-6} M phosphate whereas plants older than four weeks were able to take up phosphate at a sufficient rate to meet their requirements from concentrations between 0.04 and 0.4×10^{-6} M. This and other work has shown that the concentration required by plants is very low and differs between species, their stage of growth and growth rate. Although there can be no universally applicable threshold concentration below which plants will suffer deficiency, in flowing nutrient solutions young plants need about 10^{-6} to 10^{-5} M phosphate $(0.03-0.3 \text{ mg P l}^{-1})$ and older plants require about 10^{-6} M phosphate $(0.03 \text{ mg P l}^{-1})$ or less. These are concentrations at the root surface. Concentrations which have been shown by several workers to be required in bulk soil solution[59] are a little higher $(0.06-0.68 \text{ mg P l}^{-1})$, and this would be expected because uptake reduces the phosphate concentration at the root surface when plants are grown in static systems, e.g. soils.

Transfer to the root surface

In soil, phosphate must be transported through the soil solution to the root surface. This is mainly by diffusion, a process discussed more fully on pages 155–163. Mass flow of the soil solution to roots, the other major process whereby nutrients reach the root surface, normally provides only a small fraction (less than one-tenth) of the phosphate taken up by the plants because of the small concentration in solution. It can be calculated that for mass flow to provide the whole of the 0.7 to 1 kg P ha^{-1} day^{-1} required by rapidly growing crops (Table 3.4) transpiring at 5 mm day^{-1}, the soil solution must have a phosphate concentration of about 5 to 7×10^{-4} M, which is attained only in some heavily manured soils.

Diffusion is a slow process, and with phosphate the diffusive flux is particularly slow because of small concentration gradients. Because the flux is slow most of the phosphate taken up by plants diffuses from within a very short distance (up to about 5 mm in 1 week) of the root. In order to achieve sufficient uptake of phosphate, plant roots should therefore ramify well through the soil, creating as close contact as possible with the sources of phosphate. Root hairs increase the effective diameter of the root cylinder and so increase the volume of the phosphate feeding zone.[60] Hyphae from mycorrhizal associ-

57 Asher, C. J. in *CRC Handbook Series in Nutrition and Food*, Vol. 3 (ed Miloslav Recheigl, Jr.), p. 575, 1978.
58 Breeze, V. G. *et al.*, *J. exp. Bot.*, 1984, **35**, 1210. For results with lettuce and cabbage see Temple-Smith, A. B. and Menary, A. C., *Aust. J. Pl. Physiol.*, 1977, **4**, 505.
59 Kamprath, E. J. and Watson, M. E. in *The Role of Phosphorus in Agriculture* (eds F. E. Khasawneh *et al.*), p. 433, Am. Soc. Agron., Madison, Wisc., 1980.
60 Nye, P. H. and Tinker, P. B., *Solute Movement in the Soil-Root System*, p. 145, Blackwell, 1977; Barber, S. A., *Soil Nutrient Bioavailability*, p. 170, Wiley-Interscience, 1984.

ations with roots have a similar effect. For good phosphate supply to plants the structural condition of the soil should allow the development of a high density of roots, or of mycorrhizal hyphae, and the soil should not become dry because then diffusion becomes very slow. It follows that both soil and plant characteristics determine the diffusive flux of phosphate to the root surface.

The effect of soil texture on the supply of phosphate to plants has been demonstrated by Olsen and Watanabe.[61] They solved the diffusion equation giving the uptake of phosphate Q by a plant root during a time t in the form

$$Q = aB(C_o - C_r)f(Dt/Ba^2)$$

where a is the radius of the root, B the buffer capacity of the soil, D the diffusion coefficient of phosphate in the soil, C_o and C_r the phosphate concentrations in the bulk solution of the soil and at the root surface respectively, and for $f(Dt/Ba^2)$ see ref. 61. Table 21.3 gives their calculated values of C_o, in three soils, which are needed by a maize root if it is to take up 31 μg P g^{-1} root per day, which is a typical figure for a fast-growing maize crop. The table also gives the amount of water-soluble phosphate that must be added to the soil, and the amount of phosphate extractable by sodium bicarbonate (available phosphate) that each soil must contain to give the necessary value of C_o. It is interesting to note that the required level of available phosphate is less dependent on texture than the amount of phosphate that must be added and the necessary concentration of phosphate at the root surface. Thus clay soils may need a higher rate of phosphate manuring and appear to be more phosphate deficient than sands or silty soils, because they are more strongly buffered. For the same reason clay soils do not need as high concentrations of phosphate in solution as sandy soils.

Table 21.3 Effect of soil texture on phosphate concentration, C_o, required in the soil solution to meet the phosphorus requirement of a rapidly growing maize crop

Soil	Diffusion coefficient ($cm^2 s^{-1}$)	Buffer capacity*	C_o, μM Actual	C_o, μM Necessary	P to be added (mg/100 g soil)	Available P necessary (mg/100 g soil)
Clay	5.40×10^{-7}	261	1.0	3.5	4.05	2.20
Silty clay loam	3.23×10^{-7}	86.9	2.9	9.2	2.90	2.18
Fine sandy loam	1.12×10^{-7}	28.0	7.1	27.2	2.06	1.96

* Dimensionless.

61 Olsen, S. R. and Watanabe, F. S., *Soil Sci.*, 1970, **110**, 318.

Effect of pH

The pH of the soil also affects phosphate uptake. In nutrient solutions free of aluminium the pH at which phosphate uptake is greatest has been reported as 5.5 for sunflower and 4.3 for clover.[62] At higher pH values uptake decreases, as does the concentration of $H_2PO_4^-$ above pH 5 (see ref. 39). One reason that has been given for the effect of pH is that $H_2PO_4^-$ reacts more strongly with its binding site (DPNH) in root cells than HPO_4^{2-} reacts with its own binding site (cytochrome b).[63] However, to reach the sites the ions diffuse between and within cells of the root cortex whose walls are negatively charged, and an alternative explanation is that the HPO_4^{2-} ion is excluded from some diffusion pathways because of its double negative charge. A simple relation cannot however be expected between phosphate uptake and the pH of the external solution because the external pH will differ from that at the cellular level. In their work with clover Dunlop and Bowling[62] found that the external pH affected the electrical potential difference between the root and the outside solution, and that this was closely related to phosphate uptake. They suggested that phosphate entry to roots is coupled to the re-entry of protons previously extruded by an ATP pump. Phosphate entry is therefore favoured by a low pH (down to 4) and decreases as the pH value is increased.

In soils the main effect of pH on phosphate uptake arises indirectly. If the solution concentration of phosphate is controlled by the solubility of octa-calcium phosphate or apatite, in soils at equilibrium with pH above 7 the concentration is 10^{-6} M or less. At lower pH values, and to some extent also above pH 7, the concentration is controlled by adsorption, which is not uniquely related to pH. It is generally believed that phosphate concentration is greatest at a soil pH of 6 to 6.5, but the evidence is by no means clear. For example, Fig. 21.6 shows lowest solution concentrations in four soils, under non-equilibrium conditions, at pH 6 to 6.5.[64] There are also many reports of smaller, as well as higher availability of phosphate to plants, following lime additions to an acid soil. The reduction of phosphate solubility (and availability) after liming an acid soil probably results from high phosphate adsorption on to freshly precipitated aluminium-hydroxyl polymers. It has been suggested that when these polymers have been aged, e.g. by drying, adsorption is less, so that more phosphate remains in solution; hence acid soils which have a high phosphate adsorption capacity should be allowed to dry after liming and before phosphate fertilizer is added.[65] The increase in phosphate availability after liming, which has also been observed, can be the

62 For sunflower see Blamey, F. P. C. *et al.*, *Proc. 9th Int. Pl. Nutr. Coll.*, 1982, **1**, 66. For clover see Dunlop, J. and Bowling, D. J. F., *J. exp. Bot.*, 1978, **29**, 1147.
63 Hagen, C. A. and Hopkins, H. T., *Pl. Physiol.*, 1955, **30**, 193.
64 Murrmann, R. P. and Peech, M., *Soil Sci. Soc. Am. Proc.*, 1969, **33**, 205. For summary of early analyses see Wild, A., *Trans. 5th Int. Congr. Soil Sci.*, 1954, **2**, 500.
65 See reviews by Haynes, R. J., *Pl. Soil*, 1982, **68**, 289; *Adv. Agron.*, 1984, **37**, 249.

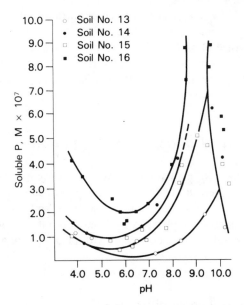

Fig. 21.6 Relation between phosphate concentration and pH of solution for four soils shaken with graded amounts of hydrochloric acid or calcium hydroxide solution.

result of less adsorption as the pH is increased, as occurs because of ligand exchange on oxide surfaces,[66] or follow from greater mineralization of soil organic phosphorus.

In measuring adsorption a distinction needs to be made between exchangeable and non-exchangeable phosphate.[67] In a laboratory experiment using three Oxisols from Brazil and Colombia liming to pH 6.2 consistently decreased the amount of non-exchangeable phosphate, measured by isotopic exchange, after phosphate had been added to the soils. This might explain the positive interaction observed in this work between lime and phosphate on yields of soya beans in the field; together they gave a greater yield response than the sum of the responses from separate applications.

Another indirect effect of soil pH, and hence an effect of liming, is that aluminium taken up by plants growing in acid soils is itself toxic (see Ch. 25). Aluminium interacts with phosphate in plants (p. 872) because aluminium is precipitated as Al–OH polymers in the root cortex where it adsorbs phosphate and prevents or reduces its transport to the stems and leaves.[68] Aluminium also blocks the numerous biochemical reactions in plant cells that involve the transfer of phosphate ions, for example in oxidative processes.

66 White, R. E., ref. 48; *Proc. 3rd Int. Congr. Phosphorus Compounds*, 1983, 53.
67 Le Mare, P. H. and Goedert, W. J., *Proc. 3rd Int. Congr. Phosphorus Compounds*, 1983, 65.
68 Clarkson, D. T., *Pl. Soil*, 1967, **27**, 347. See Haynes (ref. 65) for later work.

Raising the soil pH by liming lowers the concentration of soluble aluminium and its uptake by plants, and so has a beneficial effect on plant metabolic processes that involve phosphate. Although there are many exceptions which were referred to earlier, the beneficial effect of liming on the phosphate nutrition of plants seems most generally to arise because it reduces the uptake of aluminium by plant roots.

Replenishment of phosphate in solution

Uptake of phosphate by plants would cause its concentration in solution to fall rapidly to zero, were it not for mineralization of soil organic phosphorus compounds, the desorption from mineral surfaces and the solubilization of calcium, and possibly other, phosphates. There are many reports[69] that phosphate held on mineral surfaces correlates with plant uptake and it seems safe to assume that in fertilized soil desorption of phosphate is the principal means by which the concentration in solution is maintained.

Isotopically exchangeable phosphate

Four methods have been used to measure the amount of phosphate which can be desorbed from soil surfaces.[70] Radioisotope methods have been developed from the work of McAuliffe and his colleagues. Based on the reaction:

$$^{31}P_{surface} + {}^{32}P_{solution} \rightleftharpoons {}^{32}P_{surface} + {}^{31}P_{solution}$$

at equilibrium one may write:

$$\frac{^{32}P_{surface}}{^{31}P_{surface}} = \frac{^{32}P_{solution}}{^{31}P_{solution}} \qquad [21.1]$$

In these equations ^{31}P and ^{32}P represent orthophophate ions in which the phosphorus atom is either the naturally occurring isotope, ^{31}P, or the radioisotope, ^{32}P. Using equation [21.1], the $^{31}P_{surface}$ is calculated from the measured concentrations of ^{31}P and ^{32}P in solution at equilibrium, and the $^{32}P_{surface}$, which is given by the difference between the ^{32}P added and that left in solution. A convenient form of equation [21.1] is

$$^{31}P_{surface} = (y - y_t)x_t/y_t = \left(\frac{y}{y_t} - 1\right) x_t \qquad [21.2]$$

where y and y_t are respectively the amount of ^{32}P added and the amount left in solution after time t, both expressed in decay units, e.g. counts per minute,

69 See Mattingly, G. E. G., *MAFF Tech. Bull.*, No. 13, 1, HMSO, 1965.
70 Russell R. S. *et al., J. Soil Sci.*, 1957, **8**, 248.   For review see Larsen, S., *Adv. Agron.*, 1967, **19**, 151.

and x_t is the amount of ^{31}P in solution after time t per unit weight of soil.

The exchange between ^{32}P and surface exchangeable ^{31}P is at first rapid but becomes progressively slower. It is therefore necessary to specify the time of the exchange reaction. A practical difficulty is that in unfertilized soils x_t might be too small to measure accurately. The isotope can then be added with a 'carrier' of ^{31}P, but there is then adsorption of the added ^{31}P concurrently with isotopic exchange. White has shown how to calculate surface exchangeable phosphate under these conditions.[71]

Measurement of isotopically exchangeable phosphate (E value) is made by shaking a small amount of soil (about 1 g) for 24 hours with a solution of 0.01 M $CaCl_2$ (or other solution to approximate to that found in soil), after which the ^{32}P is added. The soil suspension is shaken for a further 24 hours, centrifuged, and ^{32}P and ^{31}P in the solution are measured.[72]

The amount of phosphate which undergoes isotopic exchange depends on the experimental conditions. The effect of increasing the time of shaking between the soil and the ^{32}P is to decrease the specific activity of the solution curvilinearly, so that after long shaking times equilibrium is approached but not reached. For practical purposes the times of exchange can be chosen to give rapidly exchanging phosphate (e.g. after 24 hours) and the 'total' isotopically exchangeable phosphate (e.g. after 150 or 170 hours), the times being those used by Talibudeen.[73] An example of this fractionation is given in Table 21.4.

Exchangeable phosphate is affected by the presence of organic anions, by the soil : water ratio if this is greater than about 2 g/100 cm³, and by tempera-

Table 21.4 Labile phosphate in some Rothamsted soils measured by isotopic exchange for 24 and 150 hours; P_T = total soil phosphate

Soil description*	mg P/100 g				
	P_{24}	P_{150}	P_T	P_{24}/P_{150}	P_{150}/P_T
1 Exhaustion land: nil	2.9	4.8	42.6	0.61	0.11
2 FYM	7.7	11.0	65.1	0.70	0.17
3 Superphosphate	6.7	9.2	67.8	0.73	0.14
4 Broadbalk: nil	2.6	4.5	60.5	0.58	0.08
5 $(NH_4)_2SO_4$ only	2.7	7.6	64.5	0.36	0.12
6 Superphosphate	18.7	24.1	115.2	0.78	0.21

* Soil 1 no fertilizer at any time.
  Soil 2 FYM up to 1901.
  Soil 3 superphosphate up to 1901.
  Soil 4 no fertilizer at any time.
  Soil 5 ammonium sulphate annually since 1843.
  Soil 6 superphosphate annually since 1843.

71 White, R. E., *Phosphorus in Agric.*, 1976, **67**, 9.
72 Le Mare, P. H., *J. Soil Sci.*, 1982, **33**, 691.
73 Talibudeen, O., *J. Soil Sci.*, 1958, **9**, 120.

ture, but with no difference between NaCl, KCl and $CaCl_2$ at the same ionic strength.[74]

A method due to Larsen[75] measures the plant uptake of ^{32}P and ^{31}P and gives the L value. ^{32}P, usually with a carrier, is thoroughly mixed with the soil, plants are grown for a period of several days or weeks and their ^{32}P and ^{31}P contents are measured. The principle, and the method of calculation, are similar to those for the E value. The calculated value also includes phosphate in the soil solution, though its contribution is negligible. A method due to Fried and Dean gives the A value.[76] It differs conceptually from that of E and L value measurements, in that the plant has access to soil phosphate (unlabelled) and to fertilizer phosphate (^{32}P labelled), which are not mixed together. Assuming that the two sources are used in direct proportion to the amounts available, the A value (soil available phosphate) is calculated from the uptake of unlabelled phosphate. The value obtained depends on the conditions of the experiment, for example on the amount of fertilizer phosphate which is added.

The final method uses anion exchange resins[77] in the chloride or bicarbonate form. The adsorption of phosphate by the resin from the soil solution results in desorption from the soil, and at the end of the extraction period the phosphate content of the resin is measured. The amount of phosphate extracted depends on the period of extraction and on the ratio of resin to soil.

The desorption reaction

As referred to above, adsorption of phosphate consists of a fast and slow reaction[78] and the desorption process behaves similarly. Fast desorption, which occurs by ligand exchange, appears to be complete in a few hours if the soil is finely ground and a sink, such as anion exchange resin, is provided for the desorbed phosphate. The rate of the slow reaction may depend on slightly soluble calcium phosphates being dissolved, on low diffusion rates through porous minerals, or a desorption requiring high activation energy. Desorption occurs in the presence of such ligands as hydroxyl, citrate, oxalate, acetate and tartrate some of which can be produced by the rhizosphere population. Desorption of phosphate in the vicinity of plant roots may, however, be more influenced by local pH changes than by the microbial production of organic anions.[79] As described elsewhere (p. 146), a high ratio

74 Mattingly, G. E. G. and Talibudeen, O., *Rothamsted Ann. Rept. for 1960*, p. 246.
75 Larsen, S., *Pl. Soil.*, 1952, **4**, 1.
76 Fried, M. and Dean, L. A., *Soil Sci.*, 1952, **73**, 263;   Fried, M., *8th Int. Congr. Soil Sci., Bucharest*, 1964, **4**, 29 discusses differences between L and A values.
77 Amer, F. *et al.*, *Pl. Soil*, 1955, **6**, 391. Zunino, H *et al.*, *Soil Sci.*, 1972, **114**, 404; Sibbesen, E., *Pl. Soil*, 1977, **46**, 665.
78 Barrow, N. J., *J. Soil Sci.*, 1978, **29**, 447, *Soil Sci.*, 1974, **118**, 380.
79 Tinker, P. B. in *The Role of Phosphorus in Agriculture* (eds F. E. Khasawneh *et al.*), p. 617, Am. Soc. Agron., Madison, Wisc., 1980.

of cations to anions taken up by plants lowers the pH in the root environment and a low ratio raises it. The effect of a change of pH on phosphate desorption depends on the soil properties. There are reports that phosphate concentration is increased by raising the pH,[80] which would be expected because of the effect of OH^- as a displacing ligand, but there are also reports that it is decreased, possibly because raising the pH to about 6 increases the hydrolysis and polymerization of aluminium on to which phosphate ions are adsorbed.[65]

Under intensive cropping over long periods the quantity of isotopically exchangeable phosphate becomes more important than the concentration in solution. This is illustrated in Fig. 21.7 which gives results of a pot experiment. The yields of grass were the same after 40 days whatever the quantity of exchangeable phosphate in the soil, but after 60 days they diverged so that yields became directly related to exchangeable phosphate. The solution concentration affected the initial but not later yields.

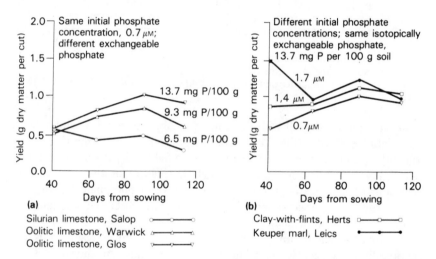

Fig. 21.7 Effects of different initial phosphate concentrations in soil solutions, and different levels of isotopically exchangeable P in calcareous soils on rate of growth of ryegrass. (From Mattingly, G. E. G., MAFF Tech. Bull., No. 13, 1, HMSO, 1965.)

Phosphate which is slowly available

Phosphate which is adsorbed by soil generally becomes more slowly exchangeable the longer the period of incubation or contact between soil and the phosphate, as shown by the work of Talibudeen referred to earlier (Table 21.4). With three soils from Brazil, isotopically exchangeable and resin-

80 Barrow, N. J., *J. Soil Sci.*, 1984, **35**, 283, reported increase and decrease of phosphate sorption by different soils as the pH was raised. Grinsted, M. J. *et al.*, *New Phytol.*, 1982, **91**, 19, found increased desorption as the soil pH was lowered from 6.2 to 4.5.

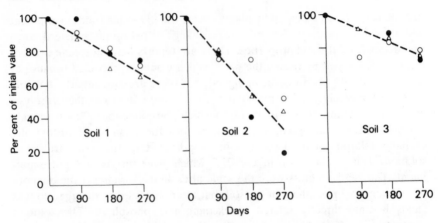

Fig. 21.8 Decrease of labile phosphate in three soils from Brazil after reaction periods of 90, 180 and 270 days with monocalcium phosphate (250 μg P g⁻¹ soil). Labile P is expressed as a per cent of the value found after a reaction period of 0 days, measured as P uptake by sorghum plants (●), P extracted by anion exchange resin (○), and as isotopically exchangeable P (△).
Soils 1 and 3 = Ultisols, soil 2 = Oxisol.
(From Cabala-Rosand, P. and Wild, A., *Pl. Soil*, 1982, **65**, 351.)

extractable phosphate fell with the time of incubation (at 23 °C and a water potential of −33 kPa) (Fig. 21.8). The steepest fall was with the Oxisol, for which the isotopically exchangeable phosphate fell to 42 per cent of its initial value after incubation for 180 days, but the reason for the difference in behaviour of the soils is not known. The uptake of phosphate by sorghum plants grown on the same soils decreased similarly after the soils and the phosphate had been incubated.

Phosphate which is not extractable by anion exchange resin, or is not isotopically exchangeable, may still be used by plants. As shown in Table 21.5, there was a substantial release of phosphate during a 10-month cropping period with ryegrass. In this experiment labile phosphate was measured by two methods, namely, by extracting 1 g soil with 2 g anion exchange resin (chloride form) in 20 ml water, and by shaking 1.5 g soil for 24 h with 0.01 M CaCl₂ containing ³²P. It is not known how much P mineralized from soil organic matter during the cropping period was taken up by the ryegrass.

The Hoosfield (Exhaustion Land) plots at Rothamsted also show that the decrease in isotopically exchangeable phosphate over the period 1958–74 was much less than the uptake of phosphate by the crops (barley) over this period (Table 21.6). The isotopically exchangeable phosphate was measured over a 168–hour period. Soil samples were taken only at a depth of 0 to 23 cm and it is uncertain how much of the phosphate taken up by the crops came from below that depth. It is probable, however, that most of the applied phosphate

Table 21.5 Uptake of phosphate by ryegrass over a 10-month cropping period in pots compared with changes in labile phosphate. All nutrients except phosphate added. (mg P kg^{-1} soil) (Data from Owusu-Bennoah (private communication))

| Soil | Uptake by ryegrass | Resin extractable P | | | |
		Initial	Final	ΔP	Release
Sonning	86.3	78.0	61.8	16.2	70.1
Hamble	66.0	63.5	33.5	30.0	36.0

| Soil | Uptake by ryegrass | Isotopically exchangeable P | | | |
		Initial	Final	ΔP	Release
Sonning	86.3	75.5	55.4	20.1	66.2
Hamble	66.0	50.6	32.1	18.5	47.5

remained in the top 23 cm because other measurements at Rothamsted have shown very little downward movement of phosphate. About 45 per cent of the phosphate residues in the Hoosfield plots remained isotopically exchangeable and the explanation for the data in Table 21.6 is probably that much of the phosphate taken up by the crops came from slowly exchangeable phosphate which accumulated from applications of farmyard manure and phosphate fertilizer in the period 1856–1901. On the unmanured plots the uptake by crops (76 kg P ha^{-1}) was little greater than the decrease in isotopically exchangeable phosphate (60 kg P ha^{-1}), probably because there was no slow release from any previous application of phosphate.

Another reason why the quantity of surface-held P which is isotopically exchangeable is not a measure of the amount available to plants, is that the solution concentration may limit uptake. This was recognized in early work by Russell and his co-workers (see ref. 70) which showed a close relation for any one soil between P uptake by plants and isotopically exchangeable P, but the relation broke down when more than one soil was included in the

Table 21.6 Comparison (kg P ha^{-1}) between phosphate uptake by crops (1958–74) and changes in isotopically exchangeable phosphate on plots which received manures 1856–1901 and then only nitrogen fertilizer; Hoosfield (Exhaustion Land), Rothamsted. (See Johnston, A. E. and Poulton, P. R., 1977, *Rothamsted Ann. Rept. for 1976*, Part 2, 53)

| Treatment 1856–1901 | P removed in crops 1958–74 | Isotopically exch. P | | | Bicarbonate sol. P | | |
		1958	1974	Decrease	1958	1974	Decrease
None	76	127	67	60	25	8	17
FYM	205	295	201	94	79	37	42
PK	171	240	174	66	63	26	37

comparison. The importance of the phosphate concentration in the soil solution in determining phosphate uptake has been discussed earlier.

Indices of available phosphate in soil

Soil chemists for a long time have been concerned with advising farmers how much fertilizer it is economically justifiable to give a crop on a particular field and have devoted a great deal of time devising simple and reliable methods to help them.

Soil extraction methods

Initially chemists looked for a chemical that would dissolve the same amount of phosphate from the soil as would the plant roots,[81] ignoring the fact that different plants extract different amounts of phosphate from the same soil. It was later realized that any standardized chemical extraction technique which placed soils in the order of crop responsiveness to phosphate was all that was needed. This led to the present procedure of correlating the amount, or the logarithm of the amount, with the response to a phosphate fertilizer of a crop grown on a range of soils usually representative of a region or a country. The crops may be grown either in field trials or in pots in a greenhouse, and the method is chosen that gives the largest correlation coefficient. When the experiment is done with crops in the field, which is required for advice to farmers, the correlation coefficient rarely exceeds 0.7. This method establishes the probable critical level below which phosphate fertilizer should be added, and the amount of phosphate extracted from soil is an *index* of availability.

It will be clear from Fig. 21.5 and the discussion of the processes of supply of phosphate to the root surface that no chemical method can define the amount of phosphate in soil which is available to crops. The uptake of phosphate depends on the extensiveness of the root system, on the number of root hairs it carries and on whether it is infected with VA mycorrhiza. The root can also increase the solubilization of soil phosphate by changing the rhizosphere pH, and, in the case of white lupin and perhaps other plant species, by secretion of metal complexing acids such as citric.[82]

Methods of analysis which give an index of the available phosphate in soils are in most demand for annual crops, many of which need a good supply early in their growth period. The principle underlying the methods in common use involves determining the amount of phosphate a soil sample, usually air-dried and ground to pass a 2 mm sieve, will release to a mild

---

81 Dyer, B., *Trans. Chem. Soc.*, 1894, **65**, 115; *Phil. Trans. R. Soc.*, 1901, **B194**, 235.
82 Gardner, W. K. *et al.*, *Pl. Soil*, 1982, **68**, 19, 33; *ibid.* 1983, **70**, 107.

extractant using short extraction times and a small volume of extractant per unit weight of soil. The extractants used include 0.5 M sodium bicarbonate, probably the most generally useful for soils that are not very acid, dilute mineral acids such as hydrochloric, dilute hydroxy acids or their acid salts, such as lactic,[83] dilute citric acid and acid fluoride solutions.[84] For research purposes isotopically exchangeable phosphate can be measured, but this method is too laborious for routine analysis of large numbers of samples. Extraction with an anion exchange resin is also used.[85]

All these methods have inherent limitations. This is shown in Table 21.7 for the response of sugar beet to phosphate in eastern England. The crop is only grown on soils that are not far from neutral and usually of light texture, but the table shows that the crop may not respond to phosphate in soils low in available phosphate and may respond on soils that are high. It is probable that much of this scatter in response is due to factors other than the availability of phosphate, so it would not be appreciably reduced by any other single determination.

Table 21.7 The frequency of response of sugar beet to phosphate fertilizers in groups of experiments arranged by soil analysis. Responses to 55 kg P ha^{-1} in kg sugar ha^{-1}. (From Warren, R. G. and Cooke, G. W., *J. agric. Sci.*, 1962, **59**, 269)

			Number of centres with response			
P soluble in 0.3 M HCl	*Number of experiments*	*Mean response*	*Over 340*	*330–180*	*170–0*	*Depression*
0.2–2.5	53	490	27	14	6	6
2.7–4.7	54	190	13	11	15	15
4.8–8.0	54	90	6	15	18	15
Over 8.0	55	90	7	14	11	23

Note: Dilute HCl was the most efficient of the extractants tested and although the method is no longer used the results illustrate the variable relationship between crop response to phosphate fertilizer and soil analysis. Sodium bicarbonate was not used in this series of experiments.

At a single site and under the controlled conditions of a field experiment, the correlation between soil analysis and crop yields can be close, as shown in Fig. 21.9 for the relationship between sodium-bicarbonate-soluble phosphate and yields of potatoes and sugar beet at Saxmundham, in eastern England. The plots contained a range of amounts of sodium-bicarbonate-

83 Egner, H., *et al.*, *Kungl. Lantbr. Hogsk. Ann.*, 1960, **26**, 204.
84 Bray, R. H. and Kurtz, L. T., *Soil Sci.*, 1945, **59**, 39. For review of methods see Kamprath, E. J. and Watson, M. E. in *The Role of Phosphorus in Agriculture* (eds F. E. Khasawneh *et al.*), p. 433, Am. Soc. Agron., Madison, Wisc., 1980.
85 Amer, F. *et al.*, *Pl. Soil*, 1955, **6**, 391; *Sibbesen, E., op. cit.*, 1977, **46**, 665.

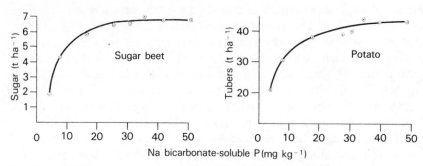

Fig. 21.9   Relationship between sodium-bicarbonate-soluble phosphate and yield of sugar (sugar beet) and potatoes at one site (Saxmundham). (From Johnston, A. E. *et al.*, *J. agric. Sci.*, 1986, **106**, 155.)

soluble phosphate as a result of different rates of application of farmyard manure and soluble phosphate fertilizer during the previous 70 years. Although varying widely between years, this method of analysis accounted on average for 84 per cent and 73 per cent of the within-year variance of the yields of potatoes and sugar beet respectively.

The experience in Britain is that extraction of soil with sodium bicarbonate is generally the best method on which to base advice to farmers on fertilizer use.[86] In this method, 5 cm^3 soil ($<$ 2 mm) is extracted for 30 minutes at 20 °C with 100 cm^3 0.5 M NaHCO$_3$ solution (pH adjusted to 8.5). After filtration the phosphate concentration of the solution is measured colorimetrically, and the result is expressed as mg per litre of soil. Advice to farmers is based on the P index (Table 3.7). For example cereals require only a maintenance application of phosphate fertilizer if the P index is 2 or greater, the recommended application* being 40, 60 and 80 kg P$_2$O$_5$ ha^{-1} for an expected grain yield of 5, 7.5 and 10 t ha^{-1}. These rates match the removal in the grain (see Table 3.1). The same application rates are recommended with a P index of 1, but at index 0 the recommended rates are 90, 110 and 130 kg P$_2$O$_5$ ha^{-1}. Potatoes are more responsive to phosphate and recommended rates for most soils are 350, 300, 250 and 200 kg P$_2$O$_5$ ha^{-1} at index 0, 1, 2 and 3 respectively, and 100–200 kg ha^{-1} at higher indices. When comparing these values of the P index with those used in other countries it should be noted that the values will depend on the conditions used for the extraction, and especially on the ratio of soil to extracting solution.[87] Also, the interpretation of the index will depend on the expected crop yield, on the sensitivity to phosphate deficiency of the crop to be grown, and on the soil properties.

86 Olsen, S. R. *et al.*, *US Dept. Agric. Circ.*, No. 939, 1954; Olson, R. A. *et al.*, *Agron. J.*, 1954, **46**, 175.   For use in Britain see Hooper, L. J., *Fertil. Soc. Proc.*, No. 118, 1970;   Mattingly, G. E. G., *Chem. Ind.*, 1980, 690.
87 Colwell, J. D., *Aust. J. exp. Agric. Anim. Husb.*, 1963, **3**, 190, for a variation of the standard method.

* 1 kg P$_2$O$_5$ = 0.44 kg P (approx.)

Other advisory methods

A method which is different in principle is to determine the amount of phosphate which must be added to the soil to raise the phosphate concentration of its solution to a sufficient level.[88] The determination is made by shaking soil samples in 0.01 M $CaCl_2$ (1 g : 10 cm³) containing graduated concentrations of monocalcium phosphate. The recommended shaking period depends on the soil. A six-day period, shaking gently for two periods of 30 minutes each day was used by Fox and Kamprath.[89] Examples of adsorption curves using this method are given in Fig. 21.10. The required concentration varies with the crop that is to be grown, and also with the soil as shown in Table 21.8. The method is particularly useful when advising on the use of phosphate fertilizers in a region where the soils vary greatly in the amounts of phosphate they adsorb. It does not, however, take into account the exchangeable phosphate on the soil surfaces. The critical test of the method will be its success in predicting the amount of fertilizer required by a crop as measured by field experimentation.

Empirical methods using extractants such as sodium bicarbonate are used

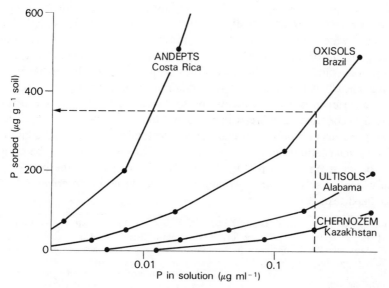

Fig. 21.10 Phosphate sorption curves for selected soils. Note log scale. (From Fox, R. L. in *Chemistry in the Soil Environment* (eds R. H. Dowdy *et al.*), p. 223, Am. Soc. Agron. Spec. Publ., No. 40, 1981.)

88 Beckwith R. S., *Aust. J. exp. Agric. Anim. Husb.*, 1965, **5**, 52; Ozanne, P. G. and Shaw, T. C., *Trans. 9th Int. Congr. Soil Sci.*, 1968, **2**, 273; Fox, R. L. *et al.*, *op. cit.* 301.

89 Fox, R. L. and Kamprath, E. J., *Soil Sci. Soc. Am. Proc.*, 1970, **34**, 902.

Table 21.8 Concentration of phosphate in solution associated with 95% of maximum yield of selected crops grown in Hawaii. (From Fox, R. L. in *Chemistry in the Soil Environment* (eds R. H. Dowdy *et al.*), p. 223, Am. Soc. Agron. Spec. Publ., Madison, No. 40, 1981).

Crop	Soil series	*Approx. concentration ($mg\ P\ l^{-1}$) in soil solution for yield indicated*	
		75% of max.	*95% of max.*
Cassava (*Manihot esculenta*)	Halii	0.003	0.005
Peanut (*Arachis hypogaea*)	Halii	0.003	0.01
Cabbage (*Brassica oleracea*)	Kula	0.012	0.04
Corn (*Zea mays*)	Halii	0.005	0.05
	Wahiawa	0.01	0.06
Sorghum bicolor (grain)	Honokaa	0.015	0.06
Soyabean (*Glycine max*)	Wahiawa	0.025	0.20
	Halii	0.025	0.20
Tomato (*Lycopersicon esculentum*)	Kula	0.05	0.20
Head lettuce (*Lactuca sativa*)	Kula	0.10	0.30

in most countries, and with experience in interpreting the amounts of phosphate extracted they are usually satisfactory. Various other methods have been used which appear to suit local conditions. For example, in the tropics where the rate of mineralization of soil organic matter is high, the organic phosphorus content of soil has sometimes been found useful,[90] as has water-soluble silica.[91] Another method is to determine the crop response to phosphate fertilizer in field experiments and to divide the region into sub-regions where an economic response is likely or not likely.[92]

The per cent phosphorus in plant leaves or total plant material is another common diagnostic technique. The method is not simple because the critical value depends on the crop that is being grown, its age and the part of the plant which is analysed. The recommended procedures are described in more specialized texts.[93]

90 Friend, M. T. and Birch, H. F., *J. agric. Sci.*, 1960, **54**, 341. Nye, P. H. and Bertheux, M. H., *J. agric. Sci.*, 1957, **49**, 141. For general discussion of the evaluation of the P status of tropical soils see Ch. 9 in Sanchez, P. A., *Properties and Management of Soils in the Tropics*, Wiley, 1976.

91 Birch, H. F., *J. agric. Sci.*, 1953, **43**, 229, 329; Garberg, P. K., *E. Afr. agric. For. J.*, 1970, **35**, 396.

92 Examples for northern Nigeria given by Goldsworthy, P. R. and Heathcote, R. G., *Emp. J. expl. Agric.*, 1963, **31**, 351 (groundnuts); *Expl. Agric.*, 1967, **3**, 29 (sorghum), 263 (maize); Scaife, M. A., *J. agric. Sci.*, 1968, **70**, 209 (maize) for Tanzania.

93 Walsh, L. M. and Beaton, J. D., *Soil Testing and Plant Analysis*, Soil Sci. Soc. Am., 1980, for methods and their interpretation. For UK *The Analysis of Agricultural Materials*, *MAFF Ref. Book* **RB 427**, 1986, gives methods, and *MAFF Reference Book*, **209**, HMSO gives interpretation. Methods reviewed by Kamprath, E. J. and Watson, M. E. in *The Role of Phosphorus in Agriculture* (eds F. E. Khasawneh *et al.*), p. 433, Am. Soc. Agron Madison, Wisc., 1980.

Phosphate fertilizers

The main sources of phosphate for fertilizers are rock phosphates of marine or igneous origin. Also used are deposits of guano, mainly from sea birds and to a less extent from bats. The most important commercial source is rock phosphate of marine origin, which may contain up to 80 per cent apatite usually with isomorphous substitution, and which occurs in large deposits in Morocco, Florida, North Carolina and elsewhere. Igneous and metamorphic rocks are mined for apatite in Russia (Kola) and S. Africa, and there are important guano deposits in S. Pacific islands.[94] Compared with soluble phosphates, these rock phosphates are only used directly as fertilizers on a small scale. Their use is mainly for perennial crops growing on acid soils because at pH 7 or above they will only maintain a phosphate concentration of less than 10^{-6} M. Their general formula is $Ca_{10-a-b} Na_a Mg_b (PO_4)_{6-x} (CO_3)_x F_{2+y}$, where y is about $0.4x$. They are called carbonate apatite or francolite. The availability of their phosphate to plants varies widely between sources and increases with the amount of substitution in the apatite lattice. Isomorphous substitution of CO_3^{2-} for PO_4^{3-} decreases the unit cell size (X-ray diffraction is used to give the a-axis dimension), decreases the crystallite size and increases the specific surface area of the apatite aggregates. The review by Khasawneh and Doll[95] should be consulted for further information on the chemical and agronomic evaluation of rock phosphates.

Fertilizers in common use

The most commonly used phosphate fertilizers are single superphosphate (18–20 per cent P_2O_5, 8–9 per cent P) containing 30–35 per cent monocalcium phosphate and 65–70 per cent gypsum, triple superphosphate (46 per cent P_2O_5, 20 per cent P) which is about 85 per cent monocalcium phosphate with a small amount of gypsum, and monoammonium and diammonium phosphates. All are water soluble. The less soluble dicalcium phosphate, which maintains a phosphate concentration of about 10^{-3} M in solution, is used in some European countries.[96] The fertilizers are made from rock phosphate and the manufacturing processes are outlined in Fig. 21.11.

Basic slag is a by-product of older methods used in the steel industry, and little is now made. It has a variable composition, and correspondingly variable agricultural value,[97] but the most useful contains calcium silicophosphate and calcium silicate, and the less useful crystalline fluor-apatites. It is a liming

94 Cathcart, J. B. in *The Role of Phosphorus in Agriculture* (see ref. 93), p. 1.
95 Khasawneh, F. E. and Doll, E. C., *Adv. Agron.*, 1978, **30**, 159;   Lehr, J. R. in *The Role of Phosphorus in Agriculture* (see ref. 93), p. 81.
96 Hoare, J. in *The Role of Phosphorus in Agriculture* (see ref. 93), p. 121; Young, R. D. and Davis, C. H., *op. cit.*, 195.
97 Brown, G. G. and Thatcher, K. F. J., *Fertil. Soc. Proc.*, No. 96, 1967.

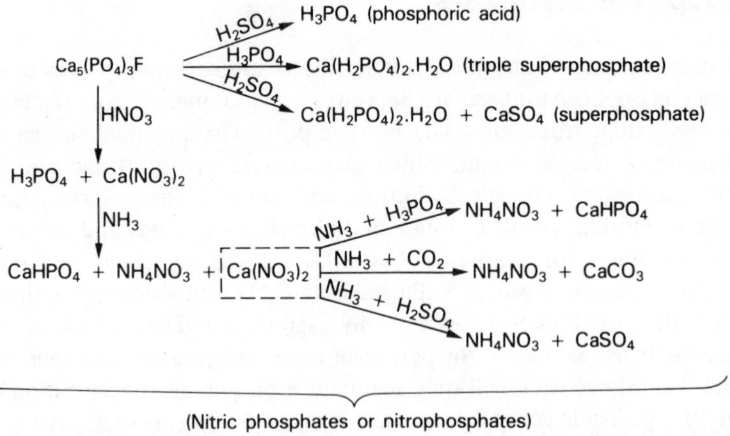

$$Ca_5(PO_4)_3F$$

with $H_2SO_4 \longrightarrow H_3PO_4$ (phosphoric acid)

with $H_3PO_4 \longrightarrow Ca(H_2PO_4)_2.H_2O$ (triple superphosphate)

with $H_2SO_4 \longrightarrow Ca(H_2PO_4)_2.H_2O + CaSO_4$ (superphosphate)

$$HNO_3$$

$$H_3PO_4 + Ca(NO_3)_2$$

$$NH_3$$

$$CaHPO_4 + NH_4NO_3 + Ca(NO_3)_2$$

$NH_3 + H_3PO_4 \longrightarrow NH_4NO_3 + CaHPO_4$

$NH_3 + CO_2 \longrightarrow NH_4NO_3 + CaCO_3$

$NH_3 + H_2SO_4 \longrightarrow NH_4NO_3 + CaSO_4$

(Nitric phosphates or nitrophosphates)

Fig. 21.11   Outline of processes used to convert rock phosphate to phosphate
fertilizers. Equations are not stoichiometric and reactions involving
fluoride are omitted. (From G. E. G. Mattingly., *Education in
Chemistry*, 1979, **16**, 41.)

material as well as a phosphate fertilizer, for the calcium silicate hydrolyses
readily to calcium hydroxide and silica.

The only common phosphate fertilizers now in use are listed in Table 3.6.
Because of transport costs, the trend has been towards the use of concentrated fertilizers. Triple (concentrated) superphosphate is now more widely
used than single (ordinary) superphosphate, and both diammonium and
monoammonium phosphate are used in fertilizers which supply two or more
nutrients. The water-soluble phosphates are usually manufactured and sold
as granules which are easier to handle than powder. Assessment of the value
of phosphate fertilizers is based mainly on their solubility found by prescribed
methods, which in many countries are defined by legislation. In Britain and
other EEC countries the superphosphates are evaluated by solubility in water
and in neutral ammonium citrate, and for basic slag 2 per cent citric acid is
used.[98] In Britain total phosphate and phosphate soluble in 2 per cent formic
acid are used to assess rock phosphates whereas in the United States the
official (AOAC) method uses neutral ammonium citrate.[99] Fineness of
grinding of rock phosphate is required to be stated in many countries. Alkaline ammonium citrate distinguishes between apatite and dicalcium phosphate
and in some European countries it is used as an official method of analysis.[100]

98 *The Fertilisers (Sampling and Analyses) Regulations*, 1978. Statutory Instruments 1978,
No. 1108, London: HMSO.

99 AOAC, 1980, *Official Methods of Analysis* (13th edn), Association of Official
Analytical Chemists, Arlington, Virginia.

100 For reviews see Jacob, K. D. and Hill, W. L. in *Soil and Fertilizer Phosphorus*, Agron.
Monog., No. IV (eds W. H. Pierre and A. G. Norman), p. 299, 1953;   Mattingly,
G. E. G. and Talibudeen, O. in *Topics in Phosphorus Chemistry*, Vol. 4 (eds M.
Grayson and E. J. Griffith), p. 157, Interscience, 1967.

The agricultural evaluation of phosphate fertilizers depends on the crop and soil on which they are to be used, as well as on their solubility and granule size.[101] Table 21.9 illustrates some of these points.[102] It shows that swedes make better use of rock phosphate than potatoes, and that for both crops the rock phosphate becomes of decreasing value as the soil pH rises towards neutrality. The difference between the crops might arise because the higher growth rate of potatoes requires a higher rate of uptake of phosphate and hence a higher concentration in solution. Alternatively, swedes might increase the solubility of rock phosphate in the way that rape increases the solubility of soil phosphate through acidification of the rhizosphere.[103].

The greater effectiveness of rock phosphate on acid than on neutral or calcareous soils has been observed generally, although if the soil is very acid toxicity of aluminium may limit the crop response. Rock phosphates dissolve under acid conditions to give in solution Ca^{2+}, PO_4^{3-} and CO_3^{2-} (which are protonated to $H_2PO_4^-$, HCO_3^- and H_2CO_3) and F^-, and other ions that are presented in the apatite lattice. Adsorption of these ions by the soil removes them from solution and leads to greater dissolution. Thus by acting as a sink the soil affects the rate at which the rock phosphate dissolves. As would be expected for a fertilizer of low solubility, rock phosphate is more reactive when it is finely ground, though there appears to be no benefit in grinding it to less than 100 mesh (150 μm).[104] Rock phosphates are not used directly as fertilizers in large quantities. Where there are local deposits of acceptable citrate solubility they are useful for grassland and tree crops, having a price advantage over water-soluble fertilizer, and saving foreign currency if sulphur has to be imported for the manufacture of superphosphate. For these reasons there is a resurgence of interest in their use, which has been fostered by research at the International Fertilizer Development Center[105] and the Tennessee Valley Authority.

The choice between water-soluble phosphates depends mainly on price per unit weight of phosphate at the farm gate. Because of transport costs this gives an advantage to the more concentrated forms, but where sulphur deficiency may occur single superphosphate is preferred to triple superphosphate because of its higher content of sulphur, present in the fertilizer as calcium sulphate. Ammonium phosphates may be more valuable than monocalcium phosphate on some soils. This might be because: (i) more phosphate is taken up when nitrogen is supplied as NH_4^+ than as NO_3^-, as found in culture

101 For example of effect of granule size see Bouldin, D. R. *et al.*, *J. agric. Fd Chem.*, 1960, **8**, 470.
102 Cooke, G. W., *J. agric. Sci.*, 1956, **48**, 74.
103 Grinsted, M. J. *et al.*, *New Phytol.*, 1982, **91**, 19; Hedley, M. J. *et al.*, *op. cit.*, 31, 45.
104 Cooke, G. W., *Emp. J. expl. Agric.*, 1956, **24**, 295; Jones, E., *J. Soil Sci.*, 1975, **26**, 66.
105 Hammond, L. L. and Leon, L. A., *3rd Int. Congr. Phosphorus Compounds*, 1983, 503.

Table 21.9 The effect of soil and crop on the relative values of phosphate fertilizer. Field experiments in the United Kingdom 1951–53; kg P from superphosphate to give the same response as 100 kg P from the fertilizer

Soils	Swedes		
	Very acid <pH 5.5	*Acid pH 5.5–6.5*	*Neutral >pH 6.5*
No. of expts	10	22	3
Dicalcium phosphate	97	85	95
Silico-phosphate	90	84	52
Gafsa rock phosphate	91	86	12
	Potatoes		
	Very acid <pH 5.5	*Acid pH 5.5–6.5*	*Neutral >pH 6.5*
No. of expts	10	15	9
Dicalcium phosphate	122	62	84
Silico-phosphate	92	56	30
Gafsa rock phosphate	34	37	4

solutions[106], or (ii) the solution diffusing from ammonium phosphate granules is much less acid than from monocalcium phosphate and so will dissolve little if any aluminium (see next section); or (iii) nitrification of the ammonium ions will cause acidification and in neutral or calcareous soils this will increase phosphate solubility.[107]

The reaction between water-soluble phosphate fertilizers and the soil

The reaction between the soil and solutions containing phosphate has usually been investigated by shaking a soil suspension with solutions of phosphate and determining the amount of phosphate which is adsorbed. Fertilizer is applied as powder or granules and there is a different set of reactions, the

106 Arnozis, P. A. and Cogliatti, D. H., *3rd Int. Congr. Phosphorus Compounds*, 1983, 611.
107 Terman, G. L., *Fertil. Soc. Proc.*, No. 123, 1971. See also Mattingly, G. E. G. and Talibudeen, O. (ref. 48) for a discussion of the effects of form of nitrogen on phosphate availability.

first being rapid absorption of water from the soil. Phosphate dissolves in the water to give a saturated or nearly saturated solution, and then diffuses from the granules or powder into the soil solution. If the fertilizer is monocalcium phosphate, it dissolves to give a solution with a pH of about 1, which is between 3 and 4.5 M in phosphate and approximately 1 M in calcium, leaving a residue of dicalcium phosphate.[108] If the phosphate is applied as ammonium phosphate, the saturated solution has a pH between 4 and 8, depending on the ratio of ammonium to phosphate.[109] The solution diffusing from a granule of monocalcium phosphate is very reactive, and, if the soil is non-calcareous, will dissolve iron and aluminium compounds in the soil which react with the phosphoric acid to give precipitates of potassium-containing phosphates, such as the aluminium-containing taranakite $(H_6K_3Al_5(PO_4)_8.18H_2O$ and the iron-aluminium phosphate $H_8K(Al,Fe)_3(PO_4)_6.6H_2O$, as well as amorphous aluminium phosphate and variscite. It is probable that in many temperate soils taranakite is the first compound precipitated, and only after all the exchangeable potassium has been used up are the simple aluminium phosphates formed;[110] it is also probable that the taranakite and other aluminium phosphates are precipitated in preference to ferric phosphates, unless the soil contains much amorphous ferric hydroxide.

The distance the phosphoric acid diffuses from a fertilizer granule depends on the amount of water-soluble phosphate in the granule and on the granule size. Sample and Taylor[111] found that for 6 mm granules containing 70 per cent of water-soluble monocalcium phosphate in a silt loam soil, the diffusion zone had a radius of about 17 mm after three weeks. This diffusion zone can be traced quite easily by autoradiography if some ^{32}P is incorporated into the phosphate, as has been shown by Bouldin and Black,[112] although they also found that the phosphate was not always precipitated uniformly in a spherical shell, but was sometimes concentrated in rings or in a series of spots.

The consequence of this movement of phosphate from fertilizer granules into the soil by diffusion is that, for a period after the phosphate has been added, there is great variability in the phosphate concentration from point to point in the soil, being much higher near each granule than in the bulk of the soil. It is from these zones of high concentration that plant roots can most easily take up phosphate. This extremely patchy distribution can be expected to last a long time in an undisturbed soil because of the slow diffusion of phosphate in soil.

108 Lindsay, W. L. and Stephenson, H. F., *Soil Sci. Soc. Am. Proc.*, 1959, **23**, 12, 18.
109 Huffman, E. O., *Fertil. Soc. Proc.*, No. 71, 1962; *J. Agric. Fd Chem.*, 1963, **11**, 182.
110 Taylor, A. W. and Gurney, E. L., *Soil Sci. Soc. Am. Proc.*, 1965, **29**, 18. For fuller discussion of these reactions see Sample, E. C. *et al.*, in *The Role of Phosphorus in Agriculture* (see ref. 93), p. 263.
111 Sample, E. C. and Taylor, A. W., *Soil Sci. Am. Proc.*, 1964, **28**, 296; Terman, G. L., *Fertil. Soc. Proc.*, No. 123, 1971.
112 Bouldin, D. R. and Black, C. A., *Soil Sci. Soc. Am. Proc.*, 1954, **18**, 255.

There has been much discussion in the past about whether adding a soluble phosphate fertilizer, and particularly superphosphate, to a soil makes it more acid. Its long-continued use on plots at Rothamsted and Woburn has not had any measurable effect on the soil pH, though a small acidifying effect might have been expected as the fertilizer may contain some free acid. In so far as the monocalcium phosphate reacts with iron or aluminium hydroxide surfaces, this should set free the calcium so increasing the calcium status of the soil, but this amount is only about 4.5 kg/100 kg with single superphosphate and about 12 kg/100 kg with triple superphosphate. On the other hand, in neutral soils, if the monocalcium phosphate is converted to hydroxyapatite, this will remove about 15 kg of exchangeable calcium for every 100 kg of single superphosphate that is so converted, so one would expect to find the pH of the soil reduced, particularly if it is rather poorly buffered. Evidence for acidification is that ten annual applications of 80 kg P ha^{-1} as triple superphosphate reduced the pH of a loam soil from 7.04 to 6.62 in the top 15 cm, with smaller reductions down to 45 cm.[113]

Phosphate fertilizers have been shown to increase soil acidity when their use allows soil organic matter to increase. On some very phosphate-deficient soils in Australia, the use of superphosphate on the pastures allows more carbon to be converted into humus, with the concurrent locking up of fertilizer phosphate in organic forms and the creation of humic carboxylic acids. In some soils in New South Wales, it was found that every 100 kg ha^{-1} of superphosphate added, containing about 7.5 kg P, increased the humus content by about 2 t ha^{-1} which would need about 90 kg Ca ha^{-1} to bring its pH to neutrality, yet the superphosphate only contained about 22 kg Ca ha^{-1}, most of which was present as gypsum.[114] As a consequence the pH of the soils fell by about 0.05 units for every 100 kg ha^{-1} of superphosphate added.

There are reports[115] that phosphate fertilizers often, though not always, increase the uptake of manganese by plants, but the explanation is not straightforward. The effect would be expected because the acid solution that diffuses through soil from a band of monocalcium phosphate has a high concentration of manganese.[116] It has, however, been shown by Le Mare[117] that although addition of phosphate increased the uptake of manganese by cotton plants grown in soil, the increase was large only for additions at high ratios of P : Ca, and uptake was reduced at P : Ca of 1 : 1 or less. It appears

113 Alben, A. O., *Soil Sci.*, 1961, **92**, 212.
114 Donald, C. M. and Williams, C. H., *Aust. J. agric. Res.*, 1954, **5**, 664; 1957, **8**, 179.
115 Page, E. R. *et al.*, *Pl. Soil*, 1963, **19**, 255; Larsen, S., *Pl. Soil*, 1964, **21**, 37. See also review by Adams, F. in *The Role of Phosphorus in Agriculture* (ref. 93), 655.
116 Lindsay, W. L. and Stephenson, H. F., *Soil Sci. Soc. Am. Proc.*, 1959, **23**, 12.
117 Le Mare, P. H., *Pl. Soil*, 1977, **47**, 593, 607, 621.

from this work that there is an interaction between calcium, phosphorus and manganese within the plant which is independent of any effects of soil acidification. Of practical interest is the observation in a field experiment on an acid soil (pH 5.8) high in manganese in Uganda[118] that small additions of triple superphosphate (12.6 kg P ha^{-1}) depressed the dry matter yield and phosphate uptake by cotton, which implies that the fertilizer induced a toxicity which might have been due to manganese. Higher applications increased crop yields. Phosphate in fertilizer granules may therefore increase manganese uptake through soil acidification, but the effect on the plant may depend on the relative uptake of phosphate and calcium.

Phosphate fertilizers have also been reported to induce zinc deficiency in plants on soils which have a marginally sufficient supply of zinc.[119] Again the effect is not well understood, but it probably arises from an imbalance between zinc and phosphate in plant metabolism rather than from the effect of phosphate on zinc solubility in soil, although the mechanism is not yet known (see also p. 798).

Downward movement of phosphate in soil

An important consequence of the reaction of fertilizer phosphate with soil is that the phosphate moves very slowly from where it has been placed, unless the soil is cultivated or otherwise disturbed. The equilibrium soil solution of moderately fertile soils is about 5×10^{-6} M P, so that if 30 cm water drains through the soil annually, it will carry with it only about 0.5 kg P ha^{-1}. Even this amount will not normally leach out of the soil because some will be absorbed by the subsoil. Average concentrations in drainage water from agricultural land have been reported to be about 5×10^{-6} M from sandy soils, 2×10^{-6} M or less from clay soils, but to be much higher from organic soils.[120] Thus very little phosphate is leached out of soils under most conditions although sands are an exception. There can, however, be loss of recently broadcast fertilizer in surface run-off[121] and by erosion.[122] Loss of topsoil by erosion can be a particularly serious source of loss because of its generally high content of phosphate.

118 Le Mare, P. H., *J. agric. Sci.*, 1968, **70**, 265, 271, 281.
119 See review by Adams, F. (ref. 115).
120 Cooke, G. W. in *Agriculture and Water Quality, MAFF Tech. Bull.*, No. 32, p. 5, 1976.
121 Ryden, J. C. *et al.*, *Adv. Agron.*, 1973, **25**, 1.
122 Taylor, A. W. and Kilmer, V. J. in *The Role of Phosphorus in Agriculture*, (see ref. 93), p. 545.

Downward movement of phosphate is greater when organic manures are added. This is illustrated in Table 21.10 which gives the total phosphate content of plots on Barnfield at Rothamsted Experimental Station which has been in root crops for most years since 1845. A little phosphate has moved below 30 cm on the plots receiving all their phosphate as superphosphate, but appreciably more has moved below this depth on plots receiving farmyard manure or both superphosphate and farmyard manure. The greater downward movement of phosphate on the farmyard manure plots might be a result of movement of organically held phosphate in the soil solution.[123] It might also be due to the greater activity of soil fauna, e.g. earthworms, on plots treated with organic manures, and hence greater incorporation of plant residues below the soil surface (see Ch. 16).

Table 21.10 The movement of phosphate down the soil profile under arable cropping. Rothamsted, Barnfield; permanent mangolds. Superphosphate and manure added annually since 1843. Total phosphate in soil (mg P kg^{-1}) in 1958. (From Johnston, A. E. in *Agriculture and Water Quality, MAFF Tech. Bull.*, No. 32, 111, London: HMSO, 1976)

Depth (cm)	No P no K	P no K	PK	FYM	FYM + P
0–23	777	1350	1295	1376	1971
23–30	464	541	524	649	781
30–46	413	446	448	525	579
46–53	401	396	397	442	411

Note: Phosphate application as single superphosphate at 34 kg P ha^{-1} (115 applications) and FYM at 35 t ha^{-1}, containing about 40 kg P ha^{-1} (102 applications).

Uptake of fertilizer phosphate by crops

An annual crop usually takes up about 5 to 25 per cent of the phosphate supplied by a single application of a water-soluble phosphate fertilizer.[124] The recovery has usually been calculated as the difference in uptake between crops grown on unfertilized and fertilized plots, which leaves uncertainty as to whether the fertilizer has stimulated root growth and so increased the uptake of soil phosphate. Using ^{32}P-labelled fertilizer to remove this uncertainty Mattingly found recoveries of 8–33 per cent from applications of 5.5

123 Wild, A., *Z. PflErnähr.* Düng, 1959, **84**, 220.
124 Smith, A. M. and Simpson, K., *J. Sci. Fd Agric.*, 1950, **1**, 208; 1956, **7**, 754; Cooke, G. W., *Fertil. Soc. Proc.*, No. 92, 1966. Higher recoveries have been reported by Child, R. *et al.*, *Emp. J. expl. Agric.*, 1955, **23**, 220.

to 27.5 kg P ha^{-1} to root crops, barley and ryegrass grown on soils at Rothamsted.[125]

Recoveries of fertilizer phosphate depend on the conditions of the experiment and a wide range of recoveries can be expected. Low recovery can be a consequence of: (i) adding the fertilizer to a soil of high phosphate status or adding an amount considerably in excess of the crop requirement; (ii) the crop having a low potential for response because growth is limited by some factor other than phosphate supply; (iii) placing the phosphate in the soil which remains dry or is inaccessible to roots for some other reason; (iv) most of the phosphate being adsorbed by the soil so that the concentration in solution remains very low; and (v) the adsorbed or precipitated phosphate being converted into non-labile form.

Placement of phosphate close to the seed, as is often done for cereals by combine drilling the fertilizer and seed, has often proved beneficial. In early experiments in Britain about half as much phosphate was needed with combine drilling as was necessary with broadcast applications.[126] Combine drilling provides seedlings with the high concentration of phosphate in the soil solution which they need for rapid growth. Placement of phosphate in a band in the soil, however, gave a smaller response than broadcast fertilizer at the start of an experiment in Brazil although a greater response after the

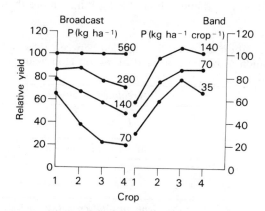

Fig. 21.12 Relative yields of four crops of corn as influenced by rates and method of application. Broadcast applications applied to Crop 1 only, band applications applied to each crop. Reference yield 560 kg P ha^{-1} broadcast. (From Yost, R. S. *et al.*, *Soil Sci. Soc. Am. J.*, 1979, **43**, 338.)

125 Mattingly, G. E. G., *Rothamsted Ann. Rept. for 1956*, p. 48, 1957.
126 Lewis, A. H., *J. agric. Sci.*, 1941, **31**, 295. For later examples see Reith, J. W. S., *Emp. J. expl. Agric.*, 1959, **27**, 300; Devine, J. R. and Holmes, M. R. J., *Exp. Husb.*, 1964, 11.

first crop.[127] The experiment was on a very high phosphate adsorbing Oxisol to which broadcast applications of 70, 140, 280 or 560 kg P ha^{-1} as single superphosphate were applied to the first crop. Banded applications of 35, 70 or 140 kg P ha^{-1} were made annually at 5 cm below the seed. The results, some of which are shown in Fig. 21.12, indicated that the best method was to combine initially broadcast and maintenance band applications.

Residual value

Phosphate from a fertilizer application not used by a crop continues to be of value to succeeding crops, although the uptake each year is usually less than in the first year. The effect can be very long lasting, as shown by the Hoosfield Exhaustion Plots on a slightly calcareous soil at Rothamsted. Some of the plots received about 30 kg P ha^{-1} as superphosphate annually from 1856 to 1901 and some others about 35 t farmyard manure ha^{-1} from 1876 to 1901, and since then they have only received nitrogen fertilizers. In the five years 1970–74 barley took up an extra 6.7 kg P ha^{-1} each year from the superphosphate plots and 10.2 kg ha^{-1} from the manure plots, compared with the uptake from the plots which had not received any phosphate (Table 21.11). The phosphate plots received about 1410 kg P ha^{-1} between 1856 and 1901, and in 1970 they contained about 400 kg P ha^{-1} more than the unman-

Table 21.11   Residual effect of fertilizer applied between 1856 and 1901 on yield and composition of barley crop in 1970–74, on Hoosfield Exhaustion Plots, Rothamsted

	Treatment 1856–1901			
	No P No K	P	PK*	Farmyard manure[†]
	Mean yield (t ha^{-1} a^{-1})			
Grain	1.83	3.76	4.5	4.75
Straw	1.38	2.62	3.12	3.11
	Nutrients in total crop (kg ha^{-1} a^{-1})			
P	4.2	10.9	13.0	14.4
K	15.3	23.9	40.7	34.2

* PK 1856–75, P alone 1878–1901.   † 1876–1901.

127 Lathwell, D. J., *Cornell Inst. Agric. Bull.*, 33, 1979;   Yost, R. S., *Soil Sci. Soc. Am. J.* 1979, **43**, 338.

ured, so the land still contained 28 per cent of the phosphate that had been added, and the barley is taking up about 1.7 per cent of this phosphate each year.[128]

The residual value of phosphate in this experiment has also been measured by comparing the response of different crops to fresh superphosphate on the no-residues and on the with-residues plots. Table 21.12 shows that the residual value of the phosphate was almost enough to give a full crop of barley and sugar beet, being equivalent to about 30 kg P ha^{-1} as fresh superphosphate, but the residual phosphate was equivalent to a fresh application of only about 6 kg P ha^{-1} for potatoes, though this figure may be exceptionally low.[129] In this experiment on Hoosfield, the three crops gave effectively the same yield on the plots receiving 56 kg P ha^{-1} as fresh superphosphate whether or not they had received phosphate between 1856 and 1901. This result is not always found at Rothamsted. On other fields crop yields on plots receiving 56 kg P ha^{-1}, or even 85 kg P ha^{-1}, as fresh superphosphate are less on those plots which have not received any phosphate fertilizer in the past, than on those which have received annual dressings. Figure 21.13 illustrates this for potatoes on Agdell field at Rothamsted. Two plots on this field have received no phosphate fertilizer since 1848 and two plots received 37 kg P ha^{-1} once every four years. The phosphate response curves indicate that with phosphate residues potatoes give higher yields than without residues, whatever amount of fresh phosphate fertilizer is applied.

Table 21.12 Crop yields (t ha^{-1}) in 1957 and 1958 on Hoosfield Exhaustion Plots, Rothamsted. Comparison can be made between: (a) plots that received phosphate in 1856–1901 (total of 1410 kg P ha^{-1}) and those that did not; and (b) the effects of 56 kg P ha^{-1} as superphosphate in each of the two test years on the 'no residues' and the 'with residues' plots

Previous history	*Potatoes (tubers)*		*Sugar beet (sugar)*		*Barley (grain)*	
	No P	*P*	*No P*	*P*	*No P*	*P*
No phosphate given	12.8	32.6	3.84	5.72	2.03	3.50
Phosphate given 1856–1901	21.1	32.6	5.67	6.00	3.11	3.48
Fresh superphosphate equivalent of residues (kg P ha^{-1})	6		27		34	

128 Johnston, A. E. and Poulton, P. R., *Rothamsted Ann. Rept. for 1976*, Part 2, p. 53.
129 Johnston, A. E. *et al.*, *Rothamsted Ann. Rept. for 1969*, Part 2, p. 39.

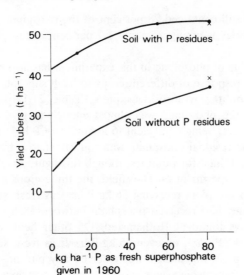

Fig. 21.13 Response of potatoes to fresh phosphate and residual phosphate. Agdell field, Rothamsted 1960. 56 kg ha^{-1} P given as superphosphate in 1959(x).

Table 21.13 Residual effect of superphosphate on Hoosfield, Rothamsted (21 seasons); 60 kg P ha^{-1} given as superphosphate once every five years

	Yield No P	*Response in year of application*	*Response to phosphate added years previously*			
			1	*2*	*3*	*4*
Potatoes t ha^{-1}						
Low nitrogen	11.45	14.5	7.5	8.3	7.5	6.3
High nitrogen	12.10	18.3	12.1	10.8	9.5	9.5
Barley grain kg ha^{-1}	3080	390	160	240	240	160
Ryegrass hay kg ha^{-1}	2120	320	300	240	140	140
Wheat grain kg ha^{-1}	3550	50	−60	10	−20	40
Soluble P in the plots manured as above						
Soluble in NaHCO$_3$ (ppm)	8	24	19	17	15	15
Soluble in 0.01 M CaCl$_2$ (10^{-6} M)	0.11	0.70	0.55	0.45	0.35	0.30

In another experiment at Rothamsted the residual value of 60 kg P ha^{-1}, added as single superphosphate once every five years, was measured in a four-course rotation. The soil was slightly calcareous. The results in Table 21.13 show that potatoes, barley and ryegrass which gave first year responses, also continued to give yield responses to phosphate applied up to four years previously.[130] The table also shows that the superphosphate applications increased both the bicarbonate-soluble phosphate and the phosphate concentration in the soil solution four years after an application. In the same experiment Gafsa rock phosphate gave almost no crop response and bicarbonate-soluble phosphate remained almost unchanged.

These results are typical of those found for many others soils. Thus Larsen[131] found that a single dressing in the field of 250 or 500 kg P ha^{-1} as granular triple superphosphate mixed with 23 different soils became non-exchangeable (measured by L value) according to a first-order reaction. In the 23 soils the half-life was between one and six years, and in half the soils it was between two and three years. There was a tendency for the half-life to be longer on acid than on neutral soils; a peat was also used and for this the half-life was 56 years. For a first-order decay the logarithm of the proportion of phosphate that remains effective gives a linear plot against time. This linear relation which was found by Larsen with English soils, was also found by Arndt and McIntyre[132] in a field experiment in northern Australia, but only for a three- to four-year period, the loss of effectiveness then becoming less rapid.

For the purpose of compensation to farmers in Great Britain for fertilizer residues, a water-soluble phosphate fertilizer is assumed to have lost half its value after one year, three-quarters of its value after two years, and all its value after three years. The results given here, and a number of others from several experimental husbandry farms in England and Wales,[133] suggest that this convention favours the incoming tenant on well-farmed land because of the long-lasting effect of regular phosphate additions, although the figures for the first two years often appear to be reasonable.

Many soils in the tropics behave similarly, in that applications of superphosphate supplying 10–20 kg P ha^{-1} have residual effects lasting a number of years.[134] This is illustrated in Table 21.14 for an experiment on a soil derived from basement complex rocks in the Machakos district of Kenya. In this experiment the effect of 40 kg P ha^{-1} was still appreciable five years after

130 Warren, R. G., *Fertil. Soc. Proc.*, No. 37, 1956.
131 Larsen, S. *et al.*, *J. Soil Sci.*, 1965, **16**, 141.
132 Arndt, W. and McIntyre, G. A., *Aust. J. agric. Res.*, 1963, **14**, 785.
133 Williams, J. H. *et al.*, *MAFF Tech. Bull.*, No. 20, 42, 1971.
134 Greenwood, M., *Emp. J. expl. Agric.*, 1951, **19**, 225; Boswinkle, E., *op. cit.*, 1961, **29**, 136. See reviews by Russell, E. W., *Fertil. Soc. Proc.*, No. 101, 1968; Jones, M. J. and Wild, A., *Soils of the West African Savanna*, Commonw. Bur. Soils Tech. Comm., No. 55, 1975.

Table 21.14   Residual effect of superphosphate on a tropical red earth, Makaveti, Kenya. Soil from basement complex. Yield and responses to superphosphate (18% P) given in 1952 in kg ha^{-1}. Pearl millet: *Pennisetum typhoides*; beans: *Phaseolus vulgaris*

Year crop	1952 Millet	1953 Beans	1954 Maize	1954 Millet	1955 Millet	1956 Millet	1957 Beans
Control yield, no P	405	760	1920	410	1250	785	785
Response to:							
20 kg P ha^{-1}	525	160	590	120	110	85	40
40 kg P ha^{-1}	720	250	785	175	180	125	125

application. Evidence from field experiments shows that it is quite wrong to generalize that soils in the tropics give low phosphate residual values.

There are, however, soils throughout the world on which even heavy applications of phosphate rapidly lose their ·fertilizing effect.[135] These are usually high in allophane, or contain aluminium or perhaps ferric hydroxide with large surface areas. Such soils are sometimes said to 'fix' phosphate or to have a 'high phosphate fixing' capacity. The term fixation was originally used to imply a loss of availability of fertilizer phosphate to crop plants although it is now frequently used to describe phosphate adsorption, whether or not adsorption is accompanied by loss of availability. Because of confusion over its meaning the term should only be used if its method of measurement is defined.

Rapid loss of fertilizer effect was observed on two soils in Scotland, one derived from a basic igneous and one from a granitic till. An application of 250 kg P ha^{-1} as superphosphate was equivalent to only 60 kg P ha^{-1} for turnips one year after application and had too small an effect to be measured three years after.[136] Younge and Plucknett[137] have given an even more extreme example from Hawaii where an application of 650 kg P ha^{-1} as superphosphate was needed for the grass *Digitaria decumbens* and the pasture legume *Desmodium intortum* to give their maximum yield, and its effectiveness decreased rapidly during the subsequent four years. An application of 1300 kg P ha^{-1}, however, maintained maximum yields for at least six years.

---

135 For general review see Sanchez, P. A. and Uehara, G. in *The Role of Phosphorus in Agriculture* (see ref. 93), p. 471. Barrow, N. J., *op. cit.*, p. 333 discusses the evaluation of phosphate residues.
136 Williams, E. G., *J. Sci. Fd Agric.*, 1950, **1**, 244.
137 Younge, O. R. and Plucknett, D. L., *Soil Sci. Soc. Am. Proc.*, 1966, **30**, 653; for another example see Kamprath, E. J., *Agron. J.*, 1967, **59**, 25.

On soils such as these, it may be more economical to supply some of the phosphate in relatively insoluble forms such as rock phosphate, fused serpentine phosphate or a high temperature calcium magnesium phosphate rather than solely as a water soluble fertilizer.[138]

It was mentioned, at the beginning of this section that an annual crop will usually take up about 5 to 25 per cent of an application of phosphate fertilizer. The recoveries decrease in subsequent years, as shown by the results in Table 21.15 for seven grassland experiments in England. The recoveries were similar on acid and neutral soils. High recovery of phosphate was reported by Goedert,[139] in an experiment in the Brazilian cerrado on a Red Latosol with high phosphate buffer capacity. By the tenth crop of maize the yield from an initial broadcast application of 70 kg P ha^{-1} as superphosphate had dropped almost to that of the control, but the ten crops had taken up 45 kg P ha^{-1}. As the yield on the control plots in the first year was only about 10 per cent of the maximum yield and the soil contained little adsorbed phosphate, the implication is that a high percentage of the phosphate was recovered, even though the soil adsorbed large amounts of phosphate, and large applications of phosphate were needed to achieve maximum yields. It therefore appears that a distinction needs to be made between soils which have a high capacity for adsorbing phosphate, but retain it so that it can still be used by plants, and those in which adsorbed phosphate undergoes a

Table 21.15 Upate of phosphate by hay and grass in successive years after phosphate manuring. Per cent of added P recovered in hay or mowings. 55 kg P ha^{-1} added as superphosphate. (Figures taken from Russell, E. J., *Min. Agric. Bull.*, No. 28, 3rd edn, 1939, and from *Basic Slag Rpts to the Min. Agric.*, 1933 and 1934)

Year	2 neutral soils	3 acid soils	2 soils grass mown*
Application	7.7	8.3	13.7
1 year after	5.7	5.0	9.4
2 years after	1.8	3.3	5.8
3 years after	1.5	2.3	2.1
Total recovery	16.7	18.9	31.0

* One acid and one neutral soil.

138 See Mainstone, B. J., *Emp. J. expl. Agric.*, 1963, **31**, 175; McLachlan, K. D. and Norman, B. W., *Aust. J. exp. Agric. Anim. Husb.*, 1969, **9**, 38, 341.
139 Goedert, W. J., *J. Soil Sci.*, 1983, **34**, 405.

further slow reaction to become less available. What is usually lacking in agronomic experiments with phosphate fertilizers is ancillary measurement of the slow and fast reactions of phosphate in soil (reactions 5 and 6 in Fig. 21.5).

22

Potassium, sodium, calcium, magnesium, sulphur, silicon

Potassium*

Large amounts of potassium are taken up by crops which attain high yields. The amount taken up annually by leafy crops such as grass swards cut three or four times and yielding 10 to 15 t dry matter per hectare is 400–500 kg K ha^{-1}, by a good crop of potatoes it is about 300 kg K ha^{-1}, and for a cereal crop yielding 5 to 10 t ha^{-1} grain the amount is 200–300 kg K ha^{-1}. These amounts are very much greater than the uptake of most other nutrients and may equal or exceed the uptake of nitrogen. The total potassium content of soil is, however, relatively large compared with even the greatest potassium uptake by crops. For example many soils in Britain contain about 1 per cent K which corresponds to about 20 t ha^{-1} to 15 cm depth, yet even soils with the highest contents might be unable to supply enough to meet the potassium requirements of a crop. As discussed for phosphate in the previous chapter, in soils where there is an insufficient supply to the crop the limitation is not the total amount present, but the concentration at the root surface and its replenishment. The concepts of nutrient availability which were described for phosphate therefore apply in general terms to potassium.

* For recent review articles see *Potassium in Agriculture* (ed. R. D. Munson), Am. Soc. Agron., Madison, 1985.

Soil minerals containing potassium

The total content of potassium in soils varies widely from less than 0.01 per cent to about 4 per cent, and is commonly about 1 per cent (Table 22.1). The potassium content of rocks also varies widely, average values being in the range 1 to 2 per cent.[1] The two groups of potassium-bearing minerals which occur in rocks and soils in largest amounts are the micas muscovite $(KAl_2(AlSi_3O_{10})(OH)_2)$ and biotite $(K(Mg,Fe)_3(AlSi_3O_{10})(OH)_2)$, and the feldspars orthoclase and microcline $(KAlSi_3O_8)$. These minerals occur extensively in sialic rocks (rich in silicon and aluminium) such as granite, but are present in smaller amounts in mafic rocks (rich in magnesium and iron), which therefore have lower total potassium contents. The potassium content of sedimentary rocks varies with their content of clay minerals and especially illite. Sandstones have generally low or very low potassium contents, clays and shales may contain 2–4 per cent K, and in calcareous rocks the potassium content varies with their clay content.

Table 22.1 Total potassium content of soils

	Per cent K	*Ref.*
Australia	<0.01 to >3.0	1
United States	0.83 (ave.)	2
England	0.2–4	3
Northern Nigeria	0.06–4	4
New Zealand	0.05–1.8	5
Malaysia	0.1–1.0	5

1 Williams, C. H. and Raupach, M. in *Soils: An Australian Viewpoint*, p. 777, CSIRO/Acad. Press, 1983.
2 Black, C. A., *Soil–Plant Relationships* (2nd edn), Wiley, 1968.
3 Arnold, P. W. and Close, B. M., *J. agric. Sci.*, 1961, **57**, 295.
4 Wild, A., *Expl. Agric.*, 1971, **7**, 257.
5 Cooke, G. W., *The Control of Soil Fertility*, Crosby Lockwood, 1967.

Weathering processes bring about the breakdown of micas and feldspars so that the minerals are reduced both in particle size and amount.[2] Potassium is released and leaching removes soluble weathering products, which reduces the potassium content of the soil. In weakly weathered soils the potassium

1 Wedepohl, K. H. (ed), *Handbook of Geochemistry*, Vol. II/2, Ch. 19, 1978, for survey of rock analyses. For total K in soils see Ure, A. M. and Berrow, M. L. in *Environmental Chemistry*, Vol. 2 (Sen. Reporter H. J. M. Bowen), R. Soc. Chem., Burlington House, London, 1982.
2 For weathering processes see Arnold, P. W., *J. Sci. Fd Agric.*, 1960, **11**, 285. For the weathering of K-bearing minerals see Rich, C. I. in *Potassium in the Soil*, Vol. 15, Int. Pot. Inst., 1972.

content is similar to that in the parent rock, and primary minerals remain in the sand and silt fractions; but after long, continued weathering and leaching the potassium content in all particle size fractions becomes low. For intermediate stages there is relatively more potassium in the clay than in the sand and silt fractions although coarse grains of stable minerals such as muscovite might persist.[3]

Feldspars weather more readily than micas of a similar particle size and biotite weathers more readily than muscovite. The minerals are decomposed by hydrolysis and by the action of organic acids which act as proton donors and complexing agents. Products of weathering may include both 2 : 1 and 1 : 1 clay minerals, ferric and aluminium oxides and hydroxides, silicic acid and potassium and other ions, including Na^+ and Mg^{2+}, which are released into solution and may subsequently be retained as exchangeable ions or leached out of the weathering zone.

Weathering by proton exchange is an important process for the release of potassium from micas. Protons replace K^+ on the mineral surface and in the interlayer space, especially at the edges of the crystals. The H^+ in turn may be replaced by $Al(OH)_n^{(3-n)+}$ to produce vermiculite- or chlorite-like minerals in the clay fraction. These degraded micas have expanded interlayer space from which more K^+ may be released. But, as discussed later, if the concentration of potassium in the soil solution is increased, e.g. when potassium fertilizer is applied, K^+ may move back into the expanded interlayer space and become fixed, partially reversing the weathering process.

Reactions of potassium ions in soils

As applies to other cations, potassium ions are held partly in the soil solution and partly at the negatively charged surfaces of clays and organic matter, and they undergo exchange reactions as described in Chapter 7. Additionally, and as discussed below, potassium ions can be held tightly inside clay minerals, when they are often described as 'fixed'.

For the exchange reactions, the relationship between potassium which is exchangeable and that which is in solution is given by the Gapon equation (see Ch. 7). If potassium is added to, or removed from, a sample of soil which is calcareous, neutral or only slightly acid, a convenient form of the equation[4] is:

$$\Delta K = K' \, a_K/a_{Ca}^{1/2}$$

3 Salmon, R. C., *Rhodesian J. agric. Res.*, 1964, **2**, 85 gives the distribution of potassium in the particle size fractions of soils from Zimbabwe.
4 Beckett, P. H. T., *J. Soil Sci.*, 1964, **15**, 1, 9. For applications see Tinker, P. B., *op. cit.* **15**, 24, 35; Addiscott, T. M., *J. agric. Sci.*, 1970, **74**, 131; Wild, A., *Expl. Agric.*, 1971, **7**, 257; Sinclair, A. H., *J. Soil Sci.*, 1979, **30**, 757.

where ΔK is the amount of potassium adsorbed or desorbed by the soil, K' is an equilibrium constant, and a_K and a_{Ca} represent the active concentrations of potassium and calcium in solution. The ratio $a_K/a_{Ca}^{1/2}$ defines the intensity, I, of potassium in solution; this form of expression should be used when comparing different soils. If, for a single soil sample, the change of potassium adsorbed or desorbed is small, the change of calcium concentration or activity in solution is also small, and can usually be neglected; under these conditions the ratio is sometimes replaced by a_K or by the solution concentration, C_K.

The relation between ΔK and $a_K/a_{Ca}^{1/2}$ is known as the quantity/intensity (Q/I) relationship for soil potassium. It is useful in characterizing the exchangeable and solution potassium in terms of three parameters: (i) the intensity (or concentration) of potassium in the equilibrium soil solution, which is the value where $\Delta K = 0$; (ii) the buffer capacity of the soil for potassium, which is the slope of the part of the Q/I curve where $\Delta K = 0$; and (iii) the quantity of potassium which can be readily desorbed, which is obtained by extrapolation of the curve to zero intensity (or concentration).

At low values of the activity ratio the slope of the Q/I curve usually increases sharply, indicating that under these conditions the potassium is held much more tightly than the rest of the potassium. It is probably held in holes or wedge-shaped spaces situated between broken edges of clay sheets into which the potassium ions just fit, for neither a well-dispersed montmorillonite clay[5] nor peat[6] holds tightly bound potassium.

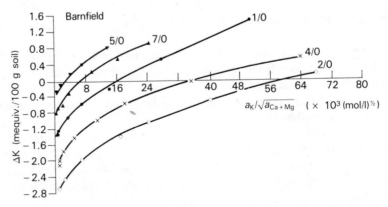

Fig. 22.1 Displacement of the Q/I curve on soil from Barnfield, Rothamsted, with addition and removal of potassium. Key: 5/0 P only; 7/0 P, Na, Mg; 1/0 FYM; 4/0 P, K, Na, Mg; 2/0 FYM, P, K. Annual rates (1903–67): 33 kg P ha^{-1}, 228 kg K ha^{-1}, 88 kg Na ha^{-1}, 22.4 kg Mg ha^{-1}, 35 t FYM ha^{-1}; no N to plots. (From Addiscott, T. M., *J. agric. Sci.*, 1970, **74**, 131.)

5 Beckett, P. H. T. and Nafady, M. H. M., *J. Soil Sci.*, 1967, **18**, 263.
6 Salmon, R. C., *J. Soil Sci.*, 1964, **15**, 273.

There is much evidence[7] that the buffer capacity is a characteristic of a soil, additions or removals of potassium giving curves which can be superimposed by upward or downward displacement, as is illustrated in Fig. 22.1. The evidence of Beckett and Nafady[5] is that organic matter has much less effect than clay minerals on the buffer capacity of soils for potassium, and they have shown that oxidation of soils by hydrogen peroxide usually increases the potassium buffer capacity by unblocking exchange sites.

Potassium release and fixation

The content of exchangeable potassium in soils is usually defined as that extracted from a sample of the soil with a solution of an ammonium salt, commonly 1 M. Ammonium and potassium ions are about the same size (0.29 and 0.27 nm diameter respectively) so that determinations by this method are not influenced by changes in size of the wedge-shaped exchange sites of the interlayer space of 2 : 1 clay minerals. A solution of an ammonium salt is used because it gives a measure of the amount of soil potassium which is readily extracted, and the amount is almost independent of the period of extraction. It is, however, an incomplete measure of the amount of potassium which can pass into solution. In soil solutions the dominant cation is usually Ca^{2+} which, although a smaller ion (0.20 nm diameter), is bigger than K^+ when hydrated. Hence, by exchange it enlarges the interlayer space, so releasing more potassium. Further, as potassium is removed from the soil solution by plant uptake, it continues to be released from the clay minerals by cation (including proton) exchange and a gradient is created which leads to diffusion of K^+ through the structure of the clay particles to their surface. The reverse reaction occurs when potassium is added to the soil solution, as by fertilizer addition. Some potassium ions will then become 'fixed' in the interlayer space, i.e. they are not extracted by an ammonium salt under defined conditions, but may nevertheless be used by plants. These reactions are shown diagrammatically in Fig. 22.2.

The relation between K^+ in clay minerals and in solution has been studied best with flakes of well-crystallized micas and vermiculite. When the crystals are placed in a suitable salt solution, K^+ present in the interlayer space of the crystals is slowly released from the broken edges, and the spacing of the interlayers is increased. This will occur with many salt solutions, ammonium salt solutions being the main exception. The stronger the binding between K^+ and the adjacent silicate layers the lower must be the concentration of potassium in the bathing solution before potassium begins to be released from between the layers. Thus if the bathing solution is 1 M NaCl, the potassium concentration in the solution must be below 10^{-5} M for potassium to be

7 Addiscott, T. M. (ref. 4). Beckett, P. H. T. and Clement, C. R., *J. Soil Sci.*, 1973, **24**, 82.

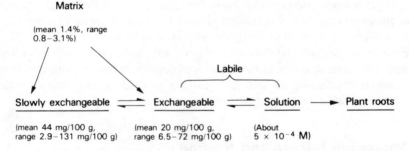

Fig. 22.2   Representation of soil potassium; amounts are those reported for 20 English soils. Matrix potassium is that present in feldspars and micas. Slowly exchangeable potassium is that released in a greenhouse experiment in which ryegrass was grown for 200 to 600 days, and exchangeable potassium is that initially present. The value for solution potassium is that of a fertile arable soil. (Adapted from data of Arnold, P. W. and Close, B. M., *J. agric. Sci.*, 1961, **57**, 295.)

released from muscovite, but only below 2×10^{-4} M from biotite.[8] The release is slow because of the large activation energy of the reaction, which is about 50 kJ mol^{-1} for biotite, and because the depth of penetration of the replacing cation increases only with the square root of the time of contact.

The micaceous clays found in soils are less ordered and as a consequence the forces between the interlayer potassium ions and the adjacent layers are weaker. Thus, the critical potassium concentration in the soil solution to bring about release of potassium may be about 5×10^{-3} M, about 500 times greater than for muscovite. This is probably because potassium ions do not occupy all the interlayer positions. Especially near the weathered edges of the clay particles the interlayer spaces are wider and are occupied by larger ions, which increases the rate of diffusion of ions into and out of the clay particles.[9]

This process of slow exchange in a solution of low potassium concentration is partly reversible but the time scales of the forward and reverse reactions are very different. If a 2 : 1 clay that has been depleted of non-exchangeable potassium is put into a more concentrated potassium solution, a proportion of the potassium will be converted into a non-exchangeable form in a matter of minutes,[10] whereas desorption of non-exchangeable potassium is always slow. This latter is illustrated in Fig. 22.3 for two Broadbalk soils,[11] one of which has received potassium fertilizers for most years since 1843 and the

---

8 Newman, A. C. D., *J. Soil Sci.*, 1969, **20**, 357.

9 Martin, J. C. *et al.*, *Soil Sci. Soc. Am. Proc.*, 1946, **10**, 94;   Barshad, I., *Soil Sci.*, 1954, **77**, 463.

10 Quoted by Arnold, P. W., *Fertil. Soc. Proc.*, No. 115, 1970.

11 Matthews, B. C. and Beckett, P. H. T., *J. agric. Sci.*, 1962, **58**, 59.

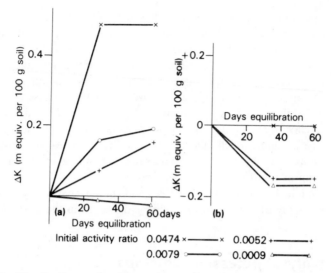

Fig. 22.3 The fixation and release of potassium from two soils on Broadbalk field, Rothamsted, when shaken in solutions of different activity ratios. (a) Plot 11, N P fertilizer, no K; $AR_0 = 0.0015$. (b) Plot 13, N P K fertilizer: $AR_0 = 0.0145$.

other has received none. It shows that if the soil is put in a solution whose potassium concentration is larger or smaller than its equilibrium value in the soil, potassium will either be taken up from the solution or released to the solution. In the graph the potassium–calcium activity ratios in the solution have been plotted rather than potassium concentrations, though the potassium concentration is the more important for uptake by plants.

Various methods have been proposed for determining the amount of soil potassium which, although non-exchangeable, or fixed, as defined earlier, can fairly readily be converted to the exchangeable form. No absolute method is possible because the potassium ions can be held in a variety of positions, ranging from those from which it can fairly readily be released to those held as tightly as between muscovite layers or within the structure of feldspars; the very tightly held potassium is referred to as matrix potassium in Fig. 22.2. Further, crops vary in their ability to reduce the potassium ion concentration in the soil solution (see p. 756), so no useful definition can be given for the crop-available fixed potassium. However, hydrogen-saturated and calcium-saturated exchange resins can be a useful means of measuring the less strongly fixed potassium.[12] Other methods are referred to later.

12 Haagsma, T. and Miller, M. H., *Soil Sci. Soc. Am. Proc.*, 1963, **27**, 153; Brown, T. E. and Matthews, B. C., *Canad. J. Soil Sci.*, 1962, **42**, 266; Goulding, K. W. T., *J. agric. Sci.*, 1984, **103**, 265.

The proportion of potassium ions added to a soil that is converted to the fixed form is increased by drying the soil. Alternate wetting and drying in the root zone of a crop tends to increase the amount of fertilizer potassium that is converted to the fixed form. The fixation of fertilizer potassium also takes place more readily in neutral than in acid soils, and liming an acid soil increases its ability to fix potassium.[13] This is probably because of the presence in acid soils of hydroxy-aluminium polymers as 'islands' or interlayers within the clay minerals. On liming to pH 6 or above the aluminium is replaced by calcium which allows the interlayer spaces to expand and contract, thus releasing or fixing potassium depending on the potassium concentration in solution, as described above.

The relation between exchangeable and non-exchangeable potassium affects the supply of potassium to crops, as discussed in the next section.

Availability of potassium to crops

The immediate source of potassium for crops is that in the soil solution. The provision of potassium to a crop therefore depends on the potassium ion concentration close to the root surface, on the rate at which potassium ions are transported through the soil solution to the root surface, on the replenishment of the solution by desorption from exchange sites on the soil colloids, and on the extent to which roots ramify through the soil. Root ramification is included because it affects the length of the transport path of potassium in the soil solution. The requirements are the same as for phosphate discussed in Chapter 21 where it was emphasized that rate of uptake by plants depends to a large extent on plant growth rate.

Potassium in soil solution

Solution and sand culture experiments have shown that the rate of potassium uptake by roots of rapidly growing crops depends on the potassium concentration at their absorbing surfaces and that it is not very sensitive to quite large changes in the concentration of other cations, such as calcium, magnesium or sodium in the solution.[14] Further, while potassium moves to the root largely by diffusion (see below) calcium moves largely by mass flow, so the ratio of these two ions in solution close to the root is different from that in the bulk soil solution. The minimum concentration of potassium required at the root surface can be measured by the flowing solution culture technique. This concentration depends on the plant species, the growth stage and the

13 York, E. T. *et al., Soil Sci.*, 1953, **76**, 379, 481; 1954, **77**, 53.
14 Wild, A. *et al., Soil Sci.*, 1969, **108**, 432 give examples with ryegrass and flax.

potential growth rate under the prevailing environmental conditions. For example, using this technique the potential growth rates of barley and fodder radish were maintained for the first 14 days after germination by concentrations of 2×10^{-5} M K, and for ryegrass 1×10^{-6} M K.[15] After 14 days the growth rates of all three species were the same at 1×10^{-6} M K as at higher concentrations.

In soils, the concentration at the root surface is usually limited by diffusion and, as a consequence, higher concentrations are required in the soil solution. Some crops require a concentration of 5×10^{-3} M K in the bulk soil solution, for others 5×10^{-4} M K is adequate, whereas 5×10^{-5} M K is inadequate for all crops.[16] As an example of the amounts these concentrations represent, a soil which holds 6 cm of solution of 5×10^{-4} M K in the top 30 cm, contains about 12 kg K ha^{-1} in solution. The range of potassium concentrations found in soil solutions is wide; values below 10^{-5} M have been reported, but are more usually about 10^{-3} to 10^{-4} M if the soil is used for intensive agriculture. The small amount in solution has two implications. First, to provide over 200 kg K ha^{-1}, as required by a good annual crop, the soil solution is itself an inadequate source and potassium must be desorbed from the soil surfaces. Second, if the crop is transpiring at a rate of 3 mm day^{-1} the mass flow of solution to the roots will provide the crop with 0.6 kg K day^{-1} if the solution is 5×10^{-4} M. This is much less than is required by a rapidly growing crop (Table 3.4). Diffusion would therefore be expected to be the dominant process for transporting potassium to the root surface (see Ch. 4) and this conclusion is supported by measurements from field and pot experiments.[17] From a pot experiment with leeks it was calculated that about 10 per cent of the K was taken up by mass flow (more strictly by apparent mass flow) and 90 per cent by diffusion.[18]

Exchangeable and non-exchangeable potassium

Although the soil solution is the immediate source of potassium used by plants, replenishment is needed if the supply is to be maintained. The potassium ions which are held on the surfaces of soil particles readily exchange with other cations and replenish the solution. Exchangeable potassium is therefore often a fairly satisfactory index of the quantity of potassium which plants can take up. This is illustrated by Fig. 22.4 which shows a good

15 Woodhouse, P. J. *et al.*, *J. exp. Bot.*, 1978, **29**, 885. See also Asher, C. J. and Ozanne, P. G., *Soil Sci.*, 1967, **103**, 155.
16 Haworth, F. and Cleaver, T. J., *J. Sci. Fd Agric.*, 1963, **14**, 264; 1965, **16**, 600. For winter wheat compare Barraclough, P. B. and Kent, R., *Rothamsted Ann. Rept for 1984*, p. 183.
17 Barber, S. A. *et al.*, *Trans. Int. Congr. Soil Sci.*, Comm. IV and V, 1962, 121.
18 Brewster, J. L. and Tinker, P. B., *Soil Sci. Soc. Am. Proc.*, 1970, **34**, 421.

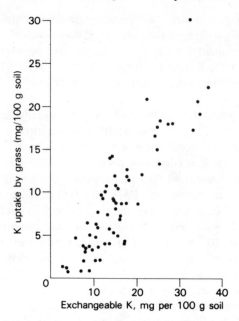

Fig. 22.4 Relationship between exchangeable K content of soils and potassium
uptake by perennial ryegrass in the greenhouse (approximately 36
days' cropping). (From Arnold, P. W., *Fertil. Soc. Proc.*, No. 72,
1962.)

correlation between short-term uptake by ryegrass and the exchangeable K
measured by extraction with neutral 1 M ammonium acetate. As pointed out
by Arnold, for all 64 soils from 50 sites the correlation coefficient was 0.88,
but more than half the soils had exchangeable potassium contents between
10 and 20 mg K/100 g soil, and for this group the correlation coefficient was
only 0.35. One reason for this partial failure of the exchangeable potassium
to account for uptake is that during cropping there is release of potassium
which is not exchangeable with NH_4^+ ions in a neutral solution.

The prediction of potassium availability is improved by inclusion of the
readily released non-exchangeable potassium.[19] This release of non-exchangeable potassium can be measured by growing crops on the soil and subtracting
the fall of exchangeable K during the cropping period from potassium taken
up by the crop. It is convenient to grow grass as the test crop because it can
be cut at intervals without replanting. Figure 22.5 gives some typical curves
for the cumulative uptake of potassium by ryegrass from four British soils,
chosen because they all have about the same clay content.[20] Three soils
readily released potassium in amounts equal to about three times the fall in

19 See Goulding, ref. 12.
20 From Arnold, P. W. and Close, B. M., *J. agric. Sci.*, 1961, **57**, 295.

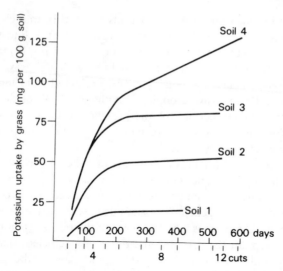

Fig. 22.5 Potassium released to ryegrass by four English soils each with about 20% clay.

	Soil 1	Soil 2	Soil 3	Soil 4
% K in clay fraction	1.96	2.66	1.69	1.78
Exchangeable K, mg per 100 g soil	9.0	14.0	33.5	72.0

exchangeable potassium, and then much more slowly, at a rate of between 3 and 10 mg/100 g soil per year.

Some soils continue to lose potassium at an appreciable rate for a very long time under intensive cropping. Soil 4 in Fig. 22.5 is an example from a temperate region: its 'slow' release rate is 35 mg/100 g annually. This soil is derived from the Upper Greensand and contains the potassium zeolite clinoptilolite in its coarse clay and fine silt fraction,[21] which weathers readily to release potassium. Some boulder clays in eastern England behave in the same way, but in these soils the potassium is released from illitic minerals.[22] Thus, the soil on Harwood's field, Saxmundham, Suffolk still supplies over 100 kg K ha^{-1} annually to crops, though it has been in continuous arable for nearly 60 years without receiving any potassium fertilizer.[23] In the same way potassium silicates, such as biotites and andesine, will readily weather under humid tropical conditions to release large amounts of potassium. In Sumatra, soils containing these minerals liberated 200 to 250 kg K ha^{-1} annu-

21 Talibudeen, O. and Weir, A. H., *J. Soil Sci.*, 1972, **23**, 456.
22 Goulding, K. W. T. and Loveland, P. J., *J. Soil Sci.*, 1986, **37**, 555.
23 Trist, P. J. O. and Boyd, D. A., *J. agric. Sci.*, 1966, **66**, 327. For release of non-exchangeable potassium on Hoosfield, Rothamsted see Johnston, A. E. and Poulton, P. R., *Rothamsted Ann. Rept. for 1976*, Part 2, p. 53.

ally to sisal and sugar cane over long periods[24] and in Hawaii elephant grass (*Pennisetum purpureum*) took up over 4 t K ha^{-1} over a four-year period without the exchangeable potassium in the top metre of soil being affected.[25] However, not all soils in tropical regions behave in this way. Only small amounts of non-exchangeable potassium were released from most northern Nigerian soils under intensive cropping.[26] Although many of these soils had high potassium contents because of the presence of large amounts of mica in the sand fraction, the clays were mostly kaolinitic and their potassium contents were consequently low.

The results in Fig. 22.5 indicate that part of the non-exchangeable potassium is readily released and taken up by roots and part is only slowly released. The distinction is not sharp, but is more easily observed by plotting the release of non-exchangeable potassium against the square root of time.[27] As shown in Fig. 22.6 there is a first step where the release is more rapid than in the second step. In the first step the diffusion coefficient that was calculated for these soils was 1.4 to 3.1 $\times$ 10^{-7} cm^2 s^{-1}, and in the second step it was 10^{-20} to 10^{-22} cm^2 s^{-1}. Release in the first step was probably from potassium fixed close to the edges of the clay minerals, and the amount was closely

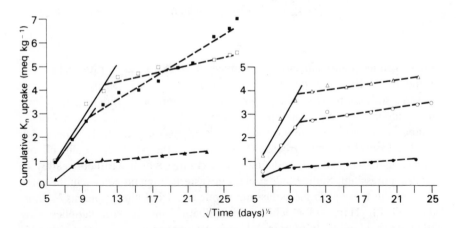

Fig. 22.6 Relationship between cumulative K$_n$ uptake by ryegrass (tops and roots) and $\sqrt{}$time for six representative Scottish soils; K$_n$ is potassium initially non-exchangeable.
▲ Loanends; ● Maryfield; △ Logie Newton;
○ Shaggart; ■ Auchnagatt; □ Auchindownie.
(From Sinclair, A. H., *J. Soil Sci.*, 1979, **30**, 775.)

24 van der Marel, H. W., *Soil Sci.*, 1947, **64**, 445.
25 Ayres, A. S., *Soil Sci. Soc. Am. Proc.*, 1946, **11**, 175.
26 See Wild, ref. 4.
27 Addiscott, T. M., *J. agric. Sci.*, 1970, **74**, 123; with Talibudeen, O., *op. cit.*, 1971, **76**, 411; with Johnston, A. E., *op. cit.*, 1975, **84**, 513.

correlated with exchangeable plus solution potassium. The high diffusion coefficient implies that the ions were diffusing through comparatively wide spaces, whereas the very low diffusion coefficient in the second step is similar to that found in micas and represents diffusion from the core of the mineral. For comparison, the diffusion coefficient in aqueous solutions at 25 °C is 2×10^{-5} cm^2 s^{-1}. Extraction with a calcium resin will distinguish between these two sources of non-exchangeable potassium.[28]

In the cropping experiments illustrated in Figs 22.5 and 22.6, uptake of non-exchangeable potassium from bulk soil was measured. Depletion of solution and exchangeable potassium is, however, much greater very close to the root, and especially within the root hair cylinder than it is further away, and the depletion is rapid. For example, using two [43]K-labelled soils it has been shown that within 0.7 mm of maize roots, which were 2.5 days old, about twice as much potassium was released from the soil as was present as exchangeable potassium before depletion.[29] On the same soils the potassium inflow with five plant species was directly related to the volume of the root hair cylinder.

To summarize: the release of non-exchangeable potassium depends on both the crop species and the soil. The properties of the crop that are important are growth rate, which determines demand for potassium, root density and volume of the root hair cylinder. The soil properties include the amount of clay, potassium content or mineralogy of the clay, the amount of previously applied potassium which had been fixed, and ped size,[30] which affects root distribution.

Soil analysis and potassium manuring

Many methods of soil analysis have been used to provide an index of the amount of potassium in soil that is available to a crop, and hence to advise farmers on the amount of fertilizer potassium to apply.[31] The usual method is to determine that which is exchangeable by extracting the soil with an ammonium salt. Ammonium nitrate (1 M) and neutral ammonium acetate (1 M) are most commonly used, the former being more convenient for flame analysis. The use of an ammonium salt has the advantage over other extractants in giving a well-defined value for exchangeable potassium in soils. Although there is no generally accepted chemical method for determining the

28 Talibudeen, O. *et al.*, *J. Soil Sci.*, 1978, **29**, 207; Goulding, ref. 12.
29 Jungk, A. *et al.*, *Proc. 9th Int. Pl. Nutr. Conf.*, 1982, **1**, 250.
30 For calculations see Heming, S. D. and Rowell, D. L., *J. Soil Sci.*, 1985, **36**, 45, 61.
31 Knudsen, D. *et al.*, in *Methods of Soil Analysis* (2nd edn) (ed. A. L. Page), Part 2, p. 225, *Agron. Monog.*, No. 9, Am. Soc. Agron., Madison, Wisc., 1982. See also Deist, J. and Talibudeen, O., *J. Soil Sci.*, 1967, **18**, 125. For use of potassium isotopes see Mercer, E. R. and Gibbs, A. R., *Trans. Comm. II and IV. Int. Soc. Soil Sci., Aberdeen*, 1966, 233; Sumner, M. E. and Bolt, G. H., *Soil Sci. Soc. Am. Proc.*, 1962, **26**, 541.

readily released non-exchangeable potassium, there are many reports[32] of the use of hot 1 M HNO_3 or HCl, Na tetraphenylboron, electro-ultrafiltration, and cation exchange resins.

In England and Wales, exchangeable potassium is determined by extracting soils with 1 M ammonium nitrate, and the results are expressed on soil volume.[33] The indices from these determinations are given in Table 3.7. The advice that is given to a farmer is based on the K index of his soil, the crop to be grown, the expected yield, and on the maintenance of the potassium balance of the soil. For example, at a K index of 1 an application of potassium fertilizer at 37 kg K ha^{-1} would be advised for a wheat crop expected to yield 7.5 t ha^{-1} grain if the straw is to be incorporated or burned. If the straw is to be removed from the field, the application would be increased to 75 kg ha^{-1}, and it would be further increased for an expected yield of 10 t ha^{-1} with the straw removed. For maincrop potatoes the advised application is 249 kg ha^{-1} at index 1 decreasing to 83 kg ha^{-1} if the index is over 3.

Crops differ, however, in their response to potassium fertilizers and in their response to residues from earlier applications. This can be illustrated by some results obtained on the Exhaustion Plots at Rothamsted.[34] Some of these plots received about 100 kg K ha^{-1} annually from 1857 to 1874 and then 150 kg K ha^{-1} from 1876 to 1901, and others received no potassium. After 1901 no further potassium has been given and the land has been cropped to cereals most years. In 1957 and 1958 several crops were grown on these plots with different levels of fresh potassium fertilizer. Table 22.2 gives some of the results of this experiment. It shows that the potassium residues left in the soil since manuring ceased in 1901 were adequate to supply the potassium needs of swedes and sugar beet, and were more effective than the 63 kg given as fertilizer to the swedes or the 125 kg given to the beet on the plots which had not received any potassium since 1857. Potatoes, on the other hand, gave a larger response to 125 kg of fresh potassium than to the residues, but a still larger response to both the residues and the fresh fertilizer; the evidence from the response curves suggests that this difference would persist if still higher levels of fresh potassium fertilizer had been given.

The reasons for this difference in crop behaviour have not been fully explored. As mentioned earlier, crop plants grown in solution culture differ

32 Haylock, O. F., *Trans. 6th Int. Congr. Soil Sci.*, 1956 B, 403; Moss, P. and Coulter, J. K., *J. Soil Sci.*, 1964, **15**, 284; Németh, K. (ed), *Application of Electro-ultrafiltration.(EUF) in Agricultural Production*, Nijhoff, The Hague, 1982; *idem, Pl. Soil*, 1982, **64**, no. 1. For review see Martin, H. W. and Sparks, D. L., *Comm. Soil Sci. Pl. Anal.*, 1985, **16**, 133.

33 *The Analysis of Agricultural Methods, MAFF Ref. Book RB427* (2nd edn), London, HMSO, 1981.

34 Warren, R. G. and Johnston, A. E., *Fertil. Soc. Proc.*, No. 72, 1962, and see *Rothamsted Ann. Rept.*, 1954, p. 59 for a similar result with timothy (*Phleum pratense*).

Table 22.2 Crop responses to potassium fertilizer residues and fresh additions of potassium fertilizer; Rothamsted, Hoosfield Exhaustion Plots. Rate of K fertilizer: barley and swedes 63 kg ha^{-1}, potatoes and sugar beet 125 kg ha^{-1}. Figures in brackets = uptake of K by crop in kg ha^{-1}.

Yields in t ha^{-1}	Yield without residues or fertilizer	Response to		
		Residue	Fertilizer without residue	Fertilizer with residues
Barley: grain	3.24 (43)	0.20 (16)	0.23 (13)	0.23 (20)
Swedes: roots	37.6 (88)	9.1 (47)	2.5 (20)	10.0 (54)
Potatoes: tubers	17.1 (38)	10.5 (34)	14.2 (58)	19.6 (81)
Sugar beet: roots	31.2 (121)	6.0 (83)	4.8 (32)	2.5 (82)

in the minimum concentration of potassium they require at the root surface to achieve their potential growth rates. It has also been shown that potatoes grown in soil can take up little potassium once the activity ratio ($a_K/a_{Ca}^{1/2}$) falls below about 3.2×10^{-4} M$^{1/2}$, while ryegrass can take up potassium down to 6.7×10^{-5} M$^{1/2}$. One reason for the difference between crops may lie in their rate of demand as determined by their growth rate. Also important is their root proliferation, production of root hairs and infection with mycorrhiza, each of which affects the length of the diffusive pathway and hence the potassium concentration at the root surface. In terms of the potassium potential, expressed as $\Delta G = RT \ln a_K/\sqrt{a_{Ca+Mg}}$, the values at which crops ceased to take up potassium in pot experiments were -23.4 kJ/equiv. for ryegrass, -20.9 kJ/equiv. for lettuce, and for potatoes it changed from -17.4 kJ/equiv. to -20.5 kJ/equiv. as the plants developed. Thus more potassium was initially available to ryegrass than to lettuce or potatoes.[35]

The results in Table 22.2 may be compared with those for residual phosphate fertilizer illustrated in Table 21.12. The results with potatoes show that it takes time to build up the potassium status of the soil, presumably because it is difficult to distribute a single dressing of fertilizer evenly through the depth of cultivation. Potassium in the subsoil has been found to be important under some conditions.[36] A single heavy application of potassium fertilizer can, however, have the undesirable effect of increasing the osmotic pressure of the soil solution, due to the high concentration of the associated anion, usually chloride, which can harm the growth of young plants during spells of dry weather (see Ch. 27 for effects of chloride on plants). It is for this reason that large applications of farmyard manure are more valuable than fertilizers for building up the relatively high concentration of potassium ions in the

35 Addiscott, T. M., *Rothamsted Ann. Rept. for 1969*, p. 63, 1970; and see also, with Johnston, A. E., *J. agric. Sci.*, 1971, **76**, 539, 553.
36 McEwen, J. and Johnston, A. E., *J. agric. Sci.*, 1979, **92**, 695.

solution needed by many young vegetable crops, for example, if they are to make rapid early growth.[37]

Crops differ in the proportion of the potassium added as fertilizer that they take up, when all other nutrients are present in ample supply. Grass will take up about 80 per cent of added potassium, whether the application is large or small, as will be shown in the next section. At Rothamsted, kale and potatoes will take up to 50 per cent of the added potassium, as does wheat on the Broadbalk plots receiving a balanced fertilizer although, in general, the cereals take up less.

Maintenance of the potassium status of soils

The grass crop can have a very great influence on the potassium status of the soil, for it is a very strong extractor of potassium. Grass therefore competes successfully with clover in grass–clover swards for a limited potassium supply, perhaps because it has a much more extensive root system in the surface soil; so it is only possible to have a vigorous clover component in a sward if the soil potassium status is adequate. Grass can take up very large amounts of potassium from soils receiving a high rate of potassium fertilizer, as is illustrated by results in Table 22.3 from a four-year experiment at the Animal and Grassland Research Station, Hurley, in which the ryegrass was cut and removed.[38] As the rate of addition of potassium fertilizer was increased, not only did the yield of the grass increase, but so also did its potassium content, so that an annual dressing of 250 kg K ha^{-1} (i.e. 1000 kg ha^{-1} in four years) was inadequate to keep the soil in potassium balance.

Table 22.3 Uptake of soil potassium by ryegrass when manured with potassium fertilizer. Nitrogen as nitrochalk (ammonium nitrate/calcium carbonate), potassium as potassium chloride, half applied in March, half in June. Totals for 4 years of the ley

Potassium added as fertilizer (kg ha^{-1})	Potassium removed in grass (kg ha^{-1})	Decrease in soil potassium (kg ha^{-1})	Exchangeable potassium at end of experiment (mg/100 g)	Yield of grass dry matter (t ha^{-1})	% potassium in dry matter
0	660	660	6.8	36	1.9
250	840	590	6.8	38	2.2
500	1030	530	6.6	39	2.6
1000	1290	290	9.9	40	3.2

37 Haworth, F. and Cleaver, T. J., *J. hort. Sci.*, 1961, **36**, 202; 1963, **38**, 40; 1966, **41**, 299.
38 Hopper, M. J. and Clement, C. R., *Trans. Comm. II and IV Int. Soc. Soil Sci.*, Aberdeen, 1967, 237.

Young grass contains a little over 2 per cent K in the dry matter at maximum growth rates[39]; if the grass is allowed to mature and is grown for hay the requirement falls below 2 per cent K. If there is a good supply of potassium in the soil, or if large applications of potassium fertilizer are made, these concentrations in the herbage are exceeded. As shown in Table 22.3 the annual fertilizer addition at 250 kg K ha^{-1} resulted in herbage containing 3.2 per cent K, and the total uptake exceeded that added in fertilizer by 290 kg K ha^{-1} over the four-year period. This luxury uptake, that is, uptake in excess of the crop requirements, occurs commonly with potassium. The drain on the soil potassium was greater when no fertilizer potassium was added.

The supply of potassium in this soil is represented diagrammatically in Fig. 22.7. It shows that the supply will be reduced within three years to the rate at which non-exchangeable potassium becomes available. This rate is less than half of that required for the potential yield of grass. During the course of the experiment the exchangeable potassium in the soil fell from about 20 mg/100 g at the start to 6.8 mg/100 g at the end on the plots receiving no potassium fertilizer.

Because grass readily takes up more potassium than it needs, it is uneconomic to try to maintain the potassium status of a soil under intensively managed grass which is cut for drying or silage. This has the corollary that the following crop will probably need an application of potassium fertilizer if its yield is not to be limited by a shortage of available potassium. If, on the other hand, the grass is grazed most of the potassium is returned via the excreta, and depletion of exchangeable potassium in the soil is much less.

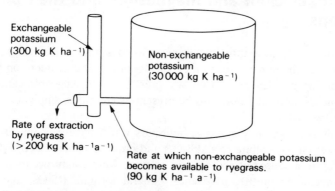

Exchangeable potassium (300 kg K ha^{-1})

Non-exchangeable potassium (30 000 kg K ha^{-1})

Rate of extraction by ryegrass (> 200 kg K ha^{-1}a^{-1})

Rate at which non-exchangeable potassium becomes available to ryegrass. (90 kg K ha^{-1} a^{-1})

Fig. 22.7 Diagrammatic representation of soil potassium and rate of extraction by ryegrass yielding 10 t dry matter ha^{-1} a^{-1}.

39 Clement, C. R. and Hopper, M. J., *NAAS Quart. Rev.*, No. 79, 101. For physiological requirements see Leigh, R. A. and Wyn Jones, R. G., *J. Sci. Fd. Agric.*, 1984, **35**, 292; *idem, New Phytol.*, 1984, **97**, 1.

Uneven distribution of potassium will, however, occur under grazing because of the voiding of excreta in patches. There is some loss from cattle which are housed for milking, but the potassium will be conserved in liquid slurry or farmyard manure and can be returned to the field.

Large amounts of potassium are removed when the whole crop is harvested. In cereal crops at harvest, 50 per cent or more of the potassium is present in the straw (Table 3.1) and will be removed if the straw is sold off the farm, but will be returned to the soil if the straw is either burned, used as a mulch or incorporated. Leafy crops such as spinach generally contain about 2 per cent K in their dry matter and large amounts are removed when they are harvested. A crop of 250 t ha^{-1} of tomatoes, which is achieved under glasshouse conditions in England, removes about 1300 kg K ha^{-1} in the fruit and foliage.[40]

From this account it will be seen that potassium fertilizer is usually required for high crop yields, but the requirement depends on previous cropping and on the levels of exchangeable and readily released non-exchangeable potassium in soil. For agricultural crops potassium is usually applied as the chloride containing 40, 50 or 60 per cent K_2O and a variable amount of sodium chloride, or less commonly as kainite which also supplies magnesium and sodium. Chloride in high concentration can harm plant growth and applications are therefore sometimes split. Potassium sulphate or nitrate are also used, especially in horticulture and for tobacco. For agricultural crops generally their higher cost is not justified.[41]

Sodium, calcium and magnesium and their uptake by crops

As for potassium, these three metals occur as constituents of silicates in rocks and remain in the mineral fraction of soils to an extent dependent on the rate of weathering and leaching. They are present as exchangeable cations and in the soil solution, and may also be present as chlorides, sulphates, carbonates or other salts.

The silicates include plagioclase feldspars which have a chemical composition between pure albite ($NaAlSi_3O_8$) and pure anorthite ($CaAl_2Si_2O_8$) and are therefore a source of sodium and calcium. Magnesium occurs in several silicates including biotite, pyroxenes, general formula $RSiO_3$, and amphiboles, general formula $R_7(Si_4O_{11})_2(OH)_2$ where R can include Mg, Fe^{2+}, Ca, Na and other metal ions. All these primary silicates are more readily weath-

40 Winsor, G. W. *Proc. Fertil. Soc.*, No. 103, 1968.
41 For further information on potassium fertilizers see Cooke, G. W., *Fertilizing for Maximum Yield* (3rd edn), 1982; Zehler, E. *et al.*, *Int. Pot. Inst. Res. Topics*, No. 9, 1981.

ered than orthoclase and muscovite, two of the main silicates which contain potassium. Magnesium differs from sodium and calcium in being a lattice constituent of a mica (biotite). Although the structure of the mica is changed on weathering, magnesium remains a lattice constituent of the weathering products if these include 2 : 1 layer lattice clay minerals (hydrous mica, vermiculite and montmorillonite). The non-clay fraction of soils which are moderately weathered and leached therefore contains only small amounts of sodium, calcium and magnesium compared with potassium. The clay fraction is usually similar except that it may contain large amounts of magnesium as a lattice constituent of the clay minerals. These differences are shown in Table 22.4 which refers to six English non-calcareous soils which have been acid washed to remove exchangeable cations. Soils that are highly weathered and leached progressively lose their sodium and calcium silicates, and ultimately their 2 : 1 layer lattice clay minerals. It is therefore common to find only small reserves of sodium and calcium to replenish the exchangeable ions when these are absorbed by plants or lost by leaching. The loss of silicates containing calcium is the more serious because it is an essential nutrient and little is added in rainfall.[42]

The amounts of sodium, calcium and magnesium taken up by some crop plants are given in Tables 3.1 and 3.2. For calcium and magnesium there is evidence that ions which are in solution or exchangeable are the source used by crops. Thus, using ^{45}Ca it has been shown that calcium which is isotopically exchangeable, and which is about equal in amount to the exchangeable calcium, is the principal source for crops.[43] For magnesium, using a range of soils from Britain, the amount of magnesium taken up by ryegrass in a pot experiment was very closely correlated with the fall in exchangeable magnesium.[44] The amounts were about equal, but a small amount of non-exchange-

Table 22.4 Composition of different fractions of six English soils (content of element in meq/100 g of fraction). (From *Rothamsted Ann. Rpt.*, 1965, 61)

	K	Na	Ca	Mg
Coarse sand	2–15	1–4	0.5–6	0.5–3
Fine sand	20–30	15–25	4–10	0.5–6
Silt	35–55	15–35	3–10	10–30
Clay	35–45	5–10	0–5	45–65

42 Goulding, K. W. T. *et al.*, *Water, Air and Soil Pollut.*, 1986, **29**, 27.
43 Newbould, P. and Russell, R. S., *Pl. Soil*, 1963, **18**, 239; Newbould, P., *J. Sci. Fd. Agric.*, 1963, **14**, 311. See also Davies, D. E. *et al.*, *Soil Sci.*, 1953, **76**, 153 for similar results with some American soils.
44 Salmon, R. C. and Arnold, P. W., *J. agric. Sci.*, 1963, **61**, 421. For similar results with some German soils see Michael, G. and Schilling, G., *Z. PflErnähr. Düng.*, 1957, **79**, 31.

able magnesium (up to 10 per cent of the total uptake) might have been taken up by the crop. It seems safe to assume that the same applies to sodium, and that for all three cations the immediate source used by plants is the soil solution and this is replenished by desorption of exchangeable cations. Some analyses of soil solutions are given in Table 22.5. They are weighted towards soils from arid regions and the solutions therefore contain higher concentrations of cations than found in leached soils.

Table 22.5 Concentrations of calcium, magnesium and sulphur in some soil solutions extracted at about field capacity. (Data from Reisenauer, H. M. in *Environmental Biology* (eds P. L. Altman and D. S. Dittmer), p. 507, Fed. Am. Soc. Exp. Biol., 1966. See also Barber, S. A., *Soil Nutrient Bioavailability*, Wiley-Interscience, 1984)

Ca		Mg		SO_4-S	
$mg\ l^{-1}$	% of samples	$mg\ l^{-1}$	% of samples	$mg\ l^{-1}$	% of samples
0–50	23.1	0–25	9.2	0–25	16.5
51–100	54.6	26–50	21.4	26–50	40.1
101–200	8.1	51–100	38.6	51–100	38.1
201–500	8.1	101–200	25.2	101–200	3.2
501–1000	5.5	201–1000	3.3	201–1000	1.5
No. of samples	979		337		693

Sodium

Although sodium is not, in a strict sense, an essential element for crop plants (see Ch. 3) it is beneficial to crops such as sugar beet, mangolds and some brassicas. Some of these crops, sugar beet for example, have a definite sodium demand if they are to give their maximum yield, and this demand is independent of, and possibly greater than, the potassium demand. This is illustrated in Table 22.6 which summarizes the results of 42 fertilizer experiments with sugar beet[45] in the United Kingdom. Exchangeable sodium is weakly held by all soils and, together with that in the soil solution is the principal source for annual crops.[46] Sugar beet responds to sodium fertilizers only if the soil contains less than about 0.05 meq/100 g exchangeable sodium.

45 Tinker, P. B., *J. agric. Sci.*, 1965, **65**, 207. See also Draycott, A. P. *et al.*, *op. cit.*, 1970, **74**, 567; Holmes, J. C. *et al.*, *Exp. Husb.*, 1961, **6**, 1.
46 Tinker, P. B., *Trans*. Comm. II and IV, Int. Soc. Soil Sci., Aberdeen, 1967, 223.

Table 22.6   Effect of sodium and potassium fertilizers on the yield of sugar beet; 42 experiments 1959–62. Na: 0, 100 and 200 kg Na ha^{-1} as agricultural salt. K: 0, 125 and 250 kg K ha^{-1} as potassium chloride. Yield of sugar in t ha^{-1}

	Na_0	Na_{100}	Na_{200}
K_0	6.43	7.01	7.21
K_{125}	6.82	7.21	7.24
K_{250}	6.97	7.20	7.31

Sodium forms a large proportion of the exchangeable cations only in alkaline and saline soils, and these soils will usually also contain sodium salts (see Ch. 27). The sodium content of soils of humid regions depends largely on their proximity to the oceans and the extent of leaching. In sea water the ionic ratio of Na to Cl is 0.86 and in the absence of industrially produced Cl this is about the ratio found in rain-water. A formula proposed by Hutton[47] for the concentration of chloride in rain-water is $y = 0.99\ d^{-1/4} - 0.23$, where $y = $ meq l^{-1} and $d = $ km from oceans. This formula will also apply approximately to sodium deposition, implying a wide range of values. A review of Australian data[48] shows deposition rates from about 1 kg ha^{-1} at inland stations to about 190 kg ha^{-1} at exposed coastal sites. Recent values of sodium deposition in rainfall in Britain are 10 to 50 kg ha^{-1} a^{-1} on the west coast, 30 kg ha^{-1} a^{-1} on the east coast and 10 to 30 kg ha^{-1} a^{-1} inland.[49]

Calcium

In nearly all soils which are neutral or only slightly acid calcium ions occupy most of the exchange sites on clays and organic matter. Calcium is present as calcite ($CaCO_3$) in soils formed from limestones and other calcareous parent materials, and in soils which have been recently limed. It is present as gypsum ($CaSO_4.2H_2O$) in some soils of arid and semi-arid regions. In humid regions calcium is leached out of the soil and as the soil becomes acid its place on cation exchange sites is taken by aluminium (see Ch. 25). Under acid conditions the ratio of exchangeable Ca to exchangeable Mg narrows and may become less than 1, probably because of the slow release of Mg from silicates.

47 Hutton, J. T., *Search*, 1976, **7**, 207. See also Hutton, J. T., *Trans. 9th Int. Congr. Soil Sci., Adelaide*, **IV**, 1968, 313;   Hutton, J. T. and Leslie, T. I., *Aust. J. agric. Res.*, 1958, **9**, 492. See also Eriksson, E., *Tellus*, 1952, **4**, 280.
48 Isbell, R. F. *et al.*, in *Soils: an Australian viewpoint*, p. 107, CSIRO/Acad. Press, 1983.
49 Reynolds, B. *et al.*, *Int. J. environ. Studies*, 1984, **22**, 195, and ref. 42.

Most soils also contain enough calcium for crop growth if the crop is not affected by acidity. But most crops with a high calcium demand are also sensitive to the relatively high concentrations of aluminium or manganese ions typically present in acid soils, so these soils must have their pH raised, by adding calcium carbonate or lime, if the crop is to grow well. There are, however, some crops which are tolerant of acid soils, whose yield may be limited by the calcium supply. Thus, potatoes may have their yield limited by lack of calcium on some acid soils in the United Kingdom, as may cotton and groundnuts on some tropical acid soils. The roots of some legumes have a high calcium demand during the process of root infection by *Rhizobia*, a demand which can be met by pelleting the seed with calcium carbonate. With non-legumes, calcium deficiency *per se* has been reported in soils of the Brazilian cerrado containing concentrations of exchangeable calcium of 0.05 meq/100 g, or less.[50]

In certain exceptional soils, such as those derived from serpentine, the magnesium to calcium ratio in the exchange complex may be so high that agricultural crops suffer from a true calcium deficiency; but these soils tend to be very infertile as they often also contain high levels of available nickel and other heavy metals.[51]

Magnesium

Magnesium is present as the carbonate in soils containing dolomite $((CaMg)CO_3)$ and is leached like calcium from soils of humid regions. It is released from magnesium silicates as already noted. As an exchangeable ion in neutral and slightly acid soils, magnesium is usually second in abundance to calcium. Magnesium ions resemble calcium ions in their behaviour in ion exchange reactions, but there is evidence that they become non-exchangeable when the pH of soils with pH-dependent charge is raised by liming.[52] There is no agreed explanation.

A deficiency of magnesium has been reported at some time in all the common crops, but it is more widespread in horticultural than agricultural crops, probably because of greater use of potassium fertilizers. In the 1950s the extent of deficiency led to its being described as a major nutrient,[53] although this is not the general experience.

Responses to an application of magnesium are normally found on three types of soil: sandy soils low in exchangeable magnesium; a few acid soils low

50 Ritchey, K. D. *et al., Soil Sci.*, 1982, **133**, 378.
51 For a short symposium see *Ecology*, 1945, **35**, 259;   Walker, R. B. *et al., Pl. Physiol.*, 1955, **30**, 214.
52 Sumner, M. E. *et al., Commun. Soil Sci. Pl. Anal.*, 1978, **9**, 995;   Chan, K. Y. *et al., Soil Sci. Soc. Am. J.*, 1979, **43**, 301.
53 Jacob, A. *Magnesium, The Fifth Major Nutrient*, 1958.

in exchangeable magnesium which have only received dressings of a calcitic limestone[54]; and, due to the competition for uptake between potassium and magnesium, some soils very high in potassium.[55] European experience indicates that crop response to a magnesium fertilizer, which is usually a dolomitic limestone on acid soils or a calcined kieserite (magnesium sulphate) on neutral soils, is unlikely if the soil contains more than 0.4 meq Mg/100 g, and many crops do not respond even if the soil contains only half this amount.[56] Thus, Tinker,[57] working with sugar beet in East Anglia, found responses to magnesium on soils with exchange capacities between 5 and 100 meq/100 g, if they contained less than 0.2 meq Mg, or alternatively if the equilibrium concentration of magnesium in a 0.01 M calcium chloride extract was less than 3×10^{-4} M. Below this concentration Tinker found a fair correlation between response and concentration. He used applications of 110 kg Mg ha^{-1} but considered that about 60 kg ha^{-1} would probably be as effective; and the mean response obtained, on responsive soils, was about 7 per cent, which was about the same as for potassium or sodium additions.

The effect of the exchangeable magnesium content of the soil on the magnesium content of the crop is particularly important for grass, because of the tendency of grasses with a low content to induce hypomagnesaemia ('staggers') in dairy cattle. This occurs in early spring and there is evidence that ammonium is involved, probably due to its competitive effect on the uptake of magnesium.[58] Salmon[59] investigated the relation between the magnesium content of ryegrass grown in pots and the activity of magnesium ions in the soil solution. At a constant level of soil K he found that the per cent magnesium in the ryegrass was proportional to $\sqrt{a_{Mg}}/a_{Mg+Ca}$ in the soil solution. With a range of K levels the per cent magnesium was inversely proportional to $a_K/\sqrt{a_{Ca+Mg}}$. Thus, if the magnesium content of the herbage is low, it is necessary to increase the magnesium ion activity, or the exchange-

54 For examples see Chambers, W. E. and Gardner, H. W., *J. Soil Sci.*, 1951, **2**, 246; Bender, W. H. and Eisenmenger, W. C., *Soil Sci.*, 1941, **52**, 297.

55 Batey, T., *MAFF Tech. Bull.*, No. 14, 143, 1967 advised maintenance of a ratio of exchangeable K : exchangeable Mg of about 2 : 1.

56 Reith, J. W. S., *J. Sci. Fd Agric.*, 1963, **14**, 417; Holmes, M. R. J., *J. agric. Sci.*, 1962, **58**, 281; Birch, J. A. *et al.*, *J. Sci. Fd Agric.*, 1966, **17**, 76; For Woburn results see Bolton, J. with Penny, A., *J. agric. Sci.*, 1968, **70**, 303; with Slope, D. B., *op. cit.*, 1971, **77**, 253. For response of maize in soils of southern Nigeria containing 0.4 meq/100 g, or less, exchangeable Mg see Kayode, G. O., *J. agric. Sci.*, 1985, **104**, 643.

57 Tinker, P. B., *J. agric. Sci.*, 1967, **68**, 205. For more recent data see Draycott, A. P. and Durrant, M. J., *op. cit.*, 1970, **75**, 137; 1971, **77**, 61.

58 The competitive effect was shown by Cox, W. J. and Reisenauer, H. M., *Pl. Soil*, 1973, **38**, 363 and *Agron. J.*, 1977, **69**, 868 in nutrient solutions and by Claassen, M. E. and Wilcox, G. E., *Agron. J.*, 1974, **66**, 521 with maize grown in two soils. For review of hypomagnesaemia see Mayland, H. F. and Grunes, D. L. in *Grass Tetany* (eds V. V. Rendig and D. L. Grunes), p. 123, Am. Soc. Agron., Madison, 1979.

59 Salmon, R. C., *Soil Sci.*, 1964, **98**, 213. See also Arnold, P. W., *MAFF Tech. Bull.*, No. 14, 1967, HMSO.

able magnesium, four-fold to double the content, which means that very heavy applications of magnesium fertilizer, of the order of 350–1700 kg Mg ha^{-1}, are needed on low-magnesium pastures.[60]

Other elements

The uptake by crops of other Group I and II cations, such as rubidium, caesium, strontium, barium and radium[61] has received some attention. Although they are not plant nutrients ^{137}Cs and ^{90}Sr can enter the soil in fall-out from nuclear explosions and so be taken up by crops. The general experience is that crop roots take up potassium and rubidium, or calcium and strontium, in much the same ratio as their activities in the soil solution, although there may be differences in the translocation of these ions after uptake; caesium and barium are taken up less strongly. There are exceptions because some crops are known to accumulate strontium, and others barium and radium, compared with calcium, but these are usually restricted to plants adapted to high strontium and barium soils. Since soils hold exchangeable barium and caesium more strongly than calcium and potassium, and strontium somewhat more strongly than calcium, the ratio of Ca : Sr : Ba in the crop is about 50 : 2 : 1 that on the exchange complex; and the ratio of K : Rb : Cs in the plant is about 25 : 1 : 1 that on the exchange complex.[62] In soils the ratio of exchangeable Ca to exchangeable Sr is usually in the range 100 to 800, and to exchangeable Ba 100 to 10 000.[63]

Sulphur

Annual crops giving moderate to good yields take up about 15 to 50 kg S ha^{-1} which is similar to their uptake of phosphorus[64] (Table 3.1). Legumes and brassica crops have high requirements for sulphur, but deficiency in tree crops has rarely been observed.[65] Sulphur deficiency in crops has been reported from most parts of the world except coastal regions, which receive sulphur in sea spray, and those close to urban and industrial centres

60 Wolton, K. M., *NAAS Quart. Rev.*, 1963, **14**, 122.
61 Smith, K. A., *Pl. Soil*, 1971, **34**, 369, 643 (radium and barium);   Robinson, W. O., Whetstone, R. R. and Edgington, G., *US Dept. Agric., Tech. Bull.*, No. 1013, 1950.
62 Menzel, R. G., *Soil Sci.*, 1954, **77**, 419. Russell, R. S., *Ann. Rev. Pl. Physiol.*, 1963, **14**, 271.
63 Bowen, H. J. M. and Dymond, J. A., *Proc. R. Soc.*, 1955, **144B**, 355.
64 See Terman, G. L., *Atmospheric Sulphur–the agronomic aspects, Sulphur Inst. Tech. Bull.*, No. 23, 1978. For review articles see *Sulfur in Agriculture* (ed. M. A. Tabatabai), Agron. Monog. No. 27, Am. Soc. Agron., Madison, Wisc., 1986.
65 For early recognition of deficiency in young tea bushes see Storey, H. H. and Leach, R., *Ann. appl. Biol.*, 1933, **20**, 23.

which receive sufficient sulphur from the atmosphere.[66] Because of reductions in the emission of SO_2 into the atmosphere from fossil fuels since the mid-1970s, and the increasing use of fertilizers low in sulphur, the occurrence of deficiency can be expected to increase.[67]

Forms of sulphur in soil

The total sulphur content of igneous and metamorphic rocks is usually in the range 300–700 μg g^{-1}, highest contents occurring in mafic rocks.[68] Sulphur occurs as the sulphide in igneous rocks, particularly basic igneous, and as these weather the sulphur is released as sulphate ions. Because sulphate is present in sea water (average concentration 28 $\times$ 10^{-3} M) many coastal alluvial soils contain sulphates or sulphides, and sedimentary rocks formed in coastal marshes are rich in ferrous sulphide or calcium sulphate, as for example, the Lower Lias and Gault clays in Britain. Sulphate is also present as an impurity, presumably as a co-precipitate in some limestones, calcium carbonate concretions and coral.[69] Total sulphur concentrations in soils therefore vary over a very wide range. They have been reported[70] to be as low as 20–30 μg S kg^{-1} in sandy soils in England and Australia to over 3–5 per cent in reclaimed tidal marshes in the Netherlands, S. Carolina, USA and elsewhere. In reclaiming soils that contain sulphide, oxidation to sulphate gives rise to severe acidity unless sufficient calcium carbonate is present (see Ch. 26 for a discussion of acid sulphate soils).

In well-aerated soils sulphur is present mainly as sulphate ions and as a component of soil organic matter. In soils of arid regions where sulphate concentrations are high, gypsum, $(CaSO_42H_2O)$ is commonly present. Sulphate ions are present in the soil solution and are adsorbed on to clay minerals and iron and aluminium oxides, especially under acid conditions[71] (see also Ch. 7). As mentioned above, in anaerobic soils sulphur can be present as the sulphide of iron and possibly of other metals including zinc.[72]

66 Martin, A., *Environ. Pollut.* (Ser. B), 1980, **1**, 177.
67 Bristow, A. W. and Garwood, E. A., *J. agric. Sci.*, 1984, **103**, 463.
68 Wedepohl, K. H., *Handbook of Geochemistry*, Vol. II–2; Ch. 16, Springer-Verlag, Berlin, 1978.
69 Williams, C. H. and Steinbergs, A., *Pl. Soil*, 1962, **17**, 279.
70 Whitehead, D. C., *Soils Fertil.*, 1964, **27**, 1. See also Dent, D., *J. Soil Sci.*, 1980, **31**, 87, for S contents of some acid sulphate soils in New Zealand. Williams, C. H. and Raupach, M. in *Soils: an Australian viewpoint*, p. 781, CSIRO/Acad. Press, 1983; general review by Ure, A. M. and Berrow, M. L. in *Environmental Chemistry*, Vol. 2 (Sen. Reporter H. J. M. Bowen), Royal Soc. Chem., Burlington House, London, 1982.
71 Harward, M. E. and Reisenauer, H. M. *Soil Sci.*, 1966, **101**, 326. Barrow, N. J., *Soil Sci.*, 1967, **104**, 342; Williams, C. H. in *Sulphur in Australasian Agriculture* (ed K. D. McLachlan), p. 21, Sydney Univ. Press, 1975.
72 Kittrick, J. A., *Soil Sci. Soc. Am. J.*, 1976, **40**, 314.

Sulphur is also present in soil organic matter, but its chemical forms are poorly understood. The fractions that have been recognized are that reduced by Raney nickel, which probably consists of the amino acids cystine and methionine, and the remainder which can be divided between that which is reduced by hydriodic acid (probably ester sulphate) and that which is not reduced (believed to be carbon bonded). A wide range of analyses has been reported. In Australian, Canadian and United States soils the fraction reduced by Raney nickel accounted for between 11 and 58 per cent of the total sulphur content. The hydriodic-acid-reducible sulphur may range between 30 and 70 per cent of the organic sulphur content.[73]

Sulphur from the atmosphere

Plants derive some of their sulphur from the atmosphere. For example, Olsen[74] found that cotton plants grown in nutrient solutions labelled with $^{35}SO_4$ obtained about 30 per cent of their sulphur from the atmosphere. There is also evidence from experiments in growth chambers that exposure of ryegrass plants to 50 μg SO_2 m^{-3} air alleviates the effect of an inadequate supply of sulphur from the soil.[75]

The atmosphere contains sulphur compounds both as gases and in particulate, or aerosol, form. The predominant gas is sulphur dioxide. In areas near sea coasts the aerosols contain sodium and other inorganic sulphates derived from sea spray. In areas receiving air from industrial and urban centres, the aerosols contain sulphate, predominantly ammonium sulphate, and as free acid emitted on burning sulphur-containing fuels. The deposition of sulphur in rainfall is therefore greatest near industrial sources where it can exceed 100 kg ha^{-1} a^{-1}. At Rothamsted[76] the annual deposition in rainfall increased from 7.8 kg ha^{-1} in 1881–87 to 20.6 in 1969–73, and more recently to 25–35 kg ha^{-1}. Background values at rural sites free of atmospheric pollution in Nigeria[77] and New Zealand are about 1 kg S ha^{-1} a^{-1}.

Dry deposition is believed to be mainly as SO_2 and to a much less extent as particulate (or aerosol) sulphate, their respective air concentrations in rural Britain in the mid-1970s being estimated[78] as 35–40 and 5–30 μg m^{-3}.

73 Freney, J. R. in *Soil Biochemistry* (eds A. D. McLaren and G. H. Peterson), p. 227, Dekker, 1967, gives analyses in a review of soil sulphur compounds and their transformations. See also Williams, C. H. in *Sulphur in Australasian Agriculture* (ed K. D. McLachlan), p. 21, Sydney Univ. Press, 1975. For more recent review of oxidation of S compounds see Wainwright, M., *Adv. Agron.*, 1984, **37**, 349.
74 Olsen, R. A., *Soil Sci.*, 1957, **84**, 107.
75 Lockyer, D. R. *et al.*, *J. exp. Bot.*, 1976, **27**, 397.
76 Goulding, K. W. T. and Poulton, P. R., *Soil Use Management*, 1985, **1**, 6.
77 Bromfield, A. R., *Tellus*, 1974, **26**, 92.
78 *Department of the Environment Central Unit on Environmental Pollution*, paper No. 7, 1976, HMSO. This paper reviews effects of atmospheric SO_2 on plants.

The amount of dry deposition depends on the nature of the surface exposed to the air and on the concentration of sulphur compounds in the air. Accession is usually expressed as a deposition velocity (V_g) which equals rate of deposition of SO_2 (μg m^{-2} s^{-1})/concentration of SO_2 in air (μg m^{-3}). This gives V_g in m s^{-1}, more usually expressed in cm s^{-1}, the values being obtained from the concentration in the air and rate of deposition. Values for plants and soils vary from >1 to about 3 cm s^{-1}. They are higher for wet soil than dry, and generally higher for plants than for soils. A concentration of SO_2–S of 30 μg m^{-3} and V_g of 1 (cm s^{-1}) corresponds to an annual rate of dry deposition of 48 kg S ha^{-1}.

The average total S deposition in Britain has been estimated as 70 kg ha^{-1} a^{-1}, of which 20 and 50 kg ha^{-1} a^{-1} are from wet and dry deposition respectively.[78] At upland sites in England and Scotland that are distant from industrial centres, the total deposition is less, probably about 10 to 20 kg S ha^{-1} a^{-1}. Because the sulphur is largely present in acids such as sulphuric, and in acid-forming compounds such as sulphur dioxide and ammonium sulphate, the deposition of sulphur, except close to the oceans, is related to acid deposition ('acid rain').[79]

Sulphur which is deposited on the soil from the atmosphere is removed by crops and can be lost by leaching and erosion, but little is returned to the atmosphere directly. There is, for example no evidence of a flux of SO_2 from soils or plants to the atmosphere except through burning. Although loss of hydrogen sulphide and volatile organic sulphur compounds can occur from soils, plants and manures, present evidence is that loss of hydrogen sulphide from soils is small, even from those soils in which biological reduction of sulphate to sulphide has occurred,[80] probably because of its reaction in soil to form FeS and other metal sulphides. Volatile organic sulphur compounds, including dimethyl sulphide, dimethyl disulphide, methyl mercaptan and carbonyl sulphide account for the loss of some sulphur from soils, and probably more than hydrogen sulphide, but there are very few estimates of amounts.[81]

Sulphur deficiency in crops

The amounts of sulphur in wet and dry deposition are sufficient to meet crop

79 Fowler, D. *et al.*, *Soil Use Management*, 1985, **1**, 3; also ref. 49.
80 Freney, J. R. and Williams, C. H. in *The Global Biogeochemical Sulphur Cycle* (eds M. V. Ivanov and J. R. Freney) p. 129, Wiley (Scope 19) and other chapters, 1983; Bremner, J. M. and Banwart, W. L., *Soil Biol. Biochem.*, 1976, **8**, 79. In early work Eriksson, E., *J. geophys. Res.*, 1963, **68**, 4001 and others suggested loss of S from soils as H_2S.
81 Bremner, J. M. and Steele, C. G., *Adv. Microbial. Ecol.*, 1978, **2**, 155; Adams, D. D. *et al.*, *J. Air Pollut. Control Assoc.*, 1979, **29**, 380; Farwell, S. O. *et al.*, *Soil Biol. Biochem.*, 1979, **11**, 411.

requirements near industrial and urban centres and near sea coasts. Deficiency in crops is usually limited to areas where the wet and dry deposition is less than 5 to 10 kg ha^{-1} of S and where soil sulphate contents are low, as in strongly leached soils. There are reports of crop response to sulphate applications from many countries, deficiencies in parts of the United States, Australia and New Zealand being particularly severe.[82]

The nutritional quality of cereal grain is reduced in sulphur-deficient plants because there is less synthesis of cysteine and methionine and of other essential amino acids. Sulphur-deficient wheat grain gives a dough which is tough and has a low extensibility, and gives bread of low quality.[83] The growth of ruminant animals is adversely affected on sulphur-deficient feed; for example, the average daily gain of lambs grazing a clover–grass pasture increased when the pasture received sulphate,[84] 90 per cent of the maximum average daily gain being obtained when the herbage contained 0.19 per cent S.

The immediate source of soil sulphur for crops is the sulphate present in the soil solution, and this is replenished by desorption from soil clays and hydrous oxides. The combined amount is measured by extracting soil samples with a solution of KH_2PO_4 or $Ca(H_2PO_4)_2$, which is the method commonly used to establish the level of soil sulphur availability.[85] A value of 10 to 13 mg S kg^{-1} soil extracted by 0.016 M $Ca(H_2PO_4)_2$ has been given as the lower limit to avoid sulphur deficiency in the field,[86] but the method has not always been found effective, one reason being that it does not include the organic sulphur that becomes available.

The principal reservoir of sulphur in soils of the humid regions is that contained in organic matter, which commonly has a N : S ratio of 10 : 1 to 10 : 2, a ratio lower than that in most plants. Mineralization of soil organic sulphur appears, however, not to be simply related to the N : S ratio, nor to the amount of nitrogen which is mineralized.[87] Nevertheless, if plant residues with a low sulphur content are mixed with soil, immobilization of sulphate occurs. A content of 0.15 per cent in the dry matter of straw is about the minimum required for its decomposition not to reduce the soil sulphate

82 For occurrence of deficiency in Australia and New Zealand see *Sulphur in Australasian Agriculture* (ed K. D. McLachlan), Sydney Univ. Press, 1975. See also *Sulphur in the Tropics*, 1979, publ. by Sulphur Institute and Int. Fert. Dev. Centre, Muscle Shoals, Alabama, and ref. 64. For a recent report of deficiency under field conditions in Scotland see Scott, N. M. *et al.*, *J. agric. Sci.*, 1984, **103**, 699.

83 Wrigley, C. W. *et al.*, *Sulphur in Agric.*, 1984, **8**, 2.

84 Jones, M. B. *et al.*, *Agron. J.*, 1982, **74**, 775. For a review on the requirements for wool growth see Downes A. M. *et al.*, in *Sulphur in Australasian Agriculture* (ed K. D. McLachlan), p. 117, Sydney Univ. Press, 1975.

85 Ensminger, L. E., *Soil Sci. Soc. Am. Proc.*, 1954, **18**, 259; Fox, R. L. *et al.*, *op. cit.*, 1964, **28**, 243; Tabatabai, M. A. in *Methods of Soil Analysis* (2nd edn) (ed A. L. Page), Part 2, p. 501, Am. Soc. Agron., Wisc., 1982.

86 Andrew, C. S. in *Sulphur in Australasian Agriculture* (ed K. D. McLachlan), p. 196; Sydney Univ. Press, 1975.

87 Haque, I. and Walmsley, D., *Pl. Soil*, 1972, **37**, 255.

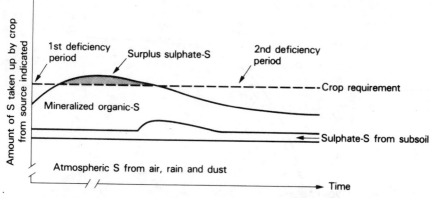

Fig. 22.8 Diagrammatic representation of the changes of sources of S to crops, from clearance of natural fallow through the cropping cycle (not to scale) where no S is supplied. (From Bromfield, A. R., *J. agric. Sci.*, 1972, **78**, 465.)

status,[88] and contents as low as one-third of this can be found in wheat straw grown on some sulphur-deficient soils. This is probably the reason for sulphur deficiency occurring in the first crop after a grass fallow which has been reported in W. Africa.[89] The schematic illustration in Fig. 22.8 is of three phases following cultivation of such a fallow. First there is initial immobilization of soil sulphate giving rise to a deficiency in a crop. This is followed by a period when mineralization of the soil organic matter provides enough sulphate to meet crop requirements.[90] Finally there is a second period of deficiency when the organic matter content has fallen too low for it to provide sufficient sulphate to meet crop requirements.

A sign of sulphur deficiency in a crop is a general chlorosis of the leaves, but these visual symptoms are not specific to sulphur deficiency. Chemical analysis of plant leaves gives a more reliable diagnosis. In general, the dry matter of young leaves should contain more than 0.15–0.2 per cent S, though the threshold value depends on the crop and decreases with the age of the leaf. Also used is the ratio of protein N to protein S, or more commonly the ratio of total N to total S. Some workers[91] prefer to use plant SO_4-S expressed

88 Stewart, B. A. *et al.*, *Soil Sci.*, *Soc. Am. Proc.*, 1966, **30**, 355. See also Barrow, N. J., *Aust. J. agric. Res.*, 1960, **11**, 317, 960; 1961, **12**, 306, who used C/S ratios.
89 Nye, P. H. and Greenland, D. J., *The Soil under Shifting Cultivation, Commonw. Bur. Soils Tech. Comm.*, No. 51, 97, 1960.
90 For evidence of uptake of sulphate by plants after mineralization of soil organic sulphur see Fox, R. L. *et al.*, *Soil Sci. Soc. Am. Proc.*, 1964, **28**, 243; Nearpass, D. C. *et al.*, *Soil Sci. Soc. Am. Proc.*, 1961, **25**, 287. For a review see Ensminger, L. E. and Freney, J. R., *Soil Sci.*, 1966, **101**, 283.
91 Eaton, F. M. in *Diagnostic Criteria for Plants and Soils* (ed H. D. Chapman), p. 444, Univ. Calif,. 1966. Scott, N. M. *et al.*, *J. Sci. Fd Agric.*, 1983, **34**, 357.

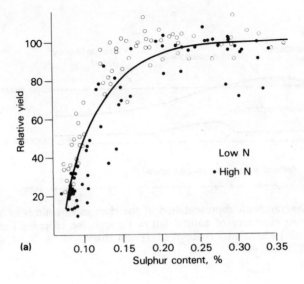

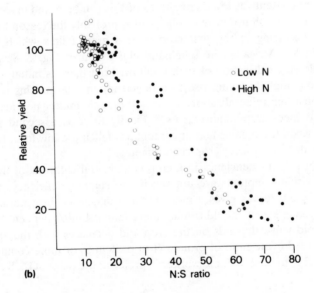

Fig. 22.9 Relationships between relative yield of shoots of ryegrass and sulphur in plants: (a) total S, (b) N : S ratio. Plants grown on 16 soils and results are for four harvests. Relative yield: (yield without added S/yield with added S) × 100. (From Jones, L. H. P. *et al., Soil Sci.,* 1972, **114**, 104. © Williams & Wilkins.)

as a percentage of total plant S, values greater than 20 indicating an adequate supply. For grasses and cereals a ratio in the leaves of total N to total S greater than 14 may indicate sulphur deficiency.[92] These relationships are illustrated in Fig. 22.9 which shows results from an experiment with soils from England and Wales in which ryegrass was grown in growth chambers, the air being filtered to remove gaseous S compounds. Yield reductions of 10 per cent and 5 per cent occurred at 0.17 per cent S and 0.20 per cent S respectively, and the N : S ratio was 14 or less in healthy plants.

Sulphur in fertilizers

Sulphur-deficient soils were often not recognized in the past because many were also phosphate deficient and received single superphosphate, which contains between 11 and 13 per cent S in addition to its 8 per cent P. Again, the common form of nitrogen fertilizer was sulphate of ammonia, which contains 24 per cent S. Thus, crops requiring both nitrogen and phosphate would have received a substantial amount of sulphate. Sulphur deficiency may become more widespread, even close to urban and industrial areas, because of less accession of sulphur from the atmosphere and the trend towards fertilizers low in sulphur.

If sulphur is the only deficient nutrient, an application can be made of calcium sulphate, which contains between 13 and 23 per cent S, depending on its purity and degree of hydration. In the United States, liquid fertilizers containing sulphur and nitrogen are sometimes used, for example ammonium polysulphide in aqueous ammonia and ammonium thiosulphate in aqueous solution.[93] When applied to soil ammonium polysulphide decomposes to release colloidal, elemental sulphur and sulphide whilst ammonium thiosulphate releases elemental sulphur and sulphate ions. Finely divided elemental sulphur may also be used in aqueous suspension or mixed with solid fertilizers such as triple superphosphate.[94] The successful use of elemental sulphur depends on the presence of sulphur-oxidizing micro-organisms in the soil, about which little is known for sulphur-deficient soils. Virtolins and Swaby[95] found that about one-third of the Australian soils they examined did not oxidize sulphur, and of those which did the dominant oxidizing organisms depended on the pH of the soil. The rate of oxidation depends on the particle size of the sulphur and is sensitive to the phosphate status of the soil[96], its

92 Dijkshoorn, W. and van Wijk, A. L., *Pl. Soil*, 1967, **26**, 129.
93 Coleman, R., *Soil Sci.*, 1966, **101**, 230; Parr, J. F. and Papendick, R. I., *Soil Sci.*, 1966, **101**, 336; Parr, J. F. and Giordano, P. M., *Soil Sci.*, 1968, **106**, 448.
94 Tisdale, S. L., Nelson, W. L. and Beaton, J. D., *Soil Fertility and Fertilizers* (4th edn), Macmillan, 1985.
95 Virtolins, M. I. and Swaby, R. J., *Aust. J. Soil Res.*, 1969, **7**, 171. For an earlier review see Starkey, R. L., *Soil Sci.*, 1966, **101**, 297.
96 Bloomfield, C., *Soil Sci.*, 1967, **103**, 219.

temperature, aeration and moisture content, and often has a fairly sharp maximum at about field capacity.[97] In some sulphur-deficient soils it may take an appreciable time for sulphur-oxidizing bacteria to build up to an appreciable population. The most reliable way of correcting sulphur deficiency is therefore by using gypsum, or by using single superphosphate if there are deficiencies of both sulphur and phosphorus.

Sulphate-containing fertilizers generally have little long-lasting residual effect, but they may do so on acid soils which adsorb sulphate or where there is little loss by leaching.[98] Fertilizers containing elemental sulphur are usually somewhat slower-acting and longer-lasting than those containing gypsum, because the sulphur itself is insoluble.[99] Greenwood[100] in northern Nigeria found that an application of 112 kg ha^{-1} of gypsum improved the yield of groundnuts for two years after it was applied with a small effect in the third year. Walker and Adams[101] in New Zealand give examples of 45 kg S ha^{-1}, given as gypsum to a grass–clover ley repeatedly cut and harvested, increasing the yield and protein content of the herbage four years after application, although by this time the clover component, which had earlier been stimulated, had returned to its original poor condition. By contrast, on an acid soil in northern Nigeria sulphate was found to accumulate when it was added in ammonium sulphate, in single superphosphate or in farmyard

Table 22.7   Total and residual sulphur in the plots of a 19-year-old fertilizer experiment at Samaru, Nigeria. (From Bromfield, A. R., *J. agric. Sci.*, 1972, **78**, 465)

		kg S ha^{-1} (0–180 cm depth)			
Treatment no.	*Annual application to soil (kg ha^{-1})*	*Total in soil*	*Residual (by diff.)*	*Applied in 19 yrs.*	*Residual/ applied*
1	Nil	389	–	0	–
2	5000 FYM	613	224	?	–
3	112 AS	776	387	511	0.76
4	112 AS + 112 SS	1102	713	788	0.90
5	112 AS + 112 SS + 5000 FYM	1359	970	788+?	–

FYM = Farmyard manure; AS = Ammonium sulphate; SS = Single superphosphate.

97 For reviews see Attoe, O. J. and Olson, R. A., *Soil Sci.*, 1966, **101**, 317; Wainwright, M. (ref. 73).
98 Jones, M. B., *et al.*, *Soil Sci. Soc. Am. Proc.*, 1968, **32**, 535;   Bromfield, A. R., *J. agric. Sci.*, 1972, **78**, 465; *idem, Expl. Agric.*, 1973, **9**, 55.
99 Jones, M. B. and Ruckman, J. E., *Agron. J.*, 1966, **58**, 409.
100 Greenwood, M., *Emp. J. expl. Agric.*, 1951, **19**, 225.
101 Walker, T. W. and Adams, A. F. R., *Pl. Soil*, 1958, **10**, 176.   For review of sulphate residual values see McLachlan, K. D. in *Sulphur in Australasian Agriculture* (ed K. D. McLachlan), p. 58, Sydney Univ. Press, 1975.

manure, as illustrated in Table 22.7. The mean annual rainfall of about 1100 mm gave through-drainage and the land was cropped every year. When land is put down to grass sulphate added in fertilizer accumulates in the soil organic matter as this increases. The effect is similar to that of added phosphate referred to on p. 700.

Silicon

Silicon is present in soils in large amounts in the forms of silica and aluminosilicates. The most common form is quartz, and this is the dominant mineral in most sandy soils. Quartz is crystalline and stable to weathering, particularly when present in sand-sized or coarser particles, but it weathers fairly easily when in clay-sized particles. The quartz is derived from the parent material and is rarely, if ever, formed in the weathering profile. Silica is also present in an hydrated form as opal, normally derived from plant leaves or diatoms. The amorphous forms in plants are known as opal phytoliths. They may accumulate in soils, and particularly in grassland soils because grass leaves contain a relatively large amount of silica.[102] The phytoliths show very clearly the fine structure of the fibrils which form the cell wall. Under some circumstances silica can be deposited between soil particles to give a silica-rich pan or silcrete which is a feature of some soils in dry regions.

Silicon is also present in the soil solution as silicic acid $Si(OH)_4$. Since the first dissociation constant of this acid has a pK of 9.6, the silicate anion is not present in any appreciable concentration in the solution of normal agricultural soils. The silicic acid is in equilibrium with a solid phase, for if the solution is removed and the soil is then left moist, the silicic acid concentration in the soil solution will return to its original value.[103] In equilibrium with amorphous silica, the concentration is 2 mM. Jones and Handreck[104] found, however, that the silicic acid concentration in the solution of a number of Australian soils varied from 0.10 mM for a lateritic (krasnozem) soil to 1.1 mM for a black clay Vertisol. Further, although the solubility of silicic acid in the pure system is independent of pH over the range 2 to 9, they found that the concentration decreased from 1.0 to 0.4 mM in an Australian soil when its pH was raised from 5.5 to 7.2 by liming. This compares with concentrations of 0.15 to 0.5 mM in the drainage water for different plots on Broadbalk field at Rothamsted, where the soil has a pH of about 7.2.[105]

102 Smithson, F., *J. Soil Sci.*, 1958, **9**, 148.
103 Alexander, G. B. *et al.*, *J. phys. Chem.*, 1954, **58**, 453.
104 Jones, L. H. P. and Handreck, K. A., *Pl. Soil*, 1965, **23**, 79.
105 Voelcker, A., quoted by Hall, A. D. in *The Book of the Rothamsted Experiments*, 1905 and 1919.

The reason the silicic acid concentration in the soil solution is less than would be expected from the solubility of amorphous silica is that it is adsorbed on the surfaces of iron and aluminium hydrous oxides by ligand exchange, as explained in Chapter 7. Thus the greater the surface area of the hydrous oxides, and the higher the pH up to 9.5, the more strongly will the soil adsorb silicic acid, and the lower will be its concentration in the soil solution.[106]

These processes may have certain consequences. In the first place, water-logging is likely to increase water-soluble silicic acid in soil, presumably due to its release from sesquioxide surfaces. In the second, extracting agents which remove free sesquioxides from a soil would be expected to release silicic acid, and this is almost universally observed. Another implication is that quartz would not be expected to occur in soils having a large surface area of hydrous sesquioxides, for these surfaces maintain a soluble silicic acid concentration less than that of the equilibrium solution of quartz (0.10 mм). Unless the quartz is coated with sesquioxide which prevents it dissolving, silica will be transferred from quartz to the sesquioxide surfaces. This may explain the low quartz content reported for a krasnozem in Australia compared with that for a Vertisol.[104]

Silicon and plant growth

Crop plants take up different amounts of silicic acid depending on the soil and on the species. Rice has a pronounced ability to take it up and the vegetative parts commonly contain 8 to 14 per cent SiO_2 in the dry matter. Other cereals and grasses normally contain 2 to 4 per cent SiO_2 whereas dicotyledonous species, including legumes, contain only 0.1 to 0.4 per cent. In several hundred samples of grass and foliage of bushes and trees belonging to 47 families and growing in the drier areas of Kenya, the SiO_2 content of grasses varied from 1 to over 10 per cent of the dry matter, with contents of between 2 and 5 per cent being common, while most other plants had silica contents below 1 per cent and many below 0.2 per cent.[107] Wheat and oat crops commonly remove between 20 and 30 kg SiO_2 t^{-1} while clover, beans and root crops remove only 2 kg or less per tonne.

Plant roots absorb silicon as silicic acid. It is possible that the amount taken up by some gramineous crops is equal to the amount of silicic acid present in the water the roots absorb, so that the greater the amount of water tran-spired, the greater their uptake of silica.[108] Dicotyledonous crops on the other

106 McKeague, J. A. and Cline, M. G., *Canad. J. Soil Sci.*, 1963, **43**, 83;   Beckwith, R. S. and Reeve, R., *Aust. J. Soil Sci.*, 1963, **1**, 157; 1964, **2**, 33.   For a review see McKeague, J. A. and Cline, M. G., *Adv. Agron.*, 1963, **15**, 339.
107 Dougall, H. W. *et al.*, *E. Afr. Wildlife J.*, 1964, **2**, 86.
108 Jones, L. H. P. and Handreck, K. A., *Pl. Soil*, 1965, **23**, 79; *Aust. J. biol. Res.*, 1967, **20**, 483; *Adv. Agron.*, 1967, **19**, 107 for review.

hand take up much less silicic acid than is present in the water their roots take up, often only about 5 per cent, and Jones and Handreck consider that the dicotyledonous crop crimson clover (*T. incarnatum*) translocated to its shoots a constant proportion of the silica that reached its roots; again, the greater the transpiration the greater the silica content of the shoots.

The general validity of these observations is uncertain because it has been shown by Barber and Shone[109] with barley that enzyme inhibitors such as sodium azide and dinitrophenol cause a reduction in silicic acid uptake without causing a corresponding reduction in transpiration. They also found that broad beans (*Vicia faba*) had a silica concentration in their sap over four times greater than in the external solution and they concluded that active transport across the root must have occurred. Further, rice will take up more silicic acid than is carried to the roots by mass flow.[110] The physiological reasons for these marked differences between species are not known.

Silicon plays an important part in the growth of plants, but it cannot yet be considered an essential element for any crop plant if the criteria referred to in Chapter 3 are used as the test of essentiality; it is essential for diatoms and for *Equisetum*.[111] The experimental difficulty in establishing whether silicon is an essential micronutrient for other plants is in removing all silicon from the growth medium. Further experiments might show it to be essential. It is, however, clear that an adequate supply of silicon in the form of silicic acid is needed by grasses and cereals to give a good yield, for it increases the strength and rigidity of their cell walls. Thus it helps rice leaves to have a more upright habit under conditions of high nitrogen manuring, which may increase the rate of photosynthesis per unit area of land.[112] It increases the oxidizing power at the surface of rice roots, probably by increasing the rigidity of the walls of the aerenchyma or gas channels within the plant. It is also essential for a good seed set in some varieties of rice and for fruit development in tomatoes.[113] Silica also increases the tolerance of some crops, especially the cereals, to high levels of available soil manganese for reasons that are not fully understood. It prevents the manganese in the leaf becoming concentrated in a number of spots which would become necrotic,[114] and because it allows a greater oxygen supply to the root surface of paddy rice, it ensures a more rapid oxidation of manganese just outside the root.

109 Barber, D. A. and Shone, M. G. T., *J. exp. Bot.*, 1966, **17**, 569.
110 Okuda, A. and Takahashi, E., *The Mineral Nutrition of Rice*, Ch. 10, IRRI, Philippines, 1965.
111 Chen, C. H. and Lewin, J., *Canad. J. Bot.*, 1969, **47**, 125; Werner, D. and Roth, R. in *Inorganic Plant Nutrition* (eds A. Läuchli and R. L. Bieleski), *Encycl. Pl. Physiol.*, Vol. 15B, p. 682, Springer-Verlag, 1983.
112 Yoshida, S. *et al.*, *Pl. Soil*, 1969, **31**, 48.
113 Miyake, Y. and Takahashi, E., *Soil Sci. Pl. Nutr.*, 1978, **24**, 175.
114 Vlamis, J. and Williams, D. E., *Pl. Soil*, 1967, **27**, 131; *Pl. Physiol.*, 1957, **32**, 404; Horst, W. J. and Marschner, H., *Pl. Soil*, 1978, **50**, 287; Vorm van der, P. D. R. and Diest van A., *Pl. Soil*, 1979, **52**, 19.

An adequate supply of silica to the cereals thickens those cell walls in which it is deposited, and this may have a number of desirable consequences. The silica reduces the tendency of a cereal to wilt during the initial stages of drought, probably because of the reduced permeability to water or water vapour of the walls of the leaf epidermal cells. There is also evidence that plants adequately supplied with silica have increased resistance to some pests and diseases. Thus, an adequate silica content may increase the resistance of some cereals to powdery mildew (*Erysiphe graminis*)[115] and of rice to blast (*Pyricularia oryzae*),[116] and to some stem borers such as *Chilo suppressalis*,[117] of sorghums to central shoot fly (*Atherigone indica*)[118] and wheat to hessian fly (*Mayetiola destructor*).[119]

Silicon in fertilizers

Fertilizers containing silica as soluble silicates, or calcium silicate slags can increase the pH of the soil and may also increase the silicic acid concentration in the soil solution. Silicate fertilizers can also increase the availability of soil

Table 22.8   Effect of silicate on the growth of barley; Hoosfield, Rothamsted

	Yield of grain ($t\ ha^{-1}$)					
	1862–1891		1932–61		1964–66	
	Without silicate	*With silicate*	*Without silicate*	*With silicate*	*Without silicate*	*With silicate*
Nitrogen alone	1.85	2.24	1.59	2.07	1.63	2.54
Complete less phosphate	1.92	2.45	1.85	2.16	2.66	4.03
Nitrogen plus phosphate	2.77	2.89	2.54	2.71	4.06	3.80
Complete	2.76	2.98	2.64	2.89	4.85	4.94

Complete: 48 kg N ha^{-1} as sodium nitrate, 32 kg P ha^{-1} as single superphosphate, 92 kg K ha^{-1} as potassium sulphate, 112 kg of magnesium sulphate, 112 kg sodium sulphate. In 1964–66 96 kg N ha^{-1} given in place of 48 kg ha^{-1}.

115 Lowig, E., *Pflanzenbau*, 1937, **13**, 362. Wagner, F., *Phytopathology*, 1940, **12**, 427.
116 Volk, R. J. *et al.*, *Phytopathology*, 1958, **48**, 179.
117 Djamin, A. and Pathak, M. D., *J. econ. Entomol.*, 1967, **60**, 347.
118 Ponnaiya, B. W. X., *J. Madras Univ.*, 1951, **21B**, 203.
119 Miller, B. S. *et al.*, *J. econ. Entomol.*, 1960, **53**, 995.

phosphate to the crop,[120] presumably by displacing phosphate absorbed on sesquioxide surfaces.[121] This is illustrated in Table 22.8 for barley on Hoosfield at Rothamsted, which shows that an application of 450 kg ha^{-1} of sodium silicate annually is still increasing the yield of no-phosphate plots after a century of use. The effect is unlikely to be due to the sodium in the silicate as the source of nitrogen is sodium nitrate, and the plots receiving potassium also receive 112 kg ha^{-1} of sodium sulphate. It is probably another example showing that the concentrations of water-soluble silicic acid and of phosphate are not two independent quantities in soils, but are closely linked since their solubilities are controlled by sesquioxide surfaces.

120 Fisher, R. A., *J. agric. Sci.*, 1929, **19**, 132;   Lemmermann, O., *Ztschr Pflanz. Düng.*, 1929, **13A**, 28.
121 Obihara, C. H. and Russell, E. W., *J. Soil Sci.*, 1972, **23**, 105.

23

Micronutrients and toxic elements

The elements now regarded as essential micronutrients for plants are iron, manganese, copper, zinc, boron, molybdenum and chlorine (but see Ch. 3 for the status of cobalt and silicon). These, together with a large number of other elements, are usually present in plants in small concentrations and were termed 'trace elements' when their concentrations could only be reported as traces. Modern methods allow measurements to be made of concentrations of ng g^{-1} or less, but the term 'trace element' remains useful if imprecise. This chapter provides an overview of the micronutrients and of most of the elements which can be toxic to plants (the toxicity of aluminium is discussed in Ch. 25, and toxicities associated with saline soils in Ch. 27). Reference is also made to some of the elements important to animals. The essential elements for plants and animals are listed in Table 23.1.

Crops are usually grown to feed animals or human beings and it is desirable that their composition is suitable for the consumer. Elements required by animals should therefore be present at sufficient concentrations.[1] For

1 For reviews of the essential elements for animals see Underwood, E. J., *Trace Elements in Human and Animal Nutrition*, Acad. Press, 1977; Mertz, W., *Science*, 1981, **213**, 1332; for a useful general review see Adriano, D. C., *Trace Elements in the Terrestrial Environment*, Springer-Verlag, 1986.

Table 23.1 The essential elements. (Adapted from ref. 11)

Plants	Animals	Typical soil content	
		Mean	Range
Major nutrients			
Carbon	Carbon	2.0%	0.7 – 50%
Hydrogen	Hydrogen		
Oxygen	Oxygen		
Nitrogen	Nitrogen	0.2%	<0.002 – >2.5%
Phosphorus	Phosphorus	0.04%	0.002 – 0.6%
Potassium	Potassium	1.8%	0.005 – 7.9%
Sulphur	Sulphur	433	3 – 8 200
Calcium	Calcium	2.0%	0.01 – 32%
Magnesium	Magnesium	0.83%	0.005 – 16%
*	Sodium	1.1%	<0.005 – 10%
m	Chlorine	485	18 – 806
Micronutrients			
Boron	*	38	0.9 – 1 000
Chlorine	M	485	18 – 806
*	Cobalt	12	0.3 – 200
*	Chromium	84	0.9 – 1 500
Copper	Copper	26	2.5 – 60
*	Fluorine	270	6 – 7 070
*	Iodine	7	<0.09 – 80
Iron	Iron	3.2%	0.01 – 21%
Manganese	Manganese	761	<1 – 18 300
Molybdenum	Molybdenum	1.9	0.07 – 5
*	Nickel	34	0.1 – 1 523
*	Silicon (1)	60%	50 – 70%
*	Selenium	0.41	0.03 – 2
*	Tin	5.8	0.1 – 40
*	Vanadium	108	3 – 1 000
Zinc	Zinc	60	1.5 – 2 000

Note: M = major nutrient; m = micronutrient; * = essentiality not established;
(1) represented as dioxide. Units of measurement are mg kg^{-1} except where % is
indicated. For definition of essentiality of elements in plants see Chapter 3.

example, selenium is not known to be essential for plants, but if the plant
is intended for animal consumption it is necessary that it should contain an
adequate concentration of this micronutrient.

The micronutrients and most of the non-essential trace elements are known
to have undesirable effects on plant and animal growth if present in excess
concentration in soils. Plants sometimes have their growth affected by excess
manganese or aluminium in acid soils and by nickel, cobalt and chromium
in acid soils derived from ultramafic igneous rocks. In addition there is

growing concern that several toxic elements are accumulating in some soils as a consequence of pollution by man's industrial or urban activities, or because of the use of metal-contaminated sewage sludge. Whenever an element or its compounds are heated, dissolved or pulverized it becomes environmentally labile and may escape from the working environment and ultimately accumulate in soils.

Much research has been carried out on the 'heavy' metals, especially cadmium, copper, lead, mercury and zinc. The term 'heavy' is not rigorously defined but most authors use it to describe metallic elements having a density greater than 6000 kg m^{-3}. Some of the more important heavy metals are listed in Table 23.2. As well as metal pollution, contamination of land and crops by arsenic and fluorine can also present problems locally.

Table 23.2   Some heavy metals with their densities and average concentrations in crustal rocks. (Adapted from Davies, 1980, p. 290, ref. 12 and Mason and Moore, 1982, ref. 2)

Element	Density (kg m^{-3}) (× 10^3)	Mean content crustal rocks (mg kg^{-1})
Ag	10.5	0.1
Au	19.3	0.005
Bi	9.8	0.17
Cd	8.7	0.2
Cr	7.2	100
Co	8.9	25
Cu	8.9	55
Fe	7.9	60 000
Hg	13.6	0.08
La	6.2	25
Mn	7.4	950
Pb	11.3	13
Mo	10.2	1.5
Ni	8.9	75
Pt	21.5	0.05
Tl	11.9	0.45
Th	11.7	9.6
Sn	7.3	2
U	19.1	2.7
V	6.1	135
W	19.3	1.5
Zn	7.1	70
Zr	6.5	165

Sources of trace elements in soils

The trace element content of a soil depends initially on the nature of its parent material. As the soil matures and ages leaching and nutrient cycling through plants will cause some elements to be concentrated in specific soil horizons while others will be lost progressively in drainage water. Concurrently, elements will be gained by the soil through pollution, from natural accretions such as volcanic emissions or deposited dusts, and from drainage water received from soils at higher elevations. An understanding of how rocks differ in trace element composition[2] is a first step in understanding how various soils differ in their composition.

Trace elements do not generally form minerals in their own right but become incorporated in silicate crystals. This incorporation is largely controlled by their valencies and ionic radii. One ion can substitute for another if the radius of the larger ion does not differ from the smaller by more than 15 per cent and their ionic charges do not differ by more than one. The ionic radii of magnesium and iron(II) are 66 pm (pm = 10^{-12} m) and 74 pm respectively, and they can be replaced by cobalt(II) (72 pm), chromium(III) (63 pm), nickel (69 pm), zinc (74 pm) and, to some extent, by copper(II) (72 pm). Consequently, soils which form on basic rocks are generally well endowed with micronutrients and crop deficiencies are rare. Indeed, toxicities may occur where soils form on ultrabasic rocks such as serpentine since they can contain high concentrations of chromium and nickel. In contrast, acid igneous rocks are generally poor in some of the micronutrients so that, for example, cobalt deficiencies in stock are common where soils are formed from granite or rhyolite. The situation is complicated when discrete sulphide phases separate out since many of the metallic micronutrients are preferentially bound to sulphur atoms rather than oxygen. They can therefore become unexpectedly enriched in some rocks. Locally, sulphide ore minerals may be found near the earth's surface and as they weather they may cause the ore metals, especially copper, lead and zinc, to accumulate in soil and, in turn, lead to the development of specialized metal-tolerant floral associations. Flora which are tolerant of high soil concentrations of chromium and nickel are found over serpentine rocks in some areas.

Agriculturally, sedimentary rocks are far more important than igneous rocks since, although they represent only 15 per cent of the crustal volume, they occur spread over the igneous basement and amount to approximately 75 per cent of the earth's surface. In sedimentary rocks, substitution in silicate lattices of primary minerals is less important than the partitioning of elements which occurred during earlier weathering cycles. As the primary minerals

---

2 For chemical principles of rock composition see Goldschmidt, V. M., *Geochemistry*, Oxford, 1958; Mason, B. and Moore, C. B., *Principles of Geochemistry*, Wiley, 1982; Krauskopf, K. B., *Introduction to Geochemistry*, McGraw-Hill, 1979.

decompose during weathering the alkaline earths and alkali metals tend to remain in solution and some of the metallic micronutrients pass into the lattices of the secondary or clay minerals. Others become adsorbed on to clay-sized particles, are incorporated into humified organic matter by complexation or separate as precipitates following changes in redox potential.

Trace element concentrations in sandstones are generally low since these rocks are frequently dominated by quartz, although the matrix may be formed from other resistant minerals, and it is the matrix which carries the micronutrients which eventually become available to plants. Consequently, micronutrient deficiencies in plants often occur where soils have developed on arenaceous sediments and other light-textured parent materials such as aeolian sands, fluvioglacial sands, cover sands and coarse glacial outwash deposits. Where carbonate sediments are very pure the only significant source of trace elements in the soil may be from extraneous materials such as loess. But many carbonate rocks are impure and their trace element contents may be boosted by iron/manganese oxides and clays. In between the two extremes of sandstone and limestone are the shales which generally have a satisfactory trace element composition but they are so variable in composition that broad guidelines may be of little help.[3]

These geochemical generalizations are useful when considering young soils where composition is broadly similar to that of the parent material. Little general guidance is available for old, deeply weathered soils rich in iron, manganese and aluminium oxides and the published literature tends to be contradictory. In Ghana, a survey of 19 profiles representing most of the major soils of the forest zone led to the conclusion that total trace element content was related to the underlying geological formations.[4] In contrast, in soils, probably older or more leached, from Queensland, Tasmania and central Australia, the concentrations of trace elements in soils and parent materials were not close enough for the soil content to be predicted from the rock composition.[5]

As silicate minerals decompose, trace elements are released and enter the soil solution. Thereafter their fate depends on a number of factors. Immediately, they may precipitate or remain in solution and the likelihood of this happening can be predicted by considering the ratio of ion charge/radius (in nm), which is often called the ionic potential (IP). Elements with values above 95 form soluble oxyanions and include B, $Cr(VI)$, $Mo(VI)$, Si and W. Those with values below 30 form soluble cations and include Cd, Co, Cu, $Fe(II)$, Pb, $Mn(II)$, Hg, Ni, Ag, Sr, Sn and Zn. These elements may become 'trapped' in precipitates of compounds of elements with IP values in the range 30 to 95. This is a group of elements which tend to accumulate in weathering

3 Mitchell, R. L. and Burridge, J. C., *Phil. Trans. R. Soc. Lond.*, 1979, **B288**, 15.
4 Burridge, J. C. and Ahn, P. M., *J. Soil Sci.*, 1965, **16**, 296.
5 Oertel, A. C., *J. Soil Sci.*, 1961, **12**, 119; Wells, N., *op. cit.*, 1960, **11**, 409.

residues and includes Cr(III), Fe(III), Mn(III), Mn(IV), Mo(IV), V(III) and V(V). Iron and manganese dominate the group and in most soils these elements precipitate as hydrous oxides, either in concretionary forms or as coatings on soil peds or within the soil capillary network. They exert a chemical control on the activities of other ions far greater than might be supposed from their concentrations.[6] The hydrous oxides act as sinks for other trace elements through several mechanisms.[7] As they form, other trace metals can be occluded in the oxide precipitate. Nickel and cobalt are associated with manganese oxides, and copper and zinc with both manganese and iron oxides. These oxides tend to have high adsorption affinities which increase with pH. The trace cations are readily adsorbed on to the oxide surfaces after which they may enter the lattice through solid state diffusion. Super-imposed on these mechanisms is the possibility that the hydrous oxides can dissolve and then precipitate again in response to changes in the redox potential or pH of the soil.

Plants exert a strong influence on the distribution of trace elements in soil. Under natural vegetation roots absorb nutrient elements from a large volume of soil which are then returned to the surface through leaf fall. As leaves decompose their trace elements are incorporated into humus through complex formation or are released to the soil solution and are then adsorbed on to humus exchange sites. In consequence soil profiles are usually charac-terized by an accumulation of many trace elements in their surface layers.[8]

Table 23.3 Approximate amounts of some trace elements and major elements removed by crops

	Removed by average crop (kg ha^{-1})
Cobalt	0.001
Molybdenum	0.01
Copper	0.1
Boron	0.2
Zinc	0.2
Manganese	0.5
Iron	0.5
Magnesium	20
Phosphorus	20
Potassium	100

6 Jenne, E. A. in *Trace Inorganics in Water, Adv. Chem.*, Ser. 73, 1986. Am. Chem. Soc., Washington.
7 Taylor, R. M. and Giles, J. B., *J. Soil Sci.*, 1970, **21**, 203; Kalbasi, M. and Racz, G. J., *Canad. J. Soil Sci.* 1978, **58**, 61.
8 Swain, D. J. and Mitchell, R. L., *J. Soil Sci.*, 1960, **11**, 347; Wright, J. R. *et al.*, *Soil Sci. Soc. Am. Proc.*, 1955, **19**, 340.

Subsequently, leaching, lessivage and cheluviation will transport these elements down the profile, either to be lost in the drainage water or to be trapped in the sesquioxide-rich (B_s) horizons of podzols or in the clay-rich (B_t) horizon of 'sols lessivés'.

Micronutrient deficiencies may occur when podzolic soils are reclaimed for agricultural use. When crops are harvested the biogeochemical cycling of elements is disrupted and the loss of micronutrients in the harvested plants may eventually lead to deficiency problems.[9] Table 23.3 gives the amounts of several nutrient elements which may be removed from the soil by crops.[10] As crop yields have continued to increase in recent decades micronutrient problems have become more common in areas where supply was already marginal; in this respect copper is notable in Great Britain and zinc in the USA.

Availability of micronutrients to plants

The total content of the macronutrients in soil is of little practical significance in plant nutrition. To some extent this is less true for the micronutrients where, for example, analyses of total copper or zinc can be used to assess the likelihood of plant deficiencies or toxicities. Table 23.1 contains generalized information on the ranges and mean concentrations in soils of the essential elements.[11] Little has yet been done to put the presentation of such data on a firm statistical footing and most compilations merely summarize the elemental values recorded in the literature. Such summaries may well be biased by an inherent research interest in unusually high or unusually low concentrations. In recent years there has been a movement towards reporting trace element concentrations in terms of probability ranges of carefully defined parent materials.[12]

Indices of plant-available micronutrients may be obtained using soil extractants. The extractants are generally dilute (<1 M) solutions of mineral or organic acids, simple salts or organic and inorganic complexing agents. They are used in three ways. Historically the earliest, and still the most common, use is either to identify soils where a yield response may be expected if a micronutrient fertilizer is applied, or to confirm a micronutrient deficiency following field observations of growth abnormalities. The second use is to

9 Mitchell, R. L., *Neth. J. agric. Sci.*, 1974, **22**, 295.

10 Mitchell, R. L., *J. R. agric. Soc. Engl.*, 1963, **124**, 75.

11 Ure, A. M. and Berrow, M. L., *Environmental Chemistry*, 2. Specialist Periodical Reports (ed H. J. M. Bowen), R. Soc. Chem., London, 1982.

12 Davies, B. E., *Geoderma*, 1983, **29**, 67. Information on the use of extractants will be found in Ch. 13, pp. 289–317 of *Micronutrients in Agriculture* (eds J. J. Mortvedt, P. M. Giordano and W. L. Lindsay), Soil Sci. Soc. Am., Madison, Wisc., 1972. See also *Applied Soil Trace Elements* (ed B. E. Davies), Wiley, 1980.

attempt to predict crop uptake of trace elements, and this is especially important where plants are growing on soils contaminated by elements such as lead or arsenic which are toxic to animals. A third use is in the chemical fractionation of soil to identify and evaluate the major chemical pools of micronutrients and toxic metals. The same extractants are used for all three purposes and none is based on any firm theoretical foundation. Yet, despite the empiricism involved in their use they have proved of practical assistance both in advisory work and in laboratory studies. Some of the more commonly used reagents are listed in Table 23.4 together with an indication of which chemical fraction they are thought to extract. The order in the table is one of increasing vigour of attack on the soil so that a solution which extracts, say, metals from the humus in soil will also extract from the soil solution and from the exchangeable pool.

The immediate source of micronutrients for the plant is the soil solution[13] but there is a paucity of information concerning its composition especially for micronutrients. Whereas the soil solution contains calcium, magnesium and potassium at concentrations typically in the range 10^{-4} to 10^{-2} M, the micronutrients tend to be present at the micromolar level. For example, in

Table 23.4 Some common diagnostic soil extractants and the chemical forms they are thought to extract

Soil fraction	Common extractant
Soil solution	H_2O; 0.01 M $CaCl_2$
Readily exchangeable	0.5 M CH_3COONH_4; 0.2 M $MgSO_4$ 0.1 M NH_4Cl; 1 M NH_4NO_3
Specifically sorbed	0.5 M CH_3COOH; 0.1 M HCl; 0.1 M HNO_3
Organically bound	0.05 M EDTA*; 0.05 M EDDHA† 0.005 M DTPA + 0.1 M TEA + 0.01 M $CaCl_2^\ddagger$
Hydrous oxide bound	1 M CH_3COONH_4 + 0.002 M $C_6H_6O_2$** 0.2 M $(COO)_2 (NH_4)_2$ + 0.15 M $(COOH)_2$ at pH 3.3
Residual	HF; mixtures of hot, concentrated acids, namely: HNO_3 + $HClO_4$; HNO_3; HNO_3 + HCl; also fusion mixtures

* Ethylene diamine tetraacetic acid, disodium or diammonium salt.
† Ethylene diamine di-(*o*-hydroxyphenyl acetic acid).
‡ Diethylene triaminepentaacetic acid and triethanolamine with calcium chloride at pH 7.3.
** Hydroquinone (1:4 dihydroxybenzene).

13 For methods of obtaining soil solutions see: Davies, B. E. and Davies, R. I., *Nature, Lond.*, 1963, **198**, 216; Whelan, B. R. and Barrow, N. J., *J. environ. Qual.*, 1980, **9**, 315.

saturation extracts of Californian soils[14] the following median concentrations ($mg\ l^{-1}$) were measured: Co <0.01, Cu 0.03, Fe 0.03, Mn <0.01, Mo <0.01, Zn 0.04. Micronutrient ions rarely exist in the soil solution as simple forms since they are liable to complex formation with both inorganic and organic ligands. Table 23.5 lists the principal chemical species predicted in aerobic soils in acid and alkaline conditions and the ordering from left to right is roughly that of decreasing concentration.[15] Copper is particularly prone to organic complexation. In displaced solutions from 20 calcareous soils in Colorado 98–99 per cent of the copper was in the form of an organic complex.[16] Much information on the speciation of trace elements has come from studies of river and sea water[17] and recently there has been considerable progress in writing computer programs (see Ch. 7) which will predict the main species present in very dilute solutions.[18]

Plant roots take up their nutrients from the soil solution but as they grow they alter their immediate environment, the rhizosphere, as discussed in Chapter 17. As a result, the chemistry of micronutrients at the root–soil interface is poorly understood. As for the macronutrients, the trace elements are transferred to plant roots by mass flow, or they may diffuse in response to a concentration gradient caused by depletion at the root surface (Ch. 4).[19]

Table 23.5 Principal chemical species of trace metals in acid and alkaline soil solutions (oxic conditions) (Adapted from Sposito, 1983, p. 150, ref. 15)

Metal	*Principal species*	
	Acid soils	*Alkaline soils*
Mn (II)	Mn^{2+}, $MnSO_4^0$, Org*	Mn^{2+}, $MnSO_4^0$, $MnCO_3^0$, $MnHCO_3^+$
Fe(II)	Fe^{2+}, $FeSO_4^0$, $FeH_2PO_4^+$	$FeCO_3^0$, Fe^{2+}, $FeHCO_3^+$, $FeSO_4^0$
Ni(II)	Ni^{2+}, $NiSO_4^0$, $NiHCO_3^+$, Org*	$NiCO_3^0$, $NiHCO_3^+$, Ni^{2+}
Cu(II)	Org*, Cu^{2+}	$CuCO_3^0$, Org*
Zn(II)	Zn^{2+}, $ZnSO_4^0$	$ZnHCO_3^+$, $ZnCO_3^0$, Zn^{2+}, $ZnSO_4^0$
Cd(II)	Cd^{2+}, $CdSO_4^0$, $CdCl^+$	Cd^{2+}, $CdCl^+$, $CdSO_4^0$, $CdHCO_3^+$
Pb(II)	Pb^{2+}, Org*, $PbSO_4^0$, $PbHCO_3^+$	$PbCO_3^0$, $PbHCO_3^+$, $Pb(CO_3)_2^{2-}$, $PbOH^+$

* Organic complexes (e.g., fulvic acid complexes).

14 Bradford, G. R. *et al.*, *Soil Sci.*, 1971, **112**, 225.
15 Sposito, G. in *Applied Environmental Geochemistry* (ed I. Thornton), Acad. Press, 1983.
16 Hodgson, J. F. *et al.*, *Soil Sci.*, *Soc. Am. Proc.*, 1966, **30**, 723.
17 Mantoura, R. F. C. *et al.*, *Estuarine and Coastal Marine Sci.*, 1978, **6**, 387.
18 Sposito, G., *Environ. Sci. Tech.*, 1981, **15**, 396.

Plants differ in their ability to absorb micronutrients even when grown in the same soil. In an experiment in Finland[20] root and grain crops, grasses and legumes were grown during two seasons at nine sites. The least varying micronutrient among the crops was copper where the highest mean content was only four times that of the lowest. For other micronutrients the respective differences were iron and zinc seven-fold, boron twenty-one-fold, manganese thirty-five-fold and molybdenum forty-six-fold. Although some species absorbed generally more micronutrients than others, their absorption ability seldom applied to all six elements investigated. There were also considerable differences between the micronutrient contents of various parts of plants. In addition to differences between species, different cultivars of the same species can vary substantially in their ability to absorb micronutrients or trace elements.[21]

Chemistry of essential plant micronutrients in soil

This part of the chapter gives an account of the individual plant micronutrients. The discussion concentrates on the occurrence of each element in the soil and the conditions which lead to deficiencies. The symptoms of deficiency and metabolic functions of each element in the plant are given in Chapter 3.

Iron

Iron is one of the most ubiquitous elements in rocks and soils. Soil colour and, hence, the visible differentiation of profiles into horizons is often related to the form and composition of re-deposited iron hydroxides and oxides, the concentrations of which are measured in per cent. The soil chemistry of iron is therefore that of a major element, yet in plant and animal nutrition it is a micronutrient and deficiencies of iron occur in crops in many countries. The total iron content of soil varies from less than 0.1 per cent in some leached sands to over 30 per cent in enriched horizons and plinthite. Freshly precipitated iron is present in forms amorphous to X-rays as $Fe_2O_3.nH_2O$, as the poorly crystalline ferrihydrite, and also as the crystalline hydrated oxides (FeOOH), goethite and lepidocrocite. The unhydrated crystalline oxides

19 *Applied Soil Trace Elements* (ed B. E. Davies), p. 287, Wiley, 1980. See also Wilkinson, H. F. in *Micronutrients in Agriculture* (eds J. J. Mortvedt, P. M. Giordano and W. L. Lindsay, p. 139), Soil Sci. Soc. Am., Madison, Wisc., 1972.
20 Yläranta, T. and Sillanpää, M., *Annales Agriculturae Fenniae*, 1984, **23**, 158.
21 Crews, H. M. and Davies, B. E., *J. agric. Sci.*, 1985, **105**, 591.

which are found in soil are haematite and maghemite (both Fe_2O_3) and magnetite, Fe_3O_4. In soils of temperate regions the main stable forms are goethite and lepidocrocite, while in the tropics the Oxisols contain haematite and maghemite. Iron also occurs[22] in primary silicates, such as micas and ferromagnesian minerals, in 2 : 1 layer lattice clay minerals and in close association with soil organic matter.

The apparent discrepancy between the ubiquitous occurrence of iron and its role as a micronutrient is explained by its very low solubility in soil. Its soil and biological chemistry is complex and the subject of long-standing and continuing research. The controlling reaction in the soil solution is

$$Fe^{3+} + 3OH^- \rightleftharpoons Fe(OH)_3$$

which has an equilibrium constant of about 10^{-39}, and the solubility of Fe(III) decreases 1000-fold for each pH unit increase. The solubility of the precipitate is therefore very low under normal soil conditions. Besides Fe^{3+} the hydrolysed species $Fe(OH)_2^+$ and $Fe(OH)^{2+}$ are also present in solution.

Redox potential largely controls the solubility of iron in soil (see Ch. 26), because both ferrous and ferric ions can exist in the soil solution and the former are more soluble. Aerobic soils usually have redox potentials (*Eh*) between 0.3 and 0.8 V and in anaerobic soils they tend to lie between 0.2 and -0.4 V. In aerobic soils below pH 5 Fe^{2+} is the stable species. Even in aerobic soils reduction can occur in wet soil aggregates. Many plants, including rice, which can grow in waterlogged soils, leak oxygen from their roots and thereby minimize the soluble iron concentration of the rhizosphere and reduce the likelihood of iron toxicity.

Calculated concentrations of ferrous and ferric iron in aerobic soil solutions have been given by Rowell.[23] The concentration of iron required for satisfactory growth in nutrient solutions has been reported[24] as $>10^{-7}$ M and in flowing nutrient solutions 10^{-6} M has been found adequate, although the threshold concentration is probably much less. The concentration of iron required in the soil solution to allow optimum growth rates is not known. Insufficient amounts would exist in the solutions of normal agricultural soils if the concentrations depended on the inorganic chemistry of Fe(II) and Fe(III). But iron readily forms organic complexes which can raise its solution concentration, and complexing and chelating ligands are available from tree canopy drip, leaf litter, mineralizing humus and root exudates. Micro-organisms also mediate in changing the metal's oxidation state. At the soil–

22 Brown, G. *et al.* in *The Chemistry of Soil Constituents* (eds D. J. Greenland and M. H. B. Hayes), p. 29, Wiley, 1978. Oades, J. M., *Outlook on Agriculture*, 1964, **41**, 143.

23 Rowell, D. L. in *The Chemistry of Soil Processes* (eds D. J. Greenland and M. H. B. Hayes), Wiley, 1981.

24 Lindsay, W. L. and Schwab, A. P., *J. Pl. Nutr.*, 1982, **5**, 821.

root interface it is probable that Fe^{3+} is decomplexed and reduced to Fe^{2+} before absorption.[25]

The importance of iron in plant nutrition was established last century by Gris[26] who recognized its role in causing yellow leaves in grape vines growing on calcareous soils. He found that the condition could be alleviated by spraying with soluble iron salts. Plants differ in their ability to absorb iron from soils, particularly from calcareous soils, and it is probably this difference which underlies the division of plants into calcifuges and calcicoles. Calcifuges, including most species of *Rhododendron* and *Erica*, display symptoms of 'lime-induced chlorosis' when growing on neutral or calcareous soils. Crops which are most likely to be affected by lime-induced chlorosis include deciduous fruit trees, grape vines, citrus fruit trees and, albeit to a lesser extent, sugar beet, soyabeans, maize and tomatoes. The situation is aggravated by a high concentration of the bicarbonate ion in the soil solution and chlorosis is common where irrigation water has a high bicarbonate content. It has been reported that high soil moisture content, poor aeration and low temperatures can also induce the condition.[27]

The iron content of normal plant tissue varies according to species but is usually within the range 50 to 250 mg Fe kg^{-1} dry matter. Total iron content is not a reliable indicator, but values below 50 mg kg^{-1} may indicate deficiency. The P : Fe ratio has also been regarded as crucial because it may affect the Fe(II) concentration in plant tissues and therefore the activity of the enzyme aconitase.[28] Other metals may induce iron deficiency. There is a well-established antagonism between manganese and iron,[29] and competition for uptake by heavy metals, including copper, nickel and zinc, can induce iron deficiency on metal-contaminated land.

Because iron deficiency in plants is primarily due to its low solubility especially in calcareous soils, an application of simple iron salts to the soil is usually ineffective. Availability can be improved by putting down a fruit orchard to grass which is cut and the cuttings left on the surface. Large quantities of organic manures can also be effective. Only a few iron chelates are effective when applied to the soil, and of these Fe–EDDHA, iron ethylene diamine di-(*o*- hydroxyphenyl acetate), is one of the most commonly used. Mature pear and apple trees need about 60 g per tree when applied to the soil. Alternatively, the leaves of the crop may be sprayed with a solution of Fe–EDTA, iron ethylene diamine tetraacetate, applied at a rate of 1 kg in 1000 l water per hectare. Applications are repeated every two weeks.

25 Uren, N. C., *J. Pl. Nutr.*, 1982, **5**, 515; Olsen, R. A. *et al.*, op. cit., 1982, **5**, 433.
26 Gris, E., *Compt. rend.*, 1843, **17**, 679. For a review of iron in calcareous soils see Chen, Y. and Barak, P., *Adv. Agron.*, 1982, **35**, 217.
27 Wallace, A., *J. Pl. Nutr.*, 1982, **5**, 277; Vose, P. B., *op. cit.*, 233.
28 DeKock, P. and Morrison, R. I., *Biochem J.*, 1958, **70**, 266, 272.
29 Somers, I. I. and Shive, J. W., *Pl. Physiol.*, 1942, **17**, 582.

Manganese

Like iron, manganese deficiencies do not normally arise from an absolute lack of the element but because soil conditions restrict the supply to the root. The total manganese content of soils commonly varies between 300 and 500 μg g^{-1}, but it can vary over a much wider range (Table 23.1). The element can be present as Mn(II) in solution, on the exchange complex and in non-stoichiometric oxides in which Mn(III) and Mn(IV) may also be present.

Soils with different pH values have widely different concentrations of manganese in solution. Thus the concentration in displaced solutions from 21 surface soils in the USA ranged from 10^{-5} M at pH 4 to 2×10^{-7} M at pH 7 to 8.[30] Soil pH has a similar effect on the level of exchangeable manganese.[31] The solution and exchangeable manganese is probably present mainly as $Mn(H_2O)_6^{2+}$ under acid conditions.* For example, it was found that 75 to 90 per cent of the total manganese in solution was present as the free ion in soils with pH 5.5 to 7.0[32]; when the pH is raised an increasing proportion is complexed with humic and fulvic acids.[33]

The concentration of divalent ions in the soil solution (which is probably the form absorbed by plant roots) is the result of two opposing reactions, namely oxidation to manganese oxides of very low solubility and their reduction. The oxidation is due to micro-organisms and its rate is relatively slow between pH 5 and 6 but increases markedly as the pH is raised to 7.5. The oxides are reduced by organic matter, or some biological processes involving microbial or root products, and the rate increases with increasing acidity. The reaction may be represented by:

$$MnO_2 + 4H^+ + 2e \rightarrow Mn^{2+} + 2H_2O$$

The redox potential for the Mn^{2+}–MnO_2 couple has been used to calculate the concentration of manganese in solution. While the calculated concentrations differ somewhat from the observed concentrations, calculation shows that they decrease by a factor of 10^4 for each unit rise in pH in the range of 4 to 7.

The manganese oxides often occur as visible concretions, or as thin black films on the surface of soil peds. In 23 Australian soils the commonest manganese minerals were identified as birnessite $M.Mn^{2+} Mn^{4+} (O.OH)_2$ where M is an alkali or alkaline earth metal, and lithiophorite $Li_2Al_8M_4.Mn_{10}^{4+}O_{35}.14H_2O$ where M is divalent manganese, cobalt or

30 Geering, H. R. *et al.*, *Soil Sci. Soc. Am. Proc.*, 1969, **33**, 81.
31 Steenbjerg, F., *Tidsskr. Planteavl.* 1933, **39**, 401.
32 Sanders, J. R., *J. Soil Sci.*, 1983, **34**, 315.
33 McBride, M. B., *Soil Sci.*, 1978, **126**, 200; *Soil Sci. Soc. Am. J.*, 1982, **46**, 1137.

* The aquo-ions will not be referred to in later sections.

nickel.[34] Hollandite $Ba(Fe^{3+}, Mn^{4+})_8O_{16}$ was found in very few soils and pyrolusite MnO_2 in only one. The crystallite size was extremely small (0.1–0.02 μm). Because of the high surface area of these oxides they contribute significantly to the surface activity of the soil. They are noted for their strong adsorption of cobalt, nickel and copper,[35] and about 80 per cent of the cobalt in some soils has been found to be associated with birnessite.[36]

Manganese differs from other micronutrients in that both deficiency and toxicity are widespread in agricultural practice. Toxicity symptoms are likely to occur in acid soils as a result of the reduction of manganese oxides (see also Chs 25 and 26). Sterilizing and air-drying a soil can also increase the solubility of manganese, and as the pH is not altered one explanation is that the manganese-oxidizing micro-organisms are destroyed.

The essentiality of manganese for several species of plants was established by McHargue[37] in 1922. Manganese deficiency symptoms have been described in most crops (see Ch. 3). Deficiencies occur rarely on soils with pH values below 6.5 and only on certain soils with higher pH. Liming is one of the most common causes of deficiency, and occurs when old pastures are ploughed and limed or marshland soils are reclaimed. An example of the effect of liming an acid soil on the uptake of manganese by wheat is given in Table 23.6.[38] Deficiency of manganese is sometimes associated with peats, particularly those overlying calcareous parent material or after they have been drained and limed; it is rarely observed on soils low in organic matter developed on chalk or limestone. Although manganese is more available where soil drainage is restricted, plant root systems may develop poorly and deficiency symptoms can then appear during a drought period. These problems become exacerbated when soils are already low in manganese, such as reclaimed (and limed) podzols or strongly leached light-textured soils over sandy parent

Table 23.6 Effect of raising the soil pH on the manganese uptake by winter wheat. Chalk applied 1934; wheat grown 1945/46. Soil: medium loam on boulder clay, Oaklands, St Albans, England

Ground chalk applied (t ha^{-1})	pH of soil	Yield of wheat grain plus straw (t ha^{-1})	Uptake of Mn (kg ha^{-1})
0	4.3	2.78	0.62
4.8	4.7	4.12	0.61
9.4	5.3	5.25	0.57
14.1	5.8	5.65	0.48
18.8	6.5	5.78	0.38

34 Taylor, R. M. *et al., Aust. J. Soil Res.*, 1964, **2**, 235.
35 McKenzie, R. M., *Aust. J. Soil Res.*, 1967, **5**, 235.
36 Taylor, R. M. and McKenzie, R. M., *Aust. J. Soil Res.*, 1966, **4**, 29.
37 McHargue, J. S., *J. Am. Chem. Soc.*, 1922, **44**, 1592.
38 Chambers, W. E. and Gardner, H. W., *J. Soil Sci.*, 1951, **2**, 246.

materials. In soils where the concentration of divalent manganese is low, an adequate supply depends on the ease with which manganese oxides can be reduced close to the roots. When readily reducible oxides were added to deficient soils with pH 7.0 to 8.0, oats, a susceptible crop, grew satisfactorily.[39] Crop roots probably play an active role in promoting the dissolution of manganese oxides through the excretion of short-lived reducing compounds.[40] Manganese deficiency has been alleviated experimentally by compacting soil[41] and it has been observed that healthy plant growth occurred in the tracks of wheeled vehicles when the rest of the crop was deficient.[42] Compaction may increase the contact between roots and manganese oxides.

Manganese interacts with other elements in soils and plants especially iron. Both are greatly affected by soil redox conditions and an excess of one can decrease uptake of the other and cause deficiency. Other elements which have been found to affect the manganese nutrition of plants include phosphorus (see p. 732), and calcium and magnesium which can decrease uptake. Silicon has been found to repress toxicity symptoms in rice and other cereals.[43]

It has been noted earlier that a portion of the higher oxides can be solubilized by mild reducing agents and these have been used in diagnostic soil testing. A widely used extractant is 0.1 per cent hydroquinone in neutral molar ammonium acetate in which <25 mg Mn kg^{-1} soil is a critical level. But the widely differing susceptibilities of crops to manganese deficiencies make the method only a rough guide. In a comparison of analytical methods using 63 soils of contrasting type ammonium acetate, with correction for soil pH, appeared to be the most satisfactory.[44]

Control of deficiencies through soil treatment is rarely effective except in the very short term. Manganese sulphate may be broadcast at high application rates, e.g. at 125–250 kg ha^{-1}, or manganous oxide can be pelleted with the seed or banded with an acid-forming fertilizer. Foliar sprays of manganese sulphate or chelated manganese may be used when symptoms of deficiency are first observed.

Copper

Deficiency of copper in crops has been reported from many countries.[45] The

39 Jones, L. H. P. and Leeper, G. W., *Pl. Soil*, 1951, **3**, 141, 154.
40 Bromfield, S. M., *Pl. Soil*, 1958, **9**, 325; *ibid.*, 1958, **10**, 147; Jauregui, M. A. and Reisenauer, H. M., *Soil Sci. Soc. Am. J.*, 1982, **46**, 314.
41 Passioura, J. B. and Leeper, G. W., *Nature, Lond.*, 1963, **200**, 29.
42 Needham, P. in *Diagnosis of Mineral Disorders in Plants* (eds C. Bould, E. J. Hewitt and P. Needham), HMSO, London, 1983.
43 Cheng, B. T. and Ouellette, G. J., *Soils Fertil.*, 1971, **34**, 589.
44 Browman, M. G. *et al.*, *J. agric. Sci.* 1969, **72**, 335.
45 Much information on copper is contained in *Copper in Soils and Plants* (eds J. F. Loneragan, A. D. Robson and R. D. Graham), Acad. Press, 1981.

total copper content of soils (Table 23.1) commonly ranges from 2.5 to 60 μg Cu g^{-1} but values up to several hundred μg g^{-1} are found in contaminated soils. Copper deficiencies in crops may occur if the content of total soil copper is very low, for example when well-developed podzols are brought into agricultural production, or in soils over inherently low copper parent materials such as granites, sandstones or sandy glacial deposits. In Scotland, 'wither tip' in cereals has been reported[46] where soils which have developed in fluvioglacial sands derived from Devonian sandstones have total copper contents of less than 6 μg g^{-1}.

Deficiencies are generally a consequence of soil conditions reducing the availability of an otherwise marginal supply. In soils copper occurs as Cu(II). At pH values of about 7 and above, Cu(OH)$^+$ is present in significant concentrations and may enter into soil adsorption reactions.[47] Lindsay suggests[48] that the solubility of soil copper is controlled by that of cupric ferrite and that the equilibrium for soil copper can be represented by the equation

$$\log \text{Cu}^{2+} = 2.8 + 2\text{pH}$$

but it seems more likely that most soil copper exists in chemisorbed or occluded forms in hydrous oxides of iron, aluminium or manganese and complexed with organic matter.

Copper readily forms organic complexes which are the predominant species in the soil solution. In peats of pH 3.5, 60–90 per cent of the copper in aqueous extracts was found to be organically complexed, and raising the pH to 6.0 increased the degree of complexation to >98 per cent.[49] These complexes are very stable and Schnitzer[50] has reported that of nine divalent ions, copper formed the most stable complex with fulvic acid, involving carboxy and phenolic groups. Complexation of copper by organic matter may control its availability, for peat binds even low concentrations very tightly. Copper deficiencies following the reclamation of peat bogs through drainage and liming are common.[51]

The importance of organic matter in the soil chemistry of copper is underlined by data from fractionation studies. One differential extraction scheme[52]

46 McBratney, A. B. *et al.*, *Agronomie*, 1982, **2**, 969;   Purves, D. and Ragg, J. M., *J. Soil Sci.*, 1962, **13**, 241.
47 Ellis, B. G. and Knezek, B. D. in *Micronutrients in Agriculture* (eds J. J. Mortvedt, P. M. Giordano and W. L. Lindsay), Soil Sci. Soc. Am., Madison, Wisc., 1972. See also Sanders, J. R. and Bloomfield, C., *J. Soil Sci.*, 1980, **31**, 53 and Mattigood, S. V. and Sposito, G., *Soil Sci. Soc. Am. J.*, 1977, **41**, 1092.
48 A full account of the equilibrium chemistry of elements in soil is given by Lindsay, W. L., *Chemical Equilibria in Soils*, Wiley, 1979.
49 Hodgson, J. F. *et al.*, *Soil Sci. Soc. Am. Proc.*, 1965, **29**, 665, and see ref. 16.
50 Schnitzer, M., *Soil Sci. Soc. Am. Proc.*, 1969, **33**, 75.
51 Pizer, N. H. *et al.*, *J. agric. Sci.*, 1966, **66**, 303;   Tills, A. R. and Alloway, B. J., *J. agric. Sci.*, 1981, **97**, 473.
52 McLaren, R. G. and Crawford, D. V., *J. Soil Sci.*, 1973, **24**, 172.

attempted to distinguish five fractions, namely, soil solution and exchangeable copper, copper weakly bound to specific sites, organically bound copper, copper occluded by oxides and residual copper. There was an association between manganese oxides and copper. Soil solution copper was controlled by equilibria involving specifically adsorbed copper, but the bulk of the plant-available reserves was believed to reside in the organically bound fraction. Another study of the adsorption of copper by soil materials at low equilibrium solution concentrations[53] demonstrated that iron and manganese oxides and humic acid adsorbed the greatest amounts, and that within the range of normal agricultural soils pH had little effect on the solution concentration of copper. A study of the form of copper in solutions of contaminated soils demonstrated that 28 per cent was associated with the organic fraction, and the manganese and iron oxides were of nearly equal importance.[54]

Copper has been known to be essential for plants since 1931.[55] Its many roles in plant metabolism have been discussed in Chapter 3. Soil analysis is useful in predicting or diagnosing copper deficiencies. Extractants currently favoured are formulated from chelating agents[56] and extract exchangeable ions and complexes with organic matter. In England and Wales the disodium salt of EDTA at 0.05 M is used for advisory purposes. Concentrations of <1.6 mg Cu l^{-1} soil indicate the need for preventative treatment and concentrations between 1.7 and 2.4 mg Cu l^{-1} suggest possible deficiency on organic and peat soils. In the USA a mixed reagent[57] of triethanolamine (TEA), diethylene triamine pentaacetic acid (DTPA) and $CaCl_2$ has been widely adopted. Copper deficiencies are associated with soil concentrations <0.2 mg kg^{-1}. A comparison of several soil extractants on English soils indicated that EDTA was more reliable than DTPA.[58]

Copper deficiency in crops has been reported in many countries.[45] It can be severe when peats are drained especially if their surface becomes dry. It occurs on thin organic soils over chalk and on many strongly weathered soils in Australia.

Copper deficiency can be corrected either by adding a copper salt to the soil or spraying it on the crop itself. If added to the soil, the recommended application can vary 100-fold depending on the soil. Applications as high as 70 kg Cu ha^{-1} as copper sulphate have been recommended in the Florida Everglades,[59] 9 kg ha^{-1} on the peats and fens of eastern England,[60] about 3–15 kg ha^{-1} on calcareous sandy soils of South Australia,[61] and 0.5 kg ha^{-1}

53 McLaren, R. G. *et al.*, *J. Soil Sci.*, 1981, **32**, 247.
54 Hickey, M. G. and Kittrick, J. A. *J. environ. Qual.*, 1984, **13**, 372.
55 Lipman, C. B. and Mackinney, G., *Pl. Physiol.*, 1931, **6**, 593; Sommer, A. L., *Pl. Physiol.*, 1931, **6**, 339.
56 Viro, P. J., *Soil Sci.*, 1955, **70**, 459.
57 Lindsay, W. L. and Norvell, W. A., *Soil Sci. Soc. Am. J.*, 1978, **42**, 421.
58 Tills, A. R. and Alloway, B. J., *J. Sci. Fd, Agric.*, 1983, **34**, 1190.
59 Bryan, O. C., *J. Am. Soc. Agron.*, 1929, **21**, 923.
60 Pizer, N. H. *et al.*, *J. agric. Sci.*, 1966, **66**, 303.
61 Riceman, D. S. and Donald, C. M., *J. Dept. Agric. S. Aust.*, 1939, **42**, 959.

on some sandy soils in Western Australia on which 2.5 kg ha^{-1} is harmful.[62] Copper sulphate or copper oxychloride is used as a crop spray at a rate of about 1 to 1.8 kg ha^{-1}, and both soil application and spray might be required in the early years of cropping after peat reclamation. A soil application can last several years.[63]

Copper sulphate combined with lime is the common fungicide Bordeaux mixture. Use of this may have prevented deficiencies in some areas but it can also lead to a build-up of copper in the soil. Toxicities have been observed in grape-growing areas of France, and the accumulation of copper in soils, litter and vegetation has been described for the coffee-growing areas of Kenya.[64]

Copper essentiality for animals was demonstrated before that for plants, but it is involved in similar oxidative systems. Increasing attention is now being paid to subclinical effects in animals where metabolic changes precede the appearance of overt deficiency.[65] These can arise where pasture has a low copper content even though the grass does not itself appear deficient. An economically important deficiency is 'swayback' in newborn lambs[66] and, as discussed later, excess herbage molybdenum will induce copper deficiency in animals. Thus, maintaining an adequate copper content of plants goes beyond avoiding crop deficiencies.

Zinc

A wide range of total zinc concentrations is found in soils (Table 23.1) but values from 50 to 300 μg g^{-1} are most common in agricultural soils. Zinc deficiency in crops can arise where there is a low total content in soils, or more commonly where soil conditions reduce its availability. Severe deficiencies have been reported from Australia,[67] the USA[68] and many other countries.

In solutions of acid soils zinc occurs mainly as Zn^{2+}. For example in three acid soils (pH 4.8–5.7) 71.0–76.0 per cent of the zinc in solution was estimated[69] to be present as Zn^{2+}. At higher pH values $Zn(OH)^+$, $Zn(HCO_3)^+$ and $ZnCO_3^0$ are present in concentrations which depend on the pH and partial pressure of carbon dioxide.

Organic complexes of zinc are less important in the soil solution than for

62 Wild, A. S. and Teakle, L. J. H., *J. Dept. Agric. W. Aust.*, 1942, **19** (ser. 2), 71, 242.
63 Reith, J. W. S., *J. agric. Sci.*, 1968, **70**, 39, and see ref. 60.
64 Lepp, N. W. *et al.*, *Pl. Soil*, 1984, **77**, 263.
65 Mills, C. F., *Trans. R. Soc. Lond.*, 1979, **B288**, 51.
66 Alloway, B. J., *J. agric. Sci.*, 1973, **80**, 521, and see ref. 1.
67 Donald, C. M. and Prescott, J. A. in *Trace Elements in Soil–Plant–Animal Systems* (eds D. J. D. Nicholas and A. R. Egan), Acad. Press, 1975.
68 Berger, K. C., *J. agri. Fd Chem.*, 1961, **10**, 178.
69 Riley, D. and Barber, S. A., *Soil Sci. Soc. Am. Proc.*, 1971, **35**, 301.

copper. In soil solutions extracted from 20 calcareous soils collected in Colorado, USA, rarely more than 75 per cent of the zinc was organically complexed.[16] The results from chemical fractionation studies confirm a reduced role for organic matter in the soil chemistry of zinc. In four contaminated soils about 39 per cent of the total zinc was associated with iron and manganese oxides and about 28 per cent with the carbonate fraction, but less than 10 per cent was organically bound.[54] Some work has been done at comparatively high rates of addition where $Zn(OH)_2$ and, in calcareous soils, $ZnCO_3$ can be precipitated. In the presence of $NH_4H_2PO_4$, zinc phosphate has been identified.[70] More usually the concentration of zinc in the soil solution is controlled, at least in dilute solution, by adsorption/desorption reactions.

The solubility of soil zinc decreases as the pH increases. In one study the total concentration of zinc in the soil solution decreased about 100-fold as the pH of the soil increased from 4.4 to 7.5.[71] In another study which investigated the effects of pH and redox potential,[72] greater amounts of zinc were extracted (by sodium acetate and potassium pyrophosphate) at low E_h and pH than at high E_h and pH, and complexation by organic matter was greater in the soil of low E_h.

The essentiality of zinc for higher plants was first proposed in 1914 by Maze and then demonstrated by Sommer and Lipman for barley and sunflower.[73] The management practices that may induce zinc deficiency include the use of phosphate fertilizers which may lead to a decrease in the uptake of zinc,[74] or, more likely, to reduced transport in the plant. Some phosphatic fertilizers, however, contain zinc as an impurity. Liming may induce deficiency because of the control of zinc solubility by pH, as mentioned earlier. Also, high rates of application of nitrogen fertilizers, by increasing total crop growth, may increase zinc requirements beyond the available supply. Zinc deficiency can be controlled by the use of sprays or zinc fertilizers. Zinc is phytotoxic and care must be taken not to exceed optimum soil levels. Zinc deficiencies can be predicted by extracting soil using the same organic complexing compounds as are used for copper. The use of a mixed solution of EDTA and ammonium carbonate has been proposed as an extractant for neutral and calcareous soils.

Molybdenum

The total molybdenum content of soils (Table 23.1) is usually in the range 0.07 to 5 $\mu g\ g^{-1}$, but values of 100 $\mu g\ g^{-1}$ have been reported for soil over black shales. In the soil solution molybdic acid, H_2MoO_4, is dissociated over the whole range of soil pH, and MoO_4^{2-} is the predominant species above

70 Kalbasi, M. *et al., Soil Sci.,* 1978, **125**, 55.
71 Jeffery, J. J. and Uren, N. C., *Aust. J. Soil Res.,* 1983, **21**, 479.
72 Sims, J. L. and Patrick, W. H., *Soil Sci. Soc. Am. J.,* 1978, **42**, 258.
73 Sommer, A. L. and Lipman, C. B., *Pl. Physiol.,* 1926, **1**, 231.
74 See Sumner, M. E. and Farina, M. P. W., *Adv. Soil Sci.,* 1986, **5**, 201.

pH 4.3. The concentration of molybdenum in the soil solution is very small and in saturation extracts from 11 soils in the USA with pH ranging from 5.0 to 7.8 it varied[75] between 3.0 and 8.1 μg dm^{-3}. The concentration in solution in one soil increased linearly from 1.3 to about 80 μg dm^{-3} as the pH was raised from 5.0 to 7.0.[76]

The immediate reservoir of molybdate in solution is that adsorbed on soil surfaces. It has been shown that hydrated ferric oxide has a much stronger affinity than aluminium oxide for molybdate, and that the 1 : 1 clay minerals, halloysite and kaolinite, are poorer adsorbents than aluminium oxide.[77] Adsorption decreases sharply as pH is raised above 4, which is consistent with ligand exchange of molybdate with exposed Fe–OH and Al–OH groups[78] (see Ch. 7). The importance of iron oxides is shown both by highly significant positive correlations between the amount of iron and molybdenum dissolved by extracting agents[79] and by the accumulation of molybdenum in concretionary ferric oxide.[80]

On average, the uptake by a crop plant growing on the one soil will increase by two- to three-fold for each unit rise in pH from 5.0 to 7.0. Larger responses to liming have been reported, for example when red clover was grown on a soil of pH 5.4 the uptake increased by as much as six-fold on liming to pH 6.4.[81] Although it is generally possible to cure deficiency by liming to pH 6.5 or 7.0, there are large areas of soil, particularly in Australia, New Zealand and the USA where it is necessary, and often cheaper, to supply molybdenum to the crop (see Colour Plate 4). Deficiency can normally be corrected in legumes and non-legumes by applications of sodium or ammonium molybdate applied to the soil at 50 to 250 g Mo ha^{-1}, and such an application will usually last for several years.[82] Alternatively, the seed may be treated by soaking in a solution of sodium molybdate, adjusted so that the addition to the soil is equivalent to 50 to 100 g Mo ha^{-1}. A foliar spray may be used to correct a molybdenum deficiency should it persist after soil or seed treatment.

It is important that molybdenum dressings given to forage plants be kept

75 Lavy, T. L. and Barber, S. A., *Soil Sci. Soc. Am. Proc.*, 1964, **28**, 93. For a review on the chemistry and availability of molybdenum in soils, see Gupta, U. C. and Lipsett, J., *Adv. Agron.*, 1981, **34**, 73.

76 Reisenauer, H. M. *et al.*, *Soil Sci. Soc. Am. Proc.*, 1962, **26**, 23.

77 Jones, L. H. P., *J. Soil Sci.*, 1957, **8**, 313.

78 Hingston, F. J. *et al.*, *J. Soil Sci.*, 1972, **23**, 177.

79 Grigg, J. L., *New Zealand J. agric. Res.*, 1960, **3**, 69;   Trobisch, S. and Schilling, G., *Chem. d. Erde*, 1963, **23**, 91;   Karimian, N. and Cox, F. R., *Soil Sci. Soc. Am. J.*, 1978, **42**, 757.

80 Wells, N., *New Zealand J. Sci. Tech.*, 1956, **37**, 482;   Norrish, K. in *Trace Elements in Soil–Plant–Animal Systems* (eds D. J. D. Nicholas and A. R. Egan), Acad. Press, 1975.

81 Mitchell, R. L. and Burridge, J. C., *Phil. Trans. R. Soc.*, London, 1979, **B288**, 15. For a review of soil and fertilizer treatments see Anderson, A. J., *Adv. Agron.*, 1956, **8**, 164.

82 Scott, R. S. *et al.*, *New Zealand J. agric. Res.*, 1963, **6**, 538.

low because the molybdenum content of the plant may become so high that it reduces the availability of copper to animals, especially cattle and sheep. In some neutral or alkaline soils, forage plants may take up abnormally high amounts of molybdenum, sometimes reaching 120 μg g^{-1} dry weight without affecting the growth of the plant.[83] Such pastures are called 'teart'. The soils involved are usually poorly drained clays fairly high in organic matter such as those derived from the lower Lias clay in the west of England. In the western States of the USA, the soils producing forage plants with high molybdenum contents are similarly poorly drained and high in organic matter but they are usually clay loams formed in granite alluvium. Toxic effects of molybdenum in ruminants usually arise when the content in forage is in the range 10 to 20 μg g^{-1} dry weight but they have been found[84] with contents as low as 4 μg g^{-1}.

It may be more difficult to decrease the molybdenum contents of forage plants growing on soils with naturally high supplies than to correct deficiency. Lowering the pH is seldom practical because the soils are generally highly buffered, but as sulphate is known to reduce the uptake of molybdate,[85] applications of sulphate-containing fertilizer may help. If drainage and aeration are improved the uptake of molybdenum can be expected to decrease.[86]

A commonly used procedure for assessing available molybdenum in acid soils is extraction with acid ammonium oxalate buffered at pH 3,[87] a reagent originally suggested by Tamm for dissolving amorphous ferric and aluminium oxides. Anion exchange resins have been used with some success in assessing available molybdenum.[88]

Boron

The total boron content of soil[89] is of no agricultural significance since it is often accounted for by minerals of the tourmaline group, which are common in granites and other end members of the magmatic crystallization series, and

83 Kubota, J. and Allaway, W. H. in *Micronutrients in Agriculture* (eds J. J. Mortvedt, P. M. Giordano and W. L. Lindsay), Soil Sci. Soc. Am., Madison, Wisc., 1972.
84 Matrone, G. in *Trace Element Metabolism in Animals* (ed C. F. Mills), E & S Livingstone, 1970.
85 Reisenauer, H. M., *Soil Sci. Soc. Am. Proc.*, 1963, **27**, 553.
86 Mitchell, R. L. in *Trace Elements in Soils and Crops*, MAFF Tech. Bull., No. 21, HMSO, 1971; Kubota, J. *et al.*, *Soil Sci. Soc. Am. Proc.*, 1963, **27**, 679.
87 Reisenauer, H. M. in *Methods of Soil Analysis* (ed-in-chief C. A. Black), p. 1050, Monog. No. 9, Am. Soc. Agron., 1965 (1st edn), but see: Robinson, W. O. and Edgington, G., *Soil Sci.*, 1954, **77**, 237; Williams, C. H. and Moore, C. W. E., *Aust. J. agric. Res.*, 1952, **3**, 343; Kubota, J. and Cary, E. E. in *Methods of Soil Analysis* (ed. A. L. Page), p. 485, Monog. No. 9, Am. Soc. Agron., 1982 (2nd edn).
88 Bhella, H. S. and Dawson, M. D., *Soil Sci. Soc. Am. Proc.*, 1972, **36**, 177; Jarrell, W. M. and Dawson, M. D., *Soil Sci. Soc. Am. J.*, 1978, **42**, 412.
89 Fleming, G. A. in *Applied Soil Trace Elements* (ed B. E. Davies), p. 155, Wiley, 1980.

are very resistant to weathering. The chemistry of boron in soil is still poorly understood. It is probably present in the soil solution as boric acid, H_3BO_3. This is a weak acid and its dissociation constant from the reaction

$$H_3BO_3 + H_2O \rightleftharpoons B(OH)_4^- + H^+$$

is about 10^{-9}. Below pH 7 the predominant form is H_3BO_3, but as the pH increases above 7 the concentration of $B(OH)_4^-$ increases. This ion is adsorbed by hydrous iron and aluminium oxides and clay minerals by ligand exchange[90] (see Ch. 7); hence adsorption increases at higher pH values, although competition with OH^- above pH 9 causes a decrease.[91] Of the clay minerals illite is the most reactive. There is evidence that some of the boron in soil is associated with organic matter,[92] and although it has been suggested that the reaction with organic matter is by condensation with diol groups,[93] there is not yet any direct proof for this reaction.

Boron is one of the micronutrients which most commonly limits the yield and quality of crops. Deficiencies have been reported in many root crops, such as swedes, turnips, sugar beet and, to a lesser extent, potatoes, legumes, including lucerne, red clover and soyabeans, and many fruit and forest trees. Warington[94] demonstrated a requirement for boron by beans and other species and her work was confirmed by Sommer and Lipman.[73]

Several factors control the availability of boron. Although it is adsorbed to an increasing extent as the pH is increased above 7, there is only weak adsorption at lower pH values. Boron is therefore leached fairly readily, resulting in deficiency in regions of high rainfall. Deficiency can be induced by application of high rates of lime to acid sands.[95] It is often pronounced in dry summers which might be due to low solution concentrations of boron in the subsoil.

The most common extractant for boron is boiling water[96] and it is agreed in many countries that concentrations <1 mg B kg^{-1} soil imply the likelihood of deficiencies in susceptible crops. It has been proposed[97] that extraction of soils with 0.01 M $CaCl_2$ + 0.05 M mannitol could be used to predict the response of wheat to boron in Australia.

90 Hingston, F. J., *Aust. J. Soil Res.*, 1964, **2**, 83; Bingham, F. T. and Page, A. L., *Soil Sci. Soc. Am. Proc.*, 1971, **35**, 892; Rhoades, J. D. *et al.*, *op. cit.*, 1970, **34**, 938; Singh, M., *Geoderma*, 1971, **5**, 209.
91 Keren, R. *et al.*, *Soil Sci. Soc. Am. J.*, 1981, **45**, 45; Mezuman, U. and Keren, R., *op. cit.*, 1981, **45**, 722.
92 Gupta, U. C., *Soil Sci. Soc. Am. Proc.*, 1968, **32**, 45; Ellis, B. G. and Knezek, B. D. in *Micronutrients in Agriculture* (eds J. L. Mortvedt, P. M. Giordano and W. L. Lindsay), p. 59, Soil Sci. Soc. Am., Madison, Wisc., 1972.
93 Parks, W. L. and White, J. L., *Soil Sci. Soc. Am. Proc.*, 1952, **16**, 298.
94 Warington, K., *Ann. Bot.*, 1923, **37**, 629.
95 For an example see Walsh, T. and Golden, J. D., *Trans. 2nd and 4th Comm., Int. Soc. Soil Sci.*, 1952, **2**, 167.
96 Berger, K. C. and Truog, E., *J. Am. Soc. Agron.*, 1940, **32**, 297.
97 Cartwright, B. *et al.*, *Aust. J. Soil Res.*, 1983, **21**, 321.

Boron

Boron deficiency can be corrected by applying a boron compound to the soil or foliar spraying. For deficient soils the usual method is to apply borax ($Na_2B_4O_7.10H_2O$) at rates of 10 to 20 kg ha^{-1} (1.2 to 2.4 kg ha^{-1} as B). To facilitate spreading of such small amounts the borax may be applied as a spray to the soil or, more usually, it is incorporated in a compound fertilizer. Borax is, however, only sparingly soluble in water and proprietary complex borates e.g. 'Solubur' containing 20 per cent B, are available which are more soluble and easier to apply either to the soil or crop. For foliar sprays, rates of 0.5 to 1.0 kg as B are used. Care must be exercised since the margin between deficiency and toxicity is narrow. Excess boron in irrigation water is troublesome in some areas (see Ch. 27).

Chlorine

Chlorine was shown to be essential for plants in 1954[98] but there is no evidence for field deficiencies. There might be adsorption in soils with positive charges, but in most soils the chlorine is present as Cl^- in the soil solution and is therefore readily leached. Near oceans the supply in rainwater is sufficient for crops, while in dry, continental interiors imperfect profile leaching should ensure that sufficient is retained in the soil.

Chemistry of other plant micronutrients in soil

Cobalt

Cobalt is essential for dinitrogen fixation by *Rhizobium* in legume nodules.[99] There is no evidence that higher plants have a direct requirement for cobalt. However, unexplained growth responses to the addition of cobalt have been observed in wheat and non-nodulated clover when grown in sterile nutrient solutions.[100] In practice, cobalt has greatest economic significance in relation to animal nutrition. Essentiality for ruminant animals was first established in Australia in 1935[101] and it was later shown that cobalt is a component of the vitamin B_{12} (cobalamine) molecule. The deficiency occurs in many parts of the world and is known by a variety of names such as 'coast disease' in

98 Broyer, T. C. *et al.*, *Pl. Physiol.*, 1954, **29**, 526.
99 Reisenauer, H. M., *Nature, Lond.*, 1960, **186**, 375.
100 Wilson, S. B. and Nicholas, D. J. D., *Phytochem.*, 1967, **6**, 1057.
101 Underwood, E. J. and Filmer, J. F., *Aust. Vet. J.*, 1935, **11**, 84.

southern Australia, 'pining' in Britain, 'bush-sickness' in New Zealand and 'salt sickness' in Florida.

Cobalt deficiency occurs on soils formed from acidic igneous rocks or arenaceous sediments which are low in total cobalt. Podzols may also be low in total cobalt because of leaching. In the soil solution cobalt is present as Co^{2+} and $Co(OH)^+$ and, like manganese, forms a weak complex with fulvic acid. Adsorption by goethite,[102] α-MnO_2[103] and soil increases with increasing pH. Availability decreases as pH rises and liming may therefore markedly reduce plant uptake. Cobalt is particularly associated with manganese oxides in soil. It has been demonstrated[104] that cobalt is specifically adsorbed by birnessite, lithiophorite and hollandite separated from soils, and by synthetic manganese oxides. Cobalt may replace Mn^{3+} in the oxide lattices. In a pot experiment when synthetic birnessite was added to soils the uptake of cobalt by subterranean clover was markedly reduced.[105] In Scotland, the availability of cobalt to plants is increased under conditions of poor soil drainage, and this has been ascribed to the reduction and dissolution of manganese hydroxides and oxides.

Uptake can be sufficient for legumes to fix dinitrogen, but the herbage may not contain enough cobalt to meet the requirements of ruminants. This requirement is for about 0.08 μg Co g^{-1} in the dry matter. Deficiency can be rectified by applying about 100 to 200 g ha^{-1} of Co as the sulphate, applied as a crop spray or fertilizer. Where cobalt deficiency occurs over a wide area, as in Australia and New Zealand, manufacturers incorporate cobalt sulphate in the fertilizer, for example, superphosphate. Alternatively, cobalt can be incorporated in salt licks.

Sodium

Sodium has been demonstrated as essential for the bladder saltbush *Atriplex vesicaria*,[106] an important fodder plant in the arid areas of Australia. The ubiquity of sodium in the environment required experimentation in filtered air in order to exclude dust before a requirement could be established. Deficient leaves contained 350 μg Na g^{-1} dry matter compared with 9100 μg g^{-1} in healthy leaves. Its interaction with potassium in the nutrition of sugar beet and other crops is discussed in Chapter 22. Although sodium is not required by pasture plants, their tissues frequently contain much less than is required by ruminant animals, and hence the common use of salt licks.

102 Forbes, E. A. *et al.*, *J. Soil Sci.*, 1976, **27**, 154.
103 Loganathan, P. *et al.*, *Soil Sci. Soc. Am. J.*, 1977, **41**, 57.
104 McKenzie, R. M., *Aust. J. Soil Res.*, 1967, **5**, 235; *ibid*, 1970, **8**, 97.
105 McKenzie, R. M., *Aust. J. Soil Res.*, 1978, **16**, 209.
106 Brownell, P. F. and Wood, J. G., *Nature, Lond.*, 1957, **179**, 635; Brownell, P. F., *Pl. Physiol.*, 1965, **40**, 460.

Nickel

Nickel is not yet accepted as an essential element for plants (Ch. 3). It has, however, been shown that when soyabeans were deprived of nickel, urea accumulated in toxic concentrations in the leaf tip leading to necrotic lesions.[107] This occurred regardless of whether the plants were supplied with inorganic nitrogen or were dependent on dinitrogen fixation. Nickel deprivation resulted in delayed nodulation and in a reduction of early growth. The condition was relieved by supplying 1 μg Ni l^{-1} nutrient solution. Nickel, like cobalt, substitutes for iron in ferromagnesian minerals. It is not known to what extent nickel is incorporated into clay minerals in normal soils but it is concentrated in hydrous oxides of iron and manganese.[108] The main concern with this element is the occasional report of nickel toxicity in some plants growing on soils developed from ultrabasic rocks. In studies based on serpentine soils in Aberdeenshire and on sand culture it has been shown that oats are particularly sensitive to nickel toxicity, and the symptoms are sufficiently characteristic for diagnostic purposes.[109] Liming alleviates nickel toxicity, an effect partly due to raising pH but largely to increasing the supply of calcium.[110] There is also evidence that high soil magnesium and low calcium are involved in causing the toxicity symptoms observed in plants growing over some serpentines. The results from centrifuged soil solutions derived from Scottish and Zimbabwean serpentine soils suggest, however, that the Mg : Ca ratio is less important than a high nickel concentration.[111]

Some animal micronutrients

Of the micronutrients required by animals, copper, molybdenum and cobalt have been mentioned above while discussing plant problems. In addition, there are several elements which are not required by plants but which are essential for farm animals and human beings, and some will be discussed next.

107 Eskew, D. L. *et al.*, *Science*, 1983, **222**, 621;   Dixon, N. E. *et al.*, *J. Am. Chem. Soc.*, 1975, **97**, 4131.
108 LeRiche, H. H. and Weir, A. H., *J. Soil Sci.*, 1963, **14**, 225;   Taylor, R. M. and McKenzie, R. M., *Aust. J. Soil Res.*, 1966, **4**, 29;   Norrish, K. (ref. 80).
109 Hewitt, E. J. in *Diagnosis of Mineral Disorders in Plants* (eds. C. Bould, E. J. Hewitt and P. Needham), HMSO, London, 1983.
110 Halstead, R. L., *Canad. J. Soil Sci.*, 1968, **48**, 301.
111 Proctor, J. *et al.*, *Nature, Lond.*, 1981, **294**, 245.

Chromium

It is now well established that chromium is required by animals for glucose metabolism, but an essential role in plant nutrition has not been demonstrated. Chromium accumulates with iron in weathering residues, and iron-rich Oxisols tend also to be high in chromium. The soil chemistry of chromium is poorly understood. It has been reported that when solutions of Cr(vi) as potassium chromate were added to soils of pH 4.7 to 7.4, the chromium was rapidly converted to insoluble forms which had properties of a mixed hydrous oxide of Cr(iii) and Fe(iii).[112] It has been suggested that the reduction depends on the presence of organic compounds, as electron donors.[113] Cr(iii) appears to be the stable form in the soil solution,[114] although present at very low concentrations. Thus, although Cr(vi) is much more toxic to plants than Cr(iii) it will be reduced to Cr(iii) after addition to aerobic soils. The rate of reduction is slower in alkaline than in acid soils, but even in alkaline soils Cr(vi) will be reduced to insoluble forms of Cr(iii) within one season.[112] The mechanisms involved in the uptake and translocation of chromium in plants are not understood, largely because of uncertainty about the ionic species present in different systems. There is, however, evidence that Cr(vi) is reduced to Cr(iii) between the root surface and the shoots and that irrespective of the form in which it is supplied most of the chromium is retained in the roots. In most studies it has been found that the levels of chromium in plants are less than 1 or 2 μg g^{-1} dry matter even when growth is reduced by toxicity, but certain species which have become adapted to soils developed on ultrabasic rocks accumulate and are tolerant to both chromium and nickel.[115]

Iodine

Iodine is needed to form the thyroid hormone thyroxine and deficiency causes the disease goitre. This is partly due to a lack of iodine in food and feeds but the condition is exacerbated by goitrogenic substances in plants especially brassicas. Soil contents range from traces to about 80 μg g^{-1} and Whitehead[116] has reported that British soils have iodine contents of between 2.7 and 36.9 μg g^{-1}. Iodine, as the iodide ion, can substitute for the hydroxyl ion in micas and in ferric hydroxides. Soils of high clay content and organic matter retain iodine, and it has been found that the adsorption of iodide is due in part to organic matter and in part to iron and aluminium oxides, with

112 Carey, E. E. *et al., J. Agric. Fd. Chem.*, 1977, **25**, 300, 305.
113 Grove, J. H. and Ellis, B. G., *Soil Sci. Soc. Am. J.*, 1980, **44**, 238.
114 Bartlett, R. J. and Kimble, J. M., *J. environ. Qual.*, 1976, **5**, 379.
115 Gambi, O. V. *et al., Acta Oecologia Oecon. Planta*, 1982, **3**, 291; Peterson, P. J., *Phil. Trans. R. Soc. Lond.*, 1979, **B288**, 169.
116 Whitehead, D. C., *Environ. Int.*, 1984, **10**, 321.

the oxides becoming increasingly important as soil pH decreases. In the soil solution iodide (I^- or I_3^-) is probably the predominant species in humid temperate areas, although iodate (IO_3^-) may be the predominant species in alkaline soils in arid areas.[89] Iodide is absorbed by plants more rapidly than iodate. The major source in soils may be oceanic salts rather than parent rock, and goitre is more prevalent where soils are very young or in areas remote from the sea.

Selenium

Selenium provides an example of a close relationship between the soil and the health of animals grazing on the pasture or range. It has not been found to be essential for plants but both toxicity and deficiency are now recognized in animals. In the USA in 1934 it was discovered that excessive concentrations of selenium in certain plants growing in some areas of the Plains and Rocky Mountain States caused a disease in animals, the symptoms being loss of hair and sloughing of hooves. The disease usually occurred when plants contained 5 μg Se g^{-1} dry matter or higher concentrations. Subsequently it was established that the soils producing such concentrations were formed from Permo-Cretaceous sedimentary rocks.

Some plants selectively accumulate selenium and contents as high as 0.9 per cent in the dry matter have been reported: these include about 24 species and varieties of the genus *Astragalus*, but native grasses and crops growing nearby usually contain less than 5 μg Se g^{-1}. The soils on which these native grasses and the selenium accumulators grow are alkaline and often contain $CaSO_4$ and $CaCO_3$.

In recent years, attention has switched from toxicity to deficiency which is probably of far greater economic significance. Selenium has now been recognized as an essential element for several animal species. In ruminants, a deficiency of dietary selenium results in degeneration of the heart and skeletal muscles ('white muscle disease'). This disorder is found if the selenium content of the feed is as low[117] as 0.02 to 0.04 μg g^{-1}. Because of the requirement for selenium by animals, considerable attention has been given to the distribution of soils and plants low in selenium in Britain and elsewhere.[118]

The conditions which control the chemical form of selenium in solution and in the solid phases of soil have been studied by Geering.[119] Depending on

117 Oldfield, J. E. in *Geochemistry and the Environment*, Vol. 1, p. 57, Natl. Acad. Sci., Washington, 1974. For a general account of selenium see Gissel-Nielsen, G. *et al.*, *Adv. Agron.*, 1984, **37**, 398.
118 Anderson, P. H. *et al.*, *Vet. Record*, 1979, 7 March, 235; Carlstrom, G. *et al.*, *Swedish J. agric. Res.*, 1979, **9**, 43.
119 Geering, H. R. *et al.*, *Soil Sci. Soc. Am. Proc.*, 1968, **32**, 35.

the pH and redox potential of the soil, it may exist in different oxidation states, namely: +6, selenate (SeO_4^{2-}); +4, selenite (SeO_3^{2-}); 0, elemental selenium (Se^o); and possibly −2, selenide (Se^{2-}). Added elemental selenium is relatively stable, but is oxidized very slowly by microbiological and non-biological processes at an increasing rate with increasing pH. In acid soils (pH 4.5 to 6.5) selenium is present predominantly as selenite (SeO_3^{2-}) adsorbed on hydrous ferric oxides and its concentration in solution is very low, whereas under alkaline, (pH 7.5 to 8.5) well-aerated conditions it is present as selenate (SeO_4^{2-}) and much more soluble and available to plants.

Considerable attention has been given to possible measures for adjusting the concentration of selenium in plants growing on acid soils with low contents. Additions of selenite to soils have been shown to be the best way of increasing uptake to adequate but non-toxic levels.[120] Because of its low solubility and availability the efficiency of selenite application is relatively low, but the application of selenate would involve the risk of producing plants containing toxic levels of selenium. Applications of selenite must, however, be carefully regulated because of the narrow margin between deficient and toxic levels in the plant.

Trace elements and environmental problems

It has been noted earlier that high concentrations of some micronutrients can adversely affect crops and animals. In addition, problems may occur where several non-essential trace elements are present in soil in elevated concentrations either from natural sources or through pollution; in recent years these elements have received attention because of their role in human health.

Much of the literature on toxic trace elements focuses on air and water pollution but foods can be an important source of these elements. In some countries the amount of these elements permitted in food is controlled by law and, for example, there are guidelines on permissible levels issued by the World Health Organization. Trace element pollution is therefore a subject of increasing importance to the soil scientist. In addition to elements considered earlier where toxicities were noted, mercury, arsenic, fluorine and lead can, under special circumstances, cause anxiety for public health because of high contents in or on plants. There is also concern about the accumulation of a range of metals in soil from the use of town sewage, and toxicities in the neighbourhood of mining and smelting operations.

120 Carey, E. E. and Allaway, W. H., *Agron. J.*, 1973, **65**, 922.

Mercury

Although inorganic mercury salts have long been known to be toxic, there has been increased and widespread concern following the deaths of many people at Minamata, Japan, due to the release of mercury in industrial effluent. The toxin was methyl mercury which formed in the sediments and subsequently accumulated in fish. This problem did not involve soils and plants.

The normal content of soil mercury is <100 ng g^{-1} but concentrations as high as 15 μg g^{-1} have been reported in polluted soils near lead mines.[121] The chlor-alkali industry, in which mercury is used in the electrolytic dissociation of brine, has been one of the most important sources of mercury pollution.[122] Coal combustion releases mercury to the environment,[123] and volcanic activity is also an important natural source of mercury emissions.

Mercury may be added to soil in sewage sludge. It has been used for several decades as organic and inorganic mercury compounds in pesticides used in agriculture, but the amounts used are relatively small. The fate of mercury in soils is complex as it may exist in three oxidation states, namely the elemental form Hg°, mercurous ion Hg_2^{2+}, and mercuric ion Hg^{2-}, depending on redox potential, pH and the presence of reactive groups. In aerobic soils, many added organic and inorganic compounds of mercury are converted to elemental mercury according to the equilibrium

$$Hg_2^{2+} \rightleftharpoons Hg^{2+} + Hg^{\circ}$$

The potential rate of volatilization of the elemental mercury is very much greater than the amounts normally applied in practice.[124] Cationic as well as anionic forms, e.g. $HgCl_3^-$ and $HgCl_4^{2-}$, can be adsorbed by hydrous oxides of iron and manganese and organic matter. The molecular species $HgCl_2$ and Hg_2Cl_4 and some mercury-containing pesticides may also be adsorbed. In anaerobic soils mercury might be precipitated as the very sparingly soluble sulphide, and it is possible that the soluble monomethyl or the insoluble dimethyl form, which is volatile, also forms. Plants take up ionic forms of mercury through their roots, but only a small proportion passes into the shoots and they may also absorb elemental mercury through their leaves.[125] In solution culture as little as 10 ng l^{-1} as Hg produces toxicity in plants.[126]

121 Davies, B. E., *Geoderma*, 1976, **16**, 183;   Warren, H. V., *Oikos*, 1969, **20**, 537;   Shacklette, H. T. *et al.*, *Mercury in the Environment*, US Geological Survey Circular, No. 644, Washington DC, 1971.

122 Lodenius, M. and Lodenius, E., *Tulisalo, Bull. environ. Contam. Toxicol.*, 1984, **32**, 439.

123 Crocket, A. B. and Kinnison, R. R., *Environ. Sci. Tech.*, 1979, **13**, 712.

124 Graham-Bryce, I. J. in *The Chemistry of Soil Processes* (eds D. J. Greenland and M. H. B. Hayes), p. 621, Wiley, 1981.

125 Lagerwerff, J. V. in *Micronutrients in Agriculture* (eds J. J. Mortvedt, P. M. Giordano and W. L. Lindsay), p. 593, Soil Sci. Soc. Am., Madison, Wisc., 1972;   Beanford, W. *et al.*, *Physiol. Pl.*, 1977, **39**, 261.

126 Mhatre, G. N. and Chaphekar, S. B., *Water, Air Soil Pollut.*, 1984, **21**, 1.

Arsenic

Arsenic is listed as a micronutrient required by animals, but while some arsenical compounds may be added to poultry and other animal diets for antibiotic purposes, it has long been known to be toxic to animals and human beings if ingested in large amounts. Normal soil concentrations are usually in the range 5–50 μg g^{-1}. Arsenic is associated with copper in sulphide ores, for example as a sulpho-arsenide, $Cu_3As_4S_4$. High concentrations may therefore be found in mine waste near old copper mines where grasses, e.g. *Agrostis tenuis* may have evolved an arsenic tolerance.[127] Soil arsenic concentrations may also be raised near copper smelters. Arsenic compounds, including lead and calcium arsenates and copper aceto-arsenite, have been extensively used as pesticides, and after many years of application arsenic may accumulate in soils to levels that reduce crop yields.

In soils arsenic is present as arsenate, AsO_4^{2-}, and there is evidence that its behaviour resembles the behaviour of phosphate. For example, arsenate is adsorbed by ligand exchange on hydrous iron and aluminium oxides.[128] Studies of soil based on electron microprobe analysis have shown a very close association of arsenic with goethite.[129] The retention in soils of arsenic by ferric hydroxide is consistent with the well-known use of ferric hydroxide as an antidote in cases of arsenic poisoning. Calcium arsenates may be precipitated in calcareous soils, and in anaerobic soils arsenate may be reduced to arsenites and possibly arsine, AsH_3.

As there are no recorded instances where naturally occurring levels of arsenic are the cause of reduced crop yields and the use of arsenic in pesticides is decreasing, the main agronomic problem is to reduce uptake from contaminated soils. Some applications of ferrous sulphate, aluminium sulphate or lime have been found to improve crop yields on contaminated soils. However, in view of the similar behaviour of arsenate and phosphate it may be difficult to develop soil treatments that will reduce the availability of arsenate without causing a concomitant decrease in the availability of phosphate.

Fluorine

Fluorine is not considered to be essential for plants but it has long been known to have beneficial effects on teeth and bone structure, and Schwarz has demonstrated its essentiality for animals.[130] High levels of fluoride can be

127 Porter, E. K. and Peterson, P. J., *Environ. Pollut.*, 1977, **14**, 255.
128 Hingston, F. J. *et al.*, *Adv. Chem. Ser.*, 1968, **70**, 82; *Disc. Faraday Soc.*, 1971, **52**, 334.
129 Norrish, K. (see ref. 80).
130 Schwarz, K., *Federation Proc.*, 1974, **33**, 1748.

toxic to both plants and animals, and toxicity due to the deposition of air-borne fumes of hydrogen fluoride and particulate fluorides on the foliage of plants has been an important practical problem.

Fluorine contents of soil (Table 23.1) are variable. Fluoride ions partially replace hydroxyl ions in minerals such as mica so that some granitic rocks have high contents of fluorine. Volcanic ash fall-out can lead to a localized fluoride enrichment.[131] Fluoride pollution may arise near brickworks and aluminium smelters, and mine tailings can contain very high total fluoride concentrations when fluorspar, CaF_2, is present, although water-soluble fluoride levels are very low. In Derbyshire, England, mine waste contained[132] up to 17 per cent total fluorine but only 5.1 μg water-soluble F g^{-1}. Rock phosphates may contain fluorapatite, and usually contain some fluorine: phosphoritic limestones may contain 3 per cent F. Superphosphate, which is manufactured from rock phosphate, has been reported as containing as much as 2 per cent F in the USA.

Although soils may contain appreciable amounts of fluoride it is not taken up by plants in amounts that are toxic to animals. In limed soils or those with adequate calcium for normal plant growth, any added fluoride reacts with calcium to form CaF_2 and with other soil constituents, e.g. aluminium oxides and clays, and is mostly unavailable to plants. The substantial amounts of fluoride added in phosphatic fertilizers to soils have not significantly increased the fluoride content of plants grown on these soils. Apparently the concentration of fluoride in solution in soils is controlled by calcium fluoride and aluminium silico-fluoride, both of which are insoluble, and possibly by fluorapatite. The fluoride ion is a strong ligand and replaces hydroxyl ions on the surface of soil minerals.[133] Most plants contain 0.10–10 μg g^{-1} dry matter, but the common tea plant *Camellia sinesis* contains elevated levels, sometimes reaching 800 μg g^{-1}.

Lead

The most general cause of the contamination of soil with lead is from the combustion of petrol containing tetramethyl and tetraethyl lead, the product of combustion being particulate lead bromochloride (PbBrCl). After emission, the very small particles are dispersed widely by the wind while coarser particles are deposited on soil and plant surfaces within 30 to 50 m of the road. The amounts of lead deposited decline exponentially with distance. There is evidence that the lead bromochloride is converted mainly to lead

131 Taves, D. R., *Science*, 1980, **210**, 1352.
132 Cooke, J. A. *et al.*, *Environ. Pollut.*, 1976, **11**, 9.
133 Hingston, F. J. *et al.*, *J. Soil Sci.*, 1972, **23**, 177. For solution relationships of fluoride minerals, see Elrashidi, M. A. and Lindsay, W. L., *Soil Sci. Soc. Am. J.*, 1985, **49**, 1133.

sulphate and occasionally to lead oxide (PbO) or lead sulphide (PbS), and it has been suggested that these conversions occur either during transport or after deposition on the soil.[134] The solubility of lead sulphate is extremely low and, although plant uptake may increase slightly as a result of deposition on the soil, most of the additional supply is retained in the roots.[135] Only about one-half of the lead deposited on the leaves can be washed off by rain, or removed by washing when vegetable foods are prepared in the kitchen.[136] The problem of ingesting food contaminated by particulate lead becomes more serious near or in towns and cities where both the road network and traffic are denser than in open country. Although market garden crops may be affected to some extent, the contamination of vegetables grown in urban gardens or municipal allotments is of much greater concern.[137]

It has been shown with several species that a large proportion of the lead taken up from solution culture is associated with the roots.[138] Liming a soil decreases uptake and studies in which air-borne lead is not a factor generally show that additions of lead to soils, whether as soluble salts or in sewage, do not appreciably increase uptake. Lead is retained in soils by adsorption on hydrous oxides, notably ferric hydroxide, and adsorption increases sharply with increasing pH. In calcareous soils lead may be precipitated as the carbonate. Retention in soils may be also partly due to organic matter.

Heavy metals from mineralization, mining and smelting

High concentrations of trace elements in soils and plants are used by geologists to prospect for metal ores and much of our knowledge of the rarer elements in soil comes from geochemical research. Metal ore bodies in the rocks below the soil profile contribute to its parent material and thereby raise the soil trace element content in their immediate vicinity. The resulting anomaly can be identified and located by sampling and chemical and statistical analysis. But the major impact on the soil arises when ores are mined, processed and smelted or refined. Losses of elements to the environment are possible at all stages of processing between separating ore from its host rock to the final product. The agencies of dispersal are air, water and gravity.

Atmospheric dispersal is in smelter smoke and similar emissions. Particulates drop to the ground at a rate determined by their particle sizes and densities and gaseous emissions are washed out by rain. The consequence is that

134 Olson, K. W. and Skogerboe, R. K., *Environ. Sci. Tech.*, 1975, **9**, 227.
135 Jones, L. H. P. *et al.*, *Pl. Soil*, 1973, **38**, 403.
136 Davies, B. E. and Holmes, P. L., *J. agric. Sci.*, 1972, **79**, 479.
137 Davies, B. E. *et al.*, *J. agric. Sci.*, 1979, **93**, 749.
138 Broyer, T. C. *et al.*, *Pl. Soil*, 1972, **36**, 301; see also ref. 135.

smelters are usually surrounded by an area of soil with high concentrations of trace metals which decrease exponentially away from the stack. Some modern smelters are in remote rural areas where forest ecosystems are polluted but others are located in farmland and near major cities. Soil samples collected along transects from the modern lead–zinc smelter at Avonmouth, near Bristol, England, contained 600 μg Pb g^{-1}, and 32 μg Cd g^{-1} within 0.3 km of the works. At 7 km the elemental concentrations were 192 μg Pb g^{-1} and 3.6 μg Cd g^{-1} and background levels had not been reached at 14 km.[139]

Running water spreads pollutants either as solutes or in the suspended sediment load, and they may be deposited on the floodplain. The quality of effluent water from mines and processing plant is now strictly controlled but this was not always so and floodplain soils in the old lead mining areas of Britain are still contaminated by lead, zinc, copper, cadmium and other metals.[140]

Gravitational transport, together with air and water erosion, leads to contamination of agricultural land downslope of old mine waste heaps.

The three processes complement and reinforce each other so that soil contamination in metal mining districts can be widespread and soil lead and zinc concentrations as high as 20 000 μg g^{-1} and cadmium contents up to 1000 μg g^{-1} have been recorded.[141]

Sewage sludge

Sewage wastes, both solid and liquid, are increasingly applied to agricultural land. They contain heavy metals in varying concentrations (Table 23.7) depending on the nature of the local industry and the proportion of domestic and industrial wastes. The solid waste, or sludge, is sometimes regularly applied at 25 t dry matter ha^{-1}. As a result, the constituent heavy metals will accumulate in the soil.[142] Among the metals, chromium, iron, lead and mercury pose relatively little hazard to plants or animals as they are converted to forms of very low solubility and availability to plants. Manganese may constitute a problem when added to acid soils in which the increased level of soluble manganese could cause phytotoxicity. Although

139 Burkitt, A. *et al.*, *Nature, Lond.*, 1972, **238**, 327. The *Wolfson Geochemical Atlas of England and Wales*, Oxford Univ. Press, 1978, gives regional distribution of 19 elements.
140 Davies, B. E. in *Applied Environmental Geochemistry* (ed I. Thornton), Acad. Press, 1983.
141 Colbourn, P. and Thornton, I., *J. Soil Sci.*, 1978, **29**, 513; Davies, B. E. and Roberts, L. J., *Water, Air Soil Pollut.*, 1978, **9**, 507.
142 For an account of the chemistry of heavy metals in soil from sewage and other sources see Jones, L. H. P. and Jarvis, S. C. in *The Chemistry of Soil Processes* (eds D. J. Greenland and M. H. B. Hayes), p. 593, Wiley, 1981.

Table 23.7 Concentrations of heavy metals (μg g^{-1} dry matter) in 42 sewage sludges from England and Wales. (From Berrow, M. L. and Webber, J., *J. Sci. Fd Agric.*, 1972, **23**, 93.)

Metal	Median	Range
Cd	—	<60–1 500
Co	12	2–260
Cr	250	40–8 800
Cu	800	200–8 000
Fe	21 000	6 000–62 000
Mn	400	150–2 500
Mo	5	2–30
Ni	80	20–5 300
Pb	700	120–3 000
Sn	120	40–700
Zn	3 000	700–49 000

molydenum is not phytotoxic at elevated levels, repeated applications of sludges containing large amounts might cause problems in animal health if the soils are of high pH. The elements of particular concern are copper, zinc, nickel and cadmium. There are reports of toxicity in certain crops caused by applications of the first three of these elements (copper, zinc and nickel) when applied with sewage sludge, and it is considered that cadmium has accumulated in some plants to levels that may be hazardous to humans.[143]

Copper, zinc, nickel and cadmium ions are strongly bound by all except sandy soils, but they are taken up by plants in greater quantities from sludge-treated soils than from untreated soils. The amounts taken up might be excessive, depending on the crop species or cultivar, the soil and the composition of the sludge. Soils with high contents of adsorbents, including organic matter, and with high pH involve less risk than acid soils with low contents of adsorbents.

The behaviour of copper, zinc and nickel in soil has been discussed earlier. The chemical behaviour of cadmium in soils resembles that of zinc, but it may be more mobile in acid soils. Although cadmium is adsorbed by ligand exchange on goethite, its affinity for the surface is less than that of zinc.[102] As cadmium has a high affinity for calcium carbonate surfaces it has been suggested that chemisorption controls the concentration of cadmium in solution in calcareous soils.[144]

The uptake of cadmium by crop plants from a sludge-treated soil is usually substantially greater than from an untreated soil but phytotoxicity is apparently rare. Zinc, which is usually present in the largest amounts in sludge, and copper and nickel are responsible either singly or in combination for

143 Naylor, L. M. and Lochr, R. C., *Environ. Sci. Tech.*, 1981, **15**, 881.
144 McBride, M. B., *Soil Sci. Soc. Am. J.*, 1980, **44**, 26.

phytotoxic effects in crops. Their relative toxicity differs with the crop species and soil but it has been suggested that copper is twice as toxic as the same weight of zinc and nickel is eight times as toxic as zinc. This has been the basis of the formulation called the 'zinc equivalent' which is used to define the maximum amounts of sludge which may be safely applied, provided the pH of the soil is not less than 6.5.[142]

If the elements have accumulated in the soil in phytotoxic amounts there are few management options. One is to increase the soil pH by liming which reduces the solubility and availability of the heavy metals. Another option is to choose a crop tolerant of high metal concentration. To minimize the health hazard a cereal rather than a vegetable crop should be grown because the metals tend to be present in lower concentrations in fruits and seeds than in leaves. For example, it has been reported that the ratio of the concentration in maize leaves to the concentration in the grain was about 20 : 1 for lead, 10 : 1 for cadmium, chromium and mercury and 1 : 1 for copper and nickel.[145]

145 Hinesly, T. D. *et al.*, *Compost Sci.*, 1972, **13**, 26.

24

Soil survey, soil classification and land evaluation

Soil varies across the earth's surface. The variation occurs both laterally and vertically and is a result of the complex interactions of the processes of soil formation (Ch. 5) and the environmental conditions, in particular climate, parent material and topography. These processes have continued over varying periods of time, with the consequence that while some soils may be millions of years old, well developed and in equilibrium with their environment, others showing incipient soil development might be only a few years old.

An important feature of the soil therefore is that it is not static, but instead is a dynamic natural body interacting with its environment, and as a consequence changing through time and in response to environmental changes. The result of these complex interactions is that the soil shows considerable variation at different times in the development of the landscape but, of far greater importance, it varies from place to place at any one time.

This might lead the observer to draw the conclusion that at no two places on the earth's surface are there two identical soils. While, at the extreme, this may be the case, it is more generally accepted that it is possible to group soils together into classes, within which many properties of the soils may be considered similar. The purpose of soil survey and soil classification is to group soils into these classes. Land evaluation seeks to take the classes of soil survey and soil classification and, together with information on other aspects of land, assess the potential of land for specific purposes.[1]

Purposes of soil survey and classification

The usual outcome of a soil survey is a map which distinguishes areas of broadly similar soils. Soils will be grouped together on the map if they have similar properties, but also, depending on the scale of the mapping, if they are close together. An area of land mapped as one soil, may therefore include soils with different properties. In contrast, soil classification endeavours to group together similar soils chiefly on the basis of their properties, their soil forming processes or, now less commonly their soil-forming factors. Ideally in a soil class there will be very few inclusions which do not readily satisfy most of the requirements of that class. It is important that users of soil surveys and systems of classification are aware of the possible differences which may exist between the classes distinguished, particularly where the same class names are used in both.

Both soil survey and soil classification simplify the complex distribution of soils by providing a framework within which inter-relationships can be established although their limitations should be recognised. An agricultural example is that soil classes might be related to nitrogen requirements of a crop; soils of Class A require x units of nitrogen per hectare to achieve a given wheat yield, soils of Class B require y units to achieve similar yields. Further, if the soil map portrays two areas of land which are predominantly Class A and Class B respectively, it will be possible to manage them differently to achieve optimum yields. It is probable, however, that within both areas of land there will be soils with different nitrogen requirements. If the farmer used only the soil map he would manage each area as if it were uniform.

Soil classification is not new, for in the twelfth century Yahya Ibn Mohammed, writing in his compendium entitled *The Book of Agriculture*,[2] noted that 'the first step in agriculture is in recognising the soil and knowing how to differentiate between the good and the poor one'. While this may have been the state of knowledge in the twelfth century, little new infor-

---

1 See for example Purnell, M. F. in *Proc. Conf. Classification and Management of Tropical Soils*, (K. T. Joseph, ed), Malaysia Soil Science Society, 1980.
2 Quoted by Johnson, W. M. in *Soil Resource Data for Agricultural Development*, p. 3 (ed L. D. Swindale), 1978.

mation became available for many years on the nature and distribution of soils, nor on the agricultural properties of the soils, their potential for producing crops or requirements for modification in order to improve or sustain yields. Much of our knowledge of soil distribution and soil–plant relationships has been gathered since the start of the nineteenth century.

Soil horizons

In many of the early studies the soil was considered to consist chiefly of the top few centimetres, little or no attention being paid to the underlying layers. As the knowledge concerning soils and soil development has increased, so has the awareness of the need to consider these underlying layers. The situation that prevails today is that we identify three broad groups of layers or horizons, labelled from the surface downwards, A, B and C, with a fourth group relating to the litter layers or organic layers at the surface. These horizons are described in Table 24.1.

Soil classification

It has already been indicated that there is an enormous diversity of soils and soil properties across the earth's surface. Given such complexity it is necessary to group soils into classes. A soil is placed in a class which accommodates a group of soils with properties within a defined range. As discussed below, the problem has usually been to decide on the properties that define the class, and indeed whether to base the classification on soil properties alone or to include environmental conditions, which then requires a theory of soil genesis. The advantage of a good classification scheme is that it allows relationships to be identified between individual soils and between classes of soils, which is of value in the organization of information.

This classification framework provides a basis for the prediction of individual properties and behaviour from a knowledge of the class characteristics. Thus, soil classification is essential as a means of communication among soil scientists. It is especially important when predicting the behaviour of soils, identifying their most appropriate use, estimating their productivity, and extrapolating the knowledge gained at one location to other often relatively little known locations.

While it is obvious that the organization of our knowledge by means of a soil classification is an important step forward in our understanding of the

Table 24.1   Nomenclature of soil horizons

1 Litter Layer and Organic Horizons

L Fresh litter in which there has been little alteration of the original plant structure

F Partially decomposed or comminuted litter remaining from earlier years, although in situations where decomposition takes place rapidly there may be a thin F layer for part of the year only. Some plant structure is visible

H Well-decomposed litter in which original plant structures cannot be seen. There may be some mixing with mineral matter

O Peaty horizons accumulated under wet conditions and saturated for at least 30 consecutive days in most years or have been artificially drained

2 Mineral Horizon

A Mineral soil horizon formed at or near the surface, characterized by incorporation of humified organic matter intimately associated with the mineral fraction

Ah : uncultivated A horizon
Ap : cultivated A horizon
Ag : gleyed A horizon

E Subsurface mineral horizon underlying the A horizon that is lighter in colour and contains less organic matter, sesquioxides and/or clay than the horizon below

Ea : the eluviated 'bleached' horizon of podzols (Spodosols)
Eb : lighter coloured horizon of *lessivé* soils
Eg : gleyed E horizon

B Mineral horizons differentiated from adjacent horizons by colour and structure. It usually underlies an A or E horizon and is either a zone of alteration, a zone of illuvial concentration of clay, iron, aluminium, or humus, singly or in combination, or a zone from which soluble materials (e.g. carbonates) are removed

Bfe : sharply defined iron pan
Bs : sesquioxide enriched, often ochreous coloured
Bt : B horizon enriched with translocated clay
Bw : B horizon showing evidence of alteration by weathering, and leaching, or reorganization *in situ*
Bg : gleyed B horizon

C Unconsolidated or weakly consolidated mineral horizon which retains evidence of rock structure and lacks features which would result from pedological alteration

soil, there remains the often problematical decision of which properties of the soil are to be used to determine and identify the soil classes. As our knowledge of soils has increased, so has the range of properties upon which we might base a soil classification.

Initially the soil classes were based on qualitative assessments or, at best, broad quantitative categories. For example in the late eighteenth and nineteenth centuries several county agricultural surveys were undertaken in England under the auspices of the Agricultural Boards which often included a simple map and classification of the soils. The survey in Norfolk was typical[3]; the soil classes were broad and related to the texture of the surface soil and its ease of cultivation. For example Young distinguished light sand, rich loam and various loams. The Norfolk survey had only five soil classes, but it is perhaps worth noting the broad correspondence between the groups he distinguished, their boundaries on the accompanying map, and the broad groups and boundaries distinguished by the Soil Survey of England and Wales[4] when mapping this area. These early surveys by Young and others were important markers in the development of soil classification and soil survey, particularly so as they drew close relationships between the soil types and agricultural activities in the areas surveyed.

Early work in Russia

Probably the most important contribution to soil classification, and to a degree soil survey, is the work in Russia of Dokuchaiev and co-workers in the second half of the nineteenth century.[5] Dokuchaiev produced a framework for the classification of soils, based upon the notion that the soil was a dynamic natural body interacting with its environment. He drew upon these interactions as a basis for a classification of soils in which he identified important soil-forming factors which he linked genetically to soil distribution.

In Dokuchaiev's framework climate was considered the dominant soil-forming factor. He identified broad relationships between the latitudinal climatic belts of the continental land masses and soil belts, and implicitly the corresponding broad vegetational belts. Soils in this climatic group he described as Normal, but they are more commonly known as Zonal soils. He identified a second class of soils where the climatic influence on soil development was considered subsidiary to that of one of the other soil-forming factors, for example where parent material was the dominant soil-forming

---

3 Young, A., *General View of the Agriculture of the County of Norfolk*, 1804.

4 Hodge, C. A. H. *et al.*, *Soils of Eastern England*, Soil Survey of England and Wales, 1984.

5 Sibertsiev, N. reproduced in Finkl, C. W., *Benchmark Papers in Soil Science – Soil classification*, p. 15, 1982, discusses the early work of Dokuchaiev and others in Russia in the late nineteenth century.

factor, as with many calcareous parent materials. This class of soils Doku-chaiev described as Transitional, more widely known as Intrazonal. A third class of soils was identified where soil development was restricted either through insufficient time or where some terrestrial process such as inundation by a river or the sea had interrupted the process. This class of soils described originally as Abnormal is more widely known as Azonal. Thus at the highest level of the classification there were three broad classes, Zonal, Intrazonal and Azonal. These classes were subdivided, for example within the Zonal soil order there were subdivisions by climate into suborders, including tundra soils, and steppe soils.

With modifications this framework provided the basis for most national soil classifications throughout the world in the first half of the twentieth century. There were of course many local classifications of a few fields or tens of square kilometres, where the soils might be classified on locally relevant properties, such as the texture and tilth of the Norfolk survey of 1804.

Development in USA

In the United States of America, USDA produced a classification in 1938 which clearly had many similarities with earlier zonal soil classifications.[6] At the highest level there was the same division into Zonal, Intrazonal and Azonal. The second level of this system with nine suborders identified broad climatic groupings such as soils of the Cold Zone within the Zonal soil order, and soils influenced by other soil-forming factors, e.g. calcimorphic soils within the Intrazonal order. No suborder was identified within the Azonal order. The third level of this system, comprising 36 great groups, was the major part of this system which provided a 'gallery of conceptual classes'.

In essence these were complex modal concepts comprising soil profile, soil environments and pedogenetic attributes. The names of these great soil groups were the names widely used throughout much of the middle of the twentieth century by soil scientists, and are names which still are perhaps most familiar to a broad spectrum of soil scientists, names such as Yellow Podzolic soils, Chernozem soils and Rendzina. In many respects this was a classification of environments rather than of soils, and it was very difficult to allocate soils to classes or distinguish different classes without substantial environmental information.

6 Baldwin, M. *et al.*, in *Soils and Men*, Yearbook of Agriculture 1938, p. 979, USDA, 1938.

Revision of USDA system leading to Soil Taxonomy

Following the publication of this 1938 USDA system of soil classification there developed a degree of dissatisfaction with many of the national soil classification schemes. Dissatisfaction was wide ranging but included such criticisms as the lack of precision in class boundaries,[7] the varied criteria used in classification including environmental factors, morphology and soil-forming processes. The 1938 classification and others which followed[8] were defined only by their central concepts. Little attention was given to class limits nor to the problems of allocation when soils lay distant from these central concepts. In particular, users of these classifications often found it difficult to allocate a new soil to a class.

The 1938 system was revised over the following 20 years[9] and by 1960, partly as a result of criticisms contained in these revisions and partly as a result of the developments in the methodology of soil classification, USDA produced a comprehensive soil classification, widely known as the Seventh Approximation.[10] This system which arose in part from published revisions and partly from a series of consultations, was extensively used for soil survey and classification in the USA by 1965 and was itself superseded by the publication of *Soil Taxonomy*.[11] Both the 1960 and 1975 versions of the classification have six categorical levels and are strongly hierarchical. *Soil Taxonomy* comprises a hierarchy of classes, a series of class definitions with clearly stated numerical limits to most, a defined language of descriptive criteria and a new vocabulary for the class names. With the development of a number of soil keys for each of the Orders it is possible to allocate an unknown soil entity to its proper class.[12]

The six levels of the classification are Order, Suborder, Great Group, Subgroup, Family and Series. At the highest level the ten Orders are each assigned a new name, discarding earlier more familiar names; for example many soils previously identified as Podzols would be allocated in the new system to the order Spodosols. The 10 Orders are:

1 Entisols   Weakly developed mineral soils
2 Vertisols   Cracking clay soils
3 Inceptisols Moderately developed soils of humid regions

7 Leeper, G. W., The classification of soils: an Australian approach, *Trans. 5th Int. Congr. Soil Sci.*, 1954, **4**, 217.
8 See for example Stephens, C. G., *Manual of Australian Soils*, CSIRO, 1953.
9 Among the several authoritative comments during this time were: Riecken, F. F. and Smith, G. D., *Soil Sci.*, 1949, **67**, 107 who dealt with the lower levels; Thorp, J. and Smith, G. D., *Soil Sci.*, 1949, **67**, 117 who discussed the higher levels.
10 Soil Survey Staff, *Soil Classification: a comprehensive system*, 7th Approximation, USDA, 1960.
11 Soil Survey Staff, *Soil Taxonomy*, USDA Handbook, No. 436, 1975.
12 A number of flow diagrams have been produced by New Zealand Soil Bureau to assist in soil identification, e.g. Thomas, R. F., Blakemore, L. C. and Kinloch, D. I., *New Zealand Soil Sci. Rept.*, 1980, No. 39.

4	Aridisols	Soils of deserts and semi-deserts
5	Mollisols	Base-rich soils with organic (non-peaty) surface horizons
6	Spodosols	Podzols
7	Alfisols	Soils with clay accumulation in the B horizon and moderately high base saturation
8	Ultisols	Soils with clay accumulation in the B horizon and low base saturation
9	Oxisols	Well-leached soils of the humid tropics
10	Histosols	Soils rich in organic materials

Throughout the system much of the previous terminology is discarded, and allocation to classes is chiefly based on morphological or measurable properties, which, for the most part, are associated with the operation of soil-forming processes or are indicators of genetic factors. At the highest levels most divisions are based upon the presence or absence of diagnostic horizons. Six diagnostic surface horizons (surface horizons are called epipedons) are identified chiefly by the nature and amount of organic matter incorporated in the layer. The subsurface diagnostic horizons, of which there are 17, relate to the pedogenetic processes and are the horizons which would be expected to result from the operation of such a process, for example, the argillic horizon is an illuvial horizon in which clay has accumulated to a significant extent.

Soil Taxonomy was devised to assist in the making and interpreting of soil maps, yet this is its major drawback, and indeed of any other strictly hierarchical system in which classification descends from the general to the specific. Soil survey often consists of identifying and mapping soil series, the lowest level of the system. Ideally the definition of a soil series should be developed by the soil surveyor to fit the local conditions of profile variation and boundaries between adjacent units. If, however, a soil series is part of the fixed hierarchy, the range of variability within the unit cannot cross the boundaries established by the definition of classes higher in the system. Indeed, the definition of all the classes above it must apply to the soil series. A possible consequence of this structure is that the surveyor may disregard what he sees in the landscape and compromise by defining what he sees in terms of the classification. This may lead to inaccuracies and lack of precision at the series level, the level of the classification which we would expect to have the greatest precision.

Although there have been many criticisms of the system of soil classification presented in *Soil Taxonomy*, it is now widely used outside the United States of America, and it is therefore imperative that soil scientists become familiar with its structure and terminology. Smith writing during the early years of the Seventh Approximation[13] listed the advantages of the new system of classification as follows:

13 Smith, G. D., *Soil Sci.*, 1963, **96**, 6.

1　It permits classification of soils rather than soil-forming processes.
2　It focuses on the soil rather than on related sciences such as geology and climatology.
3　It permits the classification of soils of unknown genesis – only knowledge of soil properties is needed.
4　It permits greater uniformity of classification as applied by a large number of soil scientists; differences in interpretation of how soil was formed do not influence its classification under this scheme.

A number of soil scientists have, however, criticized the system.[14]

FAO system

Even with the widespread use of *Soil Taxonomy* beyond the boundaries of the USA, it has not become universally accepted as a framework for the classification of all soils. In 1974 FAO/UNESCO produced a legend[15] to accompany the 1 : 5 000 000 sheets of the *Soil Map of the World*. While in most publications the legend is specifically described as not being a classification, it does appear in many respects to be a soil classification. It is unique in that by the very nature of the project to which it is attached it is established in a global context. Because of the diversity of information available on a worldwide scale and the international compromise found to be necessary for worldwide agreement, the criteria used in the classification are of necessity very general.

The system has two categories of which the higher with 26 classes is roughly equivalent to the Suborders and Great Group levels in the US system (Table 24.2). Classes are defined by the use of diagnostic horizons, many of which are broadly similar to, although generally not as precisely defined, as those described in *Soil Taxonomy*. Terminology in the system is of mixed origin, and the main units have a mixture of 'traditional' names such as Podzol, Solonetz and Chernozem; names taken from the United States and Canadian systems[16] such as Gleysol, Vertisol and Histosol; and newly coined names such as Luvisol and Acrisol.

14　See for example Webster, R., *J. Soil Sci.*, 1968, **19**, 354;　Duchaufour, P., *J. Soil Sci.*, 1963, **14**, 149.
15　Dudal, R. *et al.*, *Soil Map of the World*, Vol. 1. *Legend*, FAO–UNESCO, 1974; Landon, J. R., *Booker Tropical Soil Manual*, Longman, 1984, provides a simplified flow diagram representation of FAO–UNESCO Soil Legend and a summary of the important features of the Legend.
16　Canada Soil Survey Committee, *The Canadian System of Soil Classification*, No. 1646, 1978.

Table 24.2 Soil classes used in FAO/UNESCO *Soil Map of the World*, with *Soil Taxonomy* equivalents

FAO classes	Brief description	USDA equivalent
Acrisols	Acid low base status soils. More strongly leached than Luvisols, but insufficiently leached for ferralsols. Tend to be reddish	Ultisols (part)
Andosols	Soils derived from recent volcanic deposits	Andepts
Arenosols	Coarse, weakly developed soils with an identifiable B horizon. Clay content <15%	Psamments (part)
Cambisols	Earths with Cambic B horizon	Inceptisols (part)
Chernozems	Black earths of the steppes	Borolls (part)
Ferralsols	Strongly weathered soils of the humid tropics, with Oxic horizon	Oxisols
Fluvisols	Recent alluvial soils	Fluvents
Gleysols	Hydromorphic soils	Aquic suborders (part)
Greyzems	Grey forest soils	Borolls (part)
Histosols	Organic soils; peats	Histosols
Kastanozem	Chestnut steppe soils – similar to Chernozem, but with shallower 'erdefication', and with carbonate/gypsum horizons	Ustolls
Lithosols	Soils of <10 cm depth over hard rock	Lithic subgroups
Luvisols	Soils having argillic B horizons, with high base status	Alfisols (part)
Nitosols	Tropical soils with argillic B horizon	Some Ultisols and Alfisols
Phaeozems	Prairie soils – paler than Chernozem. Chernozem–Kastanozem intergrade	Udolls (part)
Planosols	Soils with albic E with hydromorphic properties	No equivalent
Podzoluvisols	Soils intermediate between podzols and luvisols	Glossic. Great groups of Alfisols
Podzols	Soils with spodic B horizon	Spodosols (part)
Rankers	Shallow soils with umbric A horizon	Lithic Haplumbrets
Regosols	Weakly developed soils from unconsolidated materials	Orthents Psamments
Rendzina	Shallow calcareous soils with mollic A horizon	Rendolls
Solonchaks	Saline soils	Salic great group
Solonetz	Alkali soils with natric B horizon	Natric great groups
Vertisols	Dark, montmorillonite- rich, cracking soils	Vertisols
Xerosols	Semi-desert soils with weak ochric A horizon (0.5 to 1.0% OM)	Mollic Aridisol
Yermosols	Desert soils with very weak ochric A horizon (<0.5% OM)	Typic Aridisol

Other national systems

With some hundred years or more of varied national and international attempts at soil classification there is still no universally accepted system. The systems of FAO/UNESCO and USDA's *Soil Taxonomy* have a degree of worldwide acceptance and use, but many countries still consider it necessary to establish their own national systems. While there may be some common elements and broad similarities between the national and international schemes, there are often important and in some cases fundamental differences. For example in England and Wales and Denmark,[17] profile drainage is considered of primary importance at the highest level of the classification, whereas in *Soil Taxonomy* the moisture regime is chiefly incorporated at the suborder level or below. It is argued by the national soil survey organizations that such a difference in the relative importance of drainage is necessary, given the importance of within-profile waterlogging in the two countries and the substantial influence this has on agricultural activity. Attempts to cross-reference national and international soil classification schemes have been made with only a moderate degree of success.[18]

Criticism of classification systems

In addition to these differences of emphasis there have also been widespread criticisms of almost all soil classifications. In particular, *Soil Taxonomy* and its predecessor Seventh Approximation, which in many respects were a revolution in soil classification, have been the focus of much criticism. While one of the aims of the American classifications was to establish class boundaries precisely, some have considered such precision restrictive and detrimental to the smooth working of the classification.[19] Other criticisms have included paying too little attention to soil genesis,[20] while other workers have suggested too much attention is paid to soil genesis.[21] *Soil Taxonomy* and indeed the large proportion of modern soil classifications are strongly hierarchical in structure on the assumption that soils are best classified in a rigid hierarchical structure. The imposition of a hierarchical structure to soil classification has been widely criticized[22] but even where attempts were made to

17 England and Wales: Avery, B. W., *Soil Survey Tech. Monog.*, No. 14, 1980; Denmark: Madsen, H. B., *Pedologie*, 1983, **33**, 171.
18 See for example Ragg, J. M. and Clayden, B., *Soil Survey Technical Monog.*, No. 3, 1973.
19 Webster, R., *J. Soil Sci.*, 1968, **19**, 354, objected to the American classification on the basis that exactly defined mutually exclusive classes lead to inconsistency or absurdity.
20 See for example Duchaufour, P., *J. Soil Sci.*, 1963, **14**, 149; Raeside, J. D., *Bull. Int. Soc. Soil Sci.*, 1961, **19**, 20.
21 See for example Mulcahy, M. J. and Humphries, A. W., *Soils Fertil.*, 1967, **30**, 1; Webster, R., *J. Soil Sci.*, 1968, **19**, 354.
22 The problems of hierarchies are discussed by Avery, B. W., Systematics Association Publication No. 8, *The Soil Ecosystem*, p. 9, 1969.

propose non-hierarchical structures the final classifications were hierarchical.[23] Given these widely differing viewpoints on the relative importance of criteria for classification it seems likely that the goal of a universally accepted soil classification is unachievable. Writing in 1956 Leeper summarized the situation which, some 30 years on, still prevails when he stated: 'When scientists discuss methods of analysing a solution for traces of phosphate, they are practical, reasonable and unemotional. When the same men discuss the classification of soils these virtues are liable to evaporate.'[24]

Soil survey

Soil survey has been commonly described as the art and science of producing soil maps, the suggestion being that soil survey involves both subjective and objective components. The aim of soil survey often presented is 'to describe, map and classify the soils of an area'. A good soil survey must be both practical in its purpose and scientific in its construction. With the development of soil survey and survey techniques, and the demands of greater returns for the investment, users now often expect an interpretation of the soil and soil pattern. The specific nature of this interpretation and the constraints this may impose on the survey will be returned to below. It should be stressed, however, that soil surveys are not just inventories answering questions about 'what and how much?', but also indicate where soils occur and how they are associated.

To be of any value the survey must enable the user to make more precise statements about the mapped subdivisions than is possible about the region as a whole. The straightforward aims of 'describe, map and classify', highlight one of the major dilemmas facing the soil surveyor: the link between soil survey and soil classification. In many situations the surveyor may find himself attempting to map the soils of an area with his classification predetermined, yet the process of soil surveying and the production of a soil map is one in which the soil surveyor accumulates information as he proceeds. Based upon this accumulated information he will formulate his soil classes, which he will endeavour to delineate on his map. The pre-establishment of a classification may restrict the soil surveyor in his task and possibly lead to an inferior product.

23 Avery, B. W., *Trans. Proc. 9th Int. Congr. Soil Sci.*, 1968, **4**, 169, outlines the advantages of a non-hierarchical system, but the system produced by Avery for the Soil Survey of England and Wales is strongly hierarchical.
24 Leeper, G. W., *J. Soil Sci.*, **7**, 1956, 59.

Mapping units

The unit of classification is the taxonomic unit, within which soils with properties within a given range will be found. In most modern soil classifications allocation to a taxonomic unit is solely with reference to soil properties. It is different for a map unit. Two very different soils which occur close together in an intricate pattern may be placed together in the same map unit, although they may be identified as belonging to different taxonomic units. The problem is compounded because it is common practice to use the same terms, for example soil series in detailed soil surveys, to describe both the taxonomic and map unit. Much confusion has arisen among users of soil classification and soil survey because of the failure to distinguish clearly between the two types of unit. It is often the case that the units of classification do not have to be represented on a map: they are units resulting from comparisons of soils and can be described in a monograph.

The interface between classification and mapping poses problems when endeavouring to establish limits for the range of characteristics used to define the units. Taxonomically the range can be as narrow as desired, but very narrow ranges would create a large number of units and give rise to many difficulties in mapping them. To enable a viable linkage to be established between the classification and the survey, the limits must be *wide* enough to permit reasonable uniformity over an area of practical size, but additionally the limits must be *narrow* enough to keep the unit as taxonomically homogeneous as possible.

Once limits have been established, boundaries between soil series (or other units) may be delineated on a map. As the soil is a continuum, it is impracticable to draw exact boundaries between one soil and another. Usually each delineated area contains small segments of other soils. Where the pattern of soils is complex and there are two or more dominant soil classes, the unit will be mapped as a compound map unit or complex. There are no compound taxonomic units.

The aim of soil survey is therefore to provide a basis by which information about soil distribution can be communicated. This information will chiefly relate to soil classes, but may in more detailed soil surveys relate to particular soil properties. Soil survey must answer the following questions about soil classes if it is to be of value to its users:

1 What classes are present and in what proportions do they occur?
2 What is the soil class at any site of interest in the area?
3 Where can soil of a particular class be found in an area?

Routine soil survey is generally able to provide this information. Often, however, the users of soil survey require more specific information about soil properties, for example:

1   The properties of the soil classes and the proportion of the area occupied by soil classes with particular properties, or within particular ranges of one or more properties.
2   The properties of a soil at any site of interest in the area.
3   The locality of soil with given properties, or within a given range of a property or properties.

If soil surveys are examined few will be able to provide this information for more than a small number of soil properties.

General purpose surveys

Most soil surveys are general purpose, in that they aim to describe, map and classify profile classes. In general purpose survey, the surveyor will not know whether the user will use the soil map to identify soils suitable for wheat, a roadline, a route for a below-ground gas pipeline, or other purposes. If any natural soil classes exist, the surveyor is likely to define and map them with a minimum of arbitrary subdivisions. In general purpose survey the soil surveyor will aim to produce a map portraying natural classes from which he or an associate may make specific interpretations. This is done partly because surveys are undertaken for areas sufficiently large for there to be more than one kind of important land use and also many users of the land with varied interests and requirements. Only when soil surveys are produced for small areas at large scale is it likely that single property maps will be produced in which the surveyor may discard his natural subdivisions and use arbitrarily defined soil classes which relate to the single purpose at hand. Interpretation of the general purpose soil map and classification often depends on specific information being obtained outside the remit of the survey.

The precision of the information required from the survey depends primarily on the nature of the land use decisions to be made. Irrigation farmers require a soil map to show fine differences of texture, drainage and slope, and for management purposes the mapping units must have a high degree of uniformity. In contrast, those concerned with extensive stock grazing are unlikely to modify land management practices for any but large differences in soil units. They can make do with maps showing broadly defined soil units. Failure to recognize these contrasting demands has often led to mapping at uniform intensities, when intensity of survey should have varied with information needs. Choosing a compromise intensity for general purpose survey may satisfy the needs of no users.

Scale of survey

The scale at which the soil survey is undertaken and that of the final map

Table 24.3 Scale, nature and purpose of soil survey

Scale	1:500 000	1:250 000	1:100 000	1:50 000	1:25 000	1:10 000	1:5000	1: 2500
Nature	Reconnaissance		Semi-detailed		Detailed		Intensive	
Purpose		Resource Inventory		Project Inventory				
					Feasibility Study			
						Development Survey		
							Management Survey	

depends upon the time and funds available, the purpose of the survey, and the intricacy of the patterns to be represented. There is a hierarchy of soil maps and soil surveys, where mapping procedures and purpose are related to map scale. The methods used in the survey vary considerably depending upon scale, from reconnaissance surveys at scales below 1 : 100 000, where much use is made of air photograph interpretation and only limited use of ground survey, to intensive surveys at scales larger than 1 : 10 000 where many field observations are made, often in some form of grid or detailed traverse. The purpose of the surveys will also vary considerably with map scale. For example reconnaissance surveys at 1 : 100 000 or less often provide only general information of resource distribution; intensive surveys may provide information to suggest possible management practices, such as advice on amounts of irrigation water and fertilizer applications. Table 24.3 gives a general summary of survey scale, nature and purpose.[25]

Land evaluation

The soil survey and soil map provide information about the distribution and properties of the soils in an area; they do not show whether the soil can be successfully used for a particular management system, for example for irrigation. If such decisions are expected from a survey, the soil resource information is not the end point of the survey: some form of interpretation is required.[26]

Because soil is among the most stable attributes of the land, yet flexible in its response to man and offering the prospect of improvement, information from the soil survey is often used to evaluate alternative land uses or the establishment of a set of priorities. This interpretation and evaluation is likely to be based upon a wide range of factors of which those relating to soil are an important component. They may also include natural resource information on terrain, climate, geological formations and natural vegetation, socio-economic information including infrastructure and market availability, and information on the technology of the use of the resource. For example if the evaluation is with respect to agronomic potential of an area, the information required might include the range of crops which can be grown, yields likely to be achieved, fertilizer requirements and the need for irrigation.

Soil surveys provide the sole economically viable means of obtaining infor-

25 See Young, A., *Tropical Soils and Soil Survey*, Cambridge Univ. Press, 1976; *FAO Soils Bull.*, No. 42, 1979.
26 Woode, P. R., *Soil Surv. Land Eval.*, 1981, **1**, 1, discusses the need for more refined interpretation based on soil information in Zambia. Beek, K. J., *Soil Surv. Land Eval.*, 1981, **1**, 6, 18 gives a brief historical discussion of the development from soil survey interpretation to land evaluation.

mation about the soil component of land, and therefore form an important component in the overall process of land evaluation. Land evaluation may be single purpose, in which estimates of potential for one use are made, for example forestry; multi-purpose, where a number of single purpose evaluations are combined, or general purpose in which there is a comparison of the capabilities for various land use alternatives.

Suitability or capability?

In discussions of land evaluation reference is frequently made to capability and suitability. Capability refers to evaluations with respect to broad land use types, for example some of the most widely used land capability schemes refer to the agricultural capability of the land, they do not refer to specific crops or practices. In identifying land capable of supporting general agricultural use, appropriate uses for land not capable of supporting such practices may be suggested, for example it might have capability for forestry or recreation.

Suitability refers, in contrast, to one clearly defined, reasonably homogeneous purpose or practice. This may be a major kind of land use such as irrigated agriculture, rainfed agriculture, grazing of livestock, forestry, recreation or urban development; or a more specific land use such as rainfed arable farming of cowpea, irrigated rice production, or pulpwood production from *Acacia* trees. The suitability assessment generally identifies sites on the basis of the positive features associated with the success of the particular land use. In contrast, capability is generally somewhat less precise and is often defined in terms of the limitations which restrict or prevent the activities under consideration.

Land capability classification

Land capability classification has been found by many users to be a most useful derivative of routine soil survey. The approaches to land capability classification have their roots in the early part of the twentieth century, but were first formalized in 1961 by USDA.[27] The USDA scheme was intended as a means of grouping soil mapping units, taking into account other features of land, chiefly slope angle, climate and frequency of flooding. The main concept of the scheme is that of limitation, which includes permanent limitations, for example slope angle, soil depth and climate, and temporary limitations, which can be ameliorated by land management, such as nutrient status, and in some circumstances poor drainage.

27 Klingebiel, A. A. and Montgomery, P. H., *Land Capability Classification*, USDA Agriculture Handbook, No. 210, 1961.

The original USDA classification allocates land to one of eight classes on the basis of these limitations, and assumes a moderately high level of management. Land which is allocated to any class has the potential for the use specified for that class and for all classes below it. Capability classes I to IV are identified as suitable for arable agriculture with increasing limitations from I to IV. Classes V–VIII are identified as not suitable for arable agriculture. Apart from the assumption of a 'moderately high level of management' no other social or economic factors are considered in the evaluation.

In most cases where this system, or a modification of it, has been used there are two further levels of the classification: capability subclass and capability unit. The capability subclass is defined as land which has the same kind of limitation. In the original USDA scheme four kinds of limitation were recognized; e = erosion hazard, w = excess water, s = soil root zone limitations (e.g. depth or stoniness) and c = climatic limitations. Consequently subclass IIs consists of land placed in Class II on account of soil root zone limitations, which would include shallowness or stoniness within the rooting zone. In later modifications of the system the nature of the soil root limitation is identified.

The third level, a capability unit, is a group of soil map units that have the same potential, limitations and management responses. Units are shown in arabic numbers, e.g. IIs-1, IIs-2, etc. Soils within a capability unit can be used for the same crops, require similar conservation treatment and other management practices, and have comparable productive potential. This amalgamation of soil map units may lead to the combination of pedologically distinct soil map unit units into one capability unit which will require broadly similar soil management to sustain arable agriculture.

Use in Britain

The USDA *Land Capability Classification* is simple and easy to present. It has been found relatively simple to adapt it to any physical environment and to any level of farming technology, with the consequence that there are several land capability classifications in use which owe their origins to the USDA 1961 scheme. In the United Kingdom a similar system with seven classes (omitting class V of USDA) was introduced in 1969,[28] with classes 1 to 4 embracing land suitable for arable crops and other uses, classes 5 and 6 include land generally not suitable for arable crops but useful for grazing and forestry, and class 7 includes land not suitable for any of these uses. As with the original scheme, subclasses were identified on the basis of limiting factors, that is, soil (s), wetness (w), climate (c), gradient (g) and erosion (e).

Some of these land classes are illustrated in Plates 24.1 to 24.3. Plate 24.1a

28 Bibby, J. S. and Mackney, D., *Land Use Capability Classification*, Soil Survey Tech. Monog. No. 1, 1969.

(a)

(b)

Plate 24.1   Examples of high quality arable land:
(a) Class 1 land.
(b) Class 2 land, downgraded from class 1 because the soil is shallow. In the background are moderately steep slopes (class 4g) which restrict cultivations.

(a)

(b)

Plate 24.2 Examples of class 3 quality land.
(a) Coarse sandy soil in an area where summer droughts are common and where stones hinder cultivation.
(b) Improved and unimproved wet fine-textured soils. These are placed in class 3ws irrespective of differences in present land use.

Plate 24.3 Inherently low quality land. Fine-textured soils with poor drainage (class 4w) receive run-off and seepage water from the adjacent steep slopes (class 6g). The gently sloping upland surface has shallow stony soils (class 5c). West edge of the Longmynd, Salop.

illustrates class 1 land – level, well drained, of medium texture and well suited to a wide range of arable crops. Plate 24.1b illustrates otherwise good, level, well-drained arable land but downgraded because it has rather shallow soil over chalk, which limits its range of cropping. The slopes in the background are labelled 4g because they are too steep for cultivation. Plate 24.2a illustrates well-drained arable land downgraded because the soil is gravelly coarse sand. It is thus liable to drought during the summer and the stones seriously affect cultivations and cropping. Plate 24.2b illustrates pasture land on a poorly drained, fine textured soil, unsuitable for arable farming and graded 3ws; as the photograph shows, the grading is the same for the enclosed improved pasture as for the unenclosed rough grazing, that is, it does not depend on the level of farming being practised. Plate 24.3 is a photograph of an upland area in the English Midlands, where the bottom land is graded 4w because although it is gently undulating it consists of fine textured soils which are wet due to seepage and run-off water from the escarpment slope. The escarpment itself is too steep to justify improvement and is graded 6g. The gently sloping upland surface has shallow, stony soils, and because it is sufficiently above the level of the lowland to have cooler, wetter and more sunless climate and hence a shorter growing season, it is graded 5c.

The scheme has been widely used by the Soil Survey of England and Wales and the Soil Survey of Scotland. Many of the soil maps and accompanying memoirs having additionally a land use capability map and description of the map units.[29] A scheme was introduced in 1982 for use with the soil mapping programme then in progress in Scotland,[30] where interpretative maps of land capability for agriculture accompany soil maps at a scale of 1 : 250 000. The 1982 version of the classification incorporates advances in the methods of assessment, with more subdivisions of classes 3 to 6. The two divisions in classes 3 and 4 are based on increasing restrictions to arable cropping, principally climate, in particular the reliability of suitable weather conditions and interactions between soil properties and climatic features. Qualities of land such as workability and droughtiness are particularly affected. Class 5 has three subdivisions based on potential for successful reclamation and class 6 has three based on the value of existing vegetation for grazing purposes. While broadly similar to the 1961 USDA scheme, the modifications incorporated into the 1982 classification have produced a classification which is a good discriminator of agricultural land potential.

Productivity and parametric indices

The *Land Capability Classification* and related systems are category systems in which land is grouped into a small number of discrete, ranked categories according to the limiting values of a number of soil and site properties. In allocations of this sort there often arise problems at the class boundaries: for example 'is this land low class 2 or high class 3?' An alternative to this allocation procedure is to endeavour to place land on a continuous scale of assessment. If the information is available this might be done in terms of the likely yield of land as in the Productivity Index of USDA.[31] More

---

29 See for example the series of seven maps produced by the Soil Survey of Scotland in 1982 at 1 : 250 000. Maps of soil and land capability of agriculture were produced.

30 Bibby, J. S. *et al.*, *Land Capability Classification for Agriculture*, Soil Survey of Scotland, 1982.

31 See for example, Odell, R. T. and Oschwald, W. R., *Circ. Univ. Ill. Coll. Agric. Co-op. Ext. Serv.*, No. 1016, 1970.

recently USDA has suggested ranking land by a general Soil Potential Index taking into account not only the effective yield of a crop, but also indices relating to the costs of corrective measures or the costs due to continuing limitations.[32]

Other systems do not include yield but combine various land properties that are believed to influence yield in an arithmetical formula. This approach is generally described as parametric. There are various approaches involving additions, multiplications, or some more complex functions or combinations of factors. Inputs to the productivity ratings may be direct values, such as soil depth or the values can be rated on a scale, usually 0–100. A straightforward example of the multiplicative form of these equations is given by Riquier and co-workers[33]:

Productivity Index = H. D. P. T. N. (or S). O. A. M,
  where H = soil moisture, D = drainage, P = effective depth, T = texture/structure, N = base saturation, S = soluble salt concentration, O = organic matter content, A = cation exchange capacity/nature of clay, M = mineral reserves.

These indices have been found to have local rather than universal applicability, and while they may appear to offer an arithmetical objectivity, it must be remembered that a number of subjective if not arbitrary decisions have to be taken prior to the computation of the index. These decisions include the selection of soil and land characteristics, their ranking and scaling. The final result will vary, in some cases substantially, maybe even from high to low, depending upon the criteria selected and their ranking. A further criticism levelled at many productivity indices is that in their attempts to become more comprehensive and 'objective', the number of parameters incorporated has tended to increase. A consequence of this increase in the number of parameters is that the formulation of management practices based on the parameters also becomes increasingly complex and at times impossible.

A recent development in the sphere of parametric systems is that of the *Fertility Capability Classification*.[34] This system, to date only extensively applied in South America, is parametric. A wide range of qualities and characteristics is used as the basis for an additive parametric system. The system identifies three categories of parameters: 'Type' and 'Substrate Type' refer to the broad textural class of the topsoil and subsoil. A third level, 'Modifiers' of which there are 15, relate to specific soil properties which might have a substantial influence on crop performance. The system was developed to make information available when soils are mapped using *Soil Taxonomy*

---

32 *USDA Soil Potential Ratings*, National Soils Handbook Notice. No. 31, 1978.
33 Riquier, J. *et al., A New System of Soil Appraisal in Terms of Actual and Potential Productivity*, FAO, Rome, 1970.
34 Sanchez, P. A. *et al., Geoderma*, 1982, **27**, 283.

or the legend of the FAO–UNESCO *Soil Map of the World*. This system provides classes which indicate the main soil fertility constraints for crop production. Tests have shown it to be useful with a variety of crops and farming systems in South America and the USA.

Land suitability

Classifications of land according to capability have generally assumed a moderately high level of management. This assumption, and the feature that land is often classified on the negative principles of limitations, have led to widespread criticism of land capability classification. An important alternative to land capability is land suitability which is land evaluation for a particular purpose or set of purposes. These purposes may be major kinds of land use such as irrigated agriculture, forestry or recreation, or more specific land uses such as rainfed arable farming of maize. Suitability is assessed, classified and presented separately for each kind of use. This form of evaluation is not new; the compendium of Yahya Ibn Mohammed of the twelfth century, referred to earlier, discussed the identification of good and bad soil for particular crop cultivations.

The FAO framework for land evaluation

The *Framework for Land Evaluation*[35] provides a standard set of principles and concepts on which national or regional land evaluation systems can be constructed. The Framework assesses the performance of land when used for specific purposes. It involves the execution and interpretation of surveys and studies of landform, soils, vegetation, climate and other aspects of land in order to identify and make a comparison between alternative kinds of land use in locally applicable terms.

In essence the Framework aims to answer the following questions:

1   What uses of the land are physically possible?
2   What uses of the land are economically and socially relevant?
3   What changes are desirable and feasible?
4   What are the advantages and the comparative disadvantages of the present and potential uses?

The broad scope of the Framework allows a number of possible approaches to data collection and land evaluation. Three broad approaches widely used are: qualitative land suitability classification, quantitative physical evaluation, and economic land suitability classification. These vary in the manner in

35 *A Framework for Land Evaluation*, FAO Soils Bull., No. 32, 1976.

which the results are expressed and in the expression of the limits between land suitability classes.[36]

The FAO Framework provides a general basis for land evaluation at a variety of scales. For most projects it will be necessary to quantify the relationships with more detailed specifications. The type of survey will determine the nature of the evaluation; for example in reconnaissance surveys at scales less than 1 : 100 000 the concern will be with major kinds of land use, such as rainfed agriculture, irrigated agriculture, forestry and recreation. These surveys will, in general, be qualitative in nature and incorporate only broad scale background information on economic and social conditions. As the scale of survey increases so the inclusions of information such as capital and labour intensity, technological level, infrastructure, land tenure and size of holdings will also increase, and the final assessment of suitability will be more specific.

The Framework's structure is very flexible and reasonably compatible with other systems. There are four levels or categories of the classification: land suitability orders, classes, subclasses and units.

1 *Suitability orders* separate land into 'suitable' (S) and not suitable (N) for the use under consideration. In most cases the land is classed as not suitable for one of the following reasons: (i) proposed land use is technically impracticable (e.g. soil is too stony); (ii) proposed land use is economically unprofitable; (iii) proposed land use is environmentally unsuitable (e.g. land use would result in considerable soil loss). A third order, S_c, 'conditionally suitable', is separated as a 'phase' of the order suitable, where suitability is with respect to a restricted crop range, or where the land is considered only temporarily not suitable pending further study of the nature of the problem.

2 *Suitability classes* reflect the degrees of suitability within orders, and are numbered according to decreasing degrees of suitability. (If three classes are recognized within the order suitable, as is often the case, the following names are widely used: S1 highly suitable, S2 moderately suitable, S3 marginally suitable.) There is however no limit to the number of classes; the norm is three, and the generally agreed maximum is five. Within the 'not suitable' order, N1 is used to indicate currently not suitable, N2 permanently not suitable under the present land management system.

3 *Suitability subclasses* All classes with the exception of S1 may be subdivided into subclasses according to the nature of the major limitation or limitations, for example moisture deficiency or erosion hazard. The subclasses and limitations are indicated by lower case letters placed after the class symbol. There is no limit to the number of subclasses, but they should be kept to a minimum. The symbols used are not yet standardized,

---

36 See for example *Guidelines: land evaluation for rainfed agriculture, FAO Soils Bull.*, No. 52, 1983.

but there is some agreement on the use of 'm' for moisture deficiency, 'e' for erosion hazard, 'd' for drainage limitation, 'n' for soil nutrient availability, 'r' for rooting conditions, 's' for excess salts, 'x' for toxicities and 'f' for flood hazard.

4   *Suitability units* are subdivisions of subclasses that differ from each other in detailed aspects of their production characterization or management requirements. Suitability units are most use at the farm planning level where the differences identified would lead to implementation of different farming practices.

In the procedures of land evaluation, land utilization types form the core of the Framework.[37] A land utilization type consists of a set of technical specifications for a crop or a number of crops in a given physical, economic and social setting. This may be the current environment, or a future setting modified by major land improvements. Land utilization types are defined in terms of attributes which include: crops grown, market orientation, capital intensity, labour intensity, power sources, technical knowledge, technology employed, infrastructure requirements, size and configuration of land holdings, land tenure and income levels.

Major kinds of land use such as rainfed agriculture, or more specifically its major subdivisions: annual crops, perennial field crops, tree and shrub crops and wetland rice, are possibly of use at the general reconnaissance level of survey. Even at the reconnaissance scale, however, it is usually more useful to base land evaluation on land utilization types defined for individual crops or cropping systems under broadly specified management levels.

The process of assessment of land suitability for a particular crop or combination of crops involves matching the requirements of a particular land use activity with the properties (defined as land qualities)[38] of the particular land unit, leading to the first approximation of land suitability classes. It may then be necessary to consider the effects of proposed land improvement, or ways in which the land qualities can be favourably altered by means of capital works. Following this first stage, the next step is to pursue an economic and social analysis. The results of these two stages are then brought together in the land suitability classification.

The most common form of presentation of the suitability information is in the form of maps, in which the mapped land units are portrayed with land suitability types. The land suitability maps can be presented as a single map with a tabular legend, or as a series of suitability maps, one for each kind

37 See for example Beek, K. J., *The Concept of Land Utilization Types*, FAO Soils Bull., No. 22, 103, 1974;   Beek, K. J., *Land Utilization Types in Land Evaluation*, FAO Soils Bull., No. 29, 87, 1975.

38 A list of land qualities is given in *FAO, Soils Bull.*, No. 52 (ref. 36). It should be noted that there are similarities between some of these land qualities and the modifiers of the fertility capability classification of Sanchez, P. A. *et al.* (ref. 34).

Table 24.4 Example of tabular legend to a land suitability map[38]

Land units	Area (ha)	Land utilization types				
		A	B	C	D	E
		Improved traditional farming				
		Maize	*Sorghum*	*Rice*	*Mechanical cereal cultivation*	*Coffee intermediate input*
1 Abuna plains	17 000	S2w/D/S1	S1	N2m	S1	N2m
2 Bendo plains	1 500	S1	S1	N2m/I/S2	S2q	N2m
3 Canji flood plain	2 700	N1w/D/S2	N2w	S1	S3q	N2w
4 Dipithawa hills	9 900	S3e/T/S2	S3e/T/S2	N2m	N1q	S2m
5 Escoco hill	15 500	N2e	N2e	N2m	N2q	N1m

Land suitability subclass symbols: e = erosion hazard, m = moisture availability, w = oxygen availability, q = potential for mechanization.
Land improvements: D = subsoil drainage, I = irrigation, T = terracing.

of land use. In the first of these alternatives the legend is the key feature and would have the form shown in Table 24.4.

The legend gives information on the area of each land unit and the subclass of each land utilization type. In addition, where the suitability of land could be upgraded by specified land improvement, these are indicated and the resultant potential land suitability is also given.

The evaluation of land suitability offers an important contribution to programmes of land development, particularly in less developed areas. The development of the *Framework of Land Evaluation* is an important contribution to the establishment of broadly comparable schemes for many countries. The Framework might be considered as the skeleton of land evaluation, which allows additions to be made according to the requirements of the particular survey.

Agrotechnology transfer

In 1974 the Development Support Bureau of the US Agency for International Development established the Benchmark Soils Project (BSP) to investigate the feasibility of agrotechnology transfer within the tropics and subtropics based on the classification of soils using *Soil Taxonomy*.[39] Detailed soil survey was undertaken in the selected study areas and soils were classified on the basis of soil and climatic information at the family level. The aim was that once mapped and classified, the soil survey would act as a bridge that allowed the transfer of knowledge, gained by research or by experience of cultivators, from one place to all other places where it was applicable. The experience gained on one soil family in location A could be transferred to the same soil family in location B.

The study was undertaken on three soils at locations within the tropics and subtropics. While soil series would normally be identified in routine soil survey, the level of comparison chosen was the family. This, the second lowest taxon in the system, is narrowly defined and is contrived for the explicit purpose of providing groupings of soils that are reasonably homogeneous in characteristics important to the growth of plants. The rationale is that soils belonging to the same family should have essentially the same management requirements, analogous response to soil manipulation, and similar potential for crop production. The family thus stratifies the soil population into pragmatic groups. The three soil families studied were:

1   Thixotropic isothermic Hydric Dystrandept.
2   Clayey kaolinitic isohyperthermic Tropeptic Eutrustox.
3   Clayey kaolinitic isohyperthermic Typic Paleudult.

39 Beinroth, F. M. *et al.*, *Adv. Agron.*, 1980, **33**, 303.

The feature of the soil classification system is that these short descriptions give condensed statements of the accumulated knowledge about the unique characteristics and behaviour of the soils based on detailed soil survey. Experimental sites were established in six countries in the tropics and subtropics with three locations for Hydric Dystrandept, three locations for Tropeptic Eutrustox and two locations for Typic Paleudult. At each of the sites the hypothesis concerning the transfer of agrotechnology was investigated using on-site experimentation. The conclusion from the project was that it was possible to use the soil family category of *Soil Taxonomy* for the purpose of transferring agrotechnological information and experience.

Following BSP an alternative scheme has begun to be investigated in IBSNAT (International Benchmark Sites Network for Agrotechnology Transfer).[40] BSP was based to some degree on site experimentation, IBSNAT in part follows on from BSP but is endeavouring to investigate the transfer of agrotechnology by means of systems analysis and crop simulation.

If these projects of agrotechnology transfer prove to be successful, the stock of soil survey and soil classification will be raised substantially, as they will be confirmed as providing a basis for the extrapolation of detailed information on the suitability of soil and soil management practices for a wide range of agronomic activities.

40 See for example *Agrotechnology Transfer*, 1985, **1**, also IBSNAT Prog. Rept., 1982–85, Univ. Hawaii.

25

Soil acidity and alkalinity

The adjustment of the pH of acid soils to a value suitable for subsequent crop production should be one of the first steps in soil management, and the history of the management of soil acidity is a long and interesting one (Ch. 1). The extension of agricultural production into large areas of the tropics has brought into sharp focus the differences in management required for soils and crops in these regions compared with temperate regions.[1] Over recent years environmental concern regarding the acidifying effects of 'acid rain' and the effects of planting coniferous forests on areas previously under grass or deciduous woodland have also stimulated interest in the problems of soil acidity.

The worldwide distribution of acid and alkaline soils is closely related to climate and to parent material. Leaching leads to acid soils, but with decreasing rainfall first $CaCO_3$ and then Na_2CO_3 accumulate giving calcareous and saline sodic soils. This relationship is affected by the type of parent material and by the other soil-forming factors (Ch. 5).

1 Sanchez, P. A., *Properties and Management of Soils in the Tropics*, Wiley, 1976.

Soil pH

The pH of a solution is the negative logarithm of the hydrogen ion activity (pH $= -\log$ (H$^+$) where (H$^+$) $= \gamma$[H$^+$], γ is the activity coefficient of H$^+$ and [H$^+$] is the concentration of the ion in solution in moles per litre). The range of pH values found in soil solutions is between 2 and 10.5, i.e. the activity of hydrogen ions ranges between 10^{-2} and 3×10^{-11} mol l^{-1}.

Because hydrogen is present in aqueous solution as a cation, soil pH has to be considered in terms of exchangeable H$^+$ as well as solution H$^+$, and it does not have the precise meaning given above for a solution. The pH value of the soil solution decreases through the diffuse layer close to a negatively charged surface. For many purposes it is convenient to think of soil pH as the pH of the solution in the pores of a moist soil (away from the diffuse layer region) since this solution is in contact with the surfaces of roots, and many chemical reactions in the soil are governed by this solution. However, roots are also in contact with exchange surfaces, they affect the pH around them, and the weathering of clays is dependent on the presence of H$^+$ on the surfaces of the minerals. It is therefore accepted that soil pH is defined by the method used for its measurement, summarized as follows.

A glass electrode and calomel electrode, with the voltage difference between the two measured by a millivoltmeter, are first calibrated in buffer solutions of known pH. In order to make electrical contact in a soil, a paste or suspension is prepared. Conveniently for routine analysis a 1 : 2.5 (g soil : cm^3 water) suspension is used and the pH measured in the stirred suspension after a few minutes' mixing, although other standard procedures can be adopted. For some purposes it may be useful to mix the soil and water and then separate them by filtering or centrifuging before measuring the pH in the solution. This gives a precise measurement in the solution.

The suspension effect

The addition of water to a soil changes the concentration of H$^+$ in the soil solution and in an acid soil the pH rises. However, if the soil has a negative charge, on dilution more divalent ions in the system become adsorbed (see Ch. 7). If Ca^{2+} and H$^+$ are the only cations present, the ratio of Ca^{2+} : H$^+$ on the exchange surfaces will increase as water is added and H$^+$ will be desorbed into solution counteracting to some extent the dilution by water. The significance of the exchangeable H$^+$ can be seen if a soil suspension is allowed to settle and the pH measured in the supernatant solution and then in the settled soil paste: the pH of the latter is about 0.2 pH units lower.

In calcareous soils a further complication is present. The pH of these soils is governed by the reactions between calcium carbonate, water and CO$_2$, and addition of water causes CaCO$_3$ to dissolve. At first the pH rises as dissolution proceeds and then falls again as microbial respiration increases the concentration of CO_2 dissolved in the water. Therefore the pH of these soils as measured changes with time, and often less reproducible values are obtained than for acid soils. Where Na_2CO_3 is present and the soils hold exchangeable Na^+, addition of water causes Na^+ to dissociate from the surface with a consequent rise in pH (p. 897).

The salt effect

In an attempt to avoid these dilution problems, a dilute electrolyte solution (0.01 M $CaCl_2$ or 0.1 M KCl) is sometimes used to prepare the suspension. For acid or neutral soils, this causes a depression of pH by about 0.5 units compared to that in water (ΔpH is negative). The added cation displaces H^+ into the solution and the dilution effect is counteracted. The lower pH is closer to the value in the soil solution and therefore probably more representative of the soil's pH value. It should be noted that where a soil carries a net positive charge, the addition of salt to the soil suspension causes the pH to rise (ΔpH is positive). In this case the Cl^- displaces exchangeable OH^-, although the mechanism is not a simple exchange. The ΔpH value is then an indication of the net charge on the soil. For a soil at its point of zero charge, ΔpH $= 0$ (Ch. 7).

The addition of $CaCl_2$ to calcareous soils involves adding the 'common ion' Ca, which influences the dissolution of $CaCO_3$. A high Ca^{2+} concentration in solution means that a low carbonate concentration will develop, again leading to a lower pH value than in water. However, the extent of the lowering is unpredictable since equilibrium is not attained in the short times used for routine measurements.

The high pH of sodic soils is the result of Na_2CO_3 being present. $CaCO_3$ is also usually present in these soils and the addition of $CaCl_2$ to the suspensions will tend to cause the precipitation of $CaCO_3$ as the solubility product of $CaCO_3$ will be exceeded. There may therefore be a slight fall in pH. Since the routine analysis of these soils generally entails the preparation of a saturated soil paste using water, the measurement of pH is normally made in the paste, with the value depending primarily on the concentration of Na_2CO_3 present.

Usefulness of a soil pH value

Although soil pH can have no precise value nor unambiguous meaning, a knowledge of its value is useful, and it is often the first measurement to be made in soil analysis. Both in terms of plant nutrition and for understanding the chemical properties of soils the pH value is needed. Provided care is taken to use standard procedures for its measurement, comparisons between soils can be made with confidence. The standard conditions are quoted as a soil : solution ratio (weight : volume) and the electrolyte solution is described (e.g. 1 : 2.5 in 0.01 M $CaCl_2$).

The nature of soil acidity

The effects of soil pH on plant growth are partly through effects of pH as such on root function and partly through its effects on soil properties, the predominant characteristic of acid soils being the presence of aluminium in soluble and exchangeable forms.

Aluminium in soil solutions

Aluminium in soil solutions can most simply be considered to depend on (i) the solubility of gibbsite; (ii) the hydrolysis products (ion species) of Al in solution, both monomeric and polymerized; and (iii) the interaction of Al with soluble organic ligands. Figure 25.1 gives the theoretical activities of ion species in equilibrium with gibbsite.[2] It has been derived from the following relationships:

$$K_{sp} \text{ for gibbsite} = (Al^{3+}) (OH^-)^3 = 10^{-33.78}$$

For the first hydrolysis step

$$Al^{3+} + H_2O = AlOH^{2+} + H^+$$

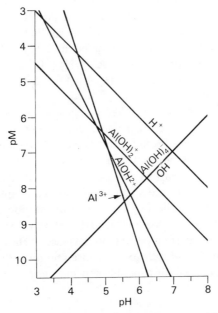

Fig. 25.1 Activities of aluminium species (M) in equilibrium with gibbsite. pM = −log activity of M in mol l^{-1}.

2 Smith, R. W. and Hem, J. D., *US. Geol. Surv. Water Supply*, Paper 1827D, 1, 1972.

which has a hydrolysis constant

$$K = \frac{(AlOH^{2+})\,(H^+)}{(Al^{3+})} = 10^{-5.0}$$

and there is good agreement regarding the value of the constant. At pH 5 $(AlOH^{2+}) = (Al^{3+})$, at pH 4 $(AlOH^{2+}) = 0.1(Al^{3+})$ and at pH 3 $(AlOH^{2+}) = 0.01(Al^{3+})$.

The second step, $Al^{3+} + 2H_2O = Al(OH)_2^+ + 2H^+$ has a constant

$$K = \frac{(Al(OH)_2^+)(H^+)^2}{(Al^{3+})} = 10^{-8.6}$$

At pH 5 $(Al(OH)_2^+) = 25\,(Al^{3+})$, at pH 4 $(Al(OH)_2^+) = 0.25\,(Al^{3+})$ and at pH 3 $(Al(OH)_2^+)) = 0.0025\,(Al^{3+})$. However, there is poor agreement regarding the hydrolysis constant partly because of difficulties in the measurement of the ion species, but mainly because of the tendency for polymeric ions to form slowly which makes equilibrium studies very difficult. In alkaline solutions $Al(OH)_4^-$ is present and the equilibrium reaction $Al(OH)_3 + H_2O = Al(OH)_4^- + H^+$, has an equilibrium constant $K = (Al(OH)_4^-)(H^+)$ with values between 10^{-13} and 10^{-14} depending on the form of the aluminium hydroxide.

Aluminium present in solution occurs in octahedral coordination with OH or H_2O. Thus Al^{3+} is present as $Al^{3+}\,6H_2O$ and $AlOH^{2+}$ as $AlOH^{2+}\,5H_2O$. Polymerization of these octahedra occurs in all but very acid solutions when the average number of hydroxyl ions bound to each Al ion rises above 1. The simplest example is the dimeric $Al_2(OH)_2(H_2O)_8^{4+}$ formed from two $AlOH(H_2O)_5^{2+}$ ions with two shared OH ions (Fig. 25.2).[3] This dimer carries two charges per Al ion.

More complex polymers form but their composition is not well established. One probable structure is an hexagonal ring of octahedral aluminiums $[Al_6(OH)_{15}(H_2O)_9]^{3+}$ linked through pairs of shared OH ions, but with extra OH in the structure. This polymer carries 0.5 charges per Al ion. Further linking together of these hexagons may occur as the structure moves towards that of gibbsite, and the polymers can be considered as intermediates in its formation. As the pH and the OH : Al ratio in solution rise, polymers of increasing complexity are formed. The OH : Al ratio within the polymer increases and the charge carried per Al ion decreases.[4] However, equilibrium may rarely occur and it is difficult to be precise regarding the nature of these polymers. The presence of silica in solution may add to the complexity of the system.[5]

3 Hem, J. D. and Roberson, C. F., *US Geol. Surv. Water Supply*, Paper 1827A, 1, 1967.
4 Hsu, P. H. in *Minerals in Soil Environment* (eds J. B. Dixon and S. B. Weed), p. 99, Soil Sci. Soc. Am., Wisc., 1977; Hsu, P. H. and Bates, T. F., *Min. Mag.*, 1964, **33**, 749.
5 Shin-Ichiro Wada and Koji Wada, *J. Soil Sci.*, 1980, **31**, 457.

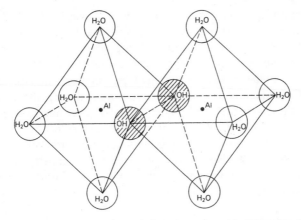

Fig. 25.2   Schematic representation of dimeric cation $Al_2(OH)_2(H_2O)_8^{4+}$.

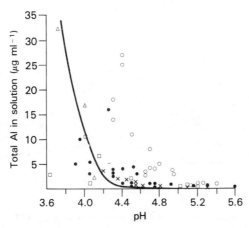

Fig. 25.3   The effect of pH on soil solution aluminium concentrations. The full line is the total solution aluminium in equilibrium with gibbsite at an ionic strength of 0.01.[6]

    x Two Scottish soils derived from basis igneous till and} acid soils
       river alluvium.[7]   } including
    ● Oxisols and Ultisols from Puerto Rico.[8]   } lime
    ○ Ultisols from southeast USA.[8]   } treatments
    □ Iron podzol on granitic till and an acid brown forest soil on glacial sand[9] including lime treatments.
    △ An upland soil from Sierra Leone,[10] pH decreases with horizon depth.

  6 See ref. 2.
  7 Bache, B. W. and Crooke, W. M., *Pl. Soil*, 1981, **61**, 365.
  8 Pearson, R. W., *Soil Acidity and Liming in the Humid Tropics, Cornell Int. agric. Bull.*, No. 30, 1975.
  9 Bache, B. W. and Sharp, G. S., *J. Soil Sci.*, 1976, **27**, 167.
10 Rhodes, E. R. and Lindsay, W. L., *J. Soil Sci.*, 1978, **29**, 324.

The significance of the above principles is shown in Fig. 25.3 where the total concentration of Al in soil solutions is related to pH for a range of soils. The trend is similar to that expected for gibbsite although most of the solutions contain more Al than predicted except at low pH.[11]

A similar solubility line is obtained for kaolinite in equilibrium with quartz.[12] Consequently there is disagreement regarding the dominance of gibbsite in these systems[13] with mineralogical and organic matter differences being important. The significance of organic Al complexes in these solutions is not known. Polymerized Al may be a significant fraction of the total in solution in some soils (Fig. 25.4),[14] although S. C. Jarvis (priv. comm.) has failed to find significant amounts in other soils. These forms of Al would raise the total concentrations.

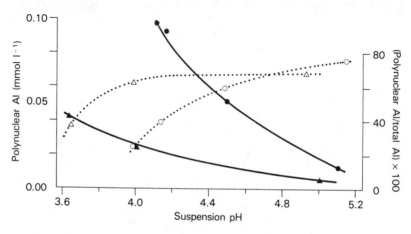

Fig. 25.4 Polynuclear Al in solution as a function of pH for 0.01 M CaCl₂ suspensions.
Full symbols and left-hand ordinate indicate concentration; open symbols and right-hand ordinate indicate polynuclear as a percentage of total Al in solution.
●, ○ Iron podzol (Countesswells series);
▲, △ Acid brown forest soil (Cawdor series).

11 See also Bache, B. W., *J. Soil Sci.*, 1974, **25**, 320; Bloom, P. R., *et al., Soil Sci. Soc. Am. J.*, 1979, **43**, 488.
12 Rhodes, E. R. and Lindsay, L. W., *J. Soil Sci.*, 1978, **29**, 324.
13 May, H. M. *et al., Geochim. Cosmochim. Acta*, 1979, **43**, 861; Bloom, P. R. *et al., Soil Sci. Soc. Am. J.*, 1979, **43**, 488.
14 Bache, B. W. and Sharp, G. S., *J. Soil Sci.*, 1976, **27**, 167.

Acidity on soil surfaces

Exchangeable hydrogen and aluminium

The traditional method of extracting exchangeable cations from soils using molar ammonium acetate buffered at pH 7 will cause any Al ions initially present to be precipitated as hydroxide. For this reason unbuffered 1 M KCl solution is best used for acid soils (10 g soil + 100 cm³ solution, leached through the soil over a period of 2 hours). This extracts the exchangeable H and Al along with other cations (predominantly Mg, Ca) at the natural soil pH. The exchangeable acidity (H + Al) is determined in the extract by titration with NaOH using phenolphthalein. Sodium fluoride added to this solution gives the following reaction:

$$Al(OH)_3 + 6NaF = 3NaOH + Na_3AlF_6$$

The NaOH released is equivalent to the amount of Al in the extract and can be titrated with HCl using the same indicator. This gives a measure of exchangeable Al in meq per 100 g soil, and the exchangeable H can be calculated by difference. Alternatively, the Al in the extract can be measured using atomic absorption spectrophotometry (mg or mmol per 100 g soil). The two methods together give the charge per Al ion.

The effects of pH on exchangeable Al are shown in Fig. 25.5.[15] Aluminium forms a large proportion of the total cations in acid soils, but little is extracted above pH 5.5. The nature of the exchangeable Al has not been examined in detail, but the average charge per Al is between 2 and 3, decreasing as pH rises, presumably the result of a mixture of the monomeric species Al^{3+},

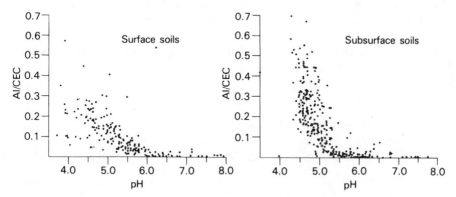

Fig. 25.5 Exchangeable Al as a proportion of CEC at pH 8.2 for soils from Virginia. (Reprinted with permission from the publisher.)

15 Thomas, G. W. and Hargrove, W. in *Soil Acidity and Liming* 2nd edn. (ed F. Adams), Ch. 1, *Agronomy*, No. 12, Am. Soc. Agron., Madison, 1984.

$AlOH^{2+}$ and $Al(OH)_2^+$ and polymeric forms.[16] There is a wide range of Al : CEC ratios present at a given pH for this set of soils. Some soils have very small amounts of soluble Al compounds present so giving limited amounts of exchangeable Al. Variability may also be the result of determining cation exchange capacity (CEC) at pH 8.2. For many soils this value is larger than the CEC at the soil's pH value, and low ratios result (p. 265).

There is a difference in behaviour between topsoils and subsoils, which is presumably an effect of organic matter. This seems to depress the ratios at low pH which may reflect the pH-dependent charge on the organic matter. Greater amounts of exchangeable Al between pH 5.5 and 6.0 may be the result of organic Al complexes both on surfaces and in solution.[17] Chelated Al in solution is, however, not as toxic as free Al, which may partly account for the lower critical pH values for plant growth in organic soils and the significance of organic matter additions in the management of soil acidity.[18] Lower total amounts of Al in organic soils may also lower the critical pH values.

Non-exchangeable acidity

Non-exchangeable acidity contributes to titratable or total acidity (note that there has been some conflict in the literature over the use of terms).[19] It may be a large proportion of the total acidity in soils rich in organic matter, oxides or allophane. Total acidity is measured by titrating a soil suspension up to a high pH (normally about pH 8), the amount of acidity in the soil being equivalent to the NaOH used. Alternatively, a triethanolamine–barium chloride buffer at pH 8 is shaken with the soil, and the amount of acidity reacting with the buffer is determined by titrating a fresh sample of buffer with acid down to the measured pH of the soil + buffer. The amount of acidity determined is dependent on the time of contact between the NaOH (or buffer) and the soil.

The non-exchangeable acidity is in the form of protons which can be released from surfaces as the pH rises, or as non-exchangeable Al. It has no direct effect on plant growth but has to be neutralized as the pH is raised. There are three main sources of protons:

1   Polymeric aluminium hydroxides which carry positive charge but are not exchangeable. As the pH rises coordinated water molecules donate protons to the soil solution and $Al(OH)_3$ is eventually produced although

16 Bache, B. W. and Sharp, G. S., *J. Soil Sci.*, 1976, **27**, 167.
17 Schnitzer, M. and Skinner, S. M., *Soil Sci.*, 1963, **96**, 86;   Evans, C. E. and Kamprath, E. J., *Soil Sci. Soc. Am. Proc.*, 1970, **34**, 893;   Bloom, P. R. *et al., op. cit.,* 1979, **43**, 488.
18 Bloom, P. R. *et al.*, (see ref. 17).
19 See ref. 15.

where present as interlayers in smectites and vermiculites, a charge of +0.3 per Al remains.[20]

2　Reactive OH groups on the surfaces of hydrated iron and aluminium oxides and on the edges of clay particles, particularly the 1 : 1 lattice clays. As the pH rises, the donation of protons to solution causes the surface to become more negative (Ch. 7).

3　Carboxyl, hydroxyl and amino groups on organic matter which dissociate as pH rises. Dissociated groups in acid soils are largely satisfied by aluminium which is difficult to exchange, but reacts with added base.[21]

The effects of the components of acidity can be seen in the shapes of titration or buffer curves of soils. These are graphs showing the changes in pH as base or acid is added to soil suspension. The slope of the curve is inversely related to the buffer capacity of the soil system which is its capacity to withstand pH

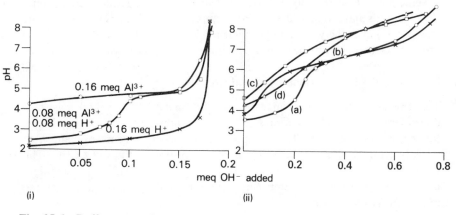

(i)　　　　　　　　　　　　　　　　(ii)

Fig. 25.6　Buffer curves for

(i) HCl and $AlCl_3$ solutions. Initial amounts of H^+ and Al^{3+} are given on the curves.

(ii) (a) fresh and (b) aged acid washed Wyoming Bentonite. 1 g samples, exchange capacity = 0.8 meq.
(c) Oxisol (Vacaria), Brazil. 1 g sample.
(d) Entisol (Pouso Redondo), Brazil. 1 g sample.

The soils have the following properties: all values are meq g^{-1}.

	Oxisol	Entisol
Exchangeable Al (1 M KCl)	0.042	0.072
Sum of exchangeable cations (1 M KCl)	0.068	0.112
Titratable acidity (pH 8)	0.48	0.48
Non-exchangeable acidity (pH 8)	0.438	0.408

20 Barnhisel, R. I. in *Minerals in Soil Environments* (eds J. B. Dixon and S. B. Weed), Ch. 10, Soil Sci. Soc. Am., Madison, Wisc., 1977.
21 See ref. 15. This Al is however exchangeable with La; Jarvis, S. C., *J. Soil Sci.*, 1986, **37**, 211.

changes as OH^- or H^+ is added. Buffer capacity is defined as $b = \dfrac{d[HS]}{d[pH]}$ where [HS] is the concentration of acid soil, i.e. of proton donating groups. It is measured by the amount of strong base consumed per unit weight (or volume) of soil in raising the equilibrium soil solution by one pH unit.[22] Examples of buffer curves are shown in Fig. 25.6. In the solutions effectively all the HCl is titrated before the $AlCl_3$ and this two-step curve can be seen for the acid washed bentonite which, when fresh, appears to have about 0.2 meq exchangeable H^+ in an exchange capacity of 0.8 meq. On ageing, acid attack of the mineral occurs with a reduction in the amount of H^+ to about 0.04 meq. The titratable acidity of the bentonite is almost equal to its CEC indicating that there is only a small amount of non-exchangeable acidity present. In the soils, the amounts of exchangeable acidity (predominantly Al) are small compared to the titratable acidity, indicating large amounts of non-exchangeable acidity.

Some soils are extremely well buffered at low pH. In Assam, India, some acid tea soils have received very large applications of ammonium sulphate (see p. 863) over the years and yet their pH has remained at about 4.5.[23] The protons released by nitrification are removed by reaction with soil minerals.

Percentage Al and "base" saturation

The presence of non-exchangeable acidity means that a soil has pH-dependent charge and a varying CEC, and it is important that the conditions for the measurement of CEC are specified, especially because exchangeable Al is often quoted as per cent aluminium saturation of the soil's CEC. The CEC for temperate soils is normally determined at pH 7 with 1 M ammonium acetate. Since these soils often carry predominantly permanent charge the procedure is satisfactory.

For soils containing appreciable amounts of non-exchangeable acidity, the CEC, determined as the sum of the exchangeable cations extracted in 1 M KCl (Σ Ca + Mg + Al + H with other cations being present in negligible amounts), is known as the effective cation exchange capacity or ECEC. For these soils aluminium saturation is lower when expressed relative to the CEC at pH 7 compared to the ECEC and does not give a true measure of the condition of the exchange surfaces (nor the best index of solution Al, see p. 871). For these soils "base" saturation will also be lower when expressed relative to the CEC at pH 7, although the value in this case is normally used as an indication of the 'deficit' of bases relative to a limed neutral soil. The term "base" applies to the less acidic cations Ca, Mg, K and Na to distinguish them from the more acidic Al, Mn and H. Both aluminium and base saturation are broadly related to pH. Per cent base saturation is about 25 at pH 5 and 75 at pH 6. Exchangeable Al does not occur above about pH 5.5–6.0,

22 Nye, P. H., *J. Soil Sci.*, 1972, **23**, 82.
23 Gokhale, N. G. and Bhattacharyya, N. G., *Emp. J. expl Agric.*, 1958, **86**, 309.

and so a 25 per cent deficit of bases at pH 6 is due partly to H^+ but may also reflect an increase in CEC as the pH is raised from 6 to the pH of the CEC determination.

The development of acidity

One of the most striking chemical properties of soils is their variability in terms of pH with values ranging from 2 for acid sulphate soils to above 10 for sodic soils. As for all other soil properties, the pH which develops depends on the soil-forming factors: parent material, climate, topography, biological activity, management, and time.

Changes in acidity within a profile can be quantified through the hydrogen ion budget which considers all inputs and losses of protons in solution and on exchange surfaces. Apart from inputs from the atmosphere and losses in leachates there is an input of protons whenever a process releases anions without releasing an equivalent (charge) amount of cations, or removes more cations than anions. A charge balance must be maintained and the deficit of positive charge is made up by protons. The converse also applies that there is a loss of protons (release of hydroxyl) whenever a process releases more cations than anions or removes more anions than cations. A net increase of protons will cause the displacement of other cations from exchange sites, normally Ca + Mg, and these may be removed by leaching.

In the discussion which follows, aspects of acidity development will be considered separately, before bringing them together in the form of H^+ budgets.

Parent material

Rain-water, in the absence of oxides of sulphur and nitrogen, is a dilute carbonic acid solution of pH 5.65. Dissolution of CO_2 produced by soil respiration lowers the pH further (see also Fig. 25.12). Soil minerals dissolve in this water and the resulting solution will have a pH which depends on the composition of the minerals. Alkaline solutions form from carbonates and hydroxides. If these are not present then neutral or only slightly alkaline solutions develop. Stumm and Morgan[24] have presented the chemical principles and Table 25.1 gives examples of the composition of waters draining from various rocks. Aluminium is not listed: it is obviously present only at very low concentrations at these pH values and remains as a component of the mineral residues. There is no tendency to develop acidity simply on the basis of mineral dissolution.

Soils have cation exchange capacities associated with both mineral and organic constituents and the cations released during dissolution will compete

24 Stumm, W. and Morgan, J. J., *Aquatic Chemistry*, Wiley-Intersci., 1970.

for these exchange sites. The divalent ions compete strongly with H^+. Parent materials and minerals can be considered as basic or acidic depending on their ratio Al : (Ca + Mg + K), because although the dissolution of the minerals is often incongruent (not complete in one step), through a weathering sequence this ratio will determine the input of acidic and basic cations to the soil system.

Leaching

Where there is a net downward movement of water through the surface layers of a parent material, concentrations of ions in the water will be lower than those given in Table 25.1 because dissolution is time dependent. The pH, however, will probably be similar to that shown until, progressively, the more soluble components of the minerals are removed, the pH begins to fall and aluminium in the residue begins to dissolve. Thus the ratio of Al : Ca + Mg + K in the solution will increase, with a parallel increase in the ratio on exchange surfaces.

Table 25.1 The composition of waters draining from a range of types of rocks (−log activity). (Reprinted by permission of John Wiley & Sons, Inc.)

Type of rocks being drained	Quartzite stream	Granite	Sandstone	Shale	Limestone	Dolomite
				ground waters		
pH	6.6	7.0	8.0	7.3	7.0	7.9
pNa	4.6	3.4	3.3	2.6	3.0	3.5
pK	5.1	4.0	4.0	4.2	3.7	—
pCa	4.3	3.5	3.0	2.5	2.7	2.8
pMg	5.1	3.8	3.5	2.5	3.4	2.8
pH_4SiO_4	4.2	3.2	3.9	3.5	3.7	3.4
$pHCO_3$	4.0	2.9	2.6	2.1	2.3	2.2
pCl	5.8	4.0	3.7	4.0	3.2	3.3
pSO_4	4.7	4.2	3.2	2.2	3.4	4.7

Soluble organic compounds will also be present in the leaching waters, the source being leaves, litter and soil organic matter. The pH values of water passing out of the horizons of a podzol are shown in Table 25.2.[25] The inflow is more acid than pure water in equilibrium with atmospheric CO_2, and the organic layers lower the pH further. Aluminium is released into solution ln the B horizon, but is precipated when weathering minerals in the C horizon raise the pH to 6.

25 Bache, B. W., *Phil. Trans. R. Soc. Lond.*, 1984, **305**, 393. For an example of measurements in an Inceptisol see Van Miegroet, H. and Cole, D. W., *Soil Sci. Soc. Am. J.*, 1985, **49**, 1274.

Table 25.2 Change in water composition by interaction with horizons of a podzol soil profile

Stage/soil horizon	pH	H	Ca + Mg	Al	Na	SO$_4$	Cl	HCO$_3$
					(meq l^{-1})			
Inflow	4.0	0.10	0.15	—	0.20	0.25	0.20	—
Humic horizon	3.5							
Exch (Ca, Mg) = 1 meq/100 g								
Total acidity (pH 8) = 100								
Intermediate 1	3.6	0.25	—	—	0.20	0.25	0.20	—
B horizon	4.5							
Exch. (Ca, Mg) = 0.05								
Exch. Al = 1.5 meq/100 g								
Total acidity = 5.0								
Intermediate 2a	4.5	0.03	—	0.22	0.20	0.25	0.20	—
Intermediate 2b	4.5	0.03	—	0.07	0.20	0.10	0.20	—
C horizon	6.0							
Decomposing minerals								
Outflow	6.0	0.001	0.30	—	0.25	0.20	0.20	0.15

Table 25.3   The effect of tree species and parent material on soil pH

	Fresh leaf	Litter	Humic layer	50 cm
Sand over Chalky Boulder Clay West Tofts, Norfolk				
Alnus incania	6.61	6.42	5.79	7.65
Betula alba	6.51	6.28	5.97	7.73
Pinus nigra	6.48	5.06	5.25	7.48
Larix leptolepis	6.40	5.18	5.20	7.71
Pseudotsuga taxifolia	6.44	5.33	5.06	7.51
Sandy loam over Old Red Sandstone Abbotswood, Forest of Dean				
Larix decidica	—	4.49	3.99	4.36
Pinus nigra	—	4.68	3.69	4.49
Quercus robur	—	4.64	5.03	4.57
Pseudotsuga taxifolia	—	4.28	4.16	4.39

Table 25.3 shows the pH values produced by various types of tree litter.[26] There are differences between species, but more important is the difference in parent material in controlling the acidity of the litter. Tree species, initial soil conditions, management and the proportion of the tree removed at felling are all important factors influencing the long-term impact of forests on soil acidity.[27] The development of acidity results in losses of calcium and magnesium from the soil from both mineral weathering and release from cation exchange surfaces. Cations are lost along with non-adsorbed anions.

Respiration and mineralization are important processes which develop acidity and cause cation loss. The processes involved are (i) CO_2 produced by respiration which dissolves to give H_2CO_3 with dissociation to H^+ and HCO_3^-, the latter being leached along with some H^+ and exchanged Ca^{2+} and Mg^{2+} and (ii) mineralization of organic matter to produce nitrate, sulphate and phosphate. The production of NO_3^- involves the oxidation of NH_4^+ with two hydrogen ions produced per NH_4^+ oxidized. Phosphate is strongly adsorbed, but all the nitrate is free to be leached, as is some of the sulphate depending on the adsorption characteristics of the soil. The losses of nitrate will be discussed further in relation to fertilizer use. Flushes of mineralization with its associated acidity production occur after deforestation, after ploughing of grassland, and especially in autumn when warm dry soils are rewetted, and in spring when soil temperature rises. Acidity development is

26 Ovington, J. D., *J. Ecol.*, 1953, **41**, 13.
27 Hornung, M., *Soil Use Management*, 1985, **1**, 24.

reduced if leaching losses of calcium, magnesium and nitrate released into soil solution are minimized by deep rooted crops.

The extent of displacement of Ca^{2+} and Mg^{2+} from exchange surfaces by H^+ and Al^{3+} will depend on the base saturation, Ca^{2+} and Mg^{2+} losses decreasing as acidity develops. Estimates based on field measurements in Table 25.4[28] show that for British soils the loss is approximately halved for each pH unit decrease. These values are approximately correct for soils from Rothamsted and Woburn (Fig. 25.7) and a model of Ca + Mg loss has been developed.[29] The values include the effects of fertilizer discussed later.

Table 25.4 Postulated annual mean losses of calcium (expressed as $CaCO_3$) by leaching from soils of pH 5.0–8.0 with 25 cm of through drainage

pH	kg $CaCO_3$ ha^{-1}*
5.0	118
5.5	168
6.0	235
6.5	336
7.0	471
7.5	672
8.0	942

* Divide by 50 to obtain the equivalent H^+ addition to the soil for comparison with the H^+ budget, p. 865.

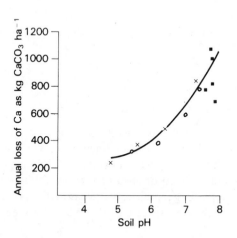

Fig. 25.7 Losses of $CaCO_3$ related to mean pH at Rothamsted (■ Broadbalk; x Sawyers Field) and Woburn (o). The line gives losses predicted by the model.

28 Gasser, J. K. R., *Exp. Husb.*, 1973, **25**, 86.
29 Bolton, J., *J. agric. Sci.*, 1977, **89**, 81.

Displacement of calcium and magnesium ions from soils is also dependent on the ratio of permanent to variable charge.[30] The weak acid groups of the variable charge surfaces release Ca^{2+} more readily than the permanent charge, and organic soils therefore release their Ca^{2+} and Mg^{2+} easily.

Crop removal of calcium and magnesium

Under natural vegetation, elements are cycled and to a large extent held against leaching losses (Table 25.5).[31] In contrast, a wheat crop may contain a total of 25 kg calcium and 8 kg magnesium per ha of which about one-fifth is in the roots when grown on a soil containing 10 t exchangeable Ca and 0.5 t exchangeable Mg per ha to 1 m depth.[32] Thus crop production involves removal of calcium, magnesium and potassium which have to be replaced. Amounts removed[33] range from about 1 kg Ca t^{-1} fresh weight for grain and potatoes to 40 kg t^{-1} for lucerne. On intensive grassland removals can be as much as the amounts lost by leaching given in Table 25.4.

Table 25.5   Calcium and magnesium distribution (kg ha^{-1}) in adjacent beech and spruce forest ecosystems on an acid brown forest soil at Kongalund, southern Sweden

	Beech		*Spruce*	
	Ca	*Mg*	*Ca*	*Mg*
Yearly litter fall	32	4	20	3
Surface litter	34	5	48	8
Exch. in soil	175	38	150	31
Total in plants above ground	603	105	459	69
Total below ground	42	14	45	11

The plant root as a source of acidity

There seems little doubt that most plants growing under natural conditions cause soil pH to rise at the root surface. Analyses of 62 plant species growing in soil[34] showed that they contained an average of 2.5 meq absorbed cations $(K+Na+Mg+Ca)$ and 3.6 meq of absorbed anions $(NO_3+SO_4+H_2PO_4+Cl)$ per g dry matter of tops. In general, roots release HCO_3^- or OH^- to maintain electrical neutrality at the root surface $(3.6 - 2.5 = 1.1$ meq HCO_3^- per g dry

30 Amedee, G. and Peech, M., *Soil Sci.*, 1976, **121**, 259;   Wiklander, L. and Anderson, A., *Geoderma*, 1972, **7**, 159.
31 Nihlgard, B., *Oikos*, 1972, **23**, 69.
32 Gregory, P. J. *et al.*, *J. agric. Sci.*, 1979, **93**, 485, 495.
33 *Lime and Liming, MAFF Ref. Book*, 35, HMSO, London, 1981.
34 Cunningham, R. K., *J. agric. Sci.*, 1964, **63**, 109.

matter).[35] When plants are supplied with NH_4^+ rather than NO_3^- then the pH in the root zone falls.[36] Flowing solution culture studies have also shown that NO_3^- fed plants raise pH and NH_4^+ fed plants lower it. Nitrate normally constitutes more than half the total absorbed anions (1.8 meq g^{-1} dry matter). If this amount of N is taken up as NH_4^+ instead of as NO_3^-, then cations $= 4.3$ and anions 1.8, giving a net acid release of $4.3 - 1.8 = 2.5$ meq H^+ g^{-1} dry matter. For legumes, where N is fixed symbiotically and little is taken up either as NO_3^- or NH_4^+, the net effect is $2.5 - 1.8 = 0.7$. meq H^+ released per g dry matter. Other values are 0.8 meq g^{-1} for lucerne[37] and 1.08 for soyabean.[38] Efflux rates of H^+ for lucerne, red clover and white clover in flowing solution culture decreased from about 200 to 50 μg H^+ g^{-1} dry shoot per day over a 75-day growth period.[39] Nye[40] has used a flux of 3×10^{-12} mol cm^{-2} s^{-1} of HCO_3^- for NO_3^- fed non-legumes and an

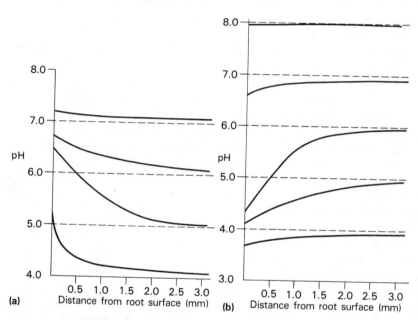

(a)   (b)

Fig. 25.8 Effect of initial soil pH (dotted lines) on pH profiles following (a) HCO_3^- and (b) H^+ release at the root surface.

35 Nye, P. H., *J. Soil Sci.*, 1968, **19**, 205.
36 Miller, M. H. *et al.*, *Agron. J.*, 1970, **62**, 524; Riley, D. and Barber, S. A., *Soil Sci. Soc. Am. Proc.*, 1971, **35**, 301; Smiley, R. W., *Soil Sci. Soc. Am. Proc.*, 1974, **38**, 759.
37 Nyatsanga, T. and Pierre, W. H., *Agron. J.*, 1973, **65**, 936.
38 Israel, D. W. and Jackson, W. A. in *Mineral Nutrition of Legumes in Tropical and Sub-tropical Soil* (eds C. S. Andrew and F. J. Kamprath), p. 113, CSIRO, Melbourne, 1978.
39 Jarvis, S. C. and Hatch, D. J., *Ann. Bot.*, 1985, **55**, 41.
40 Nye, P. H., *Pl. Soil*, 1981, **61**, 7.

equal flux of H^+ for legumes to calculate the likely changes in pH around roots in a weakly buffered soil. The calculated profiles are shown in Fig. 25.8 for the following soil conditions: root radius = 0.02 cm; buffer capacity = 10^{-6} mol cm^{-3} pH $^{-1}$, volumetric water content $\theta = 0.2$, tortuosity factor f = 0.1, P_{CO_2} = 0.01 atmosphere.

The changes in the root zone are at a maximum when the soil pH is about 5.3 because at this pH diffusion of acidity is slow. The latter also has the effect of decreasing the thickness of the zone influenced by the root, the zone being about 1 mm thick. It is from this thin zone that most of the uptake of micronutrients and phosphate probably occurs, and in which there is a large microbial population. At soil pH 8, changes in pH are small because HCO_3^- concentrations are high in the soil solution: the implications of the presence of calcium carbonate in a soil at pH 8 are not considered here. It seems likely that the gradients of CO_2 concentration in the rhizosphere will have a more significant effect on pH in alkaline soils (see p. 892). Increased buffer capacity increases the time it takes to establish a given profile, and increased water content reduces the pH changes in the rhizosphere because the ions can then diffuse more rapidly.

These predictions have not yet been tested experimentally because of the difficulties of measuring pH at points so close to the root surface. They are based on experimental determinations using soil blocks and resin paper saturated with HCO_3^- as a source of alkalinity.[41] However, measurements with dense rape root systems[42] and with both legumes and non-legumes in pots[43] have confirmed the general principles even though rape plants fed on NO_3^- took up more cations than anions and acidified the soil. Changes around single onion roots have been measured[44] giving average pH values in small volumes of soil, and around rape roots growing in agar with bromocresol purple as a pH indicator.[45] Marschner and Romheld[46] grew plants in soil, which was then impregnated with agar and indicator, and pH micro-electrodes measured pH values close to the roots (Colour Plate 5). Compared to the bulk soil (pH 6.0) maize roots changed the rhizosphere pH to 7.5 with NO_3^- but to 4.0 with NH_4^+, but even with NO_3^- fed plants, acidification occurred around the apical parts of primary roots and around lateral roots. Chickpea decreased the pH to 4.5 even with NO_3^-. White clover when supplied with NO_3^- increased the pH to 6.5, but when dependent on symbiotically fixed N, the pH decreased to 4.5.

41 Ramzan, M. and Nye, P. H., *J. Soil Sci.*, 1978, **29**, 184; Farr, E. *et al.*, *J. Soil Sci.*, 1970, **21**, 1.
42 Grinsted, M. J. *et al.*, *New Phytol.*, 1982, **91**, 19; Hedley, M. J. *et al.*, *New Phytol.*, 1982, **91**, 31.
43 Jarvis, S. C. and Robson, A. D., *Aust. J. agric. Res.*, 1983, **34**, 341, 355.
44 Mitsios, I. and Rowell, D. L., *J. Soil Sci.*, 1987, **38**, 53, 65.
45 Moorby, H. *et al.*, *Pl. Soil*, 1985, **84**, 403.
46 Marschner, H. and Romheld, V., *Z. Pflanzenphysiol.*, 1983, **111**, 241.

Fertilizer use

The use of ammonium sulphate as fertilizer to supply nitrogen to crops has a strongly acidifying influence on soil. Gasser[47] reviewed the effects of fertilizer use in terms of calcium and magnesium loss and showed that the major problem was associated with the use of ammonium salts. NH_4^+ is rapidly nitrified in neutral or slightly acid soils, the reaction being

$$(NH_4)_2SO_4 + 4O_2 = 2NO_3^- + 4H^+ + 2H_2O + SO_4^{2-}$$

If the protons produced displace Ca^{2+} and Mg^{2+} from exchange surfaces, and these are lost by leaching, 7.2 kg calcium carbonate is required per kg N applied to replace the calcium lost and to neutralize the H^+ produced. Field measurements suggest that between 4 and 5 kg calcium carbonate is required. The reasons for the difference between the theoretical and field value have not been examined in detail. An H^+ budget is not yet available for a fertilized site partly because it depends on a knowledge of the nitrogen budget of the soil which is still not fully measurable. However, the following may all be involved:

1 NH_4^+ may be volatilized as NH_3 so contributing only one H^+ per $NH_4^+ - N$.
2 Nitrate may be denitrified, and the removal of an anion from solution must result in the formation of HCO_3^- or OH^- to maintain neutrality.
3 Uptake of NO_3^- by the plant may cause HCO_3^- or OH^- to be released (note that uptake of some NH_4^+ will reduce the acidity produced by nitrification but will increase the cation : anion ratio in the plant and the net acidity effect is independent of the ratio of $NH_4^+ : NO_3^-$ taken up).
4 H^+ may react with soil minerals or organic matter releasing cations.

If ammonium nitrate is used as the fertilizer the nitrification reaction is

$$NH_4NO_3 + 2O_2 = 2NO_3^- + 2H^+ + H_2O$$

In this case 3.6 kg $CaCO_3$ per kg N applied is the calculated amount required to neutralize the H^+. Again it is found that the acidity which develops requires less $CaCO_3$ than this (2–2.5 kg). In England and Wales the Agricultural Development and Advisory Service[48] considers that 1.4–3.9 kg is required per kg N used, this range applying to both ammonium nitrate and ammonium sulphate. Urea has half the acidifying effect of NH_4^+ per kg N.[49]

Potassium fertilizers are normally chloride or sulphate salts, and assuming that potassium is either taken up by plants or enters the soil reserves, the Cl^- or SO_4^{2-} will carry away Ca^{2+} and Mg^{2+} if they are leached, and an equivalent of 1.1 kg calcium carbonate is required to neutralize the acidity from 1 kg K (1.2 kg K_2O).

47 Gasser, J. K. R., *Exp. Husb.*, 1973, **25**, 86.
48 *Lime and Liming, MAFF Ref. Book*, 34, HMSO, London, 1981.
49 Court, M. N. *et al.*, *J. Soil Sci.*, 1964, **15**, 42, 49.

Phosphate fertilizers have a negligible affect on pH since the phosphate is strongly held in the soil or is taken up by the crop. If the phosphate is supplied as mono- or diammonium phosphate, as in most compound ferti- lizers, the ammonium will contribute acidity in the way discussed above. Basic slag and many rock phosphates raise soil pH.

Acid rain

There is much discussion and argument regarding the effects of the burning of fossil fuels on the acidity of rain-water and subsequent effects of this on soils, streams and lakes,[50] and on the plant and animal life associated with these ecosystems. Reviews of the environmental aspects are available.[51]

Acid deposition involves H^+ (with NO_3^- or SO_4^{2-}) in the rain-water, dis- solved NH_4^+ which produces acidity when it is nitrified, and dry deposition of SO_2 which produces acidity when it dissolves and is oxidized in the soil. The distribution of these inputs for Britain has been examined by the Warren Spring Laboratory.[52] Total inputs through wet and dry deposition are equiv- alent to between 1 and 4 kg H^+ ha^{-1} a^{-1} of which up to about 1 kg is present as H^+ in the rain. Rainfall pH is therefore not a good indicator of total acid deposition. The latter may be small compared to total acidification in the soil, for calculations summarized in Table 25.4 show that at pH 6.5, arable soils lose 336 kg $CaCO_3$ ha^{-1} a^{-1}, equivalent to about 7 kg H^+ (20 kg H^+ at pH 8). At Rothamsted, H^+ deposition is 0.1 – 0.2 kg ha^{-1} a^{-1}, and pH of rain has risen over the last 15 years.[53] Similar input figures are available for the Netherlands[54] and lower figures for the USA.[55]

Proton budgets

Van Breemen[56] has brought together information from many sources to give the proton budgets or balance sheets shown in Table 25.6. External sources of protons are atmospheric inputs, and from mineralization of organic N.[57] Internal sources are carbonic acid, net assimilation of cations by the biomass with associated H^+ release, and release of anions with an equivalent amount

50 See for example Driscoll, C. T. and Newton, R. M., *Environ. Sci. Technol.*, 1985, **19**, 1018.
51 Hutchinson, T. C. and Havas, M., *Effects of Acid Precipitation on Terrestrial Ecosys- tems*, Plenum, 1980; Environmental Resources, *Acid Rain*, Graham and Trotman, London, 1984.
52 Barrett, C. F. *et al.*, *Acid Deposition in the United Kingdom*, Warren Spring Labora- tory, Stevenage, 1983.
53 Goulding, K. W. T. and Poulton, P. R., *Soil Use Management* 1985, **1**, 6.
54 Vermeulen, A. J., *Environ. Sci. Tech.*, 1978, **12**, 1017.
55 Counc. Agric. Sci. Tech. Rept., No. 100, *Acid Precipitation in Relation to Agriculture and Aquatic Biology*, 1984.
56 Van Breemen, N. *et al.*, *Nature, Lond.*, 1984, **307**, 599.
57 The importance of nitrification is discussed by Van Miegroet, H. and Cole, D. W., *Soil Sci. Soc. Am. J.*, 1985, **49**, 1274.

Table 25.6 Proton fluxes, rates of soil acidification and external/internal proton ratios of different ecosystems

Ecosystem	Soil, vegetation	H^+ sources, kg ha^{-1} a^{-1}		H sinks, kg ha^{-1} a^{-1}		EIPR[†]	ΔANC[‡]
		External	Internal	Internal	Drainage		
Low rate of acidification Hubbard's Brook Experimental Forest, New Hampshire	Haplorthod pH 3.6–5 hardwoods	1.4	1.2	2.4	0.1	1.17	−2.1
Intermediate rate of acidification Hackfort B, Netherlands	Dystrochrept pH 4 Quercus	4.3	1.9	5.7	0.1	2.26	−4.8
High rate of acidification Castricum, Netherlands	Udipsamment Quercus	1.6	12.6	15.1	0	0.13	−14.3
Deforested system Hubbard's Brook Experimental Forest	Haplorthod pH 3.6–5	12.2*	2.9	13.3	0.6	0.34	−2.2

* Of this 11.2 is from mineralization of surface organic matter and only 1.0 from atmospheric inputs.
[†] External: internal proton ratio.
[‡] Change in acid neutralizing capacity. Mineral weathering involves both sinks and sources of protons. The figure given is the net H^+ assimilated by the minerals. (Reprinted by permission from *Nature*, 1984, **307**, 599 © Macmillan Journals Ltd.)

of H^+ from the weathering of minerals. Internal sinks for protons are immobilization of NO_3^-, net anion assimilation by the biomass, and mineral weathering with cation release. The external sink is drainage water.

To assess the relative importance of external inputs to total sources in the ecosystem, the external: internal proton ratio (EIPR) is listed which is the ratio of atmospheric inputs of mineral and potentially mineral acids to internal sources. The acid neutralizing capacity (ANC) of the inorganic fraction of the soil is its capacity to assimilate protons, and soil acidification is defined as a decrease in ANC as minerals are weathered and is generally, but not always, accompanied by a decrease in pH. If the soil neutralizes the proton inputs then Ca^{2+}, Mg^{2+}, Na^+ and K^+ are released and exported from the soil by vegetation uptake or drainage water. If there are large H^+ inputs, low amounts of these cations which can be released by weathering, or slow dissolution relative to retention time of water in the soil, then neutralization of inputs is incomplete. The pH will then decrease and aluminium dissolution will occur.

A distinction needs to be drawn between buffer capacity and ANC. The former is the amount of H^+ required to lower the soil pH by one unit where the H^+ reacts chemically with mineral and organic matter. It is a ratio of a capacity to an intensity factor. ΔANC relates to H^+ reacting only with the mineral fraction and is a change in a capacity factor unrelated to pH.

In neutral soils rich in weatherable minerals, acidification is rapid (about 10 kg ha^{-1} a^{-1}) and is mainly associated with deprotonation of H_2CO_3. In acidic soils, acidification is less than about 2 kg ha^{-1} a^{-1} resulting from cation assimilation and organic acid deprotonation. Atmospheric deposition is between 1 and 5 kg ha^{-1} a^{-1} in north eastern USA and in central and north

Table 25.7   Acidity added to soils from various sources

	Source	H addition or equivalent (kg H ha^{-1} a^{-1})
Natural	CO_2 in soil pH > 6.5 (calc. soils)	7.2–12.8
	Org. acids in acid soils and from vegetation	0.1–0.7
Atmosphere	Wet deposition	0.3–>1.0
	Dry deposition	>0.3–>2.4
	NH_3 and NH_4^+ oxidation	0.7
Land use	Cation excess in vegetation	0.5–2*
	NH_4^+ oxidation (agric. soils) and leaching	4–6
	Oxidation of N and S from organic matter and leaching	0–10[†]

* Cation excess calculated for all ions except NH_4 and NO_3.
[†] High values only after clearance of vegetation. Acid sulphate soils are not included.

eastern Europe, and this exceeds internally generated H^+ in soils with low or intermediate rates of soil acidification. At EIPR values below 0.3, ecosystems seem able to assimilate proton inputs and acidification proceeds largely through the dissolution and leaching of cations. Values above 0.5 indicate significant dissolution of aluminium which is lost in drainage.

A summary of amounts of acidity associated with different components of the system is given in Table 25.7.[58] Nitrification of ammonium fertilizers and NO_3^- leaching are significant sources of acidity in agricultural soils, and high H^+ inputs occur after deforestation and ploughing of grassland.

Effects of acidity on plant growth

Effects of H+

The wide range of pH values found in natural soils is reflected in variations in the vegetation or crops they can support. It is not clear how far these differences are due to the sensitivity of the plant roots to H^+ or OH^- in soil or soil solution, and how far secondary effects are involved. The limited information for non-legumes in solution culture with pH between 3 and 9 suggests that root growth and function can be directly affected at pH 5 and below depending on species.[59] The injurious effects at pH 4 to 5 can be offset by additional Ca but not at pH 3.[60] Between pH 5 and 8 growth is satisfactory but at pH 9 there can be direct effects of OH^- or HCO_3^- on the absorption of phosphate and molybdate.

Effects of aluminium

Beneficial effects. Although aluminium is not known to be an essential element, it seems that low concentrations of Al stimulate the growth of some plants: examples are lucerne and red clover, 7 μM,[61] *Lolium perenne* and other grasses, 186 μM,[62] maize, 18 μM and tea.[63] The mechanisms are not clear but are almost certainly indirect.

Symptoms of aluminium toxicity An example of a crop failure due to acidity is shown in Plate 25.1. The first effects of aluminium to be observed are a shortening and thickening of the roots (Plate 25.2).[64] Roots become brown in

58 Rowell, D. L. and Wild, A., *Soil Use Management*, 1985, **1**, 32.
59 Arnon, D. I. and Johnson, C. M., *Pl. Physiol.*, 1942, **17**, 525.
60 Sutton, C. D. and Hallsworth, E. G., *Pl. Soil*, 1958, **9**, 305.
61 Macleod, L. B. and Jackson, L. P., *Canad. J. Soil Sci.*, 1965, **45**, 221.
62 Hackett, C., *J. Ecol.*, 1967, **55**, 831.
63 Paterson, J. W., Ph. D. thesis, Pennsylvania State Univ., *Diss. Abstr.*, **25**, 1965, 6142; Chenery, E. M., *Pl. Soil*, 1955, **6**, 174.
64 Wind, G. P., *Neth. J. agric. Sci.*, 1957, **15**, 259.

Plate 25.1 Barley failure on acid soil (pH 4.5). The area in the foreground is the headland to the north of the Exhaustion Land at Rothamsted Experimental Station. It did not receive chalk when the rest of the area was treated over 200 years ago and represents the acid condition of the unlimed clay with flints soil. (*Rothamsted Experimental Station Annual Report*, 1976, Part II, p. 69.)

colour and branching is reduced. As aluminium concentrations increase in soil or solution, symptoms occur similar to those of phosphorus or calcium deficiency. Roots become more susceptible to disease, and certain physiological plant diseases are thought to be caused by excess aluminium, for example bronzing of rice.[65] The similarity of the effects of aluminium excess to phosphorus or calcium deficiency make it difficult to identify the cause of reduced plant growth in acid soils.

Aluminium taken up by plants tends to accumulate in the roots and is not easily translocated to the shoots,[66] and much of the aluminium seems to be retained in the cell walls of the root.[67] Colour Plate 6 shows micrographs of maize roots influenced by aluminium.[68] The primary sites of aluminium

65 Ota, Y., quoted by Foy, C. D. in *The Plant Root and its Environment* (ed E. W. Carson), p. 601, Univ. Virginia Press, Charlottesville, 1974.

66 Foy, C. D. *et al.*, *Agron. J.*, 1967, **31**, 513.

67 Clarkson, D. T., *Pl. Soil*, 1967, **27**, 347.

68 Bennet, R. J. *et al.*, *S. Afr. J. Pl. Soil*, 1985, **2**, 1, 8; *S. Afr. J. Bot.*, **51**, 355, 363. These papers also give results of ultrastructural studies, showing that Al interferes with membrane transport.

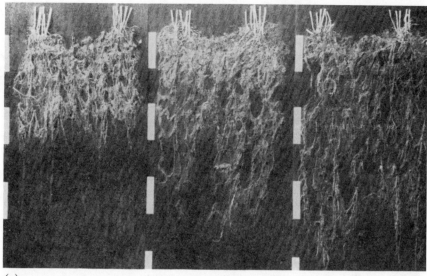

(a)

(b)

Plate 25.2 (a) Root growth of spring wheat in a field experiment on an untreated and two treated soils. From left to right: subsoils (>15 cm) with pH 3.1, 3.5 and 4.5. Each block on the scale is 10 cm.

(b) Barley roots grown in limed soil (left) and acid soil (right). (From Dale Ritchey.)

uptake are the peripheral cells of the root tip and the mucilaginous secretions surrounding the root, and entry is limited to the outer cortical cells. Increased plant tolerance seems to be associated with the decreased uptake of the aluminium,[69] and decreased translocation to tops.[70] Toxic levels of aluminium in the plant cause many physiological and biochemical changes which seem to be linked to increases in viscosity of protoplasm in root cells and decreases in permeability of root membranes to ions and water. These effects may be due to cross-linking between adjacent protein molecules, a result of the polyvalency of Al, although other trivalent ions do not always produce similar effects.[71] Aluminium also interferes with cell division, respiration, nitrogen metabolism and glucose phosphorylation.[72] These effects often occur along-side changes in uptake of phosphate, calcium, magnesium and iron.

Plant growth and solution composition   Many plants are very sensitive to aluminium in solution. Solution experiments show that plant species and varieties vary in their sensitivity, and that the critical concentration for a given plant varies depending on solution pH, calcium concentration and the experimental technique. Thus it is not possible to give exact values, but critical levels seem to vary from about 1 mM for aluminium tolerant plants to about 1 μM for aluminium sensitive plants.[73]

The variations in results are largely removed if sensitivity of roots is related to aluminium activity. Thus, cotton root growth is restricted above about 1 μM activity of Al^{3+} for soil solutions of varying composition[74] and coffee seedlings above about 0.4 μM Al^{3+} activity.[75] For wheat, $AlOH^{2+}$ may be the aluminium species most directly related to growth.[76] Blamey *et al.* added $NaOH$ and NaH_2PO_4 to nutrient culture solutions and showed that the toxic effect of aluminium on soyabean root growth was ameliorated by the reduction in monomeric Al species by polymerization or precipitation. A close curvilinear relationship was observed (Fig. 25.9)[77] between relative root length and the sum of the activities of monomeric Al in solution irrespective of total Al concentration, the concentration of polymeric Al, ionic strength,

69 Rorison, I. H. in *Nutrition of the Legumes* (ed E. C. Hallsworth), p. 43, Butterworths, 1958.

70 Foy, C. D. *et al., Agron. J.*, 1969, **61**, 505.

71 Clarkson, D. T. and Sanderson, J., *Planta*, 1969, **89**, 136.

72 Clarkson, D. T. in *Ecological Aspects of the Mineral Nutrition of Plants* (ed I. H. Rorison), p. 381, Symposium of the British Ecological Society, No. 9, Blackwell, 1969. Also the papers by Bennet, R. J. *et al.*, 1985, see ref. 68.

73 Clarkson, D. T., *J. Ecol.*, 1966, **54**, 167; Jarvis, S. C. and Hatch, D. J., *J. exp. Bot.*, 1985, **36**, 1075.

74 Adams, F. and Lund, C. F., *Soil Sci.*, 1966, **101**, 193.

75 Pavan, M. A. and Bingham, F. T., *Soil Sci. Am. J.*, 1982, **46**, 993; Pavan, M. A. *et al., op. cit.*, 1982, **46**, 1201.

76 Kerridge, R. C. *et al., Agron. J.*, 1971, **63**, 586; Moore, D. P. in *The Plant Root and its Environment* (ed. E. W. Carson), p. 135, Univ. Virginia Press, Charlottesville, 1974.

77 Blamey, F. P. C. *et al., Soil Sci.*, 1983, **136**, 197.

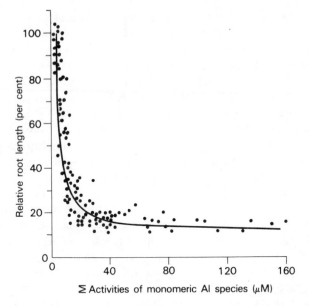

Fig. 25.9 Effect of monomeric Al species in nutrient culture on growth of soyabean roots. (Reprinted with permission © Williams & Wilkins.)

solution nutrient composition, phosphate concentration or pH. Soyabean root growth was severely reduced with monomeric aluminium activities of 10 μM.

Relationships between solution aluminium and soil acidity Because aluminium activity in soil solution is not easy to measure, effects of acidity are usually discussed either in terms of pH or exchangeable Al. Increases in solution aluminium are associated with decreased pH and increased amounts of exchangeable Al. Possibly the best index of the activity of aluminium in soil solution is the Al saturation of the CEC measured at the pH of the soil (p. 854) since this takes account of variations in CEC between soils. Plant tolerance can be expressed in terms of a critical Al saturation and raising this to a satisfactory level for a crop is a basis for determining lime requirements[78] (p. 885). Measurements on some soils of the coastal plain of southeast USA showed that aluminium concentration in soil solution was not a good indicator of aluminium toxicity possibly because non-available organically bound aluminium was included in the solution measurement. This may be the reason why aluminium in cultivated soils was more toxic than equal amounts in woodland sites; Al saturation was not a good indicator of toxicity for these soils.[79]

78 Cochrane, T. T. *et al.*, *Trop. Agric., Trin.*, 1980, **57**, 133.
79 Adams, F. and Hathcock, P. J., *Soil Sci. Soc. Am. J.*, 1984, **48**, 1305.

Aluminium–phosphate interactions   Some direct effects of aluminium on plant metabolism have already been discussed, but aluminium also binds phosphate on root surfaces, cell walls and in the free space of plant roots,[80] thus preventing phosphate from being available for root metabolism. An example of this is shown in Plate 25.3 using electron microprobe analysis of maize roots.[81] The aluminium is precipitated on the cell walls and the distribution of phosphorus is closely similar, suggesting for these high P and Al concentrations that a precipitate of aluminium phosphate is formed. Increased amounts of exchangeable aluminium may therefore lead to increased amounts of phosphate being inactivated at the root surface, so leading to phosphate deficiencies. This helps to explain why symptoms of aluminium toxicity can be very similar to those of phosphate deficiency. This is unlikely to be the full explanation, since root growth is also restricted by aluminium either directly through its effect on metabolism or indirectly because of the shortage of utilizable phosphate. Because phosphate is relatively immobile in soils its uptake is related to root length[82] and any restriction in root growth will result in reduced uptake. An example of reduced root growth in acid soils is shown in Plate 25.2a.

Phosphate uptake may also be reduced as a result of the effects of aluminium on root demand. For two Brazilian wheat varieties inflow (uptake rate per unit length) of phosphate increased by about five times as a result of liming, even though the bicarbonate-extractable phosphate was unchanged.[83] Inflow could be increased by a better growth of root hairs or increased effectiveness of mycorrhizas. The effects of acidity on the latter are variable.[84]

It has often been thought that because liming increases the uptake of phosphate by plants, the phosphate availability in the soil is increased. If uptake is the measure of availability then this is correct, but it does not follow that the chemical status of phosphate in the soil has been raised or that diffusion of phosphate to the root is made easier. The literature suggests that changes are generally small and variable.[85] Supply is affected by the following properties which may be changed by liming: (i) concentration of phosphate in soil solution[86] (ii) amounts of extractable phosphate[87] (iii) changes in buffer capacity[88] (iv) rhizosphere effects[89] and (v) mineralization of organic phos-

80 Rorison, I. H., *New Phytol.*, 1965, **64**, 23;   Clarkson, D. T., *Pl. Physiol.*, 1966, **41**, 165; *idem*, *Pl. Soil*, 1967, **27**, 347;   Miranda, L. N. de and Rowell, D. L., unpublished.
81 Rasmussan, H. P., *Planta*, 1968, **81**, 28.
82 Baldwin, J. P., *J. Soil Sci.*, 1975, **26**, 195.
83 Miranda, L. N. de and Rowell, D. L., *Pl. Soil*, 1987, in press.
84 Graw, D., *New Phytol.*, 1979, **82**, 687.
85 Haynes, R. J., *Pl. Soil*, 1982, **68**, 289;   Haynes, R. J. and Swift, R. S., *J. Soil Sci.*, 1985, **36**, 513.
86 Holford, I. C. R., *Aust. J. Soil Res.*, 1985, **23**, 75.
87 See refs. 83, 84 and 86.
88 See ref. 86.
89 Hedley, M. J. *et al.*, *New Phytol.*, 1982, **91**, 45.

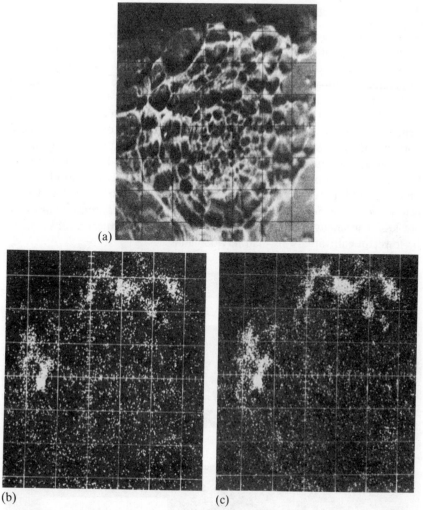

Plate 25.3 Electron probe X-ray analysis of a lateral root of maize grown in Hoagland's solution with 2 mM P and 10^{-3} M $AlCl_3$ at pH 3.2. One grid represents 40 μm.
(a) Back-scattered electron image, cross-section.
(b) Aluminium X-ray oscillogram of tissue in (a).
(c) Phosphorus X-ray oscillogram of tissue in (a).

phorus compounds[90] (see also Ch. 21). Liming results in a marked stimulus of biological activity, and in grassland soils with high organic matter contents mineralization of organic nitrogen and phosphorus compounds may be the most important initial effect.

90 Awan, A. B., *Soil Sci. Soc. Am. Proc.*, 1964, **28**, 672; Lucas, L. N. and Blue, W. G., *Trop. Agric., Trin.*, 1972, **49**, 287; Dalal, R. C., *Adv. Agron.*, 1977, **29**, 85.

Deficiencies of calcium and magnesium

Although aluminium toxicity is the major problem in acid soils, exchangeable Ca and Mg are often present in very small amounts and may be deficient. Plate 25.4 shows the effects of Ca and Al on sorghum. Increases in root elongation of tobacco both from addition of Ca and from removal of exchangeable Al have been found[91] in an Ultisol with pH 4.2 and 0.4 meq Ca per 100 g. Some Hawaiian soils are deficient in calcium and liming is of benefit for sugar cane primarily because of the addition of calcium.[92] The B horizons of some Brazilian Oxisols with about 0.2 meq Ca per 100 g and no exchangeable Al are deficient in calcium.[93] Normal root development results from the addition of 0.15 meq Ca per 100 g applied either as chloride, phosphate or carbonate, with magnesium carbonate having no effect. Magnesium can, however, be deficient in some soils.[94] Soil solution concentrations of Ca are less than about 0.5 mM in deficient soils, and this critical level seems to be lower in cultivated soils than in woodlands.[95]

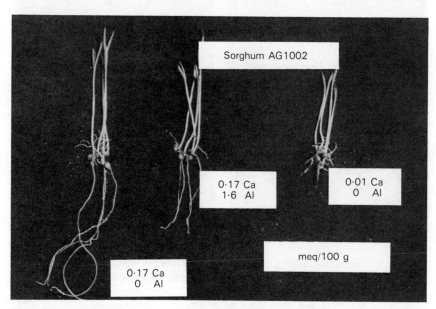

Plate 25.4   The effects of exchangeable aluminium and calcium on the growth of sorghum in soil. (From Dale Ritchey.)

91 Abruna, F. R: *et al., Soil Sci. Soc. Am. Proc.*, 1970, **34**, 629.
92 Ayers, A. S. *et al., Soil Sci. Soc. Am. Proc.*, 1965, **29**, 387.
93 Ritchey, K. D. *et al., Soil Sci.*, 1982, **133**, 378.
94 Mikklesson, D. S. *et al., Int. Rice Res. Inst. Bull.*, No. 29, 1963, quoted by Sanchez, P. A., *Properties and Management of Soils in the Tropics*, Wiley, 1976.
95 Adams, F. and Hathcock, P. J., *Soil Sci. Soc. Am. J.*, 1984, **48**, 1305.

Manganese toxicity

Toxicities of manganese can occur below pH 5.5–6.0. Yield responses to liming have been correlated with a decrease in toxic levels of manganese in tobacco leaves in a Puerto Rican Oxisol.[96] Adequate calcium is important to counter the effects of excess manganese[97] because the translocation of Ca is inhibited by Mn.[98] Thus liming will not only reduce the level of exchangeable Mn in soils and its uptake, but will also increase Ca uptake, the combined effect removing the manganese toxicity. Phosphorus also interacts with manganese (Ch. 21).

Examples of crop response to liming

Experiments carried out in the humid tropics have shown that serious limitations to crop yield can sometimes be removed by liming the acid soils which occur under these climatic conditions. A field experiment with maize in Ghana (Table 25.8) shows large responses to both lime and P in two seasons.[99] In general the largest responses to lime occurred without added fertilizer and the largest responses to phosphate occurred without added lime.

Table 25.8 Influence of lime and phosphate on the yield of maize grain grown in 1976 and 1977 on an Oxisol in Ghana

P_2O_5	Lime (t ha^{-1})			
	O	*1*	*2*	*4*
(kg ha^{-1})		Yield (t ha^{-1})		
1976				
0	1.05	1.38	2.23	3.08
25	2.63	2.79	2.70	3.01
50	2.92	3.10	2.99	3.42
75	2.90	3.13	3.20	3.06
100	2.83	3.44	3.61	3.33
1977				
0	0.71	0.91	1.52	1.63
25	1.51	1.96	1.73	1.61
50	1.55	1.91	1.88	2.55
75	2.16	1.41	1.85	2.74
100	1.43	1.78	2.16	2.33

96 Abruna, F. R. *et al., Soil Sci. Soc. Am. Proc.*, 1970, **34**, 629.
97 Vose, P. B. and Jones, D. G., *Pl. Soil*, 1963, **18**, 372; Le Mare, P. H., *op. cit.*, 1977, **47**, 621.
98 Horst, W. J. and Marschner, H., *Z. Pflanzen physiol.*, 1978, **87**, 137.
99 Lathwell, D. J., *Crop Response to Liming of Ultisols and Oxisols, Cornell Int. agric. Bull.*, No. 35, 1979.

Subsoil acidity was high, and root growth below 15 cm was probably severely restricted. This may have reduced yields in the drier 1977 season.

The interaction of liming and fertilizer applications on crop yields has been reported from Rothamsted and Woburn in southern England.[100] Four lime treatments were given and the plots received either complete nutrients (NPK Mg), or P, K, or Mg was omitted. Responses of barley to lime were small above pH 5.7, and the effect of omitting phosphate was always larger on soils with low pH. Potatoes were largely unaffected between pH 4.4 and 6.7, and omitting phosphate or magnesium had the largest effects at low pH. Oat yields were largely unaffected by pH when soils were adequately fertilized. Only on the Woburn soil, which contains little magnesium, was there an appreciable response to magnesium even on the well-limed soils. For grass crops, an experiment in Devon[101] showed that an *Agrostis*/ryegrass sward was affected by acidity below pH 5 but when reseeded to S23 ryegrass there was a response between pH 4.5 and 6.5. Subsequently *Agrostis* began to re-establish and pH again had an effect.[102]

Tolerance to acidity: differences between species and varieties

Species differ in their tolerance to acidity and aluminium and varietal differences are also important. The following reasons have been suggested for differences in tolerance:[103]

1   Differences in root morphology: the roots of some species are not apparently injured in acid conditions.
2   Some aluminium tolerant species and varieties have the ability to increase the pH of the root zone.
3   Translocation of aluminium from roots to shoots is reduced in some tolerant plants whilst in others, notably tea, tolerance is associated with accumulation in the shoots.
4   Aluminium in the roots of tolerant plants does not inhibit the uptake and translocation of calcium, magnesium and phosphate.

A comparison between two Brazilian wheat varieties showed that tolerance

100 Johnston, A. E. and Whinham, W. N., *The Use of Lime in Agricultural Soils., Fertil. Soc. Proc.*, No. 189, 1980;   Bolton, J., *Rothamsted Ann. Rept. for 1970*, Part 2, 98, 1971;   Bolton, J., *J. agric. Sci.*, 1977, **89**, 87.
101 Johnston, A. E. and Whinham, W. N. (see ref. 100).
102 For further examples of the effects of liming on yield see Gardner, H. W. and Garner, H. V., *The Use of Lime in British Agriculture*, Spon, London, 1957;   Adams, F. (ed), *Soil Acidity and Liming*, 2nd Edn., Agron. Monog. No. 12, Am. Soc. Agron., Madison, Wisc., 1984.
103 Foy, C. D. in *The Plant Root and its Environment* (ed. E. W. Carson), p. 602, Univ. Virginia Press, Charlottesville, 1974.

involved a mechanism for excluding aluminium from roots.[104] The tolerant variety with a larger root length had an aluminium inflow ten times smaller than the sensitive one, but also had a phosphate inflow four times smaller. The inflow of phosphate per unit inflow of aluminium (a mol ratio) was three times greater for the tolerant variety.

For acid soils in the tropics, Table 25.9 lists crops which are suitable for growth with only small amounts of added lime.[105] Using tolerant varieties of these crops makes it possible to manage even very acid land successfully and plant breeding may produce even more tolerant varieties.

Table 25.9 Crop and pasture species suitable for acid soils following minimum lime applications. (Adapted from Sanchez, P. A., *Properties and Management of Soils in the Tropics*, Wiley, 1976)

Lime application ($t\ ha^{-1}$)	Al Saturation (%)	pH	Crops (using tolerant varieties)
0	about 80	< 4.3	
0.25 to 0.5	68 to 75	4.5 to 4.7	Upland rice, cassava, mango, cashew, citrus, pineapple, *Stylosanthes*, *Desmodium*, kudzu, *Centrosema*, molasses grass, jaragua, *Brachiaria decumbens*, *Paspalum plicatulum*
0.5 to 1.0	45 to 58	4.7 to 5.0	Cowpeas, plantains (?)
1.0 to 2.0	31 to 45	5.0 to 5.3	Maize, black beans

For temperate regions, Table 25.10 gives critical pH values.[106] Growth is apparently depressed for some crops above pH 5.5 even though little or no exchangeable Al is present. It is not clear what the growth limiting factors are at these pH values. Non-exchangeable Al may influence the root and pH may be lower in the root zone than in the bulk soil. Crops with higher critical pH values are more sensitive to phosphate or potassium supply, and if phosphate and potassium levels of the soil are high enough, liming has little effect on the yields of most farm crops in slightly acid soil.[107] This is more likely to be true in regions with cool moist summers than in

104 Miranda, L. N. de. and Rowell, D. L., (see ref. 83).
105 Sanchez, P. A., *Properties and Management of Soils in the Tropics*, Wiley, 1976.
   Spain, J. M. *et al.* in *Soil Management in Tropical America* (eds E. Bornemisza and A. Alvarado), p. 300, North Carolina State Univ., 1975.
106 *Lime and Liming, MAFF Ref. Book*, 35, HMSO, London, 1981.
107 Aslander, A., *Soil Sci.*, 1952, **74**, 181, 436.

regions with warm drier summers, for example, in the west rather than in the east of England.

The critical pH values in Table 25.10 can only be approximate because effects of acidity are not directly related to hydrogen ion concentrations. Sugar beet and potatoes, for example, both have an appreciable calcium demand but potatoes have a much greater tolerance to aluminium than beet, while beet is more tolerant of manganese. Similarly, barley is like beet in being sensitive to aluminium and relatively tolerant to manganese, but it has a much lower calcium demand. Oats have a low calcium demand and are tolerant to both aluminium and manganese.

Table 25.10   Soil pH below which growth may be restricted on mineral soils

Arable crops				Grasses and legumes	
Barley	5.9	Oats	5.3	Clover, Alsike	5.7
Bean	6.0	Pea	5.9	Clover, red	5.9
Beet, sugar	5.9	Potato	4.9	Clover, white	5.6
Kale	5.4	Rape	5.6	Clover, wild white	4.7
Linseed	5.4	Rye	4.9	Cocksfoot	5.3
Maize	5.5	Swede	5.4	Fescues	4.7
Mangel	5.8	Turnip	5.4	Lucerne	6.2
Mustard	5.4	Wheat	5.5	Ryegrass	4.7
				Sainfoin	6.2
				Timothy	5.3
				Trefoil	6.1
				Vetches	5.9

Tables 25.9 and 25.10 do not show varietal differences, but these can be significant. *Trifolium repens* is an interesting example of quite extreme varietal differences, for certain varieties of white clover indigenous to acid soils in Wales are typically calcifuge in habit, while those indigenous to calcareous soils in Hampshire are calcicoles.[108]

The critical pH values listed are averages, but there is much heterogeneity in the field and this may be a reason for some high values. Over a field local areas of lower pH may be present which will depress the average yield. Sugar beet is a good example of this effect, for general experience of growers in eastern England and Sweden is that yields tend to be highest on fields with pH above 7. This may simply be a way of ensuring that there are no areas in the field with a pH below 5.9. Examples of pH variation are given in Table 25.11 for the pH plots from the farm of the North of Scotland College of Agriculture.[109] All fields were less than 4 ha, and samples were taken to

108 Snaydon, R. W., *J. Ecol.*, 1962, **50**, 133, 439.
109 Farr, E., *J. Sci. Fd Agric.*, 1972, **23**, 1089.

Table 25.11 pH variation at different sites in the same field

Field	Number of samples	pH range	Mean pH	SD of individual observations
1	12	5.7–6.4	6.1	0.20
2	19	4.8–5.9	5.4	0.29
3	27	5.7–6.6	6.1	0.22
4	33	5.6–6.6	6.1	0.24
5	24	4.6–6.2	5.1	0.37

about 20 cm depth. The pH values of soil samples taken from the same sites at different times during the year exhibited a similar degree of variation.

Differences between plant species in their tolerance to acidity are clearly seen in meadows and permanent pastures where many species of plants grow together in competition with each other. Plants which tolerate acidity a little better than their neighbours will spread at their expense and come to dominate the flora of acid soils, even though they do not necessarily thrive best in acid soils in the absence of competition. These effects are illustrated in the floristic composition of the long-continued fertilizer plots on Park Grass field at Rothamsted. Soil acidity can have a further important effect, for pastures on soils which are too acid for the larger species of earthworm to flourish will have a surface mat of dead vegetation, since there is no longer any mechanism by which the dead herbage can be mixed with the soil. Liming the soil to allow earthworms to dominate the soil fauna alters the composition of the herbage, encourages grass species that are more palatable to domestic stock and causes the soil to lose its slightly mor-like character and become more mull-like. The use of basic slag on hill pastures in the United Kingdom, being both a liming material and a phosphatic fertilizer, causes coarse grasses, such as species of *Nardus* and *Agrostis*, to be replaced by more nutritious species, such as ryegrass and *Poa*.

The pH of a soil can have a further effect on crop production, for it can affect the suitability of the soil as the home for soil-borne disease organisms. Thus, the actinomycete causing scab in potatoes, *Streptomyces scabies*, is much better adapted to a neutral than to an acid soil, possibly because it is sensitive to manganese. Thus, liming a soil tends to encourage scab although it can be rectified to some extent by applying manganese sulphate to the soil. Also, the fungus *Plasmodiophora brassicae*, which causes club-root or finger-and-toe in the cultivated Brassica crops (swedes, turnips, cabbage) tolerates acidity better than its host plant, and is therefore more likely to be injurious on acid soils. 'Take all' (*Ophiobolus graminis*) in wheat and barley is a further example. Thus, liming the soil will render it less suited to the pathogen, and so increase the resistance of these crops.

Subsoil acidity

Liming and its effects on plant growth are normally considered in relation
to changes occurring in the cultivated soil layer, but crops grow roots down
to much greater depths from which water and nutrients can be taken up.
Thus, subsoil acidity may cause severe reductions in crop production under
drought conditions. Field data have only become available in recent years and
have been summarized by Bouldin. Experiments in Planaltina, Brazil,
involved the incorporation of lime to two depths. Changes in per cent Al
saturation are shown in Fig. 25.10[110] for five years following these treatments.
In the topsoil the decrease in Al saturation after application is followed by
a slow increase in acidity over the period of the experiment. Lime applied

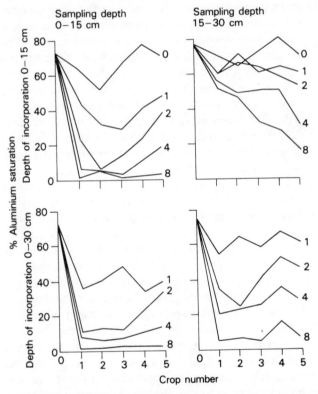

Fig. 25.10 Percentage aluminium saturation of soil samples taken at two depths
following each of five crops at Planaltina, Brazil. Lime was applied
before the first crop was planted and incorporated either 0–15 cm or
0–30 cm deep. Numbers to the right of lines are t ha^{-1}.

110 Bouldin, D. R., *The Influence of Subsoil Acidity on Crop Yield Potential, Cornell Int.
agric. Bull.*, No. 34, 1979.

in the top layer moves slowly into the lower layer. In other experiments in Brazil[111] subsoil acidity has been reduced by surface liming under a wide range of conditions after a few years. Applications of fertilizer nitrogen, potassium chloride and superphosphate (which included gypsum) enhanced the effects in the subsoil. Bouldin considers that these fertilizers contribute anions which are necessary for the leaching of Ca, and this points to a displacement of Al from the subsoil rather than a precipitation of aluminium hydroxide. On the other hand, high rates of calcium nitrate application (450 kg ha^{-1}) on a sandy loam raised the pH by almost 1 unit down to 75 cm,[112] an effect attributed to the excess of anion over cation uptake at the root surface. Higher rates of application had no extra effect presumably because the crop did not take up the extra nitrogen.

The effects of the lime treatments at Planaltina on the cumulative yields of the first four consecutive crops are shown in Fig. 25.11.[110] Differences in yield were not seen subsequently. The yield differences are likely to be the result of water use from the subsoil, for leaf water potentials measured in the third maize crop showed much higher water stress in the maize grown on the soil limed only to 15 cm. The importance of subsoil acidity in controlling water availability has also been shown for cotton.[113]

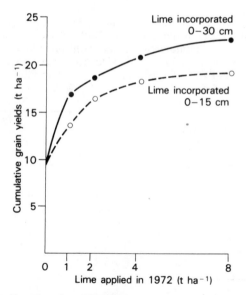

Fig. 25.11 Cumulated grain yields of four consecutive crops (three maize + one sorghum) at Planaltina, Brazil, plotted against amounts of lime applied before the first crop was planted.

111 Ritchey, K. D. *et al.*, *Agron. J.*, 1980, **72**, 40.
112 Adams, F. and Pearson, R. W., *Soil Sci. Soc. Am. Proc.*, 1969, **33**, 737.
113 Adams, F. *et al.*, *Agron. J.*, 1967, **59**, 453.

Gypsum may be useful as a means of reducing acidity in subsoils, used either alone or as phospho-gypsum. The sulphate allows the Ca to move down in solution. In laboratory columns and in field plots gypsum caused a decrease in Al saturation and an increase in pH in the subsoil of a Brazilian Oxisol.[114] In similar experiments aluminium was not leached out of the subsoil but was apparently precipitated in a form that was insoluble in water or KCl.[109] Calcium sulphate, unlike calcium carbonate, is not a sink for protons but the adsorption of sulphate on to oxide surfaces releases OH^-, so raising the pH by between 0.1 and 0.8 pH units.[115] However, $AlSO_4^+$ ion pairs are also formed, so shifting the balance of Al–OH ion species with release of OH^-. Calcium chloride additions do not raise the pH. The application of lime and gypsum together greatly increases the amount of calcium passing into the subsoil as a result of decreased sulphate adsorption in the topsoil at the higher pH values.[116] Increased leaching of potassium and magnesium results from gypsum applications and increased inputs of these nutrients may be required.

Restrictions to root growth in acid subsoils may also be alleviated by adequate phosphate fertilization.[117] Phosphorus taken up from the topsoil is translocated to roots in the subsoil where it reduces the aluminium toxicity

Table 25.12 The effect of surface applications of calcium carbonate on the surface soil and subsoil pH of old grassland. Rothamsted Park Grass plots receive 470 kg ha^{-1} a^{-1} of ammonium sulphate (which could give a loss of about 450 kg CaCO$_3$ ha^{-1} a^{-1}). Lime is applied once every 4 years at 4000 kg CaCO$_3$ ha^{-1} on Plot 9 and 7910 kg CaCO$_3$ ha^{-1} on Plot 18. Soil pH 1:5 in water. (From Warren, R. G. and Johnston, A. E., *Rothamsted Ann. Rpt.*, 1963, 240)

	Plot 9		Plot 18	
Soil	Surface soil 0–22 cm	Subsoil 22–45 cm	Surface soil 0–22 cm	Subsoil 22–45 cm
Unlimed portion				
1923	4.0	4.8	4.5	5.7
1959	3.8	4.3	4.1	4.4
Limed portion				
1923	4.5	5.2	4.7	5.6
1959	5.3	5.2	7.1	6.6

114 Ritchey, K. D. *et al.*, *Agron. J.*, 1980, **72**, 40. See also Oates, K. M. and Caldwell, A. G., *Soil Sci. Soc. Am. J.*, 1985, **49**, 915.
115 Pavan, M. A. *et al.*, *Soil Sci. Soc. Am. J.*, 1984, **48**, 33.
116 Couto, W. *et al.*, *Soil Sci.*, 1979, **127**, 106 Liming, however, does not always reduce sulphate retention. Hue, N. V. *et al.*, *Soil Sci. Soc. Am. J.*, 1985, **49**, 1196.
117 Miranda, L. N. de and Rowell, D. L., see ref. 83.

by precipitation at the root surface.

Surface applications of lime are made to permanent grassland and movement down the profile is very slow. Table 25.12 shows the effects observed on the Park Grass plots at Rothamsted. During 36 years, the effects of lime were seen only in the subsoil in Plot 18 which received the higher lime applications. Apart from leaching of calcium bicarbonate into the subsoil, biological activity may be involved in carrying surface soil down the profile.

Determination of lime requirements

Traditionally, lime requirement has been considered as the amount of lime required to raise the pH of a soil to a certain value, often about 6.5 in temperate regions. With the rapid development of tropical regions for crop production, and increased understanding of the importance of aluminium and manganese toxicities as limitations to growth, the term is now used as the amount of lime required to alleviate restrictions to crop yield. The removal of exchangeable aluminium, or its reduction to non-toxic levels, is a central consideration.

All laboratory methods for determining lime requirements must eventually be calibrated against crop yields from field experiments. It is difficult to obtain precise relationships between field trials and laboratory determinations because of natural soil variation, uneven lime application, degree of mixing between lime and soil, and losses of lime by leaching before it has reacted with the soil. It is therefore questionable whether highly accurate laboratory methods are justified for advisory work. Because of the problems, there has been a tendency to calibrate laboratory methods with incubation experiments in which soil samples are mixed with increasing amounts of lime, stored moist to facilitate reaction and soil pH measured. This calibrates the laboratory method under ideal conditions which is then used for advisory purposes making use of field experiments and local knowledge. In the USA,[118] the conversion of a laboratory determined lime requirement into a field recommendation is made assuming that there is a standard weight of soil in the cultivated layer, and that about two-thirds of the lime reacts with the soil. In Britain, the Agricultural Development and Advisory Service (ADAS) calculate a field recommendation based on uniform mixing with soil in a 20 cm layer (3300 t dry soil ha^{-1}).

Early methods for the determination of lime requirement included titration of the soil with $Ca(OH)_2$ which allowed soil pH to be raised to any chosen value[119] and, second, the back-titration of excess $Ca(HCO_3)_2$ solution

118 Adams, A. in *Modifying the Root Environment to Reduce Crop Stress* (eds G. F. Arkin and H. M. Taylor), p. 269, *Am. Soc. agric. Eng. Monog.*, No. 4, 1981.

119 Abruna, F. and Vicente, J., *Univ. Puerto Rico J. Agric.*, 1955, **39**, 41, quoted by Adams, A., 1981 (ref. 118).

after reaction with the soil which raises its pH to about 6.3.[120] These have been replaced by methods based on that devised by Woodruff[121] which are easier for routine analysis. The principle is that addition of acid will lower the pH of a buffer solution and the pH change depends on the amount of acid added. For simplicity of use a buffer is chosen which shows an approximately linear decrease in pH with added acid. After calibration of the buffer, the decrease in pH after mixing with the soil sample is measured. This allows the calculation of the amount of acidity in the soil which reacts when the soil pH is raised to a target value between 5.5 and 7 and an equivalent amount of lime is required to neutralize it. In England ADAS[122] use a modification of this method and calculate a recommendation to raise soil pH to 7, so aiming to maintain pH above 6.5 for arable crops. For grass and fruit crops, pH 6 is the aim and for potatoes 5.5.

Problems occurred when this method was used for soils containing large amounts of exchangeable Al and the method of Shoemaker, McLean and Pratt (the SMP method) was developed using a modified buffer.[123] The decrease in pH in the buffer on reaction with soil was calibrated in experiments in which varying amounts of $CaCO_3$ were incubated with soils. Lime requirements were then found to raise soil pH to 6.0, 6.4 or 6.8. These lime requirements may only be suitable for soils of similar acidic properties and CEC to those used in the original calibration,[124] and double buffer methods[125] have been developed to try to make the methods more generally applicable. Several methods have been tested against incubation experiments for 70 acid mineral soils.[126] Lime requirements of less than 10 meq $CaCO_3$ per 100 g soil to attain pH 6.5 were accurately predicted by the methods. For soils requiring between 10 and 40 meq, the predictions were between 69 and 90 per cent of the actual requirements, because the larger the requirement, the lower the final buffer plus soil pH, and so less soil acidity is neutralized. The authors presented regression equations to correct for the proportion of the soil acidity determined by the buffer. By this means the Woodruff and SMP single buffer methods were as accurate as the double buffer methods and so were recommended for routine purposes. ADAS[127] use a simple modification of the Woodruff method to avoid this problem. If the final buffer plus soil pH is below 6.0, then the measurement is repeated using a double strength buffer giving an acceptable final pH.

For the highly weathered soils of the tropics a different approach has

120 Hutchinson, H. B. and MacLennan, K., *Chem. News*, 1914, **110**, 61.
121 Woodruff, C. M., *Soil Sci.*, 1948, **66**, 53.
122 *The Analysis of Agricultural Materials*, *MAFF* Ref. Book, 427, HMSO, London, 1986.
123 Shoemaker, H. E. *et al.*, *Soil Sci. Soc. Am. Proc.*, 1961, **25**, 274.
124 Adams, A. in *Modifying the Root Environment to Reduce Crop Stress* (eds G. F. Arkin and H. M. Taylor), p. 269, *Am. Soc. agric. Eng. Monog.*, No. 4, 1981.
125 Yaan, T. L., *Soil Sci. Soc. Am. J.*, 1976, **40**, 800;   McLean, E. O. *et al.*, *op. cit.*, 1978, **42**, 311.
126 Tran, T. S. and Van Lierop, W., *Soil Sci.*, 1981, **131**, 178.
127 See ref. 122.

emerged over the last 20 years because of evidence which shows that: (i) for many crops of these areas, high pH values are not required and are in some cases harmful[128] and (ii) because aluminium is the toxic element in most acid soils, lime should be used to reduce exchangeable Al to a level which is non-toxic for the crop being grown. Thus exchangeable Al could serve as the single criterion for determining lime requirements for these soils.[129] Experiments in Natal used a tropical grass *Sorghum sudanense* growing in eight Oxisols in the glasshouse.[130] The lime requirements determined by the SMP method were calibrated against pot experiments and exchangeable Al was measured in these soils before and after liming. Table 25.13 gives the lime requirements, and shows that there is close agreement between the lime needed to give maximum yield and that required to reduce the exchangeable Al to a low level: for this crop it was 0.2 meq per 100 g. On average these required values were about one-quarter of the values given by the SMP method (pH 6).

Where moderate levels of exchangeable Al can be tolerated (up to 15 per cent Al saturation) lime requirements (LR) can be based on the equation:[131]

LR as meq $CaCO_3$ per 100 g soil = 1.5 × meq exch. Al per 100 g soil

The exchangeable Al is being neutralized by added lime, with the factor 1.5 inserted to allow for the neutralization of exchangeable hydrogen and non-exchangeable acidity, much of which may be organically bound Al which can be exchanged with La. The required level of exchangeable Al after liming depends on the crop species and variety being grown.[132] The best indicator of this tolerance level may be per cent Al saturation although in the Natal soils exchangeable Al was a better indicator than Al saturation. From the above equation, the following has been proposed:[133]

LR as meq $CaCO_3$ per 100 g = 1.5 [Initial exch. Al − Residual exch. Al]

where the residual exchangeable Al is the amount present in meq per 100 g at the residual per cent Al saturation, i.e. the level tolerated by the crop. This gives

$$LR = 1.5 \left[\text{Initial exch. Al} - \frac{\text{Residual per cent Al} \times \text{CEC}}{100} \right]$$

128 Kamprath, E. J., *Proc. Soil Crop Sci. Soc. Florida*, 1971, **31**, 200. Magnesium stress has been implicated as a cause of yield loss resulting from over liming, Grove, J. H. and Sumner, M. E., *Soil Sci. Soc. Am. J.*, 1985, **49**, 1192.
129 Kamprath, E. J., *Soil Sci. Soc. Am. Proc.*, 1970, **34**, 252.
130 Reeve, N. G. and Sumner, M. E., *Soil Sci. Soc. Am. Proc.*, 1970, **34**, 595.
131 Kamprath, E. J., *Soil Sci. Soc. Am. Proc.*, 1970, **34**, 252; Oates, K. M. and Kamprath, E. J., *Soil Sci. Soc. Am. J.*, 1983, **47**, 690.
132 Wright, M. J. (ed), *Plant Adaption to Mineral Stress in Problem Soils*, Cornell Univ., Ithaca, 1976, chapters by Rhue, R. D. and Grogen, C. P., p. 297 and by Spain, J. M., p. 213.
133 Cochrane, T. T. *et al.*, *Trop. Agric.*, *Trin.*, 1980, **57**, 133.

Table 25.13   Soil properties and lime requirements of eight Natal Oxisols with *Sorghum sudanense* as the experimental crop

Soil	pH 0.02 M CaCl₂	CEC meq/100 g 1 M KCl	Exch. cations meq/100 g			Lime requirements (t CaCO₃ ha⁻¹)		
			Ca	Mg	Al	max. yield†	Exch Al*	pH 6 (SMP)‡
Griffith	4.4	5.5	1.1	1.1	2.0	4.0	4.0	13.0
Clovelly	4.4	6.4	2.1	1.7	2.0	3.0	2.8	10.0
Lidgetton	4.5	7.3	2.3	2.0	1.7	3.0	3.0	10.0
Farmhill	4.6	5.6	2.2	1.3	0.4	1.0	0.6	9.0
Hutton	4.7	6.0	1.0	0.9	2.7	4.0	4.6	10.0
Farmingham A	4.6	4.5	0.8	0.6	2.1	3.0	2.8	8.0
Farmingham B	5.0	6.7	3.7	2.2	0.0	0.0	0.0	9.0
Balmoral	4.7	3.7	1.7	1.6	1.4	1.0	1.0	6.0

† Lime requirement to give maximum yield.
* Lime requirement to reduce exch. Al to 0.2 meq/100 g.
‡ Lime requirement determined by the SMP method to raise pH to 6.

and where the soil analysis is carried out by leaching the soil with 1 M KCl the equation can be written

$$LR = 1.5 \left[\text{Initial exch. Al} - \frac{\text{Residual per cent Al (exch. Al + Ca + Mg)}}{100} \right]$$

The lime requirement estimated from this equation agreed closely for four Ultisols from North Carolina with that measured by incubation. It is, however, unlikely that the factor 1.5 will apply to all soils because of differences in the amount of non-exchangeable acidity neutralized when exchangeable Al is reduced. Tables showing the tolerance of plant species to Al are not available, but maize appears to tolerate up to 70 per cent Al saturation compared to 30 per cent for soyabean.[134] Upland rice, cassava, cowpea, groundnut and many pasture species are tolerant also and varietal differences occur in wheat, maize, sorghum, rice and beans.[135]

Liming materials

The rate at which an application of lime reacts with soil and raises the pH depends on the composition and fineness of grinding of the lime. Lime is a general term, for it includes calcium oxide, calcium hydroxide and calcium carbonate, which is typically in the form of chalk or limestone although it may be an industrial waste product, such as that produced by sugar beet factories. Some blast furnace slags, high in calcium silicates, are also liming materials, for the calcium silicate hydrolyses in the soil to calcium hydroxide and silica. Calcium oxide and hydroxide are marketed as fine powders, and 56 kg of the oxide or 74 kg of the hydroxide are equivalent to 100 kg of the carbonate. Most limestones are calcitic, that is, most of the carbonate is present as calcium carbonate, although the content of calcium carbonate differs for limestones from different sources. Some limestones are dolomitic, with a proportion of their carbonate present as magnesium carbonate, and these are particularly valuable on acid soils low in magnesium. The acid neutralizing ability of liming materials is expressed as a neutralizing value, relative to the effect of CaO. For example, pure $CaCO_3$ has a neutralizing value of 56, i.e. 100 kg $CaCO_3$ has the same effect as 56 kg CaO.

Natural limestones must be crushed to facilitate reaction with the soil. If a chalk is used, crushing to a fine powder is technically simple and the process cheap. But it is more expensive to crush hard limestones, and the finer they are ground, the more expensive the process. On the other hand, the finer they are ground, to within certain limits, the more quickly and completely they will react with the soil when properly incorporated. Limestones crushed

134 Evans, C. E. and Kamprath, E. J., *Soil Sci. Soc. Am. Proc.*, 1970, **34**, 893.
135 Wright, M. J. (ed) (see ref. 132).

to pass a 100 mesh sieve (aperture 0.15 mm) are almost as reactive as the oxide and hydroxide. Limestone crushed to pass a 20 mesh sieve (aperture 0.84 mm), but to be retained by 40 mesh (aperture 0.42 mm), takes considerably longer to exert its full effect on the soil pH, although in general there is little difference between the effect of the coarse and fine materials after a year.[136] Limestone coarser than 2 mm is, however, relatively inefficient as a liming material. Natural limestones often contain some silica, and it is possible that the more siliceous, harder limestones may need to be crushed rather finer, for the silica coatings on the limestone nodules protect their surfaces against dissolution.[137] Details of liming materials and regulations for their use are available for the UK[138] and for the USA.[139]

The nature of soil alkalinity

Soil pH values above 7 are associated with the presence of carbonates in soil and soil solution. Figure 25.12a shows the concentrations of carbonate species in solution in eqilibrium with the atmospheric concentration of CO_2 (a partial pressure of 0.0003).[140] The point P marks the condition of pure water with dissolved CO_2 at this partial pressure: the water has a pH of 5.65. Values higher or lower than this are associated with the presence of base or acid added to the solution. Below pH 6, aqueous CO_2 and H_2CO_3 are predominant. Between pH 6 and pH 9, HCO_3^- is predominant with the concentration increasing ten-fold for each pH unit rise, and above pH 10, CO_3^{2-} becomes predominant. The total dissolved carbonate species are present in very much greater concentrations than OH^- at all pH values.

The partial pressure of CO_2 in the soil atmosphere is normally higher than that in air as a result of soil respiration, and may be up to about 100 times greater (see Ch. 9). Figure 25.12b shows the concentrations of the carbonate species for a partial pressure of 0.03. These concentrations are raised 100 times by the higher CO_2 concentration and the equilibrium pH in pure water is now 4.65. Thus soil solution pH values above or below 4.65 are due to the presence of base or acid. Figure 25.12 allows the concentrations of the carbonate species in soil solution to be found if the pH is known. Soil pH is easily measured and so solution composition can be obtained.

136 Reith, J. W. S. and Williams, E. G., *Emp. J. expl. Agric.*, 1949, **17**, 265.
137 Lahav, N. and Bolt, G. H., *Soil Sci.*, 1964, **97**, 293.
138 *Lime and Liming, MAFF Ref. Book*, 35, HMSO, London, 1981.
139 Adams, F., (ed) *Soil Acidity and Liming*, 2nd Edn., *Agron. Monog.* No. 12, Am. Soc. Agron., Madison, Wisc., 1984.
140 Stumm, W. and Morgan, J. J., *Aquatic Chemistry*, Wiley, 1970. A new carbonate species ($H_3C_2O_6^-$) has been reported by Covington, A. K., *Chem. Soc. Rev.*, 1985, **14**, 265. It may be as much as 30 per cent of the total carbonate in solution at pH 6.5.

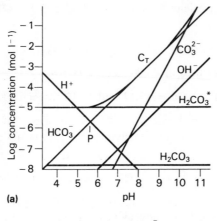

(a)

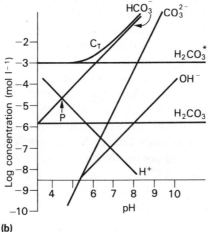

(b)

Fig. 25.12 Carbonate species in pure water in equilibrium with
(a) atmospheric CO_2. Partial pressure $P_{CO_2} = 10^{-3.5}$ atmosphere
(b) $P_{CO_2} = 10^{-1.5}$ atmosphere
$[H_2CO_3^*] = [CO_2 \text{ aq}] + [H_2CO_3]$
$C_T = [H_2CO_3^*] + [HCO_3^-] + [CO_3^{2-}]$

Calcareous soils develop pH values depending on the equilibria between $CaCO_3$, H_2O and CO_2 and for calcite, the most commonly occurring carbonate mineral, the following applies:

$$CaCO_3 = Ca^{2+} + CO_3^{2-} \quad K_{sp} = (Ca)(CO_3),$$
where parentheses indicate activities.

For an open system where calcite and water are equilibrated with a known CO_2 partial pressure

$$2pH - pCa^{2+} = 18.08 - pK_{sp} - \log P_{CO_2}$$

Two values of K_{sp} for calcite are 3.02×10^{-9},[141] and 3.3×10^{-9}.[142] Using the latter we obtain

$$2pH - pCa = 9.60 - \log P_{CO_2}$$

which for partial pressures of 0.0003 and 0.03 gives

$$2pH - pCa = 13.12 \text{ and } 11.12 \text{ respectively.}$$

In the above equation the activity of Ca is required. To avoid converting the calcium concentration into activities the following equation can be used[143]:

$$2pH + \log [Ca] = 9.60 - \log P_{CO_2} + 1.43 \sqrt{\mu}$$

where μ is the ionic strength of the solution and [Ca] is the molar concentration. Ionic strength can be conveniently calculated from the Debye–Hückel equation assuming that the solution is $Ca(HCO_3)_2$ and that $[HCO_3] = 2[Ca]$, or by measuring the electrical conductivity (EC) of the solution and using the relationship μ(mol per litre) $= 0.013$ EC mS cm^{-1}.[144] Figure 25.13 gives the values found from these equations. Soil solution Ca concentrations are around 10^{-3} M in calcareous soils and pH values for equilibrium systems lie between 7 and 8.5. (For calcite/H$_2$O, pH $= 8.4$ at $P_{CO_2} = 0.0003$.)

A calcareous subsoil can have a pH greater than 8.5 because if water, initially in equilibrium with the carbon dioxide in the atmosphere, percolates through a calcareous soil in which little carbon dioxide is being produced, the soil approximates to a closed system. When pure water initially in equilibrium with the atmosphere is equilibrated with calcite with no contact with the atmosphere, the pH rises to 9.9, pCa $= 3.9$ and [Ca] $= 1.4 \times 10^{-4}$ M.[145] pH

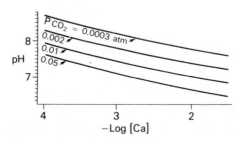

Fig. 25.13 Relationship between pH, $-\log[Ca]$ and P_{CO_2} for solutions in equilibrium with calcite.

141 Christ, C. L. *et al.*, *J. Res. US Geol. Surv.*, 1974, **2**, 175.
142 O'Connor, G. A. and Cadena, F. C., *Soil Sci.*, 1975, **102**, 182.
143 Yaalon, D. H., *Trans. 5th Int. Congr. Soil Sci.*, 1954, **2**, 356.
144 Griffin, R. A. and Jurinak, J. J., *Soil Sci.*, 1973, **116**, 26.
145 Stumm, W. and Morgan, J. J., *Aquatic Chemistry*, Wiley, 1970.

values between 8.8 and 9.8 have been measured for leachates from calcareous soils in the laboratory[146] and high pH values have also been recorded in suspensions.[147]

Soil carbonates

Calcite is the most common form of carbonate in soils but aragonite ($CaCO_3$, $K_{sp} = 10^{-8.22}$), dolomite ($CaCO_3\ MgCO_3$, $K_{sp} = (Ca)(Mg)(CO_3)^2 = 10^{-16.7}$) and magnesite ($MgCO_3$, $K_{sp} = 10^{-4.9}$) also occur. Aragonite readily weathers to calcite.[148] Calcite normally occurs in soils in an impure form, the main contaminant being magnesium within the crystal, and a variety of surface contaminants. This gives a range of values for the ion activity product $(Ca)(CO_3)$ around that of pure calcite.[149] Some calcareous soils, in which the solution is supersaturated with respect to calcite,[150] contain calcites low in magnesium. These mineral particles, after separation from the soil, gave the ion activity product expected for calcite. Supersaturation in the soil solutions may therefore result from the presence of calcium silicates more soluble than calcite, with the kinetics of Ca silicate dissolution and $CaCO_3$ precipitation controlling the ion activity products. The rate of the carbonate reactions[151] will depend on surface area: this is difficult to determine precisely because of its low value, but a few m^2 per g is normal for soils.[152] Argon adsorption can be used to determine these small areas.

The solubility of the low magnesium calcites has been shown to depend on composition of the solid phase which may be considered as a solid solution[153] and the solubility product is then expressed as $K_{sp} = (Ca)^{1-x}\ (Mg)^x\ (CO_3)$ where x is the mole fraction of $MgCO_3$ in the calcite.[154]

146 Turner, R. C. *et al., Canad. J. Soil Sci.*, 1958, **38**, 94.
147 Dowdell, R. J. *et al., J. Soil Sci.*, 1984, **35**, 169.
148 Dixon, J. B. and Weed, S. B. (eds), *Minerals in Soil Environments, Soil Sci. Soc. Am.*, Madison, Wisc., 1977.
149 Olsen, S. R. and Watanabe, F. S., *Soil Sci.*, 1959, **88**, 123; Schinas, S. and Rowell, D. L., *J. Soil Sci.*, 1977, **28**, 351; Hasset, J. J. and Jurinak, J. J., *Soil Sci. Soc. Am. Proc.*, 1971, **35**, 403.
150 Suarez, D. L. and Rhoades, J. D., *Soil Sci. Soc. Am. J.*, 1982, **46**, 716.
151 Morse, J. W. in *Carbonates: Mineralogy and Chemistry* (ed R. J. Reeder), p. 227, Min. Soc. Am., 1983; Amrhein, C., *Soil Sci. Soc. Am. J.*, 1985, **49**, 1393.
152 Lahav, N. and Bolt, G. H., *Nature, Lond.*, 1963, **200**, 1343; Abedi, M. J. and Talibudeen, O., *J. Soil Sci.* 1974, **25**, 357.
153 Mackenzie, F. T. *et al.* in *Carbonates: Mineralogy and Chemistry* (ed R. J. Reeder), p. 97, Min. Soc. Am., 1983.
154 Thornsten, D. C. and Plummet, L. N., *Am. J. Sci.*, 1977, **277**, 1202.

Variations in pH in calcareous soils

Soils containing free calcium carbonate have pH values measured in water by conventional methods of between 7.0 and 8.4. The conventional methods do not allow time for the solution to come to equilibrium with the carbonates and the partial pressure of CO_2 is not controlled. However, differences between samples measured at equilibrium are also seen in the conventional measurements. The differences in pH between the soil samples do not seem to be related to the amount of carbonate present, but reflect more closely the solubility of the carbonates.[155] Even though pH and the related solution composition are not directly related to carbonate content, the latter may be important to plants. The ability of the soil to maintain a high pH and HCO_3^- concentration close to roots will depend both on carbonate solubility and on rate of dissolution. The latter will be influenced by surface area which may increase with carbonate content.

A further source of variation in soil pH will be the plant root itself, through the production of CO_2 and the release of H^+ or HCO_3^-. The effects have been discussed earlier (p. 862): at pH values above 7, the soil has a relatively high acidity diffusion coefficient and so pH gradients are predicted to be very small for normal inputs of H^+ or HCO_3^- by the root. Release of CO_2 by the root may, however, have a more significant effect. Because of the greater solubility of CO_2 than O_2 in soil water, small changes in partial pressure of CO_2 lead to relatively large concentration gradients and therefore rapid diffusion. Even where the O_2 partial pressure falls from 0.21 in an air-filled pore to near zero at the root surface through water-saturated soil (the maximum possible difference), the partial pressure of the CO_2 will only increase to about 0.01 at the root surface compared to 0.0003 in the air filled pore.[156] This appears to be a small change, but the implications in terms of pH are significant as shown in Fig. 25.13, reducing the pH by about 1 unit.

Plant growth in calcareous soils

Plants can be roughly divided into three groups: those that grow well on calcareous soils called calcicoles; those that that do not grow well on such soils called calcifuges; and those that are tolerant of a wide range of pH. The growth of crops on calcareous soils presents few management problems compared to acid soils primarily because although there are nutrient deficiencies on calcareous soils, there are toxicities on acid soils. In acid soils the toxicities have to be removed by changing soil pH, whereas in calcareous soils nutrients can be supplemented via the soil or through foliar sprays. In both types of soil, however, choice of suitable crop or of crop variety is important.

155 Schinas, S. and Rowell, D. L., *J. Soil Sci.*, 1977, **28**, 351.
156 Greenwood, D. J., *J. Soil Sci.*, 1970, **21**, 314.

Chlorosis

It is not known if the bicarbonate ion has any direct effect on plant growth. The concentrations of HCO_3^- can become high at pH 8.5 (10^{-3}–10^{-2} M). One of the main difficulties in conventional solution culture experiments which attempt to isolate the effect of bicarbonate is that any change in bicarbonate concentration in solution is normally associated with changes in other solution properties such as pH, Ca concentration, form of phosphate ions, and possible precipitation of micronutrient metals. In soils there are even more difficulties because of associated changes in the equilibria between surfaces and solution.

In solution culture the normal effect of high bicarbonate concentrations is the development of iron chlorosis in sensitive plants. The reason for this has been reported as reduced absorption of iron by the root, accumulation of iron in the roots with reduced translocation to stems and leaves, inhibition of respiration, depression of protein synthesis and oxidase activity, inhibition of cytochrome oxidase activity and other effects.[157]

In soils, increased bicarbonate concentration in solution is associated with increased pH or increased CO_2 concentration. Increased pH will reduce the solubility of iron and manganese and so may induce chlorosis. However, it seems that under these conditions although the iron and manganese content of plants may not be much changed, the ability of sensitive plants to use these nutrients is affected. The sections on Fe and Mn as micronutrients give further information (Chs. 3 and 23). It is also known that an increased phosphate supply may cause iron chlorosis to develop possibly associated with decreased citrate levels in roots.[158]

Iron chlorosis most commonly occurs on highly calcareous soils or shallow soils over chalk. However, crops growing on apparently similar calcareous soils often show very different degrees of chlorosis. The growth of tomatoes, a moderately sensitive plant, is closely related to the solubility of the carbonates in soils[159] (Fig. 25.14) and since pH depends on this solubility, pH is a good indicator of the likelihood of chlorosis for this plant. Yield is less clearly related to the amount of $CaCO_3$ in the soil, although particle size and surface area may be important in controlling the ability of the soil to buffer pH at a high value at the root surface. 'Active $CaCO_3$' has been proposed as an index of the likelihood of chlorosis.[160] Its measurement involves shaking a soil with ammonium oxalate solution and measuring the amount of calcium oxalate precipitated as a coating on the soil carbonates under standard

157 Gauch, H. G. and Wadleigh, C. H., *Bot. Gaz.*, 1951, **112**, 259; Porter, L. K. and Thorne, D. W., *Soil Sci.*, 1955, **79**, 373; Brown, J. C., *Adv. Agron.*, 1961, **13**, 329; Chen, Y. and Barak, P., *op. cit.*, 1982, **35**, 217.
158 Brown, J. C., *Adv. Agron.*, 1961, **13**, 329.
159 Schinas, S. and Rowell, D. L., *J. Soil Sci.*, 1977, **28**, 351.
160 Drouineau, G., *Ann. Agron.*, 1942, **2**, 441; Yaalon, D. H., *Pl. Soil*, 1957, **8**, 275.

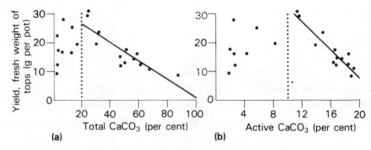

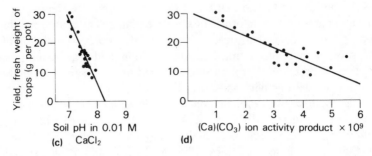

Fig. 25.14   The relationships between yield of tomato in a glasshouse experiment and (a) total $CaCO_3$, (b) active $CaCO_3$, (c) pH in 0.01 M $CaCl_2$ and (d) (Ca) (CO_3) ion activity product for 22 soils on calcareous parent materials in southern England.

conditions. It is a measure of the extent and reactivity of the carbonate surfaces and is broadly related to total carbonate. Any management procedure which increases the soil pH should be avoided where sensitive crops are grown: deep ploughing which brings up carbonate of higher solubility to increase the amount in the surface layers may well add to problems of chlorosis.

Phosphate supply and utilization

Phosphorus availability is often low in calcareous soils and the reasons for this are discussed in Chapter 21. The effects of large calcium carbonate contents have been demonstrated using the Woburn soil, treated with up to 26 per cent calcium carbonate and stored in a moist aerobic state, giving pH values ranging from 5.75 to 7.7 in 0.01 M $CaCl_2$. For pH values above 7.0, the soil solution was close to equilibrium with octacalcium phosphate[161]. The calcium phosphate potential was reduced as the calcium carbonate content and pH increased, and the H_2PO_4 concentration decreased from 10^{-5} to

161 Webber, M. D. and Mattingly, G. E. G., *J. Soil Sci.*, 1970, **21**, 111, 121.

10^{-6} M in a soil that initially had 70 mg kg^{-1} bicarbonate-extractable phosphate. On the other hand, calcareous soils become less buffered with respect to phosphate as the calcium carbonate increases because the carbonate, which contributes weakly absorbing sites, will dilute the soil material containing the hydrous oxides, which have strongly absorbing sites.[162] The rate of phosphate diffusion to roots will be increased by a reduction in buffering so tending to increase phosphate availability.

Adsorption of added phosphate in calcareous soils will depend on the numbers of strongly adsorbing and weakly adsorbing sites and the extent to which these are filled before the fertilizer is applied. A range of calcareous soils[163] had 140–345 μg P g^{-1} high energy adsorption capacity, well related to dithionite soluble Fe, and 400–663 μg P g^{-1} low energy adsorption capacity which was well related to the CaCO$_3$ surface area but not to the per cent CaCO$_3$. In fertilizing such soils it should be necessary only to saturate the high energy sites because once these are filled, soil solution concentration rises rapidly with added phosphate. If this is correct, soils with high carbonate contents may require less phosphate to raise phosphate availability to acceptable levels.

The utilization of phosphate in calcareous soils depends on plant tolerance to alkalinity. Wheat is unaffected by high carbonate levels provided phosphate is available in adequate amounts, whereas tomato is sensitive even with an adequate phosphate supply. The high bicarbonate concentration may directly affect the uptake of phosphate, or there may be an indirect effect through iron or manganese deficiencies.[164] Increased phosphate supply may intensify the chlorosis.

Other effects

Increased potassium[165] and nitrogen supply[166] may intensify the development of iron chlorosis in plants on calcareous soil, but it is not known whether these effects are significant under field conditions.

Of the arable crops grown on calcareous soils, field peas seem to be the most intolerant, with Mn deficiency commonly occurring unless foliar sprays are used. Copper deficiency in cereals also occurs on shallow chalk soils where extractable copper is low and is related to per cent organic matter content rather than per cent CaCO$_3$ or depth of soil over chalk. In arid and

162 Holford, I. C. R. and Mattingly, G. E. G., *J. Soil Sci.*, 1975, **26**, 407; Ryan, J. *et al.*, *Soil Sci. Soc. Am. J.*, 1985, **49**, 1215.

163 Holford, I. C. R. and Mattingly, G. E. G., *J. Soil Sci.*, 1975, **26**, 407. Phosphate sorption on to CaCO$_3$ reagents is well related to their surface area. Amer, F. *et al.*, *Soil Sci. Soc. Am. J.*, 1985, **49**, 1137.

164 Thorne, D. W., *Soil Sci. Soc. Am. Proc.*, 1946, **11**, 391; Brown, J. C., *Soil Sci.*, 1968, **105**, 159; Singh, M. and Dahiya, S. S., *Pl. Soil*, 1976, **44**, 511.

165 Hewitt, E. J. in *Plant Physiology*, Vol. 3 (ed F. C. Steward), Acad. Press, 1963.

166 Thorne, D. W. *et al.*, *Soil Sci. Soc. Am. Proc.*, 1950, **15**, 254.

semi-arid regions, zinc deficiency often occurs on calcareous soils[167] and in soils from similar parent material is related to the amount of $CaCO_3$, but the relationship is not seen for soils from different parent materials.[168] The nature and particle size of the carbonates seems to be important and plant uptake of zinc added to these soils is inversely related to the amount of clay-sized carbonate present.

Sodic soils

Soils containing sodium carbonate have pH values between 7 and 10.5 directly related to the amount of sodium carbonate present, this being a soluble salt and is completely dissolved in a moist soil. Using the principle of electroneutrality for a Na_2CO_3 solution[169] with concentrations in moles per litre,

$$[H] + [Na] = [OH] + [HCO_3] + 2\,[CO_3]$$

and

$$pH + p[Na] = 7.85 - \log P_{CO_2} - 0.51\sqrt{\mu}$$

In soil solutions other soluble salts may be present, e.g sodium sulphate and sodium chloride, and p[Na] in the above equation is not the total Na in solution but that which is not balanced by Cl^- or SO_4^{2-}. From the first equation [Na] is approximately equal to $[HCO_3] + 2[CO_3]$ between pH 6 and 10, and this is the alkalinity of the solution which can be substituted for pNa to give

$$\log \{2\,[CO_3] + [HCO_3]\} - pH = \log P_{CO_2} - 7.85 + 0.51\sqrt{\mu}$$

The alkalinity can be determined by titration of the solution with HCl and methyl orange. Figure 25.15 shows the data and theoretical lines for three partial pressures of CO_2[170] for solutions and soil and calcite suspensions with added Na_2CO_3, Na_2SO_4 and NaCl. The divergence from theory at the low P_{CO_2} level may be the result of slow approach to equilibrium.

The development of alkalinity may be the result of several processes:

1 Rock weathering may produce bicarbonates of Na, Ca and Mg. Evapotranspiration of the solution causes the precipitation of calcium and magnesium carbonates leaving Na^+ and HCO_3^- in solution. Subsequent further drying causes precipitation of sodium bicarbonate, often as hydrated crystals.

167 Thorne, D. W., *Adv. Agron.*, 1957, **9**, 31.
168 Navrot, J. and Ravikovitch, S., *Soil Sci.*, 1969, **105**, 184; ibid., 1969, **108**, 30.
169 Ponnamperuma, F. N., *Soil Sci.*, 1967, **103**, 90; Ponnamperuma, F. N. *et al.*, *Soil Sci. Soc. Am. Proc.*, 1969, **33**, 239.
170 Mashhady, A. S. and Rowell, D. L., *J. Soil Sci.*, 1978, **29**, 65.

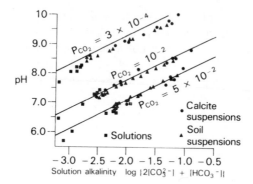

Fig. 25.15 The relationship between alkalinity and pH for solutions, and soil and calcite suspensions at three partial pressures of CO_2.

2 Irrigation water or groundwaters containing $NaHCO_3$ may enter the soil and accumulate.[171]

3 Saline waters (NaCl) may enter the soil and build up a high exchangeable sodium percentage displacing calcium and magnesium. Subsequent leaching with winter rainfall or good quality irrigation water causes the dissociation and hydrolysis of the exchangeable Na[172]:

$$Na \text{ clay} + H_2O \rightarrow H^+ \text{ clay} + Na^+ + OH^-$$
$$Na^+ + OH^- + CO_2 \rightarrow Na^+ + HCO_3^-$$

Acid attack of the clay occurs with the release of magnesium, silicon and aluminium into solution, the aluminium precipitating as hydroxide. With sufficient leaching the $NaHCO_3$ is removed and an acid soil and solution remains. Calcium-saturated exchange surfaces also dissociate during leaching[173] but to a much smaller extent than for Na due to exchange competition between Ca^{2+} and H^+.

Plant growth in sodic soils

The main limitations to plant growth in saline and sodic soils are discussed in Chapter 27. These include the general effects of increased osmotic stress on soil water supply, specific toxic effects of boron, sodium and chloride and the effects of high pH in reducing the availability of phosphate, iron, manganese and zinc.

Many sodic soils are also calcareous, but the pH is related primarily to the amount of $NaHCO_3$ in the soil solution. Reduced availability of nutrients as

171 Van Beek, C. G. E. M. and Van Breemen, N., *J. Soil Sci.*, 1973, **24**, 129.
172 Brown, D. S. and Miller, J. D., *Soil Sci. Soc. Am. Proc.*, 1971, **35**, 705;   Shainberg, I., *op. cit.*, 1973, **37**, 689;   Schramm, L. L. and Kwak, J. C. T., *Soil Sci.*, 1984, **137**, 1.
173 Frenkel, H. *et al.*, *Soil Sci. Soc. Am. J.*, 1983, **47**, 649;   Mashhady, A. S. and Rowell, D. L., *J. Soil Sci.*, 1978, **29**, 65.

a result of these high pH values may be considered in general to be an intensification of the effects discussed for calcareous soils. However iron and manganese deficiencies are often not found even for sensitive crops such as tomatoes and tree fruits, despite high pH values.[174] Soluble iron–organic complexes may be important as a means of supplying iron to crops under these conditions.[175]

174 Wallace, A. and Lunt, O. R., *Proc. Am. Soc. hort. Sci.*, 1960, **75**, 819.
175 D'Yakanova, K. U. D., *Soviet Soil Sci.*, 1962, 692;   Mashhady, A. S. and Rowell, D. L. *J. Soil Sci.*, 1978, **29**, 367.

26

Flooded and poorly drained soils

Chemical principles of reduction processes

Flooding a soil containing decomposable organic matter causes the onset of anaerobic or partially anaerobic conditions because the soil micro-organisms, in decomposing the organic matter, will use up any free oxygen dissolved in the soil water much faster than atmospheric oxygen can diffuse into the wet soil. This shortage of oxygen will cause some species of bacteria to carry out chemical reductions which may affect plant growth.[1]

Organisms obtain the energy they need for their vital processes through a series of chemical reactions involving the transfer of electrons from substances which serve as sources of energy to substances which may become products of respiration. For example, glucose is a source of electrons when it is oxidized to pyruvic acid in the first stage of its breakdown to carbon dioxide:

---

1 For a more detailed discussion of microbial metabolism in waterlogged soils, see Yoshida, T. in *Soils and Rice*, IRRI, 1978.

$$C_6H_{12}O_6 = 2CH_3COCOOH + 4H^+ + 4e^-$$

If the organisms are respiring aerobically the final electron sink is oxygen, which accepts electrons and combines with hydrogen ions to give water, so aerobic respiration involves the reduction of molecular oxygen to water:

$$O_2 + 4H^+ + 4e^- \rightarrow 2H_2O$$

In the absence of free oxygen a number of other substances can accept electrons and take part in a reduction reaction. Some ions containing oxygen, such as nitrate and sulphate, can accept electrons and lose oxygen, thus:

$$NO_3^- + 2H^+ + 2e^- \rightarrow NO_2^- + H_2O$$

and:

$$2NO_2^- + 8H^+ + 6e^- \rightarrow N_2 + 4H_2O$$

Some high valency cations will accept electrons and become reduced to a lower valency state. For example

$$Fe^{3+} + e^- \rightarrow Fe^{2+}$$
$$Mn^{4+} + 2e^- \rightarrow Mn^{2+}$$

Finally, the hydrogen ion itself can accept an electron to become hydrogen gas:

$$2H^+ + 2e^- \rightarrow H_2$$

A second consequence of bacterial activity under conditions of oxygen deficiency is that the organic compounds are no longer fully oxidized to carbon dioxide and water, but instead intermediate products are produced, such as simple fatty acids, hydroxy-carboxylic and polycarboxylic acids, alcohols and ketones. Some of these can reduce ferric oxides, bringing the iron into solution as a ferrous chelate. These organic compounds are, in turn, further decomposed with the production of carbon dioxide, methane and other hydrocarbons, and sometimes hydrogen gas. Thus, flooded soils may contain both inorganic and organic compounds in reduced forms. These will donate electrons and may absorb oxygen. The greater the tendency to donate electrons, the stronger will be the reducing agent.[2] The tendency of a solution to donate electrons to a reducible substance or to accept electrons from an oxidizable substance can be measured by its oxidation–reduction or redox potential, E_h (volts). The more strongly reducing a substance, the lower is

2 For a review of changes taking place in waterlogged soils, see Ponnamperuma, F. N., *Adv. Agron.*, 1972, **24**, 29. A useful description of the chemistry of these changes is in Morris, J. G., *A Biologist's Physical Chemistry* (2nd edn), Arnold, 1974. The application of this chemistry to soils can be found in Rowell, D. L. in *The Chemistry of Soil Processes* (eds D. J. Greenland and M. H. B. Hayes), p. 401, Wiley, 1981.

its potential. The E_h of a solution is the voltage difference between a platinum electrode and a hydrogen reference electrode but for convenience a calomel reference electrode is used to obtain a measured voltage difference[3] where

$$E_h = \text{E measured} + 0.248 \text{ V}$$

because the voltage difference between a hydrogen and a calomel electrode is 0.248 V. Transfer of protons normally occurs with the transfer of electrons in an oxidation–reduction reaction and so the redox potential depends on the pH of the solution. Thus, E_h in a well-aerated solution falls linearly by 59 mV for each pH unit rise in solution, and the relationship between E_h and pH is given in Table 26.1. This relationship holds reasonably well in aerated soils.

In the same way as one can carry out an acid–base titration, and define the buffer capacity of a solution as the amount of hydrogen ions needed to give a unit reduction in pH (p. 854), so one can define the poise of a solution as the amount of electrons, as a reducing substance, that must be added to give a unit reduction in the redox potential. A solution which can accept a large amount of acid for a small change in pH is said to be well buffered, and a solution that can accept a large amount of a reducing substance for a small change in redox potential is said to be well poised.

The various reduction reactions that can take place in a soil when it becomes anaerobic depend on the redox potential at which they are most strongly poised – those poised at a high redox potential going to completion before those poised at an appreciably lower potential become important. These potentials are also pH dependent, but the effect of a unit change in pH on the potential is not the same for all reduction reactions, and for simple inorganic reductions can vary from -59 mV to -177 mV per unit increase of pH (Table 26.1), and in soils may be as high as -232 mV.[4] This means that the relative order in which a series of reductions take place when the soil becomes anaerobic depends on the pH of the soil, if they are poised at about the same potential.

The principal inorganic reductions that poise a soil when it becomes anaerobic are nitrate to nitrite, manganic salts and manganese dioxide to manganous ions, ferric hydroxide to ferrous ions, hydrogen ions to hydrogen gas and sulphate to sulphite and sulphide. The redox potentials of these reductions at maximum poise are given in Table 26.1, both at pH 5.0 and pH 7.0. Redox potentials in aerobic soils are often around 600 mV, well below the value for O_2/H_2O in the table. This may be due to a reduced partial pressure of oxygen, heterogeneity and poor poise. Some of the reductions are reversible in the soil system, such as those for iron and manganese. However, nitrite and sulphite are reduced more easily than nitrate and

3 Ponnamperuma, F. N., *Adv. Agron.*, 1972, **24**, 29; Bohn, H. L., *Soil Sci. Soc. Am. Proc.*, 1968, **32**, 211; Liu Zhi-Guang and Yu Tian-Ren, *J. Soil Sci.*, 1984, **35**, 469.
4 Patrick, W. H., *Trans. 7th Int. Congr. Soil Sci.*, 1960, **2**, 494.

Table 26.1   Oxidation–reduction potentials of typical soil systems

System	Oxidation–reduction potential in mV, 25°C	
	At pH 5	At pH 7
$O_2 + 4H^+ + 4e^- = 2H_2O$ $E_h = 1.23 + 0.015 \log P_{O_2} - 0.059 \text{ pH}$	934	816
$NO_3^- + 2H^+ + 2e^- = NO_2^- + H_2O$ $E_h = 0.83 = 0.0295 \log NO_2^-/NO_3^- - 0.059 \text{ pH}$	539	421
$MnO_2 + 4H^+ + 2e^- = Mn^{2+} + 2H_2O$ $E_h = 1.23 - 0.0295 \log Mn^{2+} - 0.119 \text{ pH}$	634	396
$Fe(OH)_3 + 3H^+ + e^- = Fe^{2+} + 3H_2O$ $E_h = 1.06 - 0.059 \log Fe^{2+} - 0.177 \text{ pH}$	172	-182
$SO_4^{2-} + 10H^+ + 8e^- = H_2S + 4H_2O$ $E_h = 0.30 - 0.0074 \log H_2S/SO_4^{2-} - 0.075 \text{ pH}$	-67	-215
$CO_2 + 8H^+ + 8e^- = CH_4 + 2H_2O$ $E_h = 0.17 - 0.059 \log P_{CH_4}/P_{CO_2} - 0.059 \text{ pH}$	-126	-244
$2H^+ + 2e^- = H_2$ $E_h = 0.00 - 0.059 \text{ pH}$	-295	-413

sulphate, the former to dinitrogen or nitrous oxide, which are lost from the system as gases, and the latter to sulphide, which, in the absence of ferrous iron, may also be lost as gaseous hydrogen sulphide. Therefore nitrate reduction always, and sulphate reduction sometimes, are irreversible processes.

The reduction of nitrate and ferric iron can be carried out by a number of different bacteria, many of which are capable of bringing about both these reductions. Of 71 bacteria isolated from three gleyed subsoils,[5] all but three could reduce nitrate to nitrite, but only about half could carry the reduction to nitrous oxide or nitrogen gas. Most of the bacteria were either pseudo-monads or bacilli. On the other hand, only few bacteria appear capable of reducing sulphate, and these belong to the genus *Desulphovibrio*. They are most active in neutral soils and reduction occurs slowly outside the pH range 6.5 to 8.5.[6]

Some of the oxidation–reduction systems given in Table 26.1 are only partially applicable to field soils, because the compounds present in the soil are not pure. For example, ferric hydroxide nearly always contains impurities, such as manganese, because it is formed from the oxidation of ferrous iron under acid conditions and will absorb some divalent manganese under

5 Ottow, J. C. G. and Glathe, H., *Soil Biol. Biochem.*, 1971, **3**, 43.
6 Connell, W. E. and Patrick, W. H., *Science*, 1968, **159**, 86.

conditions too reducing for the manganous ion to be oxidized. Manganese dioxide often contains fewer impurities because when the redox potential has risen sufficiently high to oxidize the manganous ion, most other oxidizable cations have been oxidized.[7] However, no single form of manganese oxide dominates in soils and each is poised at a different potential.[8] Thus, iron and manganese precipitates have oxidation–reduction potentials spanning a fairly wide range, so that, in the field, some ferrous ions will be present before all the higher valency manganese compounds have been reduced.[9]

In soils containing iron, the ferric–ferrous reaction is dominant in poising the system. The ferric ion concentration in soil solutions is very low, due to the insolubility of ferric hydroxide or hydrated oxide, but the concentration of ferrous ions can be sufficiently high for plant growth to be affected. The normal reversible oxidation–reduction reaction is probably between the metastable $Fe(OH)_3$ and the metastable ferroso-ferric hydroxide $Fe_3(OH)_8$.[10] The relationship between the redox potential, the pH of the solution and the ferrous ion activity is:

$$E_h = 1.06 - 0.059 \log (Fe^{2+}) - 0.18 \text{ pH} \qquad [26.1]$$

if the pH is too low for the ferroso-ferric hydroxide to be formed. This equation holds for many soils containing moderate amounts of iron and organic matter and the term $E_h + 0.18$ pH has been proposed as a measure of the severity of reducing conditions.[11] For some rice soils this is above 1.34 for aerated soils, between 1.34 and 1.15 for soils which are suitable for rice, but below 1.15 when very severe reducing conditions prevail. Since the reduction process removes hydrogen ions from the solution, the pH of acid soils rises, and once it has risen sufficiently high, ferroso-ferric hydroxide begins to precipitate, provided there is sufficient ferrous iron present, and controls the concentration of ferrous ions in solution. This hydroxide gives the soils a typical green–grey colour. The soil solutions should now conform to the following relationships:

$$Fe_3(OH)_8 + 8H^+ + 2e^- = 3Fe^{2+} + 8H_2O$$
$$E_h = 1.37 - 0.088 \log (Fe^{2+}) - 0.24 \text{ pH} \qquad [26.2]$$

and $3Fe(OH)_3 + H^+ + e^- = Fe_3(OH)_8 + H_2O$
$$E_h = 0.43 + 0.059 \text{ pH} \qquad [26.3]$$

7 For examples see Collins, J. F. and Buol, S. W., *Soil Sci.*, 1970, **110**, 111, 157.
8 Ponnamperuma, F. N. *et al.*, *Soil Sci.*, 1969, **108**, 48.
9 Valera, C. I. and Alexander, M., *Pl. Soil*, 1961, **15**, 268. Jordan, J. H. *et al.*, *Soil Sci.*, 1967, **104**, 129.
10 Ponnamperuma, F. N. and others, *Soil Sci.*, 1966, **101**, 421; ibid., 1967, **103**, 374; *Int. Rice Res. Inst., Ann. Rept. for 1965*, p. 126.
11 Jeffery, J. W. O., *J. Soil Sci.*, 1961, **12**, 172.

which combined with equation [26.2] gives

$$\log (Fe^{2+}) = 10.6 - 2 \text{ pH}. \qquad\qquad [26.4]$$

In equation [26.2] the constant falls within the range 1.34 to 1.39, the lower values being for soils low in iron and high in manganese. The constant in equation [26.3] falls within the range 0.41 to 0.55 for different soils, the higher values being found for soils low in iron but high in manganese.

Soils, however, contain carbon dioxide dissolved in the soil solution, and, if the conditions are suitable, either ferrous carbonate will be precipitated, or the carbon dioxide in the solution, through its effect on pH, will alter the concentration of ferrous ions in equilibrium with $Fe_3(OH)_8$, if this is present. In the presence of sufficient iron, $Fe_3(OH)_8$ seems to control the Fe^{2+} concentrations in the solution, with the pH being related to the carbon dioxide in the solution through the equation[12] $\text{pH} = 6.12 - 0.57 \, P_{CO_2}$, where P_{CO_2} is the partial pressure of CO_2.

It is interesting to note that if a soil contains free calcium carbonate, the relationship between the soil pH and the CO_2 is:

$$\text{pH} = 6.03 - 0.67 \log P_{CO_2}$$

so the value of pH of a soil containing adequate iron for $Fe_3(OH)_8$ to be formed and of a soil containing free calcium carbonate are almost the same, and decreases from about pH 7.2 to 6.7 as the CO_2 partial pressure rises from 0.01 to 0.1 atmospheres. Thus the pH of both acid ferruginous soils and calcareous soils tends to stabilize at about 6.7–7.0 after flooding.[13]

The behaviour of manganese oxides in the presence of quartz sand under reducing conditions has been compared with that of 16 Philippine rice soils all containing manganese.[14] The results suggest that initially MnO_2 is reduced partly to Mn^{2+} and partly to Mn_2O_3, and later Mn_2O_3 and any Mn_3O_4 are reduced to Mn^{2+}. The redox potentials are all lower, for a given activity of manganous ions in solution, than expected from the theoretical equations, pointing to uncertainties regarding the composition of the oxides involved in the equilibria (see Ch. 23 for an account of Mn minerals found in soils). Although the formation of manganous carbonate is probably the mechanism for removing manganous ions from solution as the pH rises during reduction, it is unlikely that the pH of the soil is ever controlled by this compound.[15]

The sulphate–sulphide reduction which is brought about by species of the bacteria *Desulphovibrio* is of great importance in anaerobic soils, because hydrogen sulphide is extremely toxic to many plants, and a concentration of 10^{-6} M can affect the functioning or growth of roots. However in the presence

12 Ponnamperuma, F. N. and others, *Soil Sci.*, 1966, **101**, 421; *ibid.*, 1967, **103**, 90, 374; *Soil Sci. Soc. Am. Proc.*, 1969, **33**, 239.
13 For an example see Mukhopadhyay, A. *et al.*, *Soil Sci.*, 1967, **104**, 107.
14 Loy, T. A. and Tianco, E. M., *Soil Sci.*, 1969, **108**, 48.
15 Mukhopadhyay, A. *et al.*, *Soil Sci.*, 1967, **104**, 107.

of ferrous ions, the sparingly soluble ferrous sulphide is produced, which will only permit a hydrogen sulphide concentration of 10^{-8} M in neutral soils when sulphide production is active. If the soil is well supplied with active iron, that is, ferric iron that can be easily reduced, it is usually well poised at a redox potential too high for sulphide formation,[16] unless large amounts of easily decomposable organic material have been added.[17] Zinc sulphide may also be produced but apparently not copper or manganese sulphide.[18] As a result, the availability of Zn decreases, sulphide needs to be oxidized to sulphate for the plants to obtain the sulphur they need and loss of sulphate by leaching is prevented.

The reduction products from organic compounds

The reduction of decomposable organic compounds in flooded soils differs from inorganic reductions in that a whole sequence of reduction products are likely to be involved. If carbohydrate-rich materials are added to a flooded soil, the first products are typically gases and a whole range of fatty and hydroxy acids. The latter are, in turn, reduced eventually to other gaseous products.[19]

The principal organic acids produced are acetic, with smaller quantities of proprionic, butyric, lactic, valeric, fumaric and succinic acids.[20] Their total concentration in the soil can exceed 10^{-2} M for several weeks after flooding soil containing much readily decomposable plant residues.[21] Since 10^{-2} M acetic acid is toxic to the roots of many plants, and butyric and some of the other acids are toxic at 10^{-4} M, conditions very unfavourable to plant growth will be present during this period.[22] If this soil is to be used for rice, either these acids must be washed out of the soil by flooding and leaching, or the land must be left wet long enough for their concentration to fall to a low level before the crop is planted. It is probable that the undissociated acids are toxic rather than their anions, so these problems are more serious on soils with pH values below about 5.5.[23]

16 Aomine, S., *Soil Sci.*, 1962, **94**, 6.
17 Bloomfield, C., *J. Soil Sci.*, 1969, **20**, 207.
18 *Int. Rice Res. Inst., Ann. Rept. for 1972*, p. 204.
19 IRRI, *Organic Matter and Rice*, 1984. Chapters by Watanabe, I., p. 237, Neue, H. U. and Scharpenseel, H. W., p. 311 and Tsutsuki, K., p. 329; Ponnamperuma, F. N. in *Flooding and Plant Growth* (ed T. T. Kozlowski), p. 9, Acad. Press, 1984; Yoshida, T. in *Soils and Rice*, p. 445, IRRI, 1978; Lynch, J. M., *Crit. Rev. Microbiol.*, 1976, **5**, 67.
20 Stevenson, F. J. in *Soil Biochemistry* (eds A. D. McLaren and G. H. Peterson), Arnold, 1967; *Int. Rice Res. Inst. Ann. Rept. for 1969*, p. 135.
21 Wang, T. S. C. *et al.*, *Soil Sci.*, 1967, **104**, 138; Mitsui, S. *et al.*, *5th Int. Congr. Soil Sci.*, 1950, **2**, 364.
22 *Int. Rice Res. Inst., Ann. Rept. for 1969*, p. 135.
23 *Int. Rice Res. Inst., Ann. Rept. for 1965*, p. 62.

When the soil is first flooded and decomposition begins, a number of gases are given off. Apart from nitrogen and nitrous oxide produced from the reduction of nitrate, hydrogen gas[24] and a range of low molecular weight hydrocarbons are also produced. These include methane, ethane, propane and n- and isobutane and ethylene, propylene and butene-1, but no acetylenic compounds. The production of these hydrocarbons only occurs in the initial stages of flooding, and typically ceases after a few days. Their rate of production roughly parallels that of nitrous oxide, showing that it only takes place under rather mild reducing conditions. Table 26.2 gives the amounts of the hydrocarbons, up to the propane group, for six British soils.[25] The butane group is not included because the technique used was not suited to their quantitative measurement, though they were probably present in quantities comparable with the C_3 hydrocarbons.

Table 26.2 Production of hydrocarbons in soils during 10 days under anaerobic conditions; production in $\mu g\ kg^{-1}$ soil

Soil	% organic matter	Methane CH_4	Ethylene C_2H_4	Ethane C_2H_6	Propylene C_3H_6	Propane C_3H_8
Sand	1.4	0.3	0.6	0.1	0.1	1.7
Sandy loam	3.9	3.1	5.5	0.5	1.0	1.6
Gault clay	5.0	12.5	7.6	0.6	1.0	1.6
Loam from basalt	9.8	17.9	13.3	0.5	1.2	1.7

The biochemistry of the processes is not well understood.[26] Ethylene production begins within a few hours of the oxygen concentration in the soil falling below about 1 per cent, so is presumably brought about by organisms that are also active in aerated soils, that is, they are presumably facultative anaerobes. The process is in part carried out by enzymes not dependent on cellular activity, for sterilizing the soil with gamma radiation only reduced production by 50 per cent compared with a reduction by 90 per cent if sterilized by autoclaving.

Ethylene is the only one of these hydrocarbons to have a marked effect on root development of many crops. The saturated hydrocarbons are almost without effect, and the effects of propylene and butene are less marked than those of ethylene, and they occur in lower concentrations in the soil. The concentration of ethylene in the soil will be given here in terms of the

24 Bell, R. G., *Soil Biol. Biochem.*, 1969, **1**, 105.
25 Smith, K. A. and Restall, S. W.F., *J. Soil Sci.*, 1971, **22**, 430. For an earlier account see Russell, R. S., *Nature, Lond.*, 1969, **222**, 769. The effects of short-term waterlogging have been examined by Orchard, P. W. *et al.*, 1984. *Pl. Soil*, **81**, 119; *ibid.*, 1985, **88**, 407, 421.
26 Lynch, J. M., *J. Gen. Microbiol.*, 1974, **80**, 187; *ibid.*, **83**, 407; Lynch, J. M., *Crit. Revs. Microbiol.*, 1976, **5**, 67.

concentration of ethylene in the gas phase that is equilibrium with the solution: a concentration of 1 ppm is in equilibrium with a soil solution containing 1.5×10^{-7} g litre^{-1} (5.3×10^{-9} M) ethylene at 20 °C. Plant roots differ widely in their reaction to ethylene. Thus the seminal roots of tobacco and tomato will have their rate of elongation reduced by 75 per cent if the soil atmosphere contains 1 ppm of ethylene, and this concentration will reduce the rate of elongation of barley roots by 60 per cent, rye roots by 25 per cent, but will not affect the elongation of several varieties of rice. Even 0.3 ppm will reduce the rate of growth of barley roots by 50 per cent, though this concentration is without effect on rye.[27] Wheat and oats are intermediate between rye and barley. The sensitivity of these plants to ethylene is in line with field experience, for tomatoes and tobacco are known to be very intolerant of anaerobic conditions and rye and wheat are known to be more tolerant than barley (see also Ch. 4).

Ethylene will persist in poorly drained soils in the field for periods of weeks, particularly in the subsoil. Figure 26.1 gives the ethylene concentration in the surface and subsoil of a pasture soil in a poorly drained Oxford clay in southern England,[28] and it shows quite clearly that the ethylene concentration is high enough, over an appreciable period of time, to influence the root development of the crop. It is probable that the ethylene content of many poorly drained soils is a more common source of damage to crop growth than lack of oxygen or high carbon dioxide content in the soil. It is likely that two of the most useful parameters to measure the degree of temporary anaerobicity in a soil will be the ethylene and the nitrous oxide content of the soil water, although *in situ* redox potential measurements may be useful.[29]

Three other points about the effect of ethylene should be noted. It affects roots growing in aerated soils, so the damage it does in a poorly drained soil is not confined to the roots present in anaerobic soil volumes. It causes epinasty (down curving of leaves while still turgid due to more rapid growth on the upper side) and other deformities in plants, probably due to preventing the root from deactivating indolyl acetic acid produced in the tops,[30] and epinasty can be mistaken for wilting if the observer is not looking out for the difference. It can presumably build up very quickly in soils of moderate organic matter content in warm weather when the soil surface is temporarily sealed by heavy rain giving a surface crust or if irrigation water stands on the surface of the soil. Field observers have often noted what they considered to be the crop wilting under these conditions, but the collapse of

27 Smith, K. A. and Restall, S. W. F., *J. Soil Sci.*, 1971, **22**, 430.
28 Dowdell, R. J. and Smith, K. A., *Soil Biol. Biochem.*, 1972, **4**, 325; Burford, J. R., *J. environ. Qual.*, 1975, **4**, 55.
29 Colbourn, P., 1984, *AFRC Letcombe Laboratory Ann. Rept.*, 1983, p. 26; Conto, W. et al., *Soil Sci. Soc. Am. J.*, 1985, **49**, 1245.
30 Phillips, D. J., *Ann. Bot.*, 1964, **28**, 17, 37.

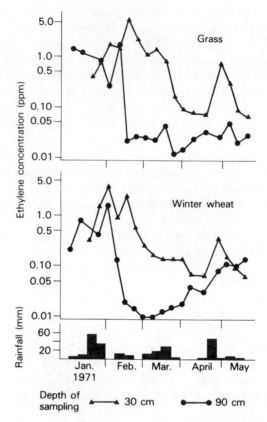

Fig. 26.1   Mean concentration of ethylene in the soil atmosphere of an Oxford clay soil under grass and under winter wheat.

the leaf from the upright position may be due to an epinastic effect.

Plant roots are directly affected by lack of oxygen.[31] As with soil which contains microsites of anaerobiosis in the interior of peds,[32] so roots also contain these microsites in their meristems even under aerobic conditions. These zones extend when oxygen concentrations fall. If the plants are intolerant to these conditions, an acceleration of glycolysis is initiated which increases the production of ethanol to toxic levels. The internal production of ethanol is more toxic than that produced by bacteria, and causes dissolution of membranes. Tissues are so badly damaged that even on a return

31 See also Cannell, R. Q. in *Applied Biology* (ed T. H. Coaker), Vol. II, p. 1., Acad. Press, 1977.
32 Currie, J. A., *J. Sci. Fd Agric.*, 1962, **13**, 380;   Greenwood, D. J. and Goodman, D., *op. cit.*, 1964, **15**, 579;   Sexstone A. J. *et al., Soil Sci., Soc. Am. J.*, 1985, **49**, 645; Myrold, D. D. and Tiedje, J. M., *op. cit.*, 651.

to aerobic conditions, anaerobic production of ethanol continues. The initiation of this toxic reaction is thought to explain the sensitivity of many plants to short periods of waterlogging, although the role of ethanol has been questioned by some workers.[33] In flood-tolerant plants there is a minimal acceleration of glycolysis and the flow of carbon is directed away from ethanol to a range of other metabolic products.

The gases given off after the first few days of active reduction are predominantly carbon dioxide and methane. Methane is therefore produced by two quite different processes – an early-stage process as just described, in which only small amounts of methane are produced, and a late-stage process in which an appreciable proportion of the carbon in the original organic matter is converted into this compound. This methane is produced by a number of strict anaerobes, most of which can only utilize simple organic compounds such as the fatty and hydroxy acids produced by the earlier group of organisms or by anaerobes capable of decomposing cellulose. Methane is, therefore, liberated over an appreciable period of time in flooded paddy soils. In some Indian paddy fields the gases leaving the reducing layer were predominantly methane, with some hydrogen and carbon dioxide.[34] However, no methane escaped from the soil, for the methane that was produced was oxidized by bacteria to carbon dioxide in the soil surface, or in the water on the surface, and this was then used by the algal films in photosynthesis. The principal gases escaping were therefore oxygen and possibly nitrogen.

Changes caused by flooding a dry soil

These changes will depend on the supply of decomposable organic matter, the various substances present in the soil which can be reduced, and the temperature, because reducing conditions are brought about almost entirely by microbial activity. Thus if a soil low in organic matter is flooded, there will be little change in its redox potential. Low soil temperatures only allow slow activity, so the redox potentials again will only fall gradually, and the higher the temperature, up to a certain limit, the more rapid its fall. Soils high in nitrate, or soils to which nitrate is added, maintain a redox potential of between +400 to +200 mV until all the nitrate is reduced, which may only take a few weeks, and then the potential may fall rapidly.[35] The potential in

33 Jackson, M. B. *et al., Pl. Cell Environ.*, 1982, **5**, 163.

34 Harrison, W. H. and Aiyer, P. A. S., *India Dept. Agric. Mem. Chem. Ser.*, 1913, **3**, 65; 1914, **4**, 1; 1920, **5**, 181. For later work see De, P. K. and Digar, S., *J. agric. Sci.*, 1954, **44**, 129; 1955, **45**, 280.

35 Couto, W. *et al., Soil Sci. Soc. Am., J.*, 1985, **49**, 1245; Ponnamperuma, F. N. and Castro, R. V., *Trans. 8th Int. Congr. Soil Sci.*, 1964, **3**, 379; Takkar, P. N., *Soil Sci.*, 1969, **108**, 108.

soils well supplied with active manganese and moderately supplied with organic matter falls fairly rapidly to about $+100$ mV and then over several months to about -50 mV. The potential in soils reasonably well supplied with iron and organic matter falls to about -50 mV and then slowly over a period of months to -200 mV. The potential in soils low in nitrate, manganese and iron but well supplied with organic matter will fall rapidly to about -250 mV and hydrogen sulphide is likely to be present after a short time provided the pH has risen to about neutrality. In soils low in organic matter the potential will only fall slowly, and may be still positive after several months of waterlogging. Under very active reducing conditions, brought about by the incorporation of decomposable organic matter, and in the presence of soluble sulphates, a proportion of the sulphide produced is in the form of hydrogen sulphide,[36] even in soils well supplied with ferric iron, due to the mobilization of ferric iron being the rate-limiting process for the formation of ferrous sulphide.

In some soils the redox potential will show a more complex change, rapidly falling to a low value directly after waterlogging, then rising rapidly to a value well below its initial value, and finally falling slowly. The fast fall is due to high bacterial activity before the ferric–ferrous system becomes operative, for if the iron is only present in crystalline forms, the amount of reducible iron is controlled by the rate of solution of ferric iron. This effect would be expected to be most noticeable when a dryland soil is flooded for the first time, because the iron in soils regularly flooded and drained usually remains in a form which is readily reduced.[37] In some soils nitrate remains much longer than expected, and instead of all the nitrate-nitrogen being lost as gas in a few weeks, up to one-half of this nitrogen may become incorporated in the organic matter.[38] The cause of this difference in behaviour is not known.

The pH of an acid soil usually rises after being waterlogged and strongly reducing conditions are often rapidly built up. As already mentioned on p. 904, the pH will stabilize at about 6.5 or just above, if sufficient iron is present, but the rise will be delayed if the temperature, the amount of reducible substances, or the amount of ferric iron is too low to produce sufficient ferrous iron for the $Fe_3(OH)_8$–$FeCO_3$–CO_2 buffer to become operative. It frequently happens in soils being poised by the ferric–ferrous reduction that the theoretical relation between the redox potential of the soil, its pH and the concentration of ferrous iron in the solution does not hold. This is because the redox potential is not uniform throughout the soil, but is much lower on the surface of the bacterial cells, which are the source of the electrons causing the reduction, than in the soil solution a little way from these cells. This results in the redox potential of the solution extracted from the

36 Bloomfield, C., *J. Soil Sci.*, 1969, **20**, 207;   Connell, W. E. and Patrick, W. H., *Soil Sci. Am. Proc.*, 1969, **33**, 71;   Ayotade, K. A., *Pl. Soil*, 1977, **46**, 381.
37 Bloomfield, C., *J. Soil Sci.*, 1969, **20**, 207.
38 *Int. Rice Res. Inst., Ann. Rept. for 1966*, p. 128.

soil being higher than the redox potential of the soil itself. The theoretical relations should be valid for the solution provided the chemical composition of the solid phases controlling the redox potential has been correctly identified, and the solution is free of bacterial cells. It must also be remembered that some of the ferrous iron in solution is likely to be present as a chelate and not as free ions, particularly in the early stages of reduction. The total ferrous iron is therefore not a correct measure of the free ferrous activity required in the theoretical relationship. This mosaic of actively reducing spots with large redox potential gradients into the solution may involve a range of reductions and oxidations taking place outside the cells involving systems that cannot yet always be predicted. As a consequence, the effect of a change of the pH of the soil on the redox potential cannot be predicted.[39] The concentration of ferrous iron in the soil solution often rises rapidly to a maximum and then falls as reduction proceeds, Fig. 26.2.[40] The cause of the fall is prob-

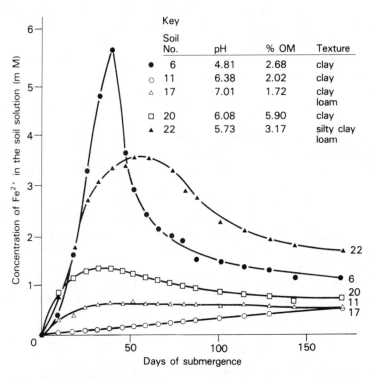

Fig. 26.2   Changes in the ferrous iron concentration in the soil solution during prolonged flooding.

39 Ponnamperuma, F. N. in *Soils and Rice*, p. 421, IRRI, 1978.
40 Taken from Ponnamperuma, F. N. in *The Mineral Nutrition of the Rice Plant*, p. 315, IRRI, 1965.

ably the precipitation of $Fe_3(OH)_8$ brought about by a rise of pH.

Flooding a soil, by causing the formation of manganous and ferrous ions which can take part in cation exchange,[41] will cause exchangeable calcium, magnesium and potassium to come into solution. The reduction and dissolution of ferric oxides also typically releases phosphate into solution,[42] for phosphate is often strongly adsorbed on active ferric hydroxides. A rise in pH will increase the negative charge on these surfaces which will also cause phosphate to be released. Flooding often releases silicate[43] for the same reason. If water slowly percolates through the soil, it will remove the ions, and this may impoverish it faster than normal percolation under aerobic conditions. The ferrous and manganous ions present in the percolate may be precipitated lower down the profile if oxidizing conditions are present in the subsoil. Oxidizing conditions will often persist in subsoils of land which is periodically flooded because reduction processes can only be active when there is microbial activity, that is, under conditions where there is a supply of decomposable organic matter.[44] Thus, the ferrous iron, which can be oxidized fairly easily, is converted into ferric hydroxide near the interface between the reducing and oxidizing regions, and the manganous ions, which need stronger oxidizing conditions to be converted into manganese dioxide, are precipitated below the iron hydroxide.

If a soil is subjected to several drying and flooding cycles without the addition of any organic matter, E_h falls increasingly slowly with time after each inundation, because the amount of readily reducible organic matter present at each inundation decreases rapidly in contrast to the behaviour of well-drained soils that are wetted but not flooded. This slower build-up of reducing conditions after each inundation means that if nitrate, for example, is added to the soil after it has been through several cycles, it will be reduced more slowly, and so will last for a longer time in the flooded soil.[45]

The formation of different reduction products in waterlogged soils containing decomposable organic matter can be summarized as follows. Molecular oxygen disappears at an E_h (pH 7) of about 550 mV, nitrate is lost at about 400 mV, and manganous and ferrous appear below 400 mV. Sulphide is formed below 0 mV and methane below about −200 mV. The E_h value for a given change seems to vary between soils for reasons not fully

41 Ponnamperuma, F. N. in *Flooding and Plant Growth* (ed T. T. Kozlowski), p. 9, Acad. Press, 1984.

42 See for example Gasser, J. K. R. and Bloomfield, C., *J. Soil Sci.*, 1955, **6**, 219; Patrick, W. H., *Trans. 8th Int. Congr. Soil Sci.*, 1964, **4**, 605.

43 Mortimer, C. H., *J. Ecol.*, 1941, **29**, 280.

44 Couto, W. *et al.*, *Soil Sci. Soc. Am. J.*, 1985, **49**, 1245.

45 Patrick, W. H. and Wyatt, R., *Soil Sci. Soc. Am. Proc.*, 1964, **28**, 647.

understood. Thus, nitrate loss has been reported at 500–600 mV,[46] 340 mV[47] and 200 mV.[48]

Effect of adding decomposable organic matter to flooded soils

If organic matter is added to a flooded soil, by the puddling in of green weeds, for example, or the ploughing in of a green manure crop before flooding, there is usually a flush of microbial activity. The extent of this flush depends on how easily the organic matter can be decomposed. Initially, there is a fall in pH and redox potential, a rise in the partial pressure of carbon dioxide, and a rise in products of reduction such as simple organic acids, followed by production of methane, hydrogen and other gases. The higher the soil temperature the more marked are these effects, and in hot climates this flush is over after about two weeks. The severe reducing conditions raise the concentrations of ferrous and manganous ions in the soil solution very rapidly, reaching a maximum in about four weeks if the temperatures are

Table 26.3 Effect of adding rice straw to a flooded soil on its ferrous iron status. (From *Int. Rice Res. Inst. Ann. Rept.*, p. 61, 1963)

Weeks incubated	Without rice straw			With 5% rice straw		
	Extractable Fe^{2+} meq/100 g soil	Fe^{2+} in solution 10^{-4} M	Complexed Fe^{2+} in solution 10^{-4} M	Extractable Fe^{2+} meq/100 g soil	Fe^{2+} in solution 10^{-4} M	Complexed Fe^{2+} in solution 10^{-4} M
Acid soil, initial pH 4.81, organic matter content 2.68%						
0	0.68	0.18	–	1.5	0.13	–
4	80	9.0	3.4	100	15.8	4.9
8	78	2.2	–	93	7.7	4.5
12	80	1.6	–	96	5.2	1.1
Neutral soil, initial pH 6.64, organic matter content 2.02%						
0	0.01	0.03	–	0.36	0.02	–
4	1.35	0.05	–	37	2.2	1.4
8	0.79	0.09	–	33	1.6	1.4
12	1.35	0.13	–	34	0.9	–

46 Takai, V. and Kamura, T., *Folia Microbiol.*, 1966, **11**, 304.
47 Pearsall, W. H. and Mortimer, C. H., *J. Ecol.*, 1939, **27**, 483;   Patrick, W. H., *Trans. 7th Int. Congr. Soil Sci.*, 1960, **2**, 494.
48 Bell, R. G., *Soil Biol. Biochem.*, 1969, **1**, 105.

high, as is shown in Table 26.3 for ferrous iron. It may take 10 to 12 weeks for the soil conditions to reach a fairly steady state.

The bacteria carrying out the decomposition release nitrogen surplus to their requirements as ammonium ions, sulphur as sulphate, hydrogen sulphide or mercaptans and phosphorus as phosphate. The amount of nitrogen released during decomposition differs from that in aerobic conditions in that the nitrogen factor (N immobilized per 100 g material) is lower for anaerobic than aerobic decomposition. Thus, for rice straw decomposing under aerobic conditions a nitrogen factor of 0.54 was found after six months, compared to 0.3 under mildly anaerobic conditions, or 0.07 under strictly anaerobic conditions.[49] The reason for this difference is that anaerobes obtain much less energy than aerobes per unit of carbohydrate decomposed, for only half the carbon they metabolize is lost as carbon dioxide, the other half being lost as methane. Organic matter of an appreciably lower N content can therefore be added to a flooded soil than can be added to a well-drained soil before the decomposition begins to remove ammonium ions from the solution. For example, a critical value for rice straw not to decrease the grain yield of rice was 0.55 per cent N[50] whereas that for dryland crops is between 1.5 and 1.7 per cent. Although these low-nitrogen organic compounds will release ammonium ions, they decompose slowly under anaerobic conditions, so only substances with higher N percentages will release appreciable amounts of ammonium over periods of a few weeks.

Once a flooded soil has come to a stable condition, it is not easy to disturb its stability as long as there is no drainage and the soil remains flooded. If fresh organic matter is carefully incorporated in the soil, with minimum disturbance, it decomposes very slowly with only small changes in redox potential, and correspondingly if nitrate is added it is only slowly reduced. However, if the soil is drained and reflooded, there will be a rapid build-up in the rate of decomposition and of reducing conditions.[51]

Redox conditions can change very rapidly over short distances in flooded soils or very wet soils in the field. Thus, small pieces of organic matter being actively decomposed by the soil bacteria in an anaerobic volume of soil can be surrounded by a thin zone where very strong reducing conditions have been set up, even though there may be a pore containing oxygen only a few millimetres away. Analysis of soil air from wet soils with very restricted air space will not necessarily be of any value in indicating the conditions prevailing in the bulk of the soil. An analysis of the soil water will give a much more useful picture, yet even this will give only an average picture of the conditions prevailing in the volume being sampled. Thus, a bulk sample

49 Acharya, C. N., *Biochem. J.*, 1935, **29**, 1116.
50 Williams, W. A. *et al.*, *Pl. Soil*, 1968, **28**, 49.
51 Ponnamperuma, F. N., *Int. Rice Res. Inst.*, *Ann. Rept. for 1966*, p. 113.

of water extracted from a British wet heath contained both small amounts of dissolved oxygen (2 ppm) and appreciable amounts of hydrogen sulphide (10 to 20 ppm).[52]

Adaptation of plants to anaerobic conditions

The various processes by which some plant species can overcome the effects of poor aeration or of reducing conditions can be studied in plants adapted to moor or marshland conditions.[53] A redox potential of about 320 mV at pH 5 is a useful dividing line between oxidizing and reducing soils,[54] so that plants adapted to soils with a lower redox potential than this have developed some protective or adaptive mechanism in their roots to allow them to function as organs for the extraction of nutrients and water from the soil. Plants adapted to conditions of poor soil aeration develop a system of interconnected internal air spaces (aerenchma) through which oxygen from the atmosphere can diffuse into the root, and into the soil immediately outside, and this rate of diffusion can be measured electrometrically.[55] This allows the uptake of ions by the root to take place as an aerobic process.

The seedlings of some crops are adapted in this way though the rate of oxygen transport becomes much less as they grow (see Ch. 4).[56] Paddy rice varieties are well adapted for this mode of transport, as are many marsh plants, though most varieties of rice have lost this adaptation at the onset of flowering. It is possible that some of the carbon dioxide that accumulates in the soil near the roots may diffuse from the soil through the plant to the atmosphere, for the carbon dioxide content of the air in the plant tissue increases as the carbon dioxide concentration outside the roots increases.[57]

A consequence of oxygen diffusing out of plant roots growing in flooded soils is that they are often surrounded by cylinders of soil having a dark brown colour, due to the conditions being sufficiently oxidizing for the formation of ferric hydroxide and manganese dioxide. In addition to the atmospheric oxygen diffusing out of the root, there may also be a supply of oxygen

52 Sheikh, K. H., *J. Ecol.*, 1969, **57**, 727.
53 Sheikh, K. H. and Rutter, A. J., *J. Ecol.*, 1969, **57**, 713. See also Jones, H. E. and Etherington, J. P., *J. Ecol.*, 1970, **58**, 487 for a comparative study of the survival of *Erica* species in waterlogged soils.
54 Pearsall, W. H., *J. Ecol.*, 1938, **26**, 180, 194, 298; Pearsall, W. H. and Mortimer, C. H., *op. cit.*, 1939, **27**, 483.
55 See for example Armstrong, W., *J. Soil Sci.*, 1967, **18**, 27; Sexstone, A. J. *et al.*, *Soil Sci. Soc. Am. J.*, 1985, **49**, 645.
56 Greenwood, D. J., *New Phytol.*, 1967, **66**, 337; Greenwood, D. J. and Goodman, D., *op. cit.*, 1971, **70**, 85, for some vegetable crops; Jensen, C. R. *et al.*, *Science*, 1964, **144**, 550 for maize; Barber, D. A. *et al.*, *J. exp. Bot.*, 1962, **13**, 397 for barley.
57 Webster, J. R., *J. Ecol.*, 1962, **50**, 619.

produced enzymically outside the root surface from the decomposition of glycollic acid to carbon dioxide through glyoxalate, oxalate and formate, with the liberation of hydrogen peroxide.[58] The cylinder may extend for a few millimetres from the root surface during the active growth of the root, but it breaks up as soon as the oxidizing conditions cease.[59] The root surface typically remains white if it is surrounded by an oxidized sheath but, if the outward rate of oxygen diffusion is too low for this, it may still be adequate to cause ferric hydroxide to be precipitated on the epidermal cell walls of the root, or within the intercellular spaces around the cortical cells; in either case the root surface is stained brown.[60] Marsh plants differ considerably in their ability to tolerate hydrogen sulphide in the soil outside their roots, probably because of differences in the rate at which oxygen can diffuse out of the roots. In part this may be due to their ability to precipitate ferric hydroxide in the intercellular spaces in the cortical and epidermal layers which can absorb the hydrogen sulphide before it reaches the inside of the root cells. The xeromorphic features of many marsh plants suggests that low rates of water uptake will reduce the need for detoxification of reduced ions at the root surface.[61]

The soil conditions prevailing in a paddy field

The principal agricultural use of flooded soils is for the production of swamp or paddy rice, which is a very important crop of the tropics and subtropics. Paddy soils are not necessarily soils with a water table at or near the surface. Many have a deep water table but their surface soil is made nearly impermeable either by being puddled when wet or by having a compacted layer in the subsurface. Nor are they normally flooded throughout the year, for many are only flooded during the greater part of the growing season. Thus, a typical paddy soil is characterized by alternate periods of reducing conditions, when it is cropped to rice, and oxidizing conditions during the dry season, when it may either be allowed to dry out with its natural weed population or be planted with a dry-season crop. The depth to which oxidizing conditions prevail before the land is next flooded depends on the depth of the water table, which may be very shallow for fields on the flood plains of rivers, but may increase in depth as the fields extend up the valley sides.

58 Mitsui, S. in *Mineral Nutrition of the Rice Plant*, IRRI., 1965; Armstrong, W. *Physiol. Plant.*, 1967, **20**, 920.
59 Bidwell, O. W. *et al., Trans. 9th Int. Congr. Soil Sci.*, 1986, **4**, 683.
60 Armstrong, W. and Boatman, D. J., *J. Ecol.*, 1967, **55**, 101.
61 Crawford, R. M. M., *J. Ecol.*, 1966, **54**, 406;   McManmon, M. and Crawford, R. M. M., *New Phytol.*, 1971, **70**, 299.

A typical soil profile in a fertile paddy soil with a deep gley horizon during the middle of the growing season is as follows[62]:

1 There is a layer of water on the soil surface in which there is a growth, sometimes copious, of algae, which keep this water strongly oxygenated.

2 There is a thin surface film of oxidized soil (E_h between 400 and 600 mV) brown or yellow–brown in colour, which receives its oxygen from the surface water film, and may be several millimetres thick.[63]

3 Below this is a thicker layer of reduced soil about 15 cm deep (E_h between 0 and 100 mV), typically grey or blue–grey in colour. The boundary between the oxidized and reduced layer may be fairly sharp, it need not be horizontal, and it may be marked by a thin brown film of ferric hydroxide.[64] As will be described later, this layer can be very heterogeneous.

4 Below this is an oxidized layer, again often with a sharp boundary, where indurated ferric hydroxide may occur below which may be an enrichment of black manganese dioxide. This layer probably remains oxidized during the period of flooding because of low microbial activity and an adequate amount of poising material to prevent serious reducing conditions.

5 This lower oxidized layer often overlies a permanently gleyed subsoil.

A healthy young rice plant has its roots entirely in the reducing layer and they are typically either surrounded by a brown sheath of ferric hydroxide or stained brown. This results in the reducing layer containing pockets of soil that are oxidized. Further, fresh organic matter has usually been incorporated in this layer when the soil was prepared for the crop. It often contains channels up which the gases escaped that were produced during the fermentation of this organic matter, and down which oxygenated water can diffuse or percolate, resulting in the soil bounding these channels also becoming oxidized. Thus, this reduced layer in a fertile paddy rice soil is characterized by having a mosaic of reducing and oxidizing conditions.[65] As the rice plant develops, the root system in the reducing layer loses its ability to transmit oxygen, and a new root system develops in the thin oxidized surface layer of the soil. This becomes the active system from heading or flowering time onwards.[66]

62 Aomine, S., *Soil Sci.*, 1962, **94**, 6; Koenigs, F. F. R., *Trans. 4th Int. Congr. Soil Sci.*, 1950, **1**, 297; de Gee, J. C., *op. cit.*, 300; Grant, C. J. in *Mineral Nutrition of the Rice Plant*, IRRI, 1965.

63 Oxygen concentrations in saturated soil have been mapped by Sexstone, A. J. *et al.*, *Soil Sci. Soc. Am. J.*, 1985, **49**, 645.

64 For laboratory studies on the formation of iron and manganese enriched layers, see Collins, J. F. and Buol, S. W., *Soil Sci.*, 1970, **110**, 111, 157.

65 For an experimental demonstration of this, see *Int. Rice Res. Inst. Ann. Rept. for 1966*, p. 110.

66 Alberda, T., *Pl. Soil*, 1953, **5**, 1.

The reducing layer in soils less suited to paddy rice typically differs from that in a good soil in that it is more uniform in colour, being grey to the depth of cultivation or disturbance and often bleached below this. The whole layer often overlies a layer indurated with ferric hydroxide and perhaps silica. Such soils are called degraded by the Japanese workers. They are characterized by being low in iron and manganese, so are typically found on parent materials derived from siliceous marine deposits, and from granites and rhyolites. The reducing layer in these soils will have a redox potential sufficiently low for hydrogen sulphide to be formed. The rice roots are greyish white and the oxygen diffusion rate from the root to the soil is very low or is zero, due to reducing substances diffusing to the root surface faster than the roots can release oxygen. Ferrous iron that may be close to the root surface will not become oxidized. These undesirable conditions can sometimes be mitigated by draining the surface water off the soil for a period, one or more times during the first part of the growing season, which allows the crop to dry the soil with the temporary development of oxidizing conditions.[67] The reducing conditions may also encourage the loss of soluble silica and ferrous iron, and ferrous iron can also be lost as colloidal particles of ferrous sulphide stabilized by hydrogen sulphide. Both of these may be precipitated in the indurated layer underlying the reducing layer. The oxidized layer below this may be very poorly developed, particularly in poorly drained soils or in some degraded soils. The layer may be grey with flecks of orange or brown, instead of being brown to red with flecks of grey. If it is well structured, the ped faces may be coated with a thin ferric hydroxide film though the interior of the peds remain reduced.

The management of rice soils

Most of the world's rice is grown on puddled soils*. Puddling is wet cultivation of land that mixes soil and water to produce a soft, slowly permeable layer. It is achieved by ploughing and harrowing the saturated soil. A clay content exceeding 20 per cent, dominance of smectite clays and exchangeable sodium facilitate puddling. Soils with a high content of organic matter, oxides of iron and aluminium, 1 : 1 clays or allophane puddle less readily.

A paddy soil before flooding usually carries either a crop of weeds or a green manure crop, and this is incorporated into the soil, either before or after flooding, so that very active microbial decomposition is usually taking place in the soil during the first few weeks after flooding. It is common practice therefore not to transplant or sow rice for two to three weeks after flooding, to allow the first flush of decomposition to pass, so that the rice

67 Strickland, R. W., *Aust. J. exp. Agric. Anim. Husb.*, 1968, **8**, 212.

* For a review on puddled rice soils, see Sharma, P. K. and De Datta, S. K., *Adv. Soil Sci.*, 1986, **5**, 139.

roots will be in a relatively favourable environment for growth from the beginning. If the soil is well poised, the roots should continue to grow and function effectively.

Rice growing on degraded soils can suffer from a number of physiological diseases,[68] known by local names, which rarely have very specific symptoms because of the variety of unfavourable soil conditions which can induce them. If the principal trouble is due to hydrogen sulphide or simple fatty acids, the damage can be reduced by ensuring as little fresh organic matter as possible is incorporated in the soil, and flooding the soil some time before the crop is planted. These degraded soils are often very acid, and some contain sufficient iron for the ferrous concentration in the reducing layer to rise to toxic levels. This can be ameliorated by liming, which raises soil pH and reduces the ferrous concentration. It is possible that the toxic effect of a high ferrous concentration can be alleviated, at least to some extent, by the use of a potassium fertilizer, for this increases the potassium and reduces the iron uptake by the plant.[69] Since these soils may be low in available silica, silicate slags are often the preferred liming material. A simple method to distinguish between suitable and degraded soils is to determine the concentration of ferrous iron in the reducing layer, for it should lie between 10^{-3} and 10^{-4} M. If it is too low sulphides may become important and if too high iron toxicity may occur.[70]

The choice of date at which the soil is puddled, and when the crop is planted, can be of great importance, particularly in regions where there is a marked seasonal change of temperature.[71] If the soil temperature is below 15 °C decomposition goes relatively slowly, and the rise in soil pH becomes delayed. If fertilizers are not being used, the rate of release of nitrogen from the organic matter and the rate of build-up of available phosphate are low, although these effects can be mitigated by proper fertilizer use. However, if much organic matter is incorporated in cool weather, there may be a flush of decomposition, with harmful strong reducing conditions occurring in mid-summer, when the temperature has risen appreciably and the crop is well established but still susceptible.

Drainage and water control in paddy soils

Although paddy rice is grown in flooded soils, there is considerable exper-imental evidence that a certain downward movement of water through the soil

68 Baba, I. *et al.*, in *Mineral Nutrition of the Rice Plant*, 1965, IRRI, 1965; Mitsui, S. *et al.*, *Trans. 5th Int. Congr. Soil Sci.*, 1954, **2**, 364;   Hollis, J. P., *Louisiana Agric. Exp. Stn. Bull.*, 1967, 614; *Int. Rice Res. Inst. Ann. Rept.*, 1965, **45**.
69 Tanaka, A. and Tadano, T., *Potash Rev.*, 1972, subject 9, suite 21.
70 Jeffery, J. W. O., *J. Soil Sci.*, 1961, **12**, 172, 317.
71 *Int. Rice Res. Inst. Ann. Rept. for 1967*, p. 122.

is desirable. This may be particularly important early in the season if there is a considerable amount of organic matter decomposing. The percolating water carries with it soluble organic substances, which can reduce the soil around the channels down which the water percolates, often giving a skin of grey-coloured soil. Some of these substances are probably fatty acids and other compounds liable to harm the roots of the young rice plant if their concentration becomes too high. In parts of Japan tile drainage is practised to maintain an adequate rate of percolation. Green manuring and the use of composts is also valuable in some areas, probably because they help to maintain soil structure in the initial stages of flooding which allows adequate percolation. Drainage may also decrease the partial pressure of CO_2 in the soil solution, so raising the pH and decreasing the ferrous concentration. Drainage rates of about 0.5 to 1.0 cm day^{-1} are probably adequate.[72] They will not involve too large a waste of water, nor will they impoverish the soil by excessive leaching of soluble ions. However, over-drainage of calcareous or sodic soils can be undesirable, for by lowering the CO_2 concentration in the soil water, the pH may rise sufficiently to cause loss of yield through too low an uptake of ferrous iron.

The function of the water in which paddy rice is grown is to maintain adequate reducing conditions in the soil to help control weeds and to prevent water stress. Its depth need only be sufficient to ensure that it covers the whole field. Thus, level fields require only shallow depths of water, but since it is never possible to have an absolutely level field, it is usual to aim for an average depth of 5 to 7.5 cm on well-levelled fields. The depth has no effect on yield provided it is less than 10 cm.[73] Evaporation of standing water with little movement of water from the field leads to accumulation of carbonates, water-soluble salts and silica, and in some instances high boron concentrations and alkalinity with low zinc availability. On the other hand, irrigation water can be a substantial source of potassium and sulphur. Assuming that 1 m of water is used per crop, 1 ppm of an element in the water corresponds to 10 kg ha^{-1} per crop added to the soil. Flooding may be particularly important in mid-season from panicle initiation to heading, as there may be a much higher ferrous iron demand in the reproductive than in the vegetative phase of growth.[74] Paddy rice can, however, be grown without flooding on a number of soils, and some of the high yielding IRRI rices appear to give very satisfactory yields on aerated soils in which the water deficit is always kept small. This may be, in part, because such soils will always contain anaerobic regions possibly inside peds.

72 *Int. Rice Res. Inst. Ann. Rept. for 1967*, p. 134.
73 Wickham, T. N. and Sen, L. N. in *Soils and Rice*, p. 649, IRRI, 1978.
74 *Int. Rice Res. Inst. Ann. Rept. for 1966*, p. 117;   Sharma, P. K. and De Datta, S. K., *Soil Sci. Soc. Am. J.*, 1985, **49**, 1451.

Nutrition and manuring of paddy rice

Unmanured rice obtains its nitrogen from five sources: from the ammonium produced in the reduction zone by decomposition of organic matter; from nitrate produced by decomposition of organic matter in the surface oxidizing layer of the soil; from nitrogen fixed by blue–green algae and other photo-synthetic micro-organisms[75] in the water films above the soil, and from nitrogen fixed by heterotrophic nitrogen fixing organisms in the soil or the water film,[76] and in the rhizosphere.[77] These sources are discussed in Chapter 19.

The nitrogen fixed by the algal films may be transferred to the rice as nitrate from mid-season onwards after the film has built up strongly and after the rice has developed a surface root system. In good paddy soils these sources of nitrogen are adequate to maintain rice yields at a modest level almost indefinitely, in contrast to dryland soils in which crop yields fall to very low levels under continuous cropping. In unfertilized wetland rice fields, biological nitrogen fixation is the major source of soil nitrogen. Meas-urements at the International Rice Research Institute in the Philippines (IRRI) revealed fixation of 63 kg N ha^{-1} in a period of 17 weeks.[78] In a continuous rice experiment after 31 successive crops (three per year) on a field receiving no fertilizer, the yields stabilized at 3–4 t ha^{-1} per season.[79] The nitrogen removed by the crop is about 60–80 kg ha^{-1} per season, or 180–240 kg ha^{-1} a^{-1}, with no decrease in soil nitrogen over the period.

Additional nitrogen must be used if high yields of rice are wanted.[80] Leguminous green manures ploughed in at the beginning of the season have limited value because if too much readily decomposable green matter is incorporated in the soil, reducing conditions develop which are too severe for the young rice roots. The use of nitrogen fertilizers requires care[81] due to the rapid transitions that can occur between aerobic conditions, when ammonium ions are converted to nitrate, and reducing conditions, when the nitrate is almost wholly denitrified. Thus, an ammonium fertilizer or urea broadcast on the surface of a flooded soil is likely to be converted to nitrate in this surface layer. In so far as this nitrate can seep down into the reduced layer through channels of oxidized soil, or the rice plant has developed a root system in the soil surface, it can be taken up by the plant, for rice takes up

75 Kobayashi, M. *et al.*, *Soil Sci.*, 1967, **104**, 113.
76 MacRae, I. C. and Castro, T. F., *Soil Sci.*, 1967, **103**, 277; Magdoft, F. R. and Bouldin, D. R., *Pl. Soil*, 1970, **33**, 49.
77 Yoshida, T. and Ancajas, R. R., *Soil Sci. Soc. Am. Proc.*, 1971, **35**, 156.
78 Yoshida, T. and Ancajas, R. R., *Soil Sci. Soc. Am. Proc.*, 1973, **37**, 42.
79 De Datta, S. K., *Principles and Practice of Rice Production*, Wiley, 1981; Westcott, M. P. and Mikkelsen, D. S., *Soil Sci. Soc. Am. J.*, 1985, **49**, 1470.
80 De Datta, S. K. and Magnaye, C. P., *Soils Fertil.*, 1969, **32**, 103. See also ref. 79.
81 Savant, N. K. and De Datta, S. K., *Adv. Agron.*, 1983, **35**, 241; Fillery, I. R. P. *et al.*, *Soil Sci. Soc. Am. J.*, 1984, **48**, 196, 203, 914. See also ref. 79.

nitrate at least as easily as ammonium.[82] However, in many paddy soils the reducing conditions in the root zone are sufficiently severe that any nitrate entering it, or any nitrate present before the reducing conditions set in, is likely to be denitrified. Therefore adding an ammonium fertilizer to a damp soil one or two weeks before it is flooded, or broadcasting it on the soil surface, is likely to be an inefficient method. In some soils, possibly of low biological activity, a proportion of the nitrate persists for several weeks, and becomes incorporated in the soil organic matter.[83] The ideal method of use early in the season is to place an ammonium or urea fertilizer or inject anhydrous ammonia in the zone that will become the reducing zone, normally about 10 cm deep, before the land is flooded.[84] If nitrate fertilizers must be used, they should only be applied as a top dressing in mid-season, because the active rice roots are then in the oxidizing layer in the soil surface.

High-yielding varieties of rice require a high level of nitrogen in their early stages of growth, but this must be applied with care, otherwise it may cause such a vigorous growth of green algae that the young plant is smothered. It is common practice not to give all the nitrogen at this stage, particularly if a high application is to be used; some is applied later in the season as a top dressing at about the time of panicle initiation, when the rice roots are near the soil surface. High rates of nitrogen can only be used, however, if weed growth can be suppressed, which normally involves both good management and the use of appropriate herbicides. The maximum rate of nitrogen that can be used profitably, and the response of the crop to nitrogen, depends on the level of solar radiation in regions where water and temperature are adequate, particularly during the period from panicle initiation to harvest. Thus at IRRI[85] their short stiff-strawed variety IR8 only gave a response of 15 kg of grain per kg N during the rainy cloudy season, but 30 kg during the dry sunny season. The most profitable dressings of nitrogen were in the two seasons about 60 and 120 kg, and the yields about 6.5 and 9.5 t ha^{-1}, respectively.

The two most commonly used nitrogen fertilizers are ammonium sulphate and urea. The continued use of high levels of ammonium sulphate causes acidity, as in well-drained soils, and may give accumulation of sulphides, so urea is probably the better fertilizer unless anhydrous ammonia can be used. There is interest in the use of slow-acting fertilizers and nitrification inhibitors, particularly in Japan. The Japanese workers have evidence that under their conditions high levels of ammonium nitrogen (100 to 150 kg ha^{-1}) will affect rice growth adversely, so that if the nitrogen is to be given in one

82 Malavolta, E., *Pl. Physiol.*, 1954, **29**, 98.
83 MacRae, I. C. *et al.*, *Soil Sci.*, 1968, **105**, 327.
84 Simsiman, G. W. *et al.*, *J. agric. Sci.*, 1967, **69**, 189; Aleksic, Z. *et al.*, *Pl. Soil*, 1968, **29**, 338.
85 De Datta, S. K. *et al.*, *Agron. J.*, 1968, **60**, 643; Tanaka, A., *et al.*, *Proc. 9th Int. Congr. Soil Sci.*, 1968, **4**, 1.

application rather than in split dressings, slow-acting nitrogen fertilizers are required. These are made either by bonding pellets of urea in a bitumen or plastic base so it can diffuse only slowly into the soil solution, or by using insoluble substituted ureas such as guanyl urea, $NH_2C(NH_2)_2NHCONH_2$, or crotylidene diurea which only slowly hydrolyse to urea.[86] Large granules of urea may also be used. If urea is to be broadcast on the soil surface after the rice has been planted, mixing a nitrification inhibitor with the urea will help to ensure that the ammonium produced in the oxidizing layer of the soil will remain in that form, and not be nitrified and lost if it seeps into the reducing layer. Harmful effects of high ammonium applications have not been found on the fine textured soils at IRRI, wheras the Japanese observations were for coarse textured soils.

Phosphatic fertilizers are not required so often on paddy as on dryland soils, probably because the phosphate can be held on ferric oxide surfaces during dry periods and released when the iron is reduced. Single superphosphate is, like ammonium sulphate, often undesirable because of the calcium sulphate it contains. Rock and bone phosphates are probably useful on a wider range of paddy than of dryland soils. This is probably because soluble organic compounds produced in the reducing layer increase the solubility of the phosphate through their ability to chelate calcium. The Japanese have also used a fused magnesium serpentine phosphate made by sintering a serpentine with a phosphate rock. This contains about 8.5 per cent P, and 20 per cent SiO_2 as hydrolysable silicates. It has a liming action and supplies silicate to the crop.

Scientists at IRRI using 160 Philippine wetland rice soils have found that those dominated by halloysite, kaolinite or amorphous material in the clay fraction have low available phosphorus contents (Olsen P value < 5 mg P kg^{-1}); those with these clays in minor amounts ranged from 5 to 10 mg P kg^{-1} indicating moderate concentrations of available phosphorus; and none of the soils above 10 mg P kg^{-1} contained these minerals in the clay fraction. Because of adsorption and other causes less than 10 per cent of phosphorus applied to wetland rice is taken up by the crop in the first season. Potassium is often needed for high yields of rice and potassium chloride is a suitable fertilizer.

The sulphur nutrition of rice raises interesting problems because ferrous sulphide is so insoluble. Presumably the sulphide is oxidized in the immediate vicinity of the root to give sulphate ions. Copper sulphide, particularly in the presence of ferrous sulphide, has an extremely low solubility, so rice presumably obtains most of its copper from soluble organic complexes, as in aerobic soils.[87] The uptake of sulphur from soil sulphides is shown in Table 26.4. They are apparently partially oxidized in the soil adjacent to the growing

86 Hamamoto, M., *Fertil. Soc. Proc.*, 1966, **90**.
87 Ponnamperuma, F. N., Behaviour of minor elements in padi soils, IRRI Research Paper, Series 8, 1977.

Table 26.4 The effects of ^{35}S-labelled Na_2S, MnS, FeS, ZnS, CuS and HgS placed in the lower rooting zone of a Crowley silt loam, on the yield of the aerial portion of the rice plant, uptake of ^{35}S by the rice plant, and per cent uptake of the added ^{35}S by the rice plant

Treatment	K_{sp}	Total added ^{35}S (cpm)	Yield (g)	Uptake		% uptake of added ^{35}S
				Total (cpm)	Per 0.1 g plant material (cpm)	
$Na_2{}^{35}S$	1.0×10^{-1}	876 053	0.158	39 422	24 950	4.49
$Mn^{35}S$	1.4×10^{-15}	1 068 000	0.326	6 315	1 744	0.62
$Fe^{35}S$	3.7×10^{-19}	960 000	0.303	4 141	1 366	0.43
$Zn^{35}S$	1.2×10^{-23}	1 026 900	0.313	3 948	1 261	0.38
$Cu^{35}S$	3.5×10^{-38}	724 875	0.204	1 287	631	0.18
$Hg^{35}S*$	1.0×10^{-50}	1 410 900	0.174	0	0	0

* Rice was grown on the HgS-treated soil for 42 days as compared with 21 days for the other sulphide treatments.

roots. The degree of oxidation and the uptake are directly related to the solubility of the sulphides.[88]

Rice is one of the few agricultural crops that respond to silicate (see also Chs. 3 and 23). Silicate fertilizers are used fairly widely in Japan, and could probably be profitably used in many other areas of Southeast Asia in soils containing low amounts of available silica.[89] The commonest silicate fertilizers used in Japan are some blast furnace (not basic) slags, though the fused magnesium serpentine phosphate is also used. Both of these have useful liming properties since most of the soils low in available silica are also acid.

Acid sulphate soils

These are also known as salt marsh soils or cat clays and are found primarily in coastal, deltaic or estuarine areas of the humid tropics.[90] Before being drained they contain large amounts of ferrous disulphide (pyrite) and are 'potential acid sulphate soils' with pH values near neutral. After draining they become very acidic with pH values sometimes below 2 and may show yellow mottles of basic ferric sulphate (jarosite). The soils are laid down under anaerobic conditions and sulphate derived from sea water is reduced by *Desulphovibrio* and *Desulphotomaculum* to sulphide during the anaerobic decomposition of large amounts of organic matter.[91] If silt being deposited contains ferric iron, this will be reduced and ferrous monosulphide and iron pyrite, FeS_2, will be formed.[92] The former is rarely present in amounts exceeding 200 μg g^{-1}, but several per cent of pyrite may be present.

As the silt builds up and the sea retreats, or after drainage during reclamation, oxygen enters the soil. Ferrous monosulphide oxidizes readily under neutral conditions, whereas pyrite oxidizes only under acid conditions. Chemical and biochemical steps are involved in the oxidation of ferrous to ferric and sulphide to sulphate.[93]

If the pH is below 3.5, *Thiobacillus ferro-oxidans* is the dominant organism carrying out the oxidation, and it oxidizes ferrous iron to ferric and sulphide to elemental sulphur or sulphate.[94] It releases more ferrous iron than it oxidizes, so that if water is moving through the soil, ferrous iron will be leached out. When this acid drainage water comes into contact with air, the

88 Engler, R. M. and Patrick, W. H., *Soil Sci.*, 1975, **119**, 217.
89 Kawaguchi, K. and Kyuma, K., *Proc. 9th Int. Congr. Soil Sci.*, 1968, **4**, 19.
90 Bloomfield, C. and Coulter, J. K., *Adv. Agron.*, 1973, **25**, 265; Dost, H. (ed), *Acid Sulphate Soil*, Publ. No. 18, Vol. I and II, Int. Inst. Land Recl. Imp., Wageningen, 1973.
91 Driessen, P. M. in *Soils and Rice*, p. 763, IRRI, 1978.
92 Harmsen, G. W., *Pl. Soil*, 1954, **5**, 324.
93 Van Breemen, N. and Pons, L. J. in *Soils and Rice*, p. 739, IRRI, 1978.
94 Bloomfield, C., *J. Soil Sci.*, 1972, **23**, 1.

bacteria will oxidize it to ferric hydroxide (ochre), which will accumulate in the drainage channels. If the conditions are less acid, the pyrite will oxidize slowly in the pH range up to 5, giving elemental sulphur and ferric ions, and the sulphur will be oxidized in this range by *Thiobacillus thio-oxidans*.

Reclamation of these soils raises difficult problems. In areas such as the Dutch sea marshes, calcium carbonate was laid down with the pyrite so that on reclamation the acid produced reacts with the carbonate to give calcium sulphate. But in saline mangrove swamps on tropical coastlines there is generally no carbonate in the soil and large amounts of lime would be needed to neutralize the acidity. The pyrite may be present in large crystals which take several years to oxidize[95] so that control of acidity would be needed over this period, and it is rarely worth while to use these soils for dryland agriculture.

It is technically possible to empolder many of these marshes and use them for rice cultivation, but it can be extremely difficult to prevent acidity building up. In Sierra Leone, where much work has been done on this problem, the recommended system is to drain the land and allow the maximum oxidation of sulphide to take place during the dry season. The sulphuric acid is then washed out either by flooding or by the rains, the surface soil is limed to put some calcium back, and the crop planted. This method has the weakness that, if the soil contains much pyrite, only a portion will oxidize in the first season, so acidity will build up again in subsequent dry seasons.

The oxidation–reduction processes outlined above are of importance in land drainage. Where near-neutral peaty areas are drained ferrous hydroxide may be oxidized to ferric hydroxide in field drains. It often occurs as a raglike ochre, the result of oxidation by filamentous bacteria. If the soil is pyritic, ochre can again be deposited in the drains where ferrous and sulphide are oxidized. The practical problem of blocked drains may result.[96]

95 Hart, M. G. R., *Pl. Soil*, 1959, **11**, 215; *ibid.*, 1962, **17**, 87.
96 Bloomfield, C., *J. Soil Sci.*, 1972, **23**, 1.

27

The management of irrigated saline and sodic soils*

The effect of soluble salts on plant growth

Soluble salts can have two types of effect on the growing plant: specific effects due to particular ions they contain being harmful to the crop, and a general effect due to the raising of the osmotic pressure of the solution around the roots of the crop.

Specific effects fall into two classes: those operative at low and those at high concentrations. Of the former, only two salts are normally of importance – sodium carbonate and soluble borates. The former may be harmful in itself, but its harmful effect is more likely to be due to the consequences of the high pH it brings about. Thus, many nutrients, such as phosphate, iron, zinc and manganese, become unavailable to the plant at these high pH values on the one hand, and the soil structure tends to become water-unstable on the other, thus bringing about conditions of low water permeability, poor aeration and an almost unworkable soil.

Irrigation water containing more than 0.75 mg l^{-1} boron must be treated with caution as a concentration as high as this will affect the yield of many

---

* See Chapter 5 for a description and classifications of these soils.

sensitive crops such as citrus. Water containing 4 to 6 mg l^{-1} restricts crop-ping to boron-tolerant crops such as some varieties of sugar beet, lucerne, sorghum, cotton and the date palm.[1] Some ions have a toxic effect at high concentrations which intensifies the harmful effect of the osmotic pressure they cause. Thus many fruit trees are sensitive to high concentrations of chloride or sodium, and flax and some grasses are sensitive to high concen-trations of sulphate. Again, at equal fairly high osmotic pressures, magnesium ions are more toxic than calcium, and calcium may be more toxic than sodium, though the latter is usually in crops that can take up calcium easily but exclude sodium. A high concentration of sodium chloride may also be harmful due to its effect on nutrient uptake. Thus, it may interfere with the uptake of potassium, and some varieties of barley, when young, then need appreciably higher concentrations of potassium in the soil solution if their yield is not to be affected.[2] A number of crops show considerable genetic variability in their reaction to high salt concentrations, so it is sometimes possible to breed new varieties that are more tolerant to them.

The general effect of a high salt content in the soil is to give a dwarfed, stunted plant. This is often not apparent in the field if there are no patches of low-salt soil to act as controls, and yields can be reduced by over 20 per cent without salt damage being apparent to the farmer. As the salt content becomes higher the stunting becomes more noticeable, the leaves of the crop become dull-coloured and often bluish-green, and they become coated with a waxy deposit. Further, because many crops growing in very saline soils do not display the symptoms of wilting very clearly, considerable loss of yield can occur if irrigation is applied only when the plants are obviously wilted. The effects of salt damage have been summarized as follows[3]:

1   Physiological drought which is a direct osmotic effect.
2   Increased hydraulic resistance of roots and leaves.
3   Alteration of hormone levels so influencing growth rates.
4   Direct damage, particularly to photosynthetic mechanisms.
5   Ion competition, increasing energy use to maintain the K : Na balance.

The significance of each of these has not been fully investigated.

As the osmotic pressure in the soil solution increases (the osmotic potential becomes more negative), so does the osmotic pressure in the cell sap. The difference between these may remain the same, with the cell sap having a pressure of about 1.0 MPa higher,[4] or the osmotic pressure of the cell sap

1 For a list of boron-tolerant and boron-sensitive crops see Bernstein, L. in *Drainage for Agriculture* (ed J. van Schilfgaarde), *Agron. Monog.*, No. 17, 39, Am. Soc. Agron., Madison, 1974;   Maas, E. V., *Applied Agricultural Research*, 1986, **1**, 12.
2 For solution culture see Greenway, H., *Aust. J. biol. Sci.*, 1963, **16**, 616.
3 Poljakoff-Maybee, A. and Gale, J. (eds), *Plants in Saline Environments*, Springer, 1975.
4 Slatyer, R. O., *Aust. J. biol. Sci.*, 1961, **14**, 519; Bernstein, L., *Am. J. Bot.*, 1961, **48**, 909;   Hayward, H. E. and Wadleigh, C. H., *Adv. Agron.*, 1949, **1**, 11, found this for lucerne also.

may increase more rapidly than the osmotic pressure of the soil solution.[5] As the osmotic pressure in the external solution increases, the transpiration rate and stomatal resistance in the leaves may remain constant, as found for cotton up to external pressures of 1.0 to 1.2 MPa,[5] or it may decrease.[6] But it usually reduces the growth rate of the plant and its rate of photosynthesis, though it sometimes reduces and sometimes increases the dark respiration rate.[7]

Wadleigh and Ayers[8] showed that, for some crops at least, crop growth was controlled by the total potential of the soil water, and the decrease in growth rate due to this lowering of total potential was the same whether it was due to a lowering of the matric or the osmotic component. Figure 27.1 illustrates their results with beans (*Phaseolus vulgaris*). The plants were grown in a loam soil to which no sodium chloride, or 0.1, 0.2 or 0.4 per cent, was added, and the soil was watered when one-half, two-thirds, or nine-tenths of the available water, as measured in the salt-free soil, had been used. In the field the effect

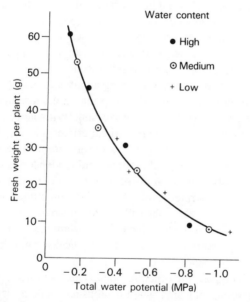

Fig. 27.1 The relation between the fresh weight of beans and the total water potential.

5 Boyer, J. S., *Pl. Physiol.*, 1965, **40**, 229.
6 Eaton, F. M., *J. agric. Res.*, 1942, **64**, 357.
7 For a review see Maas, E. V. and Nieman, R. H. in *Crop Tolerance to Suboptimal Land Conditions* (ed G. E. Jung), p. 277, Am. Soc. Agron. Spec. Publ., 1978; Shone, M. G. T. and Gale, J., *J. exp. Bot.*, 1983, **34**, 1117.
8 Wadleigh, C. H. and Ayers, A. D., *Pl. Physiol.*, 1945, **20**, 106. For the corresponding result with guayule rubber see *US Dept. Agric., Tech. Bull.*, No. 925, 1946.

of a moderate concentration of salts on a crop's growth depends on the shape of the moisture content − suction curve for the soil. Typically, for light-to medium-textured soils, much of the available water is held at relatively high matric potentials; so if the soil is non-saline, the crop can use most of the available water before its matric potential has fallen below −0.2 to −0.3 MPa. As the crop uses water the osmotic potential decreases roughly proportionately to the amount of water used, so the total potential of the water drops more rapidly in the presence than in the absence of soluble salts. However, in the field, salts are usually unevenly distributed in the soil, so that those roots growing in volumes of soil containing less salt than average will take up relatively more water than those roots growing in volumes containing more than average.[9]

Plants differ in their ability to withstand the harmful effects of salinity in the field. Early work[10] showed that plants have different abilities to extract water from soils in the wilting range, and plants better adapted to saline soils tend to have a greater ability to extract water at the drier end of this range. However, salt tolerance and drought tolerance are not necessarily related; coconut, for example, is salt tolerant but drought sensitive. The greater the salinity of the soil, the less water a crop can remove before it begins to suffer from water shortage, so that irrigated soils with an appreciable salt content need more frequent irrigations than non-saline soils.[11] Crop yields can easily be reduced unnecessarily by allowing too long an interval between irrigations; and this is liable to happen because the crop may not show signs of wilting as clearly as if it were growing in a low-salt soil. With more frequent irrigations, aeration may become limiting[12] and experiments in the Arava Valley of Israel have shown that beans respond to the injection of oxygen into the trickle irrigation water.[13] There is also some field evidence that loss of yield due to moderate salinity can be more serious on soils of low than of high fertility, and that a moderate level of salinity sometimes increases the response of a crop to fertilizers, particularly to phosphate and perhaps also to nitrogen.[14] Phosphate fertilizers have the advantage that they do not increase the osmotic pressure of the soil solution because phosphate is usually strongly adsorbed by the soil.

Salt tolerance is complex for other reasons. The tolerance of a plant may be low when it is young but high when established: lucerne is an example. The plant may survive at high salt contents, but will make very little growth.

9 For an example see Gardner, W. R., *Proc. Symp. FAO/IAEA Istanbul*, 1967, 335.
10 Briggs, L. T. and Shantz, H. L., *Bot. Gaz.*, 1912, **53**, 229.
11 Shalhevet, J. *et al.*, *Irrig. Sci.*, 1983, **4**, 83.
12 Wesseling, J. in *Drainage for Agriculture* (ed J. van Schilfgaarde), *Agron. Monog.*, No. 17, 7, Am. Soc. Agron., Madison, 1974.
13 Shalhevet, J. and Zwerman, P. J., *Soil Sci.*, 1962, **93**, 172.
14 Ravikovitch, S. and Porath, A., *Pl. Soil*, 1967, **26**, 49; Bernstein, L. *et al.*, *Agron. J.*, 1974, **66**, 412.

It only grows slowly at moderate salt contents, and hence will be of little commercial value. Again, though the plant may grow in fairly saline soils, the quality of the part harvested may be affected. Thus, cereals will grow and produce green matter in soils too saline for them to produce any grain, and sugar beet growing in saline soils produces a root low in sugar which is difficult to refine. Again, in practice, salt tolerance is often bound up with tolerance to sodicity, high pH and low calcium on the one hand, and ability to withstand prolonged waterlogging during irrigation, which is a common consequence of sodicity, on the other.

In spite of these limitations, crops can be graded into three categories of salt tolerance when grown under irrigation: tolerant, moderately tolerant and sensitive to salts. Table 27.1 gives such a grading for crops grown under irrigation in the west of the USA.[15] Dates, cotton, sugar beet and barley fall into the tolerant class, as do bermuda grass (*Cynodon dactylon*) in the warmer and tall wheatgrass (*Agropyron elongatum*) in the cooler regions; while peas, beans (both *Vicia faba* and many *Phaseolus* species) and most clovers are in the sensitive group.

Maas[16] has given extensive data on crop tolerance: taking barley (*Hordeum vulgare*) as a tolerant crop and beans (*Phaseolus vulgaris*) as a sensitive crop the conditions shown in Table 27.2 apply.

For many practical purposes it can be assumed that the osmotic potential of a solution in MPa is approximately $-0.04 \times$ the electrical conductivity in dS m^{-1} and that the salt concentration in the soil water at field capacity is

Table 27.1 The relative tolerance of crops to salts in the western states of America

Good tolerance	Moderate tolerance	Sensitive
Date palm	Pomegranate, fig, olive, grape	Pear, apple, orange, grapefruit, almond, apricot, peach
Barley	Rye, wheat, oats	Field beans
Sugar- and fodder-beet	Rice, sorghum	Green beans
Rape, kale	Maize	
Cotton	Potatoes, peas	
Bermuda and Rhodes grass	Sweet clover	White, alsike and red clover
Bird's-foot trefoil	Ryegrass	Ladino clover
	Strawberry clover	
	Lucerne	
	Cocksfoot	

15 *Diagnosis and Improvement of Saline and Alkali Soils*, US Dept. Agric. Handb., **60**, 67, 1954, now out of print, but similar tables are available in Hoffman, G. J. *et al.* in *Design and Operation of Farm Irrigation Systems* (ed M. E. Jensen), p. 145, *Am. Soc. Agric. Engrs. Monog.*, No. 3, 1980.
16 Maas, E. V. as ref. 1.

Table 27.2   Soil conditions and crop tolerance

	Barley	*Beans*
Electrical conductivity ($dS\ m^{-1}$) of the saturation extract for*		
(a) 50% yield	18	3.6
(b) 50% emergence	16–28	8.0
Maximum allowable salinity without yield reduction		
(a) electrical conductivity ($dS\ m^{-1}$) of a saturation extract	8	1.0
(b) osmotic potential (MPa) of soil water at field capacity	−0.66	−0.07
(c) salt content of the saturation extract ($mg\ g^{-1}$)	about 8	about 0.6

approximately 2 × the value for the saturation extract. Also the concentration of salt expressed as meq per litre in a solution is approximately 10 × the electrical conductivity in dS m^{-1}.

The management of irrigated soils

Quality of irrigation water

The water used for irrigation, whether taken from a river or from wells, is never pure water but always contains dissolved salts. Much of the water applied to the land will be taken up by the crop and transpired, and if the salt concentration of the water is high, most of the salts will be left behind in the soil.

The salt content of irrigation water is commonly specified in practice by its electrical conductivity, for this is a good index and easy to measure. Table 27.3 gives the composition of some river waters in the western part of the United States that are used for irrigation,[17] and it shows that their electrical conductivity varies from 0.1 to 3.2 dS m^{-1}, corresponding to concentrations between 1.5 and 38 meq l^{-1} and that sodium constitutes between 13 and 67 per cent of the total cations. The US Salinity Laboratory[17] grades the

17 Taken from *The Diagnosis and Improvement of Saline and Alkali Soils*, US Dept. Agric. Handb., **60**, 1954.

* Electrical conductivity in this chapter is given in units of siemens per metre, S m^{-1}, usually dS m^{-1}, where dS m^{-1} = mmhos cm^{-1}.

Table 27.3 Chemical composition of some river waters used for irrigation in the western United States. Concentration of ions in meq l⁻¹

River Location	Columbia Wenatchee Wash.	Rio Grande		Gila Florence Ariz.	Pecos Carlsbad N. Mex.	Humboldt Rye Patch N. Mex.
		Otowi Br. N. Mex.	El Paso Texas			
Electrical conductivity dS m⁻¹ at 25°C	0.15	0.34	1.16	1.72	3.21	1.17
Dissolved solids mg l⁻¹	78	227	754	983	2380	658
Sum of cations	1.5	3.4	11.5	16.8	38.0	11.5
Ca	0.9	1.9	4.2	3.6	17.3	1.7
Mg	0.4	0.7	1.4	2.0	9.2	1.9
Na	0.2	0.8	6.0	11.3	11.5	7.9
HCO₃	1.2	1.8	3.6	3.7	3.2	5.2
SO₄	0.2	1.5	5.0	3.3	23.1	2.2
Cl	0.1	0.1	3.1	10.0	12.0	4.5
Soluble sodium*	13	24	52	67	30	68
Sodium adsorption ratio	0.2	0.7	3.6	6.7	3.2	5.9

* as % of total soluble cations

quality of irrigation water, based on its soluble salt content into four classes (p. 939), those with an electrical conductivity below 0.25 dS m^{-1} which do not contain enough soluble salts to cause any trouble and those with conductivities between 0.25 and 0.75, between 0.75 and 2.25, and above 2.25, for which special management is needed. These limits are now considered to be low and three ranges of <0.7, 0.7–3 and >3 have been proposed[18] where the restrictions to use of the water are none, slight to moderate and severe respectively.

Management for salt control

In regions where there is sufficient rainfall at some period of the year for water to drain through the soil profile, the salts left behind from the irrigation water will be leached out. But in regions of light rainfall, these salts will accumulate in the soil. In warm arid regions, in which many of the large irrigation schemes in the world are situated, these accumulations can be very large. Thus if the irrigation water contains 500 mg l^{-1} dissolved salts – a not uncommon figure for many rivers flowing in semi-arid areas – and 50 cm of irrigation water are used for a crop, and most crops will need at least this amount, the water will add 2.5 t ha^{-1} of salts during the growing season. Thus, irrigation schemes in semi-arid areas must be managed so that these salts can be removed at regular intervals from the root zone of the crop.

As a measure of the salinity of soils, and in order to classify crop sensitivity to salt, the US Soil Salinity Laboratory uses as a criterion the electrical conductivity (EC) of the saturation extracts of soils. This is found by adding water to a soil, whilst stirring, until it is just saturated, extracting a small sample of soil solution with a suction funnel (the saturation extract) and measuring its EC. Salt sensitive crops tolerate salinity up to an EC of 4 dS m^{-1}, moderately tolerant crops up to 8 dS m^{-1} and very tolerant crops up to about 16 dS m^{-1}. If crop yields are not to suffer from salinity, the conductivity of the soil solution must be kept below the appropriate value for the crop being grown. Since the soil solution becomes more concentrated as the crop removes water from the soil, the more saline the water, the less water can the crop be allowed to remove before irrigation is reapplied; and sufficient must be reapplied to wash the concentrated soil solution below the root zone of the crop. Naturally this can only be done if the groundwater level is below the root zone and the soil profile is permeable.

The first condition for the success of any irrigation scheme is therefore that the groundwater level be kept deep enough, and in practice difficulties arise if it comes closer than 1 m from the surface. If the profile is reasonably

18 Westcott, D. W. and Ayers, R. S. in *Irrigation with Reclaimed Municipal Wastewater*, Calif. State Water Res. Control Board Rept., No. 84–1, Ch. 3, 1984; also ref. 28.

permeable, and the groundwater naturally high, pumping of the groundwater may have to be used; and if the quality of the river water is good, this pumped groundwater can be put into the irrigation ditches to increase the amount of irrigation water available. Since in many schemes land is more plentiful than water, this pumping of groundwater not only keeps existing land in cultivation but increases the area of land that can be cultivated. If the permeability of the land is not very good, it may be necessary to install a system of deep drainage ditches, sometimes with an associated tile drainage scheme.

The proportion of applied irrigation water that must be used for leaching out the residual salts from the profile is called the leaching requirement.[19] On simple theory it is the ratio of the salt concentration in the irrigation water to the maximum permitted salt concentration in the soil solution, measured as electrical conductivity. Thus, if two irrigation waters have electrical conductivities of 0.1 and 0.5 dS m^{-1}, respectively, and it is decided to keep the salt concentration in the soil solution below 4 dS m^{-1}, the leaching requirements will be 0.025 and 0.125, respectively.

In practice more water than this is commonly needed. In the first place, the leaching water does not move down uniformly through the profile, for the soil usually contains a number of cracks and channels down which some of the water will move rapidly, taking only a little of the salts within the soil crumbs or clods with it. In the second place, the simple picture assumes that the leaching water dissolves no minerals from the soil and that no chemical reactions, such as precipitation of calcium carbonate, take place between the dissolved salts and the soil. This restriction is usually of little importance if the dissolved salts are predominantly chlorides with small amounts of sulphate, but can be very important if the water contains much sodium carbonate, as will be discussed on p. 943. It is not necessary to leach out residual salts after each irrigation, and the term leaching requirement refers to the proportion of water that must be used for leaching over a period of time. The periods between leachings must be shorter, the greater the leaching requirement. Normally leaching is carried out as a pre-irrigation before the crop is sown: leaching after this is influenced by opportunity, time and crop constraints.

Plants are usually most sensitive to salts at the seedling stage, making it particularly important to keep the soil around the seeds low in salt. This is usually quite easy if the irrigation water is low in salts, but can be difficult if it is high because the soil surface is bare when the crop is sown, and evaporation of water will be taking place from the soil surface, which will cause salts to concentrate there. If the crop being grown is very high-priced, this can be prevented by covering the surface with polythene sheet. The problem

19 Rhoades, J. D. in *Drainage for Agriculture* (ed J. van Schilfgaarde), *Agron. Monog.*, No. 17, 433, Am. Soc. Agron., Madison, 1974.

can be minimized by encouraging the salts to concentrate in definite bands, and placing the seeds sufficiently far from these bands to allow the roots of the young plant to grow in a low-salt solution. One method[20] is to ridge the land and run the irrigation water along the furrows between the ridges. If the ridge is symmetrical with flat sides either coming to an apex or having a flat top, salts will concentrate in the apex or in the centre of the flat top, for this is where the upper wetting fronts from the irrigation water in the furrows will meet. This will leave the flat side or the edge of the flat top with a salt content close to that of the irrigation water, so roots from seeds planted in these positions will start growing in a low-salt solution.

Deterioration of structure in sodic soils

No irrigation scheme can succeed unless the soil profile remains permeable, and this, in turn, depends both on the proportion of the exchangeable cations held by the soil that are sodium (the exchangeable sodium percentage, ESP), and on the concentration of soluble salts in the percolating water. A soil can have a high ESP and still remain permeable if the percolating solution with which it is in equilibrium is sufficiently concentrated; but as the solution becomes more dilute, a concentration will be reached when the permeability begins to drop, due to the swelling of the soil crumbs reducing the size of the channels and pores down which the solution is percolating. As the concentration drops still further, the clay particles will begin to deflocculate and completely block these pores, so the soil becomes impermeable. Calcium-saturated soils show this swelling phenomenon to a much smaller extent, and they only deflocculate if subjected to shearing stresses.

The ESP at which these swelling and deflocculation processes become important in a given concentration of solution depends on the type of clay, the content and distribution of the sesquioxide films, and the amount and type of organic matter. Ferruginous soils, and particularly ferruginous kaolinitic soils[21] and old pasture soils[22] can have a high ESP and maintain their permeability in dilute salt solutions, but alluvial soils containing much fine sand and silt, with micaceous-type clays and small amounts of organic matter, will lose permeability in such a solution with a low ESP.

It is important to stress that there is no universal value for the minimum ESP a soil must possess for its permeability to be appreciably affected by the

20 Bernstein, L. and Fireman, M., *Soil Sci.*, 1957, **83**, 249;   Hoffman, G. J. *et al.* in *Design and Operation of Farm Irrigation Systems* (ed M. E. Jensen), p. 145, Am. Soc. Agric. Engrs., *Monog.*, No. 3, 1980.
21 El-Swaify, S. A. and Swindale, L. D., *Trans. 9th Int. Congr. Soil Sci.*, 1968, **1**, 381; *idem, Soil Sci.*, 1970; **109**, 197;   El Rayah, H. M. E. and Rowell, D. L., *J. Soil Sci.*, 1973, **24**, 137.
22 Emerson, W. W., *J. Soil Sci.*, 1954, **5**, 233.

irrigation water, for this depends on the soil, on its management and on the concentration of the salts in the percolating water.[23] An ESP of 15 has, however, been widely used, particularly in the USA, as the critical content above which the soil structure will become unstable, and while this figure is a useful guide for many irrigated soils, it must not be taken as valid for all irrigated soils. The critical figure, or rather the safe content of exchangeable sodium that will not cause sufficient swelling to affect permeability or will not give a turbid drainage water, should be determined for each soil using a solution corresponding to the lowest salt content of the irrigation water. It is likely that for low-salt waters and for alluvial soils containing micaceous and montmorillonitic clay that are high in fine sand and silt, as many such soils are, 15 per cent exchangeable sodium is a useful general guide to the probability of an unstable structure, and a content of 7.5 per cent exchangeable sodium may begin to give trouble due to swelling though not to dispersion.

An ESP of 5 is often used as the critical value in Australia, much lower than the value in the USA. This reflects the standard method used to determine instability which involves leaching with the local tap water: this water in California generally has a higher salt content than the Australian waters and so prevents changes at ESP values below 15.[24] Salt concentrations of less than 3 meq l^{-1} are critical for this range of ESP, and it is this range of concentration that can be produced by the weathering of minerals so contributing to the differences in the stability of soils.[25] A montmorillonite clay containing 12.5 per cent exchangeable sodium and 87.5 per cent exchangeable calcium behaved almost like the calcium-saturated clay provided it was subjected to no mechanical stress. The latter is important particularly at the surface of soils where stress is applied by water drop impact – the greater the stress, the lower the ESP at which structural breakdown occurs for a given water.[26] Surface crusts develop which are composed of a compacted layer (0.1 mm thick) underlain by a clay enriched layer with high bulk density (2–3 mm thick). Movement of clay often leaves loose sand on the surface. ESP values greater than 3 seem to be required before clay moves under these conditions.[27]

23 For examples see Quirk, J. P. and Schofield, R. K., *J. Soil Sci.*, 1955, **6**, 163;   Rowell, D. L. *et al.*, *op. cit.*, 1969, **20**, 176.
24 Loveday, J. (ed.), *Methods for Analysis of Irrigated Soils*, Tech. Comm., No. 54, Commonw. Bur. Soils, 1974; McIntyre, D. S., *Aust. J. Soil Res.*, 1979, **17**, 115.
25 Shainberg, I. and Letey, J., *Hilgardia*, 1984, **52**, 1;   Rowell, D. L. and Shainberg, I., *J. Soil Sci.*, 1979, **30**, 719, Suarez, D. L. and Rhoades, J. D., *Soil Sci. Soc. Am. J.*, 1982, **46**, 716;   Shainberg, I. and Gal, M., *J. Soil Sci.*, 1982, **33**, 489.
26 Oster, J. D. and Shainberg, I., *Int. Symp. on Salt Affected Soils*, 1980, 195, Karnal, India; Rowell, D. L. *et al.*, *J. Soil Sci.*, 1969, **20**, 176.
27 Morin, J. *et al.*, *J. Hydrol.*, 1981, **52**, 321;   Tarchitzky, J. *et al.*, *Geoderma*, 1984, **33**, 135;   Shainberg, I., *Adv. Soil Sci.*, 1985, **1**, 101. Agassi, M. *et al.*, *Geoderma*, 1985, **36**, 263.

The exchangeable cations held by most irrigated soils consist of calcium and magnesium, with a small proportion of exchangeable potassium and a variable proportion of exchangeable sodium. This is because irrigation water is usually low in potassium but, as shown in Table 27.3, may contain appreciable amounts of sodium salts. As irrigation water percolates through the soil, there will be a redistribution of cations between the solution and the exchange sites, and as the soil dries, the concentration of the soil solution will increase, which will alter the exchange equilibrium between the mono- and the divalent cations.

The Gapon equation, which was discussed on p. 278, states that the ratio of the exchangeable sodium to exchangeable calcium plus magnesium on a soil is proportional to the ratio of the activity of the sodium ions to the square root of the activities of the calcium plus magnesium ions in the solution. For many practical applications no great error is made if the activities are replaced by the concentrations. Exchangeable sodium percentage, ESP, is the exchangeable sodium as a percentage of the cation exchange capacity, normally determined using sodium acetate at pH 8.2. The sodium adsorption ratio, SAR, is the ratio of the sodium ion concentration to the square root of the calcium plus magnesium ion concentrations, all expressed in millimoles per litre. This differs from the activity ratios used elsewhere in this book not only because concentrations are used rather than activities, but also because the concentrations are in millimoles per litre. The ion concentration ratio $[Na]/\sqrt{[Ca+Mg]}$ with concentrations in mol l^{-1}, must be multiplied by $\sqrt{1000}$ = 31.6 to convert to SAR. If concentrations are expressed in meq l^{-1}, SAR = $[Na]/\sqrt{[Ca+Mg]/2}$, and the ESP of many irrigated soils is approximately equal to the SAR of the solution in equilibrium with the soil.[28]

The SAR of the irrigation water will influence the ESP of the soil, but the relation between the two is not straightforward, for the ESP of the soil is conditioned by the SAR of the soil solution, and this is constantly changing. When water percolates into a soil, changes occur at the interface between the water and soil solution. Cation exchange, hydrolysis of sodium and dissolution or precipitation of minerals may all change the ESP.[29] After irrigation, the solution slowly becomes more concentrated as the crop transpires water and its SAR rises. If the effect of concentration is to cause some calcium or magnesium to precipitate out as the carbonate, this will also cause it to rise still further. Again, it takes time for the equilibrium to be established between the solution and the inside of soil crumbs and peds, so that the actual average ESP of the soil is likely to lag behind changes in the SAR of the soil solution. Californian experience, as summarized in the USDA classi-

---

28 For a more detailed discussion see Ayers, R. S. and Westcott, D. W., *Water Quality for Agriculture*, Irrigation and Drainage Paper, No. 29, FAO, Rome, 1976.

29 Rowell, D. L., *Irr. Sci.*, 1985, **6**, 211; Oster, J. D. and Tanji, K. K., *J. Irrig. Drainage Eng.*, 1985, **111**, 207.

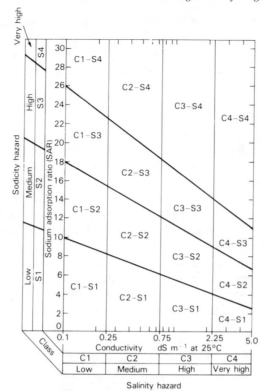

Fig. 27.2 Diagram for the classification of irrigation water.

fication of irrigation waters[30] (Fig. 27.2), has shown that the higher the total salt content of the irrigation water, the lower must be its SAR if the ESP of the soil is to remain below a given level. The reasons for this probably include buffering of the soil, the effect of varying total solution concentration on cation exchange, and the likelihood of precipitation of Ca salts at high salt concentration.

Although magnesium is equated with calcium in considering SAR and ESP values, there are indications that it may contribute to instability in some soils and the development of a 'magnesium solonetz' has been described.[31] However, results are conflicting and it may be that illitic soils are particularly sensitive to the presence of magnesium.[32] A distinction has been made[33]

30 Richards, L. A. (ed), *The Diagnosis and Improvement of Saline and Alkali Soils*. US Dept. Agric. Handb., **60**, 1954.
31 Ellis, J. H. and Caldwell, O. G., *Trans. 3rd Int. Congr. Soil Sci.*, 1935, **1**, 348.
32 Emerson, W. W. and Bakker, A. C., *Aust. J. Soil Res.*, 1973, **11**, 151; Emerson, W. W. and Chi, C. L., *op. cit.*, 1977, **15**, 255.
33 Rahman, W. A. and Rowell, D. L., *J. Soil Sci.*, 1979, **30**, 535; Vesque, C. S., *Soil Sci. Soc. Am. J.*, 1985, **49**, 1153, 1160; Tucker, B. M., *Aust. J. Soil Res.*, 1985, **23**, 405.

between a specific effect of Mg when present as an exchangeable cation, and a non-specific effect caused by the tendency for more Na to be adsorbed from a solution of given SAR when Mg is present, which occurs in soils dominated by clays other than vermiculite. Mg may have a further effect: if Mg is a product of the weathering of soil minerals, then the rate of release of cations may be reduced by the presence of solution and exchangeable Mg (the common ion effect). This may lead to lower soil solution concentrations during leaching, causing instability.[34] There is, however, little indication that the presence of high ratios of Mg : Ca (e.g. 1 : 1) will contribute to instability by more than the equivalent of 2–3 per cent exchangeable Na.

There is a group of alluvial soils for which permeability, in the normal sense of the word, is irrelevant, and these are cracking black clays high in montmorillonite – the Vertisols. They cover the Gezira in the Sudan, for example, where they are used successfully for irrigated cotton and other crops. The clays are almost impermeable but, on drying, cracks open up which may become several centimetres wide at the soil surface, as is shown in Table 27.4, and some penetrate to a depth of 0.5 m.[35] When such a soil is irrigated, the water runs down the cracks before they have time to seal up – and, in fact, they usually never seal up completely – and the water becomes distributed along the systems of smaller cracks produced by the drying of the clay by the plant roots. These soils are therefore irrigable even though they have a high ESP and have an unstable structure, and Fig. 27.3 shows that useful yields of cotton can be obtained,[36] even with an ESP of 40. However, although these clays are montmorillonitic, they may contain aluminium

Table 27.4   Types of cracks and cracking pattern of Gezira clay soils (percentage distribution)

Depth of cracks			Width of cracks			Spaces between cracks		
cm	Site		cm	Site		cm	Site	
	1	2		1	2		1	2
<16	4	1	<2.6	18	9	<21	22	39
16–35	32	45	2.6–3.5	42	25	21–40	29	53
36–55	58	19	3.6–4.5	27	22	41–60	20	7
56–85	6	21	4.6–5.5	11	28	61–80	20	1
>85	0	14	>5.5	2	16	>80	9	0

Site 1 irrigated. Site 2 not irrigated.

34 Alperovitch, N. *et al.*, *J. Soil Sci.*, 1981, **32**, 543.
35 El Abedine, A. Z. and Robinson, G. H., *Geoderma*, 1971, **5**, 229.
36 Robinson, G. H., *J. Soil Sci.*, 1971, **22**, 328.

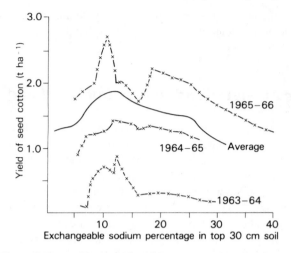

Fig. 27.3 Effect of the exchangeable sodium percentage of soil on cotton yields in the Sudan Gezira.

hydroxide which could help to stabilize the cracks to some extent. On the other hand, these soils can only be irrigated with waters low in salts because of the small amount of water and salts that drain out of the soil profile. The water of the Blue Nile is low in salts, and the salt content is kept at a fairly constant level because the crops carried off the land contain sufficient anions and cations to prevent the soil salinity from building up and affecting crop yield appreciably.

It is sometimes necessary to use a water for irrigation with an SAR which is likely to give an ESP in the soil sufficiently high for the soil permeability to fall to an unacceptably low level. The exchangeable sodium can still be kept low if the SAR of the irrigation water or of the soil water is lowered by adding gypsum (calcium sulphate) either to the irrigation water or to the soil; or sulphur itself can be added to the soil if it contains free calcium carbonate, for it will be oxidized to sulphuric acid which will react with the calcium carbonate to give calcium sulphate. The advantage of adding gypsum to the irrigation water, where this is economically feasible, is that it helps maintain the permeability of the surface soil. Calcium carbonate can also be added to a soil, if it is non-calcareous, but its solubility is much less than that of gypsum, so it has a much smaller effect on the SAR of the soil solution.

The maintenance of a high level of decomposable organic matter and the use of crops with a high root respiration rate can be a valuable means of maintaining permeability when high SAR waters must be used, again, provided there are reserves of calcium carbonate in the soil because the $CaCO_3$ is more soluble when the CO_2 concentration is increased. This will involve either the use of green manures or farmyard manure on the one hand, or of some grass and legume pastures on the other.[37]

Soluble carbonates and the control of pH and sodicity

Irrigation water usually contains free bicarbonate ions as well as sodium; and if both of these accumulate in the soil, its pH will rise to values as high as 10. This rise in pH can have two very undesirable consequences – it will increase the salt concentration necessary to keep the soil flocculated and so permeable,[38] and it may cause nutritional troubles in the crop, which may show as chloroses,[39] once it has exceeded a certain value, which for many irrigated crops is about pH 8.5. If the irrigation water has a fairly high neutral salt content, the pH of the soil does not normally rise above 8.5, so problems of high pH are typically problems with waters low in neutral salts.

Waters can contain bicarbonate and give no danger of high pH if they also contain about the same concentration of calcium and magnesium ions as bicarbonate and carbonate. Sodium bicarbonate cannot accumulate because as the soil solution becomes more concentrated due to transpiration, calcium and magnesium carbonates will precipitate out. Irrigation water in which the total bicarbonate and carbonate concentration is greater than that of the divalent ions is thus dangerous, and can only be used in long-term irrigation if precautions are taken to prevent sodium carbonate accumulating in the soil. Eaton[40] introduced the concept of the residual sodium carbonate in an irrigation water, which he defined as the excess concentration of bicarbonate and carbonate anions over the calcium and magnesium cations. He suggested that if this exceeded 2.5 meq l^{-1} the water would be unsuitable for irrigation, and if between 1.25 and 2.5 meq l^{-1} would give a serious danger of pH rising in the soil. The water of the Humboldt River, whose analysis is given in Table 27.3, is an example of water that is dangerous to use for this reason.

Bicarbonate-containing water, however, may increase the level of exchangeable sodium in the soil, even if the concentration of calcium and magnesium ions exceeds that of the bicarbonate and carbonate, because the precipitation of these cations as insoluble carbonates increases the SAR of the soil solution, and hence the level of exchangeable sodium. The reason that it only may, and not that it must, increase the SAR of the water is that

---

37 For an example from Nebraska of crops which help to maintain soil permeability to irrigation water see Mazurak, A. P. *et al.*, *Agron. J.*, 1955, **47**, 490.
38 Suarez, D. L. *et al.*, *Soil Sci. Soc. Am. J.*, 1984, **48**, 50.
39 Salem, M. A. and El Kadi, M. A., *Pl. Soil*, 1965, **23**, 377.
40 Eaton, F. M., *Soil Sci.*, 1950, **69**, 123; Wilcox, L. V. *et al.*, *op. cit.*, 1954, **77**, 259.

when a bicarbonate-containing water is percolating through a soil containing free calcium carbonate, the water will either dissolve some calcium carbonate or precipitate out calcium carbonate, depending whether the percolating solution is under- or oversaturated with respect to the solubility of the carbonate.

Whether or not a solution is over- or undersaturated with respect to calcium carbonate can be predicted from the solubility product of calcium carbonate, the dissociation constant of bicarbonate to carbonate, and the composition of the irrigation water. The degree of under- or oversaturation can be expressed as a precipitation index (PI) for that water, which is defined by

$$PI = 8.4 - pH_c$$

where pH_c is the pH which the water, with the given amounts of divalent cations and bicarbonate and carbonate anions, would have if it were in equilibrium with solid calcium carbonate. This is given by

$$pH_c = pK_2 - pK_c + p[Ca^{2+} + Mg^{2+}] + p[HCO_3^- + 2CO_3^{2-}]$$

where pK_2 is the negative logarithm of the second dissociation constant of carbonic acid, pK_c the negative logarithm of the solubility product for calcium carbonate, both corrected for the ionic strength of the percolating solution, and $p[Ca^{2+} + Mg^{2+}]$ and $p[HCO_3^- + 2CO_3^{2-}]$ are the negative logarithms of the concentrations in moles per litre in the equilibrium solution.[41] The figure 8.4 is the pH of the system calcium carbonate–carbon dioxide–water when the solution is in equilibrium with air containing 0.03 per cent carbon dioxide (see p. 890). If the precipitation index is positive, percolating the irrigation water through the soil will cause calcium and magnesium carbonates to precipitate out from the solution, thus increasing the SAR of the solution; and if it is negative, it will dissolve some carbonate, so decreasing the SAR.

It is possible to calculate the effect of this precipitation or dissolution of calcium carbonate on the SAR of the soil solution. If the leaching fraction, LF, is the proportion of applied irrigation water that leaches out of the soil, and if SAR_d is the SAR of the drainage water and SAR_i is that of the irrigation water, then

$$SAR_d = (1/\sqrt{LF}).SAR_i [1 + (8.4 - pH_c)]$$

if this is the complete description of what is happening. The leaching fraction enters as the square root since it is assumed to be equal to the ratio of the

---

41 Bower, C. A. *et al.*, *Soil Sci. Soc. Am. Proc.*, 1965, **29**, 61 give tables to help with the calculation of pH_c, and experimental results showing the accuracy of the Precipitation Index calculation.

salt concentrations in the irrigation to the drainage water, and the SAR is proportional to the square root of the salt concentration. The Californian workers have found this calculation may overestimate the SAR of the drainage water,[42] and Rhoades[43] suggested this was because some soils contain calcium silicates which slowly dissolve in the soil solution, releasing calcium ions. He finds he can allow for this by multiplying the SAR of the drainage water, as calculated from the above equation, by the factor $y^{(1 + 2LF)}$ where y is a constant, which for his soils equals 0.7; and he has published tables which give the calculated values of the leaching fraction for different values of pH_c and the SAR of the irrigation and drainage waters.

Pratt and Bair[44] showed that the empirical equality of the ESP of the soil and the SAR of the soil solution with which it is in equilibrium holds for many irrigated alluvial Californian soils even if they contain soluble carbonates, provided the pH of the soil saturation extract does not exceed 8.6, though the ESP is appreciably higher than the SAR at higher pH values. Thus provided this soil pH is less than 8.6, the ESP of the soil in equilibrium with the drainage water is approximately given by

$$ESP = \frac{y^{(1 + 2LF)}}{\sqrt{(LF)}} SAR_i (1 + PI)$$

This relation allows the computation of the leaching fraction, still uncorrected for non-uniformity of water flow through the soil, that must be used if the ESP of the soil is to be maintained below a given level when the irrigation water contains soluble carbonates. Suarez[45] has revised these equations to avoid the empirical use of the precipitation index. His equation is

$$SAR_d = \frac{Na_{iw}/LF}{\{Mg_{iw}/LF + Ca_{eq}\}^{1/2}}$$

where Na_{iw} and Mg_{iw} are the concentrations in the irrigation water and Ca_{eq} is the concentration in equilibrium with a $CaCO_3$ solid phase at the bottom of the root zone (all concentrations in mmol l^{-1}). The latter can be calculated from a knowledge of the CO_2 partial pressure in the root zone and is approximately equal to $9.33 (P_{CO_2})^{1/3}$. Suarez gives the information necessary for more accurate calculation of Ca_{eq}. Development of understanding of the solubility of Ca–Mg carbonates (p. 891) may allow refinement of the role of Mg in these systems.

Soils will sometimes contain appreciable levels of sodium carbonate formed in the soil profile. Sodic soils tend to deflocculate and so become anaerobic; if there are both decomposable organic matter and soluble sulphates in the

42 Bower, C. A. *et al.*, *Soil Sci.*, 1968, **106**, 29.
43 Rhoades, J. D., *Soil Sci. Soc. Am. Proc.*, 1968, **32**, 643, 648, 652.
44 Pratt, P. F. and Bair, F. C., *Soil Sci. Soc. Am. Proc.*, 1969, **33**, 880.
45 Suarez, D. L., *Soil Sci. Soc. Am. J.*, 1981, **45**, 469.

soil, the reducing conditions set up by the bacteria decomposing the organic matter will produce hydrogen sulphide. If Na^+ had been balancing the charge of this SO_4^{2-}, it will now be balancing bicarbonate or carbonate anions. Thus, one equivalent of sodium carbonate can be produced for each equivalent of sulphate reduced.[46]

The problems of managing soils when a high carbonate irrigation water is used are similar to those when a water with a high SAR is used. If the soil does not contain free calcium carbonate, it is usually helpful to add it to help maintain the reserves of calcium, and it may also be necessary to use gypsum or sulphur. Australian experience[47] has shown that there are advantages in adding gypsum to the irrigation water if the soils are unstable clays high in exchangeable sodium, for this helps to maintain a stable flocculated surface tilth, and a good surface permeability. Again, problems of management are greatly eased if systems of cropping can be used which allow the maintenance of a high level of microbial and root respiration, for the maintenance of a high level of carbon dioxide production will help maintain the concentration of calcium ions in the soil solution.

The reclamation of saline and sodic soils

The reclamation of land whose sole defect is that it contains too high a concentration of neutral salts is, in theory, simple, for it is only necessary to supply sufficient water over and above the transpirational needs of the crop to leach the salts out of the soil (see p. 935), and prevent the uncontrolled entry of new salts. But this can only be done if the soil is permeable, and the groundwater is well below the root zone or can be brought below the root zone by drainage. This process may, however, take time if the water that leaches out of the profile flows down a restricted number of channels or pores, because only the salts that diffuse out of the body of the soil into this percolating water will be removed. Deep saline clay soils, for example, can only be reclaimed economically if they are naturally well structured and can carry the cost of tile drainage. If the soil is permeable and the groundwater level is deep, well below the root zone of the crop, it is merely necessary to leach the salts within the root zone to below the level from which the roots take water, and to ensure that there is sufficient water available for the regular leaching of any salts that accumulate subsequently.

A common cause of salinity is, however, a high water table, and this must

46 Whittig, L. D. and Janitzky, P., *J. Soil Sci.*, 1963, **14**, 322; *ibid.*, 1964, **15**, 145; Ogata, G. and Bower, C. A., *Soil Sci. Soc. Am. Proc.*, 1965, **29**, 23.
47 Scotter, D. R. and Loveday, J., *Aust. J. Soil Res.*, 1966, **4**, 69; Davidson, J. L. and Quirk, J. P., *Aust. J. agric. Res.*, 1961, **12**, 100.

be lowered if the land is to be successfully reclaimed. The water table may be high because there is no effective drainage system, or because water is rapidly leaking out of unlined irrigation channels dug in permeable soil, or because water is seeping in from higher ground. In arid areas the groundwater is almost always saline, so a high water table almost always will cause serious salinity problems. Reclamation may involve the installation of deep drainage ditches and a drainage canal entering a river well below the irrigation scheme, it may involve lining the main irrigation canals to prevent seepage of water from them into the groundwater, or it may involve a properly designed series of cut-off drains to prevent groundwater entering the area. However, in many areas water is becoming sufficiently valuable, and power sufficiently cheap, for the groundwater level to be kept low by pumping; the groundwater being pumped into the irrigation ditches and diluted with the irrigation water.

Nevertheless, salinity control can have an unfortunate consequence. If the area being reclaimed is upstream of other irrigation schemes, and if the salts are to be run into drains discharging lower down into the river, these salts are being exported from the upstream scheme and imported into the downstream site. Since salt-affected soils may contain 10 t of salts or more per hectare, it may be necessary to limit the annual export of salts during the reclamation process. Even when the upstream scheme is not badly salinized, irrigation will always lower the quality of the water leaving the scheme, for crop transpiration increases the concentration of salts in the water, and therefore its SAR; and since concentration of the soil solution often leads to precipitation of calcium and magnesium carbonates, this again will raise the SAR.

Table 27.5 illustrates this effect for two adjacent sites on the Rio Grande in New Mexico and Texas, and it also gives the salt balance for these two schemes.[48] In the upper Mesilla area the salts are approximately in balance over the scheme, while the El Paso area is gaining about 150 000 t of salts a year. In detail the Mesilla area is losing about 40 000 t of chloride and 20 000 t of sodium annually, while the El Paso area is losing about 50 000 t of chloride but retaining half of the sodium being lost from the Mesilla scheme. Thus the two areas between them have taken about 75 per cent of the water entering the Mesilla area, have taken about half the calcium and sulphate, but have added 90 000 t of chloride to the water leaving the El Paso area.

Land affected by sodicity is usually black in colour, because the sodium carbonate present deflocculates the soil, and the humic colloids are dispersed from the clay and colour the soil much more strongly. When wet, the land usually has a fairly characteristic smell, presumably due to the activity of anaerobic bacteria.

48 Taken from Scofield, C. S., *J. agric. Res.*, 1940, **61**, 17.

Table 27.5 Annual salt and water balance in two irrigation areas of the Rio Grande (New Mexico and Texas). Water in Mt, salts and ions in kt

Area under irrigation	Mesilla area about 32 000 hectares			El Paso area about 48 000 hectares		
	Entering	Leaving	% retained	Entering	Leaving	% retained
Water	915	611	33.3	611	213	65.1
Concentration of salts in water, mg l^{-1}	660	1010		1010	2200	
Total salts	599.4	608.1	−1.5	608.1	461.8	24.1
Ca	80.6	69.7	14.5	69.7	41.3	40.7
Mg	17.6	15.5	11.9	15.5	10.2	33.9
Na + K	102.0	122.6	−20.3	122.6	111.7	9.0
$HCO_3 + CO_3$*	93.9	80.5	14.3	80.5	28.2	64.9
SO_4	233.4	208.0	10.9	208.0	108.7	47.7
Cl	71.3	111.1	−55.8	111.1	161.4	−45.3
NO_3	1.4	1.5	−3.0	1.5	0.5	65.2

* Computed as carbonate.

The principles underlying the reclamation of land rendered infertile by sodicity are: first, to ensure that drainage is adequate and that saline water is not seeping in from higher ground; and, second, to replace some of the exchangeable sodium by calcium.[49] If the soil contains gypsum, draining the land and flooding it with water is probably all that is required, although if the soil is initially impermeable careful choice of crops is required. But if the soil is low in gypsum, the primary trouble, which is either present, or which will develop unless precautions are taken, is the impermeability of the surface soil to water, and one great danger in reclamation that must be guarded against is decreasing the permeability of the soil. Soils containing much exchangeable sodium, or free sodium carbonate, will deflocculate and become impermeable to water if wetted with pure water, or with rain, whereas if they contain soluble salts, or the irrigation water has a high salinity, they may remain flocculated and permeable. Hence, the second principle in reclaiming sodic soils is to maintain a sufficiently high salt content in the soil during the process of leaching out the exchangeable sodium, and in some areas high-salt waters are used to help the leaching initially.[50] Provided the soil remains permeable, drainage, adding gypsum or sulphur,

49 Schilfgaarde, J. van (ed), *Drainage for Agriculture, Agron. Monog.*, No. 17, Am. Soc. Agron., Madison, 1974; Richards, L. A., *Diagnosis and Improvement of Saline and Alkali Soils*, US Dept. Agric. Handb., **60**, 1954.
50 For examples see Reeve, R. C. and Bower, C. A., *Soil Sci.*, 1960, **90**, 139; *Soil Sci., Soc. Am. Proc.*, 1966, **30**, 494; Muhammed, S. *et al.*, *Soil Sci.*, 1969, **108**, 249.

and flushing down the salts will remove exchangeable sodium without difficulty.

Many unused sodic soils are, however, almost impermeable to begin with, and the improvement in permeability is the primary problem.[51] Typically this is done by replacing the exchangeable sodium in the surface layer and so stabilizing it, and then deepening this stabilized layer. Adding gypsum to the surface soil, or on some lighter soils working in farmyard manure or compost and then letting it wet and dry a few times will be enough to give a few centimetres of stable permeable soil. The soil may then be flooded, provided an adequate drainage system has been installed, to allow the gypsum to wash down slowly into the subsoil, improving the permeability of every layer into which it penetrates, for rarely does a drained soil have a permeability of less than a few centimetres of water a month. Figure 27.4, taken from some experiments on an impermeable Gezira clay, in which sodium constitutes 10 per cent of the exchangeable ions, shows this effect of gypsum of increasing the permeability of the subsoil.[52] The land was flooded in April, and shortly afterwards soil water content was determined. The soil was left fallow and then flooded again in December and soil samples again taken. Without

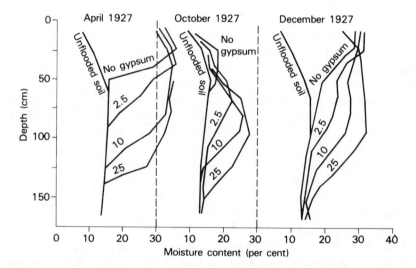

Fig. 27.4 Effect of gypsum on the permeability of soil. Percentage of moisture at different depths of unflooded soil (left-hand line) and of four flooded soils (right-hand lines) treated with different quantities of gypsum (t ha^{-1}).

51 For an American bulletin on methods of doing this, see Reitemeier, R. F. *et al.*, *US Dept. Agric. Tech. Bull.*, No. 937, 1948.
52 Greene, H., *J. agric. Sci.*, 1928, **18**, 531. See also with Snow, O. W., *J. agric. Sci.*, 1939, **29**, 1.

gypsum the water remained near the surface: with increasing applications of gypsum it penetrated deeper into the soil. The drying out affected the first 60 cm equally whatever the gypsum treatment but, as would be expected, much of the water that had penetrated below this depth remained in the subsoil. The permeability of the soil can also sometimes be increased by very deep ploughing,[53] particularly if some gypsum is ploughed in at the same time or if gypsum is present in the subsoil. This is probably most efficacious if the subsoil has a compacted layer. Deep ploughing can bring this layer up to the surface, break it up to some extent, and expose it to the weather so its structure can be mellowed by drying and wetting.

The second operation is to establish a crop on the land, either sown or of natural weeds, for the plant roots will continue to increase the permeability of the subsoil, both by extracting water from it, causing cracks to develop which will let water down quickly, and also by respiring carbon dioxide which will reduce the sodicity. The choice of crop is, however, limited, for it must be able to withstand the prolonged waterlogging necessary for washing down as much sodium as possible into the subsoil. Rice is an ideal crop, if other conditions are suitable, as it can be kept waterlogged throughout much of the season. Sweet clover (*Melilotus alba*) and strawberry clover (*Trifolium fragiferum*) are also suitable, as are many grasses, and under some conditions so is lucerne, though lucerne will not stand waterlogging as well as the others. Once these crops are established, they are encouraged to root deeply by being given heavy irrigations at as long intervals as possible. The roots take as much water as they can from the subsoil before the next irrigation, though they must not be allowed to dry the soil so much that the soil solution becomes sufficiently concentrated to harm the roots. The first crop to be taken is often ploughed in as green manure, as the plant may contain too much salt for animals, and the green manure will not only produce carbon dioxide in the soil during its decomposition, but will also slowly set free plant nutrients such as phosphate, iron, manganese and zinc, which may have low availability in sodic soils.

A good crop rotation is an excellent insurance against sodicity problems, for grass, clover and lucerne leys can all build up the structure of a soil and improve its stability; and if these leys are consumed on the farm by dairy cattle, they will be returned to the land as farmyard manure, which again has a valuable action in maintaining the permeability of the surface soil.

The most usual addition to the soil to help displace exchangeable sodium is gypsum, $CaSO_4.2H_2O$. This is more efficient than calcium carbonate because 1 litre of water will dissolve 30 meq of calcium as calcium sulphate, but at pH 8.4 only 1 meq of calcium from the carbonate, although at pH 7 it will dissolve 5 meq. In soil more Ca will dissolve than this (for gypsum 2–4

53 Kovda, V. A., *Khim. Sotsial. Zemled.*, 1941, No. 4, 31; Antipov-Karataev, N. and Zaitzev, A. A., *Dokuchaev Inst. Soils*, 1946, Anth. 14.

times more) because Ca will move on to exchange surfaces.[54] Calcium chloride is more soluble but its use will normally be uneconomic.[55] Sulphur is also used when it can be obtained cheaply. If these ameliorating substances are used with complete efficiency 10 t of gypsum correspond to 5.8 t of calcium carbonate or 1.9 t of sulphur, and if applied to a hectare of land would displace 2.2 meq of sodium per 100 g soil from the top 30 cm layer. Since it is difficult to determine how much exchangeable sodium a soil holds when in the presence of sodium salts, Schofield and Taylor[56] suggested that it is preferable to determine how much calcium the soil would need to take up to saturate its exchange complex with calcium; they developed a simple method which gives this figure to an adequate accuracy for most practical purposes.

Reclamation of soils damaged by sea water

It happens from time to time that low-lying land near the sea becomes flooded with sea water, usually due to exceptionally high tides overtopping a protective sea wall. This happened in East Anglia in 1897[57] and again over a wider area of eastern England[58] and parts of Holland and Belgium[59] in 1953. This flooding leaves large quantities of salts, principally sodium chloride, in the land which must be got rid of in such a way that the soil does not lose its structure, for the soil is now a saline sodic soil and reclamation practices described above need to be used.

After the sea water has drained away, crops may be killed by the direct effect of the residual salt. If the land is ploughed, once it has dried, a good seedbed can usually be made quite easily as the soil works and crumbles well. As rain removes the salt, the soil structure deteriorates, it becomes increasingly difficult to work, water begins to stand on the surface, and it dries into hard lumps.

It is possible to prevent deterioration by adding the requisite amount of gypsum to the soil. After the 1953 floods in eastern England, gypsum added at the rate of about 6 t ha^{-1} annually for two to three years on medium-

54 Oster, J. D. and Tanji, K. K., *J. Irrig. Drainage Eng.*, 1985, **111**, 207.
55 Magdoff, F. and Bresler, E. in *Physical Aspects of Soil Water and Salt in Ecosystems*, (eds A. Hadas *et al.*), p. 441, Ecological Studies, Vol. 4, Springer-Verlag, Berlin, 1973.
56 Schofield, R. K. and Taylor, A. W., *J. Soil Sci.*, 1961, **12**, 269.
57 See *Report on Injury to Agricultural Land on the Coast of Essex by the Inundation of Sea Water on 29 November 1897*, by Dymond, T. S. and Hughes, F., Chelmsford, 1899.
58 See *The East Coast Floods 1953*, Rept. by MAFF, London, 1962. Also for the Lindsey region, Blood, J. W., *NAAS Quart. Rev.*, 1955, **27**, 125; Heafield, T. G. and Ashley, G. D., *op. cit.*, 1956, **32**, 47.
59 For some Dutch experience of reclaiming land flooded with sea water see Beekom, C. W. C. van *et al.*, *Neth. J. agric. Sci.*, 1953, **1**, 153, 225.

textured soils kept the soils reasonably permeable and prevented the pH from rising much above 8.0; but on heavier soils dressings up to 25 t ha^{-1} were used. In the early stage of reclamation the land should be cultivated as shallow as possible, and this should be done only when the surface is dry enough to carry the tractor and implements without the soil sticking in a compact mass.

The experience in the 1953 floods showed that soils differ considerably in their reaction to flooding.[60] Soils well supplied with organic matter, such as old pastures or soils that had only been out of pasture for a few years, showed much less collapse of structure than soils that had been under intensive arable crops where the soil organic matter had fallen to a low level. Some soils also appeared to be inherently more stable than others, possibly because their structure was still being stabilized by sesquioxide cements formed during the process of formation from salt marsh. But many of the finer-textured soils, particularly those high in fine sand, silt or clay, remained in a difficult state for a number of years, and the dispersion of subsoil clay caused the tile drains to become filled with fine particles, and the sides of drainage ditches to become unstable.

The soils also differed considerably in the rate that salt was leached out of the profile, being very rapid for the sandy soils but slow for the fine sandy loams and clays. In fact some of the sandy soils could be brought back into cultivation a few months after the sea water had been drained away.

The reclamation of land from the sea poses much the same problems as reclamation after sea flooding.[61] The first problem is to install a drainage system and encourage the soil to crack, so salts can be washed down. It may then be necessary to add gypsum, though with careful management this is not always necessary if the soil is calcareous. A number of the clays formed in the shallow water under the sea may not only contain calcium carbonate, but ferrous sulphide as well. When these soils are drained and air enters, the ferrous sulphide is oxidized and the sulphuric acid set free by hydrolysis is neutralized by the calcium carbonate (p. 926). The calcium sulphate which is formed helps to improve the permeability of the profile.

60 Pizer, N. H., *Chem. Ind.*, 1966, 791.
61 Hazelden, J. *et al.*, *Saline soils in North Kent*, Soil Surv. Spec. Surv. No. 14, 1986.

Appendix 1

Units and Conversion factors

Physical quantity	S.I. unit	symbol	Factor S.I./c.g.s unit
Length	metre	m	10^2
Mass	kilogram	kg	10^3
Time	second	s	1
Temperature	Kelvin	K	
Density	kilogram metre^{-3}	kg m^{-3}	10^{-3}
Force	Newton	N	10^5
Pressure	Pascal	Pa (N m^{-2})	10
Work ⎫ Energy ⎭	Joule	J	
Power	Watt	W (J s^{-1})	10^7
Surface tension	Newton metre^{-1}	N m^{-1}	10^3
Viscosity (dynamic)		kg m^{-1}s^{-1}	10

Physical quantity	c.g.s. unit	symbol	Factor c.g.s./S.I. unit
Length	centimetre	cm	10^{-2}
Mass	gram	g	10^{-3}
Time	second	s	1
Temperature	degree Celsius	°C	
Density	gram cm^{-3}	g cm^{-3}	10^3
Force	dyne	dyne	10^{-5}
Pressure	dyne cm^{-2}	dyne cm^{-2}	10^{-1}
Work ⎫ Energy ⎭	⎰ erg ⎱ calorie	erg cal	10^{-7} 4.18
Power	erg s^{-1}	erg s^{-1}	10^{-7}
Surface tension	dyne cm^{-1}	dyne cm^{-1}	10^{-3}
Viscosity (dynamic)	poise	g cm^{-1}s^{-1}	10^{-1}

Electrical units	*unit*	*symbol*
Electric current	Ampere	A
Electric charge	Coulomb	$C = A\ s$
EMF, P.D.	Volt	$V = W\ A^{-1}$
Resistance	Ohm	$\Omega = V\ A^{-1}$
Resistivity	Ohm metre	$\Omega\ m$
Conductance	Siemen	$S = \Omega^{-1}$
Conductivity	Siemen m^{-1}	$S\ m^{-1}$
	$(dS\ m^{-1} = mmho\ cm^{-1})$	

Other units

1 acre = 4046 metre^2
1 bar = 100 kPa
1 metre head of water = 9.8 kPa
1 standard atmosphere = 101.3 kPa
1 mm mercury = 133 Pa
1 litre = $10^{-3}\ m^3$.

Concentration

Commonly concentration is expressed as mass per unit volume
e.g. $kg\ m^{-3} = g\ l^{-1} = 1\ mg\ ml^{-1} = 1\ \mu g\ \mu l^{-1}$ = approximately 1000 ppm
Molar (M) is 1 gram mol of substance dissolved to make 1 litre solution.

Equivalents

The equivalent of an ion is the molar mass in grams divided by its valency, i.e. the equivalent of a positively charged ion is the mass in grams of those ions that would replace 1 gram of hydrogen ions in a cation exchange reaction. Increasingly, use is made of the unit centimole of univalent cations per kg soil (cmol$_c$ kg^{-1} = 1 meq per 100 g soil).

Appendix 2

Botanical names of some plant species cited in this book

Common name	Botanical name
Alder	*Alnus* spp.
Almond	*Prunus amygdalus*
Apple	*Malus* spp.
Apricot	*Prunus armeniaca*
Banana	*Musa* spp.
Barley	*Hordeum vulgare*
Bean	*Vicia faba or Phaseolus* spp.
Beech	*Fagus sylvatica*
Bermuda grass	*Cynodon dactylon*
Birdsfoot trefoil	*Lotus corniculatus*
Black medic	*Medicago lupulina*
Buckwheat	*Fagopyrum spp.*
Cabbage	*Brassica oleracea* var. *capitata*
Cacao	*Threobroma cacao*
Carrot	*Daucus carota* var. *sativa*
Cashew	*Anacardium occidentale*
Cassava	*Manihot esculenta, M. utilissima*
Cauliflower	*Brassica oleracea* var. *botrytis*
Chickpea	*Cicer arietinum*
Citrus	*Citrus* spp.
Clover	*Trifolium* spp.
Clover, Alsike	*T. hybridum*
Crimson	*T. incarnatum*
Ladino	*T. repens*
Red	*T. pratense*
Strawberry	*T. fragiferum*
Subterranean	*T. subterraneum*
Sweet	*Melilotus alba*
White	*T. repens*
Cocksfoot	*Dactylis glomerata*
Coconut	*Cocus nucifera*
Coffee	*Coffea arabica*

Common name	Botanical name
Cotton	*Gossypium* spp.
Cowpea	*Vigna* spp.
Date palm	*Phoenix* spp.
Douglas fir	*Pseudotsuga menziesii*
Fig	*Ficus* spp.
Flax	*Linum* spp.
Gooseberry	*Ribes* spp.
Grape	*Vitis* spp.
Grapefruit	*Citrus paradasi*
Groundnut	*Arachis hypogaea*
Hazel	*Corylus avellana*
Hemp	*Cannabis sativa*
Horse chestnut	*Aesculus hippocastanum*
Kale	*Brassica oleracea* var. *acephala*
Leek	*Allium porrum*
Lentil	*Lens esculenta, L. culinaris*
Lettuce	*Lactuca sativa*
Linseed	*Linum* spp.
Loblolly pine	*Pinus taeda*
Lodgepole pine	*Pinus contorta*
Lucerne (alfalfa)	*Medicago sativa*
Lupin	*Lupinus* spp.
Maize (corn)	*Zea mays*
Mangel	*Beta vulgaris*
Mango	*Mangifera* spp.
Marrow	*Cucurbita pepo*
Medics	*Medicago* spp.
Millet	*Panicum* spp., *Setaria* spp., *Echinochloa* spp.
Mung bean	*Phaseolus mungo*
Mustard, white	*Sinapis alba*
Oat	*Avena sativa*
Olive	*Olea europaea*
Onion	*Allium cepa*
Orange	*Citrus sinensis, C. aurantium*
Pacific silver fir	*Abies* spp.
Parsnip	*Pastinaca sativa*
Peach	*Prunus persica*
Pear	*Pyrus communis*
Pearl millet	*Pennisetum typhoides*
Pea	*Pisum* spp.
Perennial ryegrass	*Lolium perenne*

Common name	Botanical name
Pigeon pea	*Cajanus cajan*
Pineapple	*Ananas comosus*
Plum	*Prunus domestica*
Potato	*Solanum tuberosum*
Radish	*Raphanus sativus*
Rape	*Brassica napus*
Red beet	*Beta vulgaris*
Red currant	*Ribes sativum*
Rice	*Oryza sativa*
Rye	*Secale cereale*
Ryegrass	*Lolium perenne*
Sainfoin	*Onobrychis viciifolia*
Serradella	*Ornithopus sativus*
Sesame	*Sesamum indicum*
Sitka spruce	*Picea sitchensis*
Sorghum	*Sorghum* spp.
Soyabean	*Glycine max*
Spearmint	*Mentha spicata*
Spinach	*Spinacea oleracea*
Squash	*Cucurbita maxima*
Strawberry	*Fragaria* spp.
Sugar cane	*Saccharum officinarum*
Sugar beet	*Beta vulgaris*
Sunflower	*Helianthus annuus*
Swede	*Brassica napobrassica*
Tea	*Camellia sinensis*
Timothy	*Phleum pratense*
Tobacco	*Nicotiana tobacum*
Tomato	*Lycopersicon esculentum*
Tree lupin	*Lupinus arboreus*
Turnip	*Brassica rapa*
Vetch	*Vicia sativa*
Wheat	*Triticum* spp.
Wheat grass	*Agropyron elongatum*
Willow	*Salix* spp.

Subject index

acetic acid production in soils, 905
acetylene in measurement of denitrification, 675
  nitrogen fixation, 548, 626
acid sulphate soils, 925
acidity in soils, 844 *et seq.*
  acid neutralizing capacity, 866
  buffer capacity, 866
  development, 855–64
  effect of acid rain, 864
    deforestation, 867
    fertilizers, 689–90, 863
    flooding, 901, 909, 913
    leaching, 856
    parent material, 855
    ploughing, 867
    respiration, 858
    roots, 860
  effect on actinomycetes, 458
    aluminium, 851
    bacteria to fungi ratio, 459
    calcium and magnesium, 859, 874
    decomposition of organic matter, 596
    exchangeable aluminium, 887
    manganese, 793, 875
    molybdenum, 799, 874
    nitrification, 617
    nitrogen fixation by legumes, 537, 646
    phosphate, 714–6, 872–4
    plant growth, 867–83
    plant pathogens, 879
  produced by litter, 857
  proton budgets, 864–7
  subsoil acidity, 875, 880
  tolerance of crops, 876, 878, 887
  see also liming of soils
Acrasieae, 470
Acrisols, 823, 824
actinomycetes, 458
  nitrogen fixation, 641
  producing geosmin, 458
  soil desiccation, 458
  see also Frankia
added nitrogen interaction, 688
adhesion of soil, 409

adsorption
  envelope, 254
  isotherms, 269–72
  versus precipitation, 272–3
  see also individual elements
aeration of soil,
  effect of earthworms, 513
    irrigation, 930
  effect on crop growth, 308–14
    fungi, 463
    microbial activity, 307–8
    nitrification, 615–6
  measurement, 305–7
  processes, 301
aerenchyma, 310–2
aerobic organisms, oxygen requirement, 489, 520
aggregate stability, 386–91
  effect of earthworms, 400–1
    farmyard manure, 437
    fungi, 397
    organic matter, 393–5
    plant roots, 399
    polysaccharides, 393–7
    soil conditioners, 402–3
aggregates, classification, 390–3
  internal structure, 384–6
agricultural land capability, 28, 831
agricultural surveys, 819
agrotechnology transfer, 25, 842–3
albedo of soil, 283, 288
albic horizon, 184
albite, 218
alder, nitrogen fixation, 533, 641, 654
Alfisols, 191, 195, 203, 822, 824
algae, 465–8
  effect of pH, 466
  number 467
alkalinity,
  calcareous soils, 889–91
  plant tolerance 895
  sodic soils, 896–8
  soil solution, 888
  see also sodium
allophane, 188, 254

Author index